AF352910

Mapping Abiotic Stress-Tolerance Genes in Plants

Mapping Abiotic Stress-Tolerance Genes in Plants

Special Issue Editor

Richard R.-C. Wang

MDPI • Basel • Beijing • Wuhan • Barcelona • Belgrade • Manchester • Tokyo • Cluj • Tianjin

Special Issue Editor
Richard R.-C. Wang
USDA-ARS
USA

Editorial Office
MDPI
St. Alban-Anlage 66
4052 Basel, Switzerland

This is a reprint of articles from the Special Issue published online in the open access journal *International Journal of Molecular Sciences* (ISSN 1422-0067) (available at: https://www.mdpi.com/journal/ijms/special_issues/plant_gene_abiotic).

For citation purposes, cite each article independently as indicated on the article page online and as indicated below:

LastName, A.A.; LastName, B.B.; LastName, C.C. Article Title. *Journal Name* **Year**, *Article Number*, Page Range.

ISBN 978-3-03936-114-4 (Pbk)
ISBN 978-3-03936-115-1 (PDF)

Contents

About the Special Issue Editor

Richard R.-C. Wang (Ph.D.) obtained his B.S. degree from the National Taiwan University in 1967, and M.S. and Ph.D. degrees from Rutgers, the State University of New Jersey (New Brunswick) in 1971 and 1974, respectively. He was a postdoctoral fellow at Kansas State University, cytopathologist at DeKalb Hybrid Wheat, Inc. (Wichita, KS, USA), Research Scientist at the International Plant Research Institute (San Carlos, CA, USA), before being a Research Geneticist for the USDA, Agricultural Research Services, in the Forage and Range Research Laboratory, Logan, Utah, USA. He has engaged in agricultural research over the past 47 years, authored or co-authored 150 refereed journal papers, 5 book chapters, and edited 1 symposium proceedings and 2 journal Special Issues. He is internationally recognized for his research on gene transfer from alien species into wheat, wide hybridization, genome analysis, and molecular markers in perennial Triticeae. His laboratory has hosted many visiting scientists from China, Russia, Azerbaijan, and Korea. He has been given honorary professorships at Northwest A&F University (1993) and China Agricultural University (1995). He was presented an Outstanding Alumnus Award (Fenton Graduate Alumni Award in Plant Biology) from Rutgers University (2003), a Fellow in the American Society of Agronomy (ASA, 2003), and Crop Science Society of America (CSSA, 2005). Dr. Wang is serving on the /International Journal of Molecular Sciences/ editorial board and has been a Guest Editor of two Special Issues on genes responding to biotic and abiotic stresses in plants.

International Journal of
Molecular Sciences

Editorial

Chromosomal Distribution of Genes Conferring Tolerance to Abiotic Stresses Versus That of Genes Controlling Resistance to Biotic Stresses in Plants

Richard R.-C. Wang

USDA-ARS Forage and Range Research Lab, Utah State University, Logan, UT 84322-6300, USA; richard.wang@usda.gov

Received: 29 February 2020; Accepted: 4 March 2020; Published: 6 March 2020

1. Introduction

Tolerance to abiotic stresses caused by environmental conditions can prevent yield loss in crops for sustaining agricultural productivity [1]. Resistance to biotic stresses caused by diseases and insects can prevent or reduce yield loss in crops [2]. For each crop or plant species, there are many abiotic threats, such as changes in temperature, soil salinity/alkalinity, water shortage, and soil contaminants, as well as biotic challenges from pathogens (bacteria, viruses, and fungi), insects, and nematodes. Plants need to possess genes conferring tolerance to these abiotic stresses to adapt to the changing environment, due to global climate changes, in which they are growing. Due to the coevolution of plants and stress-causing organisms [3], plants need to possess multiple resistance genes to deal with the rise of new virulence in stress-causing organisms. Plant breeders are constantly looking for new resistance genes to combat evolving organisms that pose a threat to susceptible crops. As a result, plant geneticists have identified many resistance genes in various crops, and molecular geneticists have developed molecular markers for most of those genes. Similarly, researchers are investigating plant mechanisms and underlying genetic systems involved in plant tolerance to abiotic stresses, hoping to breed crops resilient to adverse environmental conditions.

With the advent of whole-genome sequencing in many important crops, it is time to map the detailed chromosomal locations of known genes that are involved in tolerance to various abiotic stresses as well as in the resistance to biotic stresses in important plant species. In the Special Issue, "Mapping Abiotic Stress-Tolerance Genes in Plants" of International Journal of Molecular Sciences, 21 papers, including two reviews and 19 research articles, were published [4–24]. Eleven research articles [3,25–34] were published in the Special Issue "Mapping Plant Genes that Confer Resistance to Biotic Stress."

In this editorial, I firstly express my appreciation to all authors for their contribution to the two Special Issues. Secondly, I will compare the chromosomal distribution patterns of genes for the two types of stresses that plants faced (Tables 1 and 2). The evidence obtained supports my long-held hypothesis that genes conferring resistance to biotic stresses are more likely to be located in the distal portion of chromosomes than the proximal portion in order to adapt to the host-pest coevolution. On the other hand, abiotic-stress tolerance genes should have a lower ratio of distal to proximal distribution than that for biotic stresses to maintain the stability of genes regulating plant growth and development. Knowing the relationship between gene functions and their chromosomal distribution patterns, plant breeders can select the most appropriate and efficient method to improve crops for withstanding stresses and ensuring productivity and food security.

Table 1. Chromosomal distribution of genes controlling tolerance to abiotic stresses.

Plant Species	Genes	Stress	Mechanisms	Chromosome	Chromosome Arm			Reference
					distal	proximal	total	
Barley (*Hordeum vulgare*)	P-Type ATPase (*HvPAA1*) gene in a single QTL *qShCd7H*	Cadmium	Plasma membrane-localized cation-transporting ATPase.	7H	0	1	1	Wang et al. 2019 [4]
Cotton (*Gossypium hirsutum*)	ROS-network genes (*CSD1, APX1, APX2, MDAR1, GPX4-6-7, FER2, RBOH6, RBOH11, FRO5, AOX, GLR, and PER, etc.*)	Cold, heat, dehydration, salt	ROS network-mediated signal pathway.	Nine each of A and D genomes	21	15	36	Xu et al. 2019 [5]
Pear (*Pyrus pyrifolia*)	C-repeat binding factor (*PpyCBF1 to 6*) genes	Low temperature, salt, drought, and abscisic acid (ABA).	ABA-dependent and -independent pathways, ROS and antioxidant.	1, 4, 6, 7, 14 and one scoffold.	-	-	-	Ahmad et al. 2019 [6]
Rice (*Oryza sativa*)	AHA2, FRO2, IRT1, FIT, FRD3, FPN1, YSL2, VIT1, NRAMP3/4	Iron deficiency.	Iron acquisition from soil, iron transport fromroots to shoots, and iron storage in cells.	-	-	-	-	Zhang et al. 2019 [7]
Arabidopsis thaliana	Stress-Responsive NAC Transcription Factor (*LlNAC2*) of tiger lily	Cold, drought, salt stresses, and abscisic acid (ABA).	DREB1/ZFHD4/CBF-COR interaction and ABA signaling pathways.	1S (in *Arabidopsis*)	1	0	1	Yong et al. 2019a [8]
Arabidopsis thaliana	MYB related homolog (*LlMYB3*) of tiger lily	Cold, drought, and salt stresses, ABA treatment.	LlCHS2 and anthocyanin biosynthesis pathway.	5L (in *Arabidopsis*)	1	0	1	Yong et al. 2019b [9]
Soybean (*Glycine max*)	four QTLs for resistance to high-intensity UV-B irradiation (UVBR12-1, 6-1, 10-1, and 14-1)	UV-B irradiation (high light, heat, dehydration),	Possibly, actin-binding spectrin like protein interacting with membrane phosphoinositides in cellular signaling for defense.	12, 6, 10, and 14	2	2	4	Yoon et al. 2019 [10]
Woodland Strawberry (*Fragaria vesca*)	GIBBERELLIN-INSENSITIVE (GAI), REPRESSOR OF GA1-3 (RGA) and SCARECROW (SCR) protein (*FveGRAS*) genes	Cold, heat, and GA3 treatments.	Stolon formation, fruit ripening and abiotic stresses.	All 7	25	10	35	Chen et al. 2019 [11]
Arabidopsis thaliana	N-MYC Downregulated Like Proteins (NDL1, NDL2, NDL3) interacting with ANN1, SLT1, OAS-TL, ARS27A, RGS1, AGB1	Heat, cold, dehydration, DNA damage, reducing agent, increased intracellular calcium, metal ions like cadmium, nickel and cobalt, hormones.	N-MYC Downregulated Like Proteins (NDLs) interacting with G-Proteins in signal transduction in response to drought, heat, salinity and light intensity.	All 5	5	4	9	Katiyar and Mudgil 2019 [12]

Table 1. *Cont.*

Plant Species	Genes	Stress	Mechanisms	Chromosome	Chromosome Arm			Reference
					distal	proximal	total	
Soybean (*Glycine max*)	calmodulin binding transcription activator gene (*GmCAMTA*)	Drought.	Calmodulin binding Ca-CaM-CAMTA-mediated stress regulatory mechanisms.	8 out of 20 (5, 7, 8, 9, 11, 15, 17, 18)	10	5	15	Noman et al. 2019 [13]
Cotton (*Gossypium hirsutum*)	nodule inception-like protein (*GhNLP*) genes	Nitrogen deficiency	Promoters of NLP genes interact with stress-associated transcription factors and be targeted by many miRNAs.	All 26	91	14	105	Magwanga et al. 2019 [14]
Cucumber (*Cucumis sativus* L.)	GAGA-binding BASIC PENTACYSTEINE (BPC) transcription factor genes (*CsBPCs*)	Salt, drought, cold, heat, ABA, SA, JA, ETH, 2,4-D, GA.	Germination, growth and development, as well as responses to abiotic stresses and plant hormones.	3 of 7 (2, 5, 7)	3	1	4	Li et al. 2019 [15]
Carnation (*Dianthus caryophyllus*)	Heat shock transcription factors (Hsfs)	Heat, drought, cold, salt, ABA, SA.	Promoters included various cis-acting elements that were related to stress, hormones, as well as development processes, controlling reactive oxygen species homeostasis, and ABA-mediated stress signaling.	17 scaffolds	10	7	17	Li et al. 2019 [16]
Cotton (*Gossypium hirsutum*)	Histone Acetyltransferase (HAT) Gene family	Salt, drought, cold, heavy metal, DNA damage, ABA, NAA.	Affect cotton growth, fiber development, and stress adaptation by regulation of chromatin structure, activate the gene transcription implicated in various cellular processes.	8 of 26 (A-5,6,8,11 and D-5,6,10,11)	16	2	18	Imran et al. 2019 [17]
Chinese kale (*Brassica oleracea*)	multi-protein bridging factor (MBF) 1c (*BocMBF1c*)	Heat stress: cellular response to hypoxia, ethylene-activated signaling pathway, positive regulation of transcription, DNA-templated response to abscisic acid heat, and water deprivation.	BocMBF1c contains three heat shock elements (HSEs) and helix-turn-helix (HTH) domains, regulating ABRFs, SA, trehalose, and ET thermal resistance-related pathways by binding with CTAGA, including *DREB2A*.	not presented; ortholog on chromosome 3 of *Arabidopsis thaliana**	-; 0	-; 1 *	-; 1 *	Zou et al. 2019 [18]
Soybean (*Glycine max*)	Pentatricopeptide-repeat (PPR) proteins DYW subgroup genes; *GmPPR4*	Drought and salt.	Delayed leaf rolling; higher content of proline (Pro); and lower contents of H2O2, O2, and malondialdehyde (MDA); increased transcripts of several drought-inducible genes.	all 20 chromosomes; *Gm*PPR4 is on chromosome 1 distal end	143	36	179	Su et al. 2019 [19]

Table 1. *Cont.*

Plant Species	Genes	Stress	Mechanisms	Chromosome	Chromosome Arm			Reference
					distal	proximal	total	
Bread wheat (*Triticum aestivum*)	*WRKY transcription factor superfamily genes; TaWRKY13*[*]	Salt, drought, ABA, cold.	More root development, increased proline (Pro) and decreased malondialdehyde (MDA) contents.	all chromosomes except 4B and 7B; 2A	33	24; 1[*]	57	Zhou et al. 2019 [20]
oilseed rape (*Brassica napus*)	Fructose-1,6-bisphosphate aldolase (FBA) gene family (*BnaFBA*)	Salt, heat, drought, *Sclerotinia sclerotiorum* infection, and strigolactones (SLs) treatments.	Processes of glycolysis, gluconeogenesis, and Calvin cycle; Various cis-acting regulatory elements existed within the promoter regions of BnaFBA genes.	19 on 15 *B. napus* chromosomes; 3 others to 2 random chromosomes (two on the An chromosomes and one on the Cn chromosome)	7	15	22	Zhao et al. 2019 [21]
Sorghum (*Sorghum bicolor*)	stay-green QTL	Drought and heat.	N/C supply-demand, photosynthesis, water use efficiency, leaf anatomy, mineral and sugar transportation, senescence.	All 7	10	7	17	Kamal et al. 2019 [22]
Wheat (*Triticum aestivum*)				1A, 2A, 4A, 5A, 1B, 2B, 3B, 4B, 4D, 7D	10	8	18	
Rice (*Oryza sativa*)				2 to 12	9	18	27	
Maize (*Zea mays*)				1, 2, 3, 5, 6, 8, 9	12	11	23	
Barley (*Hordeum vulgare*)				All 7	4	6	10	
Soybean (*Glycine max*)	Calcium-dependent protein kinases (CDPKs) genes; *GmCDPK3*[*]	Drought and salt.	Increased proline (Pro) and chlorophyll contents and decreased malondialdehyde (MDA) content.	12 of 20 (1 to 6, 10, 11, 14, 16, 18, 19)	14; 1[*]	3	17	Wang et al. 2019 [23]
Radish (*Raphanus sativus*)	Lipoxygenases (LOXs) gene family *RsLOX*	Abiotic (drought, salinity, heat, and cold) and biotic (*Plasmodiophora brassicae* infection) stress conditions.	three tandem-clustered RsLOX genes are involved in responses to various environmental stresses via the jasmonic acid pathway.	5 of 9 (2, 5, 7, 8, 9)	5	6	11	Wang et al. [24]
Total					432	196	628	
Ratio						2.2:1		

* The chromosome position is not in Brassica but is in Arabidopsis.

Table 2. Chromosomal distribution of genes controlling resistance to biotic stresses.

Plant Species	Genes	Biotic Stress	Mechanisms	Chromosome	Chromosome Arm			Reference
					distal	proximal	total	
Glycine max	*RpsX*	Phytophthora root rot (PRR) caused by *Phytophthora sojae* (Rps).	A 144-bp insertion in the Glyma.03g027200 sequence resulted in two additional leucine-rich (LRR) encoding fragments.	3	1	0	1	Zhong et al. 2019 [25]
Oryza sativa	QTL *qFfR9* with 35.15% additive effect	Bakanae disease (BD), caused by the fungal pathogen *Fusarium fujikuroi*.	Eight genes in the QTL may be candidate genes for BD resistance.	9	1	0	1	Kang et al. 2019 [26]
Triticum aestivum	18 QTL	Karnal bunt caused by *Neovossia indica*.	QTL are associated with NBS-LRR proteins, Serine/threonine-protein kinase, Protein Kinase family protein, Kinase family protein, Receptor-like kinase, C2H2-like zinc finger protein, F-box domain containing protein, Glycosyltransferase and Transcription factor gene families.	1D, 2B, 2D, 4A, 4B, 5A, 5B, 6A, 6B, 7B, 7D	15	3	18	Gupta et al. 2019 [27]
Oryza sativa	Lesion mimic mutant (LMM) gene *LMM24*	*lmm24* exhibited enhanced resistance to rice blast fungus *Magnaporthe oryzae* and up-regulation of defense response genes.	Receptor-like cytoplasmic kinase 109 (OsRLCK109) leads to dark brown lesions in leaves and growth retardation due to enhanced ROS accumulation.	LOC_Os03g24930 on chromosome 3	0	1	1	Zhang et al. 2019 [28]
Triticum aestivum	3 QTL for stripe rust resistance	Stripe rust, caused by *Puccinia striiformis* f. sp. *tritici*.	QTL on 1B may be *Yr29 (an APR gene)*; the *minor* QTL on 2Al may be a new stripe rust resistance locus; Qyr.saas-7B could be in the same locus of QYr.nsw-7B from Tiritea.	1BL, 2AL, 7BL	3	0	3	Yang et al. 2019 [29]
Triticum aestivum	124 genomic regions associated with various diseases; several genes in those significant genomic regions had gene annotations suggesting their involvement in disease resistance.	wheat rusts (leaf; *Puccinia triticina*, stem; *P. graminis* f.sp. *tritici*, and stripe; *P. striiformis* f.sp. *tritici*) and crown rot (*Fusarium* spp.); cereal cyst nematode (*Heterodera* spp.); and Hessian fly (*Mayetiola destructor*).	Five genes were annotated as the leucine-rich repeat protein family and six genes were annotated as the F-box family protein, which were also reported to be involved in abiotic stress tolerance such as drought; Calcium-binding protein; ARM repeat superfamily protein; Elongation factor 1 alpha; Peroxidase; WAT1-related protein/EamA-like transporter family.	21 chromosomes	97	27	124	Bhatta et al. 2019 [30]
Dasypyrum villosum to *Triticum aestivum*	*Sr52*	Wheat stem rust caused by *Puccinia graminis* f. sp. *Tritici*.	Resistant to stem rust Ug99 races.	6V#3L bin FL 0.92–1.00 to 6AL.	1	0	1	Li et al. 2019 [31]

Table 2. *Cont.*

Plant Species	Genes	Biotic Stress	Mechanisms	Chromosome	Chromosome Arm			Reference
					distal	proximal	total	
Triticum aestivum	Seven significant additive QTLs for TS resistance explaining 2.98 to 23.32% of the phenotypic variation; five QTLs explaining 5.24 to 20.87% of SNB resistance	Tan Spot (induced by *Pyrenophora tritici-repentis*) and Septoria Nodorum Blotch (caused by *Parastagonospora nodorum*).	Quantitative resistance: fungus *P. tritici-repentis* isolates produce at least three host-selective toxins (HSTs), Ptr ToxA, Ptr ToxB and Ptr ToxC that interact with products of specific host sensitivity genes located on chromosome arm 5BL, 2BS., and 1AS, respectively, to cause disease.	TS (1A, 1B, 5B, 7B and 7D); SNB (1A, 5A, and 5B)	7	5	12	Singh et al. 2019 [32]
Capsicum annuum	A major QTL *qRRs-10.1*	bacterial wilt (BW), caused by *Ralstonia solanacearum*.	A cluster of five predicted R genes and three defense-related genes.	chromosome 10	0	1	1	Du et al. 2019 [33]
Aegilops searsii to *Triticum aestivum*	*Pm57*	Powdery mildew caused by *Blumeria graminis* f. sp. *tritici*.	Ten genes that are putative R genes which includes six coiled-coil nucleotide-binding site-leucine-rich repeat (CNL), three nucleotide-binding site-leucine-rich repeat (NL) and a leucine-rich receptor-like repeat (RLP) encoding proteins.	2S^s#1, fraction length 0.72–0.87	1	0	1	Dong et al. 2020 [34]
Vitis quinquangularis	Transcription Factor *VqMYB14*	bacterial flagellin peptide flg22 and harpins (glycine-rich and heat-stable proteins that are secreted through type III secretion system in gram-negative plant-pathogenic bacteria).	The promoter of *VqMYB14 is* induced by the elicitors flg22 to confer basal immunity (also called pathogen-associated molecular pattern (PAMP)-triggered immunity, PTI) and triggered by harpin to confer effector-triggered immunity (ETI). Overexpression of *VqMYB14* enhance the main stilbene contents and expression of stilbene biosynthesis genes.	chromosome 7	0	1	1	Luo et al. 2020 [3]
Total					126	38	164	
Ratio						3.3:1		

2. Chromosomal Distribution Patterns of Genes for Abiotic-Stress Tolerance vs. Biotic-Stress Resistance

Studying abiotic-stress tolerance, the authors of these 21 articles in this Special Issue covered *Hordeum vulgare, Gossypium hirsutum, Pyrus pyrifolia, Oryza sativa, Glycine max, Fragaria vesca, Cucumis sativus, Dianthus caryophyllus, Brassica oleracea, B. napus, Sorghum bicolor, Triticum aestivum, Zea mays, Raphanus sativus,* and the model plant *Arabidopsis thaliana* (Table 1). The abiotic stresses studied include cold, heat, drought, salt, iron deficiency, nitrogen deficiency, UV irradiation, DNA damage, reducing agent, phytohormones (GA, SA, JA, ABA, ethylene, 2,4-D, and NAA), and heavy metals (cadmium, nickel and cobalt). Two [6,7] of the 21 articles did not present information on the chromosomal locations of genes for abiotic-stress tolerance, and one [18] did not map the BocMBF1c gene to the target species *B. oleracea* but did locate the orthologous gene identified in *A. thaliana* to the proximal section of chromosome 3.

Many transcription factor gene families (TFs) were studied in the majority of these 21 articles [6,8,9,11,13–20,23,24]. Various putative stress-related and hormone-responsive cis-acting regulatory elements were identified in the promotor of these TFs. "The cis-regulatory sequences are linear nucleotide fragments of non-coding DNA with the main role of regulating gene expression and in turn, controls the development and physiology of an organism" [35]. Therefore, variations among members of TFs observed in those studies might account for the varying regulation of gene expression in different organs and tissues or at different developmental stages to respond to different stresses.

Among the 11 articles in the Special Issue on plant genes conferring resistance to biotic stresses [3,25–34], seven articles reported results from single resistant genes (or QTL) for crops and plant species, including soybean, rice, wheat, *Dasypyrum villosum, Aegilops searsii, Capsicum annuum,* and *Vitis quinquangularis.* The other four articles [27,29,30,32] analyzed multiple QTLs or genomic regions for one or more diseases.

For genes controlling tolerance to abiotic stresses, an averaged 2.2 to 1 ratio of distal to proximal chromosomal distribution was obtained from the 21 articles (Table 1). In comparison, the 11 articles on genes conferring resistance to biotic stresses resulted in a 3.3 to 1 ratio (Table 2). Therefore, 77% of genes conferring resistance to biotic stresses were located in the distal section of chromosomes, while 69% of those for abiotic-stress tolerance were distally located. This slightly higher number of genes in the distal section of chromosomes is advantageous for plant adaptation, because genetic variability generated from the high recombination rate in distal recombination hotspots enables plants to deal with environmental changes and new virulent pests.

Conflicts of Interest: The author declares no conflicts of interest.

References

1. Assmann, S.M. Natural variation in abiotic stress and climate change responses in *Arabidopsis*: Implications for twenty-first-century agriculture. *Int. J. Plant Sci.* **2013**, *174*, 3–26. [CrossRef]
2. Atkinson, N.J.; Urwin, P.E. The interaction of plant biotic and abiotic stresses: From genes to the field. *J. Exp. Bot.* **2012**, *63*, 3523–3543. [CrossRef] [PubMed]
3. Luo, Y.Y.; Wang, Q.Y.; Bai, R.; Li, R.X.; Chen, L.; Xu, Y.F.; Zhang, M.; Duan, D. The effect of transcription factor MYB14 on defense mechanisms in *Vitis quinquangularis*-Pingyi. *Int. J. Mol. Sci.* **2020**, *21*, 706. [CrossRef] [PubMed]
4. Wang, X.K.; Gong, X.; Cao, F.B.; Wang, Y.Z.; Zhang, G.P.; Wu, F.B. HvPAA1 encodes a P-Type ATPase, a novel gene for cadmium accumulation and tolerance in barley (*Hordeum vulgare* L.). *Int. J. Mol. Sci.* **2019**, *20*, 1732. [CrossRef] [PubMed]
5. Xu, Y.C.; Magwanga, R.O.; Cai, X.Y.; Zhou, Z.L.; Wang, X.X.; Wang, Y.H.; Zhang, Z.M.; Jin, D.S.; Guo, X.L.; Wei, Y.Y.; et al. Deep transcriptome analysis reveals reactive oxygen species (ROS) network evolution, response to abiotic stress, and regulation of fiber development in cotton. *Int. J. Mol. Sci.* **2019**, *20*, 1863. [CrossRef]
6. Ahmad, M.; Li, J.Z.; Yang, Q.S.; Jamil, W.; Teng, Y.W.; Bai, S.L. Phylogenetic, molecular, and functional characterization of PpyCBF proteins in Asian pears (*Pyrus pyrifolia*). *Int. J. Mol. Sci.* **2019**, *20*, 2074. [CrossRef] [PubMed]

7. Zhang, X.X.; Zhang, D.; Sun, W.; Wang, T.Z. The adaptive mechanism of plants to iron deficiency via iron uptake, transport, and homeostasis. *Int. J. Mol. Sci.* **2019**, *20*, 2424. [CrossRef]

8. Yong, Y.B.; Zhang, Y.; Lyu, Y.M. A stress-responsive NAC transcription factor from tiger lily (*LlNAC2*) interacts with *LlDREB1* and *LlZHFD4* and enhances various abiotic stress tolerance in Arabidopsis. *Int. J. Mol. Sci.* **2019**, *20*, 3225. [CrossRef]

9. Yong, Y.B.; Zhang, Y.; Lyu, Y.M. A MYB-related transcription factor from *Lilium lancifolium* L. (*LlMYB3*) is involved in anthocyanin biosynthesis pathway and enhances multiple abiotic stress tolerance in *Arabidopsis thaliana*. *Int. J. Mol. Sci.* **2019**, *20*, 3195. [CrossRef]

10. Yoon, M.Y.; Kim, M.Y.; Ha, J.; Lee, T.; Kim, K.D.; Lee, S.H. QTL analysis of resistance to high-intensity UV-B irradiation in soybean (*Glycine max* [L.] Merr.). *Int. J. Mol. Sci.* **2019**, *20*, 3287. [CrossRef]

11. Chen, H.; Li, H.H.; Lu, X.Q.; Chen, L.Z.; Liu, J.; Wu, H. Identification and expression analysis of GRAS transcription factors to elucidate candidate genes related to stolons, fruit ripening and abiotic stresses in woodland strawberry (*Fragaria vesca*). *Int. J. Mol. Sci.* **2019**, *20*, 4593. [CrossRef] [PubMed]

12. Katiyar, A.; Mudgil, Y. Arabidopsis NDL-AGB1 modules play role in abiotic stress and hormonal responses along with their specific functions. *Int. J. Mol. Sci.* **2019**, *20*, 4736. [CrossRef] [PubMed]

13. Noman, M.; Jameel, A.; Qiang, W.D.; Ahmad, N.; Liu, W.C.; Wang, F.W.; Li, H.Y. Overexpression of *GmCAMTA12* enhanced drought tolerance in Arabidopsis and soybean. *Int. J. Mol. Sci.* **2019**, *20*, 4849. [CrossRef] [PubMed]

14. Magwanga, R.O.; Kirungu, J.N.; Lu, P.; Cai, X.Y.; Zhou, Z.L.; Xu, Y.C.; Hou, Y.Q.; Agong, S.G.; Wang, K.B.; Liu, F. Map-based functional analysis of the *GhNLP* genes reveals their roles in enhancing tolerance to N-deficiency in cotton. *Int. J. Mol. Sci.* **2019**, *20*, 4953. [CrossRef] [PubMed]

15. Li, S.Z.; Miao, L.; Huang, B.; Gao, L.H.; He, C.X.; Yan, Y.; Wang, J.; Yu, X.C.; Li, Y.S. Genome-wide identification and characterization of cucumber BPC transcription factors and their responses to abiotic stresses and exogenous phytohormones. *Int. J. Mol. Sci.* **2019**, *20*, 5048. [CrossRef] [PubMed]

16. Li, W.; Wan, X.L.; Yu, J.Y.; Wang, K.L.; Zhang, J. Genome-wide identification, classification, and expression analysis of the *Hsf* gene family in carnation (*Dianthus caryophyllus*). *Int. J. Mol. Sci.* **2019**, *20*, 5233. [CrossRef] [PubMed]

17. Imran, M.; Shafiq, S.; Farooq, M.A.; Naeem, M.K.; Widemann, E.; Bakhsh, A.; Jensen, K.B.; Wang, R.R.-C. Comparative genome-wide analysis and expression profiling of histone acetyltransferase (HAT) gene family in response to hormonal applications, metal and abiotic stresses in cotton. *Int. J. Mol. Sci.* **2019**, *20*, 5311. [CrossRef]

18. Zou, L.F.; Yu, B.W.; Ma, X.L.; Cao, B.H.; Chen, G.J.; Chen, C.M.; Lei, J.J. Cloning and expression analysis of the *BocMBF1c* gene involved in heat tolerance in Chinese kale. *Int. J. Mol. Sci.* **2019**, *20*, 5637. [CrossRef]

19. Su, H.G.; Li, B.; Song, X.Y.; Ma, J.; Chen, J.; Zhou, Y.B.; Chen, M.; Min, D.H.; Xu, Z.S.; Ma, Y.Z. Genome-wide analysis of the DYW subgroup PPR gene family and identification of *GmPPR4* responses to drought stress. *Int. J. Mol. Sci.* **2019**, *20*, 5667. [CrossRef]

20. Zhou, S.; Zheng, W.J.; Liu, B.H.; Zheng, J.C.; Dong, F.S.; Liu, Z.F.; Wen, Z.Y.; Yang, F.; Wang, H.B.; Xu, Z.S.; et al. Characterizing the role of *TaWRKY13* in salt tolerance. *Int. J. Mol. Sci.* **2019**, *20*, 5712. [CrossRef]

21. Zhao, W.; Liu, H.F.; Zhang, L.; Hu, Z.Y.; Liu, J.; Hua, W.; Xu, S.M.; Liu, J. Genome-wide identification and characterization of *FBA* gene family in polyploid crop *Brassica napus*. *Int. J. Mol. Sci.* **2019**, *20*, 5749. [CrossRef] [PubMed]

22. Kamal, N.M.; Gorafi, Y.S.A.; Abdelrahman, M.; Abdellatef, E.; Tsujimoto, H. Stay-green trait: A prospective approach for yield potential, and drought and heats stress adaptation in globally important cereals. *Int. J. Mol. Sci.* **2019**, *20*, 5837. [CrossRef] [PubMed]

23. Wang, D.; Liu, Y.X.; Yu, Q.; Zhao, S.P.; Zhao, J.Y.; Ru, J.N.; Cao, X.Y.; Fang, Z.W.; Chen, J.; Zhou, Y.B.; et al. Functional analysis of the soybean *GmCDPK3* gene responding to drought and salt stresses. *Int. J. Mol. Sci.* **2019**, *20*, 5909. [CrossRef] [PubMed]

24. Wang, J.L.; Hu, T.H.; Wang, W.L.; Hu, H.J.; Wei, Q.Z.; Wei, X.C.; Bao, C.L. Bioinformatics analysis of the lipoxygenase gene family in radish (*Raphanus sativus*) and functional characterization in response to abiotic and biotic stresses. *Int. J. Mol. Sci.* **2019**, *20*, 6095. [CrossRef]

25. Zhong, C.; Li, Y.P.; Sun, S.L.; Duan, C.X.; Zhu, Z.D. Genetic mapping and molecular characterization of a broad-spectrum *Phytophthora sojae* resistance gene in Chinese soybean. *Int. J. Mol. Sci.* **2019**, *20*, 1809. [CrossRef]

26. Kang, D.Y.; Cheon, K.S.; Oh, J.; Oh, H.; Kim, S.L.; Kim, N.; Lee, E.; Choi, I.; Baek, J.; Kim, K.H.; et al. Rice genome resequencing reveals a major quantitative trait locus for resistance to Bakanae disease caused by *Fusarium fujikuroi*. *Int. J. Mol. Sci.* **2019**, *20*, 2598. [CrossRef]

27. Gupta, V.; He, X.Y.; Kumar, N.; Fuentes-Davila, G.; Sharma, R.K.; Dreisigacker, S.; Juliana, P.; Ataei, N.; Singh, P.K. Genome wide association study of Karnal bunt resistance in a wheat germplasm collection from Afghanistan. *Int. J. Mol. Sci.* **2019**, *20*, 3124. [CrossRef]
28. Zhang, Y.; Liu, Q.; Zhang, Y.X.; Chen, Y.Y.; Yu, N.; Cao, Y.R.; Zhan, X.D.; Cheng, S.H.; Cao, L.Y. LMM24 encodes receptor-like cytoplasmic kinase 109, which regulates cell death and defense responses in rice. *Int. J. Mol. Sci.* **2019**, *20*, 3243. [CrossRef]
29. Yang, M.Y.; Li, G.R.; Wan, H.S.; Li, L.P.; Li, J.; Yang, W.Y.; Pu, Z.J.; Yang, Z.J.; Yang, E.N. Identification of QTLs for stripe rust resistance in a recombinant inbred line population. *Int. J. Mol. Sci.* **2019**, *20*, 3410. [CrossRef]
30. Bhatta, M.; Morgounov, A.; Belamkar, V.; Wegulo, S.N.; Dababat, A.A.; Erginbas-Orakci, G.; El Bouhssini, M.; Gautam, P.; Poland, J.; Akci, N.; et al. Genome-wide association study for multiple biotic stress resistance in synthetic hexaploid wheat. *Int. J. Mol. Sci.* **2019**, *20*, 3667. [CrossRef]
31. Li, H.H.; Dong, Z.J.; Ma, C.; Tian, X.B.; Qi, Z.J.; Wu, N.; Friebe, B.; Xiang, Z.G.; Xia, Q.; Liu, W.X.; et al. Physical mapping of stem rust resistance gene *Sr52* from *Dasypyrum villosum* based on *ph1b*-Induced homoeologous recombination. *Int. J. Mol. Sci.* **2019**, *20*, 4887. [CrossRef] [PubMed]
32. Singh, P.K.; Singh, S.; Deng, Z.Y.; He, X.Y.; Kehel, Z.; Singh, R.P. Characterization of QTLs for seedling resistance to Tan Spot and Septoria Nodorum Blotch in the PBW343/Kenya Nyangumi wheat recombinant inbred lines population. *Int. J. Mol. Sci.* **2019**, *20*, 5432. [CrossRef] [PubMed]
33. Du, H.S.; Wen, C.L.; Zhang, X.F.; Xu, X.L.; Yang, J.J.; Chen, B.; Geng, S.S. Identification of a major QTL (qRRs-10.1) that confers resistance to *Ralstonia solanacearum* in pepper (*Capsicum annuum*) using SLAF-BSA and QTL mapping. *Int. J. Mol. Sci.* **2019**, *20*, 5887. [CrossRef] [PubMed]
34. Dong, Z.J.; Tian, X.B.; Ma, C.; Xia, Q.; Wang, B.L.; Chen, Q.F.; Sehgal, S.K.; Friebe, B.; Li, H.H.; Liu, W.X. Physical mapping of Pm57, a powdery mildew resistance gene derived from *Aegilops searsii*. *Int. J. Mol. Sci.* **2020**, *21*, 322. [CrossRef]
35. Wittkopp, P.J.; Kalay, G. Cis-regulatory elements: Molecular mechanisms and evolutionary processes underlying divergence. *Nat. Rev. Genet.* **2011**, *13*, 59–69. [CrossRef]

International Journal of
Molecular Sciences

Article

HvPAA1 Encodes a P-Type ATPase, a Novel Gene for Cadmium Accumulation and Tolerance in Barley (*Hordeum vulgare* L.)

Xin-Ke Wang [1], Xue Gong [2], Fangbin Cao [1], Yizhou Wang [1], Guoping Zhang [1] and Feibo Wu [1,3,*]

[1] Department of Agronomy, College of Agriculture and Biotechnology, Zijingang Campus, Zhejiang University, Hangzhou 310058, China; ke87795593@163.com (X.-K.W.); caofangbin@zju.edu.cn (F.C.); wangyizhou@zju.edu.cn (Y.W.); zhanggp@zju.edu.cn (G.Z.)

[2] School of Agriculture, Food and Wine, the University of Adelaide, Waite Campus, Adelaide 5064, Australia; redchian123@gmail.com

[3] Jiangsu Co-Innovation Center for Modern Production Technology of Grain Crops, Yangzhou University, Yangzhou 225009, China

* Correspondence: wufeibo@zju.edu.cn

Received: 1 March 2019; Accepted: 2 April 2019; Published: 8 April 2019

Abstract: The identification of gene(s) that are involved in Cd accumulation/tolerance is vital in developing crop cultivars with low Cd accumulation. We developed a doubled haploid (DH) population that was derived from a cross of Suyinmai 2 (Cd-sensitive) × Weisuobuzhi (Cd-tolerant) to conduct quantitative trait loci (QTL) mapping studies. We assessed chlorophyll content, traits that are associated with development, metal concentration, and antioxidative enzyme activity in DH population lines and parents under control and Cd stress conditions. A single QTL, designated as *qShCd7H*, was identified on chromosome 7H that was linked to shoot Cd concentration; *qShCd7H* explained 17% of the phenotypic variation. Comparative genomics, map-based cloning, and gene silencing were used in isolation, cloning, and functional characterization of the candidate gene. A novel gene *HvPAA1*, being related to shoot Cd concentration, was identified from *qShCd7H*. Sequence comparison indicated that *HvPAA1* carried seven domains with an N-glycosylation motif. *HvPAA1* is predominantly expressed in shoots. Subcellular localization verified that HvPAA1 is located in plasma membrane. The silencing of *HvPAA1* resulted in growth inhibition, greater Cd accumulation, and a significant decrease in Cd tolerance. We conclude *HvPAA1* is a novel plasma membrane-localized ATPase that contributes to Cd tolerance and accumulation in barley. The results provide us with new insights that may aid in the screening and development of Cd-tolerant and low-Cd-accumulation crops.

Keywords: barley (*Hordeum vulgare* L.); BSMV-VIGS; cadmium tolerance and accumulation; *HvPAA1*; quantitative trait loci (QTLs)

1. Introduction

Cadmium (Cd) is highly toxic to organisms, even at low concentrations. Cd has relatively high bioavailability in soil when compared with other heavy metals, and it is readily taken up by plants, which is a threat to human health via the food chain [1,2]. Therefore, it is imperative to identify the genotypes with less Cd uptake from soil and the relevant genes for developing crop cultivars with reduced Cd uptake and accumulation.

Quantitative trait loci (QTLs) that contribute to Cd accumulation have been reported in wheat [3], rice [4,5], maize [6], soybean [7], and barley [8]. A major Cd-specific QTL has been identified on the short arm of chromosome 7 in rice [4], and major QTLs that contribute to Cd tolerance during seedling stage found in *Arabidopsis* [9] and maize [6]. A major QTL that affects Cd accumulation in seeds was detected in soybean [7], and a major QTL that contributes to low Cd uptake was mapped to the 5BL chromosome in wheat [3,10]. Studies have been conducted with the goal of identifying Cd-specific genes. For example, Zhao et al. [11] performed a comparative transcriptome analysis and found that Cd stress induced the expression of general and specific genes in *Arabidopsis thaliana* roots. Villiers et al. [12] found that Cd-detoxifying proteins, such as phytochelatin synthase (PCS), antioxidative enzymes, and glutathione S-transferases (GST), were upregulated in plants that were subject to Cd stress. The loss of function of *CAX1* in *A. thaliana* resulted in higher Cd sensitivity in the presence of low concentrations of Cd [13]. Our previous study identified several transporter genes, including *HvZIP3* and *HvZIP8*, which are associated with low grain Cd accumulation [14]. Elevated expression of the heavy metal transporting ATPase 4 (HMA4), which is a P_{1B}-type ATPase, increased the Cd tolerance in plants that were subject to Cd stress resulting from the accumulation of low levels of cellular Cd in the cytoplasm [9]. Despite the identification of these genes, the molecular mechanisms underlying Cd tolerance and accumulation are still unclear at the present time.

P-type ATPases are transmembrane proteins that play a crucial role in the transport of a wide variety of cations across membranes and they are vital for ion homeostasis and heavy metal detoxification [15]. Plant P_{1B} ATPases have been classified that are expected to be involved in the transport of heavy metals. Multiple alignments between all of the P_{1B} ATPase sequences indicate that HMA1, HMA2, HMA3, and HMA4 in *Arabidopsis* likely to serve as $Zn^{2+}/Co^{2}+/Cd^{2+}/Pb^{2+}$ ATPases, while RAN1, PAA1, and HMA5 are candidate Cu^{2+}/Ag^{2+} ATPases [16]. However, *PAAs* are poorly characterized in barley, and their function, structure, and expression remain unclear.

In this study, we developed a doubled haploid (DH) barley population that was derived from a cross of Cd-tolerant and Cd-sensitive parents to identify the genes that are associated with Cd tolerance and accumulation, which were then characterized in terms of function. As a result, a novel p-type ATPase, *HvPAA1*, was identified and then cloned for the first time. Furthermore, the function of *HvPAA1* was investigated using BSMV-VIGS (barley stripe mosaic virus- virus induced gene silencing) system. The results showed that *HvPAA1* plays a vital role in the detoxification of Cd in barley.

2. Results

2.1. Transgressive Segregation was Observed for Most Investigated Traits

Table 1 summarizes the means, ranges, skewness, and kurtosis of the 17 assessed traits. In general, the values of the growth traits were reduced in the presence Cd stress relative to the control for the parental and DH lines. The coefficient variance (CV) of these parameters for the DH lines ranged between 4.2%-48.86% in the control and between 6.2%–119.43% in 10 μM Cd (Table 1). The Cd tolerance index for 11 traits revealed that Suyinmai 2 exhibited a greater reduction in growth than Weisuobuzhi (Table 1). In general, Cd decreased APX activity in shoots and roots and guaiacol peroxidase (POD) activity in shoots in both parental and DH lines. However, POD and catalase (CAT) activities in roots reflected differences in CTI between Suyinmai 2 and Weisuobuzhi. The kurtosis and the skewness for the majority of the traits examined was less than 1, with the exception of the shoot Zn concentration in the presence of 10 μM Cd, which indicated that the variations in these traits were consistent with a normal distribution.

2.2. Cd, Zn and Mn Concentrations

Concentrations of Cd, Zn, and Mn in shoots and roots were measured under both control and Cd stress conditions; however, no Cd was detected in the control. Shoot Cd concentration in the tolerant parent, Weisuobuzhi, was significantly lower than that in the sensitive parent, Suyinmai 2, in the presence of 10 μM Cd. After treatment with 10 μM Cd, a four-fold phenotypic variation in Sh_{Cd} was observed that ranged between 74.2–277.6 mg kg^{-1} with an average of 164.26 mg kg^{-1}. In contrast, root R_{Cd} was significantly higher in Weisuobuzhi than in Suyinmai 2. An approximately three-fold phenotypic variation in Cd concentration in the DH lines was observed in roots that ranged between 1161.0–2749.7 mg kg^{-1}, with an average of 1923.77 mg kg^{-1}. Cd accumulation was markedly higher in Suyinmai 2 than in Weisuobuzhi. Of note, most of the DH lines accumulated less Cd than the parental line (Weisuobuzhi; Figure 1).

Table 1. Phenotypes of the traits of growth and physiology and Cd tolerance indexes of the doubled haploid population from the cross between Suyinmai 2 and Weisuobuzhi.

	Parents		The Doubled Haploid Population (108 Lines)					
Traits	**Suyinmai 2**	**Weisuobuzhi**	**Skewness**	**Kurtosis**	**Min**	**Max**	**Mean**	**CV (%)**
Under control condition								
SDW	0.40	0.70 *	0.50	0.20	0.30	0.50	0.40	14.10
RDW	0.10	0.20 *	0.10	0.00	0.10	0.20	0.10	17.40
RL	22.90	24.80	0.60	2.50	18.50	31.10	22.60	8.50
PH	41.70	45.50	0.00	0.30	32.20	46.50	39.60	6.60
SPAD	32.00	30.30	0.80	4.20	27.60	33.60	29.80	4.20
L$_{POD}$	9.85	8.15	0.98	0.86	2.64	19.91	9.17	35.74
L$_{APX}$	3.62	24.43 *	3.64	21.79	5.04	63.30	15.07	48.86
L$_{CAT}$	0.46	0.37	0.06	−1.01	0.10	0.94	0.47	47.82
R$_{POD}$	32.59	25.56	2.16	10.95	17.90	89.45	36.37	24.44
R$_{APX}$	15.90	8.27 *	0.88	1.58	3.91	29.28	14.60	15.85
R$_{CAT}$	0.12	0.39 *	1.44	1.48	0.13	2.11	0.65	23.53
Sh$_{Mn}$	30.49	44.27 *	−0.10	−0.97	20.11	56.19	37.07	24.66
Sh$_{Zn}$	98.12	86.00 *	1.16	3.14	71.17	167.81	94.48	16.47
R$_{Mn}$	134.14	167.45 *	0.79	1.48	90.82	245.09	135.32	21.50
R$_{Zn}$	137.90	149.70	0.23	−0.12	84.99	180.93	143.31	12.81
Under 10 µM Cd condition								
SDW	0.20	0.40 *	0.60	1.10	0.20	0.40	0.30	15.30
RDW	0.10	0.10	0.20	0.20	0.00	0.10	0.10	16.30
RL	16.80	22.80 *	0.50	−0.20	16.10	23.40	19.20	8.80
PH	21.20	23.60	0.70	1.10	20.30	29.50	23.70	7.30
SPAD	27.10	29.00	−0.10	−0.50	22.80	30.10	26.70	6.20
L$_{POD}$	35.81	15.03 *	1.09	1.45	8.49	48.99	21.07	39.08
L$_{APX}$	12.79	47.77 *	0.25	1.62	6.01	65.02	20.68	60.48
L$_{CAT}$	0.30	0.12	2.65	6.79	0.10	2.78	0.47	119.43
R$_{POD}$	43.14	28.05 *	−0.35	0.047	14.35	67.61	42.89	16.35
R$_{APX}$	21.40	6.36 *	0.72	0.99	4.23	46.49	18.68	17.20
R$_{CAT}$	0.13	0.91 *	2.04	5.61	0.11	4.06	0.80	17.04
Sh$_{Cd}$	199.60	172.20 *	0.10	−0.40	98.32	213.70	156.70	15.80
Sh$_{Mn}$	30.80	45.90 *	0.50	1.70	24.20	65.10	42.10	16.20
Sh$_{Zn}$	95.20	77.30 *	2.60	16.40	61.50	173.00	86.70	15.20
R$_{Cd}$	1861.10	2235.00 *	0.10	−0.30	1161.00	2749.70	1921.50	18.20
R$_{Mn}$	127.40	188.30 *	0.80	1.10	73.20	294.40	156.10	24.90
R$_{Zn}$	158.80	169.00	0.60	2.30	76.30	248.20	157.90	15.70
Cd tolerance index (CTI)								
SDW	−36.3	−34.4	0.2	−0.7	−41.3	−0.2	−21.3	48.1
RDW	−41.5	−27.7	0.1	0.8	−43.7	−0.8	−21.9	47.8
RL	−26.5	−7.8	−0.1	0.0	−39.3	−29.0	−14.8	56.1
PH	−49.2	−48.1	0.5	−0.6	−48.9	−29.0	−40.1	10.4
SPAD	−15.5	−4.2	−0.2	0.7	−25.8	−0.4	−10.7	51.5
L$_{POD}$	263.4	84.3	0.7	−0.2	3.2	391.3	135.3	66.7
L$_{APX}$	253.9	95.6	4.7	27.9	39.1	341.8	167.6	79.2
L$_{CAT}$	−35.6	−66.5	1.2	1.6	−99.0	38.5	−59.5	51.1
R$_{POD}$	−42.5	9.8	1.8	3.5	2.6	187.9	42.5	87.4
R$_{APX}$	34.6	268.3	1.6	2.6	3.0	436.3	97.4	97.7
R$_{CAT}$	−50.4	16.5	0.4	−1.1	−99.8	−2.1	−57.0	52.3

Cd tolerance index (CTI) was calculated as CTI = (parameter under Cd stress − parameter under control)/parameter under the control) × 100%. * significant difference between the two parents Suyinmai 2 and Weisuobuzhi at the 0.05 level. SDW, shoot dry weight (g plant^{-1}); RDW, root dry weight (g plant^{-1}); RL, root length (cm); PH, plant height (cm); SPAD, SPAD value; L$_{POD}$, leaf POD activity (OD470 g^{-1} FW min^{-1}); L$_{APX}$, leaf APX activity (mmol g^{-1} FW min^{-1}); L$_{CAT}$, leaf CAT activity (mmol g^{-1} FW min^{-1}); R$_{POD}$, root POD activity; R$_{APX}$, root APX activity; R$_{CAT}$, root CAT activity. Sh$_{Cd}$, Sh$_{Mn}$, Sh$_{Zn}$, and R$_{Cd}$, R$_{Mn}$, R$_{Zn}$ represent Cd, Mn, Zn concentrations (mg kg^{-1} DW) in shoots and roots, respectively. No Cd in plants was detected under control condition. CV (%), the coefficient of variation was presented as the absolute value of the ratio of the standard deviation to the mean value.

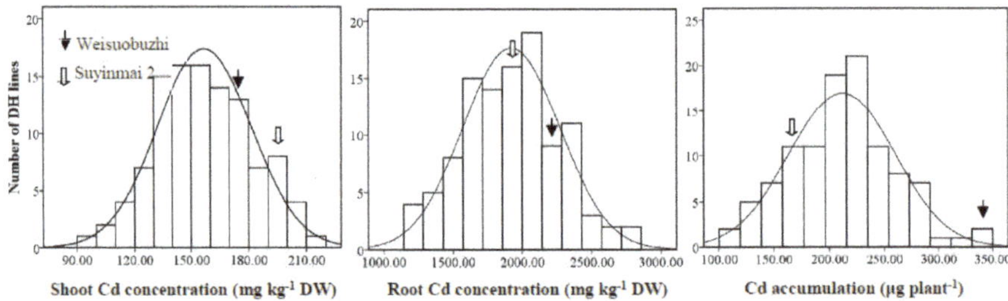

Figure 1. Frequency distribution of cadmium (Cd) concentration in shoots and roots, and Cd accumulation per plant. Parents and 108 DH lines were grown in a nutrient solution containing 10 µM $CdCl_2$ for 15 days. The data are represented as the mean of three replicates obtained from the spatial analysis. The solid arrow represents Weisuobuzhi (Cd-tolerant genotype) and the hollow arrow represents Suyinmai 2 (Cd-sensitive).

The Zn and Mn concentrations significantly differed between the parental and DH lines in both shoots and roots, and significantly higher concentrations were observed in roots than in shoots. The 10 µM Cd stress decreased the Zn concentration in shoots in Weisuobuzhi but not in Suyinmai 2. However, Cd stress resulted in an increase in Zn concentration in roots from both the parental and DH lines. In DH lines, an approximately 2.7-fold variation was observed for Sh_{Mn} and Sh_{Zn}, in comparison with 3.3-fold and 4.0-fold variation for R_{Mn} and R_{Zn}, respectively, in 10 µM Cd (Table 1).

2.3. Genetic Linkage Maps

Within the linkage map, a total of 1572 markers (1532 GBS and 40 SSR markers) were successfully assigned to the seven barley chromosomes. The total length of the map was 1134.5 cM, the length of each chromosome varied from 126.7 cM to 238.59 cM, and the number of markers per chromosome varied from 50 to 549. The average distance between two adjacent makers was 0.72 cM and it ranged between 0.45 to 2.77 cM (Figure S1).

2.4. Collocations of QTLs were Found for the Investigated Traits

Among the 17 traits that were investigated in the three conditions (control, Cd treatment and CTI), QTLs were identified for eight traits: the chloroform content presented as SPAD (Soil and Plant Analyzer Development) value, plant height, root length, leaf and root POD, leaf APX, Mn concentration in root, and Cd concentration in shoot. A total of 24 QTLs were detected from this study, 15 of which were from Cd stress; the majority of the QTLs were resolved within 10 cM (Table 2, Figure 2). The majorities of the QTLs identified during this study were for plant height and Mn concentration and it explained 10% to 17% of the phenotypic variation (Table 2). In terms of the additive effect, Weisuobuzhi's allele increased the plant height, root length, and leaf APX, regardless of QTL positions. Suyinmai 2's allele increased SPAD and POD values, regardless of plant tissues and Cd treatments. Some QTLs for plant height were mapped to the same positions under control and Cd stress on chromosome 5H and Weisuobuzhi's allele increased plant height. QTLs for leaf POD were also detected under both the control and Cd stress, but they were mapped to different chromosomal regions. The QTLs were detected on chromosomes 2H, 3H, and 5H for Mn concentration in root under Cd stress. These QTLs had a similar effect that explained around 13% of phenotypic variation. The Weisuobuzhi allele increase Mn concentration on chromosomes 2H and 3H, and the Suyinmai 2 allele increased the Mn concentration on chromosome 5H.

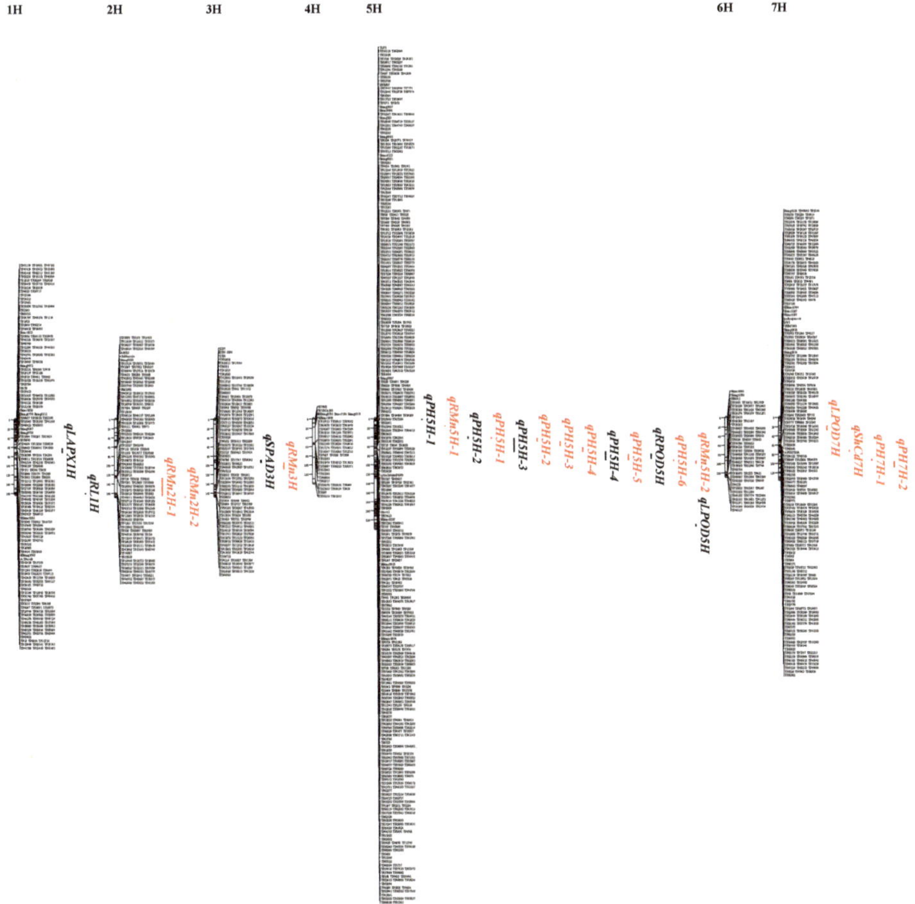

Figure 2. Quantitative trait loci (QTL) significantly associated with different parameters linked to cadmium (Cd) tolerance in the Suyinmai 2 × Weisuobuzhi doubled haploid (DH) population obtained using the multiple QTL model (MQM) mapping method. QTLs in black and red were detected in control and 10 μM Cd conditions, respectively. The one- and two-logarithm of odds (LOD) support intervals for each of the QTLs, as calculated in Mapchart, are displayed on the right side of each linkage group.

Table 2. Quantitative trait loci (QTLs) identified under control and 10 μM Cd condition.

Trait	QTL [a]	Chr. [b]	Position (cM) [c]	Interval (cM) [d]	Marker [e]	LOD [f]	R^2 [g] (%)	Add [h]
			Under control condition					
SPAD value	qSPAD3H	3H	91.88	5.88	TP41927	2.82	11.30	0.42
Plant height (cm)	qPH5H-1	5H	27.20	3.90	Bmag323	2.83	11.40	−0.90
	qPH5H-2	5H	42.46	2.81	TP33838	3.26	13.00	−0.95
	qPH5H-3	5H	64.57	29.50	TP63873	4.31	16.80	−1.06
	qPH5H-4	5H	84.09	17.50	TP21896	3.77	14.90	−1.03
Root length (cm)	qRL1H	1H	161.32	1.29	TP13	2.87	11.50	−0.66
Leaf POD (OD470 g^{-1} FW min^{-1})	qLPOD5H	5H	233.32	3.91	TP52094	2.86	11.50	1.12
Root POD (OD470 g^{-1} FW min^{-1})	qRPOD5H	5H	84.09	4.00	TP21896	2.73	10.90	3.05
Leaf APX (mmol g^{-1} FW min^{-1})	qLAPX1H	1H	70.90	9.20	EBmac0501	2.59	10.40	−0.54

Table 2. *Cont.*

Trait	QTL [a]	Chr. [b]	Position (cM) [c]	Interval (cM) [d]	Marker [e]	LOD [f]	R^2 [g] (%)	Add [h]
			Under 10 µM Cd condition					
Plant height (cm)	*qPH5H-1*	5H	37.24	4.24	TP20247	2.76	11.10	−0.58
	qPH5H-2	5H	42.46	2.81	TP33838	3.87	15.20	−0.68
	qPH5H-3	5H	43.76	5.97	TP220	3.55	14.00	−0.66
	qPH5H-4	5H	55.47	11.94	TP53164	3.55	14.10	−0.62
	qPH5H-5	5H	64.57	8.23	TP13622	3.76	14.80	−0.66
	qPH5H-6	5H	88.09	13.27	TP13479	3.80	15.00	−0.67
	qPH7H-1	7H	92.02	11.78	TP26119	3.00	12.00	−0.60
	qPH7H-2	7H	100.78	5.13	TP35770	2.81	11.50	−0.59
Leaf POD (OD470 g^{-1} FW min^{-1})	*qLPOD7H*	7H	22.50	8.80	Bmac0187	3.00	12.00	3.06
Root Mn (µg g^{-1} DW)	*qRMn2H-1*	2H	159.46	39.00	TP48811	3.19	12.70	−13.88
	qRMn2H-2	2H	168.58	2.60	TP10446	3.09	12.30	−13.65
	qRMn3H	3H	107.69	54.17	TP60843	3.46	13.70	−14.36
	qRMn5H-1	5H	15.90	1.29	HVM07	2.73	11.00	12.86
	qRMn5H-2	5H	94.60	5.11	TP27265	3.07	12.30	13.60
Shoot Cd (µg g^{-1} DW)	*qShCd7H*	7H	72.39	2.60	TP30771	4.28	17.00	10.67

[a] QTLs are named by trait and chromosome; [b] The chromosome on which the QTL is mapped; [c] The position (in cM) of the QTL on the chromosome; [d] The genetic distance between two markers; [e] The marker at the maximum logarithm of odds (LOD) score for QTL; [f] The LOD score at the QTL; [g] The percentage of explained variance of the marker at QTL; [h] Additive effect, the estimated additive effect.

2.5. A single QTL was Identified on Chromosome 7H for Shoot Cd Concentration

A single QTL for Cd concentration in shoots was identified on chromosome 7H in Cd stress conditions that had the greatest effects among all QTLs. This QTL explained 17% of phenotypic variation (Figure 3); we designated this QTL *qShCd7H*. The Suyinmai 2 allele increased the Cd concentration in shoots by 11% in comparison to the Weisuobuzhi allele (Table 2). The *qShCd7H* QTL was mapped to a 2.6 cM region between the GBS markers TP18054 and TP11089 (Table S2). The 2.6 cM region included approximately 60 genes that were annotated in the Barley Map and IBCS genome database (Table S4). As a result, AK355848.1 was cloned, sequenced, and functionally analyzed in the parents Suyinmai 2 and Weisuobuzhi, as well as the check variety Zhenong 8.

2.6. Cloning, Sequencing and Functional Analysis of A Candidate Gene Designated HvPAA1

The full length of the cloned cDNA was 2412 bp and its encoded protein contains 803 amino acids and it has a predicted molecular weight of 85.1 KDa and a theoretical pI of 6.35 (Figure S2). The parents shared the same amino acid and CDS sequence (Figure S4). There are seven important domains, including four transmembrane domains, one HMA domain, one E1-E2_ATPase domain, and one HAD domain, and, most importantly, an N-glycosylation motif (Figure 4). The predicted sequence of *HvPAA1* between the transcription start site and the termination site showed lengthy and diverse introns (Figure S5). Furthermore, the alignment result (Figure S6) showed that Weisuobuzhi and Suyinmai 2 contain four single nucleotide polymorphism sites (SNPs), all of which are located in intron. Polygenetic analysis showed that the 28 genes were clustered into three major groups; the cloned gene was a member of the group that was composed of ATP1, ATP2, HMA4, HMA5, PAA1, and PAA2. The cloned sequences of Suyinmai 2, Weisuobuzhi, and Zhenong 8 were mostly similar to those from *OsPAA1* (*Oryza sativa* L.) and *ZmPAA1* (*Zea mays* L.) and they showed 89.7%, 88.1%, and 68.3% similarity to those from *OsPAA1*, *ZmPAA1*, and *AtPAA1*, respectively (Figure S3). As a result, we designated this gene as barley *HvPAA1*.

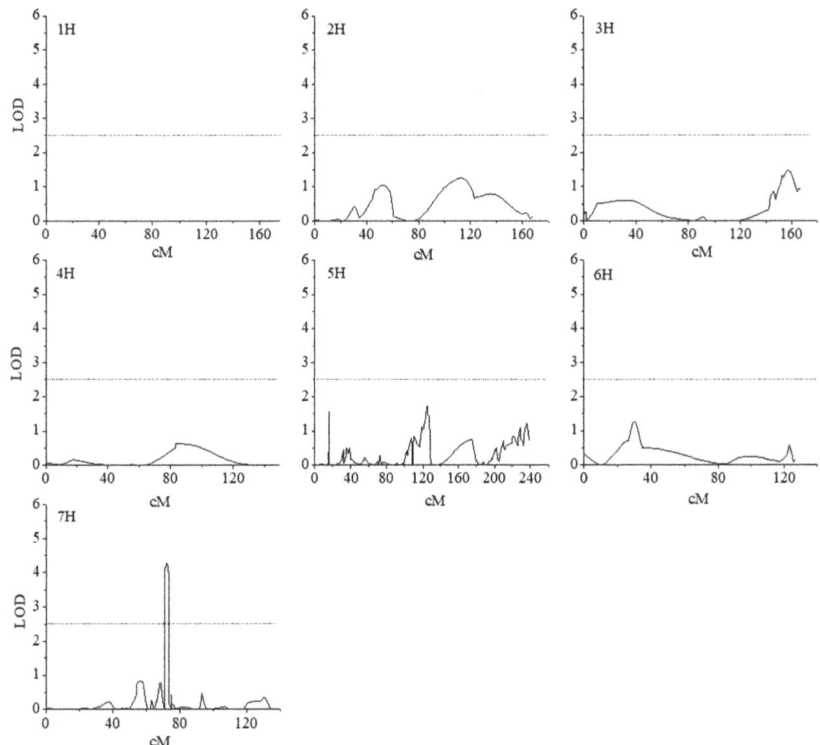

Figure 3. Logarithm of odds (LOD) score profiles for Cd tolerance in terms of shoot Cd concentration. The map positions are indicated along the abscissa. The LOD scores are indicated along the y-axis. The dashed lines represent the LOD score threshold (2.5) at a 0.05 error level for QTL detection.

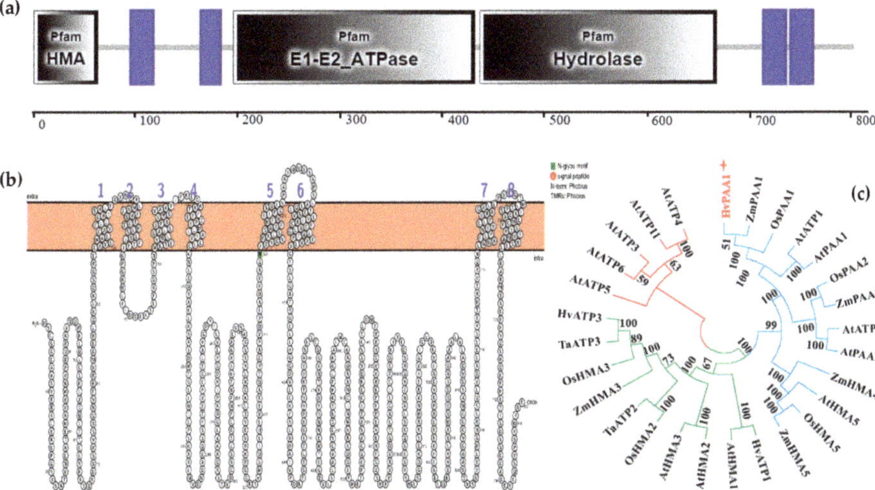

Figure 4. Protein structure and phylogenetic analysis of HvPAA1. (**a**) Amino acid sequence analysis based on the SMATR database. Some features and domains are not shown due to overlap with other annotations. (**b**) Amino acid sequence analysis based on the Protter database. (**c**) Phylogenetic tree constructed in MEGA7 using the neighbor-joining algorithm analysis method.

2.7. Expression Pattern and Subcellular Localization of HvPAA1

Exposure to 5 μM Cd significantly promoted the relative levels of *HvPAA1* transcript in leaves and stems, and the leaves were found to have higher levels than stems (Figure 5a). The relative level of *HvPAA1* transcript peaked at 3 h after Cd exposure and then sharply decreased to the same level that was observed at 0 h after 3 d and 7 d (Figure 5b). To determine the subcellular localization of HvPAA1, we transiently expressed HvPAA1–GFP (green fluoresence protein) in a nuclear marker-containing *Nicotiana benthamiana* line that expressed red fluorescent protein (RFP) that was fused to histone 2B (H2B) [17]. Based on the GFP signal, the HvPAA1 protein was localized to plasma membrane (Figure 5c).

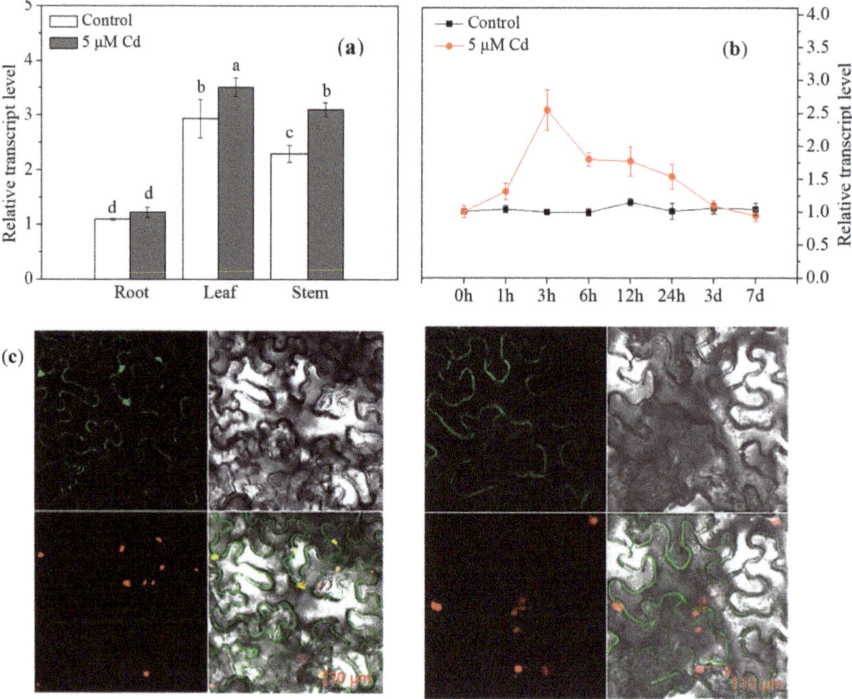

Figure 5. Tissue expression and time-dependent expression patterns of HvPAA1 and the subcellular localization of HvPAA1. (**a,b**) qRT-PCR analysis of the relative transcript levels of HvPAA1 in different tissues and at different times in Zhenong 8. (**c**) Transient expression of GFP and HvPAA1-sGFP fusion protein in tobacco. Microscopic image on left shows GFP; the image on right shows the HvPAA1-sGFP fusion protein. The results represent an average (±SE) from three independent experiments. Different letters indicate significant differences among the treatments within three sampling data according to Duncan's multiple range test with $p < 0.05$. The error bars represent the SE values.

2.8. Phenotypic Analysis and Cd Concentration in BSMV-Inoculated Lines

To clarify the function of *HvPAA1*, a BSMV-VIGS-based gene silencing technology was used to determine whether the silencing of this gene at the mRNA level affected Cd concentration in barley. Cd exposure increased the transcript level of *HvPAA1* in both mock- and BSMV:HvPAA1-inoculated Zhenong 8 plants at 1 dpi when compared with the controls (Figure 6b). However, mock-inoculated seedlings that were treated with Cd exhibited a similar expression pattern to that of the mock-inoculated seedlings at 5 and 10 dpi, which was similar to the results of the time-dependent experiment. The *HvPAA1* transcript level in leaves of BSMV:HvPAA1-inoculated seedlings was decreased by

80.4%, 76.9%, and 58.7% when compared to mock-inoculated seedlings at 1, 5, and 10 dpi, respectively. The *HvPAA1* transcript level was significantly increased in leaves from BSMV:HvPAA1-inoculated seedlings under Cd stress as compared with BSMV:HvPAA1-inoculated seedlings at 1 and 10 dpi; at 5 dpi, no significant differences were found between them.

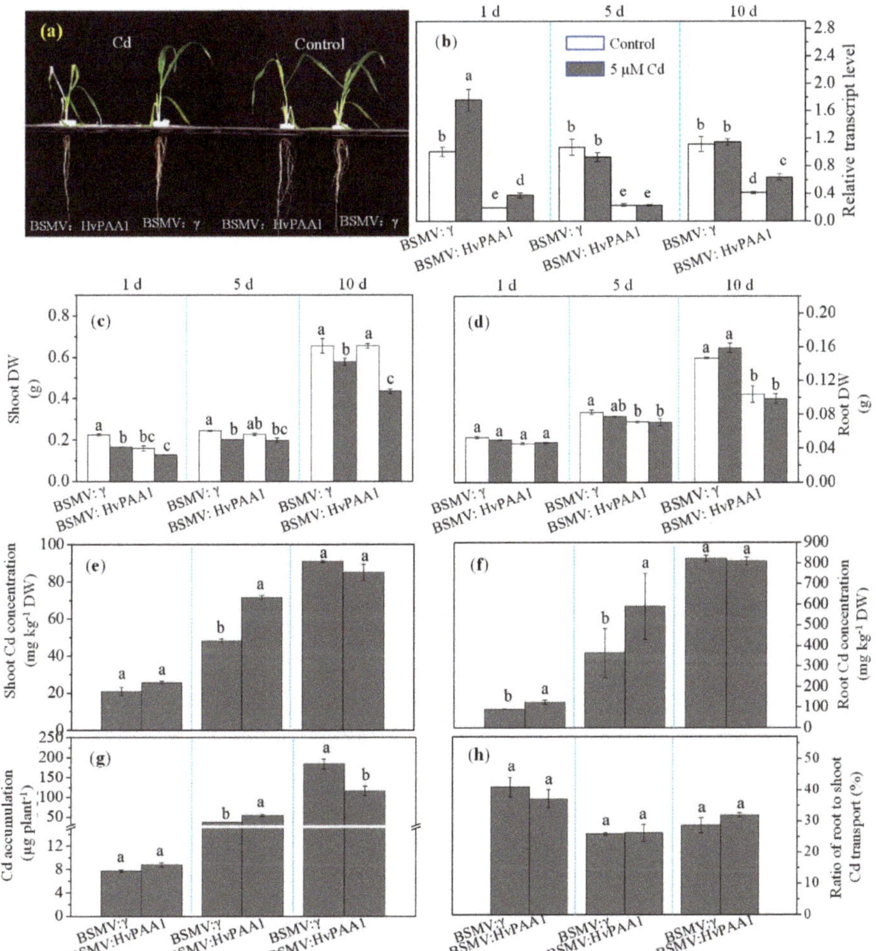

Figure 6. Functional assessment of *HvPAA1* in Zhenong 8 using BSMV-VIGS (barley stripe mosaic virus- virus induced gene silencing). (**a**) Morphology of mock- or BSMV: HvPAA1-inoculated Zhenong 8 seedlings after 10 days of exposure to 5 μM Cd. Scale bars = 6 cm. (**b**) RT-PCR analysis of the relative transcript levels of *HvPAA1* in Zhenong 8 leaves. Different letters indicate significant differences among the treatments within 3 sampling data (1, 5, 10 d after the initiation of Cd stress) according to Duncan's multiple range test with $p < 0.05$. (**c,d**) Dry weights of mock- and BSMV:HvPAA1-inoculated Zhenong 8 plants. (**e,f**) Cd concentrations of mock- and BSMV:HvPAA1-inoculated Zhenong 8 plants grown in 5 μM Cd. (**g**) Cd accumulation in mock- and BSMV:HvPAA1-inoculated Zhenong 8 plants grown in 5 μM Cd. (**h**) root-to-shoot Cd transport ratio. Error bars represent the SE values. Different letters indicate significant differences between control and Cd treatment, between each sampling date, and between the mock- and BSMV:HvPAA1 inoculated seedlings according to Duncan's multiple range tests with $p < 0.05$.

As shown in Figure 6a,c, the mock-inoculated seedlings in control showed no significant differences when compared to that in Cd stress at each of the three sampling times. In the absence of Cd, the mock- and BSMV:HvPAA1-inoculated seedlings exhibited similar root dry weights at 1 dpi and similar shoot dry weights at 5 and 10 dpi. However, the presence of 5 μM Cd resulted in decreases in the shoot dry weight, especially at 10 dpi, in BSMV:HvPAA1-inoculated plants when compared with the BSMV:HvPAA1-inoculated plants. Nevertheless, no differences were found in RDW of BSMV:HvPAA1-inoculated seedlings under Cd and control conditions.

As for Cd concentration in mock- and BSMV:HvPAA1-inoculated seedlings that were exposed to 5 μM Cd (Figure 6e,f), shoot Cd concentration showed no significant differences between the mock- and BSMV:HvPAA1-inoculated seedlings at 1 dpi, while root Cd concentration in BSMV:HvPAA1-inoculated seedlings was notably higher than in mock-inoculated seedlings. At 5 dpi, the silencing of *HvPAA1* significantly increased both the shoot and root Cd concentrations by 48.3% and 62.2%, respectively, as compared with the mock-inoculated seedlings. Meanwhile, BSMV:HvPAA1-inoculated seedlings accumulated more Cd than the mock-inoculated seedlings, which was in contrast to the result that was observed at 10 dpi (Figure 6g). This may have been due to the higher dry weight of mock-inoculated seedlings at 10 dpi. As shown in Figure 6h, the root-to-shoot transport ratio showed no significant difference between mock- and BSMV:HvPAA1-inoculated seedlings at 1, 5, and 10 dpi. These results proved that *HvPAA1* is closely associated with barley Cd tolerance and accumulation, but it has little effect on the root-to-shoot transport ratio.

3. Discussion

Cd is harmful to human health if it is incorporated into the food chain. The development of low Cd crops has become an urgent need and it relies heavily on knowledge regarding Cd uptake, accumulation, translocation, and exudation and excavating related genes. Such an understanding is vital in sustainable crop production, so as to alleviate health risks that are associated with exposure to Cd-contented food.

3.1. QTLs for Traits Investigated

In the present study, all of the detected 24 QTLs for the traits investigated were located on chromosome 1H, 2H, 3H, 5H, and 7H, respectively, which have a high density of markers. QTLs for SPAD value, root length, and leaf APX activity were only detected under control condition. As to QTLs for plant height, a total of 12 QTLs were detected, four of them were detected under the control condition located on chromosome 5H, the rest of them were detected under Cd stress, and they distributed on chromosome 5H and 7H. Furthermore, *qPH5H-2* under the control condition was overlapped with *qPH5H-2* under Cd stress, suggesting that this QTL region contains important genetic messages that are associated with plant height. This overlapped QTL was on chromosome 5H and it resolved in 2.81 cM, with the negative additive effect indicating Weisuobuzhi's allele increased plant height. For leaf POD activity, a single QTL was detected under the control and Cd condition, respectively, but the location was differed from each other. However, *qRPOD5H*, which was associated with root POD activity under the control condition, have the same chromosome location but different position as *qLPOD5H* that was detected under control condition. Five QTLs that were associated with root Mn concentration under Cd stress were found on chromosome 2H, 3H, and 5H, which can explain 62% of the phenotypic variation in total.

3.2. A major QTL qShCd7H was Detected for Cd Concentration in Shoot

Cadmium concentration could not be detected during our control experiment. A significantly larger phenotypic difference in terms of the shoot Cd concentration was observed in the parental and DH lines in the presence of Cd stress, which allowed for a single QTL, *qShCd7H*, to be identified. The *qShCd7H* QTL was associated with Cd concentration in the presence of 10 μM Cd stress, and the Suyinmai 2 allele was shown to increase the Cd concentration in shoots. During the phenotypic analysis,

Suyinmai 2 exhibited a significantly higher shoot concentration than Weisuobuzhi, which suggested that the Cd concentration during seedling stage might be controlled by a major gene with dominant effects. The influence of a single QTL with dominant effects on Cd concentration in the presence of Cd stress has been well documented in other crops [3,4,7,10]. It is worth noting that, in the presence of low amounts (0.5 μM) of Cd stress, a marker-trait association with shoot Cd concentration was reported for all barley chromosomes [8]. The *qShCd7H* QTL was mostly likely the same QTL as the QTL that was anchored by marker 3140-491, and it was detected from a soil culture, which demonstrates that *qShCd7H* could be detected in different genetic backgrounds and test environments. This is very important for the marker-assisted selection of Cd accumulation during a short-term experiment. The GBS markers, TP18054 and TP11089, which were used to identify *qShCd7H*, could be converted into local laboratory-based PCR markers.

3.3. A Novel Gene HvPAA1 is the Candidate Gene of qShCd7H for Cd Tolerance

A novel gene, *HvPAA1*, related to shoot Cd concentration, was identified from *qShCd7H*. The polygenetic analysis showed that *HvPAA1* was clustered into the *PAA1* group. The amino acid sequence of *HvPAA1* was mostly related to the *OsPAA1* (*Oryza sativa* L.). Sequence comparison between two parents demonstrated that they possess the same CDS sequence but have four SNPs in introns. Sequence analysis indicated that *HvPAA1* carried seven domains with an N-glycosylation motif, which is required for the regulation of enzymatic activity according to Migocka [18]. PAA1, also known as HMA6, is a Cu-ATPase that belongs to the heavy-metal ATPase (P_{1B}-ATPases, HMAs) subfamily of cation-transporting P-ATPases. Both Cu- and Cd stress had a significant impact on the gene expression levels of *OsHMA6*, *ZmHMA6*, and *SbHMA6*, but the influence of Cd stress on gene expression appeared to be greater than that of Cu-induced stress [19]. The *Arabidopsis AtHMA6* is primarily involved in Cu transport [20,21], suggesting that the function of *HMA6* may be relevant to its sequence and the effects of external stress during different growth stages and/or in different plant organs. Other studies have attempted to reveal the relationship between Cd and Cu, and they have found that their relationship is dependent on the relevant genotypes, concentrations, and tissues [22–24]. The role of *PAA1* (*HMA6*) in Cd tolerance has not been reported in barley previously. Thus, we consider *HvPAA1* to be the candidate gene for *qShCd7H* with regard to Cd tolerance in barley. Further studies are needed to verify the relationship between *HvPAA1* and Cu transportation in the future.

Regarding the physical position of *HvPAA1*, the EnsemblPlants and IPK database were used to investigate the position of *HvPAA1* in the barley physical map. The results indicated that this gene is located at 167 Mbp (million base pairs) on the chromosome 7H short arm approximately, while the whole chromosome size is about 680 Mbp. Mascher et al. [25] revealed a chromosome structure of the barley genome, in which has information regarding the position of centromere. According to their results, we find centromere on 7H is near to 350 Mbp. As a result, *HvPAA1* is far away from the centromere or telomere.

3.4. HvPAA1, Localized on Plasma Membrane, Contributes to Cd Tolerance and Accumulation in Barley

The tissue-specific expression pattern indicated that *HvPAA1* was mainly expressed in shoots in the absence of Cd, but that expression was induced by Cd in leaves and stems, which indicated that the expression of *HvPAA1* was upregulated by 5 μM Cd in shoots. This result was in agreement with those that were observed for the *OsHMA6* gene in rice, which showed that Cd stress significantly increased *OsHMA6* expression in shoots and stems, but not in roots [19]. The upregulation of *HvPAA1* peaked at 3 h and then dropped to the same level that was observed at 0 h at 3 and 7 dpi, which demonstrated that the Cd expression of *HvPAA1* was time-dependent and Cd only induced increases in the relative transcript level at 1 dpi. Our subcellular localization experiment showed that *HvPAA1* is located at the plasma membrane, similarly to *OsHMA6* in rice, *ZmHMA6* in maize, and *SbHMA6* in sorghum [19], but differently from *AtHMA6* in *Arabidopsis thaliana*, which is localized to the chloroplast. The differences in locations and functions of *PAA1* may be due to functional differences in different plant species.

The characterization of gene function via overexpression or silencing in transgenic plants plays an important role in producing plants with novel traits, but this approach is time consuming [26]. Virus-induced gene silencing (VIGS) is a powerful functional genomics tool that can be used for rapid gene function analysis [27]. The expression of *HvPAA1* was significantly decreased after BSMV:HvPAA1-inoculation at 1, 5, and 10 dpi; a rise in the transcript level was detected at 10 dpi, although it was significantly lower than the increase that was observed in mock-inoculated plants. Cadmium stress increased *HvPAA1* expression level in BSMV:HvPAA1-inoculatd plants at 1 and 10 dpi. The silencing of *HvPAA1* resulted in an obvious decrease in the root and shoot dry weights at 10 dpi, which indicates that *HvPAA1* was responsible for Cd tolerance in Zhenong 8 (Figure 6). Meanwhile, the silencing of *HvPAA1* also resulted in increased Cd concentration in shoots/roots at 5 dpi. The results from the three sampling times differed from each other, which may be the result of the different expression level of *HvPAA1* at 1, 5, and 10 dpi. These results all indicated that *HvPAA1* is involved in Cd tolerance in barley.

4. Materials and Methods

4.1. Plant Materials and Experimental Design

A population consisting of 108 DH lines for QTL analysis was derived from a cross between the Suyinmai 2 (Cd-sensitive) and Weisuobuzhi (Cd-tolerant) [28] via a microspore culture.

The hydroponic experiments were carried out at Zijingang Campus of Zhejiang University, Hangzhou, China, according to previously described methods [14,29]. At the second leaf stage (10 days old), uniform plants were selected and then transplanted into 35 L containers filled with 30 L of basal nutrient solution (BNS) [29]. The container was covered with a polystyrene plate with 64 evenly spaced holes (two plants per hole) and then placed in a net house. On the seventh day after transplantation, 10 μM Cd (CdCl$_2$) was added to the appropriate containers (10 μM Cd), and BNS (control) was added to the remaining containers. The experiment was laid out with a split-plot design with Cd treatment as main plot and DH line as subplot; there were three replicates for each treatment. The nutrient solution was continuously aerated with pumps and renewed every four days. The plants were collected after 15 days of treatment.

4.2. Phenotyping of the DH population Lines and Parents

After 15 days of treatment, the SPAD values (chlorophyll meter readings) were determined in the top second fully expanded leaves using a chlorophyll meter (Minolta SPAD-502). Plants were collected and separated into roots and shoots after the plant height (PH) and root length (RL) were measured. Shoot dry weight (SDW) and root dry weight (RDW) was measured, after which the shoots and roots were powdered and ashed at 550 °C for 12 h. The ash was digested with 5 mL 30% HNO$_3$ and then diluted with deionized water. Cd, Mn, and Zn concentrations in shoots (Sh$_{Cd}$, Sh$_{Mn}$, and Sh$_{Zn}$) and roots (R$_{Cd}$, R$_{Mn}$, and R$_{Zn}$) were determined using flame atomic absorption spectrometry (AA6300; Shimadzu, Tokyo, Japan).

4.3. Antioxidative Enzyme Activities

The fully expanded leaves and root samples were immediately frozen in liquid nitrogen and stored at −80 °C for the determination of guaiacol peroxidase (POD), catalase (CAT), and ascorbic acid oxidase (APX) activity in the shoots (Sh$_{POD}$, Sh$_{CAT}$, and Sh$_{APX}$) and roots (R$_{POD}$, R$_{CAT}$, and R$_{APX}$), according to Chen et al. [30] and Wu et al. [29].

Cd tolerance indexes (CTI) were calculated as a reduced (−)/increased (+) percentage of the control. i.e., CTI = [(parameter under stress − parameter under control)/parameter under control] × 100%].

4.4. Genotyping, Linkage Map Construction and QTL Mapping

Genomic DNA from each of the parental and DH lines was extracted from fresh leaves that were obtained from the seedling stage using the DNeasy Plant Mini Kit (QIAGEN, Hilden, Germany) according to the manufacturer's protocol. The DH and parental lines were genotyped using sequencing technology (GBS). The GBS library was prepared using protocols that were described by Elshire et al. [31] and Poland et al. [32]. The DNA samples were digested using PstI and MspI for complexity reduction and they were then barcoded and multiplexed. Each GBS library, which contained 96 DNA samples (96-plex), was run on a single lane of an Illumina HiSeq2000 for sequencing. The GBS raw data were analyzed using the Universal Network Enabled Analysis Kit (UNEAK) pipeline in TASSEL [33]. Heterozygous markers and those that were missing more than 20% of their data were removed. Subsequent to this procedure, 1532 GBS markers were used for linkage map construction. The population was also genotyped using simple sequence repeat (SSR) markers, which were amplified using a polymerase chain reaction (PCR) method (Tables S2 and S3). The software package JoinMap 4.0 constructed the Weisuobuzhi × Suyinmai 2 linkage map [34]. The QTL intervals for all measured data were defined within the linkage map using MapQTL 5.0 software [35]. All data were expressed as the mean of three replicates that were conducted for the purposes of the QTL analysis. Interval mapping (IM) analysis was first performed to determine the potential regions of the QTLs, after which the markers that were close to the detected QTLs (identified by IM mapping) were selected as cofactors to be used to test the multiple QTL model (MQM) [36]. A logarithm of odds (LOD) threshold > 2.5 ($\alpha = 0.05$) was the criterion that was used to confirm the presence of a QTL during the IM and MQM analysis. The detected QTLs were named according to the barley QTL nomenclature described by Szucs et al. [37].

4.5. Identification, Amplification, and Sequencing of Candidate Genes

Based on the QTL mapping study, a major QTL that was responsible for the Cd concentration in shoots was located, and the candidate genes that contribute to this QTL were identified. The positions of the flanking markers were aligned with the physical barley map using BLASTN tag sequences that were mapped against the sequence of the IBSC barley Morex cultivar in RefSeq v1.0 (http://webblast.ipk-gatersleben.de/registration/) and they are shown in Table S3. All of the genes within the QTL interval were extracted from the Barleymap database (http://floresta.eead.csic.es/barleymap). Based on gene annotation and literature studies, the AK355848.1 gene, which was annotated as "copper-transporting ATPase 1", was compared against the National Center for Biotechnology Information (NCBI) nucleotide database while using BLAST. Very little information about AK355848.1 in barley was available based on a literature search. EnsemblPlants (http://plants.ensembl.org/index.html) and IPK (https://webblast.ipk-gatersleben.de/barley_ibsc/) database investigated the chromosome position of AK355848. The sequence of AK355848.1 was then cloned in the parents Suyinmai 2 and Weisuobuzhi and it was designated as the barley *HvPAA1*. Another Cd-tolerance cultivar Zhenong 8 was introduced to this study as a check variety with high-grain-Cd-accumulation [14]. Total RNA was extracted from Suyinmai 2, Weisuobuzhi, and Zhenong 8, according to the instructions that were included with the TaKaRa MiniBST Plant RNA Extraction Kit (Takara, Japan). Supplementary Table S1 shows the primer pairs. The PCR products were connected into the pMD18-T vector (Takara, Japan) and sequenced. The SMART (http://smart.embl-heidelberg.de/) and Protter (http://wlab.ethz.ch/protter/start/) databases were used to predict the amino acid sequence of the cloned gene. The amino acid composition, CDS, protein sequence, and structure of the cloned gene were compared with those of 14 other members of the heavy-metal ATPase (P_{1B}-ATPases, HMAs) family. A phylogenetic tree was constructed in MEGA7 using the neighbor-joining algorithm analysis method that is based on the protein sequences of 28 ATPs. The full-length sequences of *HvPAA1* from two parents, including the exons and introns, were also analyzed.

4.6. Expression Patterns, Tissue and Subcellular Localization of HvPAA1

Hydroponic experiments were repeated for quantitative real-time PCR (qRT-PCR) expression analysis. The Zhenong 8 seedlings were exposed to 5 μM Cd (CdCl$_2$) for up to seven days to determine the time-dependent expression of *HvPAA1* in barley. Tissue-specific expression analysis of *HvPAA1* was conducted using roots, stems, and leaves from two-leaf-stage seedlings. Supplementary Table S1 shows the primer sequences that were used.

To investigate subcellular localization, the coding regions of *HvPAA1* were directly amplified from the full length cDNA using primers 35SGFP-HvPAA1-F and 35SGFP-HvPAA1-R (Supplementary Table S1) and then connected into the pCAMBIA 1300 vector, which contains a CaMV 35S promoter:green fluorescent protein (35S:GFP) cassette, to create a HvPAA1-GFP fusion protein. The 35S:HvPAA1-GFP fusion construct was then introduced into the *Agrobacterium tumefaciens* GV3101 strain. *Nicotiana benthamiana* containing a red nuclear histone 2B marker (H2B) [17] was infected by the *Agrobacterium*-mediated system, as previously described [38].

4.7. BSMV Inoculation and Measurement of the Relative Transcript Level, Cd Tolerance and Metal Concentration

Total RNA was extracted from Zhenong 8. A PrimeScriptTM RT reagent kit (Takara, Japan) was used to synthesis the first-strand cDNA. A 286 bp cDNA fragment from the barley phytoene desaturase gene (*HvPDS*) and a 239 bp cDNA fragment from the *HvPAA1* gene containing *NheI* sites were amplified while using oligonucleotide primers (Table S1). These two fragments were inserted in reverse into the RNAγ cDNA strand to create two cDNA clones, BSMV:HvPDS, and BSMV:HvPAA1, to ensure further gene silencing in the Zhenong 8.

Inoculation was performed according to the protocol that was described by He et al. [27]. Seven days after inoculation, the transcript levels of *HvPDS* and *HvPAA1* were verified using qRT-PCR, after which the plants were treated with 5 μM CdCl$_2$ for 10 d. Six replicates were used for each treatment.

The third leaves, which were collected at 1, 5, and 10 dpi, were used for the analysis of the transcript levels. Only plants that exhibited viral infection symptoms were chosen to serve as samples. The RNA extraction, cDNA synthesis, and qRT-PCR were performed, as described above.

The Cd tolerance of BSMV:HvPAA1 was evaluated. The roots and shoots of infected plants were separately harvested at 1, 5, and 10 dpi. SDW, RDW, Cd concentration/accumulation, and Cd root-to-shoot transport ratio were measured, as described above. All of the treatments used three replicates.

5. Conclusions

In conclusion, we developed a new mapping population that was based on the cross of Suyinmai 2 and Weisuobuzhi that utilizes 1532 GBS and 40 SSR markers. Seventeen traits were assessed in control and Cd stress conditions, and 24 QTLs were identified. A major QTL, *qShCd7H*, which is associated with Cd concentration, was mapped to chromosome 7H and it was found to explain 17% of phenotypic variation. The *qShCd7H* QTL was resolved to a 2.6 cM region between the GBS markers TP18054 and TP11089, which resulted in the cloning of a copper-transporting ATPase 1 (AK355848.1). We designated the cloned gene *HvPAA1*, which was found to be expressed mainly in shoots and localized to plasma membrane, and we observed that the silencing of *HvPAA1* leads to a decrease in biomass and increase in Cd concentration. All of this evidence confirmed that *HvPAA1* is a candidate gene for Cd concentration in *qShCd7H*. The results provide a molecular basis for understanding Cd accumulation in barley that will contribute to the development of molecular markers that can be used to quickly select low Cd accumulation varieties and to facilitate the cultivation of low-Cd barley varieties.

Supplementary Materials: Supplementary materials can be found at http://www.mdpi.com/1422-0067/20/7/1732/s1.

Author Contributions: Conceptualization, F.W. and X.G.; Methodology, F.W., G.Z. and X.G.; Formal analysis, X.-K.W. and G.X.; Investigation, X.-K.W., F.C. and Y.W; Writing—original draft preparation, X.-K.W.; Writing—review and editing, X.-K.W., F.W., G.X., Y.W. and G.Z.; Supervision, F.W.

Funding: The project was supported by the Key Research Foundation of Science and Technology Department of Zhejiang Province of China (Grant No. 2016C02050-9-7).

Conflicts of Interest: The authors declare no conflict of interest.

References

1. Clemens, S.; Ma, J.F. Toxic heavy metal and metalloid accumulation in crop plants and foods. *Annu. Rev. Plant Biol.* **2016**, *67*, 489. [CrossRef] [PubMed]
2. Wagnerm, G.J.; Donald, L.S. Accumulation of cadmium in crop plants and its consequences to human health. *Adv. Agron.* **1993**, *51*, 173–212.
3. Abuhammad, W.A.; Mamidi, S.; Kumar, A.; Pirseyedi, S.; Manthey, F.A.; Kianian, S.F.; Alamri, M.S.; Mergoum, M.; Elias, E.M. Identification and validation of a major cadmium accumulation locus and closely associated SNP markers in North Dakota durum wheat cultivars. *Mol. Breed.* **2016**, *36*, 112. [CrossRef]
4. Ishikawa, S.; Abe, T.; Kuramata, M.; Yamaguchi, M.; Ando, T.; Yamamoto, T.; Yano, M. A major quantitative trait locus for increasing cadmium-specific concentration in rice grain is located on the short arm of chromosome 7. *J. Exp. Bot.* **2010**, *61*, 923. [CrossRef] [PubMed]
5. Xue, D.; Chen, M.; Zhang, G.P. Mapping of QTLs associated with cadmium tolerance and accumulation during seedling stage in rice (*Oryza sativa* L.). *Euphytica* **2009**, *165*, 587–596. [CrossRef]
6. Zhao, X.W.; Luo, L.X.; Cao, Y.H.; Liu, Y.J.; Li, Y.H.; Wu, W.M.; Lan, Y.Z.; Jiang, Y.W.; Gao, S.B.; Zhang, Z.M.; et al. Genome-wide association analysis and QTL mapping reveal the genetic control of cadmium accumulation in maize leaf. *BMC Genom.* **2018**, *19*, 91. [CrossRef] [PubMed]
7. Benitez, E.R.; Hajika, M.; Yamada, T.; Takahashi, K.; Oki, N.; Yamada, N.; Nakamura, T.; Kanamaru, K. A major QTL controlling seed cadmium accumulation in soybean. *Crop Sci.* **2010**, *50*, 1728–1734. [CrossRef]
8. Wu, D.Z.; Sato, K.; Ma, J.F. Genome-wide association mapping of cadmium accumulation in different organs of barley. *New Phytol.* **2015**, *208*, 817–829. [CrossRef]
9. Courbot, M.; Willems, G.; Motte, P.; Arvidsson, S.; Roosens, N.; Laprade, P.S.; Verbruggen, N. A major quantitative trait locus for cadmium tolerance in *Arabidopsis halleri* colocalizes with HMA4, a gene encoding a heavy metal ATPase. *Plant Physiol.* **2007**, *144*, 1052–1065. [CrossRef] [PubMed]
10. Abbasabadi, A.O.; Kumar, A.; Pirseyedi, S.; Salsman, E.; Dobrydina, M.; Poudel, R.S.; AbuHammad, W.A.; Chao, S.M.; Faris, D.J.; Elias, E.M. Identification and validation of a new source of low grain cadmium accumulation in durum wheat. *G3-Genes Genomes Genet.* **2018**, *8*, 923–932.
11. Zhao, S.D.; Koyama, H. Comparative transcriptomic characterization of aluminium, sodium chloride, cadmium and copper rhizotoxicities in *Arabidopsis thaliana*. *BMC Plant Biol.* **2009**, *9*, 32. [CrossRef]
12. Villiers, F.; Ducruix, C.; Hugouvieux, V.; Jarno, N.; Ezan, E.; Garin, J.; Junot, C.; Bourguignon, J. Investigating the plant response to cadmium exposure by proteomic and metabolomic approaches. *Proteomics* **2011**, *11*, 1650–1663. [CrossRef]
13. Baliardini, C.; Meyer, C.L.; Salis, P.; Laprade, P.S.; Verbruggen, N. CATION EXCHANGER1 cosegregates with cadmium tolerance in the metal hyperaccumulator *Arabidopsis halleri* and plays a role in limiting oxidative stress in *Arabidopsis* spp. *Plant Physiol.* **2015**, *169*, 549–559. [CrossRef]
14. Sun, H.Y.; Chen, Z.H.; Chen, F.; Xie, L.P.; Zhang, G.P.; Vincze, E.; Wu, F.B. DNA microarray revealed and RNAi plants confirmed key genes conferring low Cd accumulation in barley grains. *BMC Plant Biol.* **2015**, *15*, 259. [CrossRef]
15. Mayerhofer, H.; Sautron, E.; Rolland, N.; Catty, P.; Berny, D.S.; Peyroula, E.P.; Ravaud, S. Structural insights into the nucleotide binding domains of the P_{1B}-type ATPases HMA6 and HMA8 from *Arabidopsis thaliana*. *PLoS ONE* **2016**, *11*, e0165666. [CrossRef]
16. Axelsen, K.B.; Palmgren, M.G. Inventory of the superfamily of P-type ion pumps in *Arabidopsis*. *Plant Physiol.* **2001**, *126*, 696–706. [CrossRef]
17. Martin, K.; Kopperud, K.; Chakrabarty, R.; Banerjee, R.; Brooks, R.; Goodin, M.M. Transient expression in *Nicotiana benthamiana* fluorescent marker lines provides enhanced definition of protein localization, movement and interactions in planta. *Plant J.* **2009**, *59*, 150–162. [CrossRef]
18. Migocka, M. Copper-transporting ATPases: The evolutionarily conserved machineries for balancing copper in living systems. *IUBMB Life* **2015**, *67*, 737–745. [CrossRef]

19. E, Z.G.; Li, T.T.; Chen, C.; Wang, L. Genome-wide survey and expression analysis of P_{1B}-ATPases in rice, maize and sorghum. *Rice Sci.* **2018**, *25*, 208–217.

20. Boutigny, S.; Sautron, E.; Finazzi, G.; Rivasseau, C.; Barrand, A.F.; Pilon, M.; Rolland, N.; Berny, D.S. HMA1 and PAA1, two chloroplast-envelope P_{1B}-ATPases, play distinct roles in chloroplast copper homeostasis. *J. Exp. Bot.* **2004**, *65*, 1529–1540. [CrossRef]

21. Shikanai, T.; Moule, P.M.; Munekage, Y.; Niyogi, K.K.; Pilon, M. PAA1, a P-type ATPase of *Arabidopsis*, functions in copper transport in chloroplasts. *Plant Cell* **2003**, *15*, 1333–1346. [CrossRef]

22. Wu, F.B.; Zhang, G.P.; Yu, J. Interaction of cadmium and four microelements for uptake and translocation in different barley genotypes. *Commun. Soil Sci. Plant* **2003**, *34*, 2003–2020. [CrossRef]

23. Wu, H.X.; Wu, F.B.; Zhang, G.P.; Bachir, D.M.L. Effect of cadmium on uptake and translocation of three microelements in cotton. *J. Plant Nutr.* **2004**, *27*, 2019–2032. [CrossRef]

24. Przedpelska-Wasowicz, E.; Polatajko, A.; Wierzbicka, M. The influence of cadmium stress on the content of mineral nutrients and metal-binding proteins in *Arabidopsis helleri*. *Water Air Soil Pollut.* **2012**, *223*, 5445–5458. [CrossRef]

25. Mascher, M.; Gundlach, H.; Himmelbach, A.; Beier, S.; Twardziok, S.O.; Wicker, T.; Radchuk, V.; Dockter, C.; Hedley, P.E.; Russell, J.; et al. A chromosome conformation capture ordered sequence of the barley genome. *Nature* **2017**, *544*, 427–433. [CrossRef]

26. Cheuk, A.; Houde, M. A new Barley Stripe Mosaic Virus allows large protein overexpression for rapid function analysis. *Plant Physiol.* **2018**, *176*, 1919–1931. [CrossRef]

27. He, X.Y.; Zeng, J.B.; Cao, F.B.; Ahmed, I.M.; Zhang, G.P.; Vincze, E.; Wu, F.B. *HvEXPB7*, a novel β-expansin gene revealed by the root hair transcriptome of Tibetan wild barley, improves root hair growth under drought stress. *J. Exp. Bot.* **2015**, *66*, 7405–7419. [CrossRef]

28. Chen, F.; Wang, F.; Zhang, G.P.; Wu, F.B. Identification of barley varieties tolerant to cadmium toxicity. *Biol. Trace Elem. Res.* **2008**, *121*, 171–179. [CrossRef]

29. Wu, F.B.; Zhang, G.P.; Dominy, P. Four barley genotypes respond differently to cadmium, lipid peroxidation and activities of antioxidant capacity. *Environ. Exp. Bot.* **2003**, *50*, 67–78. [CrossRef]

30. Chen, F.; Wang, F.; Wu, F.B.; Mao, W.H.; Zhang, G.P.; Zhou, M.X. Modulation of exogenous glutathione in antioxidant defense system against Cd stress in the two barley genotypes differing in Cd tolerance. *Plant Physiol. Biochem.* **2010**, *48*, 663–672. [CrossRef]

31. Elshire, R.J.; Glaubitz, J.C.; Sun, Q.; Poland, J.A.; Kawamoto, K.; Buckler, E.S.; Mitchell, S.E. A robust, simple genotyping-by-sequencing (GBS) approach for high diversity species. *PLoS ONE* **2011**, *6*, e19379. [CrossRef]

32. Poland, J.A.; Brown, P.J.; Sorrells, M.E.; Jannink, J.J. Development of high-density genetic maps for barley and wheat using a novel two-enzyme genotyping-by-sequencing approach. *PLoS ONE* **2012**, *7*, e32253. [CrossRef]

33. Lu, F.; Lipka, A.E.; Glaubitz, J.; Elshire, R.; Cherney, J.H.; Casler, M.D.; Buckler, E.S.; Costich, D.E. Switchgrass genomic diversity, ploidy, and evolution: Novel insights from a network-based SNP discovery protocol. *PLoS Genet.* **2013**, *9*, e1003215. [CrossRef]

34. Li, H.; Vaillancourt, R.; Mendham, N.; Zhou, M. Comparative mapping of quantitative trait loci associated with waterlogging tolerance in barley (*Hordeum vulgare* L.). *BMC Genom.* **2008**, *9*, 401. [CrossRef]

35. Van, O.J. *MapQTL®5. Software for the Mapping of Quantitative Trait Loci in Experimental Populations*; Kyazma BV: Wageningen, The Netherlands, 2004.

36. Ye, L.Z.; Huang, Y.Q.; Hu, H.L.; Dai, F.; Zhang, G.P. Identification of QTLs associated with haze active proteins in barley. *Euphytica* **2015**, *205*, 799–807. [CrossRef]

37. Szucs, P.; Blake, V.C.; Bhat, P.R.; Chao, S.A.M.; Close, T.J.; Marcos, C.A.; Muehlbauer, G.J.; Ramsay, L.; Waugh, R.; Hayes, P.M. An integrated resource for barley linkage map and malting quality QTL alignment. *Plant Genome* **2009**, *2*, 134–140. [CrossRef]

38. Xu, Y.X.; Zhang, S.N.; Guo, H.P.; Wang, S.K.; Xu, L.G.; Li, C.Y.; Qian, Q.; Chen, F.; Geisler, M.; Qi, Y.H. *OsABCB14* functions in auxin transport and iron homeostasis in rice (*Oryza sativa* L.). *Plant J.* **2014**, *79*, 106–117. [CrossRef]

International Journal of
Molecular Sciences

Article

Deep Transcriptome Analysis Reveals Reactive Oxygen Species (ROS) Network Evolution, Response to Abiotic Stress, and Regulation of Fiber Development in Cotton

Yanchao Xu [1,†], Richard Odongo Magwanga [1,2,†], Xiaoyan Cai [1,†], Zhongli Zhou [1,†], Xingxing Wang [1], Yuhong Wang [1], Zhenmei Zhang [1], Dingsha Jin [1], Xinlei Guo [1], Yangyang Wei [1,3], Zhenqing Li [1], Kunbo Wang [1,*] and Fang Liu [1,*]

[1] State Key Laboratory of Cotton Biology, Institute of Cotton Research, Chinese Academy of Agricultural Sciences (ICR, CAAS), Anyang 455000, China; xuyanchao2016@163.com (Y.X.); magwangarichard@yahoo.com (R.O.M.); caixy@cricaas.com.cn (X.C.); zhouzl@cricaas.com.cn (Z.Z.); wangxx@cricaas.com.cn (X.W.); wangyh2525377@163.com (Y.W.); xyxuezhihua@163.com (Z.Z.); jindingsha@163.com (D.J.); guoxlcaas@163.com (X.G.); weiyangyang511@126.com (Y.W.); lizhenqing2019@163.com (Z.L.)

[2] Jaramogi Oginga Odinga University of Science and Technology (JOOUST), School of Biological and Physical Sciences (SPBS), P.O BOX 210-40600, Bondo 210-40600, Kenya

[3] Biological and Food Engineering, Anyang Institute of Technology, Anyang 455000, China

[*] Correspondence: wkbcri@cricaas.com.cn (K.W.); liufang@caas.cn (F.L.); Tel.: +86-13569081788 (K.W.); +86-13949507902 (F.L.)

[†] These authors contributed equally to this work.

Received: 28 February 2019; Accepted: 8 April 2019; Published: 15 April 2019

Abstract: Reactive oxygen species (ROS) are important molecules in the plant, which are involved in many biological processes, including fiber development and adaptation to abiotic stress in cotton. We carried out transcription analysis to determine the evolution of the *ROS* genes and analyzed their expression levels in various tissues of cotton plant under abiotic stress conditions. There were 515, 260, and 261 genes of ROS network that were identified in *Gossypium hirsutum* (AD$_1$ genome), *G. arboreum* (A genome), and *G. raimondii* (D genome), respectively. The *ROS* network genes were found to be distributed in all the cotton chromosomes, but with a tendency of aggregating on either the lower or upper arms of the chromosomes. Moreover, all the cotton *ROS* network genes were grouped into 17 families as per the phylogenetic tress analysis. A total of 243 gene pairs were orthologous in *G. arboreum* and *G. raimondii*. There were 240 gene pairs that were orthologous in *G. arboreum*, *G. raimondii*, and *G. hirsutum*. The synonymous substitution value (K_s) peaks of orthologous gene pairs between the At subgenome and the A progenitor genome (*G. arboreum*), D subgenome and D progenitor genome (*G. raimondii*) were 0.004 and 0.015, respectively. The K_s peaks of ROS network orthologous gene pairs between the two progenitor genomes (A and D genomes) and two subgenomes (At and Dt subgenome) were 0.045. The majority of K_a/K_s value of orthologous gene pairs between the A, D genomes and two subgenomes of TM-1 were lower than 1.0. RNA seq. analysis and RT-qPCR validation, showed that, CSD1,2,3,5,6; FSD1,2; MSD1,2; APX3,11; FRO5.6; and RBOH6 played a major role in fiber development while CSD1, APX1, APX2, MDAR1, GPX4-6-7, FER2, RBOH6, RBOH11, and FRO5 were integral for enhancing salt stress in cotton. ROS network-mediated signal pathway enhances the mechanism of fiber development and regulation of abiotic stress in Gossypium. This study will enhance the understanding of ROS network and form the basic foundation in exploring the mechanism of ROS network-involving the fiber development and regulation of abiotic stress in cotton.

Keywords: *ROS* genes; transcription analysis; abiotic stress; fiber development

---

1. Introduction

Gossypium is one of the largest and most widely distributed genus with more than 50 species [1,2]. Cotton has a long evolution and domestication history. *Gossypium* diverged from its relatives approximately 10–15 million years ago (MYA), and evolved into three major diploid lineages, the New World clade, the African-Asian clade and the Australian clade, about 5–10 MYA [3]. The allopolyploid cottons emerged about 1–2 MYA due to an intergenomic hybridization event between A and D genomes [3,4], and were domesticated at least 4000 to 5000 years ago [5]. Both *G. arboreum* and *G. hirsutum* have a natural long spinable fiber, although, *G. raimondii* also generates fiber, its fibers are short and not spinable [6]. During human domestication history of cotton, the central focus was on the fiber quality and quantity [7]. The changes in the environment has resulted in to paradigm shift not only to explore on fiber and fiber attributes, but also on the survival strategies developed by plants, whereas plants try to adapt to various environmental conditions in the course of their evolutionary history. Thus, positive selections of genes were different during evolution and human domestication. Increasing number of genome sequencing and resequencing, mRNA sequencing and phenotype assessment of cotton [8] provided an important resource for studying potential biological mechanism related to abiotic stress tolerance in cotton.

Reactive oxygen species (ROS), including singlet oxygen ($^1O_2$), superoxide anion (O_2^-), hydrogen peroxide (H_2O_2), and hydroxyl radical ($\bullet OH$), produced by many of the metabolic pathways in aerobic cells, most likely appeared on Earth together with the first atmospheric oxygen molecules about 2.4–3.8 billion years ago and have been a constant companion of aerobic life [9]. Many biological processes are controlled by ROS, which include but are not limited to programmed cell death, biotic and abiotic stress responses, and plant growth and development [10]. However, ROS have been thought to be toxic and unwanted molecules for a long time, but now are widely known as indispensable molecule for plant lives [11]. ROS level is regulated by an intricate network composed of at least 152 genes in Arabidopsis [12–15], but ROS regulation, mechanisms and the genes involved is not clear in cotton. The ROS regulation network consists of genes which functions as inductors of enzymes which do scavenge on the ROS molecules thereby reducing their toxicity to the plant cell [12].

Cotton fiber development and abiotic stress responses are controlled by many molecules and complex pathways, including ROS [16]. The evolution of long spinnable fibers in cotton is accompanied by novel expression of genes assisting in the regulation of ROS levels [17]. A calcium sensor, *GhCaM7*, might modulate ROS production and act as a molecular link between Ca^{2+} and ROS signal pathways in early fiber development stages [18]. ROS-producing genes, including gene encoding NADPH oxidases/respiratory burst oxidase homologue (RBOH), regulated cell expansion through activation of Ca^{2+} channels, and are also involved in ROS regulation in the plant cell [19,20]. The detection of these genes found to have an integral role in the ROS regulation and fiber development, clearly shows that ROS-scavenging pathway may also be important in the regulation of fiber development. A cytosolic ascorbate peroxidase (APX) gene, *GhAPX1AT/DT*, regulating optimal H_2O_2 level, is the key mechanism regulating fiber elongation and deregulation of the increase in H_2O_2 results in shorter fiber by initiating secondary cell wall-related gene expression [21]. An ascorbate oxidase (AO) gene, *GhAO1*, controls cotton fiber elongation via the auxin-mediated signaling pathway [22]. A NAC transcription factor, *GhFSN1*, acts as a positive regulator in controlling second cell wall formation of cotton fiber [23]. Many of the abiotic stresses cause oxidative stress and genes of ROS network also involve in abiotic stress responses [24]. Furthermore, transcriptome analysis of an upland cotton, *G. hirsutum*, has provided the basis to elucidate the various mechanisms adopted by plants in dealing with salt stress [25].

In this study, genes in ROS network were identified in cotton. Evolutionary analysis proved that ROS network was a conservative and stable system during *Gossypium* evolution and domestication.

Genes of the ROS network that are involved in fiber development and abiotic stress were detected in this research work. A simple ROS signal pathway was used to describe the potential mechanism of fiber development and regulation of abiotic stress in *Gossypium*. In summary, this study will enhance the understanding of ROS network and pave the path in studying the mechanism of ROS network-involving fiber development and regulation of abiotic stress in cotton.

2. Results

2.1. ROS Network Genes Identification in Gossypium Species

A total of 515, 260, and 261 ROS candidate proteins were identified in *G. hirsutum*, *G. arboreum*, and *G. raimondii*, respectively (Table S1). The proportions of the ROS proteins obtained in the three cotton species, supports the evolution theory of the tetraploid cotton. The allotetraploid cotton, *G. hirsutum* harbored twice the number individual numbers of the ROS proteins obtained for either of the two diploid cotton species. In all the ROS proteins obtained from the three cotton species, they were further subdivided into different classes or sub families, in which the highest ROS proteins were observed for peroxidase (PER) with 279 proteins encoding the ROS genes, accounting for 26.9% of all the ROS proteins in the three cotton species. These proteins included 36 superoxide dismutase (SOD), 47 ascorbate peroxidase (APX), 24 monodehydroascorbate reductase (MDAR), 20 dehydroascorbate reductase (DHAR), 12 glutathione reductase (GR), 17 catalase (CAT), 32 glutathione peroxidase (GPX), 21 ferritin (FER), 58 respiratory burst oxidase homologue (RBOH), 27 NADPH-like oxidase (FRO), 19 Ubiquinol oxidase (AOX), 37 peroxiredoxin (PrxR), 224 thioredoxin reductase (Trx), 183 Glutaredoxin (GLR) and 279 peroxidase (PER). Beside NADPH oxidases and NADPH oxidase-like (ROS producer), all of the genes encode ROS-scavenging enzymes. There are 152 genes, so far identified in Arabidopsis. Moreover, in genome-wide identification of the *ROS* genes in Mizuna plants, a total of 32 *ROS* genes were identified out of the 22,428 transcript, in which 8258 and 14,170 transcripts were up- and down-regulated, respectively [26]. The high number of the *ROS* genes in *Gossypium*, suggested a more complex ROS network in relation to fiber development and enhancing abiotic stress tolerance.

2.2. Chromosome Mapping of the ROS Network Gene Family

The *ROS* genes were mapped and found to be widely distributed across the entire cotton chromosomes. In tetraploid cotton, *G. hirsutum*, the *ROS* genes were mapped in all the 26 chromosomes, in which the highest gene loci density were observed in chromosome At05 and its homeolog chromosome with 46 (18.11%) and 38 (14.28%) *ROS* genes, respectively, while the lowest gene loci density in chromosome At04 with only 10 *ROS* genes accounting for 1.94% of all the *ROS* genes identified in *G. hirsutum* (Figure 1A). The gene distribution among the diploid cotton species, *G. arboreum* and *G. raimondii* were found to be present in each of the 13 chromosomes, in which the highest gene loci were detected in chromosome A10 (35, 12.36%) and D09 (40, 13.89%) in *G. arboreum* and *G. raimondii*, respectively (Figure 1B,C). Moreover, the genes mapping in the individual chromosomes showed the tendencies of the genes aggregating on either upper or the lower arms of the chromosomes, similar observation have been previously observed in the distribution of the *LEA* genes in cotton, with high gene loci densities being observed in chromosomes A10 and D09 [27]. The detection of the high number of the *ROS* genes within the chromosomes showed that the three sets of chromosomes harbor significant genes with diverse roles in enhancing abiotic stress tolerance in plants. This result suggests that chromosome D09, chromosome Dt05 and its homeolog chromosome At05 play a core role in the evolution and dynamic balance in the *ROS* genes balance in cotton.

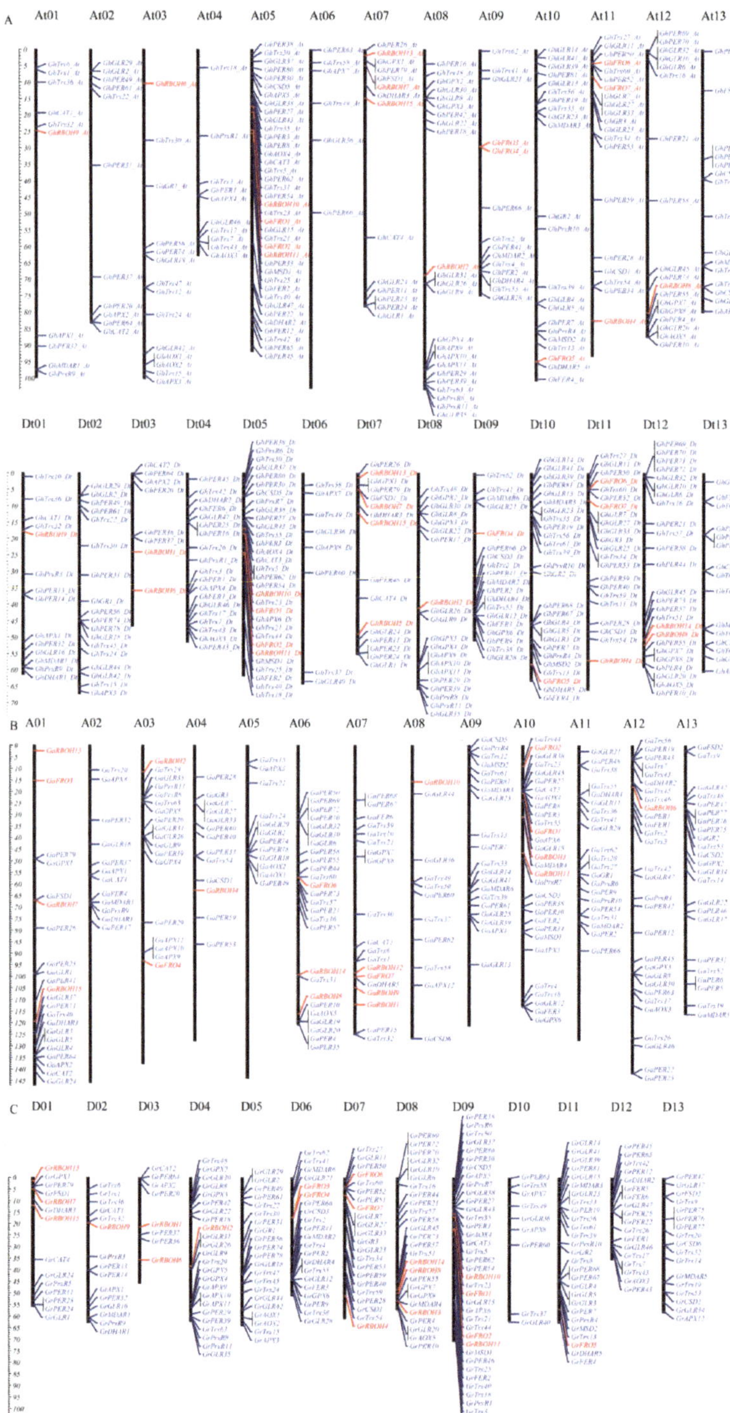

Figure 1. Distribution of reactive oxygen species (ROS) network genes in chromosomes of *G. hirsutum*, *G. arboreum* and *G. raimondii*. (**A**) Distribution of ROS network genes in chromosomes of *G. arboreum*. (**B**) Distribution of ROS network genes in chromosomes of *raimondii*. (**C**) Distribution of ROS network genes in chromosomes of *G. hirsutum*.

2.3. Phylogenetic Tree, Gene Duplication, and Synteny Analysis of the ROS Network Gene Family

In evaluating the evolution of the *ROS* network genes, the entire *ROS* network genes were aligned and phylogenetic tree constructed in order to understand the pattern and nature of their evolution. All the genes were classified into seventeen (17) different clades and each clade was made up of members of a particular family. The 17 clades were superoxide dismutase (SOD), ascorbate peroxidase (APX), monodehydroascorbate reductase (MDAR), dehydroascorbate reductase (DHAR), glutathione reductase (GR), catalase (CAT), glutathione peroxidase (GPX), ferritin (FER), respiratory burst oxidase homologue (RBOH), NADPH-like oxidase (FRO), Ubiquinol oxidase (AOX), peroxiredoxin (PrxR), thioredoxin reductase (Trx), Glutaredoxin (GLR), and peroxidase (PER) (Figure 2A). A total of 683 *ROS* network genes were found to have undergone duplications, in which 631 genes through segmental duplication while only 52 genes were tandemly duplicated (Table S1). The results were in agreement to previous findings in which a number stress responsive genes evolution were majorly governed by segmental type of gene duplication, for instance the *LEA* genes [27]. Gene duplication is considered as the main processes which do lead to the expansion of various gene families in organisms, and it is a prevailing feature in plant genomes [28,29]. Generally, a high number of gene copies in a gene family are maintained through large-scale segmental duplication or small-scale tandem duplication during evolution.

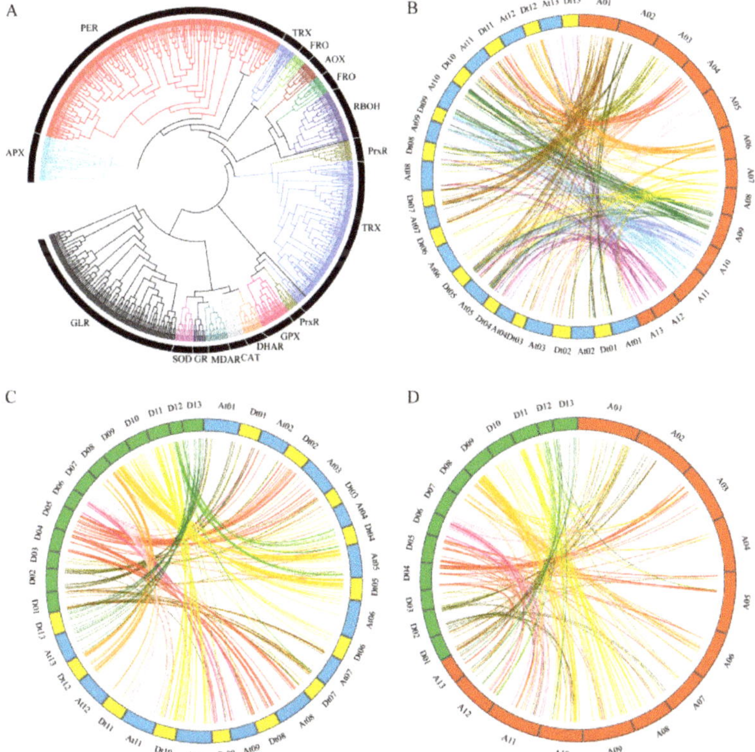

Figure 2. Phylogenetic tree and synteny analysis. (**A**) Phylogenetic tress analysis of all the cotton *ROS* network genes in *G. hirsutum*, *G. raimondii*, and *G. arboreum*. (**B**) Synteny analysis between A genome and allotetraploid AD$_1$ genome. (**C**) Synteny analysis between D genome and allotetraploid AD$_1$ genome. (**D**) Synteny analysis between A and D genomes. Red, green, yellow, and blue block means A genome, D genome, Allotetraploid cotton (TM-1) A and D subgenomes, respectively. The lines with different color mean genes of different chromosome (A and D genomes).

Synteny analysis, showed a total of 161, 166, 99, and 105 orthologous genes synteny blocks between At and A, Dt and A, At and D, and Dt and D, respectively (Figure 2B–D). It is interesting to note that more *ROS* genes were observed between the syntenic blocks among the three cotton species, for instance, at most five genes were found within the syntenic regions between At and A, and the same number of genes were also observed between Dt and A. However, at most 20 genes have been found within the syntenic regions between Dt and D (Table 1 and Table S2). The high number of *ROS* genes detected within the same syntenic regions showed that this syntenic regions have been highly conserved, and this indicates their integral role they play within the plants. This result supports the principle of asymmetric selection of genes within the genomes [8].

Table 1. The number of synteny blocks which including 1–20 gene number, respectively.

	Number of Synteny Block			
NGEB	At vs. AA	At vs. DD	Dt vs. AA	Dt vs. DD
1	136 *	63	140	58
2	19	10	19	13
3	3	11	4	8
4	2	7	2	3
5	1	7	1	4
6	0	3	0	3
7	0	1	0	2
8	0	2	0	2
9	0	0	0	2
10	0	1	0	0
11	0	0	0	2
15	0	0	0	0
16	0	0	0	1
20	0	0	0	1

NGEB: number of genes per synteny block; * for example, there are 136 synteny blocks which include one ROS gene between At subgenome and A genome.

2.4. Orthologous ROS Gene Analysis and Evolution Prediction through Synonymous (K_s) and Non-Synonymous (K_a) Rate of Substituion

Allopolyploid cotton evolved following trans-oceanic dispersal of an A genome diploid to the Americas, where the immigrant underwent hybridization, as female, with a native D genome diploid similar to modern *G. raimondii* [30]. The formation of ROS network might be earlier than that of *Gossypium* diverged from its closest relatives and Allopolyploid cottons formation. Phylogenetic analysis (Table S3) revealed that, a total of 241 paired genes, were orthologous among the three cotton species, *G. raimondii*, *G. arboreum* and *G. hirsutum*, found to be either directly or indirectly involved in the ROS network. In the analysis of the evolutionary history of the tetraploid cotton, 243 genes were orthologous between *G. arboreum* and *G. raimondii*. *G. arboreum* contained 14 genes, but not in *G. raimondii*, while only five specific genes were found to be orthologs between *G. arboreum* and At subgenome. In the D genome, *G. raimondii* had 17 genes, but not in *G. arboreum* and only five specific genes were found to be ortholog between *G. raimondii* and Dt subgenome (Figure 3A). This result suggested that 243 genes were formed evolved much earlier before the separation of the A and D cotton genomes, which are the *G. arboreum* and *G. raimondii*, respectively. Thirty-one (31) genes were found to have evolved before the separations of the two diploid cotton species, and even the emergence of the allotetraploid cotton. three genes were found to have originated from D, and introgressed into the Dt subgenome, of the allotetraploid cotton, while some element of gene loss was evident, this could have been occasioned by either loose of function or chromosomal rearrangement during the evolution of allotetraploid from the two diploid cotton species, A and D cotton genomes (Figure 3B). The synonymous substitution value (K_s) peaks of orthologous gene pairs between At/A and Dt/D was estimated at 0.004 and 0.015, respectively. The distribution of 241 putative ortholog pairs between the three cotton species/genomes (mode K_s = 0.0040–0.015) clearly indicated that the secondary peak for cotton, lies to the left of the ortholog peak, which represents a burst of gene duplications that occurred in cotton after the separation between the two lineages. Similar observation has been made in the analysis of the genome evolution of soybean, which showed that soybean genome also underwent whole genome duplication [31]. From our findings, the *ROS* genes network seemed to have occurred in tandem with the evolution of tetraploid cotton, approximately 0.8–2.8 million years ago (MYA) [2]. Both synonymous site (K_s) peaks of the ROS orthologous gene pairs between the two cotton progenitors, A and D genomes, and two subgenomes (At and Dt) was 0.045 (Figure 3C,D). Moreover, we estimated the *ROS* genes network divergence time between the A and D genomes to have occurred approximately 8.6 MYA, in the results was in agreement to the divergence time of the two diploid cotton, *G. arboreum* and *G. raimondii* which occurred 5–10 MYA [32]. Additionally, the synonymous (K_s) and non-synonymous (K_a) substitution values ratio (K_a/K_s) were less than one (1) that suggested that the *ROS* gene network evolution, was primarily governed by purifying selection, similar results were also obtained in the analysis of the 156 paralogous pairs of the *LEA* genes in cotton, in which over 99% of the orthologous gene pairs had the K_a/K_s ratio values of less than 1 [27].

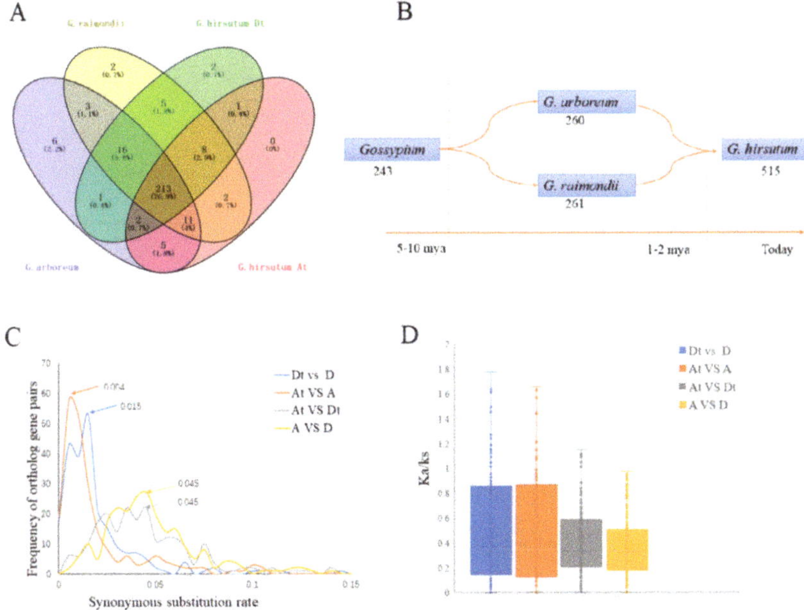

Figure 3. Orthologous gene pairs and evolution analysis of the *ROS* genes network in *G. arboreum*, *G. raimondii*, and *G. hirsutum*. (**A**) Orthologous gene pairs of the *ROS* genes network. (**B**) The orthologous genes pairs changes with a temporal depiction of phenomena that characterize polyploid evolution in Gossypium. (**C**) Distribution of K_s values for orthologous genes pairs between the A genome (*G. arboreum*), D genome (*G. raimondii*), and two subgenomes of TM-1(*G. hirsutum* At subgenome and Dt subgenome). At Dt;' t' indicates tetraploid. Peak values for each comparison are indicated with arrows. (**D**) Distribution of K_a/K_s value between the A genome, D genome and two subgenomes of TM-1.

2.5. Analysis of the Expression Levels of Homeolog and Ortholog Genes ROS Network Genes at 10 DPA and 20 DPA of Cotton Fiber Development

RNA-seq data of the public database (https://www.ncbi.nlm.nih.gov/) were downloaded and reanalyzed, which showed 192 *ROS* genes pairs were expressed during cotton fiber development (Tables S3 and S4). Moreover, we analyzed the homeolog and ortholog gene pair's expression levels at 10 and 20 DPA in the three cotton species, *G. hirsutum*, *G. raimondii*, and *G. arboreum*. It is interesting that the expression levels of genes encoding the SOD proteins, such as the CSD, FSD, and MSD proteins exhibited lower expression levels at 10 and 20 DPA in *G. raimondii* but were significantly higher in *G. arboreum* and *G. hirsutum*. The expression levels of most of the *SOD* genes were higher at 10 DPA than at 20 DPA (Figure 4A,B). The result showed that the *SOD* genes are integral at 10 DPA, being a critical stage of fiber initiation and elongation, moreover, at the stage of 10 DPA, ABA inhibits fiber cell initiation and elongation [33]. Moreover, previous studies have shown that cotton annexin proteins; AnxGh1:AAR13288, AnxGh2:AAB67993, AnxGhFx:FJ415173, and AnxGhF:AAC33305 are present in higher amount in fibers of 10 DPA wild-type plants as compared to the fuzzless-lintless mutant [34]. This result indicated that *SOD* genes were highly expressed in taxa with long fiber (*G. arboreum* and *G. hirsutum*), while were either down regulated or partially expressed in cotton species with shorter fibers, such *G. raimondii* of the D genome. Additionally, the expression level of *SOD* genes showed up regulation at fiber elongation phase (10 DPA) compared to the secondary cell wall biosynthesis/thickening phase (20 DPA). The results obtained for the expression pattern of the *ROS* genes, correlated positively to previous findings on the expression patterns of some of the genes known to be highly involved in fiber development in cotton, such as APX, MDAR, GR, GPX, FER, AOX, PrxR, Trx, GLR, and PER [21,22,35]. The *SOD* gene family and other ROS scavenging genes are believed to be involved in cotton fiber development through the change in their expression level in the course of fiber development period. On the other hand, most of the genes with functional role in the regulation of ROS, including *RBOH* and *FRO* genes were up regulated at the secondary cell wall biosynthesis phase (20 DPA) relative to fiber elongation phase (10 DPA) in *G. hirsutum*, *G. arboreum*, or *G. raimondii*. In summary, ROS level might be low at fiber elongation phase, but high at secondary cell wall biosynthesis phase in *G. hirsutum* and *G. arboreum*. Compared with *G. hirsutum* and *G. arboreum*, although the expression model of ROS production genes was similar in *G. raimondii*, the expression level of ROS removing genes was lower. Fine regulation of steady-state levels of ROS are essential for proper cell elongation [36], and long and short fiber properties were regulated by ROS removing genes, such as *SOD* gene family. Moreover, the Ascorbate peroxidase 1 *(APX1)*, has been found to be integral in the process of cotton fiber development, and that *APX* gene profiling in upland cotton, *G. hirsutum* has provided valuable information on the redox homeostasis in the process of fiber development [37]. Furthermore, when the cytosolic H_2O_2-scavenging enzyme, APX1, the whole chloroplastic H_2O_2-scavenging process stops in *A. thaliana* and the level H_2O_2 increases resulting into protein oxidation [38]. Ferric Chelate Reductase-2 *(FRO2)*, Ferric Chelate Reductase-5 *(FRO5)*, and respiratory burst oxidase homologs *(RBOH6)* orthologous gene pairs, modulation ROS level in fiber development, were identified by GWAS analysis and RNA-seq analysis [8]. Furthermore, the RBOH do modulate the ROS and in turn promote root growth in Arabidopsis [39].

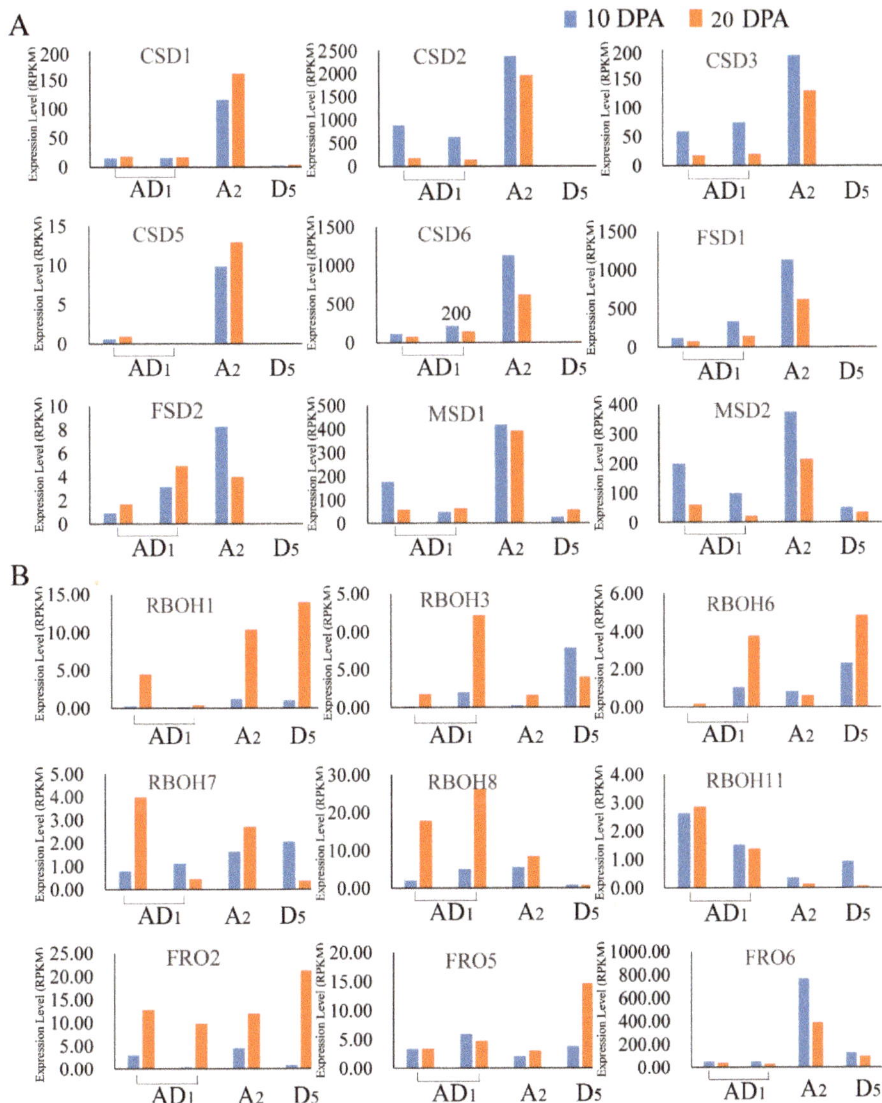

Figure 4. Expression level obtains from RNA-seq data and was show with RPKM (Reads per Kilobase of exon model per Million mapped reads). (**A**) The expression level of superoxide dismutase (SOD) genes family. (**B**) The expression of respiratory burst oxidase homologue (RBOH)and NADPH-like oxidase (FRO).

2.6. ROS Network Genes Response to Abiotic Stress

We downloaded data of ROS-gene expression levels from pubic RNA-Seq data of *G. hirsutum* (TM-1). After analyzing gene expression under cold, heat, dehydration, and salt stress treatment, a total of 135 *ROS* genes were founded to be upregulated by abiotic stresses (Table S5). Compared with the controls, 92 (68%), 79 (59%), 74 (55%), and 85 (62.9%) genes were differently expressed under cold, heat, dehydration, and salt stress, respectively. Twenty-nine genes were differentially expressed under all four different stress conditions, which included CSD, APX, MDAR, CAT, GPX, FER, RBOH,

FRO, AOX, GLR, and PER. The majority of the differentially expressed genes (DEGs) showed more expression changes under cold stress than under other abiotic stresses, suggesting that *ROS* genes were more sensitive to cold stress. ROS-producing genes were up regulated after cold, heat, dehydration, and salt treatments. FRO5 gene shown higher expression level than other ROS-producing DEGs and was upregulated after 3 and 6 h of post stress exposure. *APX1* and *MDAR1* genes, showed higher expression level than other ROS-scavenging DEGs, and were upregulated after 3 and 6 h (Figure 5). These results proved that the *ROS* gene do respond positively towards abiotic stress during the early developmental stages. *RBOH6* and *APX1* genes were highly expressed in *G. hirsutum* D subgenome than At subgenome which suggested that some gene pairs of *G. hirsutum* D subgenome and A subgenome might be playing different roles in cotton growth and development [8].

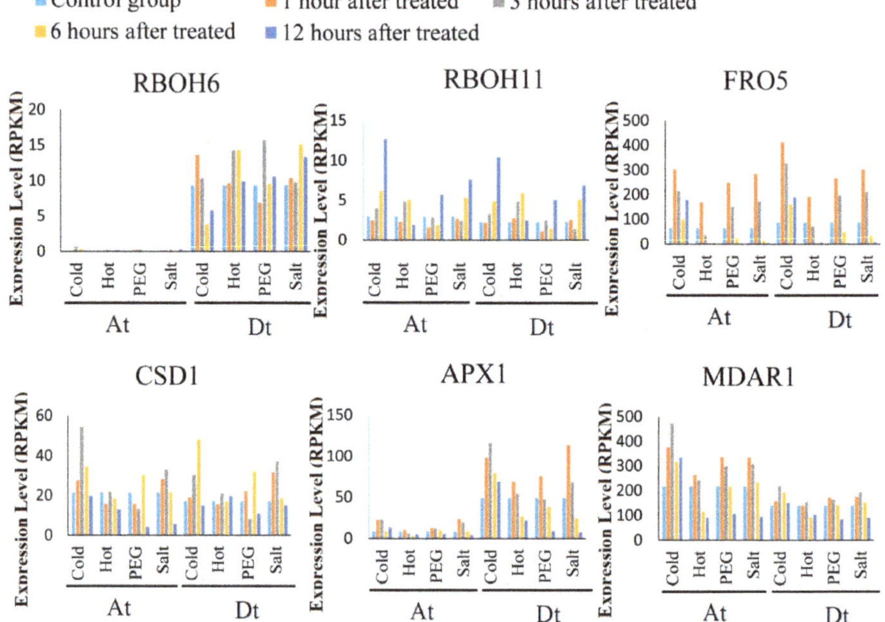

Figure 5. Expression of ROS Network genes in *G. arboreum*, *G. hirsutum*, and *G. raimondii* under abiotic stress. Expression level obtained from RNA-seq data and was show with RPKM (Reads per Kilobase of exon model per Million mapped reads). (**A**) The expression of *RBOH6/RBOH11* and *FRO5* orthologous gene pairs. (**B**) The expression level of *CSD1*, *APX1*, *and MDAR1* orthologous gene pairs.

2.7. Analysis of Morphological and Physiological Changes under Salt Stress

Four semi-wild (Latifolium 130, Latifolium 32, Latifolium 40, and Marie-galante 85) and two cultivated (CRI12 and CRI16) *Gossypium hirsutum* accessions were evaluated for phenotypic traits under salt treatment. Among them, Marie-galante 85 (M85) was highly tolerant to salinity stress, as evidenced by significantly higher relative stem length, relative root water content, relative stem water content, and lower rate of area of damaged leaves (Table 2; Figure 6). Compared with the controls, proline content was significantly enhanced under salt stress in M85, but was significantly lower in CRI12. These results suggested that M85 is highly tolerant compared to CRI12, to salt stress. The content of MDA was not significantly changed between normal condition and salt stress condition in both CRI12 and M85. However, the concentration level of MDA was relatively lower in the leaf tissues of M85 compared to the level in CRI12. Proline plays an important role in maintaining osmotic balance and protecting some important enzymes in plant cells, thus mainly acts as osmoprotectant when plants are exposed to water deficit condition, the detection of high proline content in the leaf tissues of M85, showed that the genotypes has higher adaptability to salinity compared to CRI12. Therefore, the same with salt (NaCl) stress, cotton is exposed to complex alkali-salt stress could result in osmotic stress. Malondialdehyde (MDA) is the final product of lipid peroxidation, and its content can reflect stress tolerance level of the plant [36]. However, in our study, the content of MDA was not significantly changed after salinity treatment. It suggested that alkali-salt stress has not resulted in lipid peroxidation of M85 and CRI12. Taken together, cotton regulation of complex alkali-salt stress was different from salt (NaCl) stress, and M85 had a higher tolerance to osmotic stress than CRI12. Reactive oxygen species (ROS) balance was also observed. The content of hydrogen peroxide (H_2O_2) and the activity of superoxide dismutase (SOD), peroxidase (POD), catalase (CAT), ascorbate peroxidase (APX), and glutathione reductase (GR) were measured. The content of hydrogen peroxide was significantly enhanced in root and leaf of CRI12 under alkali-salt stress. Hydrogen peroxide was also increased in leaf of M85, but significantly decreased in root of M85 under alkali-salt stress. Furthermore, we found that the activity of POD, CAT, APX, and GR was not significantly changed in roots of M85, suggesting that ROS production and removal was balanced in roots of M85. The content of hydrogen peroxide was higher in leaf than that in root. In addition, the activity of POD was significantly increased in leaf of M85, but POD, CAT, APX, GR, and SOD was not significantly affected. This result suggests that the leaf of M85 was better to remove excessive accumulation of ROS. Additionally, the activity of SOD was significantly increased under alkali-salt stress in root of M85, and the activity of POD and CAT was significantly enhanced in root of CRI12, suggesting that redox enzymes was act on different conditions, flexibly [40].

Table 2. Phenotype of six *G. hirsutum* accessions at 48 h post to complex alkali-saline stress.

Accession	RL	SL	Area of Damaged Blades			
			First Leaf *	Second Leaf *	Third Leaf *	Fourth Leaf *
Latifolium 130	0.90 [a]	0.80 [c]	0.97 [a]	0.73 [a]	0.65 [a]	0.00 [a]
Latifolium 32	0.92 [a]	0.97 [a,b]	0.17 [b]	0.03 [b]	0.00 [b]	0.00 [a]
Latifolium 40	0.87 [a]	0.89 [b,c]	0.40 [b]	0.20 [b]	0.23 [a,b]	0.03 [a]
Marie-galante 85	0.92 [a]	1.02 [a]	0.32 [b]	0.03 [b]	0.00 [b]	0.00 [a]
CRI 12	0.98 [a]	0.68 [d]	0.38 [b]	0.08 [b]	0.02 [b]	0.00 [a]
CRI 16	1.12 [a]	0.94 [a,b]	0.38 [b]	0.10 [b]	0.05 [b]	0.05 [a]

Letters following the number represents significance at 0.05 probability. Different letters a, b, c and d in a column mean statistically significant difference at the 5% level. * Represents the value of leaves damaged rate (Area of damaged leaf/Area of leaf). First leaf means that the expansion of the first true leaf.

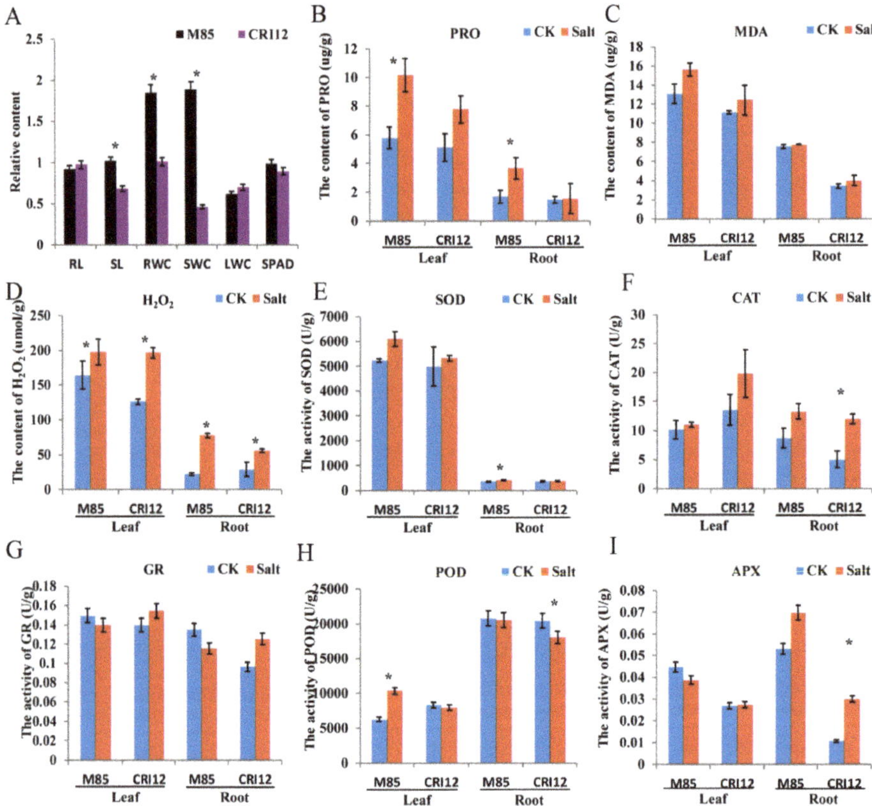

Figure 6. Complex alkali-salinity stress effects on the growth and the balance between intracellular active oxygen production and the defense system of cotton seedlings. (**A**) The phenotype of M85 and CRI12 indicated by blue and red arrows, respectively. The relative content of RL (root length), SL (stem length), RWC (root water content), SWC (stem water content), LWC (leaf water content), and SPAD (chlorophyll SPAD value) could visually show the variations among the plants. (**B**,**C**) The content of MDA (malondialdehyde) and PRO (proline) were measured at 48 h post to complex alkali-saline stress. (**D**) The content of hydrogen peroxide (H_2O_2) in leaves. (**E–I**) The enzymatic activity of superoxide dismutase (SOD), peroxidase (POD), catalase (CAT), ascorbate peroxidase (APX), and glutathione reductase (GR) were measured in leaves and roots, respectively. Error bars represent the standard deviation (SD) of three biological replicates.

2.8. RNA-Seq Analysis and ROS Network Response to Alkali-Salt Stress

After Illumina Hiseq2500 sequencing, a total of 39.62–57.16 M clean reads were generated for each sample, in which approximately 85.65%–89.61% of reads were mapped to the reference genome, and 77.79%–82.27% of clean reads were mapped to the unique locations of the reference genome (Table S6). Unique mapped reads were assembled and annotated with Cufflinks and were used for analyzing mechanism of plant response to alkali-salt stress. Based on a cutoff q-value ≤ 0.01 and fold change ≥ 2, different expressed genes (DEGs) were identified. The number of DEGs was higher at 12 h post alkali-salt stress treatment than that of 3 h in both roots and leaves in CRI12 and M85 moreover; the ROS network genes analysis in the same region by principal component analysis showed genes were aggregated at a particular region (Figure 7A,B).

The ROS balance system was found to highly enriched, the finding which was consistent with the previous studies [41]. Moreover, the majority of the DEGs were found to be enriched in response to oxidative stress (GO0006979), oxidation-reduction process (GO0055114), oxidoreductase activity (GO0016717, GO0016491, and GO0046857), and peroxidase activities (GO0004601) (Figure S1). We also observed H_2O_2 accumulation in leaves and roots under alkali-saline stress. The result shows the importance of ROS networks in acclimation for alkali-saline stress in *Gossypium hirsutum*. DEGs were also enriched in protein serine/threonine kinase activity (GO0004674). This result supports that the serine/threonine kinases, act as one of signal transduction molecular, is an essential kinase for oxidative stress responses [42]. Genes of ROS network that control ROS levels, encoded most oxidation-reduction proteins. The correlation analysis ($R^2 > 0.7$) between different cotton accessions through orthologous genes expression levels analysis were significantly higher (Figure 7C). Based on the analysis of the various *ROS* network genes, we further predicted the interlink between fiber and abiotic stress responsive *ROS* genes (Figure 8). We found that CSD1-6, FSD1-2, MSD 1-2, APX3, APX11, FRO5, FRO6, and RBOH6 have a functional role in enhancing fiber development, more specifically at 10 and 20 DPA, while CSD1, APX1, APX2, MDAR1, GPX4-6-7, FER2, RBOH6, RBOH11, and FRO5 are specifically involved in enhancing abiotic stress tolerance, more salt stress. Finally, ten genes were selected from three different libraries and validated through RT-qPCR, the expression levels were same as per the RNA sequencing data (Figure 7D).

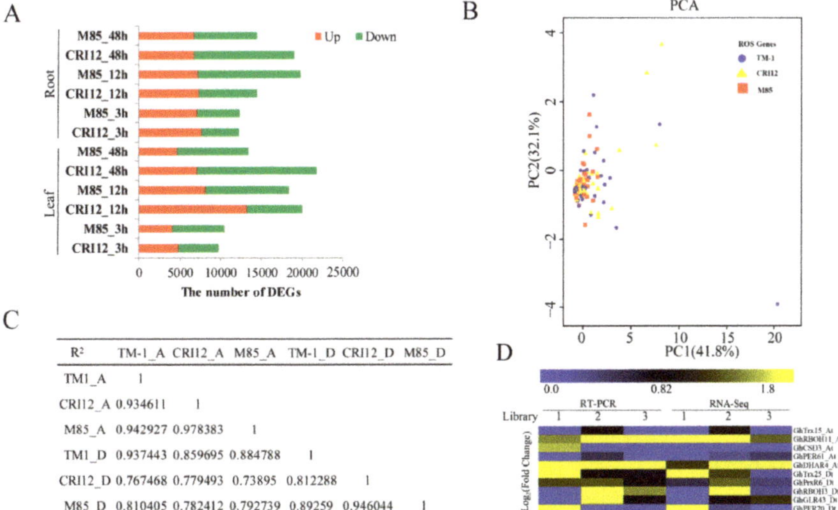

Figure 7. RNA-Seq analyses and RT-qPCR validation of the ROS network genes in M85 and CRI12 under alkali-salt stress. (**A**) The number of differentially expressed genes (DEGs) of *G. hirsutum races marie-galant* (M85) and *G. hirsutum cultivar* (CRI12) under complex alkali-saline stress at 3, 12, and 48 h. Red block is up-regulated DEGs. Green block is down-regulated DEGs. (**B**) PCA (principal component analysis) plot of the first two components for genes of ROS network. The dot color scheme represents different cotton accession (TM1, red; M85, yellow; CRI12, blue.). (**C**) Correlation coefficients of TM-1, CRI12, and M85. (**D**) RT-qPCR analysis of selected ROS genes, 1, 2, and 3 refers to the libraries, yellow-upregulated, blue: down-regulated, and black: no expression.

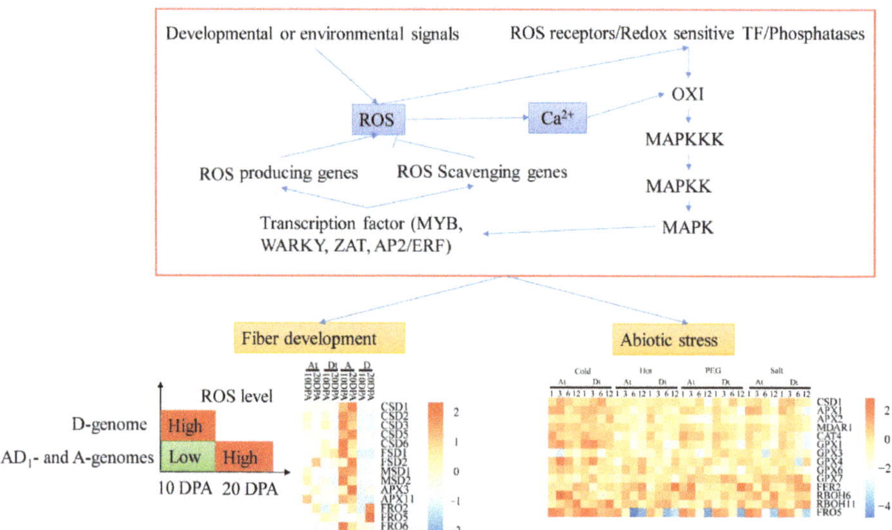

Figure 8. ROS network mediated signal pathway to illuminate the mechanism of fiber development and regulation of abiotic stress in *Gossypium*. Red box indicates modulation of ROS level by the reactive oxygen gene network of plants. ROS network regulated fiber development and abiotic stress by control ROS level, but it is not clear about how to do it. RNA-seq data of TM-1, CRI12 and M85 proved that serine/threonine kinase (OXI), and motigen-activated protein kinase (MAPK) cascades might involve in the alkali-salt stress (one of abiotic stresses) response which ROS network mediate. Under fiber development-box, heatmap of some ROS network genes indicted genes expression levels at 10 and 20 DPA (A, *G. arboreum*; D, *G. raimondii*; At, *G. hirsutum* A subgenome; Dt, *G. hirsutum* D subgenome) by R package (Pheatmap, scale = "row"). Under abiotic stress-box, heatmap indicated value of genes expression levels fold change ($\log_2$ (FC)) under cold, hot, dehydration and salt stress by Pheatmap package.

3. Discussion

In plants, many different biology processes are controlled by ROS [43]. For example, fiber development and biotic and abiotic stresses have previously been found to be partially regulated by ROS in cotton [44,45]. During 5–10 MYA, *Gossypium* split into three major diploid lineages: the New World clade (D genome), the African-Asian (A, B, E, and F genomes) clade, and the Australian clade (C, G, and K genomes) [1]. Allotetraploid Upland cotton was formed approximately 1–2 MYA [46] and domesticated for at least 4000–5000 years ago [47]. During human domestication history, the central focus was the fiber quality and quantity, but in the recent past, researchers have found that fiber and fiber qualities are not only affected by the intrinsic nature of the cotton plant, but greatly influenced by the environmental conditions, and therefore in elucidating the mechanisms of fiber development and improvement of its qualities, multidimensional approach has so far been preferred by plant breeders. The change in environmental condition, occasioned by continued human interference, rainfall has become scarce and very erratic in nature, in addition to the emergence of other forms of abiotic stresses, such as salinity, heat, and cold, among others [48]. The sessile nature of plants, makes them to be much prawn to full effects of abiotic stresses, and thus in the course of plant evolution, various survival strategies have been adopted by plants, more so the evolution of various stress responsive plants transcription factors, in which *ROS* genes are among one of them. In our study, we found that ROS network was conserved in cotton. The Evolutionary history of *ROS* genes network in cotton was similar to *Gossypium* evolution. Thus, ROS network was positively selected during cotton evolution and the result of K_a/K_s analysis also supported this view. Gene order and colinearity of allotetraploid cotton are largely conserved between the A and D subgenomes and the extant D progenitor genome (*G. raimondii*) [47], and the genes colinearity of D09 (*G. raimondii*) was conserved with Dt05 and At05 (*G. hirsutum*). Interestingly, more potential *ROS* genes were mapped to A10 (35, 12.36%), D09 (40, 13.89%), At05 (46, 18.11%), and Dt05 (38, 14.28%) in *G. arboreum*, G. raimondii, *G. hirsutum* A subgenome, and *G. hirsutum* D subgenome, respectively. It suggested that the location of *ROS* genes was conserved during the domestication process and D09, Dt05, and At05 play a core role in regulated ROS dynamic balance. Above all, ROS plays an important role in cotton growth and development, because of ROS network genes were positive selection and conserved during evolution and domestication of cotton.

At least 15 gene families (SOD, APX, MDAR, DHAR, GR, CAT, GPX, FER, RBOH, FRO, AOX, PrxR, Trx, GLR, and PER) of the ROS network were identified in *Gossypium*. SOD, APX, and *GPX* gene families were identified and analyzed. Our results are consistent with the results of *SOD* gene analyses in previous studies [37] and *SOD* genes cover the chromosome at the same locations. However, two genes were annotated as copper chaperone for superoxide dismutase in *G. arboreum* and *G. raimondii*. Compared with a previous study, the member of *GPX* genes family was different because of different genomes (NBI_*Gossypium hirsutum* vs. BGI_*Gossypium hirsutum*) used for identifying *GPX* genes. In previous studies, ROS were found to regulate development, differentiation, redox levels, stress signaling, interactions with other organisms, systemic responses, and cell death in higher plants [49–51]. Wild cottons allocate greater resources to stress response pathways, while domestication led to reprogrammed resource allocation toward increased fiber growth, possibly through modulating stress-response network [52]. Modulating reactive oxygen species (ROS) production and scavenging could be regulating cotton fiber elongation [53], abiotic stress [41].

In previous studies, different developmental or environmental signals feed into the ROS signaling network and perturb ROS homeostasis in a compartment-specific or even cell-specific manner [12]. Then, the fluctuant ROS is perceived by receptor proteins (NPR, HSF), redox-sensitive transcription factors, or direct inhibition of phosphatases [54]. Serine/threonine protein kinase (OXI1) is involved in ROS sensing and the activation of mitogen-activated-protein kinases (MAPKs) [55] that plays a central role in ROS signal network [56–58]. ROS-producing or -scavenging pathway is activated to regulate ROS levels. Different developmental, metabolic and defense pathways are activated to regulate growth and development. In cotton, several families of transcription factors, such as *MYB, WRKY, AP2/ERF*, and *NAC* have been found to be involved in regulating plant response to abiotic stress [59], and some of them are also involved in ROS signal network [60].

Cotton fiber development is a complicated biological process. Many researches demonstrate that fiber development was regulated by ROS-mediated pathway. In the previous study, 119 association signals, 71 SNP associated with yield-related traits, 45 association with fiber qualities, and three associated with resistance to Verticillium wilt, were identified through genome-wide association study, and a total of 43 *ROS* genes were found [61]. Among the 43 genes, seven genes were within the associated SNP signals region of 200 kb (±100 kb), and one of the 43 genes, the following genes were found to have a direct involvement in the regulation of fiber development, for instance *Gh_A13G1654 (GLR34)* was at the downstream, and was highly associated to signals A13:75463332 (associated with fiber elongation), A13:75463913 (associated with fiber strength), and A13:75473149 (associated with fiber length); *Gh_D05G2413 (FRO2)* was at the downstream, at the boll weight association signal D05: 24,174,044 (associated with boll weight). This finding may partially support the speculation that ROS network associated with some different plant traits. ROS is involved in regulating plant cell expansion [62], which suggests ROS may also be involved in regulating cotton fiber development [17]. High concentration of ROS in wild-cotton fiber cells inhibits fiber elongation by regulating secondary-cell-wall synthesis [8]. Moreover, the H_2O_2, involved in cell-wall relaxation, is necessary molecule for cell elongation, however, higher ROS levels may halt cell elongation [63–65]. Compared with fiber elongation (10 DPA), ROS level was regulated by ROS network, was upregulated at secondary cell wall biosynthesis (20 DPA) [21]. It is interesting that the expression levels of the *SOD* genes and some other ROS-scavenging genes were higher *G. hirsutum* and *G. arboreum*, and they are known to have relatively longer fiber compared *G. raimondii*. Moreover, *G. hirsutum* and *G. arboreum*, exhibited lower ROS-scavenging enzymes, which implied that they had the higher ROS-producing ability than *G. raimondii*. The results therefore showed that perturbed ROS levels as a result of the *SOD* genes and some other genes could perhaps have a net effect on the regulation of various stages of fiber development in *G. hirsutum*, *G. arboreum*, and *G. raimondii*. In addition, fluctuating levels in ROS levels, caused by different expression level of ROS-scavenging genes and ROS-producing genes at fiber elongation and secondary cell wall biosynthesis phases, showed the integral role played by the *ROS* genes in fiber development.

Almost all the abiotic stress factors do lead to the development of secondary stress condition such as osmotic and oxidative stresses [66]. RNA-seq data, physiological, and phenotypic index analysis in the two upland cotton genotypes, M85 and CRI12 suggested that the *ROS* gene network is partially responsible to salinity stress tolerance. The majority of the *ROS* genes were expressed under abiotic stress, which demonstrated that ROS network was also interconnected with abiotic stress tolerance in cotton [67–69]. According to the expression values (FPKM) of the ROS network genes in *G. hirsutum* TM-1, the expression levels of the homologous genes in under various abiotic stress conditions were: 92 (68%), 79 (59%), 74 (55%), and 85 (62.9%) under cold, heat, dehydration, and salt stress conditions, respectively. A total of 29 genes were differentially expressed under four different stress conditions, in which out of the six genes, three genes were the ROS-producing genes, while the other three were the ROS-scavenging genes.

4. Materials and Methods

4.1. Identification of the ROS Genes in Cotton

By against the complete genomes of *Gossypium hirsutum* (NAU-NBI Assembly v1.1 & Annotation v1.1), *G. arboreum* (BGI Assembly v2.0 & Annotation v1.0), and *G. raimondii* (JGI assembly v2.0 & annotation v2.1) which were downloaded from CottonGen database (https://www.cottongen.org), a total of 152 *ROS* genes controlling ROS levels in *Arabidopsis* [12] were used to identify *ROS* genes in cotton genome using BLASTP using the following criteria [70]: the longest transcript in each gene loci was chosen to represent that locus and the sequences were filtered out when (1) CDS length with < 150 bp, (2) CDSs with percentages of ambiguous nucleotides ('N') > 50%, (3) CDSs with internal termination codons, and (4) the CDSs with hits (BLAST identities ≥80%) to RepBase sequences (http://www.girinst.org/repbase/index.html). The conservative domains of candidate genes were obtained from Pfam website (http://pfam.xfam.org/) [71] and NCBI Web CD-Search Tool (https://www.ncbi.nlm.nih.gov/Structure/bwrpsb/bwrpsb.cgi) [72]. The biophysical characters of the encoded proteins were computed using the ExPASy ProtParam tool (http://us.expasy.org/tools/protparam.html).

4.2. Phylogenetic Tree, Orthologous Gene Pairs and Values of K_a, K_s and K_a/K_s

Full-length sequences of ROS-network proteins were first aligned with the ClustalX program (http://www.clustal.org/clustal2/) [73], then phylogenetic trees were constructed by using neighbor-joining (NJ) method with 1,000 bootstrap replicas [74] and the Poisson model by using MEGA 7.0 software (http://www.megasoftware.net). Meanwhile, the orthologous gene pairs of the ROS network in A, D genomes, At and Dt subgenomes were searched by InParanoid software (http://inparanoid.sbc.su.se/cgi-bin/index.cgi). Additionally, the evolutionary rates K_a, K_s, and K_a/K_s ratio were estimated by K_aK_s_Calculator package (https://kakscalculator.herokuapp.com/). On the basis of the synonymous substitutions per year (λ) of 2.6×10^{-9} for cotton, the divergent time of ROS orthologous gene pairs were estimated by T = $K_s/2\lambda \times 10^{-6}$ MYA [47].

4.3. Chromosomal Locations, Conserved Synteny Blocks and Gene Duplication

Chromosomal positions of the *ROS* genes were obtained from the annotation information of cotton genomes (*G. arboreum*, *G. raimondii* and *G. hirsutum*). The chromosomal distributions of genes were drafted on the cotton chromosomes according to the gene positions by Mapchart software [75]. The conserved synteny blocks between A genome and At subgenome, D genome and Dt genome in cotton were inferred using the MCSCANX program with the default parameters (http://chibba.pgml.uga.edu/mcscan2/) [76]. A local BLASTP program was run on different genomes with parameters (e^{-10}). The syntenic relationships were illustrated with the CIRCOS program (http://circos.ca/) [77]. The duplicate gene classifier program of MCSCANX software was used to identify each ROS network gene duplication with the default parameters (OVERLAP WINDOW: 5). All the ROS network genes were classified into various types of duplications.

4.4. Plant Materials and Salt Stress Treatment

G. hirsutum var. marie-galante 85 (M85) and *G. hirsutum* cv. CRI12 were selected for the salinity treatment. Semi-wild cotton marie-galante was originally collected from Yucatan, Mexico and cultivated in the National Wild Cotton Nursery (Sanya, Hainan, China). *G. hirsutum* cv. CRI12 was widely planted in china. The seeds of M85 and CRI12 were first germinated at 28 °C in a 16 h light/8 h dark cycle. Then, seedlings were planted in the normal solutions for three weeks. The similar growing seedlings were selected for salt-alkali stress. The salt-alkali treatment was imposed by simulating the actual saline condition of Xinjiang province, a region commonly known for salt stress [78]. We examined the soil salt composition and simulated a similar environment to impose salt-alkali stress, the composition was made of 2.664 mg:g^{-1} of calcium chloride ($CaCl_2$), 0.179 mg:g^{-1} of sodium hydrogen carbonate ($NaHCO_3$), 9.94 mg:g^{-1} of sodium sulphate (Na_2SO_4), 0.848 mg:g^{-1} of potassium sulphate (K_2SO_4) and finally 3.587 mg:g^{-1} of hydrated magnesium sulphate ($MgSO_4 \cdot 7H_2O$) [79].

4.5. Measurement of Morphological, Physiological and Biochemical Traits

Samples of roots and leaves that collected at 48 h after salt-alkali stress treatment were used for measuring physiological traits. And, seedlings were also used for measuring phenotype at 48 h after salt-alkali stress treatment. Malondialdehyde (MDA) and Proline (PRO), the most important protective macromolecules in response to salt stress, were measured using the kits (Nanjing Jiancheng Bioengineering Institute). Hydrogen peroxide (H_2O_2), superoxide dismutase (SOD), peroxidase (POD), ascorbate peroxidase (APX), catalase (CAT) and glutathione reductase (GR), important ROS and ROS-scavenging molecules, were measured using the corresponding assay kits (Nanjing Jiancheng Bioengineering Institute). The root length (RL), stem length (SL), roots fresh weight (RWC) and stem fresh weight (SWC) were measured after 48 h salt stress. The value of SPAD, one of photosynthesis indices, was measured using chlorophyll meter model SPAD-502.

4.6. RNA Extraction, cDNA Library Construction, and RNA-Seq

Roots and leaves were collected at 0, 3, 12 and 48 h time points after salt-alkali stress treatment. Then, collected samples were used for transcriptome sequencing. Total RNA was extracted from each cotton sample using TRlzol Reagent (Life technologies, CA, USA) according to the instruction manual. RNA integrity and concentration were checked using an Agilent 2100 Bioanalyzer (Agilent Technologies, Inc., Santa Clara, CA, USA). The mRNAs were isolated by NEBNext Poly (A) mRNA Magnetic Isolation Module (NEB, E7490) by New England BioLabs (Ipswich, MA, USA). The cDNA libraries were constructed by following the manufacturer's instructions of NEBNext Ultra RNA Library Prep Kit for Illumina (NEB, E7530) and NEBNext Multiplex Oligos for Illumina (NEB, E7500) (Illumina, San Diego, CA, USA). Briefly, the enriched mRNA was fragmented into RNAs with approximately 200 nt, which were used to synthesize the first-strand cDNA and then the second cDNA. The double-stranded cDNAs were performed end-repair/dA-tail and adaptor ligation. The suitable fragments were isolated by Agencourt AMPure XP beads (Beckman Coulter, Inc., Indianapolis, IN, USA) and enriched by PCR amplification. Finally, the constructed cDNA libraries were sequenced on a flow cell using an Illumina HiSeq™ 2500 sequencing platform.

4.7. Expression Analysis

Whole-transcriptome sequencing data for *G. hirsutum* were obtained from the NCBI Sequence Read Archive (SRA) (http://www.ncbi.nlm.nih.gov/sra) to analyze the tissue/organ-specific, stage-specific and stress-induced expression patterns of cotton *ROS* genes [47]. The details of the SRA are shown in (Tables S7 and S8). The row data of RNA-seq of M85 and CRI12 were separately analyzed and the clean reads were obtained by removing reads containing adapter, reads containing ploy-N and lower quality reads from raw data. At the same time, Q_{30} [80], GC-content and sequence duplication level of the clean data were calculated. Raw sequences were transformed into clean reads after data processing. These clean reads were then mapped to the reference genome sequence. Only reads with a perfect match or one mismatch were further analyzed and annotated based on the reference genome. Tophat2 [81] software was used to map with reference genome. Base on the reference genome, using Cufflinks software [82], mapped reads had been assembled. Quantification of gene expression levels was estimated by fragments per kilobase of transcript per million fragments mapped (FPKM). Base on mapped reads, using FPKM as the index, each gene was estimated by Cuffquant and Cufform. EBSeq software was used to identify the differential expression genes by Fold change ≥ 2 and FDR ≤ 0.01 (False Discovery Rate). FDR was corrected using Benjamini-Hochberg method by p-value. Finally based on the RNA sequence expression data, ten (10) genes were selected and their expression validated through RT-qPCR in relation to salt-alkali stress. In each, three libraries were chosen and the gene specific primers were designed by using primer premier 5 (Table S9). The RT-qPCR analysis was carried out as described by Magwanga et al [83], with GhActin as the internal control gene.

4.8. Statistical Analysis

The experimental data on phenotype and physiology were statistically analyzed using the SAS version 8.2. Graphic presentations were performed using OrginPro 8.0 program and R packages.

5. Conclusions

Due to the ever-changing environmental conditions, the available land for cultivation is declining as a result of various environmental stress factors such as drought and salinity, among others [84]. The combined effect of drought and salinity has been estimated to cause huge losses in agricultural crops as over 6% of arable lands are saline [85]. Thus non edible crops such as cotton are currently being grown in harsher conditions, in soils with high levels of salt content, but due to the importance of the cotton plant, focus is aimed of developing a more robust and salt tolerant cotton cultivars. When plants are exposed to any form stress, the ROS equilibrium shifts leading to high concentrations which results in toxicity and eventually leads to total plant death [86]. In this research work, we carried out deep transcriptome analysis of the ROS network genes, evaluated their evolution pattern and their expression levels under abiotic stress and fiber development stages. A total of 515, 260, and 261 ROS candidate proteins were identified in *G. hirsutum*, *G. arboreum*, and *G. raimondii*, respectively. The high numbers of the ROS network genes were found to have evolved through segmental gene duplication compared to tandem, a result which was in agreement to evolution pattern of various stress responsive genes such the *LEA* genes which evolved mainly through segmental type of gene duplication [27]. Moreover, RNA sequencing and RT-qPCR validation revealed that the ROS genes had a putative role in enhancing abiotic stress tolerance in cotton and in turn promote fiber development in cotton. In relation to fiber development, the following genes were found to be highly upregulated, CSD1,2,3,5,6; FSD1,2; MSD1,2; APX3,11; FRO5.6; and RBOH6 while for salt stress tolerance, CSD1, APX1, APX2, MDAR1, GPX4-6-7, FER2, RBOH6, RBOH11; and FRO5 were found to have a significant role. This research lays a significant foundation for future exploration of the *ROS* genes in developing a more stress tolerant cotton genotypes and with superior fiber quality.

Supplementary Materials: Supplementary Materials can be found at http://www.mdpi.com/1422-0067/20/8/1863/s1. Table S1. Characteristics of genes of ROS network in *Gossypium hirsutum*, *G. arboreum*, and *G. raimondii*; Table S2. Synteny blocks of *G.hirsutum* and diploid cotton (*G. arboreum* and *G. raimondii*); Table S3. Orthologous gene pairs of *G. hirsutum*, *G. arboreum*, and *G. raimondii*; Table S4. Expression levels of ROS network genes of *G. hirsutum* (TM-1), *G. arboreum*, and *G. raimondii* fiber; Table S5. Expression levels of ROS network genes of *G. hirsutum* (TM-1) leaves under abiotic stress; Table S6. Summary of the results of RNA-seq and map to the *G. hirsutum* genome; Table S7. Gene accession number and samples information of transcriptome data used in our research; Table S8. Genes of ROS network within 2 Mb (±1 Mb) of association signals; Figure S1. GO enrichment analysis of M85 and CRI12 under alkali-salt stress.

Author Contributions: F.L. and K.W. conceived and designed the study. Y.X., R.O.M., Y.W. (Yangyang Wei) and Z.Z. (Zhongli Zhou) performed the experiments. Y.X., R.O.M., X.C., Z.Z. (Zhongli Zhou), X.W., Y.W. (Yuhong Wang), Z.Z. (Zhenmei Zhang), D.J. and X.G. contributed the materials/analysis tools. Y.X. and R.O.M. wrote the manuscript. F.L., K.W., X.C., R.O.M., Z.Z. (Zhongli Zhou), Z.L. and D.J. revised the manuscript. All authors reviewed and approved the final manuscript.

Funding: This research program was financially sponsored by the National Natural Science Foundation of China (31671745, 31601352, and 31530053) and the National key research and development plan (2016YFD0100306, 2016YFD0101401).

Acknowledgments: We deeply indebted to the entire research team for their valuable effort during the period of this research work.

Conflicts of Interest: The authors declare that they have no competing interests.

Abbreviations

ROS	Reactive oxygen species
SOD	Superoxide dismutase
K_a	Synonymous
MDAR	Monodehydroascorbate reductase
K_s	Nonsynonymous
DHAR	Dehydroascorbate reductase
MYA	Million years ago
GR	Glutathione reductase
RBOH	Respiratory burst oxidase homologue
CAT	Catalase
NADPH	Nicotinamide adenine dinucleotide phosphate
GPX	Glutathione peroxidase
APX	Ascorbate peroxidase
FER	Ferritin
AO	Ascorbate oxidase
FRO	NADPH-like oxidase
AOX	Ubiquinol oxidase
PrxR	Peroxiredoxin
GLR	Glutaredoxin
Trx	Thioredoxin reductase
PER	Peroxidase
SNP	Single nucleotide polymorphism
DPA	Days post anthesis
GWAS	Genome-wide association study.

References

1. Wendel, J.F.; Flagel, L.E.; Adams, K.L. Jeans, genes, and genomes: Cotton as a model for studying polyploidy. In *Polyploidy and Genome Evolution*; Springer: Berlin/Heidelberg, Germany, 2012; pp. 181–207. ISBN 9783642314421.

2. Bradshaw, J.E. Genome Evolution and Polyploidy. In *Plant Breeding: Past, Present and Future*; Springer: Dordrecht, The Netherlands, 2016; pp. 233–269.

3. Flagel, L.E.; Wendel, J.F.; Udall, J.A. Duplicate gene evolution, homoeologous recombination, and transcriptome characterization in allopolyploid cotton. *BMC Genomics* **2012**, *13*. [CrossRef]

4. Shan, X.; Liu, Z.; Dong, Z.; Wang, Y.; Chen, Y.; Lin, X.; Long, L.; Han, F.; Dong, Y.; Liu, B. Mobilization of the active MITE transposons mPing and Pong in rice by introgression from wild rice (*Zizania latifolia* Griseb.). *Mol. Biol. Evol.* **2005**, *22*, 976–990. [CrossRef]

5. Dos Soares, L.A.A.; Fernandes, P.D.; de Lima, G.S.; Brito, M.E.B.; do Nascimento, R.; Arriel, N.H.C. Physiology and production of naturally-colored cotton under irrigation strategies using salinized water. *Pesqui. Agropecu. Bras.* **2018**, *53*, 746–755. [CrossRef]

6. Xu, Z.; Yu, J.; Kohel, R.J.; Percy, R.G.; Beavis, W.D.; Main, D.; Yu, J.Z. Distribution and evolution of cotton fiber development genes in the fibreless *Gossypium raimondii* genome. *Genomics* **2015**, *106*, 61–69. [CrossRef] [PubMed]

7. Haigler, C.H.; Betancur, L.; Stiff, M.R.; Tuttle, J.R. Cotton fiber: A powerful single-cell model for cell wall and cellulose research. *Front. Plant Sci.* **2012**, *3*. [CrossRef] [PubMed]

8. Wang, M.; Tu, L.; Lin, M.; Lin, Z.; Wang, P.; Yang, Q.; Ye, Z.; Shen, C.; Li, J.; Zhang, L.; et al. Asymmetric subgenome selection and cis-regulatory divergence during cotton domestication. *Nat. Genet.* **2017**, *49*, 579–587. [CrossRef]

9. Mittler, R. ROS Are Good. *Trends Plant Sci.* **2017**, *22*, 11–19. [CrossRef] [PubMed]

10. Mullineaux, P.; Karpinski, S. Signal transduction in response to excess light: Getting out of the chloroplast. *Curr. Opin. Plant Biol.* **2002**, *5*, 43–48. [CrossRef]

11. Ba, Q.S.; Liu, H.Z.; Wang, J.W.; Che, H.X.; Niu, N.; Wang, J.S.; Ma, S.C.; Zhang, G.S. Relationship between metabolism of reactive oxygen species and chemically induced male sterility in wheat (*Triticum aestivum* L.). *Can. J. Plant Sci.* **2013**, *93*, 675–681. [CrossRef]

12. Mittler, R.; Vanderauwera, S.; Gollery, M.; Van Breusegem, F. Reactive oxygen gene network of plants. *Trends Plant Sci.* **2004**, *9*, 490–498. [CrossRef]

13. Perata, P.; Pucciariello, C.; Novi, G.; Parlanti, S.; Banti, V. Reactive Oxygen Species-Driven Transcription in Arabidopsis under Oxygen Deprivation. *Plant Physiol.* **2012**, *159*, 184–196. [CrossRef]

14. Leymarie, J.; Vitkauskaité, G.; Hoang, H.H.; Gendreau, E.; Chazoule, V.; Meimoun, P.; Corbineau, F.; El-Maarouf-Bouteau, H.; Bailly, C. Role of reactive oxygen species in the regulation of arabidopsis seed dormancy. *Plant Cell Physiol.* **2012**, *53*, 96–106. [CrossRef] [PubMed]

15. Fu, Y.; Guo, C.; Wu, H.; Chen, C. Arginine decarboxylase ADC2 enhances salt tolerance through increasing ROS scavenging enzyme activity in *Arabidopsis thaliana*. *Plant Growth Regul.* **2017**, *83*, 253–263. [CrossRef]

16. Waszczak, C.; Carmody, M.; Kangasjärvi, J. Reactive Oxygen Species in Plant Signaling. *Annu. Rev. Plant Biol.* **2018**, *69*. [CrossRef] [PubMed]

17. Hovav, R.; Udall, J.A.; Chaudhary, B.; Hovav, E.; Flagel, L.; Hu, G.; Wendel, J.F. The evolution of spinnable cotton fiber entailed prolonged development and a novel metabolism. *PLoS Genet.* **2008**, *4*. [CrossRef]

18. Tang, W.; Tu, L.; Yang, X.; Tan, J.; Deng, F.; Hao, J.; Guo, K.; Lindsey, K.; Zhang, X. The calcium sensor GhCaM7 promotes cotton fiber elongation by modulating reactive oxygen species (ROS) production. *New Phytol.* **2014**, *202*, 509–520. [CrossRef]

19. Damiano, S.; Fusco, R.; Morano, A.; de Mizio, M.; Paternò, R.; de Rosa, A.; Spinelli, R.; Amente, S.; Frunzio, R.; Mondola, P.; et al. Reactive oxygen species regulate the levels of dual oxidase (duox1-2) in human neuroblastoma cells. *PLoS ONE* **2012**, *7*. [CrossRef]

20. Kwak, H.J.; Liu, P.; Bajrami, B.; Xu, Y.; Park, S.Y.; Nombela-Arrieta, C.; Mondal, S.; Sun, Y.; Zhu, H.; Chai, L.; et al. Myeloid Cell-Derived Reactive Oxygen Species Externally Regulate the Proliferation of Myeloid Progenitors in Emergency Granulopoiesis. *Immunity* **2015**, *42*, 159–171. [CrossRef]

21. Guo, K.; Du, X.; Tu, L.; Tang, W.; Wang, P.; Wang, M.; Liu, Z.; Zhang, X. Fibre elongation requires normal redox homeostasis modulated by cytosolic ascorbate peroxidase in cotton (*Gossypium hirsutum*). *J. Exp. Bot.* **2016**, *67*, 3289–3301. [CrossRef]

22. Li, R.; Xin, S.; Tao, C.; Jin, X.; Li, H. Cotton ascorbate oxidase promotes cell growth in cultured tobacco bright yellow-2 cells through generation of apoplast oxidation. *Int. J. Mol. Sci.* **2017**, *18*, 1346. [CrossRef]

23. Zhang, J.; Huang, G.Q.; Zou, D.; Yan, J.Q.; Li, Y.; Hu, S.; Li, X.B. The cotton (*Gossypium hirsutum*) NAC transcription factor (FSN1) as a positive regulator participates in controlling secondary cell wall biosynthesis and modification of fibers. *New Phytol.* **2018**, *217*, 625–640. [CrossRef]

24. Watkins, J.M.; Chapman, J.M.; Muday, G.K. Abscisic Acid-Induced Reactive Oxygen Species Are Modulated by Flavonols to Control Stomata Aperture. *Plant Physiol.* **2017**, *175*, 1807–1825. [CrossRef]

25. Zhang, B.; Chen, X.; Lu, X.; Shu, N.; Wang, X.; Yang, X.; Wang, S.; Wang, J.; Guo, L.; Wang, D.; et al. Transcriptome Analysis of *Gossypium hirsutum* L. Reveals Different Mechanisms among NaCl, NaOH and Na$_2$CO$_3$ Stress Tolerance. *Sci. Rep.* **2018**, *8*. [CrossRef]

26. Sugimoto, M.; Oono, Y.; Gusev, O.; Matsumoto, T.; Yazawa, T.; Levinskikh, M.A.; Sychev, V.N.; Bingham, G.E.; Wheeler, R.; Hummerick, M. Genome-wide expression analysis of reactive oxygen species gene network in Mizuna plants grown in long-term spaceflight. *BMC Plant Biol.* **2014**, *14*. [CrossRef]

27. Magwanga, R.O.; Lu, P.; Kirungu, J.N.; Lu, H.; Wang, X.; Cai, X.; Zhou, Z.; Zhang, Z.; Salih, H.; Wang, K.; et al. Characterization of the late embryogenesis abundant (LEA) proteins family and their role in drought stress tolerance in upland cotton. *BMC Genet.* **2018**, *19*. [CrossRef]

28. Flagel, L.E.; Wendel, J.F. Gene duplication and evolutionary novelty in plants. *New Phytol.* **2009**, *183*, 557–564. [CrossRef]

29. Soltis, P.S.; Burleigh, J.G.; Chanderbali, A.S.; Yoo, M.-J.; Soltis, D.E. Gene and Genome Duplications in Plants. In *Evolution after Gene Duplication*; Wiley: London, UK, 2011; pp. 269–298.

30. Wang, Q.; Wang, S.; Zhu, X.; Fang, L.; Hu, Y.; Chen, X.; Huang, X.; Du, X.; Chen, S.; Wan, Q.; et al. Genomic insights into divergence and dual domestication of cultivated allotetraploid cottons. *Genome Biol.* **2017**, *18*. [CrossRef]

31. Shoemaker, R.C.; Polzin, K.; Labate, J.; Specht, J.; Brummer, E.C.; Olson, T.; Young, N.; Concibido, V.; Wilcox, J.; Tamulonis, J.P.; et al. Genome duplication in soybean (Glycine subgenus soja). *Genetics* **1996**, *144*, 329–338. [CrossRef]

32. Small, R.L.; Wendel, J.F. Phylogeny, duplication, and intraspecific variation of Adh sequences in new world diploid cottons (*Gossypium* L., Malvaceae). *Mol. Phylogenet. Evol.* **2000**, *16*, 73–84. [CrossRef]

33. Miao, Q.; Deng, P.; Saha, S.; Jenkins, J.N.; Hsu, C.-Y.; Abdurakhmonov, I.Y.; Buriev, Z.T.; Pepper, A.; Ma, D.-P. Transcriptome Analysis of Ten-DPA Fiber in an Upland Cotton (*Gossypium hirsutum*) Line with Improved Fiber Traits from Phytochrome A1 RNAi Plants. *Am. J. Plant Sci.* **2017**, *8*, 2530–2553. [CrossRef]

34. Huang, Y.; Wang, J.; Zhang, L.; Zuo, K. A Cotton Annexin Protein AnxGb6 Regulates Fiber Elongation through Its Interaction with Actin 1. *PLoS ONE* **2013**, *8*. [CrossRef]

35. Li, H.B.; Qin, Y.M.; Pang, Y.; Song, W.Q.; Mei, W.Q.; Zhu, Y.X. A cotton ascorbate peroxidase is involved in hydrogen peroxide homeostasis during fibre cell development. *New Phytol.* **2007**, *175*, 462–471. [CrossRef]

36. Cosgrove, D.J. Enzymes and Other Agents That Enhance Cell Wall Extensibility. *Annu. Rev. Plant Physiol. Plant Mol. Biol.* **2002**, *50*, 391–417. [CrossRef]

37. Tao, C.; Jin, X.; Zhu, L.; Xie, Q.; Wang, X.; Li, H. Genome-wide investigation and expression profiling of APX gene family in *Gossypium hirsutum* provide new insights in redox homeostasis maintenance during different fiber development stages. *Mol. Genet. Genomics* **2018**, *293*, 685–697. [CrossRef]

38. Davletova, S.; Rizhsky, L.; Liang, H.; Shengqiang, Z.; Oliver, D.J.; Coutu, J.; Shulaev, V.; Schlauch, K.; Mittler, R. Cytosolic Ascorbate Peroxidase 1 Is a Central Component of the Reactive Oxygen Gene Network of Arabidopsis. *Plant Cell* **2005**, *17*, 268–281. [CrossRef]

39. Orman-Ligeza, B.; Parizot, B.; de Rycke, R.; Fernandez, A.; Himschoot, E.; Van Breusegem, F.; Bennett, M.J.; Périlleux, C.; Beeckman, T.; Draye, X. RBOH-mediated ROS production facilitates lateral root emergence in *Arabidopsis*. *Development* **2016**, *143*, 3328–3339. [CrossRef]

40. Sarwar, M.B.; Sadique, S.; Hassan, S.; Riaz, S.; Rashid, B.; Bahaeldeen Babiker Mohamed, T.H. Physio-Biochemical and Molecular Responses in Transgenic Cotton under Drought Stress. *J. Agric. Sci.* **2017**, *23*, 157–166.

41. Wei, Y.; Xu, Y.; Lu, P.; Wang, X.; Li, Z.; Cai, X.; Zhou, Z.; Wang, Y.; Zhang, Z.; Lin, Z.; et al. Salt stress responsiveness of a wild cotton species (*Gossypium klotzschianum*) based on transcriptomic analysis. *PLoS ONE* **2017**, *12*. [CrossRef]

42. Enomoto, A.; Kido, N.; Ito, M.; Takamatsu, N.; Miyagawa, K. Serine-Threonine Kinase 38 is regulated by Glycogen Synthase Kinase-3 and modulates oxidative stress-induced cell death. *Free Radic. Biol. Med.* **2012**, *52*, 507–515. [CrossRef]

43. Vaahtera, L.; Brosché, M.; Wrzaczek, M.; Kangasjärvi, J. Specificity in ROS Signaling and Transcript Signatures. *Antioxid. Redox Signal.* **2013**, *21*, 1422–1441. [CrossRef]

44. Jiang, P.; Zhang, X.; Zhu, Y.; Zhu, W.; Xie, H.; Wang, X. Metabolism of reactive oxygen species in cotton cytoplasmic male sterility and its restoration. *Plant Cell Rep.* **2007**, *26*, 1627–1634. [CrossRef]

45. Deng, M.H.; Wen, J.F.; Huo, J.L.; Zhu, H.S.; Dai, X.Z.; Zhang, Z.Q.; Zhou, H.; Zou, X.X. Relationship of metabolism of reactive oxygen species with cytoplasmic male sterility in pepper (*Capsicum annuum* L.). *Sci. Hortic.* **2012**, *134*, 232–236. [CrossRef]

46. Senchina, D.S.; Alvarez, I.; Cronn, R.C.; Liu, B.; Rong, J.; Noyes, R.D.; Paterson, A.H.; Wing, R.A.; Wilkins, T.A.; Wendel, J.F. Rate variation among nuclear genes and the age of polyploidy in Gossypium. *Mol. Biol. Evol.* **2003**, *20*, 633–643. [CrossRef]

47. Zhang, T.; Hu, Y.; Jiang, W.; Fang, L.; Guan, X.; Chen, J.; Zhang, J.; Saski, C.A.; Scheffler, B.E.; Stelly, D.M.; et al. Sequencing of allotetraploid cotton (*Gossypium hirsutum* L. acc. TM-1) provides a resource for fiber improvement. *Nat. Biotechnol.* **2015**, *33*, 531–537. [CrossRef]

48. Magwanga, R.; Lu, P.; Kirungu, J.; Diouf, L.; Dong, Q.; Hu, Y.; Cai, X.; Xu, Y.; Hou, Y.; Zhou, Z.; et al. GBS Mapping and Analysis of Genes Conserved between *Gossypium tomentosum* and *Gossypium hirsutum* Cotton Cultivars that Respond to Drought Stress at the Seedling Stage of the BC2F2 Generation. *Int. J. Mol. Sci.* **2018**, *19*, 1614. [CrossRef]

49. Simon, U.K.; Polanschütz, L.M.; Koffler, B.E.; Zechmann, B. High Resolution Imaging of Temporal and Spatial Changes of Subcellular Ascorbate, Glutathione and H2O2 Distribution during Botrytis cinerea Infection in Arabidopsis. *PLoS ONE* **2013**, *8*. [CrossRef]

50. Heino, P.; Kariola, T.; Pennanen, V.; Broberg, M.; Survila, M.; Palva, E.T.; Sipari, N.; Davidsson, P.R. Peroxidase-Generated Apoplastic ROS Impair Cuticle Integrity and Contribute to DAMP-Elicited Defenses. *Front. Plant Sci.* **2016**, *7*. [CrossRef]

51. Rossi, F.R.; Krapp, A.R.; Bisaro, F.; Maiale, S.J.; Pieckenstain, F.L.; Carrillo, N. Reactive oxygen species generated in chloroplasts contribute to tobacco leaf infection by the necrotrophic fungus Botrytis cinerea. *Plant J.* **2017**, *92*, 761–773. [CrossRef] [PubMed]

52. Yoo, M.J.; Wendel, J.F. Comparative Evolutionary and Developmental Dynamics of the Cotton (*Gossypium hirsutum*) Fiber Transcriptome. *PLoS Genet.* **2014**, *10*. [CrossRef]

53. Li, S.; Zhang, F.; Wu, S.; Jin, X.; Zhang, T.; Guo, W.; Wang, L.; Cheng, C. A Cotton Annexin Affects Fiber Elongation and Secondary Cell Wall Biosynthesis Associated with Ca^{2+} Influx, ROS Homeostasis, and Actin Filament Reorganization. *Plant Physiol.* **2016**, *171*, 1750–1770. [CrossRef]

54. Van Montagu, M.; Van Camp, W.; Vranova, E.; Villarroel, R.; Atichartpongkul, S.; Inze, D. Comprehensive analysis of gene expression in *Nicotiana tabacum* leaves acclimated to oxidative stress. *Proc. Natl. Acad. Sci. USA* **2002**, *99*, 10870–10875. [CrossRef]

55. Rentel, M.C.; Lecourieux, D.; Ouaked, F.; Usher, S.L.; Petersen, L.; Okamoto, H.; Knight, H.; Peck, S.C.; Grierson, C.S.; Hirt, H.; et al. OXI1 kinase is necessary for oxidative burst-mediated signalling in Arabidopsis. *Nature* **2004**, *427*, 858–861. [CrossRef]

56. Jia, H.; Hao, L.; Guo, X.; Liu, S.; Yan, Y.; Guo, X. A Raf-like MAPKKK gene, GhRaf19, negatively regulates tolerance to drought and salt and positively regulates resistance to cold stress by modulating reactive oxygen species in cotton. *Plant Sci.* **2016**, *252*, 267–281. [CrossRef]

57. Chu, X.; Wang, C.; Chen, X.; Lu, W.; Li, H.; Wang, X.; Hao, L.; Guo, X. The cotton WRKY gene GhWRKY41 positively regulates salt and drought stress tolerance in transgenic *Nicotiana benthamiana*. *PLoS ONE* **2015**, *10*. [CrossRef]

58. Tang, C.; Kang, Z.; Chen, S.; Deng, L.; Chang, D.; Wang, X. TaADF3, an Actin-Depolymerizing Factor, Negatively Modulates Wheat Resistance Against *Puccinia striiformis*. *Front. Plant Sci.* **2016**, *6*. [CrossRef]

59. Shang, H.; Wang, Z.; Zou, C.; Zhang, Z.; Li, W.; Li, J.; Shi, Y.; Gong, W.; Chen, T.; Liu, A.; et al. Comprehensive analysis of NAC transcription factors in diploid Gossypium: Sequence conservation and expression analysis uncover their roles during fiber development. *Sci. China Life Sci.* **2016**, *59*, 142–153. [CrossRef]

60. Miao, Y.; Wang, P.; Du, Y.; Song, C.-P.; Zhao, X. The MPK6-ERF6-ROS-Responsive cis-Acting Element7/GCC Box Complex Modulates Oxidative Gene Transcription and the Oxidative Response in Arabidopsis. *Plant Physiol.* **2013**, *161*, 1392–1408. [CrossRef]

61. Fang, L.; Wang, Q.; Hu, Y.; Jia, Y.; Chen, J.; Liu, B.; Zhang, Z.; Guan, X.; Chen, S.; Zhou, B.; et al. Genomic analyses in cotton identify signatures of selection and loci associated with fiber quality and yield traits. *Nat. Genet.* **2017**, *49*, 1089–1098. [CrossRef]

62. Apel, K.; Hirt, H. REACTIVE OXYGEN SPECIES: Metabolism, Oxidative Stress, and Signal Transduction. *Annu. Rev. Plant Biol.* **2004**, *55*, 373–399. [CrossRef] [PubMed]

63. Rodriguez, A.A. Reactive Oxygen Species in the Elongation Zone of Maize Leaves Are Necessary for Leaf Extension. *Plant Physiol.* **2002**, *129*, 1627–1632. [CrossRef] [PubMed]

64. Rodríguez, A.A.; Ramiro Lascano, H.; Bustos, D.; Taleisnik, E. Salinity-induced decrease in NADPH oxidase activity in the maize leaf blade elongation zone. *J. Plant Physiol.* **2007**, *164*, 223–230. [CrossRef]

65. Yamaguchi, M.; Sharp, R.E. Complexity and coordination of root growth at low water potentials: Recent advances from transcriptomic and proteomic analyses. *Plant Cell Environ.* **2010**, *33*, 590–603. [CrossRef]

66. Shivashankara, K.S.; Pavithra, K.C.; Geetha, G.A. Antioxidant protection mechanism during abiotic stresses. In *Abiotic Stress Physiology of Horticultural Crops*; Springer: New Delhi, India, 2016; pp. 47–70. ISBN 9788132227250.

67. Lee, S.; Seo, P.J.; Lee, H.J.; Park, C.M. A NAC transcription factor NTL4 promotes reactive oxygen species production during drought-induced leaf senescence in Arabidopsis. *Plant J.* **2012**, *70*, 831–844. [CrossRef]

68. Zhou, T.; Yang, X.; Wang, L.; Xu, J.; Zhang, X. GhTZF1 regulates drought stress responses and delays leaf senescence by inhibiting reactive oxygen species accumulation in transgenic Arabidopsis. *Plant Mol. Biol.* **2014**, *85*, 163–177. [CrossRef]

69. Mahmood, K.; El-Kereamy, A.; Kim, S.H.; Nambara, E.; Rothstein, S.J. ANAC032 positively regulates age-dependent and stress-induced senescence in arabidopsis thaliana. *Plant Cell Physiol.* **2016**, *57*, 2029–2046. [CrossRef]

70. Wang, J.; Wang, J.; Qiu, Q.; Wang, K.; Yang, H.; Abbott, R.J.; Liu, J.; Yu, L.; Ma, H.; Lu, Y.; et al. Erratum: Genomic insights into salt adaptation in a desert poplar. *Nat. Commun.* **2014**, *5*. [CrossRef]

71. Finn, R.D.; Coggill, P.; Eberhardt, R.Y.; Eddy, S.R.; Mistry, J.; Mitchell, A.L.; Potter, S.C.; Punta, M.; Qureshi, M.; Sangrador-Vegas, A.; et al. The Pfam protein families database: Towards a more sustainable future. *Nucleic Acids Res.* **2016**, *44*, D279–D285. [CrossRef]

72. Marchler-Bauer, A.; Bo, Y.; Han, L.; He, J.; Lanczycki, C.J.; Lu, S.; Chitsaz, F.; Derbyshire, M.K.; Geer, R.C.; Gonzales, N.R.; et al. CDD/SPARCLE: Functional classification of proteins via subfamily domain architectures. *Nucleic Acids Res.* **2017**, *45*, D200–D203. [CrossRef]

73. Larkin, M.A.; Blackshields, G.; Brown, N.P.; Chenna, R.; Mcgettigan, P.A.; McWilliam, H.; Valentin, F.; Wallace, I.M.; Wilm, A.; Lopez, R.; et al. Clustal W and Clustal X version 2.0. *Bioinformatics* **2007**, *23*, 2947–2948. [CrossRef]

74. Kumar, S.; Stecher, G.; Tamura, K. MEGA7: Molecular Evolutionary Genetics Analysis Version 7.0 for Bigger Datasets. *Mol. Biol. Evol.* **2016**, *33*, 1870–1874. [CrossRef]

75. Voorrips, R.E. MapChart: Software for the Graphical Presentation of Linkage Maps and QTLs. *J. Hered.* **2002**, *93*, 77–78. [CrossRef]

76. Wang, Y.; Tang, H.; Debarry, J.D.; Tan, X.; Li, J.; Wang, X.; Lee, T.H.; Jin, H.; Marler, B.; Guo, H.; et al. MCScanX: A toolkit for detection and evolutionary analysis of gene synteny and collinearity. *Nucleic Acids Res.* **2012**, *40*. [CrossRef] [PubMed]

77. Krzywinski, M.; Schein, J.; Birol, I.; Connors, J.; Gascoyne, R.; Horsman, D.; Jones, S.J.; Marra, M.A. Circos: An information aesthetic for comparative genomics. *Genome Res.* **2009**, *19*, 1639–1645. [CrossRef] [PubMed]

78. Wang, Y.; Xiao, D.; Li, Y.; Li, X. Soil salinity evolution and its relationship with dynamics of groundwater in the oasis of inland river basins: Case study from the Fubei Region of Xinjiang Province, China. *Environ. Monit. Assess.* **2008**, *140*, 291–302. [CrossRef] [PubMed]

79. Yanchao, X.; Yangyang, W.; Zhenqing, L.; Xiaoyan, C.; Yuhong, W.; Xingxing, W.; Zhenmei, Z.; Kunbo, W.; Fang, L.; Zhongli, Z. Integrated Evaluation and the Physiological and Biochemical Responses of Semi-Wild Cotton under Complex Salt-Alkali Stress. *Cotton Sci.* **2018**, 231–241. [CrossRef]

80. Ewing, B.; Hillier, L.D.; Wendl, M.C.; Green, P. Base-calling of automated sequencer traces using phred. I. Accuracy assessment. *Genome Res.* **1998**, *8*, 175–185. [CrossRef]

81. Kim, D.; Pertea, G.; Trapnell, C.; Pimentel, H.; Kelley, R.; Salzberg, S.L. *TopHat2: Accurate* alignment of transcriptomes in the presence of insertions, deletions and gene fusions. *Genome Biol.* **2013**, *14*, R36. [CrossRef] [PubMed]

Int. J. Mol. Sci. **2019**, *20*, 1863

82. Trapnell, C.; Williams, B.A.; Pertea, G.; Mortazavi, A.; Kwan, G.; van Baren, M.J.; Salzberg, S.L.; Wold, B.J.; Pachter, L. Transcript assembly and quantification by RNA-Seq reveals unannotated transcripts and isoform switching during cell differentiation. *Nat. Biotechnol.* **2010**, *28*, 511–515. [CrossRef] [PubMed]

83. Magwanga, R.O.; Lu, P.; Kirungu, J.N.; Dong, Q.; Hu, Y.; Zhou, Z.; Cai, X.; Wang, X.; Hou, Y.; Wang, K.; et al. Cotton Late Embryogenesis Abundant (LEA2) Genes Promote Root Growth and Confers Drought Stress Tolerance in Transgenic *Arabidopsis thaliana*. *G3 Genes Genomes Genet.* **2018**, *8*, 2781–2803. [CrossRef]

84. Yoshino, M. Climate Change and Its Effects on Agriculture. *J. Agric. Meteorol.* **2011**, *46*, 251–254. [CrossRef]

85. Diouf, L.; Pan, Z.; He, S.-P.; Gong, W.-F.; Jia, Y.H.; Magwanga, R.O.; Romy, K.R.E.; Rashid, H.; Kirungu, J.N.; Du, X. High-Density Linkage Map Construction and Mapping of Salt-Tolerant QTLs at Seedling Stage in Upland Cotton Using Genotyping by Sequencing (GBS). *Int. J. Mol. Sci.* **2017**, *18*, 2622. [CrossRef]

86. Kohli, S.K.; Handa, N.; Gautam, V.; Bali, S.; Sharma, A.; Khanna, K.; Arora, S.; Thukral, A.K.; Ohri, P.; Karpets, Y.V.; et al. ROS signaling in plants under heavy metal stress. In *Reactive Oxygen Species and Antioxidant Systems in Plants: Role and Regulation under Abiotic Stress*; Springer: Singapore, 2017; pp. 185–214. ISBN 9789811052545.

International Journal of
Molecular Sciences

Article

Phylogenetic, Molecular, and Functional Characterization of PpyCBF Proteins in Asian Pears (*Pyrus pyrifolia*)

Mudassar Ahmad [1,2,3,†], **Jianzhao Li** [1,2,3,†], **Qinsong Yang** [1,2,3], **Wajeeha Jamil** [1,2,3], **Yuanwen Teng** [1,2,3,*] and **Songling Bai** [1,2,3,*]

[1] Department of Horticulture, Zhejiang University, Hangzhou 310058, Zhejiang, China; ahmad_mudassar@zju.edu.cn (M.A.); 21316043@zju.edu.cn (J.L.); qsyang@zju.edu.cn (Q.Y.); 11616122@zju.edu.cn (W.J.)
[2] The Key Laboratory of Horticultural Plant Growth, Development and Quality Improvement, the Ministry of Agriculture of China, Hangzhou 310058, Zhejiang, China
[3] Zhejiang Provincial Key Laboratory of Integrative Biology of Horticultural Plants, Hangzhou 310058, Zhejiang, China
[*] Correspondence: ywteng@zju.edu.cn (Y.T.); songlingbai@zju.edu.cn (S.B.)
[†] These authors contributed equally to this work.

Received: 24 March 2019; Accepted: 24 April 2019; Published: 26 April 2019

Abstract: C-repeat binding factor/dehydration-responsive element (CBF/DRE) transcription factors (TFs) participate in a variety of adaptive mechanisms, and are involved in molecular signaling and abiotic stress tolerance in plants. In pear (*Pyrus pyrifolia*) and other rosaceous crops, the independent evolution of CBF subfamily members requires investigation to understand the possible divergent functions of these proteins. In this study, phylogenetic analysis divided six *PpyCBFs* from the Asian pear genome into three clades/subtypes, and collinearity and phylogenetic analyses suggested that *PpyCBF3* was the mother CBF. All *PpyCBFs* were found to be highly expressed in response to low temperature, salt, drought, and abscisic acid (ABA) as well as bud endodormancy, similar to *PpyCORs* (*PpyCOR47*, *PpyCOR15A*, *PpyRD29A*, and *PpyKIN*). Transcript levels of clade II *PpyCBFs* during low temperature and ABA treatments were higher than those of clades I and III. Ectopic expression of *PpyCBF2* and *PpyCBF3* in *Arabidopsis* enhanced its tolerance against abiotic stresses, especially to low temperature in the first case and salt and drought stresses in the latter, and resulted in lower reactive oxygen species (ROS) and antioxidant gene activities compared with the wild type. The increased expression of endogenous ABA-dependent and -independent genes during normal conditions in *PpyCBF2*- and *PpyCBF3*-overexpressing *Arabidopsis* lines suggested that *PpyCBFs* were involved in both ABA-dependent and -independent pathways. All *PpyCBFs*, especially the mother CBF, had high transactivation activities with 6XCCGAC binding elements. Luciferase and Y1H assays revealed the existence of phylogenetically and promoter-dependent conserved *CBF–COR* cascades in the pear. The presence of a previously identified CCGA binding site, combined with the results of mutagenesis of the CGACA binding site of the *PpyCOR15A* promoter, indicated that CGA was a core binding element of *PpyCBFs*. In conclusion, PpyCBF TFs might operate redundantly via both ABA-dependent and -independent pathways, and are strongly linked to abiotic stress signaling and responses in the Asian pear.

Keywords: asian pears; CBF; gene functions; CRT/DRE binding sites

1. Introduction

C-repeat binding factors/dehydration-responsive elements (CBFs/DREs) constitute a subfamily of the Apetala1/ethylene responsive factor (AP1/ERF) family and are characterized by the presence of one AP2 domain [1] that contains 60–70 highly conserved amino acid residues [2]. All CBFs have CBF signature motifs (PKK/RPAGRxKFxETRHP and DSAWR) that distinguish these factors from other AP1/ERF members harboring an AP2 domain [3]. This CBF motif specifically binds to the dehydration-responsive/C repeat (DRE/CRT) element (CCGAC) of downstream genes to regulate their expressions [4]. CBFs have a well-known role in cold response and acclimation in both herbaceous [5] and woody [6] plants. Studies on the poplar (*Populus trichocarpa),* eucalyptus (*Eucalyptus globulus),* grape *(Vitis vinifera),* sweet cherry *(Prunus avium),* birch (*Betula pendula),* citrus (*Citrus paradisi),* and dwarf apple *(Malus baccata),* have revealed that the cold acclimation function of CBF is highly conserved in these woody plants [7,8]. Nevertheless, several recent studies have suggested that the multiple CBF paralogs that have evolved in plants might perform different functions [9]. In this aspect, (i) CBF paralogs can influence each other's expressions. In *Arabidopsis,* for example, *AtCBF2* negatively regulates the expressions of *AtCBF1* and *AtCBF3* [10]. (ii) In addition, CBF paralogs have different tissue specificities and expression times following cold stress. For example, *PtCBF2* and *PtCBF4* in poplars were detected only in leaves, whereas *PtCBF1* and *PtCBF3* were also expressed in leaves, stems, and dormant buds [11]. A similar result has also been reported in grapes, where *Vitis CBF4* was present in mature leaves and buds, while *Vitis CBF1, CBF2,* and *CBF3* were only found in young leaves and buds [12,13]. (iii) Several *CBF* genes have also been found to be induced by other abiotic stresses (drought and salt) and molecular signals (such as abscisic acid signaling). These include *GmDREB1G-1* and *GmDREB1G-2* in soybeans [14], *VrCBF1* and *VrCBF4* in grapse [9], *MbDREB1A* in dwarf apples [15], and *AtDDF1, AtDDF2* [16], and *AtCBF4* [17] in *Arabidopsis.* (iv) Overexpressed CBF paralogs from other species conferred various levels of abiotic stress tolerance on plants. For example, overexpression of both *VrCBF1* and *VrCBF4* enhances abiotic stress tolerance in *Arabidopsis,* but *VrCBF1* is mainly responsible for drought tolerance, while *VrCBF4* confers most of the cold tolerance [9].

A core set of robustly stress-responsive plant genes, known as *COR* (cold-regulated), *RD* (responsive to dehydration), and *KIN* (cold-induced), have been identified from numerous differential screening and cloning studies over the years. Many *COR* genes contain one or more similar CRT (CCGAC) elements in their promoters, which are also found in *CRT/DRE* genes, and interestingly, they all have abiotic stress responsiveness [18]. Abiotic stress rapidly induces CBFs, which then activate various downstream cold-responsive (*COR*) genes whose products collectively increase a plant's abiotic tolerance capacity through necessary physiological and biochemical alterations [19]. The cold-stress induction of *CBF* and *COR* genes is also regulated by the circadian clock [20]. An important feature of abiotic stresses, especially low temperature, is a hyperosmotic signal that causes the phytohormone abscisic acid (ABA) to accumulate. ABA in turn provokes many adaptive responses, such as bud endodormancy, in plants [21]. Low temperatures and ABA have recently been reported to synergistically promote cold-hardiness and CBF expression in dormant grape buds [21]. These adaptive mechanisms are not only affected by ABA contents, but also by ABA signaling pathways [22]. For example, high ABA levels lead to endodormancy [23], inhibition of ABA pathways promotes germination and lateral root formation [24], while the reduction of ABA enhances water transpiration through stomatal pores [25].

Adaptive mechanisms, molecular signaling, and tolerance to abiotic stresses are also determined by many up- and downstream transcription factors of *CBF* genes. During the adaptive process of bud endodormancy in pears, for example, *PpICE3* works upstream of *PpCBF1,* while *PpCBF1, PpCBF2* and *PpCBF4* activate downstream *PpDAM1* and *PpDAM3* genes that induce endodormancy by inhibiting *PpFT2.* Meanwhile, microRNA *miR6390* degrades dormancy associated MADS (DAM) box genes to release endodormancy [22,26]. *MdMYB* and *MdHY5* in apples and *PbeNAC1* in pear have also been found to be involved in the regulation of *CBF* genes and the acquisition of abiotic stress tolerance [27–29]. In regards to molecular signals such as ABA, the PYR/RCAR–PP2C complex [30] inhibits PP2C [31] and activates SnRK2s, which not only target ABA-responsive genes (*ABF/ABI5*-type basic/region leucine zipper) [32], but also phosphorylate *ICE1* to activate CBF–COR cascades and promote plant tolerance through ABA signaling [33]. During abiotic stress, many transcription factors, i.e., COLD1, NAC, bHLH, ICE1, MYB, SnrK2, ABF, HOS1, and SIZ1, have been found to function upstream of CBFs, while

ADF, ZAT, LOS, SFR, and RAP function downstream to induce plant tolerance [34]. Consequently, CBF is the central regulator of plant adaptation and abiotic stress tolerance via both ABA-dependent and -independent pathways [15].

Pyrus germplasm resources, which are distributed worldwide, are most plentiful in China, especially in the western and southwestern mountainous areas [35,36]. Numerous genes and TFs with functions related to plant dispersal, adaptation to natural habitats, and stress tolerance had been identified and characterized in plants, including AREB/ABF, MYB, AP2/EREBP, bZIP, HSF, CBF/DREB, MYC, HB, NAC, and WRKY. Among them, the CBF/DREB subfamily occupies a major position in both herbaceous [5] and woody [6] plants. The complete CBF subfamily and the possible divergent functions of its members have never been fully studied in rosaceous groups. In this study, we identified 15 *PpyCBFs* from the pear genome database, but were unable to predict their functions through phylogenetic analysis. Hence, we tested the hypothesis to know whether all *PpyCBF* paralogs had different functions or not. We therefore selected six of the 15 *PpyCBFs* after characterization and checked their responses to abiotic stresses, ABA treatment, and bud endodormancy compared with abiotic stress-responsive *PpyCOR* genes. We also generated *PpyCBF2-* and *PpyCBF3*-overexpressing *Arabidopsis* plants and analyzed their abiotic stress tolerances, endogenous gene expressions, and ROS accumulations. After checking the binding activity of all *PpyCBFs* with the *cis*-element (CCGAC), we also studied their possible abiotic regulatory pathways and binding sites in pears.

2. Results

2.1. Identifications and Characterizations of PpyCBF Subfamily

To identify *PpyCBFs*, we first carried out a hidden Markov model search against the pear genome database. This approach identified 15 PpyCBF TFs, which were then subjected to phylogenetic analysis and further confirmation of their sequence identities and chromosomal positions. Pairwise sequence identities among isolated *PpyCBFs* were all very high, ranging from 0.271 (*PpyCBF9* and *PpyCBF10* vs. *PpyCBF12*) to 0.994 (*PpyCBF15* vs. *PpyCBF4*) (Table S1). Sequences that had an identity >0.90 and were on the same phylogenetic branch (*PpyCBFs 7,8,9,10,11,12,13,14*), incomplete (*PpyCBF12*), or on a scaffold (*PpyCBFs 7,8,10,11,13,14,15*) were eliminated from further analysis, whereas their corresponding sequences, i.e., *PpyCBFs 1–6*, were retained (Figure 1a, Table S1). To explore evolutionary relationships within the isolated subfamily, we first constructed a phylogenetic tree of sequences of similar candidates in *Pyrus (Ppy)*, *Arabidopsis (At)*, *Malus (Md)*, *Prunus (Ppe)*, *Fragaria (Fv)*, and *Vitis (Vv)*. The phylogenetic analysis distributed the *PpyCBFs* into three main clades/subtypes: *PpyCBF3* in clade I, *PpyCBFs 1,2,4* in clade II, and *PpyCBF5* and *PpyCBF6* in clade III. Interestingly, *PpyCBFs*, along with *CBFs* of other rosaceous crop species, appeared to be evolved independently of model crop *CBFs (AtCBFs 1–4)*. With the exception of *PpyCBF3*, which was clustered in clade I with *Arabidopsis CBFs*, all other *PpyCBFs* were placed in clades II and III with *MdDREBs* and *PpeDREBs* (Figure 1a). This independent evolution of *PpyCBFs* suggested their potential divergent functions and served as the impetus for our study to explore and elucidate the regulation of this family in pears.

Since *PpyCBFs* belong to the AP2/ERF family, we performed a collinearity analysis of the entire family to understand *PpyCBF* evolution and gene duplication (Figure S1a). We found 68 duplicated AP2/ERF pairs. Among them, two pairs, i.e., *Pbr013924(PpyCBF3):Pbr032764(PpyCBF5)* and *Pbr013924(PpyCBF3):Pbr021781(PpyCBF1)*, belonged to its *PpyCBF* subfamily (Figure S1b). These results suggest that clades II and III of CBFs, i.e., *PpyCBF1* and *PpyCBF5*, evolved from *PpyCBF3*, which was found in an ancestral clade with both monocot and dicot plants (Figure 1a). To examine diversification in gene structures and uncover potential conserved motifs in these selected *PpyCBFs*, we constructed another phylogenetic tree, which revealed that both duplicated *PpyCBF3* and *PpyCBF5*, and *PpyCBF2* and *PpyCBF4* had potentially similar functions. In addition, *PpyCBF5* together with *PpyCBF6* were in a sister relationship with a cluster comprising *PpyCBF1* and *PpyCBFs 2,4*, with the branch leading to these genes in turn joined to the ancestral *CBF* (Figure 1b). Regarding gene structures and conserved motifs, *PpyCBF5* was the only gene with just one intron. All the others had exonic regions (Figure 1c). Alignment of *PpyCBFs* in each phylogenetic clade revealed 10 different types of common motifs (Figure 1d). These findings indicate that *PpyCBFs* in the same clade have similar gene structures and motifs, and possibly similar functions.

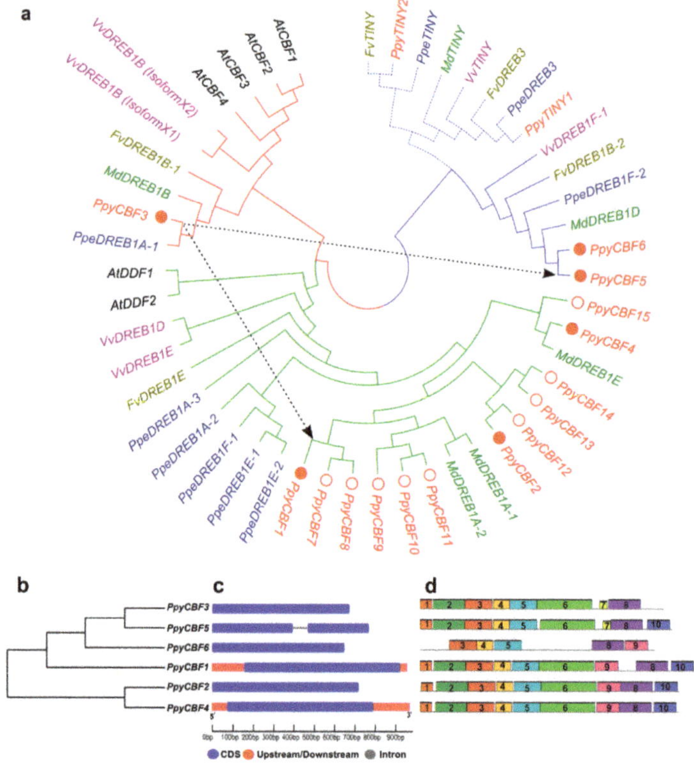

Figure 1. Identification and characterization of *PpyCBFs*. (**a**) Phylogenetic analysis of *PpyCBF* transcription factors with similar TFs of *Arabidopsis* (*At*), *Malus* (*Md*), *Prunus* (*Ppe*), *Fragaria* (*Fv*), and *Vitis* (*Vv*) species. Red, green, and blue colors indicate clades/subtypes I, II, and III of CBFs, respectively, while compact and hollow red circles indicate selected and rejected *PpyCBFs*, respectively. Arrow lines indicate the evolution of clades II and III from clade I. (**b**) Phylogenetic analysis of selected *PpyCBFs*. (**c**) Gene structure of *PpyCBFs*. Blue, black, and red lines indicate exon, intron, and upstream/downstream sections in gene structure. (**d**) Protein motif: Schematic diagrams of possible conserved motifs (1–10) in *PpyCBF* proteins, indicated by different colors.

2.2. Strong Induction of PpyCBF Transcription by Various Abiotic Stresses and ABA Treatment

To better understand the functions of *PpyCBFs*, we examined transcript levels of *PpyCBFs* in explants of *Pyrus pyrifolia* 'Dangshan Suli' subjected to different abiotic stress treatments, i.e., low temperature (4 °C), drought (15% polyethylene glycol (PEG)) and salt (200 mM NaCl), for 0, 6, 12, 24, and 48 h. qRT-PCR analysis revealed that the expressions of all six *PpyCBF* genes were induced by all abiotic stresses, but each gene responded differently to various stresses depending on its associated clade (Figure 2a). During cold treatment, expressions of *PpyCBFs* were all constant from 6 to 48 h and significantly higher than the control, with relative abundances of clade II CBFs which were much higher (~200–1600) than those of clade I and II CBFs (~2–50). During salt treatment, all *PpyCBFs* were statistically at their maximums after 12 and 48 h except for *PpyCBF4* (which peaked only at 48 h). The responses of clade I and III *PpyCBFs* were higher at early stages of salt stress than those of clade II *PpyCBFs*. Under drought conditions, *PpyCBF3* (12 h), *PpyCBF2* (24 h), *PpyCBF4* (24 h), and *PpyCBF5* (48 h) were accentuated, while *PpyCBF1* and *PpyCBF6* were downregulated. To determine whether *PpyCBFs* respond to ABA, we also tested their expressions in pear calli after 0, 3, 6, 12, and 48 h of ABA treatment (100 µM). Notably, all *PpyCBFs* had responses to ABA after 3 and 48 h. Short-term ABA exposure significantly promoted the expressions of clade II *PpyCBFs*, whereas longer exposure significantly induced the members of the other two clades *(PpyCBF3* and *PpyCBF6)*. Expression levels of clade II *PpyCBFs* were much higher than those of clades I and III. Significant downregulation of *PpyCBF3* (24 h), *PpyCBF1* (24 h), *PpyCBF5* (6 h), and *PpyCBF6* (12 h) was also observed during ABA treatment of pear calli (Figure 2a). In summary, clade I and III *PpyCBFs* exhibited higher levels of transcripts during salinity and drought treatments, whereas clade II *PpyCBF* transcripts were more abundant during low temperature and ABA stresses.

We also compared the expressions of *PpyCBFs* with those of *COR* genes (*PpyCOR47*, *PpyCOR15A*, *PpyRD29A*, and *PpyKIN*) during ABA treatment and abiotic stress. qRT-PCR analysis uncovered highly significant expressions of *PpyCORs* during cold, salt, and drought stresses, the exception being *PpyRD29A* during drought. Likewise, *PpyCORs* exhibited a highly significant, constant response throughout ABA treatment (Figure 2b). To confirm the above results and check the stress status of explants and calli, we measured expression levels of antioxidant genes (*PpySOD*, *PpyPOD*, *PpyAPX*, and *PpyCAT*) during abiotic stress and those of ABA-responsive genes (*PpyCYP707A-2*, *PpySnRK2-1* and *PpySnRK2-4*, *PpyABi5*, and *PpyPYL-2*) subjected to ABA treatment (Figure S2). The expressions of all these genes were found to be high. These results not only verify the effectiveness of the treatments, but also suggested that all *PpyCBFs* were differentially induced according to their clades during abiotic stresses and ABA treatments.

To understand the possible transcriptional regulatory cascades of *PpyCBFs*, we also analyzed their promoters. We detected numerous *cis* elements responsive to biotic and abiotic stresses, molecular signaling, and plant adaptation in promoters of *PpyCBF* transcription factors related to cold, salt, drought, oxidation, light, heavy metals, pathogens, heat, ABA, giberllic acid, and auxin, namely, ABI3/VP1, AP2/EREBP, AP2/RAV, ARF, bHLH, bZIP, ERF, GATA, MADS, MYB, MYC, NAC, TCP/PCF1, and WRKY *cis* elements (Table 1 and Table S2). We found varying degrees of differences between the types and numbers of *PpyCBF* regulatory elements. The presence of these *cis* elements suggests that ABA and stress-inducible expressions of *PpyCBFs* are transcriptionally regulated.

2.3. Increased Transcripts of PpyCBFs Induced by Low Temperature and ABA during Pear Bud Endodormancy

As inferred from the above results, all *PpyCBFs* responded to ABA and low temperature, two basic factors for the establishment of bud endodormancy. We therefore also verified the expressions of *PpyCBFs* during the endodormancy period from September to February in Asian pear cultivars 'Dangshan Suli' and 'Cuiguan' at 15-day intervals in 2016–2017 and 2017–2018. During bud endodormancy, we observed two peaks in *PpyCBF* expression, the first one related to low temperature and the other dependent on ABA. In both pear cultivars, all *PpyCBFs* had their first expression peaks on January 1–12, 2017, and January 10–11, 2018, with their maximum expressions on November 15 and October 15 of the two respective years (Figure 3). As reported in our previous study [22], below-normal maximum and minimum temperatures were observed from October 15 to November 15 during 2016–2017, with the winter season also delayed in 2016–2017 compared with 2017–2018 (November vs. October). These events ultimately affected the transcription of *CBFs* during both years. Nevertheless, *PpyCBF* transcripts in both cultivars had their second expression peaks between January 1–20, 2017, and from December 1, 2017, to January 1, 2018, with maximums observed in the middle of January and December in the two successive years. This indicated ABA-dependent responses of *PpyCBFs* during bud endodormancy (Figure 3) because, in our previous study of ABA-responsive genes, *PpyNCED1*, *PpyCYP707A-3* and *PpyCYP707A-4*, and *PpyLs 2,3,6,7,8* were at their peaks on January 1–20 during bud endodormancy [23]. Interestingly, the relative abundances of clade II *PpyCBFs* during low temperature and ABA peaks were higher than those of clades I and III during both years in both cultivars, consistent with our results discussed earlier (Figure 2a).

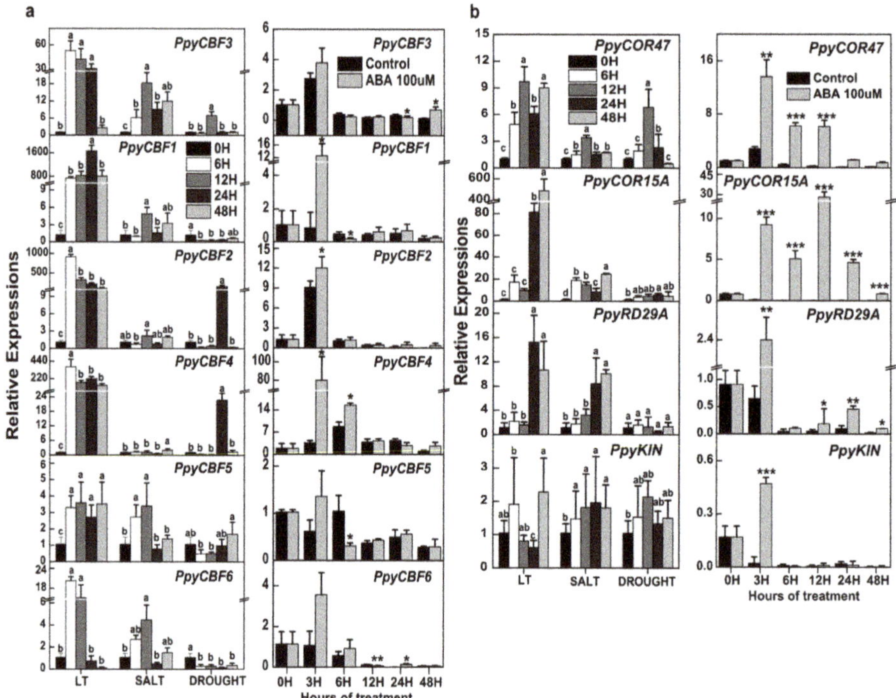

Figure 2. Relative expressions of *PpyCBFs* and *PpyCORs* during abiotic stresses and exogenous abscisic acid (ABA). (**a**) Expression analysis of *PpyCBFs* during abiotic stresses (cold, salt, and drought) and ABA according to their phylogenetic clades. (**b**) Expression analysis of *PpyCOR47, 15A, RD29A*, and *KIN* in the same samples for comparison study. Both relative expressions were normalized to *PpyActin* expression level. Error bars indicate standard errors from three biological replicates (* $p < 0.05$, ** $p < 0.01$, *** $p < 0.001$) while means with different letters had significant differences ($p < 0.05$).

To further clarify low-temperature and ABA responses of *PpyCBFs* during bud endodormancy, we rechecked the responses of the studied *PpyCORs* during pear bud endodormancy to verify their high expressions during low temperature and ABA treatments (Figure 2b). Similar to the *PpyCBFs*, all *PpyCORs* (*PpyCOR47, 15A, RD29A*, and *KIN*) had expression peaks from November 15, 2016, to December 1, 2016, and from October 1, 2017, to November 1, 2017, corresponding to a low-temperature response, and from January 1–10, 2017, and from December 12, 2017, to January 1, 2018, corresponding to an ABA response, in both cultivars, with the exception of *PpyKIN* during 2016–2017 (Figure S3). The relative abundance of *PpyCOR15A* during low temperature and ABA peaks was higher than that of other *CORs* during low-temperature and ABA treatments (Figure 2b). These results not only reveal the responses of *PpyCBFs* and *PpyCORs* during bud endodormancy but also demonstrate their obvious correlation to each other.

Table 1. Promoter analysis of all isolated *PpyCBFs*.

TFs family	Functions	*cis*-Element	Sequences	*PpyCBF3*	*PpyCBF1*	*PpyCBF2*	*PpyCBF4*	*PpyCBF5*	*PpyCBF6*
ABI3/VP1	ABA responsive	ABRE	CATGC	1	4	1	4	1	1
AP2/EREBP	Cold, drought, NaCl	CRT/DRE	CCGAC	6	4	1	4	8	3
AP2/RAV	Photoperiodism, flowering	B3	CAACA	10	8	5	7	9	8
ARF	Auxin response	SURE	GAGACA	3	2	2	2	2	1
bHLH	Iron toxicity	IRO2	CACGTGG	0	0	2	2	0	2
bZIP	ABA, NaCl, drought, heat	G-box1	CACGTG	0	1	2	2	0	3
bZIP	Salt, Pathogen	GT-1-like box	GAAAAA	3	3	7	3	4	4
ERF	Defense responses	GCC box	AGCCG	7	1	0	4	9	0
GATA	Light response	GATA box	GATA	14	16	16	11	12	15
MADS	Plant development	MIKC	CC[A/T]5	1	0	1	3	1	2
MYB like	Light response	I BOX	AAACCA	1	0	2	1	0	0
MYB/SANT	Gibberellin response	GARC	AACAAA	6	3	6	4	2	3
MYC-like bHLH	Cold stress	ICE1-like	CATTTG	1	1	4	1	2	1
NAC	Cold, drought, NaCl	NAC	CATGT	2	3	3	2	3	3
TCP/PCF1	Oxidative stress	Site 2	TGGGC	3	1	3	1	1	2
WRKY	Bacterial blight	PRE2	ACGCTG	1	0	0	0	2	0
WRKY	Bacterial blight	PRE4	TGCGCT	1	0	0	0	2	1

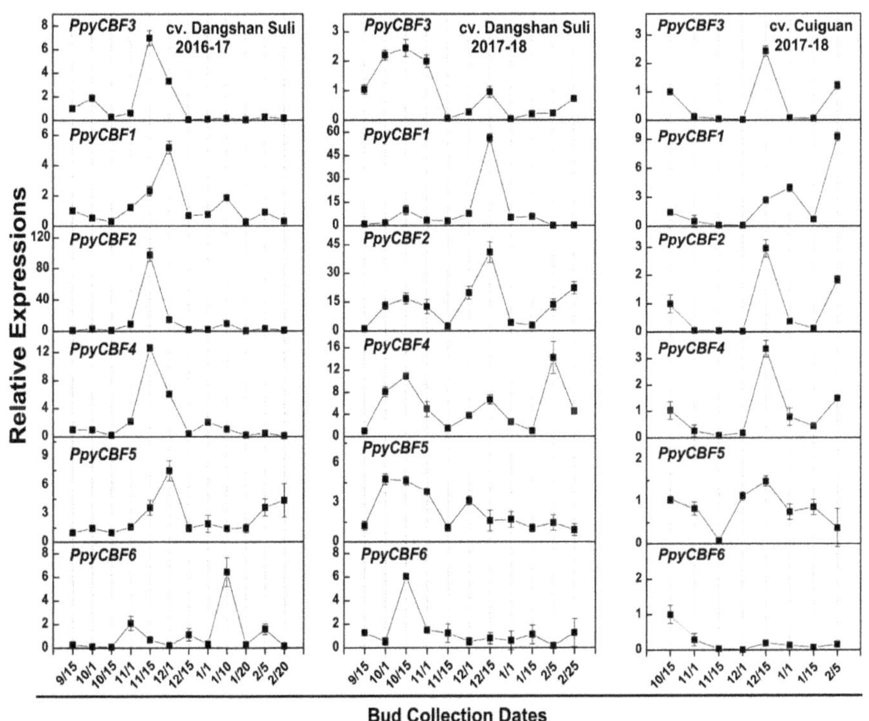

Figure 3. Relative expressions of *PpyCBFs* during bud endodormancy in *Pyrus pyrifolia* cv. 'Dangshan Suli' and 'Cuiguan' during two successive years 2016–2017 and 2017–2018. Buds were collected from September 15 to February 25 with about 15-day intervals. The data were normalized to *PpyActin* levels and the mean expression value was premeditated from four independent replicates. The standard deviation was shown by vertical bars.

2.4. Overexpressions of PpyCBF2 and PpyCBF3 Positively Regulate Abiotic Stress Tolerances in Transgenic Arabidopsis

To test whether *PpyCBFs* overexpression positively enhances abiotic stress tolerance, pCAMBIA1301 overexpression constructs of *PpyCBF2* (the most transcriptionally activated CBF) and *PpyCBF3* (the mother CBF) were transformed into *Arabidopsis*. Consistent with abiotic stress assays, phenotypes of both *PpyCBF2-ox* and *PpyCBF3-ox* transgenic lines were superior in several respects to the wild type (Figure S4a). Ectopic expression of *PpyCBF2* and *PpyCBF3* led to highly significantly increased root lengths after treatment with low temperature (1.7 and 1.3 cm, respectively), salt (1.5 and 2.1 cm), and drought (2.0 and 2.5 cm) compared with wild-type plants (0.8, 0.7, and 0.6 cm under low temperature, salinity, and drought, respectively), whereas no differences were observed among wild-type, *PpyCBF2-ox*, and *PpyCBF3-ox* plants under non-stress conditions (2.1, 2.2, and 1.9 cm, respectively) (Figure 4a). Interestingly, *PpyCBF2-ox* plants under low temperature stress and *PpyCBF3-ox* plants under salinity and drought stress had more pronounced length increases relative to the wild type, but more growth retardation was observed in all plants during low temperature stress than during salt and drought stress.

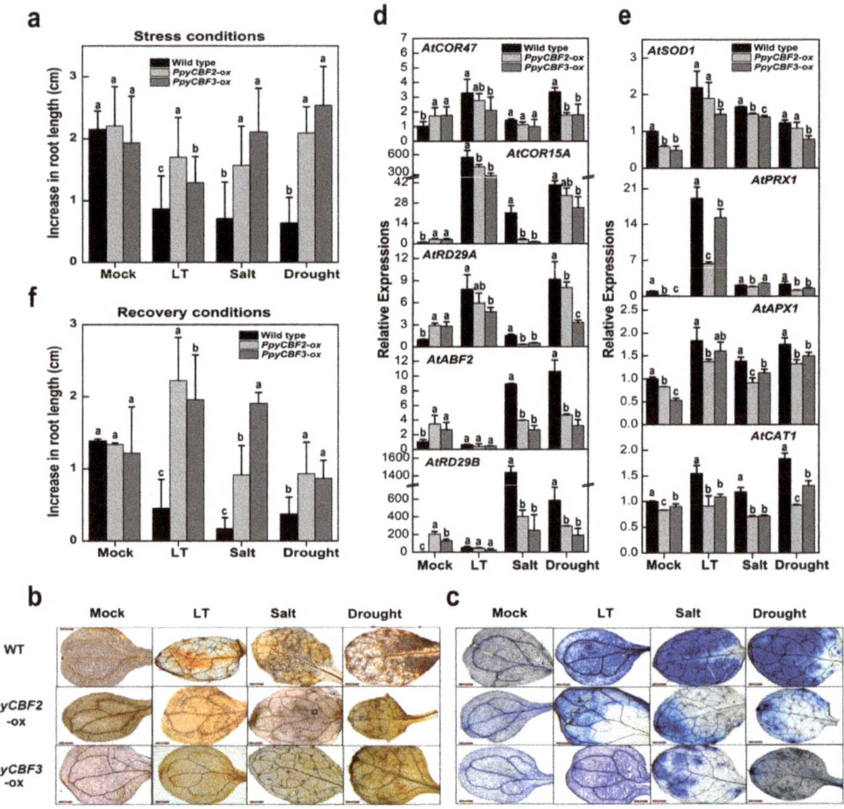

Figure 4. Overexpression analysis of *PpyCBFs* 2 and 3 in *Arabidopsis* during abiotic stresses. (**a**) Increase in root length (cm) of wild type (WT) and overexpressed lines during low temperature (LT), salt, and drought treatments by using ImageJ software. Error bars indicate standard errors from three biological replicates. (**b,c**) Diaminobenzidine (DAB) and nitroblue tetrazolium (NBT) staining of WT and overexpressed leaves after abiotic stresses to check ROS accumulation where brown and blue spots indicate the presence of H_2O_2 and $O2^{\bullet-}$ in situ while the red bar scale represent 200 μm. (**d,e**) Endogenous gene expressions of ABA-independent (*AtCOR47, AtCOR15A* and *AtRD29A*), ABA-dependent (*AtABF2* and *AtRD29B*) and antioxidant genes (*AtSOD1, AtPRX1, AtAPX1* and *AtCAT1*) in WT and overexpressed lines during control and abiotic stresses, normalized to *AtPP2A* expression levels. (**f**) Increase in root length to monitor the recovery among overexpressed and WTs *Arabidopsis* under normal conditions after abiotic stresses. Error bars indicate standard error from three biological replicates. Means with different letters had significant differences ($p < 0.05$).

To confirm the effect of *PpyCBF2-ox* and *PpyCBF3-ox* on endogenous *Arabidopsis* genes, we examined the expressions of three ABA-independent (*AtCOR47/RD17*, *AtCOR15a*, and *AtRD29A/COR78/LTI78*), two ABA-dependent (*AtABF2* and *AtRD29B*) and four antioxidant (*AtSOD1*, *AtPRX1*, *AtAPX1*, *AtCAT1*) genes. In *Arabidopsis* overexpressing either *PpyCBF2* or *PpyCBF3* under control or unstressed conditions, the ABA-dependent and -independent genes were significantly upregulated, and the antioxidant genes were downregulated (Figure 4d,e). Under each stress treatment, relative abundances of all stress-responsive and antioxidant genes were significantly lower in both overexpressing *Arabidopsis* lines, relative to the wild type (Figure 4e), while antioxidant gene expressions were higher in *PpyCBF3-ox* plants than in *PpyCBF2-ox* ones. To verify the above results, we investigated the accumulations of H_2O_2 and $O_2^{\bullet-}$ by examining diaminobenzidine (DAB) and nitroblue tetrazolium (NBT) precipitation in *PpyCBF2-ox*, *PpyCBF3-ox*, and wild-type plants. Although no differences were apparent between wild-type and overexpressing plants under control conditions, more intense brown and blue precipitates were observed under abiotic stress in leaves of wild-type plants stained with DAB and NBT, respectively.

The results of DAB and NBT staining indicate that overexpressing plants accumulated less H_2O_2 and $O_2^{\bullet-}$ during abiotic stress than the wild type (Figure 4b,c). The more pronounced activity of major H_2O_2- and $O2^{\bullet-}$-scavenging enzymes (AtPRX, AtAPX, AtCAT and AtSOD) in wild-type plants was due to the higher accumulation of these toxic molecules, whereas the higher activity of antioxidant genes in *PpyCBF3-ox* plants indicated that scavenging of accumulated ROS was more successful in *PpyCBF3-ox* than in *PpyCBF2-ox* plants (Figure 4b,c,e).

After abiotic stress treatments, both wild-type and overexpressing plants were grown under control conditions for 7 days to monitor their recovery. Almost all CBF transgenic plants exhibited more pronounced prostrate growth during recovery than wild-type ones, which were found to be under severe stress (Figure S4b). After salt stress, both overexpressing lines experienced significant growth. Following low-temperature and drought treatments, *PpyCBF2-ox* and *PpyCBF3-ox* plants had significantly longer roots than their respective wild type (Figure 4f).

2.5. PpyCBF Transcriptional Activation of 6X C-Repeat Binding Sites and Stress-Responsive Genes

To examine *PpyCBF* abiotic regulatory cascades, we first measured the CRT-dependent transactivation activities of *PpyCBFs* in dual luciferase assays. For this analysis, full-length *PpyCBFs* were inserted into a SK vector, and 6X C-repeat binding sites (CCGAC) were inserted along with a 35S promoter into a LUC vector. We found that all *PpyCBFs* had transcriptional activities with the 6X C-repeat binding sites, with the ancestral CBF (*PpyCBF3*) showing the strongest interaction with these binding sites (Figure S5).

To further investigate possible transcriptional regulatory linkages involved in pear abiotic stress pathways, dual luciferase (in vitro) and Y1H (in vivo) assays were performed with *PpyCBF* and *PpyCOR* promoters. The dual luciferase assays revealed that *PpyCBFs 1–6*, *PpyCBFs 1,2,4,5*, *PpyCBFs 1–4*, and *PpyCBF2* could significantly transactivate the promoters of *PpyCOR47*, *PpyCOR15A*, *PpyRD29A*, and *PpyKIN*, respectively. Clade II *PpyCBFs* had high transcriptional activities with *PpyCOR47, 15A*, and *RD29A*, while clade I and III *PpyCBFs* had little interaction with *PpyRD29A* (Figure 5a). In view of these results, Y1H assays were performed between *PpyCBF* genes and *PpyCOR* promoters. The Y1H results validated the direct interactions of *PpyCBFs 2,4,5* with *PpyCOR47*, *PpyCBFs 2* and *5* with *PpyCOR15A*, and *PpyCBFs 2* and *4* with *PpyRD29A* promoters, while no interactions were detected between *PpyKIN–PpyCBFs*. Interestingly, the ancestral CBF did not show any physical interaction with stress-responsive genes, while *PpyCBF2* was found to be the most active transcriptional regulator during abiotic stress signaling (Figure 5b).

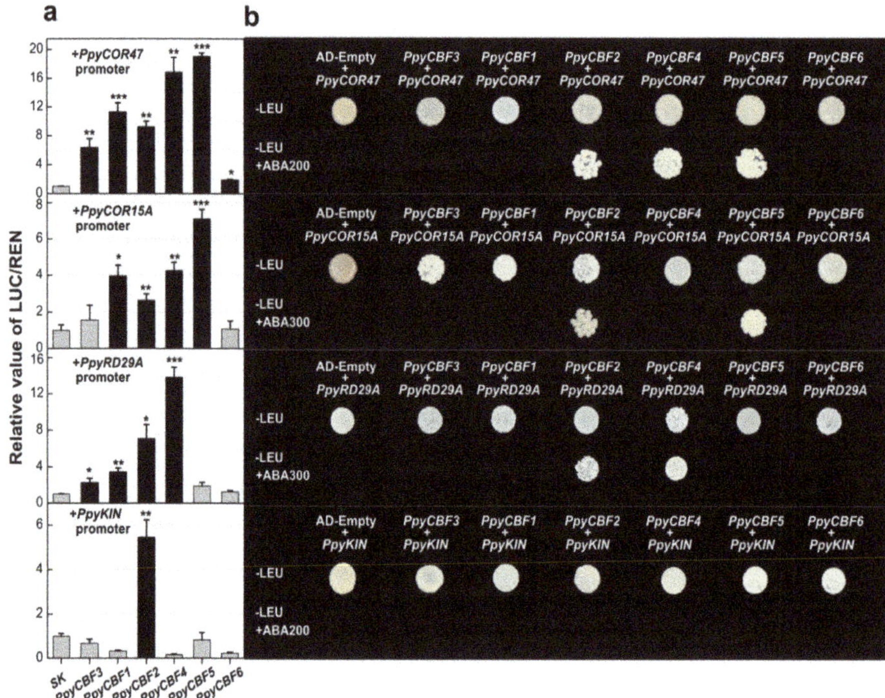

Figure 5. In vivo and in vitro regulations of *PpyCBFs* on the promoters of stress-related genes. (**a**) Dual luciferase assay to check the in vitro regulations. The ratio of firefly luciferase/renilla luciferase (LUC/REN) of the empty vector (pGreenII 0029 62-SK) plus promoter was used as calibrator (set as 1). Three independent experiments were done to verify the results. Error bars show SEs with at least four biological replicates, while asterisks show significant differences of genes SK with empty SK (* $p < 0.05$, ** $p < 0.01$, *** $p < 0.001$). (**b**) Y1H assay shows in vivo binding of *PpyCBFs* on *PpyCOR* promoters. Synthetic dropout (SD) medium without Leu and supplemented with 200 and 300 ng mL−1 ABA was used. Yeast grew on ABA-supplemented plates, indicating the possible direct interactions.

2.6. PpyCBFs Can Also Bind at the TCGAC Binding Site in the PpyCOR15A Promoter

The above findings indicate that *PpyCBFs* have transcriptional activities with 6X CCGAC binding sites. According to an analysis of *PpyCOR* promoters, however, *PpyCOR15A* had no CRT binding site in its promoter region, but had high transcriptional activities with *PpyCBFs* (Table S3). To identify the unique *PpyCBF* binding site in the *PpyCOR15A* promoter, we therefore first divided the *PpyCOR15A* promoter into four fragments. We observed both in vitro and in vivo interactions of *PpyCBFs* with fragment 2 of *PpyCOR15A* (Figure 6b,c). We identified three possible CBF-binding sites in this region, CGACA, CCGA and TCCG, and mutated them into CTTTA, CTTT and GTTG, respectively (Figure 6a). Luciferase and Y1H assays proved that the mutation at the CGACA binding site reduced the transcriptional activities and physical interactions of all *PpyCBFs* with the *PpyCOR15A* promoter present at −615 to −610 bp from the start codon. No effects on transcriptional regulation or direct interactions were observed at the second and third mutation sites. Hence, *PpyCBFs* can also bind to the TCGAC binding site, and the deletion of one cytosine from the CRT binding site did not influence its binding activity with the *PpyCOR15A* promoter in pears.

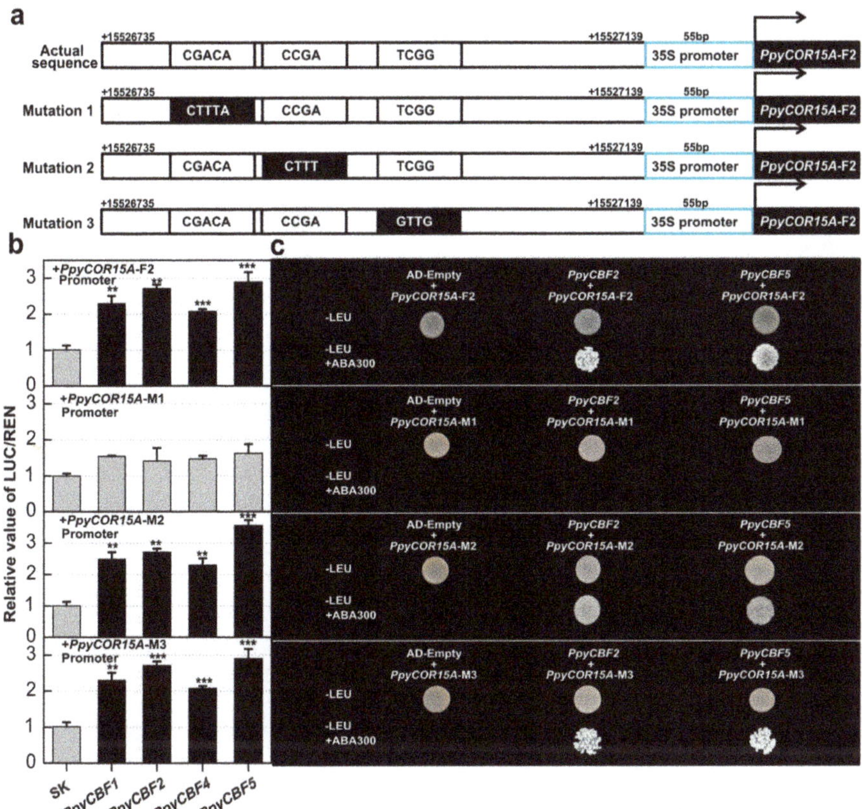

Figure 6. PpyCBFs can also bind at TCGAC binding site in the *PpyCOR15A* promoter. (**a**) Schematic diagrams of mutations at three different motif sites for *PpyCOR15A* promoters, indicated with mutation 1, 2, and 3. Possible CBF-binding sites in *PpyCOR15A* promoter are represented with white rectangles while mutations at these sites are represented by black rectangles. (**b**) Dual-luciferase assays were performed with actual and mutated promoters of the *PpyCOR15A* promoter. The ratio of LUC/REN of the empty vector (pGreenII 0029 62-SK) plus promoter was used as the calibrator (set as 1). Three independent experiments (with minimum four replicates) were performed to verify the results. Error bars show SEs with at least four biological replicates while asterisks show significant differences with empty SK (** $p < 0.01$, *** $p < 0.001$). (**c**) Y1H assay was performed to check physical interaction of PpyCBFs 2 and 5 with actual and mutated promoters of *PpyCOR15A*. Yeast grows on synthetic dropout without leucine but having Aureobasidin A 300 (SD/−leu + ABA300) indicating the possible direct interactions.

3. Discussion

In this study, we isolated 15 PpyCBF TFs from the pear genome. On the basis of sequence identity, phylogeny, conserved domain sequence (CDS) completeness, and scaffold position, however, only six *PpyCBFs* genes were selected for further study (Figure 1 and Table S1). Several CBF-specific domains, especially AP2, had strong conservations in plants, ultimately reflecting their high levels of identity [1,4]. This result explains why many identical amino acid residues and homologous groups were also found among CBFs of pears (Table S1) and other crop species, such as *Arabidopsis*, soybeans, apples, grapes, and different grasses [9,10,14,37]. Phylogenetic analysis provided evidence of independent evolution and three main PpyCBF clades/subtypes, while collinearity analyses uncovered two duplicated gene pairs (Figure 1 and Figure S1). The first clade not only contained CBFs from dicot and monocot crop species, but also the collinear gene *PpyCBF3*. The presence of *PpyCBF3* in this first clade along with genes from both monocots and dicots, and the evolutionary relationship of this clade to the other two CBF clades suggested that *PpyCBF3* might be the ancestral CBF from which all other CBFs were derived during whole-genome duplication in pears prior to their divergence from apples. This result is similar to soybeans, where the presence of orthologs from both dicot and monocot plants suggests that *GmDREB1* clade/subtype 4 genes are the ancestral genes in the GmDREB1 family [14]. Rosaceous and *Arabidopsis* crop CBFs may have evolved completely independently of one another, as CBF regulation in woody plants appears to be more complex than that in herbaceous plants [11].

As mentioned above, *PpyCBFs* were found to have different predicted functions than those of *AtCBFs*, which was corroborated by abiotic stress and bud endodormancy experiments that revealed that *PpyCBFs 1–6* were not only induced by low temperature, salt, and drought stresses, but also by exogenous and endogenous ABA (Figures 2a and 3). The predicted functions and expressions of these *PpyCBFs* were similar to those of *MbDREB1* in apples [15], *PaDREB1* in sweet cherries [38], *BrCBF* in non-heading Chinese cabbages [39], and *VviDREB1* in cowberries [40] during abiotic stress, but they were dissimilar to *AtCBFs 1–3* in *Arabidopsis*, which is only low-temperature responsive [10]. A proposed explanation for these expression changes is that cold, drought, and high salinity all cause osmotic stress [5]. In Japanese pears during bud endodormancy, we observed that the expressions of CBF/DREB4, DREB1E, DREB2, DREB2A, and DREB2D first peaked on December 24 and then suddenly declined on January 8, with a second expression peak on January 20 in both 'TH3' and 'Hengshani' cultivars [41]. We hypothesized that the first peak was low-temperature-responsive, while the second was ABA-responsive. To confirm in vivo functions of *PpyCBFs* in plants, we ectopically expressed two *PpyCBF* genes, *PpyCBF2* and *PpyCBF3*, in *Arabidopsis*. We found that plants of the two exogenous *PpyCBF-ox Arabidopsis* lines had higher resistance to low temperature (10 °C), salt (50 mM), and drought (10%) stresses than the wild type (Figure 4a), similar to results in transgenic plants overexpressing DREB1s from apples, soybeans, grapes, and cabbages [9,14,15,39]. Interestingly, overexpression of *PpyCBFs* did not cause a dwarf phenotype in transgenic *Arabidopsis* grown on Murashige–Skoog (MS) medium (Figure S4), an outcome in agreement with observations from overexpression of *MbDREB1* genes in *Arabidopsis* [15]. One notable feature of low-temperature stress and CBF overexpression is that both cause marked growth retardation resulting from the promotion of GA catabolism by two CBF-regulated isoforms (*GA2ox3* and *GA2ox6*) and subsequent accumulation of DELLA proteins [42]. Some evidence suggests that at least a few CBF paralogs have evolved to execute different functions [9], which would explain the differential responses of *PpyCBF* paralogs to various stresses observed in our study (Figure 2a). In particular, *PpyCBFs* from clade II were not only more cold-responsive during abiotic stress and bud endodormancy, but they also exhibited higher resistance in overexpressing *Arabidopsis* to cold stress compared with salt and drought stresses. In contrast, clade I and III CBFs were highly salt- and drought-responsive and were more resistant in transgenic *Arabidopsis* to these stresses (Figures 2 and 3). This situation is similar to soybeans, where the expressions of *GmDREB1* genes assigned to phylogenetic subtypes 1 and 2 were found to be induced by low-temperature, salinity, drought, and heat stresses, whereas those of subtype 4 were only induced by low temperature and salt [14].

The expression patterns of CBFs and CORs in pear are similar to those in other plant species [34]. Our qRT-PCR analysis revealed that *PpyCOR* expressions were increased not only by cold, salt, and drought stresses, but also by endogenous and exogenous ABA (Figure 2b). This result is unsurprising, as CBF-induced tolerance to cold, salt, drought, and ABA has been repeatedly correlated with increased expressions of *COR* genes [9]. Significantly higher amounts of *PpyCOR15a* and *PpyCOR47* transcripts were detected during abiotic stress, however, the reason why the expressions of *PpyRD29A* and *PpyKIN* did not follow the same trend as other *COR* genes is unclear. We note that specific information on all *COR* genes in pears are still limited. In regard to the effect of *PpyCBFs* on endogenous ABA-dependent and -independent genes, we observed significantly higher expressions of these genes under normal, unstressed conditions in *PpyCBF2-ox* and *PpyCBF3-ox* lines than in the wild type (Figure 4d). These findings suggest that *PpyCBF2* and *PpyCBF3* participate in both ABA-dependent and -independent pathways during abiotic stress signaling. Similar findings have also been reported for apples, grapes, and potatoes, where overexpressed *MbDREB1*, *VvCBF*, and *ScCBF1* significantly increase the expressions of ABA-independent (*AtCOR15a*, *AtRD29A*, *AtCOR6.6*, and *AtCOR47*) and ABA-dependent (*AtRD29B*, *AtRAB18*, *AtABI1*, and *AtABI2*) genes during normal conditions [9,15]. Interestingly, the expressions of all stress-responsive genes during abiotic stress conditions were significantly lower in overexpressing lines than the wild type, as the overexpressing lines had more resistance than the wild type because of the endogenous activation of *AtCOR* genes (Figure 4d).

Upon further investigation of transcriptional regulatory pathways of *PpyCBFs*, we uncovered their central role during abiotic stress signaling in pears (Figure 5 and Table 1). The results of our luciferase and Y1H assays indicated the existence of at least two main types of transcriptional interactions associated with CBF clades. In other words, all clade CBFs (except PpyCBF6) had interactions with *PpyCOR47* and *15A*, while clade II *PpyCBFs* had a stronger association with *PpyRD29A* compared with clades I and III. *PpyCBFs* were involved in the same CBF–COR cascades during abiotic stresses that are conserved in multiple plant species such as *Arabidopsis* and *Brachypodium*, with *AtCBF1–3* and *BdCBF1* showing interactions with *COR* genes by binding CRT/DRE (CCGAC) elements [34,37]. We also observed high transcriptional activities of all *PpyCBFs* with 6XCRT/DRE (CCGAC) binding sites. An analysis of *PpyCOR* gene promoters uncovered no CCGAC binding sites in the promoters of *PpyCOR15A*, *PpyKIN*, or *PpyRD29A* (Table S3), but we detected their strong in vivo and in vitro interactions with *PpyCBFs*. By mutating the CGAC binding site in *PpyCOR15A*, we were able to determine that *PpyCBFs* can also bind to the TCGAC binding site (Figure 6). In our previous study, we found that *PpCBF2* can also bind to the CCGA binding site in the *PpCBF4* promoter [22], which indicates that CGA is the actual core of the CBF binding site in pears.

To investigate the underlying mechanism of transcriptional regulation of *PpyCBF* expression by abiotic stress and ABA treatments, we examined the promoter regions of all *PpyCBFs* (Table 1). We found that *PpyCBF* expressions during abiotic stress are regulated by CRT/DRE, GT-1-like box, ICE1-like, NAC, and I BOX TFs, whereas during ABA treatment, ABRE and G-box1 TFs are involved. A bZIP transcription factor specifically recognizes G-box1 in promoters of ABA-responsive genes [43]. The absence of G-box1 *cis* elements and the presence of ABRE *cis* elements in *PpyCBF3* and *PpyCBF5* indicates that these genes are only regulated by the ABI3/VP1 cascade. In contrast, clade II *PpyCBFs* are regulated by both b-ZIP and ABI3 TFs, which explains why the expressions of clade II CBFs during ABA stress were relatively higher than those of *PpyCBF3* and *PpyCBF5* (Figure 2a). NAC TFs in pears are highly abiotic-stress responsive [44]. ICE-1 encoding a MYC-like basic helix–loop–helix protein that binds to Myc recognition sequences [33] and transcriptional induction of *PpCBFs* by *PpICE1s* have already been observed in pears [22]. *DREB1* genes are also negatively regulated by MYB15, an R2R3-type MYB transcription factor in *Arabidopsis* [7]. In both *Arabidopsis* and soybeans, a bZip TF recognizes GT-1-like boxes and plays a role in salt- and pathogen-induced gene expression [45]. MIKC *cis* elements in *PpyCBFs* also display a dormancy response, as the CBF–DAM regulon aids pear adaptation through bud endodormancy [22]. Given the above mentioned results, the relatively high abundance of *PpyCBFs* in the face of abiotic stress as well as exogenous and endogenous ABA, the induction of ABA-dependent and -independent genes in overexpressed *Arabidopsis* under control conditions, and the in vivo and in vitro interactions of PpyCBFs with PpyCORs and the presence of both stress- and ABA-related *cis* elements in their promoters.

4. Materials and Methods

4.1. Identification and Characterization of PpyCBFs

Protein sequences of PpyCBF subfamily members and PpyCORs were retrieved from the Pear Genome Project database (http://peargenome.njau.edu.cn/), while two databases were used to obtain *Malus* (Md), *Prunus* (Ppe), *Fragaria* (Fv), and *Vitis* (Vv) CBFs: The Genome Database for Rosaceae (GDR; http://www.rosaceae.org/) and the Plant Transcription Factor database (Plant TFDB v4.0; http://planttfdb.cbi.pku.edu.cn/). AtCBFs were downloaded from the Arabidopsis Information Resource (https://www.arabidopsis.org/). Collinear blocks of PpyCBFs and whole genomes within species were identified in MCScanX with default settings and an E-value $\leq 1 \times 10^{-10}$. After aligning all sequences in ClustalX, the resulting identity matrix was checked using BioEdit software. Phylogenetic analysis of PpyCBFs and CBFs of other crop species was performed by the neighbor-joining method with 1000 bootstrap replicates in MEGA v7.0. Gene structure and motif analyses were carried out using Gene Structure Display Server v2.0 (http://gsds.cbi.pku.edu.cn/) and MEME v5.0.4 (http://meme-suite.org/tools/meme) tools with default parameters. The PlantPan2.0 (http://plantpan2.itps.ncku.edu.tw/) database with 2000 nucleotides was used for promoter analysis.

4.2. Plant Materials and Abiotic Stress Treatments

For abiotic stress experiments, vegetative buds of Asian pear cultivar 'Dangshan Suli' were collected before bud break in March 2018. After collection, buds were washed, sterilized, and then grown in half-strength MS medium to generate pear seedlings. Seedlings of a uniform size with six to eight leaves were randomly selected for abiotic stress treatments. For the low temperature treatment, seedlings in MS medium were exposed to 4 °C, while drought and salt stress treatments were carried out by respectively adding 200 mM NaCl and 15% PEG6000 to half-strength MS medium. Samples were collected with three replicates after 0, 6, 12, 24, and 48 h of treatment. For ABA stress treatments, wild-type pear calli were placed in half-strength MS medium containing 100 µM ABA (stressed) or 100 µM absolute ethanol (Mock), and sampling was carried out with three replicates of each treatment group after 0, 3, 6, 12, 24, and 48 h. Following the abiotic stress treatments, each sample was immediately frozen in liquid nitrogen and stored at −80 °C. Plant materials and methods for study of bud endodormancy in pears were the same as those of a previously published study [44].

4.3. Analysis of Stress Tolerance of Transgenic Plants

After amplification, *PpyCBF2* and *PpyCBF3* coding sequences were cloned into a pCAMBIA 1301 vector to generate 35S::PpyCBFs constructs. The recombinant plasmids were inserted into Agrobacterium EHA105 cells and then transformed into flowering *Arabidopsis thaliana* plants by the floral dip method. After 7 days, the floral dip procedure was repeated. Following seed collection, the transgenic *Arabidopsis* plants were screened on MS medium containing 1 µg mL^{-1} of the antibiotic hygromycin. Putative transformants among the T_1 progeny, confirmed by RT-PCR using PpyCBF2- and PpyCBF3-ORF-F/R primers, were regrown using the same procedure to obtain T_3 progeny. The line of T_3 plants with the highest PpyCBF2 and PpyCBF3 abundances was selected and grown to generate T_4 progeny, which were used to assess in vivo abiotic stress tolerance. For this assessment, seeds of wild-type and overexpressed lines were germinated on MS medium for 14 days, and their seedlings were then grown for 5 days on vertical plates containing MS medium supplemented with either 50 mM NaCl (to assess salt tolerance) or 10% PEG (to assess drought tolerance). As a control, another set of seedlings were grown on MS medium with no supplement. To assess cold tolerance, seedlings on MS plates were exposed to 10 °C for 21 days. After abiotic stress treatments, all seedlings were grown under normal conditions on MS medium for 5 days to check their recovery rate. ImageJ v1.8.0 software was used to measure root lengths of wild-type and overexpressed lines under normal and abiotic stress conditions.

4.4. Histochemical Analysis of H_2O_2 and $O_2^{\bullet-}$

For histochemical analysis of H_2O_2 and $O2^{\bullet-}$, fresh diaminobenzidine (DAB) and nitroblue tetrazolium (NBT) solutions were prepared following a method reported previously [46]. Plant leaves were immersed in DAB and NBT solutions and incubated overnight at room temperature in darkness, the latter achieved by wrapping in aluminum foil. To remove chlorophyll for proper visualization, the leaves were bleached in absolute ethanol for 10 min at 95 °C in a water bath. Photographs of stained samples were taken using a Leica DMLB fluorescence microscope, where brown and blue spots respectively indicated the presence of H_2O_2 and $O_2^{\bullet-}$ in situ.

4.5. RNA Extraction and cDNA Synthesis

Total RNA was extracted from three biological replicates using a modified cetyltrimethylammonium bromide method as described in our previous study [47]. cDNA was then synthesized from 4 µg of DNA-free RNA using an iScript cDNA Synthesis kit (Bio-Rad, Foster, CA, USA) following the manufacturer's instructions. Ten-fold diluted cDNA was used as a template for qRT-PCR analysis.

4.6. qRT-PCR Analysis

qRT-PCR amplifications were performed in 15 µL reaction volumes composed of 7.5 µL SYBR Premix Ex *Taq* (TliRNaseH Plus, Takara Biotechnology (Dalian) Co., Ltd. Dalian, China), 1 µL cDNA, 0.5 µL each of forward and reverse primers, and 5.5 µL RNase-free water. The amplifications were carried out on a CFX Connect real-time PCR system (Bio-Rad, Hercules, CA, USA) according to the following protocol: 95 °C for 30 s, followed by 40 cycles of 95 °C for 5 s and 60 °C for 20 s. Melting curves were used to confirm the specificity of the qRT-PCR primers. Relative gene transcript levels were determined using the $2^{-\Delta\Delta Ct}$ method and normalized against *PpyActin* (JN684184).

4.7. Site-Directed Mutagenesis of Gene Promoters

To check possible binding sites of PpyCBFs in *PpyCOR* promoters, the predicted sites were altered by directed mutagenesis. Motif mutations were carried out using a mutagenesis system after designing specific primers for possible binding sites. Transactivation effects of PpyCBFs on mutated promoters were further examined using dual luciferase and Y1H assays.

4.8. Transient Expression and Luciferase Measurement

A dual luciferase assay was used to detect in vivo transactivation effects of transcription factors. Full-length *PpyCBF* and *PpyCOR* promoters (2000 nucleotides) were inserted into pGreenII 0029 62-SK and pGreenII 0800-LUC vectors, respectively. The dual luciferase assay was carried out with *Nicotiana benthamiana* leaves according to our previously described protocol [22]. Three independent experiments with a minimum of four replicates were performed to verify the results.

4.9. Yeast One-Hybrid Assay

Y1H assays were conducted using a Matchmaker Gold Yeast One-Hybrid System kit (Clontech, Takara, Japan) according to the instructions in the user manual. Subsequent analyses were completed as previously described [48].

4.10. Statistical Analysis

Experiments were set up according to a completely randomized design. Analysis of variance followed by Duncan's multiple range test was used to test the overall significance of differences among treatments ($p < 0.05$). Significant differences between treatments were assessed by Student's *t*-test at $p < 0.05$, $p < 0.01$, and $p < 0.001$. All data were analyzed in SPSS v25 (SPSS Inc., Chicago, IL, USA).

5. Conclusions

We identified six *PpyCBF* homologues (*PpCBF1-6*) encoding potential transcription factors in Asian pear. All *PpyCBF* members accentuated during different abiotic stresses and endo and exogenous ABA. II clade *PpyCBFs* were not only more low temperature (LT) and ABA responsive but also enhanced LT stress tolerance in overexpressed Arabidopsis as compared to I and III clades *PpyCBFs*. Ectopic expressions of *PpyCBF2* and *PpyCBF3* in Arabidopsis also increased the expressions of endogenous

ABA dependent and independent genes during normal conditions. A conversed CBF-COR regulatory cascade was also observed in pear. We conclude that *PpyCBFs* may follow both ABA-dependent and -independent stress signaling pathways during abiotic stress in pears. PpyCBF transcription factors may thus act redundantly during abiotic stress through ABA-dependent and -independent pathways. The results of our investigation, the first to differentiate the functions of the complete CBF subfamily in any rosaceous crop species, should have an important influence on the study of stress in woody species and may be applicable for the genetic engineering of different functions of transcription factors in other plant species.

Supplementary Materials: Supplementary materials can be found at http://www.mdpi.com/1422-0067/20/9/2074/s1.

Author Contributions: S.B. and Y.T. perceived and planned the study and M.A. and J.L. performed most of the experiments and analyses. M.A. and Q.Y. collected the samples and extracted total RNAs for qPCR. J.L. and W.J. helped in luciferase and Y1H assays, and data arrangements. M.A., S.B., and Y.T. wrote the manuscript. All authors read and approved the final manuscript.

Funding: This work was supported by the National Key Research and Developmental Program of China (2018YFD1000104) to S.B., National Natural Science Foundation of China (31501736) to S.B., and the Earmarked Fund for China Agriculture Research System (CARS-28) to Y.T.

Acknowledgments: We thank the Dangshan Suli Germplasm Resources Center for providing plant materials. We also say special thanks to Muhammad Ali Raza for valuable efforts and instructions in growing of transgenic *Arabidopsis*.

Conflicts of Interest: The authors declare no conflict of interest.

Abbreviations

CORs	Cold Regulons
HMM	Hidden Markov Model
MEGA	Molecular Evolutionary Genetics Analysis
TF	Transcription factor
Y1H	Yeast one hybrid
SnRK2	Snf1-Related kinase 2
CTAB	Cetyltrimethyl Ammonium Bromide

References

1. Nakano, T.; Suzuki, K.; Fujimura, T.; Shinshi, H. Genome-wide analysis of the ERF gene family in *Arabidopsis* and rice. *Plant Physiol.* **2006**, *140*, 411–432. [CrossRef] [PubMed]

2. Licausi, F.; Ohme Takagi, M.; Perata, P. APETALA 2/Ethylene Responsive Factor (AP 2/ERF) transcription factors: Mediators of stress responses and developmental programs. *New Phytol.* **2013**, *199*, 639–649. [CrossRef]

3. Jaglo, K.R.; Kleff, S.; Amundsen, K.L.; Zhang, X.; Haake, V.; Zhang, J.Z.; Deits, T.; Thomashow, M.F. Components of the *Arabidopsis* C-repeat/dehydration-responsive element binding factor cold-response pathway are conserved in *Brassica napus* and other plant species. *Plant Physiol.* **2001**, *127*, 910–917. [CrossRef]

4. Sakuma, Y.; Liu, Q.; Dubouzet, J.G.; Abe, H.; Shinozaki, K.; Yamaguchi-Shinozaki, K. DNA-binding specificity of the ERF/AP2 domain of *Arabidopsis* DREBs, transcription factors involved in dehydration-and cold-inducible gene expression. *Biochem. Biophys. Res. Commun.* **2002**, *290*, 998–1009. [CrossRef] [PubMed]

5. Thomashow, M.F. Molecular basis of plant cold acclimation: Insights gained from studying the CBF cold response pathway. *Plant Physiol.* **2010**, *154*, 571–577. [CrossRef]

6. Welling, A.; Palva, E.T. Involvement of CBF transcription factors in winter hardiness in birch. *Plant Physiol.* **2008**, *147*, 1199–1211. [CrossRef] [PubMed]

7. Agarwal, P.K.; Agarwal, P.; Reddy, M.; Sopory, S.K. Role of DREB transcription factors in abiotic and biotic stress tolerance in plants. *Plant Cell Rep.* **2006**, *25*, 1263–1274. [CrossRef]

8. Mizoi, J.; Shinozaki, K.; Yamaguchi Shinozaki, K. AP2/ERF family transcription factors in plant abiotic stress responses. *Biochim. Biophys. Acta* **2012**, *1819*, 86–96. [CrossRef]

9. Siddiqua, M.; Nassuth, A. Vitis *CBF1* and Vitis *CBF4* differ in their effect on *Arabidopsis* abiotic stress tolerance, development and gene expression. *Plant Cell Environ.* **2011**, *34*, 1345–1359. [CrossRef] [PubMed]

10. Novillo, F.; Alonso, J.M.; Ecker, J.R.; Salinas, J. *CBF2/DREB1C* is a negative regulator of *CBF1/DREB1B* and *CBF3/DREB1A* expression and plays a central role in stress tolerance in *Arabidopsis*. *Proc. Natl. Acad. Sci. USA* **2004**, *101*, 3985–3990. [CrossRef]

11. Benedict, C.; Skinner, J.S.; Meng, R.; Chang, Y.; Bhalerao, R.; Huner, N.P.; Finn, C.E.; Chen, T.H.; Hurry, V. The CBF1-dependent low temperature signalling pathway, regulon and increase in freeze tolerance are conserved in *Populus* spp. *Plant Cell Environ.* **2006**, *29*, 1259–1272. [CrossRef] [PubMed]

12. Xiao, H.; Siddiqua, M.; Braybrook, S.; Nassuth, A. Three grape *CBF/DREB1* genes respond to low temperature, drought and abscisic acid. *Plant Cell Environ.* **2006**, *29*, 1410–1421. [CrossRef] [PubMed]

13. Xiao, H.; Tattersall, E.A.; Siddiqua, M.K.; Cramer, G.R.; Nassuth, A. *CBF4* is a unique member of the CBF transcription factor family of *Vitis vinifera* and *Vitis riparia*. *Plant Cell Environ.* **2008**, *31*, 1–10. [CrossRef]

14. Kidokoro, S.; Watanabe, K.; Ohori, T.; Moriwaki, T.; Maruyama, K.; Mizoi, J.; Myint Phyu Sin Htwe, N.; Fujita, Y.; Sekita, S.; Shinozaki, K. Soybean *DREB 1/CBF* type transcription factors function in heat and drought as well as cold stress responsive gene expression. *Plant J.* **2015**, *81*, 505–518. [CrossRef] [PubMed]

15. Yang, W.; Liu, X.D.; Chi, X.J.; Wu, C.A.; Li, Y.Z.; Song, L.L.; Liu, X.M.; Wang, Y.F.; Wang, F.W.; Zhang, C. Dwarf apple *MbDREB1* enhances plant tolerance to low temperature, drought, and salt stress via both ABA-dependent and ABA-independent pathways. *Planta* **2011**, *233*, 219–229. [CrossRef] [PubMed]

16. Magome, H.; Yamaguchi, S.; Hanada, A.; Kamiya, Y.; Oda, K. dwarf and delayed-flowering 1, a novel *Arabidopsis* mutant deficient in gibberellin biosynthesis because of overexpression of a putative AP2 transcription factor. *Plant J.* **2004**, *37*, 720–729. [CrossRef]

17. Haake, V.; Cook, D.; Riechmann, J.; Pineda, O.; Thomashow, M.F.; Zhang, J.Z. Transcription factor *CBF4* is a regulator of drought adaptation in *Arabidopsis*. *Plant Physiol.* **2002**, *130*, 639–648. [CrossRef]

18. Liu, Q.; Kasuga, M.; Sakuma, Y.; Abe, H.; Miura, S.; Yamaguchi Shinozaki, K.; Shinozaki, K. Two transcription factors, *DREB1* and *DREB2*, with an EREBP/AP2 DNA binding domain separate two cellular signal transduction pathways in drought-and low-temperature-responsive gene expression, respectively, in *Arabidopsis*. *Plant Cell* **1998**, *10*, 1391–1406. [CrossRef] [PubMed]

19. Zhu, J.; Dong, C.H.; Zhu, J.K. Interplay between cold-responsive gene regulation, metabolism and RNA processing during plant cold acclimation. *Curr. Opin. Plant Biol.* **2007**, *10*, 290–295. [CrossRef]

20. Norén, L.; Kindgren, P.; Stachula, P.; Rühl, M.; Eriksson, M.E.; Hurry, V.; Strand, Å. Circadian and plastid signaling pathways are integrated to ensure correct expression of the *CBF* and *COR* genes during photoperiodic growth. *Plant Physiol.* **2016**, *171*, 1392–1406.

21. Rubio, S.; Noriega, X.; Pérez, F.J. Abscisic acid (ABA) and low temperatures synergistically increase the expression of *CBF/DREB1* transcription factors and cold-hardiness in grapevine dormant buds. *Ann. Bot.* **2018**, *20*, 1–9. [CrossRef]

22. Li, J.; Yan, X.; Yang, Q.; Ma, Y.; Yang, B.; Tian, J.; Teng, Y.; Bai, S. *PpCBFs* selectively regulate *PpDAMs* and contribute to the pear bud endodormancy process. *Plant Mol. Biol.* **2019**, 1–12. [CrossRef]

23. Li, J.; Xu, Y.; Niu, Q.; He, L.; Teng, Y.; Bai, S. Abscisic Acid (ABA) promotes the induction and maintenance of Pear (*Pyrus pyrifolia* White Pear Group) flower bud endodormancy. *Int. J. Mol. Sci.* **2018**, *19*, 310. [CrossRef]

24. Xiong, L.; Wang, R.G.; Mao, G.; Koczan, J.M. Identification of drought tolerance determinants by genetic analysis of root response to drought stress and abscisic acid. *Plant Physiol.* **2006**, *142*, 1065–1074. [CrossRef] [PubMed]

25. Hetherington, A.M. Guard cell signaling. *Cell* **2001**, *107*, 711–714. [CrossRef]

26. Niu, Q.; Li, J.; Cai, D.; Qian, M.; Jia, H.; Bai, S.; Hussain, S.; Liu, G.; Teng, Y.; Zheng, X. Dormancy-associated MADS-box genes and microRNAs jointly control dormancy transition in pear (*Pyrus pyrifolia* white pear group) flower bud. *J. Exp. Bot.* **2015**, *67*, 239–257. [CrossRef] [PubMed]

27. Xie, Y.; Chen, P.; Yan, Y.; Bao, C.; Li, X.; Wang, L.; Shen, X.; Li, H.; Liu, X.; Niu, C. An atypical R2R3 MYB transcription factor increases cold hardiness by CBF dependent and CBF independent pathways in apple. *New Phytologist.* **2018**, *218*, 201–218. [CrossRef] [PubMed]

28. An, J.; Li, R.; Qu, F.; You, C.; Wang, X.; Hao, Y. An apple *NAC* transcription factor negatively regulates cold tolerance via CBF-dependent pathway. *J. Plant Physiol.* **2018**, *221*, 74–80. [CrossRef]

29. Jin, C.; Li, K.Q.; Xu, X.Y.; Zhang, H.P.; Chen, H.X.; Chen, Y.H.; Hao, J.; Wang, Y.; Huang, X.S.; Zhang, S.L. A novel NAC transcription factor, *PbeNAC1*, of *Pyrus betulifolia* confers cold and drought tolerance via Interacting with *PbeDREBs* and activating the expression of stress-responsive genes. *Front. Plant Sci.* **2017**, *8*, 1049. [CrossRef]

30. Ma, Y.; Szostkiewicz, I.; Korte, A.; Moes, D.; Yang, Y.; Christmann, A.; Grill, E. Regulators of *PP2C* phosphatase activity function as abscisic acid sensors. *Science* **2009**, *324*, 1064–1068. [CrossRef]

31. Park, S.Y.; Fung, P.; Nishimura, N.; Jensen, D.R.; Fujii, H.; Zhao, Y.; Lumba, S.; Santiago, J.; Rodrigues, A.; Tsz fung, F.C. Abscisic acid inhibits type 2C protein phosphatases via the *PYR/PYL* family of START proteins. *Science* **2009**, *324*, 1068–1071. [CrossRef] [PubMed]

32. Yoshida, T.; Fujita, Y.; Sayama, H.; Kidokoro, S.; Maruyama, K.; Mizoi, J.; Shinozaki, K.; Yamaguchi-Shinozaki, K. *AREB1*, *AREB2*, and *ABF3* are master transcription factors that cooperatively regulate ABRE-dependent ABA signaling involved in drought stress tolerance and require ABA for full activation. *Plant J.* **2010**, *61*, 672–685. [CrossRef]

33. Zhan, X.; Zhu, J.K.; Lang, Z. Increasing freezing tolerance: Kinase regulation of *ICE1*. *Dev. Cell* **2015**, *32*, 257–258. [CrossRef]

34. Zhou, M.; Shen, C.; Wu, L.; Tang, K.; Lin, J. CBF dependent signaling pathway: A key responder to low temperature stress in plants. *Crit. Rev. Biotechnol.* **2011**, *31*, 186–192. [CrossRef] [PubMed]

35. Teng, Y. The pear industry and research in China. *Acta Hortic.* **2011**, *909*, 161–170. [CrossRef]

36. Teng, Y. Advances in the research on phylogeny of the genus *Pyrus* and the origin of pear cultivars native to East Asia. *J. Fruit Sci.* **2017**, *34*, 370–378.

37. Ryu, J.Y.; Hong, S.Y.; Jo, S.H.; Woo, J.C.; Lee, S.; Park, C.M. Molecular and functional characterization of cold-responsive C-repeat binding factors from *Brachypodium distachyon*. *BMC Plant Biol.* **2014**, *14*, 15. [CrossRef] [PubMed]

38. Kitashiba, H.; Ishizaka, T.; Isuzugawa, K.; Nishimura, K.; Suzuki, T. Expression of a sweet cherry DREB1/CBF ortholog in Arabidopsis confers salt and freezing tolerance. *J. Plant Physiol.* **2004**, *161*, 1171–1176. [CrossRef] [PubMed]

39. Jiang, F.; Wang, F.; Wu, Z.; Li, Y.; Shi, G.; Hu, J.; Hou, X. Components of the *Arabidopsis* CBF cold-response pathway are conserved in non-heading Chinese cabbage. *Plant Mol. Biol. Rep.* **2011**, *29*, 525–532. [CrossRef]

40. Wang, Q.J.; Xu, K.Y.; Tong, Z.G.; Wang, S.H.; Gao, Z.H.; Zhang, J.Y.; Zong, C.W.; Qiao, Y.S.; Zhang, Z. Characterization of a new dehydration responsive element binding factor in central arctic cowberry. *Plant Cell Tissue Organ Cult.* **2010**, *101*, 211–219. [CrossRef]

41. Takemura, Y.; Kuroki, K.; Jiang, M.; Matsumoto, K.; Tamura, F. Identification of the expressed protein and the impact of change in ascorbate peroxidase activity related to endodormancy breaking in *Pyrus pyrifolia*. *Plant Physiol. Biochem.* **2015**, *86*, 121–129. [CrossRef] [PubMed]

42. Achard, P.; Renou, J.P.; Berthomé, R.; Harberd, N.P.; Genschik, P. Plant DELLAs restrain growth and promote survival of adversity by reducing the levels of reactive oxygen species. *Curr. Biol.* **2008**, *18*, 656–660. [CrossRef] [PubMed]

43. Uno, Y.; Furihata, T.; Abe, H.; Yoshida, R.; Shinozaki, K.; Yamaguchi Shinozaki, K. Arabidopsis basic leucine zipper transcription factors involved in an abscisic acid dependent signal transduction pathway under drought and high-salinity conditions. *Proc. Natl. Acad. Sci. USA* **2000**, *97*, 11632–11637. [CrossRef] [PubMed]

44. Ahmad, M.; Yan, X.; Li, J.; Yang, Q.; Jamil, W.; Teng, Y.; Bai, S. Genome wide identification and predicted functional analyses of *NAC* transcription factors in Asian pears. *BMC Plant Biol.* **2018**, *18*, 214. [CrossRef] [PubMed]

45. Park, H.C.; Kim, M.L.; Kang, Y.H.; Jeon, J.M.; Yoo, J.H.; Kim, M.C.; Park, C.Y.; Jeong, J.C.; Moon, B.C.; Lee, J.H. Pathogen-and NaCl-induced expression of the *SCaM-4* promoter is mediated in part by a GT-1 box that interacts with a GT-1-like transcription factor. *Plant Physiol.* **2004**, *135*, 2150–2161. [CrossRef] [PubMed]

46. Khan, A.R.; Wakeel, A.; Muhammad, N.; Liu, B.; Wu, M.; Liu, Y.; Ali, I.; Zaidi, S.H.R.; Azhar, W.; Song, G. Involvement of ethylene signaling in zinc oxide nanoparticle mediated biochemical changes in *Arabidopsis thaliana* leaves. *Environ. Sci. Nano* **2019**, *6*, 341–355. [CrossRef]

47. Yang, Q.; Niu, Q.; Li, J.; Zheng, X.; Ma, Y.; Bai, S.; Teng, Y. *PpHB22*, a member of HD-Zip proteins, activates *PpDAM1* to regulate bud dormancy transition in 'Suli' pear (*Pyrus pyrifolia* White Pear Group). *Plant Physiol. Biochem.* **2018**, *127*, 355–365. [CrossRef]

48. Tao, R.; Bai, S.; Ni, J.; Yang, Q.; Zhao, Y.; Teng, Y. The blue light signal transduction pathway is involved in anthocyanin accumulation in 'Red Zaosu' pear. *Planta* **2018**, 1–12. [CrossRef] [PubMed]

International Journal of
Molecular Sciences

Review

The Adaptive Mechanism of Plants to Iron Deficiency via Iron Uptake, Transport, and Homeostasis

Xinxin Zhang [1,†], Di Zhang [1,2], Wei Sun [3] and Tianzuo Wang [1,2,*]

[1] State Key Laboratory of Vegetation and Environmental Change, Institute of Botany, The Chinese Academy of Sciences, Beijing 100093, China; zhangxinxin19870620@hotmail.com (X.Z.); zhangdi@ibcas.ac.cn (D.Z.)
[2] College of Resources and Environment, University of Chinese Academy of Sciences, Beijing 100049, China
[3] Key Laboratory of Vegetation Ecology, Ministry of Education, Institute of Grassland Science, Northeast Normal University, Changchun 130024, China; sunwei@nenu.edu.cn
* Correspondence: tzwang@ibcas.ac.cn
† Current address: State key Laboratory of Protein and Plant Gene Research, Peking-Tsinghua Center for Life Sciences, School of Advanced Agricultural Sciences and School of Life Sciences, Peking University, Beijing 100871, China.

Received: 16 April 2019; Accepted: 14 May 2019; Published: 16 May 2019

Abstract: Iron is an essential element for plant growth and development. While abundant in soil, the available Fe in soil is limited. In this regard, plants have evolved a series of mechanisms for efficient iron uptake, allowing plants to better adapt to iron deficient conditions. These mechanisms include iron acquisition from soil, iron transport from roots to shoots, and iron storage in cells. The mobilization of Fe in plants often occurs via chelating with phytosiderophores, citrate, nicotianamine, mugineic acid, or in the form of free iron ions. Recent work further elucidates that these genes' response to iron deficiency are tightly controlled at transcriptional and posttranscriptional levels to maintain iron homeostasis. Moreover, increasing evidences shed light on certain factors that are identified to be interconnected and integrated to adjust iron deficiency. In this review, we highlight the molecular and physiological bases of iron acquisition from soil to plants and transport mechanisms for tolerating iron deficiency in dicotyledonous plants and rice.

Keywords: iron deficiency; acquisition; transport; homeostasis

---

1. Introduction

Iron (Fe) is an essential micronutrient for plant growth development and plays a key role in regulating numerous cellular processes. Iron, as an important co-factor for enzymes, plays an important role in regulating plant photosynthesis, mitochondrial respiration, the synthesis and repair of nucleotides, and metal homeostasis, especially in the maintenance of structural integrity of various proteins [1]. While Fe is abundant in soil, the available Fe in soil for plants is often insufficient, particularly in calcareous soils, due to low solubility of Fe. Iron deficiency is one of the most important factors limiting crop production in the world. Plants grown in low Fe soils often exhibit chlorosis and decreased photosynthesis, leading to reduction in yield and quality of crops. To cope with this situation, plants have evolved a series of sophisticate mechanisms to adapt to iron-deficient conditions in soil. In addition, iron deficiency is a significant worldwide problem, seriously affecting over 30% of the world's population (http://www.who.int/nutrition/topics/ida/en/). Anemia as one of the severest nutritional disorders is caused by low iron in humans. Therefore, elucidation of the molecular and physiological mechanisms by which plants sense, respond, and adapt to Fe deficiency would contribute to cultivating crop varieties with high Fe efficiency.

2. Iron Acquisition from Soil to Roots

Although iron is considered as the fourth most abundant element, one-third of soil on the Earth is estimated as Fe deficient [1]. The solubility and availability of iron in soil can be affected by multiple factors, including soil pH, the redox potential, microbial processes, and the amounts of organic matter and aeration in soil [2]. As a vital cofactor for enzymes, iron takes part in distinct processes, such as facilitating various chemical reactions, modulating protein stability, hormonal regulation, and nitrogen assimilation [1]. Iron deficiency could result in interveinal chlorosis in young leaves as the result of reduced chlorophyll content. The young leaves exhibit yellow color while the veins remain green. All these ultimately lead to the reduction of yield and quality [1,3]. In addition, other nutrients have antagonistic effects on iron uptake, which can significantly reduce the yield of the crops [4].

Iron in the rhizosphere is mainly present as Fe^{3+} which is not readily accessible to plants. Different plant species have evolved different strategies for iron acquisition from soil (Figure 1). Non-graminaceous plants, such as tomato and *Arabidopsis*, known as strategy-I plants, use a reduction-based strategy, in which plasma-membrane (PM)-localized H^+-ATPases (AHAs) release the protons to increase rhizosphere acidification and promote Fe^{3+} solubility. Subsequently, the available ferric Fe^{3+} is reduced to the more soluble ferrous Fe (Fe(II)) by ferric reduction oxidases (FROs) at the apoplast [5]. The reduced ferrous ion (Fe^{2+}) is imported into root cells by the Fe^{2+}-regulated transporters such as the iron-regulated transporter (IRT1) [6,7]. Additionally, graminaceous plants, including rice, barley, and maize, known as strategy-II plants, use a chelation-based strategy to release phytosiderophores (PS). PS, as strong Fe chelators, are secreted into the rhizosphere with a high affinity for binding Fe (III) [8,9]. PS-Fe(III) is then taken up into root cells through the yellow stripe (YS) or yellow stripe-like (YSL) transporters [10].

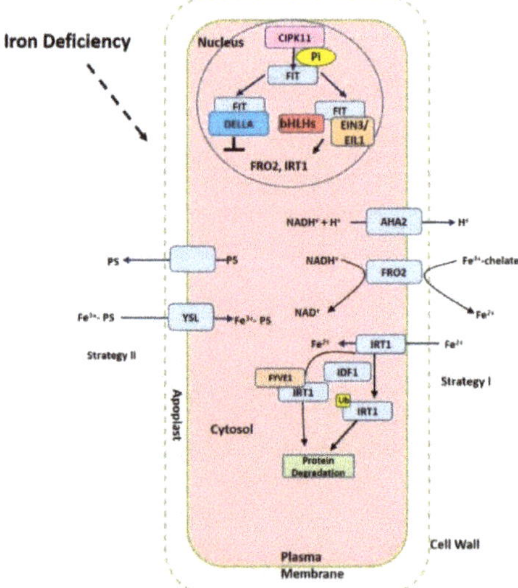

Figure 1. Summary of the iron-deficient response in plant cells. The proton ATPase AHA2, Ferric chelate reductase FRO2 (ferric reduction oxidase), Fe^{2+}-regulated transporters iron-regulated transporter (IRT1) and FER-like iron deficiency-induced transcription factor (FIT) are activated under iron starvation, respectively. AHA2 (H^+-ATPase) increases the acidification of rhizosphere to facilitate iron solubilization. FRO2 reduces ferric iron to ferrous iron that is imported into the cell via IRT1. The expression of *FRO2*, *IRT1* can be induced via FIT interaction with other transcription factors such as bHLHs and EIN3/EIL1 but prevented with DELLA.

Iron deficiency triggers the expression of many Fe uptake-associated genes. The expression of *AtAHA2* and *AtAHA7*, for example, are at higher levels under iron-deficient conditions, but *AtAHA1* is not induced by iron deficiency [11]. Twelve PM H^+-ATPases AHAs are encoded in the *Arabidopsis* genome [11]. AtAHA2 is primarily responsible for the of rhizosphere acidification of root hairs under iron deficiency. Loss function of *AtAHA2* compromised proton extrusion capacity. AHA7 is crucial for the formation of root hairs induced by iron deficiency via mediating H^+ efflux in the root hair zone. The fine-tuned regulation of root tip H^+ extrusion by PM H^+-ATPase is required for root hair formation. H^+ efflux through PM H^+-ATPase causes the acidification of the cell wall apoplast, which is crucial for the root hair initiation [11]. The loss function of *AtAHA7* contributed to a decreased frequency of root hairs [11]. However, the mechanism of *AHAs* regulation remains unknown. Recent findings indicate that cytochrome B5 reductase 1 (CBR1) is able to activate plasma membrane-localized H^+-ATPases, which is achieved by facilitating the content of unsaturated fatty acids [12]. *CBR1* expression is induced under iron-deficient conditions. CBR1 localizes to endoplasmic reticulum (ER) membrane and plays an important role in electron transfer from NADH to cytochrome b5. Then the cytochrome b5 mediates the electrons transfer to fatty acids desaturase 2 (FAD2) and fatty acids desaturase 3 (FAD3), allowing for double bonds into fatty acids. FAD2 is responsible for converting oleic acid (18:1) to linoleic acid (18:2), and FAD3 contributes to the conversion of 18:2 to linolenic acid (18:3). On the other side, 20 or 50 µM of the unsaturated fatty acids 18:2 or 18:3 can strongly activate H^+-ATPase [12]. Other compounds such as phenolics, organic acids, flavonoids, and flavins have also been implicated in the acidification–reduction strategy to uptake iron (Strategy I) [3,13–15]. These small compounds significantly promote reutilization and uptake of apoplastic iron via chelation or the reduction of iron in soil. Recently it was reported that coumarins involved in iron acquisition are secreted and essential for iron uptake under iron-limited conditions [16,17]. The plants are able to secret an array of coumarin-type compounds under different iron nutrition conditions, which facilitate Fe(III) availability [18]. The synthesis of these coumarins require Feruloyl coenzyme A 6'-hydrozylase 1 (F6'H1) enzyme [19]. ATP-BINDING CASSETTE G37 (ABCG37/PDR9) transporters contribute to the exudation of coumarins [17]. Both *F6'H1* and *PDR9* transcript expression are upregulated by iron deficiency [19,20].

Subsequently, the soluble Fe^{3+} is reduced into Fe^{2+} in root apoplast via cellular membrane localized ferric reductase oxidase 2 (FRO2). This protein has 725 amino acids with 8 transmembrane domains, containing motif for binding hemes and NADPH [21]. The electron from NADPH in the cytoplasmic side is transferred via two hemes and Flavin to the Fe^{3+} in apoplast [22]. *FRO2* is primarily expressed in roots [23]. In addition to expression in roots, *FRO2* is largely present in flowers [24]. *FRO2* transcription and post-transcription are both regulated by iron concentration, since the activity of FRO2 in *FRO2* overexpression lines is highly induced under iron deficiency [24]. In addition, iron deficiency facilitates the stability of *FRO2* mRNA [24]. A total of 50 FROs were identified in plants [25] and 8 FROs are encoded in the *Arabidopsis* genome [26]. These FROs have different tissue-specific expression patterns. *AtFRO3* and *AtFRO5* are predominantly expressed in roots, while *AtFRO6*, *AtFRO7* and *AtFRO8* gene expression primarily occur in shoots. *AtFRO1* and *AtFRO4* are present in both roots and leaves [23,27–29].

After Fe^{3+} reduced to Fe^{2+} in root rhizosphere, Fe^{2+} can be imported into cells by IRT1 with high affinity to Fe^{2+} ($K_m = 6$ µM). IRT1 is the most important root transporter for ferrous Fe uptake from the soil, while the uptake of other divalent cations (manganese, zinc, cobalt, and cadmium) can also be promoted by IRT1 [6,7,30]. IRT1 is identified in *Arabidopsis* and can rescue the defects of the *fet3fet4* mutants of yeast that are impaired in Fe uptake [6]. The expression of *IRT1* is highly induced under iron-limited conditions [6,7]. IRT1 belongs to ZIP family and consists of 347 amino acids with 8 transmembrane domains. IRT1 can also promote the uptake of and Zn^{2+} but IRT1 can transport Zn only under low pH [30,31]. IRT1 is present in early endosomes/trans-Golgi network compartments (EE/TGN). Early studies found that IRT1 degradation and recycling between EE/TGN and the plasma membrane are modulated by ubiquitination and monoubiquitin-dependent endocytosis [32]. The IRT1 protein can transport to a vacuole for degradation [32]. IRT1 degradation factor1 (IDF1), a RING-type E3 ubiquitin ligase, is found to be responsible for IRT1 ubiquitination on plasma membrane via clathrin-mediated endocytosis. Thus, Fe-deficient induced IDF1 facilitating IRT1 degradation develops a negative feedback loop to fine tune the iron homeostasis [33]. It should be noted that recent studies point to the fact that non-iron elements (Zn, Mn, and Co) are also able to regulate this trafficking of IRT1 between EE/TGN and the plasma membrane in root epidermal cells [34]. Moreover, FYVE1, a phosphatidylinositol-3-phosphate-binding protein, is also required for the recycling of IRT1 and its polar localization to outer polar domain of plasma membrane [34]. SORTING NEXIN (SNX) protein was found to co-localize with IRT1 and is also important for recycling internalized IRT1. In the *snx1* mutants, the degradation of IRT1 is enhanced [35]. Further studies reveal that there exist other transporters for iron uptake. Natural resistance associated macrophage proteins (NRAMPs) were identified as a ubiquitous family of metal efflux transporters. Quite intriguingly, NRAMP1 that acts as a transporter of manganese is also essential for low-affinity iron uptake. Pleckstrin homolog (PH) domain-containing protein AtPH1 binds phosphatidylinositol 3-phosphate (PI3P) in the late endosome, which regulates the localization of NRAMP1 to the vacuole [36].

The strategy II plants, such as rice, can secrete phytosiderophores (PS) in rhizosphere for efficiently increasing the solubility of Fe^{3+}, ultimately facilitating the available iron for root acquisition [37]. $PS-Fe^{3+}$ complexes are then imported into root epidermis cells by a specific transporter [37]. PS belong to the family of mugineic acid (MAs), such as mugineic acid (MA), 2′-deoxymugineic acid (DMA), 3-epihydroxymugineic acid (epi-HMA), and 3-epihydroxy 2′-deoxymugineic acid (epi-HDMA) [38,39]. MAs are synthesized from three S-adenosyl-methionine molecules [40]. Yellow stripe 1 (YS1) is firstly identified from maize and targeted to the plasma membrane, which is likely to responsible for transporting Fe^{3+}-PS into root cells [10]. YS1 consists of 682 amino acid with 12 transmembrane domains [3]. The transcript expression of *ZmYS1* is highly induced in both root and shoot of maize under iron-deficient condition [10,41]. Eighteen putative yellow stripe 1 (YS1)-like genes (OsYSLs) are identified in the rice genome [42].

Fe deficiency readily results in interveinal chlorosis in young leaves, ultimately reducing the yield and grain quality [43]. In order to tolerate iron deficiency, various physiological processes are induced in the root rhizosphere, including ferric reductase activity, the ratio of root and shoot, and photosynthesis. Also, root morphology is altered according to the local availability of iron and for optimizing iron uptake, such as increasing lateral root numbers, extra root hairs, and developing transfer cells to facilitate contact surface with soil [44].

3. Iron Transport Mechanism in Plants

After iron is transported to the root endodermis from epidermis via apoplastic and symplastic pathway, it needs to be transported to the above ground parts of plants through the xylem (Figure 2). The contents of organic acids, such as citrate, malate, and succinate, are elevated in xylem under iron deficient conditions [45]. The usage of various approaches, such as the theoretical calculations, high-pressure liquid chromatography (HPLC) coupled to electrospray time-of-flight mass spectrometry (HPLC-ESI-TOFMS) and inductively coupled plasma mass spectrometry (HPLC-ICP-MS), detects the natural Fe complex and provides evidence for the transport of iron in xylem to shoots which predominantly occurs as Fe^{3+}-citrate complex [46–49]. The transport of citrate and iron to the xylem is mediated by ferric reductase defective 3 (FRD3) in *Arabidopsis* and its ortholog FRDL1 in rice, which is crucial for iron translocation [50,51]. FRD3 is present only in pericycle and cells neighboring the vascular tissue [50]. *frd3* mutants exhibit severe Fe-deficient phenotype even under Fe-sufficient conditions. Less citrate and less Fe are contained in xylem sap of *frd3* mutants as compared to wild type [50]. *Osfrdl* mutants also contain reduced citrate and Fe in the xylem resembling Fe-deficiency phenotype in *frd3* mutants [52]. Therefore, it is tempted to speculate that graminaceous and nongraminaceous share the similar mechanism by which Fe is transported from root to shoot although the uptake strategies for iron are very different. Ferroportin1 (FPN1) is also responsible for loading iron into the xylem [44]. The *Arabidopsis* genome contains three FPN which have different subcellular localizations. FPN1, for example, is targeted to the plasma membrane, FPN2 on the vacuolar membrane and FPN3 on the chloroplast envelop [44,53,54]. Fe is also capable of translocation in xylem in the form of Fe-nicotianamine (NA) and Fe-MAs. NA as a non-protein amino acid is produced from S-adenosyl methionine by nicotianamine synthase (NAS) and is also the direct biochemical precursor to PS [55,56]. In rice, NA and DMA are present in xylem exudates [57,58].

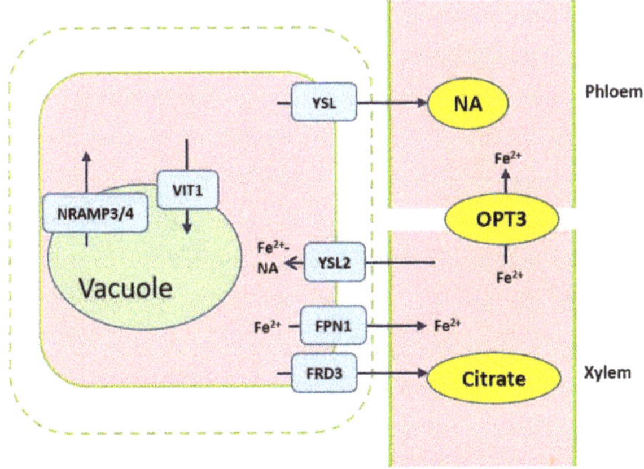

Figure 2. Overview of iron transport from roots to shoots. Ferric reductase defective 3 (FRD3) and ferroportin1 (FPN1) are responsible for importing citrate and iron into the xylem. Iron chelation with citrate or NA are translocated to shoots. Yellow stripe-like 2 (YSL2) contributes to the Fe^{2+}-NA distribution from the xylem to neighboring cells. Iron is loaded into vacuole through VIT1, while iron efflux of vacuolar occurs via NRAMP3 and NRAMP4. OPT3 mediates the Fe transport to sink tissues via the phloem.

Once the iron reaches the leaves, it must be unloaded to leaf cells from the apoplastic space. NA and DMA are also required for the phloem-based transport [59]. AtYSL1, AtYSL2, and AtYSL3, as metal-NA transporters, are involved in this process, responsible for moving iron from apoplast to symplast [60,61]. These three genes are highly expressed in vascular parenchyma cells of leaves [60,61]. AtYSL2 plays a major role in regulating the lateral distribution of iron from xylem to shoot cells in *Arabidopsis* [54,60]. Moreover, AtYSL1 and AtYSL3 appear to transport the Fe-NA chelate from senescent leaves into the inflorescences and seeds. *ysl1* and *ysl3* mutants contain reduced iron content in leaves and seeds [60,62,63]. In rice, OsYSL2 is likely to be involved in the translocation of Fe(II)-NA to shoots and seeds [42,64]. *OsYSL16* is expressed in the cells surrounding xylem and contributes to Fe(III)-MA allocation via the vascular bundle [65]. OsYSL18 also transports Fe(III)-DMA in reproductive organs and phloem of lamina joints [66]. Recent studies point to OsYSL9 which is involved in the Fe distribution in developing seeds via Fe(II)-NA and Fe(III)-DMA form [67]. Additionally, oligo peptide transporter 3 (OPT3) mediates the Fe transport to sink tissues via the phloem and recirculation in the roots in *Arabidopsis* [68]. Meanwhile, OPT3 is also found to take part in the control of iron movement out of the leaves to root or developing tissues in the form of iron ions rather than iron-ligand complexes [69,70]. Heat shock cognate protein B (HSCB) as a mitochondrial cochaperone participates in iron translocation from roots to shoots [71]. *HSCB* overexpression lines caused iron accumulation in roots but low iron levels in shoots; while *hscb* knockdown plants showed iron accumulation in shoots despite the reduced contents of iron uptake in roots [71].

4. Iron Storage in Cells

Iron mobilization in cells is essential for plant growth and development, especially under iron-deficient conditions. When transporting across cellular or intracellular membranes, ferric iron is usually reduced to ferrous iron [72]. Iron can produce cytotoxic oxygen radicals, such as hydroxyl radicals and superoxide anions [16]. Generally, the cellular iron is stored in vacuoles and is also likely to be sequestrated into ferritin, which will become available for various metabolic reactions. In *Arabidopsis* seeds, the vacuole is the major iron store containing about 50% of total iron, while ferritins play a minor role in iron storage including about 5% iron [16,73]. Ferritin is important for fine tuning the quantity of metal which is required for metabolic purposes [74]. In the vacuole of *Arabidopsis* seeds, globoids act as an important site for Fe storage [16]. However, in pea, the amount of iron-ferritin is present at about 92% of the total seed iron in embryo axis [75]. Therefore, these findings suggest that the way for iron storage in seeds may be different between different species, such as *pea* and *Arabidopsis* [73]. Plastids also act as a sink for iron in cells and appear to function in sensing and maintaining iron concentration in the plants to adapt various changes [76]. In chloroplast, ferritins represent one candidate to form the complex with Fe [76]. In *Arabidopsis*, three of ferritins are localized to chloroplasts. In addition, NA might also play a role in maintaining Fe soluble in plastids [76].

The changes of iron content in vacuole might trigger distinct responses. The vacuolar iron transporter 1 (VIT1), an orthologue of the yeast iron transporter Ca^{2+}-sensitive cross-complementer 1 (CCC1), was first identified in *Arabidopsis* [77]. AtVIT1 was found to control iron sequestration into vacuoles. Despite there being no difference in the iron content of seeds between *vit1* mutants and wild type, the iron accumulation is absent in the vacuoles of provascular cells [77]. So, what else could modulate iron mobilization efflux from vacuolar? AtNRAMP3 and AtNRAMP4 are responsible for Fe efflux from the vacuolar into the cytosol, and consequently essential for seed germination under Fe deficiency [78,79]. However, we cannot exclude other efflux transporters localized in vacuolar. In rice, the molecular mechanism underlying Fe transport in cells has also been well uncovered. OsVIT1, OsVIT2, and OsNRAMPs affect Fe translocation from the vacuole to other parts [80–83].

Ferritins, as another iron pool, are a class of universal 24-mer multi-meric, which are encoded by nuclear genes [84]. The structure of ferritins is highly conserved in eukaryotes [74]. In *Arabidopsis*, four ferritin genes (AtFer1–4) have been identified, among which FER1, FER3, and FER4 are proposed to exist in leaves while FER2 is present in seeds [74]. Recent studies found that ferritins are vital for protecting cells against oxidative stress [73]. Recently it was reported that ferritins are also involved in root system architecture regulation. Triple mutants of *fer1 fer3 fer4* exhibited altered root architecture which was caused by the alteration in the production and balance of reactive oxygen species (ROS) [85].

In addition, mitochondrion as a crucial iron sink provides available iron for the proper respiration. In rice, FRO3 and FRO8 appear to play roles in Fe^{3+} reduction in the mitochondrial membrane and mitochondrial iron transporters (MITs) are responsible for the translocation of iron from cytoplasm to mitochondrial [86]. Although the total iron content of shoots is increased in *mit* knockdown mutants as compared to wild type, the iron concentration in mitochondria is reduced, which further suggest iron is mistransported in the mitochondria of these mutants. Additionally, *mit* knockdown mutants contain a significant reduction of chlorophyll content and impair plant growth [87].

Also, chloroplast represents one of the main sinks for iron in plant cells. The iron transport across the chloroplast inner envelope also depends on reduction-based strategy. AtFRO7 as a chloroplast Fe (III) chelate reductase is targeted to the chloroplast envelope and putatively function in Fe^{3+} reduction in chloroplast. AtFRO7 is required for the survival of young seedlings under iron-deficient conditions. Under Fe-deficient conditions, loss of function of *FRO7* reduces the Fe content and hampers the reductase activity of chloroplast, leading to chlorotic appearance [29]. AtYSL6 is localized to the chloroplast envelope. Plants lacking *ATYSL4* and *ATYSL6* exhibit iron over-accumulated chloroplasts and the overexpression lines are characterized by decreased Fe content in chloroplast, suggesting that YSL4 and YSL6 take part in the release of iron from chloroplast [88]. In addition, PERMEASE IN CHLOROPLASTS1 (PIC1) as an ancient permease plays a role in chloroplast Fe uptake and maintaining Fe homeostasis. Interestingly, PIC1 was identified as the first protein involved in Fe uptake in plastid [89], which is localized to the inner envelope and contain four membrane-spanning α-helices [89]. The *pic1* mutant exhibits altered mesophyll organization, disrupted chloroplast and thylakoid development, which is consistent with Fe-deficiency phenotype [89]. Furthermore, recent findings further confirm this function of PIC1 in plastid Fe-transport using *PIC1* knockdown and overexpression lines in *Nicotiana tabacum* [90].

5. Transcriptional and Posttranscriptional Regulation of Fe-related Genes

Since Fe is vital for cellular process, a sophisticated regulatory mechanism to sense and adjust iron deficiency is essential for providing sufficient iron for plant growth and development. To avoid iron deficiency, various genes involved in iron acquisition and internal translocation are fine-tune regulated at the transcriptional and posttranscriptional level in adapt to iron deficient condition (Figure 1). Fe efficiency reactions (FER) was firstly identified in tomato and encoding a bHLH transcription factor. In this regard, FER controls the root physiology and morphology adapt to iron deficiency [91]. The basic helix-loop-helix (bHLH) FER-like iron deficiency-induced transcription factor (FIT) was identified in *Arabidopsis* and involved in iron sensing, responding, and acquisition through regulating the expression of FRO2 and IRT1 [92]. The ethylene-responsive transcription factors Ethylene Insensitive3 (EIN3) and EIN3-Like1 (EIL1) both enable interact with FIT, consequently activating FIT [93]. The activated FIT can up-regulate the transcript expression of AHA2, FRO2, and IRT1 [94–96]. Extensively, FIT activity is modulated via interaction with other proteins. The expressions of bHLH038, bHLH039, bHLH100, and bHLH101 have been reported to be increased under Fe starvation and interact with FIT [97,98]. These interactions result in the activation of FIT and consequently activate the expression of FIT target genes such as IRT1 and FRO2 [98,99]. However, the transcript expression of NRAMP3 is not influenced by the activated FIT [94]. What is more, the activity of FIT can be inhibited by the interaction of DELLA with FIT [100]. In addition, a bHLH transcription factor POPEYE (PYE) is identified which as part of an iron regulatory network is independent of FIT. PYE is capable of interacting with another PYE homologs-bHLH transcription factor IAA-Leu Resistant3 (ILR3), which regulates the iron deficiency response and are both required for maintaining iron homeostasis [101]. Under low iron conditions, PYE is expressed in the root vasculature, columella root cap, and also lateral root cap. Interestingly, its strongest expression occurs in the pericycle of the maturation zone [101]. A putative E3 ligase protein BRUTUS (BTS) can also interact with ILR3, but plays a negative role in response to iron deficiency [101]. Also, the transcription factors, MYB family members MYB10 and MYB72, are implicated in the regulation of NAS4 expression [102,103]. WRK46 not only regulates the expression of NAS but also enables to directly bind the promoter of VIT-LIKE1 via the W-boxes, thereby controlling the iron translocation [104]. YSL2 expression can be controlled by the transcription factors IDEF1 and IDEF2 in rice [105]. In rice, Fe-deficiency-inducible bHLH transcription factor OsIRO2, as the homologue of AtbHLH39, enhances the expression of YSL15 [106].

6. Function of Other Factors in the Iron Homeostasis

Although recent studies have demonstrated Fe-related genes are associated with plants response to Fe deficiency, the reality of signal network appears to be more complicated in adapting to iron deficiency. Various plant hormones, messenger molecules and kinases are implicated into this process. Auxin analogs for example can increase the activity of the root ferric chelate reductase (FCR) in bean [107–109]. In *Arabidopsis*, abscisic acid (ABA) and gibberellin have been suggested to facilitate the Fe deficient response, while cytokinin and jasmonic acid prevent this response [110–113]. ABA, for example, promotes the secretion of phenolics and also iron efflux from vacuole via up-regulation of AtNRAMP3. Further studies suggest that ABA enhances the Fe translocation from root to shoot [110]. Nitric oxide (NO) is also found to act as a component of Fe signal pathway and activate root FCR activity under iron deficiency via acting downstream of auxin in *Arabidopsis* [114]. NO plays a role in the synthesis of cell wall. Cell wall consists of pectin, cellulose, and hemicellulose. Cell walls are full of negative charges, which provide the binding sites for metal ions. Pectin is secreted into the apoplast from the symplast. Pectin methylesterase (PME) contributes to de-methylation of pectin that can increase carboxylic groups and hence provides more negative charged sites for iron in cell wall. Fe-deficiency induced NO prevents pectin methylation of cell wall and stimulates the PME activity. These together enhance the Fe retention in root apoplast. In this regard, NO limits iron translocation from root to shoot [115]. Recent evidence points to Ca^{2+} direct interrelations of Fe signal. An important signaling network in deciphering Ca^{2+} signals is formed by specific interactions of 10 calcium B-like proteins (CBLs) and 26 CBL interacting protein kinases (CIPKs) in *Arabidopsis* [116,117]. CIPK23 could be as "nutritional sensors" to sense and mediate the iron homeostasis in *Arabidopsis*. *cipk23* mutants exhibit lower activity of FCR and the regulation of FCR activity by CIPK23 is not related to the transcript expression of *FRO2, FRO3,* and *FRO5* [118]. Additionally, it has been found that CIPKs are also involved in the regulation of H^+ homeostasis. CIPK11/PKS5 suppresses the activity of the PM H^+-ATPase (AHA2) via phosphorylation which prevents the interaction between AHA2 and 14-3-3 protein, and thus inhibits the extrusion of protons (H^+) to the extracellular space [119]. Moreover, CIPK11 interacts with FIT and activates FIT via phosphorylation at Ser272, allowing for FIT-dependent Fe deficiency responses. Mutation at Ser272 of FIT affects seed iron content [120].

7. Conclusions

Iron acts as an essential element not only in plant physiological functions but also in the maintenance of various cell processes. Over the past decades, accumulating progresses have been achieved in understanding how the plants adapt to iron deficiency in soil. Cellular, biochemical, molecular, genetics, and genomic approaches facilitate a better understanding of iron uptake, transport, and utilization. However, how to observe the iron dynamics in plants, especially in different tissues and cells, is still a notable challenge. Despite a wealth of information pointing to the identities for many genes responsible for iron uptake from soil, transport from roots to shoots, storage in cells, and even their regulation at the transcription and post-transcription level, further research is clearly needed to uncover the further interconnection and integration of signaling pathways of iron deficiency into development and physiological networks. Finally, all of this information underlying the mechanism of iron uptake, transport, and homeostasis will be of great benefit to plants and human health.

Author Contributions: X.Z. wrote and designed the manuscript. T.W. read and approved the contents. D.Z. and W.S. edited the manuscript.

Funding: This work was funded by the National Key Research and Development Program of China (2018YFD0700202 and 2016YFC0500601), and the National Natural Science Foundation of China (31300231).

Conflicts of Interest: The authors declare no conflict of interest.

References

1. Mahender, A.; Swamy, B.P.M.; Anandan, A.; Ali, J. Tolerance of iron-deficient and -toxic soil conditions in rice. *Plants* **2019**, *8*, 31. [CrossRef] [PubMed]
2. Becker, M.; Asch, F. Iron toxicity in rice—Conditions and management concepts. *J. Plant Nutr. Soil Sci.* **2010**, *168*, 558–573. [CrossRef]
3. Curie, C.; Briat, J.F. Iron transport and signaling in plants. *Annu. Rev. Plant Biol.* **2003**, *54*, 183–206. [CrossRef] [PubMed]
4. Fageria, N.K.; Baligar, V.C.; Li, Y.C. The role of nutrient efficient plants in improving crop yields in the twenty first century. *J. Plant Nutr.* **2008**, *31*, 1121–1157. [CrossRef]
5. Yi, Y.; Guerinot, M.L. Genetic evidence that induction of root Fe(III) chelate reductase activity is necessary for iron uptake under iron deficiency. *Plant J.* **1996**, *10*, 835–844. [CrossRef] [PubMed]
6. Eide, D.; Broderius, M.; Fett, J.; Guerinot, M.L. A novel iron-regulated metal transporter from plants identified by functional expression in yeast. *Proc. Natl. Acad. Sci. USA* **1996**, *93*, 5624–5628. [CrossRef]

7.   Vert, G.; Grotz, N.; Dédaldéchamp, F.; Gaymard, F.; Guerinot, M.L.; Briat, J.F.; Curie, C. IRT1, an *Arabidopsis* transporter essential for iron uptake from the soil and for plant growth. *Plant Cell* **2002**, *14*, 1223–1233. [CrossRef]

8.   Takagi, S.-I. Naturally occurring iron-chelating compounds in oat- and rice-root washings. *Soil Sci. Plant Nutr.* **1976**, *22*, 423–433. [CrossRef]

9.   Takagi, S.I.; Nomoto, K.; Takemoto, T. Physiological aspect of mugineic acid, a possible phytosiderophore of graminaceous plants. *J. Plant Nutr.* **1984**, *7*, 469–477. [CrossRef]

10.   Curie, C.; Panaviene, Z.; Loulergue, C.; Dellaporta, S.L.; Briat, J.F.; Walker, E.L. Maize yellow stripe1 encodes a membrane protein directly involved in Fe(III) uptake. *Nature* **2001**, *409*, 346. [CrossRef] [PubMed]

11.   Yuan, W.; Zhang, D.; Song, T.; Xu, F.; Lin, S.; Xu, W.; Li, Q.; Zhu, Y.; Liang, J.; Zhang, J. *Arabidopsis* plasma membrane H^+-ATPase genes AHA2 and AHA7 have distinct and overlapping roles in the modulation of root tip H^+ efflux in response to low-phosphorus stress. *J. Exp. Bot.* **2017**, *68*, 1731–1741. [CrossRef]

12.   Oh, Y.J.; Kim, H.; Seo, S.H.; Hwang, B.G.; Chang, Y.S.; Lee, J.; Lee, D.W.; Sohn, E.J.; Lee, S.J.; Lee, Y.; et al. Cytochrome b5 reductase 1 triggers serial reactions that lead to iron uptake in plants. *Mol. Plant* **2016**, *9*, 501–513. [CrossRef] [PubMed]

13.   Abadía, J.; López Millán, A.F.; Rombolà, A.; Abadía, A. Organic acids and Fe deficiency: A review. *Plant Soil* **2002**, *241*, 75–86. [CrossRef]

14.   Jin, C.W.; You, G.Y.; He, Y.F.; Tang, C.; Wu, P.; Zheng, S.J. Iron deficiency-induced secretion of phenolics facilitates the reutilization of root apoplastic iron in red clover. *Plant Physiol.* **2007**, *144*, 278–285. [CrossRef]

15.   Connorton, J.M.; Balk, J.; Rodríguez Celma, J. Iron homeostasis in plants—A brief overview. *Metallomics* **2017**, *9*, 813–823. [CrossRef] [PubMed]

16.   Jeong, J.; Merkovich, A.; Clyne, M.; Connolly, E.L. Directing iron transport in dicots: Regulation of iron acquisition and translocation. *Curr. Opin. Plant Biol.* **2017**, *39*, 106–113. [CrossRef]

17.   Clemens, S.; Weber, M. The essential role of coumarin secretion for Fe acquisition from alkaline soil. *Plant Signal. Behav.* **2015**, *11*, e1114197. [CrossRef] [PubMed]

18.   Sisó Terraza, P.; Luis Villarroya, A.; Fourcroy, P.; Briat, J.F.; Abadía, A.; Gaymard, F.; Abadía, J.; Álvarez Fernández, A. Accumulation and secretion of coumarinolignans and other coumarins in *Arabidopsis thaliana* roots in response to iron deficiency at high pH. *Front. Plant Sci.* **2016**, *7*, 1711. [CrossRef] [PubMed]

19.   Fourcroy, P.; Tissot, N.; Gaymard, F.; Briat, J.F.; Dubos, C. Facilitated Fe nutrition by phenolic compounds excreted by the *Arabidopsis* ABCG37/PDR9 transporter requires the IRT1/FRO2 high affinity root Fe^{2+} transport system. *Mol. Plant* **2016**, *9*, 485–488. [CrossRef] [PubMed]

20.   Schmid, N.B.; Giehl, R.F.H.; Döll, S.; Mock, H.P.; Strehmel, N.; Scheel, D.; Kong, X.; Hider, R.C.; von Wirén, N. Feruloyl-CoA 6′-hydroxylase1-dependent coumarins mediate iron acquisition from alkaline substrates in *Arabidopsis*. *Plant Physiol.* **2014**, *164*, 160–172. [CrossRef]

21.   Robinson, N.J.; Procter, C.M.; Connolly, E.L.; Guerinot, M.L. A ferric-chelate reductase for iron uptake from soils. *Nature* **1999**, *397*, 694–697. [CrossRef] [PubMed]

22.   Schagerlöf, U.; Wilson, G.; Hebert, H.; Al Karadaghi, S.; Hägerhäll, C. Transmembrane topology of FRO2, a ferric chelate reductase from *Arabidopsis thaliana*. *Plant Mol. Biol.* **2006**, *62*, 215–221. [CrossRef] [PubMed]

23.   Mukherjee, I.; Campbell, N.H.; Ash, J.S.; Connolly, E.L. Expression profiling of the *Arabidopsis* ferric chelate reductase (FRO) gene family reveals differential regulation by iron and copper. *Planta* **2006**, *223*, 1178–1190. [CrossRef]

24.   Connolly, E.L.; Campbell, N.H.; Grotz, N.; Prichard, C.L.; Guerinot, M.L. Overexpression of the *FRO2* ferric chelate reductase confers tolerance to growth on low iron and uncovers posttranscriptional control. *Plant Physiol.* **2003**, *133*, 1102–1110. [CrossRef] [PubMed]

25.   Chang, Y.L.; Li, W.Y.; Miao, H.; Yang, S.Q.; Li, R.; Wang, X.; Li, W.Q.; Chen, K.M. Comprehensive genomic analysis and expression profiling of the NOX gene families under abiotic stresses and hormones in plants. *Genome Biol. Evol.* **2016**, *8*, 791–810. [CrossRef]

26.   Wu, H.; Li, L.; Du, J.; Yuan, Y.; Cheng, X.; Ling, H.Q. Molecular and biochemical characterization of the Fe(III) chelate reductase gene family in *Arabidopsis thaliana*. *Plant Cell Physiol.* **2005**, *46*, 1505–1514. [CrossRef]

27.   Feng, H.; An, F.; Zhang, S.; Ji, Z.; Ling, H.Q.; Zuo, J. Light-regulated, tissue-specific, and cell differentiation-specific expression of the *Arabidopsis* Fe(III)-chelate reductase gene *AtFRO6*. *Plant Physiol.* **2006**, *140*, 1345–1354. [CrossRef]

28. Li, L.-Y.; Cai, Q.-Y.; Yu, D.-S.; Guo, C.-H. Overexpression of *AtFRO6* in transgenic tobacco enhances ferric chelate reductase activity in leaves and increases tolerance to iron-deficiency chlorosis. *Mol. Biol. Rep.* **2011**, *38*, 3605–3613. [CrossRef] [PubMed]

29. Jeong, J.; Cohu, C.; Kerkeb, L.; Pilon, M.; Connolly, E.L.; Guerinot, M.L. Chloroplast Fe(III) chelate reductase activity is essential for seedling viability under iron limiting conditions. *Proc. Natl. Acad. Sci. USA* **2008**, *105*, 10619–10624. [CrossRef]

30. Korshunova, Y.O.; Eide, D.; Clark, W.G.; Guerinot, M.L.; Pakrasi, H.B. The IRT1 protein from *Arabidopsis thaliana* is a metal transporter with a broad substrate range. *Plant Mol. Biol.* **1999**, *40*, 37–44. [CrossRef] [PubMed]

31. Rogers, E.E.; Eide, D.J.; Guerinot, M.L. Altered selectivity in an *Arabidopsis* metal transporter. *Proc. Natl. Acad. Sci. USA* **2000**, *97*, 12356–12360. [CrossRef]

32. Barberon, M.; Zelazny, E.; Robert, S.; Conéjéro, G.; Curie, C.; Friml, J.; Vert, G. Monoubiquitin-dependent endocytosis of the IRON-REGULATED TRANSPORTER 1 (IRT1) transporter controls iron uptake in plants. *Proc. Natl. Acad. Sci. USA* **2011**, *108*, E450–E458. [CrossRef] [PubMed]

33. Shin, L.J.; Lo, J.C.; Chen, G.H.; Callis, J.; Fu, H.; Yeh, K.C. IRT1 degradation factor1, a RING E3 ubiquitin ligase, regulates the degradation of iron-regulated transporter1 in *Arabidopsis*. *Plant Cell* **2013**, *25*, 3039–3051. [CrossRef]

34. Barberon, M.; Dubeaux, G.; Kolb, C.; Isono, E.; Zelazny, E.; Vert, G. Polarization of IRON-REGULATED TRANSPORTER 1 (IRT1) to the plant-soil interface plays crucial role in metal homeostasis. *Proc. Natl. Acad. Sci. USA* **2014**, *111*, 8293–8298. [CrossRef]

35. Ivanov, R.; Brumbarova, T.; Blum, A.; Jantke, A.M.; Fink Straube, C.; Bauer, P. SORTING NEXIN1 is required for modulating the trafficking and stability of the *Arabidopsis* IRON-REGULATED TRANSPORTER1. *Plant Cell* **2014**, *26*, 1294–1307. [CrossRef] [PubMed]

36. Agorio, A.; Giraudat, J.; Bianchi, M.W.; Marion, J.; Espagne, C.; Castaings, L.; Lelièvre, F.; Curie, C.; Thomine, S.; Merlot, S. Phosphatidylinositol 3-phosphate-binding protein AtPH1 controls the localization of the metal transporter NRAMP1 in *Arabidopsis*. *Proc. Natl. Acad. Sci. USA* **2017**, *114*, E3354–E3363. [CrossRef] [PubMed]

37. Takemoto, T.; Nomoto, K.; Fushiya, S.; Ouchi, R.; Kusano, G.; Hikino, H.; Takagi, S.I.; Matsuura, Y.; Kakudo, M. Structure of mugineic acid, a new amino acid possessing an iron-chelating activity from roots washings of water-cultured *Hordeum vulgare* L. *Proc. Jpn. Acad.* **1978**, *54*, 469–473. [CrossRef]

38. Mori, S. Iron acquisition by plants. *Curr. Opin. Plant Biol.* **1999**, *2*, 250–253. [CrossRef]

39. Negishi, T.; Nakanishi, H.; Yazaki, J.; Kishimoto, N.; Fujii, F.; Shimbo, K.; Yamamoto, K.; Sakata, K.; Sasaki, T.; Kikuchi, S.; et al. cDNA microarray analysis of gene expression during Fe-deficiency stress in barley suggests that polar transport of vesicles is implicated in phytosiderophore secretion in Fe-deficient barley roots. *Plant J.* **2002**, *30*, 83–94. [CrossRef] [PubMed]

40. Conte, S.S.; Walker, E.L. Transporters contributing to iron trafficking in plants. *Mol. Plant* **2011**, *4*, 464–476. [CrossRef] [PubMed]

41. Roberts, L.A.; Pierson, A.J.; Panaviene, Z.; Walker, E.L. Yellow Stripe1 expanded roles for the maize iron-phytosiderophore transporter. *Plant Physiol.* **2004**, *135*, 112–120. [CrossRef]

42. Koike, S.; Inoue, H.; Mizuno, D.; Takahashi, M.; Nakanishi, H.; Mori, S.; Nishizawa, N.K. OsYSL2 is a rice metal-nicotianamine transporter that is regulated by iron and expressed in the phloem. *Plant J.* **2004**, *39*, 415–424. [CrossRef]

43. Karim, M.R. Responses of *Aerobic rice* (*Oryza sativa* L.) to iron deficiency. *J. Integr. Agric.* **2012**, *11*, 938–945.

44. Morrissey, J.; Baxter, I.R.; Lee, J.; Li, L.; Lahner, B.; Grotz, N.; Kaplan, J.; Salt, D.E.; Guerinot, M.L. The ferroportin metal efflux proteins function in iron and cobalt homeostasis in *Arabidopsis*. *Plant Cell* **2009**, *21*, 3326–3338. [CrossRef] [PubMed]

45. López-Millán, A.F.; Morales, F.; Abadía, A.; Abadía, J. Effects of iron deficiency on the composition of the leaf apoplastic fluid and xylem sap in sugar beet. Implications for iron and carbon transport. *Plant Physiol.* **2000**, *124*, 873–884. [CrossRef] [PubMed]

46. Durrett, T.P.; Gassmann, W.; Rogers, E.E. The FRD3-mediated efflux of citrate into the root vasculature is necessary for efficient iron translocation. *Plant Physiol.* **2007**, *144*, 197–205. [CrossRef] [PubMed]

47. Rellán Álvarez, R.; Giner Martínez Sierra, J.; Orduna, J.; Orera, I.; Rodríguez Castrillón, J.Á.; García Alonso, J.I.; Abadía, J.; Álvarez Fernández, A. Identification of a tri-iron(III), tri-citrate complex in the xylem sap of iron-deficient tomato resupplied with iron: New insights into plant iron long-distance transport. *Plant Cell Physiol.* **2009**, *51*, 91–102. [CrossRef] [PubMed]

48. Rellán-Álvarez, R.; Abadía, J.; Álvarez-Fernández, A. Formation of metal-nicotianamine complexes as affected by pH, ligand exchange with citrate and metal exchange. A study by electrospray ionization time-of-flight mass spectrometry. *Rapid Commun. Mass Sp.* **2008**, *22*, 1553–1562.

49. Von Wiren, N.; Klair, S.; Bansal, S.; Briat, J.F.; Khodr, H.; Shioiri, T.; Leigh, R.A.; Hider, R.C. Nicotianamine chelates both FeIII and FeII. Implications for metal transport in plants. *Plant Physiol.* **1999**, *119*, 1107–1114. [CrossRef]

50. Green, L.S.; Rogers, E.E. FRD3 controls iron localization in *Arabidopsis. Plant Physiol.* **2004**, *136*, 2523–2531. [CrossRef]

51. Yokosho, K.; Yamaji, N.; Ma, J.F. OsFRDL1 expressed in nodes is required for distribution of iron to grains in rice. *J. Exp. Bot.* **2016**, *67*, 5485–5494. [CrossRef] [PubMed]

52. Yokosho, K.; Yamaji, N.; Ueno, D.; Mitani, N.; Ma, J.F. OsFRDL1 is a citrate transporter required for efficient translocation of iron in rice. *Plant Physiol.* **2009**, *149*, 297–305. [CrossRef] [PubMed]

53. Conte, S.; Stevenson, D.; Furner, I.; Lloyd, A. Multiple antibiotic resistance in *Arabidopsis* is conferred by mutations in a chloroplast-localized transport protein. *Plant Physiol.* **2009**, *151*, 559–573. [CrossRef]

54. Schaaf, G.; Häberle, J.; von Wirén, N.; Schikora, A.; Curie, C.; Vert, G.; Briat, J.-F.; Ludewig, U. A Putative function for the *Arabidopsis* Fe–phytosiderophore transporter homolog AtYSL2 in Fe and Zn homeostasis. *Plant Cell Physiol.* **2005**, *46*, 762–774. [CrossRef]

55. Bonneau, J.; Baumann, U.; Beasley, J.; Li, Y.; Johnson, A.A.T. Identification and molecular characterization of the nicotianamine synthase gene family in bread wheat. *Plant Biotechnol. J.* **2016**, *14*, 2228–2239. [CrossRef]

56. Inoue, H.; Higuchi, K.; Takahashi, M.; Nakanishi, H.; Mori, S.; Nishizawa, N.K. Three rice nicotianamine synthase genes, *OsNAS1*, *OsNAS2*, and *OsNAS3* are expressed in cells involved in long-distance transport of iron and differentially regulated by iron. *Plant J.* **2003**, *36*, 366–381. [CrossRef]

57. Kawai, S.; Kamei, S.; Matsuda, Y.; Ando, R.; Kondo, S.; Ishizawa, A.; Alam, S. Concentrations of iron and phytosiderophores in xylem sap of iron-deficient barley plants. *J. Soil Sci. Plant Nutr.* **2001**, *47*, 265–272. [CrossRef]

58. Mori, S.; Nishizawa, N.; Hayashi, H.; Chino, M.; Yoshimura, E.; Ishihara, J. Why are young rice plants highly susceptible to iron deficiency? *Plant Soil* **1991**, *130*, 143–156. [CrossRef]

59. Curie, C.; Cassin, G.; Couch, D.; Divol, F.; Higuchi, K.; Le Jean, M.; Misson, J.; Schikora, A.; Czernic, P.; Mari, S. Metal movement within the plant: Contribution of nicotianamine and yellow stripe 1-like transporters. *Ann. Bot.* **2009**, *103*, 1–11. [CrossRef]

60. DiDonato, R.J., Jr.; Roberts, L.A.; Sanderson, T.; Eisley, R.B.; Walker, E.L. *Arabidopsis* Yellow Stripe-Like2 (YSL2): A metal-regulated gene encoding a plasma membrane transporter of nicotianamine–metal complexes. *Plant J.* **2004**, *39*, 403–414. [CrossRef]

61. Waters, B.M.; Chu, H.H.; DiDonato, R.J.; Roberts, L.A.; Eisley, R.B.; Lahner, B.; Salt, D.E.; Walker, E.L. Mutations in *Arabidopsis* Yellow Stripe-Like1 and Yellow Stripe-Like3 reveal their roles in metal ion homeostasis and loading of metal ions in seeds. *Plant Physiol.* **2006**, *141*, 1446–1458. [CrossRef]

62. Jean, M.L.; Schikora, A.; Mari, S.; Briat, J.F.; Curie, C. A loss-of-function mutation in AtYSL1 reveals its role in iron and nicotianamine seed loading. *Plant J.* **2005**, *44*, 769–782. [CrossRef] [PubMed]

63. Chu, H.-H.; Chiecko, J.; Punshon, T.; Lanzirotti, A.; Lahner, B.; Salt, D.E.; Walker, E.L. Successful reproduction requires the function of *Arabidopsis* Yellow Stripe-Like1 and Yellow Stripe-Like3 metal-nicotianamine transporters in both vegetative and reproductive structures. *Plant Physiol.* **2010**, *154*, 197–210. [CrossRef]

64. Ishimaru, Y.; Masuda, H.; Bashir, K.; Inoue, H.; Tsukamoto, T.; Takahashi, M.; Nakanishi, H.; Aoki, N.; Hirose, T.; Ohsugi, R.; et al. Rice metal-nicotianamine transporter, OsYSL2, is required for the long-distance transport of iron and manganese. *Plant J.* **2010**, *62*, 379–390. [CrossRef]

65. Kakei, Y.; Ishimaru, Y.; Kobayashi, T.; Yamakawa, T.; Nakanishi, H.; Nishizawa, N.K. OsYSL16 plays a role in the allocation of iron. *Plant Mol. Biol.* **2012**, *79*, 583–594. [CrossRef] [PubMed]

66. Aoyama, T.; Kobayashi, T.; Takahashi, M.; Nagasaka, S.; Usuda, K.; Kakei, Y.; Ishimaru, Y.; Nakanishi, H.; Mori, S.; Nishizawa, N.K. OsYSL18 is a rice iron(III)-deoxymugineic acid transporter specifically expressed in reproductive organs and phloem of lamina joints. *Plant Mol. Biol.* **2009**, *70*, 681–692. [CrossRef] [PubMed]

67. Senoura, T.; Sakashita, E.; Kobayashi, T.; Takahashi, M.; Aung, M.; Masuda, H.; Nakanishi, H.; Nishizawa, N.K. The iron-chelate transporter OsYSL9 plays a role in iron distribution in developing rice grains. *Plant Mol. Biol.* **2017**, *95*, 375–387. [CrossRef] [PubMed]

68. Zhai, Z.; Gayomba, S.R.; Jung, H.I.; Vimalakumari, N.K.; Piñeros, M.; Craft, E.; Rutzke, M.A.; Danku, J.; Lahner, B.; Punshon, T.; et al. OPT3 is a phloem-specific iron transporter that is essential for systemic iron signaling and redistribution of iron and cadmium in *Arabidopsis*. *Plant Cell* **2014**, *26*, 2249–2264. [CrossRef]

69. Mendoza-Cózatl, D.G.; Xie, Q.; Akmakjian, G.Z.; Jobe, T.O.; Patel, A.; Stacey, M.G.; Song, L.; Demoin, D.W.; Jurisson, S.S.; Stacey, G.; et al. OPT3 is a component of the iron-signaling network between leaves and roots and misregulation of OPT3 leads to an over-accumulation of cadmium in seeds. *Mol. Plant* **2014**, *7*, 1455–1469. [CrossRef] [PubMed]

70. Wintz, H.; Fox, T.; Wu, Y.-Y.; Feng, V.; Chen, W.; Chang, H.-S.; Zhu, T.; Vulpe, C. Expression profiles of *Arabidopsis thaliana* in mineral deficiencies reveal novel transporters involved in metal homeostasis. *J. Biol. Chem.* **2003**, *278*, 47644–47653. [CrossRef] [PubMed]

71. Leaden, L.; Pagani, M.A.; Balparda, M.; Busi, M.V.; Gomez-Casati, D.F. Altered levels of AtHSCB disrupts iron translocation from roots to shoots. *Plant Mol. Biol.* **2016**, *92*, 613–628. [CrossRef] [PubMed]

72. Jain, A.; Wilson, G.T.; Connolly, E.L. The diverse roles of FRO family metalloreductases in iron and copper homeostasis. *Front. Plant Sci.* **2014**, *5*, 100. [CrossRef] [PubMed]

73. Ravet, K.; Touraine, B.; Boucherez, J.; Briat, J.-F.; Gaymard, F.; Cellier, F. Ferritins control interaction between iron homeostasis and oxidative stress in *Arabidopsis*. *Plant J.* **2009**, *57*, 400–412. [CrossRef] [PubMed]

74. Briat, J.-F.; Duc, C.; Ravet, K.; Gaymard, F. Ferritins and iron storage in plants. *BBA-Gen. Subj.* **2010**, *1800*, 806–814. [CrossRef]

75. Marentes, E.; Grusak, M.A. Iron transport and storage within the seed coat and embryo of developing seeds of pea (*Pisum sativum* L.). *Seed Sci. Res.* **1998**, *8*, 367–375. [CrossRef]

76. López-Millán, A.F.; Duy, D.; Philippar, K. Chloroplast iron transport proteins—Function and impact on plant physiology. *Front. Plant Sci.* **2016**, *7*, 178. [CrossRef] [PubMed]

77. Kim, S.A.; Punshon, T.; Lanzirotti, A.; Li, L.; Alonso, J.M.; Ecker, J.R.; Kaplan, J.; Guerinot, M.L. Localization of iron in *Arabidopsis* seed requires the vacuolar membrane transporter VIT1. *Science* **2006**, *314*, 1295–1298. [CrossRef] [PubMed]

78. Lanquar, V.; Lelièvre, F.; Bolte, S.; Hamès, C.; Alcon, C.; Neumann, D.; Vansuyt, G.; Curie, C.; Schröder, A.; Krämer, U.; et al. Mobilization of vacuolar iron by AtNRAMP3 and AtNRAMP4 is essential for seed germination on low iron. *EMBO J.* **2005**, *24*, 4041–4051. [CrossRef] [PubMed]

79. Thomine, S.; Lelièvre, F.; Debarbieux, E.; Schroeder, J.I.; Barbier-Brygoo, H. AtNRAMP3, a multispecific vacuolar metal transporter involved in plant responses to iron deficiency. *Plant J.* **2003**, *34*, 685–695. [CrossRef] [PubMed]

80. Bashir, K.; Takahashi, R.; Akhtar, S.; Ishimaru, Y.; Nakanishi, H.; Nishizawa, N.K. The knockdown of *OsVIT2* and *MIT* affects iron localization in rice seed. *Rice* **2013**, *6*, 31. [CrossRef] [PubMed]

81. Bashir, K.; Hanada, K.; Shimizu, M.; Seki, M.; Nakanishi, H.; Nishizawa, N.K. Transcriptomic analysis of rice in response to iron deficiency and excess. *Rice* **2014**, *7*, 18. [CrossRef] [PubMed]

82. Zhang, Y.; Xu, Y.H.; Yi, H.Y.; Gong, J.M. Vacuolar membrane transporters OsVIT1 and OsVIT2 modulate iron translocation between flag leaves and seeds in rice. *Plant J.* **2012**, *72*, 400–410. [CrossRef]

83. Nevo, Y.; Nelson, N. The NRAMP family of metal-ion transporters. *Biochim. Biophys. Acta* **2006**, *1763*, 609–620. [CrossRef] [PubMed]

84. Harrison, P.M.; Arosio, P. The ferritins: Molecular properties, iron storage function and cellular regulation. *Biochim. Biophys. Acta* **1996**, *1275*, 161–203. [CrossRef]

85. Reyt, G.; Boudouf, S.; Boucherez, J.; Gaymard, F.; Briat, J.F. Iron-and ferritin-dependent reactive oxygen species distribution: Impact on *Arabidopsis* root system architecture. *Mol. Plant* **2015**, *8*, 439–453. [CrossRef] [PubMed]

86. Jain, A.; Connolly, E.L. Mitochondrial iron transport and homeostasis in plants. *Front. Plant Sci.* **2013**, *4*, 348. [CrossRef] [PubMed]

87. Bashir, K.; Ishimaru, Y.; Shimo, H.; Nagasaka, S.; Fujimoto, M.; Takanashi, H.; Tsutsumi, N.; An, G.; Nakanishi, H.; Nishizawa, N.K. The rice mitochondrial iron transporter is essential for plant growth. *Nat. Commun.* **2011**, *2*, 322. [CrossRef]

88. Conte, S.S.; Chu, H.H.; Rodriguez, D.C.; Punshon, T.; Vasques, K.A.; Salt, D.E.; Walker, E.L. *Arabidopsis thaliana* Yellow Stripe1-Like4 and Yellow Stripe1-Like6 localize to internal cellular membranes and are involved in metal ion homeostasis. *Front. Plant Sci.* **2013**, *4*, 283. [CrossRef] [PubMed]

89. Duy, D.; Wanner, G.; Meda, A.R.; von Wirén, N.; Soll, J.; Philippar, K. PIC1, an ancient permease in *Arabidopsis* chloroplasts, mediates iron transport. *Plant Cell* **2007**, *19*, 986–1006. [CrossRef] [PubMed]

90. Gong, X.; Guo, C.; Terachi, T.; Cai, H.; Yu, D. Tobacco PIC1 mediates iron transport and regulates chloroplast development. *Plant Mol. Biol. Rep.* **2015**, *33*, 401–413. [CrossRef]

91. Ling, H.Q.; Pich, A.; Scholz, G.; Ganal, M.W. Genetic analysis of two tomato mutants affected in the regulation of iron metabolism. *Mol. Gen. Genet.* **1996**, *252*, 87–92. [CrossRef] [PubMed]

92. Bauer, P.; Ling, H.Q.; Guerinot, M.L. FIT, the FER-LIKE IRON DEFICIENCY INDUCED TRANSCRIPTION FACTOR in *Arabidopsis*. *Plant Physiol. Biochem.* **2007**, *45*, 260–261. [CrossRef]

93. Lingam, S.; Mohrbacher, J.; Brumbarova, T.; Potuschak, T.; Fink-Straube, C.; Blondet, E.; Genschik, P.; Bauer, P. Interaction between the bHLH transcription factor FIT and ETHYLENE INSENSITIVE3/ETHYLENE INSENSITIVE3-LIKE1 reveals molecular linkage between the regulation of iron acquisition and ethylene signaling in *Arabidopsis*. *Plant Cell* **2011**, *23*, 1815–1829. [CrossRef]

94. Colangelo, E.P.; Guerinot, M.L. The essential basic Helix-Loop-Helix protein FIT1 is required for the iron deficiency response. *Plant Cell* **2004**, *16*, 3400–3412. [CrossRef]

95. Jakoby, M.; Wang, H.Y.; Reidt, W.; Weisshaar, B.; Bauer, P. FRU (BHLH029) is required for induction of iron mobilization genes in *Arabidopsis thaliana*. *FEBS Lett.* **2004**, *577*, 528–534. [CrossRef] [PubMed]

96. Yuan, Y.X.; Zhang, J.; Wang, D.W.; Ling, H.Q. AtbHLH29 of *Arabidopsis thaliana* is a functional ortholog of tomato FER involved in controlling iron acquisition in strategy I plants. *Cell Res.* **2005**, *15*, 613. [CrossRef]

97. Wang, H.-Y.; Klatte, M.; Jakoby, M.; Bäumlein, H.; Weisshaar, B.; Bauer, P. Iron deficiency-mediated stress regulation of four subgroup Ib BHLH genes in *Arabidopsis thaliana*. *Planta* **2007**, *226*, 897–908. [CrossRef] [PubMed]

98. Wang, N.; Cui, Y.; Liu, Y.; Fan, H.; Du, J.; Huang, Z.; Yuan, Y.; Wu, H.; Ling, H.Q. Requirement and functional redundancy of Ib subgroup bHLH proteins for iron deficiency responses and uptake in *Arabidopsis thaliana*. *Mol. Plant* **2013**, *6*, 503–513. [CrossRef] [PubMed]

99. Yuan, Y.; Wu, H.; Wang, N.; Li, J.; Zhao, W.; Du, J.; Wang, D.; Ling, H.Q. FIT interacts with AtbHLH38 and AtbHLH39 in regulating iron uptake gene expression for iron homeostasis in *Arabidopsis*. *Cell Res.* **2008**, *18*, 385. [CrossRef] [PubMed]

100. Wild, M.; Davière, J.M.; Regnault, T.; Sakvarelidze Achard, L.; Carrera, E.; Lopez Diaz, I.; Cayrel, A.; Dubeaux, G.; Vert, G.; Achard, P. Tissue-specific regulation of gibberellin signaling fine-tunes *Arabidopsis* iron-deficiency responses. *Dev. Cell* **2016**, *37*, 190–200. [CrossRef] [PubMed]

101. Long, T.A.; Tsukagoshi, H.; Busch, W.; Lahner, B.; Salt, D.E.; Benfey, P.N. The bHLH transcription factor POPEYE regulates response to iron deficiency in *Arabidopsis* roots. *Plant Cell* **2010**, *22*, 2219–2236. [CrossRef] [PubMed]

102. Palmer, C.M.; Hindt, M.N.; Schmidt, H.; Clemens, S.; Guerinot, M.L. MYB10 and MYB72 are required for growth under iron-limiting conditions. *PLoS Genet.* **2013**, *9*, e1003953. [CrossRef] [PubMed]

103. Zamioudis, C.; Hanson, J.; Pieterse, C.M.J. β-Glucosidase BGLU42 is a MYB72-dependent key regulator of rhizobacteria-induced systemic resistance and modulates iron deficiency responses in *Arabidopsis* roots. *New Phytol.* **2014**, *204*, 368–379. [CrossRef] [PubMed]

104. Yan, J.Y.; Li, C.X.; Sun, L.; Ren, J.Y.; Li, G.X.; Ding, Z.J.; Zheng, S.J. A WRKY transcription factor regulates fe translocation under Fe deficiency. *Plant Physiol.* **2016**, *171*, 2017–2027. [CrossRef]

105. Kobayashi, T.; Ogo, Y.; Aung, M.S.; Nozoye, T.; Itai, R.N.; Nakanishi, H.; Yamakawa, T.; Nishizawa, N.K. The spatial expression and regulation of transcription factors IDEF1 and IDEF2. *Ann. Bot.* **2010**, *105*, 1109–1117. [CrossRef] [PubMed]

106. Ogo, Y.; Nakanishi Itai, R.; Nakanishi, H.; Kobayashi, T.; Takahashi, M.; Mori, S.; Nishizawa, N.K. The rice bHLH protein OsIRO2 is an essential regulator of the genes involved in Fe uptake under Fe-deficient conditions. *Plant J.* **2007**, *51*, 366–377. [CrossRef] [PubMed]

107. Li, C.; Zhu, X.; Zhang, F.S. Role of shoot in regulation of iron deficiency responses in cucumber and bean plants. *J. Plant Nutr.* **2000**, *23*, 1809–1818. [CrossRef]

108. Li, X.; Li, C. Is ethylene involved in regulation of root ferric reductase activity of dicotyledonous species under iron deficiency? *Plant Soil* **2004**, *261*, 147–153. [CrossRef]

109. Shao, J.Z.; Tang, C.; Arakawa, Y.; Masaoka, Y. The responses of red clover (*Trifolium pratense* L.) to iron deficiency: A root Fe(III) chelate reductase. *Plant Sci.* **2003**, *164*, 679–687.

110. Lei, G.J.; Zhu, X.F.; Wang, Z.W.; Dong, F.; Dong, N.Y.; Zheng, S.J. Abscisic acid alleviates iron deficiency by promoting root iron reutilization and transport from root to shoot in *Arabidopsis*. *Plant Cell Environ.* **2014**, *37*, 852–863. [CrossRef]

111. Matsuoka, K.; Bidadi, H.; Furukawa, J.; Satoh, S.; Asahina, M.; Yamaguchi, S. Gibberellin-induced expression of Fe uptake-related genes in *Arabidopsis*. *Plant Cell Physiol.* **2013**, *55*, 87–98. [CrossRef]

112. Séguéla, M.; Briat, J.-F.; Vert, G.; Curie, C. Cytokinins negatively regulate the root iron uptake machinery in *Arabidopsis* through a growth-dependent pathway. *Plant J.* **2008**, *55*, 289–300. [CrossRef]

113. Maurer, F.; Müller, S.; Bauer, P. Suppression of Fe deficiency gene expression by jasmonate. *Plant Physiol. Biochem.* **2011**, *49*, 530–536. [CrossRef] [PubMed]

114. Chen, W.W.; Yang, J.L.; Qin, C.; Jin, C.W.; Mo, J.H.; Ye, T.; Zheng, S.J. Nitric oxide acts downstream of auxin to trigger root ferric-chelate reductase activity in response to iron deficiency in *Arabidopsis*. *Plant Physiol.* **2010**, *154*, 810–819. [CrossRef]

115. Ye, Y.Q.; Jin, C.W.; Fan, S.K.; Mao, Q.Q.; Sun, C.L.; Yu, Y.; Lin, X.Y. Elevation of NO production increases Fe immobilization in the Fe-deficiency roots apoplast by decreasing pectin methylation of cell wall. *Sci. Rep.* **2015**, *5*, 10746. [CrossRef]

116. Kudla, J.; Xu, Q.; Harter, K.; Gruissem, W.; Luan, S. Genes for calcineurin B-like proteins in *Arabidopsis* are differentially regulated by stress signals. *Proc. Natl. Acad. Sci. USA* **1999**, *96*, 4718–4723. [CrossRef] [PubMed]

117. Weinl, S.; Kudla, J. The CBL–CIPK Ca^{2+}-decoding signaling network: Function and perspectives. *New Phytol.* **2009**, *184*, 517–528. [CrossRef] [PubMed]

118. Tian, Q.; Zhang, X.; Yang, A.; Wang, T.; Zhang, W.-H. CIPK23 is involved in iron acquisition of *Arabidopsis* by affecting ferric chelate reductase activity. *Plant Sci.* **2016**, *246*, 70–79. [CrossRef]

119. Fuglsang, A.T.; Guo, Y.; Cuin, T.A.; Qiu, Q.; Song, C.; Kristiansen, K.A.; Bych, K.; Schulz, A.; Shabala, S.; Schumaker, K.S. *Arabidopsis* protein kinase PKS5 inhibits the plasma membrane H^+-ATPase by preventing interaction with 14-3-3 protein. *Plant Cell* **2007**, *19*, 1617–1634. [CrossRef] [PubMed]

120. Gratz, R.; Manishankar, P.; Ivanov, R.; Köster, P.; Mohr, I.; Trofimov, K.; Steinhorst, L.; Meiser, J.; Mai, H.-J.; Drerup, M.; et al. CIPK11-dependent phosphorylation modulates FIT activity to promote *Arabidopsis* iron acquisition in response to calcium signaling. *Dev. Cell* **2019**, *48*, 726–740. [CrossRef] [PubMed]

International Journal of
Molecular Sciences

Article

A Stress-Responsive NAC Transcription Factor from Tiger Lily (LlNAC2) Interacts with LlDREB1 and LlZHFD4 and Enhances Various Abiotic Stress Tolerance in Arabidopsis

Yubing Yong, Yue Zhang and Yingmin Lyu *

Beijing Key Laboratory of Ornamental Germplasm Innovation and Molecular Breeding, National Engineering Research Center for Floriculture, College of Landscape Architecture, Beijing Forestry University, Beijing 100083, China
* Correspondence: luyingmin@bjfu.edu.cn

Received: 11 June 2019; Accepted: 27 June 2019; Published: 30 June 2019

Abstract: Our previous studies have indicated that a partial NAC domain protein gene is strongly up-regulated by cold stress (4 °C) in tiger lily (*Lilium lancifolium*). In this study, we cloned the full-length of this *NAC* gene, *LlNAC2*, to further investigate the function of *LlNAC2* in response to various abiotic stresses and the possible involvement in stress tolerance of the tiger lily plant. *LlNAC2* was noticeably induced by cold, drought, salt stresses, and abscisic acid (ABA) treatment. Promoter analysis showed that various stress-related cis-acting regulatory elements were located in the promoter of *LlNAC2*; and the promoter was sufficient to enhance activity of GUS protein under cold, salt stresses and ABA treatment. DREB1 (dehydration-responsive binding protein1) from tiger lily (LlDREB1) was proved to be able to bind to the promoter of *LlNAC2* by yeast one-hybrid (Y1H) assay. LlNAC2 was shown to physically interact with LlDREB1 and zinc finger-homeodomain ZFHD4 from the tiger lily (LlZFHD4) by bimolecular fluorescence complementation (BiFC) assay. Overexpressing *LlNAC2* in *Arabidopsis thaliana* showed ABA hypersensitivity and enhanced tolerance to cold, drought, and salt stresses. These findings indicated LlNAC2 is involved in both DREB/CBF-COR and ABA signaling pathways to regulate stress tolerance of the tiger lily.

Keywords: NAC; DREB1; ZFHD4; interaction; abiotic stresses; lily

---

1. Introduction

Plants have evolved a series of complex mechanisms to deal with environmental stressors, such as drought, high salinity, and extreme temperatures [1]. As crucial regulatory proteins, transcription factors (TFs) can specifically bind to cis-acting regulatory elements in the promoter of target genes to up or down-regulate their transcript levels [2,3]. Multiple types of TFs have shown to play important roles in plant responses to stresses, such as AP2/EREBP [4,5], MYB [6,7], bHLH [8], WRKY [9,10], zinc finger [11,12], and NAC [13,14] families. Thus, the identification and functional characterization of TFs contribute significantly to our understanding of the transcriptional regulatory networks in plants under abiotic stresses, which are indispensable parts in developing transgenic crops with enhanced tolerance to unfavorable growth conditions as well.

As one of the largest plant-specific TF groups, NAC (NAM, ATAF1,2 and CUC2) domain proteins have been reported to be essential regulators for plant response to abiotic stresses [15,16], which contain conserved N-terminal NAC DNA binding domains and variable C-terminal regions involved in transcriptional activation [17,18]. Most of the NAC genes were demonstrated to play positive roles in plant stress resistance [19]. For instance, overexpression of eight *NAC* genes (e.g., *OsNAC9*, *OsNAC10*) enhanced the drought and salinity tolerance significantly in rice (*Oryza sativa*) [1]. In wheat

(*Triticum aestivum*), overexpression of *TaNAC2a*, *TaNAC67*, and *TaNAC47* enhanced multi-abiotic stress tolerances in transgenic plants [20,21]. In *Arabidopsis thaliana*, overexpression of *ANAC019*, *ANAC055*, and *ANAC072* genes enhanced plant tolerance to drought [22]; and *ATAF1* exhibited positive regulation of drought ABA, salt, and oxidative tolerance in the transgenic plants [23]. Also, overexpressing the *TsNAC1* (*Thellungiella halophila*), pumpkin (*Cucurbita moschata*) *CmNAC1* [24], cotton (*Gossypium hirsutum*) *GhNAC2* [25], rose (*Rosa rugosa*) *RhNAC3* [26] genes showed increased tolerance to diverse abiotic stresses in transgenic plants. In contrast, only a few *NAC* genes were found to function as negative regulators of tolerance regulation to abiotic stress, such as *NAC016* [27] and *NTL4* [28] in Arabidopsis, and *ShNAC1* (*Solanum habrochaites*) [19] and *SlSRN1* (*Solanum lycopersicum*) in tomato [29].

On the other hand, the identification of C-repeat binding factor (*CBF*) genes has been known as one of the most important milestones in the cold signaling network elucidation [30]. CBFs (CBF1/2/3), also called dehydration-responsive binding protein 1 (DREB1B/1C/1A), are from the APETALA2/ethylene responsive factor (AP2/ERF) family, which can bind to C-repeat/dehydration-responsive elements (CRT/DREs) in the promoters of certain cold-responsive genes (*COR*) and regulate their expressions and functions [31–33]. Many findings have implied that CBF can function as a master regulator of the cold signaling pathway. Interestingly, some studies reported that some NAC TFs were also involved in the DREB/CBF-COR pathway. For instance, in *Pyrus betulifolia*, PbeNAC1 induces the expression of some stress-associated genes by interacting with PbeDREB1 and PbeDREB2A [30]. GmNAC20 can bind to the promoter of *GmDREB1A* and suppress the expression of *GmDREB1C* to regulate *COR* genes and mediate cold tolerance in soybean (*Glycine max*) [33]. MaNAC1 can interact with MaCBF1 to mediate cold tolerance of banana (*Musa acuminata*) [34].

Lily is one of the most important flower crops cultivated worldwide. Most commercial cultivars of lily are sensitive to abiotic stresses, such as cold, drought, and salt. However, *Lilium lancifolium*, also called tiger lily, is one of the most widely distributed wild lilies in Asia which has strong abiotic stress resistance [35]. It suggests that tiger lily may have distinctive molecular mechanisms of abiotic stress resistance [36]. In our previous study, the unigene contig16924 coding for a partial NAC domain protein was found to be strongly up-regulated by low temperature in the tiger lily [36,37]. Thus, we characterized this *NAC* gene, *LlNAC2* in this report. *LlNAC2* was induced not only by cold stress but also by drought, salt, and exogenous ABA treatments. We also found the activity of *LlNAC2* promoter can be induced by cold, salt, and exogenous ABA treatments. In addition, A DREB1 from tiger lily (LlDREB1) was proved to bind to *LlNAC2* promoter, while LlNAC2 can physically interact with LlDREB1 and zinc finger-homeodomain ZFHD4 from tiger lily (LlZFHD4). Furthermore, the transgenic plants of *LlNAC2* showed good tolerance under cold, drought, and salt treatments, and activated the expression of many stress-responsive genes in Arabidopsis. Together, our results indicate that LlNAC2 plays a positive role in plant stress tolerance and can be a candidate gene utilized in transgenic breeding to enhance the abiotic stress tolerance of other crops.

2. Results

2.1. Cloning and Sequence Analysis of LlNAC2 Gene

LlNAC2 gene contained a complete open reading frame (ORF) of 906 bp, with 74 bp 5′ UTR, 231 bp 3′ UTR. It encodes a polypeptide protein of 302 amino acids with a calculated molecular mass of 81.11 kDa and an isoelectric point of 5.03. Amino acid analysis revealed that the LlNAC2 contained a diverse activation domain in C-terminal, and a conserved NAC domain in the N-terminal region which can be divided into five subdomains, namely A to E (Figure 1a). A phylogenetic tree based on the sequences of LlNAC2 and some other known stress-responsive NACs was constructed, which revealed that LlNAC2 was clustered closely to rice ONAC048 and Arabidopsis ATAF1, with 65% and 63% of identity (Figure 1b). We also BLAST the DNA sequence of *LlNAC2* to the whole-genome sequence of *Arabidopsis thaliana* using The Arabidopsis Information Resource (TAIR). The BLAST result showed that the highest score (bits) significant alignment of *LlNAC2* was *AT1G01720.1* (*ATAF1*), which was located in the No. 1 chromosome (Supplementary Figure S1).

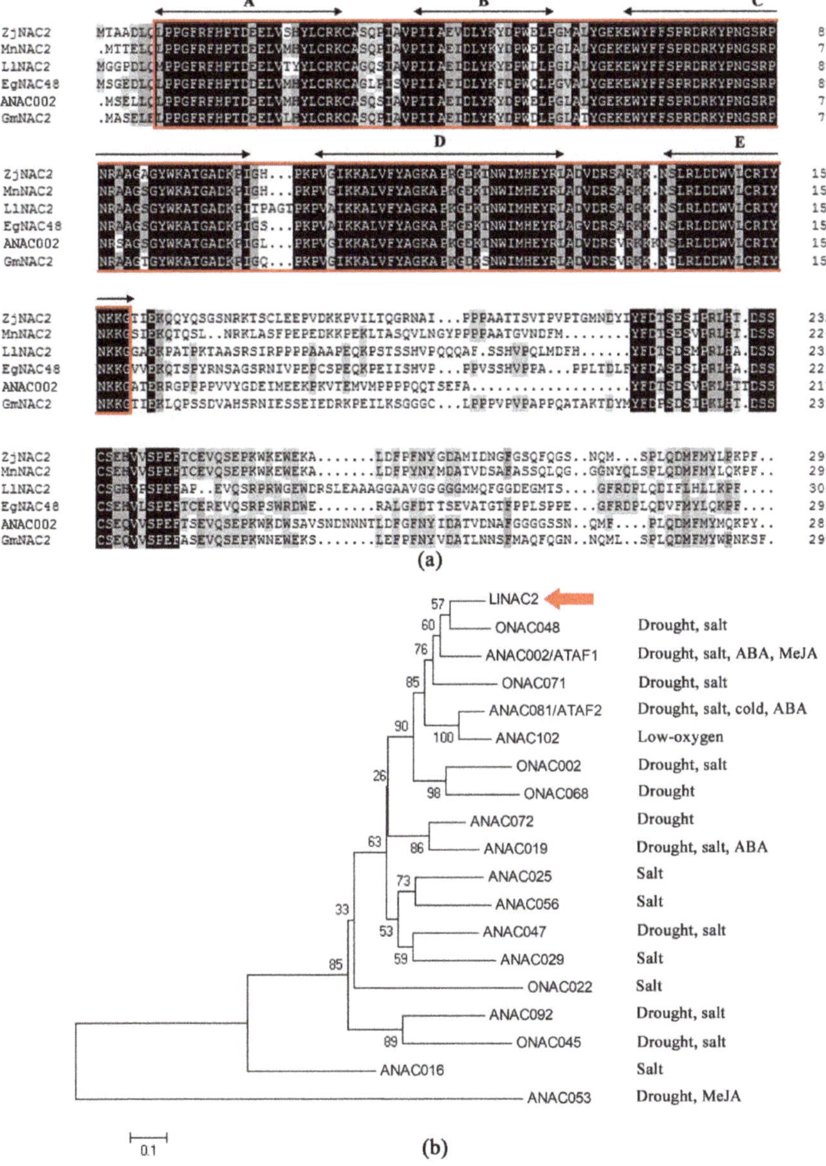

Figure 1. Characterization of LlNAC2. (**a**) Alignment of LlNAC2 with *Ziziphus jujuba* ZjNAC2, *Morus notabilis* MnNAC2, *Elaeis guineensis* EgNAC48, Arabidopsis ANAC002, and *Glycine max* GmNAC2. Black arrowed lines indicate the locations of the five highly conserved subdomains A–E. The conserved NAC domain is boxed and identical amino acids are shaded in black. (**b**) Phylogenetic tree analysis of LlNAC2 with other known stress-responsive NAC proteins. Protein sequences used in multiple sequence alignments and phylogenic tree analysis are shown in Supplementary Table S2.

2.2. Subcellular Localization and Transactivation Assay of LlNAC2

The GFP-LlNAC2 fusion construct and the GFP control in pBI121-GFP vector driven by CaMV35S promoter were transiently expressed in tobacco epidermal cells and visualized under a laser scanning confocal microscope to determine the subcellular localization of LlNAC2. Results showed the fluorescence signal from GFP alone was widely distributed throughout the cells, whereas the GFP-LlNAC2 fusion protein fluorescence signal was mainly detected in the nucleus (Figure 2a), which demonstrated that LlNAC2 is a nuclear protein.

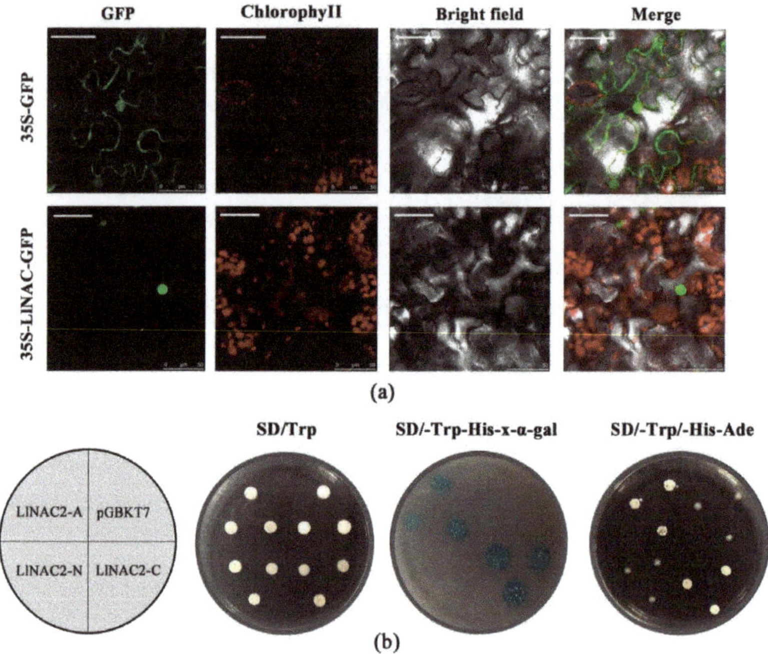

Figure 2. Nuclear localization and transactivation assay of LlNAC2. (**a**) 35S-GFP and 35S-LlNAC2-GFP fusion proteins were transiently expressed in tobacco leaves and observed under a laser scanning confocal microscope. GFP, chlorophyII, bright field, and merged images were taken (Scale bar, 50 μm). (**b**) Full-length protein (LlNAC2-A), N-terminal fragment (LlNAC2-N) and C-terminal fragment (LlNAC2-C) were inserted to the pGBKT7 vector and expressed in yeast strain Y2HGold. Transformed yeasts were dripped on the SD/-Trp, SD/-Trp-His-x-gal and SD/-Trp-His-Ade plates after being cultured for 3 days under 30 °C. The pGBDKT7 vector was used as a negative control.

The entire coding region, N-terminal, and C-terminal domain coding sequence were inserted into the pGBDKT7 vector to investigate the transcriptional activity of LlNAC2 protein. The transactivation results showed that all transformed yeast cells grew well on SD/-Trp medium (Figure 2b). The yeast strain containing the full-length LlNAC2 (LlNAC2-A) and the C-terminus of LlNAC2 (LlNAC2-C) could grow well on the selection medium SD/-Trp/-His/-Ade, while the cells with the N-terminus of LlNAC2 (LlNAC2-N) and pGBDKT7 empty vector could not grow normally (Figure 2b). Furthermore, the yeast cells that grew well on the SD/-Trp/-His-x-α-gal medium appeared blue in the presence of α-galactosidase (Figure 2b). These results indicated that LlNAC2 is a transcriptional activator, and its transactivation domain locates in the C-terminal region.

2.3. *LlNAC2 is Induced by Multiple Stresses and ABA*

The qRT-PCR analysis revealed that the *LlNAC2* gene has relative high expression level in bulb, while its expression level was low in leaf and stem (Figure 3a). Since *LlNAC2* was identified from cold-treated RNA-seq data [36], we also analyzed its expression levels under cold stress in stems, roots, and bulbs (Figure 3b). The results showed that *LlNAC2* was up-regulated in root and stem after 2 h at 4 °C treatment. However, at the same time, the expression of *LlNAC2* was strongly down-regulated and then up-regulated in bulb after 16 h at 4 °C treatment. Thus, we suppose that *LlNAC2* might play a special role in cold resistance of bulb in the tiger lily. Under ABA treatment, the expression of *LlNAC2* was significantly and rapidly up-regulated within 2 h, leading to a three-fold to four-fold increase, and then it was induced specifically again at 24 h (Figure 3c). However, treatment of tiger lily plants with cold and drought stress conditions induced the expression of *LlNAC2* until 24 h, leading to a five-fold to six-fold increase (Figure 3c,f). Similarly, salt treatment also induced the expression of *LlNAC2* slowly, showing a three-fold to five-fold increase during 12–24 h after treatment (Figure 3e). These results showed that *LlNAC2* is a stress-responsive *NAC* gene in the tiger lily.

The 1527 bp upstream of ATG start codon *LlNAC2* promoter sequence was also analyzed. Sequences of various putative stress-related and hormone-responsive cis-acting regulatory elements were identified in the *LlNAC2* promoter, including DRE/CRT (dehydration responsive element/C-repeat element) motif, LTRE (low-temperature-responsive element), MYCRS (MYC recognition site), MYBRS (MYB recognition site), ABRE (ABA responsive element), ARF (auxin response factor binding site), TGA-element and CGTCA-motif (Figure 3g, Supplementary Table S3). These cis-acting regulatory elements may be responsible for the induce expression of *LlNAC2* under stresses.

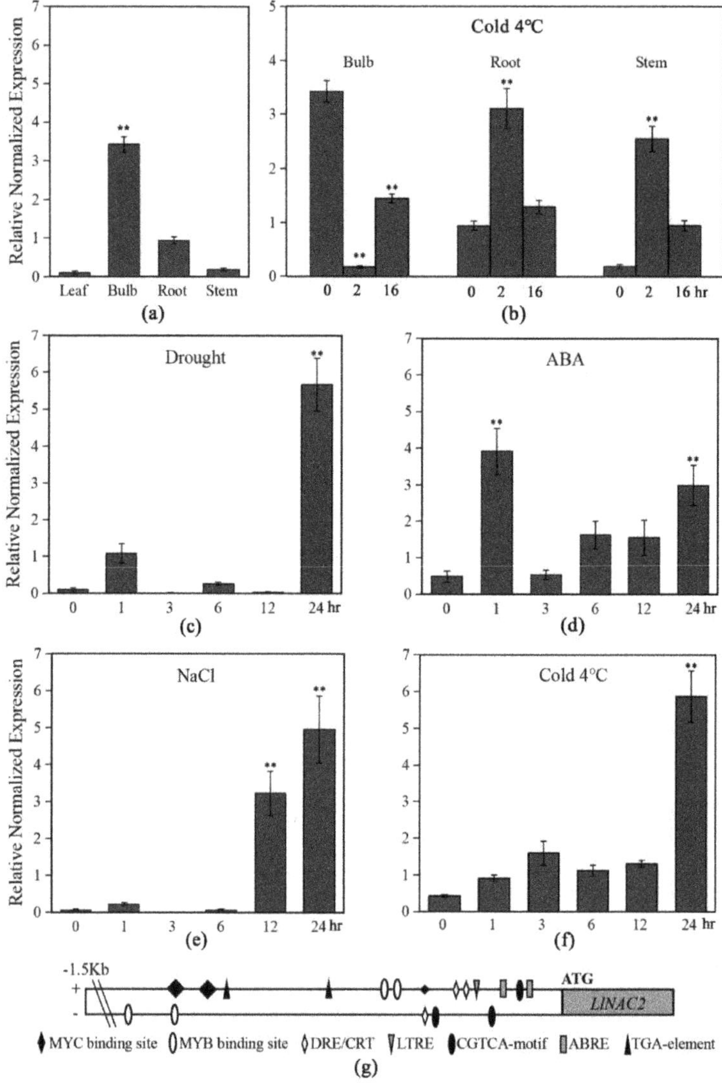

Figure 3. Expression patterns of *LlNAC2* in tiger lily seedlings under different stress treatments and the distribution of stress-related cis-elements in the *LlNAC2* promoter. Expression patterns of *LlNAC2* in leaves, bulbs, roots, and stems (**a**), and expression patterns of *LlNAC2* after cold treatments in leaves (**c**), bulbs, roots, and stems (**b**), and after ABA (**d**), NaCl (**e**), drought (**f**) treatments by qRT-PCR analysis. Transcript levels were normalized to *LlTIP1*. Three biological replications were performed. The bars show the SD. Asterisks indicate a significant difference ** $p < 0.01$ compared with the corresponding controls. (**g**) Distribution of major stress-related cis-elements in the promoter region of *LlNAC2* gene.

2.4. Promoter Activity of LlNAC2 Is Induced by Cold, Salt and ABA Treatments

To further clarify the regulatory mechanism underlying the stress-inducible expression of *LlNAC2*, the full-length *LlNAC2* promoter (proLlNAC2) was cloned and used to drive the *GUS* expression in Arabidopsis. The results showed that the untreated and drought-treated proLlNAC2 transformant (proLlNAC2-trans) Arabidopsis showed apparent lighter blue than that of the CaMV35S transformant (CaMV35S-trans) Arabidopsis plants (Figure 4a). However, the cold, salt, and ABA-treated proLlNAC2-trans showed darker blue than the untreated proLlNAC2-trans and the CaMV35S-trans Arabidopsis (Figure 4a), suggesting the expression of the *GUS* gene was increased after cold, salt stresses and ABA treatment, and the *GUS* transcript level driven by *LlNAC2* promoter is higher than that driven by CaMV35S. To further confirm these findings, *GUS* expression from the *LlNAC2* promoter was detected by qRT-PCR and fluorometric GUS assay. The expression of *GUS* gene increased significantly after 6 h or 12 h cold, salt, or ABA treatments in proLlNAC2-trans Arabidopsis plants (Figure 4b). From the GUS enzyme activity assay, we also found the same phenomenon as that shown in GUS histochemical staining results. Thus, the *LlNAC2* promoter mediated GUS activity increased under cold, salt, and ABA treatments.

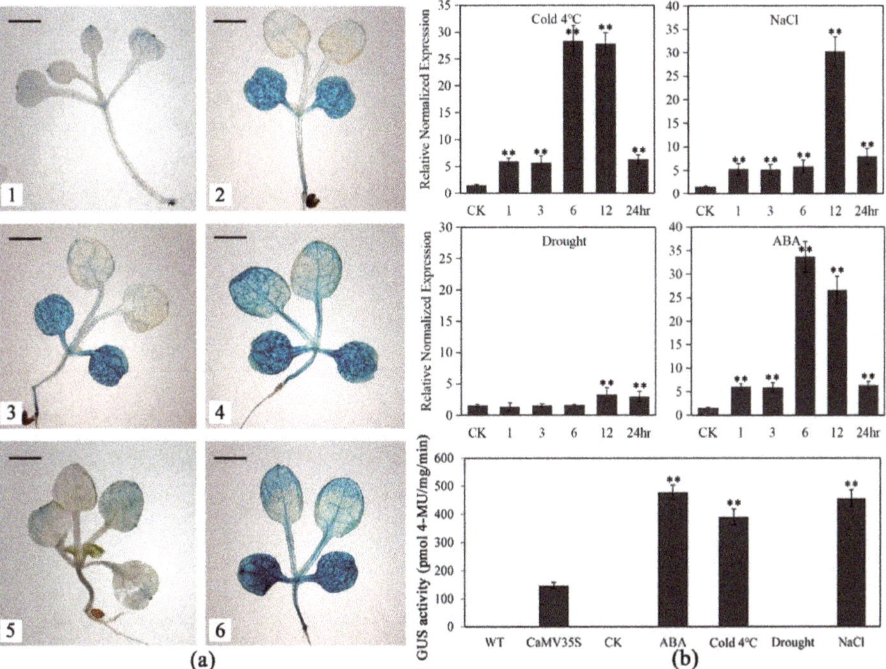

Figure 4. GUS activity mediated by *LlNAC2* promoter in transgenic Arabidopsis plants. (**a**) Beta-glucuronidase (GUS) expression in untreated (1), cold (3), NaCl (4), drought (5), and ABA-treated (6) proLlNAC2-trans and CaMV35S-trans (2) Arabidopsis plants (Scale bar, 2 mm). (**b**) The *GUS* transcript levels and enzyme activity in the leaves of the transgenic Arabidopsis under cold (4 °C), salt, drought, and ABA treatments. The transgenic plants treated for 12 h under above treatments were used for fluorometric GUS assay. GUS activity from the CaMV35S-trans, untreated proLlNAC2-trans (CK), and wild type (WT) served as controls. Twelve transgenic lines were acquired. Three biological replications were performed. The bars show the SD. Asterisks indicate a significant difference (** $p < 0.01$) compared with the corresponding controls.

2.5. LlDREB1 Can Bind to the Promoter of LlNAC2

DREBs were known to play a vital role in plant response to abiotic stresses, and some NAC TFs were found to be involved in the DREB/CBF-COR pathway in many studies [30,33,34,38] which inspired us to further explore the NAC-DREB/CBF-COR signaling cascade. In our previous study, a novel *DREB1* gene from the tiger lily (LlDREB1, NCBI accession No. KJ467618) induced by cold stress was identified [35]. Thus, we performed the Y1H assay to explore whether there is an interaction between the LlDREB1 protein and *LlNAC2* promoter. The minimal inhibitory concentration of Aureobasidin A (AbA) for bait yeast strains was found to be 150 ng/mL (Supplementary Figure S3). Yeast cells transformed with pGADT7-LlDREB1 and pAbAi-CRT/DREs grew well on SD/Leu plates with 150 ng/mL and 200 ng/mL AbA (Figure 5a). This suggested that LlDREB1 possessed the ability to bind CRT/DREs as general CBF/DREB1 TFs. The fragment (−487 to −312) of the *LlNAC2* promoter containing three CRT/DREs was cloned and its interaction with LlDREB1 was detected (Figure 5b); the result showed LlDREB1 could bind to this fragment, indicating LlDREB1 might be involved in the regulatory pathway of *LlNAC2*.

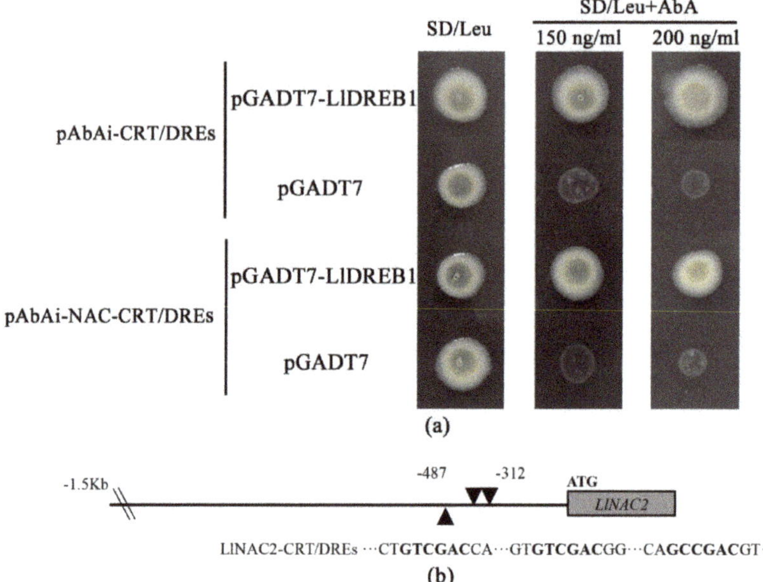

Figure 5. Yeast one-hybrid analysis of LlDREB1 binding to *LlNAC2* promoter. (**a**) Yeast one-hybrid analysis. Three tandem repeats of a CRT/DRE and the −487 to −312 region of the *LlNAC2* promoter were inserted in front of the reporter gene AUR1-C. Yeast strain Y1HGold was co-transformed with bait (pAbAi-CRT/DREs or pAbAi-NAC-CRT/DREs) and a prey (pGADT7 or pGADT7-LlDREB1) construct. Interaction between bait and prey was determined by cell growth on SD medium lacking Leu in the presence of 200 ng mL^{-1} AbA. (**b**) Schematic representation of CRT/DRE recognition elements in the *LlNAC2* promoter.

2.6. *LlNAC2 Interacts with LlDREB1 or LlZFHD4*

We have already found LlNAC2 was highly co-expressed with another cold-responsive TF zinc finger homeodomain protein LlZFHD4 [35]. To further explore the interactions between LlNAC2, LlZFHD4, and LlDREB1 proteins, BiFC assay was performed with a yellow fluorescent protein (YFP). YFP was observed in tobacco epidermal cells, transformed with vectors containing LlNAC2-pSPYNE/LlZFHD4-pSPYCE and LlNAC2-pSPYNE/LlDREB1-pSPYCE localized to the nuclei; while no YFP was detected in negative control pSPYNE173/pSPYCE (M) and LlZFHD4-pSPYNE/LlDREB1-pSPYCE (Figure 6). The result showed that LlNAC2 could interact with LlDREB1 and LlZFHD4.

Bright Field   YFP   Merge

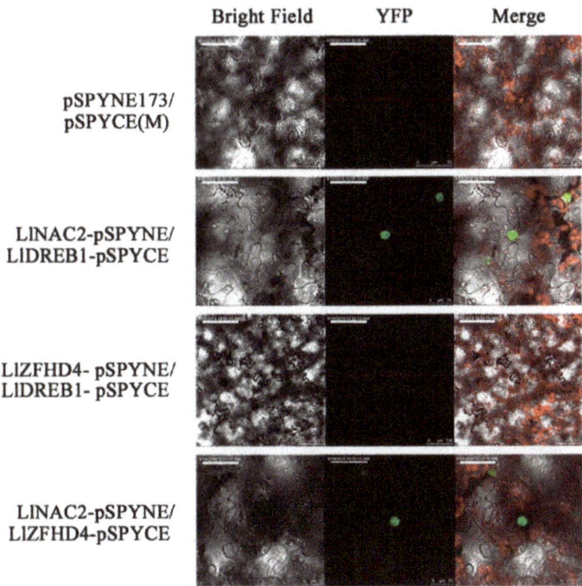

Figure 6. Bimolecular fluorescence complementation assay using tobacco epidermal cells. Negative controls were pSPYNE173/pSPYCE (M) and LlZFHD4-pSPYNE/LlDREB1-pSPYCE. Scale bars for LlZFHD4-pSPYNE/LlDREB1-pSPYCE, 100 μm; for others, 25 μm.

2.7. Overexpression of LlNAC2 in Arabidopsis Enhance Tolerance to Cold and Drought Stresses

Among 12 independent homozygous *LlNAC2* transgenic lines, LlNAC2-5 (L5) and LlNAC2-6 (L6) with relatively high expression levels (Supplementary Figure S4) were selected for further analysis.

To study the effect of *LlNAC2* overexpression on cold stress, *LlNAC2* transgenic lines and wild-type (WT) plants were grown in equal amounts of potting soil for 3 weeks under normal conditions, and cold stress was applied by being exposed to various freezing temperatures for 12 h. The results showed that all plants grew well under −2 °C and −4 °C treatment as same as under normal temperature 22 °C (Figure 7a). When the temperature was decreased to −6 °C, most of the WT plants were dead with a survival rate at around 19%, but over half of two transgenic plants survived (Figure 7a,c). At −8 °C, all WT plants were dead whereas the survival rate for transgenic plants was observed at 30%–35% (Figure 7a,c). In a further experiment, 3-week-old plants were treated at 4 °C for 3 h, and the relative electrolyte leakage and soluble sugars were measured after treatment. As a result, the transgenic plants showed lower electrolyte leakage and higher levels of soluble sugars relative to WT plants (Figure 7d,e).

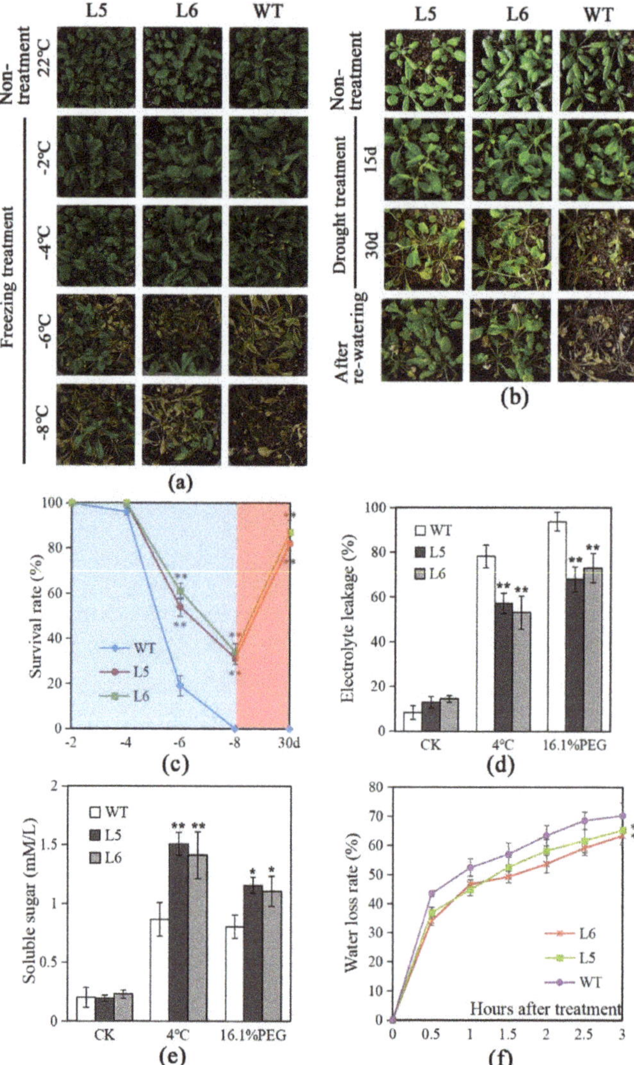

Figure 7. Overexpression of *LlNAC2* in Arabidopsis enhances plant tolerance to cold and drought stresses. Performance of WT and *LlNAC2* transgenic plants after freezing (**a**) and drought (**b**) treatments. (**c**) Survival rate of plants in (**a**) under freezing temperatures (blue region) and in (**b**) after drought treatment for 30 days (red region). Relative electrolyte leakage (**d**) and soluble sugar content (**e**) in WT and *LlNAC2* transgenic lines after 4 °C and 16.1% PEG6000 (−0.5 MPa) treatment for 3 h. (**f**) Water loss rate of leaves from WT and *LlNAC2* transgenic plants. Three biological replications were performed, each replication contains 30 plants. The bars show the SD. Asterisks indicate a significant difference $0.01 < * p < 0.05$ and $** p < 0.01$ compared with the corresponding controls.

Similarly, to study drought stress tolerance, WT plants showed visible symptoms of drought-induced damage and even death after withholding water for 30 days, while some transgenic plants remained green with expanded leaves (Figure 7b). When plants were rehydrated, about 80% of the transgenic plants recovered well, whereas the WT plants failed and ultimately died (Figure 7b,c). Additionally, after being treated with 16.1% PEG6000 (−0.5 MPa) for 3 h, transgenic plants showed lower electrolyte leakage and higher levels of soluble sugars compared to WT plants (Figure 7d,e). Also, the water-loss rates were found lower in transgenic plants after 3 h treatment (Figure 7f). These results suggest that the *LlNAC2* transgenic plants conferred tolerance to cold and drought stresses in some degree.

2.8. Overexpression of LlNAC2 in Arabidopsis Increases Seed Sensitivity to ABA and Tolerance to NaCl

Salt tolerance and ABA sensitivity of *LlNAC2* transgenic plants were also assessed. WT and two *LlNAC2* transgenic lines were placed on MS agar plates supplemented with 50 mM or 100 mM NaCl and 2 µM ABA. There were no differences in plant morphology between WT and LlNAC2 transgenic Arabidopsis on MS agar plates (Figure 8a,b). However, *LlNAC2* transgenic Arabidopsis displayed significantly higher germination ratios than WT plants following NaCl treatment according to both cotyledon greening and radicle protrusion (Figure 8a,c,d). In contrast, the germination ratio of transgenic plants seeds was remarkably lower than that of WT seeds in the MS medium containing 2 µM ABA according to cotyledon greening (Figure 8b,c). Therefore, we suggested that *LlNAC2* transgenic plants are more tolerant to salt stresses and more hypersensitive to ABA than WT plants.

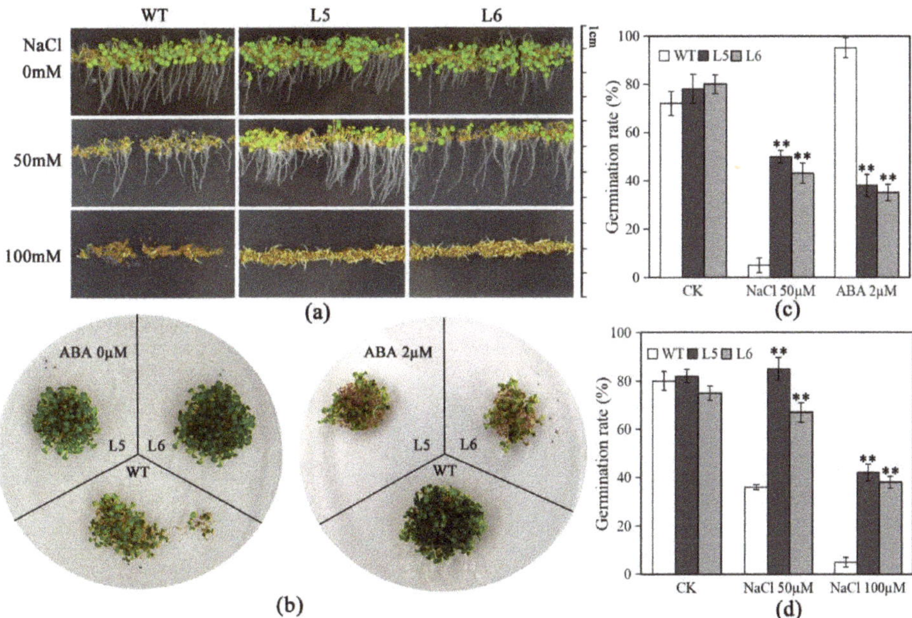

Figure 8. *LlNAC2* transgenic lines showed ABA hypersensitivity and enhanced salt tolerance. Germination of WT of Col-0 and 35S-LlNAC2 seeds on MS supplemented with (**a**) 50 mM or 100 mM NaCl and (**b**) 2 µM ABA assayed by both (**c**) cotyledon greening and (**d**) radicle protrusion. For (**a**) and (**b**), images were acquired after 7 days of incubation at 22 °C. Three biological replications were performed. The bars show the SD. Asterisks indicate a significant difference ** $p < 0.01$ compared with the corresponding controls.

2.9. Altered Expression of Stress-Related Genes in LlNAC2 Transgenic Arabidopsis Plants

LlNAC2 transgenic Arabidopsis plants showed enhanced tolerance to cold, drought, and salt stresses. Thus, we detected the transcript levels of some stress-responsive genes from Arabidopsis in the transgenic plants, including *AtRD20, AtCOR47, AtRD29A, AtRD29B, AtLEA14, AtGolS1, AtAPX2,* and *AtGSTF6* genes (NCBI accession numbers are shown in Supplementary Table S1). The qRT-PCR results showed higher transcripts of these genes accumulated in *LlNAC2* transgenic plants compared to WT plants (Figure 9). We suppose the activated expression of these stress-responsive genes may contribute to the stronger stress tolerance in *LlNAC2* transgenic plants.

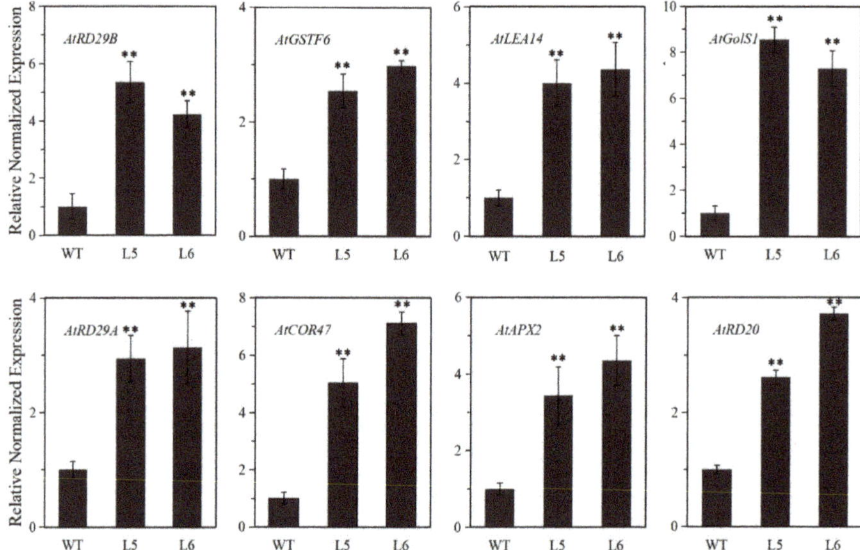

Figure 9. Transcript levels of the stress-related genes under normal condition in WT and *LlNAC2* transgenic Arabidopsis plants. Three biological replications were performed. The bars show the SD. Asterisks indicate a significant difference (** $p < 0.01$) compared with the corresponding controls.

3. Discussion

In this study, we identified a novel stress-related NAC TF gene, *LlNAC2*, from tiger lily. Sequence analysis shows that LlNAC2 has a highly conserved sequence in the N-terminal region and has high sequence identity with ATAF1 from Arabidopsis. Like most transcription factors, LlNAC2 protein is localized in the nucleus with transactivation activity in the C-terminal region. Our previous transcriptome data analysis identified a unigene contig16924 coding for partial LlNAC2, which showed significant changes in expression in the tiger lily under cold stress [35,36]. Here, we further confirmed that *LlNAC2* not only participates in cold stress response, but also responds to drought and salt stress, and its expression is sensitive to ABA signaling molecules as well. Also, the *LlNAC2* promoter is shown to be a stress-inducible promoter, and the GUS activity driven by the *LlNAC2* promoter is significantly higher than that driven by the CaMV35S promoter. Given that using a stress-inducible promoter to drive transgene expression can not only result in remarkable gains in stress tolerance, but also avoid impacting important traits in crops negatively [39], the *LlNAC2* promoter can be a candidate stress-inducible promoter utilized in transgenic breeding to enhance the stress tolerance of crops.

Many NAC domain proteins from different species were shown to function positively in regulation of plant stress tolerance [1,21,40,41]. We also found tolerance to cold, drought, and salt stresses were likely conferred by overexpressing *LlNAC2* in Arabidopsis. At the morphological level, compared to WT plants, *LlNAC2* transgenic lines showed a higher seed germination ratio under salt stress condition, and performed better growth performance and higher survival rate under drought and cold stress conditions. At the physiological level, the lower electrolyte leakage amount and higher soluble sugar level were observed in transgenic plants after dehydration and chilling treatment. Furthermore, the leaf water-loss ratios of the *LlNAC2* transgenic plants were lower in transgenic plants than in WT plants. These physiological indices changes likely resulted in enhancing tolerance to stresses at the physiological level, which were consistent with previous reports about NAC TFs in maize, rice, and rose [21,26,42]. At the gene transcription level, the *LlNAC2* gene may improve stress tolerance by regulating downstream stress-responsive genes. The results showed that higher transcript levels of 8 picked stress-responsive genes accumulated in *LlNAC2* transgenic plants. More importantly, putative NAC binding cis-elements were also reported to localize in the promoter sequences of these genes [27,43,44], indicating these genes might be transcriptionally regulated directly by LlNAC2.

On the other hand, we found overexpression of *LlNAC2* resulted in improved ABA sensitivity, which was also observed on many NAC TFs from other species, such as GmNAC20, MlNAC5, AtATAF1, ONAC022, and TaNAC47 [1,21,23,42,45]. Promoter analysis showed that there are two ABREs (YACGTGGC, Y = C/T) and eight core DPBF (Dc3 promoter-binding factor) (ACACNNG) cis-acting elements in *LlNAC2* promoter. ABREs can be recognized by ABRE-binding factors (ABRE/ABFs), which are involved in ABRE-dependent ABA signaling pathway in response to drought stress [46]; the core DPBF (Dc3 Promoter-Binding Factor) (ACACNNG) elements were also known to participate in ABA responsiveness [47,48]. The presence of these cis-elements might explain why *LlNAC2* was induced more quickly in response to ABA treatment compared with other stresses treatment in the tiger lily. Given that most of the picked genes above were also reported to be ABA-responsive genes, these results suggest that the LlNAC2 may go through the more efficient ABA signaling pathway to enhance cold, drought, and salt stress tolerance in transgenic plants.

Except for the ABA signaling pathway, NAC domain proteins have received much attention as regulators in various stress signaling pathways [49], in which some studies showed some NAC TFs were also involved in the DREB/CBF-COR pathway [1,30,33,34]. CBF/DREB1 is known to activate stress-responsive genes under abiotic stress conditions by binding to specific cis-elements such as DRE/CRTs present in their promoters [4,50]. In the promoter region of *LlNAC2*, we found there are one LTRE (CCGAC) and three CRT/DREs (GC/TCGAC) elements, which are known to be necessary and sufficient for gene transcription under cold stress [51–54]. Furthermore, a novel *DREB1* gene from the tiger lily (*LlDREB1*) induced by cold stress was identified in our earlier study [35]. We thus further analyzed the interactions between LlDREB1 and LlNAC2. Y1H assay confirmed that LlDREB1 could bind to the *LlNAC2* promoter. It provides a direct evidence for *LlNAC2* as a novel target of LlDREB1. Interestingly, BiFC assay showed that the LlNAC2 protein can interact with the LlDREB1 protein in tobacco epidermal cells. These findings indicate that the *LlNAC2* gene might be regulated by LlDREB1 TF under abiotic stress conditions and LlNAC2 TF can in turn interact with LlDREB1 to more efficiently regulate the expression of downstream stress-related genes. It may also explain why *LlNAC2* was induced more slowly but more strongly in response to cold compared with salt, drought, and ABA treatments in the tiger lily. However, in soybean, GmNAC20 was found to be an upstream protein of GmDREB1A which can directly bind to the promoter of *GmDREB1A* [1]. Thus, we suppose that there are various ways for NAC TFs participating in the DREB/CBF-COR pathway, and LlNAC2 is involved in DREB/CBF-COR pathway, to mediate stress tolerance of the tiger lily.

In Arabidopsis, an interaction between stress-inducible zinc finger homeodomain ZFHD1 and NAC TFs was detected in the report of Lam-Son Phan Tran [55]. It showed co-overexpression of *ZFHD1* and *NAC* genes, enhanced expression of *ERD1* in both Arabidopsis T87 protoplasts, and transgenic Arabidopsis plants. Also, in our previous study, *LlNAC2* was highly co-expressed with another cold-responsive TF, the zinc finger homeodomain protein 4 LlZFHD4 [36]; and here we confirmed that the LlNAC2 protein can physically interact with the LlZFHD4 protein by BiFC assay. However, further analysis is needed to confirm whether this interaction functions the same as AtZFHD1 and AtNAC TFs in Arabidopsis.

In conclusion, LlNAC2 is a nucleus-localized transcriptional activator which is regulated by cold, drought, and salt stresses and sensitive to ABA. The *LlNAC2* promoter is a stress-inducible promoter which showed higher promoter activity than CaMV35S under cold, salt, and ABA conditions. Overexpressing *LlNAC2* in Arabidopsis showed ABA hypersensitivity and enhancing tolerance of transgenic plants to freezing, dehydration, and salt conditions. We demonstrated that the *LlNAC2* gene is not only a novel direct target of LlDREB1, but the LlNAC2 protein can also interact with the LlDREB1 protein. These data indicate LlNAC2 is involved in both DREB/CBF-COR and ABA-dependent pathways to mediate stress tolerance of the tiger lily (Figure 10). Moreover, the protein interaction of the co-expressed LlNAC2 and LlZFHD4 gene was confirmed, providing another pathway involving regulator LlNAC2. Therefore, our future efforts will be focused on investigating the role of these interactive proteins in regulating expression and modulating the function of LlNAC2 under various stress conditions.

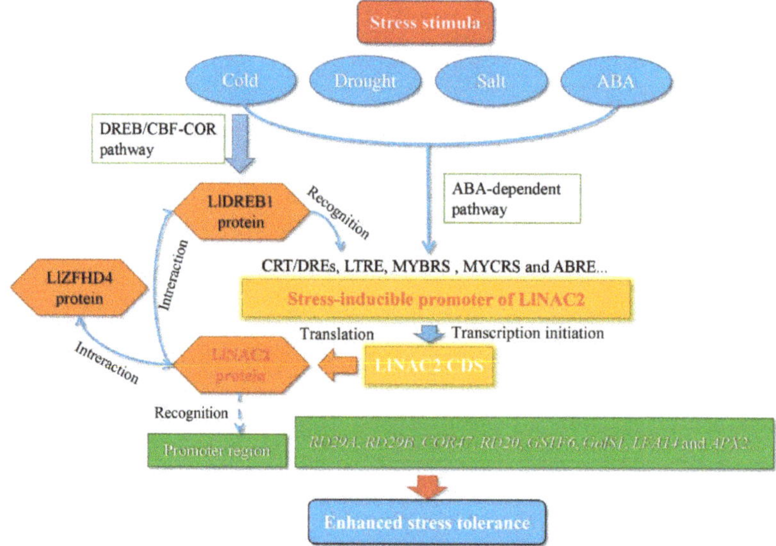

Figure 10. LlNAC2 is involved in both DREB/CBF-COR and ABA-dependent pathways to mediate stress tolerance of the tiger lily.

4. Materials and Methods

4.1. Plant Materials and Abiotic Stress Treatments

The wild tiger lily (*Lilium lancifolium*) was used as experimental material in this study. The seedlings preparation method was described in our previous study [36]. The bulbs of the tiger lily were cleaned, disinfected, and then stored at 4 °C; in March, the bulbs were box-cultivated in a greenhouse (116.3° E, 40.0° N) under controlled conditions. The model plant *Arabidopsis thaliana* Columbia-0 (Col-0) was selected for the transgenic study of *LlNAC2*. Arabidopsis plants were grown in 8 cm × 8 cm plastic pots containing a 1:1 mixture of sterile peat soil and vermiculite under controlled conditions (22/16 °C, 16 h light/8 h dark, 65% relative humidity, and 1000 lx light intensity). Seeds of *Nicotiana benthamiana* were planted and cultured under the same conditions.

For the expression analysis of *LlNAC2* in response to abiotic stress and ABA treatment, 8-week-old tiger lily seedlings were treated with 4 °C, 16.1% PEG6000 (−0.5 MPa), 100 mM NaCl and 100 μM exogenous ABA for 0, 1, 3, 6, 12, and 24 h, respectively. Leaf samples were collected and immediately frozen with liquid nitrogen and stored at −80 °C for RNA isolation.

4.2. Full-Length cDNA Cloning and Sequence Analysis of LlNAC2

The 3′-complete sequence cDNA of *LlNAC2* gene was obtained from the transcriptome data of cold-treated tiger lily leaves in our laboratory. Using the SMARTer RACE 5′/3′ Kit (Clontech, United States), we performed a 5′-rapid amplification of cDNA ends (5′-RACE) of *LlNAC2* according to the manufacturer's protocol. Two reverse primer pairs (Supplementary Table S1) were designed to amplify the 5′-complete sequence cDNA of *LlNAC2*. The 3′- and 5′-sequences were assembled by DNAMAN (version 7), resulting in the full-length cDNA of *LlNAC2* gene. The coding sequence of *LlNAC2* was amplified and cloned into pEASYT1-Blunt vector (TransGen Biotech, Beijing, China). Plasmid pEASYT1-LlNAC2 was used as a template for all experiments after sequencing. Multiple sequence alignments were performed using DNAMAN (version 7). Phylogenetic tree analysis was performed using the neighbor-joining method in MEGA5 software with 1000 replications. The NCBI accession numbers of genes used in multiple sequence alignments and phylogenetic tree analysis are shown in Supplementary Table S2. The theoretical molecular weight and isoelectric point were calculated using ExPASy (http://expasy.org/tools/protparam.html). The NAC domain region was identified with SCANPROSITE (http://www.expasy.ch/tools/scanprosite/).

4.3. RNA Isolation and Quantitative Real-Time PCR Analysis

Total RNA was isolated using an RNAisomate RNA Easyspin isolation system (Aidlab Biotech, Beijing, China). First-strand cDNA synthesis was performed using Prime Script II 1st strand cDNA Synthesis Kit (Takara, Shiga, Japan) according to the manufacturer's instructions. The qRT-PCR was performed using a Bio-Rad/CFX Connect™ Real-Time PCR Detection System (Bio-Rad, CA, USA) with SYBR® qPCR mix (Takara, Shiga, Japan) according to the manufacturer's protocol. Relative mRNA content was calculated using the $2^{-\triangle\triangle Ct}$ method against the internal reference gene encoding tiger lily tonoplast intrinsic protein 1 (LlTIP1) [35] and Arabidopsis *Atactin* gene (NCBI accession No. NM_112764). The primers used in this study were designed with Primer Premier 5 and are listed in Supplementary Table S1. All reactions were performed in three biological replicates. Student's t-test was performed for all statistical analysis in this study.

4.4. Promoter Cloning and Sequence Analysis

LlNAC2 promoter sequence was cloned from tiger lily genomic DNA using Genome Walker Universal Kits (Clontech, United States). The cis-acting motifs present in the LlNAC2 promoter were predicted using the online search tool PLACE (http://www.dna.affrc.go.jp/PLACE/) database.

4.5. Subcellular Localization of LlNAC2

By using ClonExpressII One Step Cloning Kits (Vazyme, Piscataway, NJ, United States), full-length *LlNAC2* was inserted into vector pBI121-GFP at *Xho*I and *Sal*I sites. The pBI121-LlNAC2-GFP plasmid and the pBI121-GFP empty vector were transformed into *Agrobacterium tumefaciens* GV3101 and infiltrated separately into *N. benthamiana* leaves. After infiltration, the plants were grown in a growth room under controlled conditions (22/16 °C, 16 h light/8 h dark, 65% relative humidity, and 1000 lx light intensity) for 32 h. GFP fluorescence signals were excited at 488 nm and detected under Leica TCS SP8 Confocal Laser Scanning Platform (Leica SP8, Leica, America) using a 500–530 nm emission filter.

4.6. Transcription Activation Assay of LlNAC2 in Yeast

To investigate the transcriptional activity of the LlNAC2 protein, the full-length coding sequence of *LlNAC2* and the sequence encoding the N-terminus (1–453 bp) and C-terminus (454–906 bp) were inserted into the *Eco*RI and *Bam*HI sites of the pGBKT7 vector, resulting in plasmid pGBKT7-LlNAC2-A, pGBKT7-LlNAC2-N (1–151 aa), and pGBKT7-LlNAC2-C (152–302 aa). These plasmids were transformed into the yeast strain Y2HGold separately by using Quick Easy Yeast Transformation Mix (Clontech, United States). The transformed yeast cells were incubated on SD/-Trp, SD/-Trp/-His-Ade and SD/-Trp-His-x-α-gal plates [24].

4.7. Yeast One-Hybrid (Y1H) Assays

Y1H assay was carried out using the Matchmaker™ Gold Yeast One-Hybrid System (Clontech, United States). Three tandem copies of CRT/DRE were generated by oligonucleotide synthesis and cloned into the pAbAi (bait) vector (Clontech, United States). A fragment (−312 to −487) of the *LlNAC2* promoter was amplified by PCR, and also cloned into the pAbAi (bait) vector to generate pAbAi-NAC-CRT/DREs plasmid (shown in Figure 4B). Full-length *LlDREB1* (NCBI accession No.KJ467618) was amplified and inserted into pGADT7 (prey) vector (Clontech, United States) yielding plasmid pGADT7-LlDREB1. The bait plasmids were linearized and transformed into the yeast strain Y1HGold. Positive yeast cells were then transformed with pGADT7-LlDREB1 plasmid. The DNA–protein interaction was determined based on the growth ability of the co-transformants on SD/-Leu medium with Aureobasidin A (AbA).

4.8. BiFC Assay

Full-length *LlNAC2* and zinc finger homeodomain protein *LlZFHD4* (sequence shown in Supplementary Figure S2) were cloned into the pSPYNE173 vector, *LlDREB1* and *LlZFHD4* were cloned into the pSPYCE(M) vector [56], respectively. Co-expression was executed on *N. benthamiana* leaves as described in subcellular localization assessments. After infiltration, the plants were grown under 24 h dark and then 16 h light/8 h dark for 32 h. YFPs were excited at 514 nm. Images were generated through Leica TCS SP8 Confocal Laser Scanning Platform using a 500–530 nm emission filter.

4.9. Generation of LlNAC2 Transgenic Arabidopsis

The *LlNAC2* open read frame (ORF) was cloned into the pBI121 vector under control of a CaMV35S promoter; the *LlNAC2* promoter region was inserted into the CaMV35S-GUS vector by replacing the CaMV35S promoter. The recombinant vectors and empty GUS vector were transformed into 5-week-old Arabidopsis ecotype Col-0 plants by the floral-dip method [57]. Transformed seeds were selected on MS medium containing 50 mg/L kanamycin. All transgenic lines were identified by RT-PCR; T3-generation homozygous lines were selected for gene functional analysis.

4.10. Histochemical Staining and Fluorometric GUS Assay

Histochemical staining and fluorometric GUS assay analysis for GUS activity was carried out as described before [58]. Transgenic *LlNAC2* Arabidopsis plants were treated with 4 °C, 16.1% PEG6000 (−0.5 MPa), 100 mM NaCl and 100 μM exogenous ABA for different durations before sampling. The leaves of stress-treated transgenic *LlNAC2* Arabidopsis were incubated in GUS reaction buffer (Huayueyang Biotech, Beijing, China). Photos of those stained samples were obtained by a Leica TL3000 Ergo microscope under white light. Leaves of stress-treated transgenic Arabidopsis were also used to examine the GUS gene expression level by qRT-PCR, and determine GUS enzyme activity.

4.11. Abiotic Stress Tolerance Assays and ABA Sensitivity Analysis

For cold and drought treatments, 3-week-old seedlings were kept at 4 °C for 3 h, then at −2, −4, −6 or 8 °C, respectively, for 12 h. After that, the plants were kept at 4 °C for 3 h before transferring to a normal condition at 22 °C [34]. For the drought treatment, the water intake of 3-week-old seedlings in water-saturated substrate was withheld for 30 days, followed by rehydrating the seedlings for 7 days [24].

For determining the salt tolerance and ABA sensitivity in transgenic plants, Arabidopsis seeds were cultivated on MS medium supplemented with 0 and 2 μM ABA or 50 and 100 mM NaCl, respectively, under continuous light at 22 °C in a growth chamber. The germination rate was scored on the 7th day after planting on the plates.

4.12. Measurements of Relative Electrolyte Leakage, Soluble Sugar, and Water Loss Rate

The relative electrolyte leakage, soluble sugar content, and water loss rate were evaluated following the method described previously [21,59]. The relative electrolyte leakage was evaluated by determining the relative conductivity of fresh leaves (100 mg) in solution using a conductivity detector.

The anthrone-sulfuric acid colorimetry was used for determining the soluble sugar. The water loss rate was calculated related to the initial fresh weight of the leaf samples; the samples were placed on the lab bench (20–22 °C, humidity 45–60%) and weighed at designated time points. All the measurements were performed with ten plants in triplicate.

Supplementary Materials: Supplementary Materials can be found at http://www.mdpi.com/1422-0067/20/13/3225/s1.

Author Contributions: Y.Y. designed of the study and performed the statistical analysis and molecular experiments. Y.Z. conceived of the study, and helped to draft the manuscript. Y.Y., Y.Z. and Y.L. read and approved the final manuscript.

Funding: This research was funded the China National Natural Science Foundation (grant No. 31672190, 31872138, 31071815 and 1272204) and China National Key Research & Development Project, grant number 2018YFD1000402.

Conflicts of Interest: The authors declare no conflict of interest. The funders had no role in the design of the study; in the collection, analyses, or interpretation of data; in the writing of the manuscript, or in the decision to publish the results.

Abbreviations

ABA	abscisic acid
TF	transcription factor
Y1H	yeast one-hybrid
BiFC	bimolecular fluorescence complementation
GFP	green fluorescent protein
AbA	aureobasidin A
GUS	β-glucuronidase

References

1. Hong, Y.; Zhang, H.; Huang, L.; Li, D.; Song, F. Overexpression of a Stress-Responsive NAC Transcription Factor Gene *ONAC022* Improves Drought and Salt Tolerance in Rice. *Front. Plant Sci.* **2016**, *7*, 4. [CrossRef] [PubMed]

2. Chung, P.J.; Jung, H.; Choi, Y.D.; Kim, J.K. Genome-wide analyses of direct target genes of four rice NAC-domain transcription factors involved in drought tolerance. *BMC Genom.* **2018**, *19*, 40. [CrossRef] [PubMed]

3. Liu, C.; Wang, B.; Li, Z.; Peng, Z.; Zhang, J. TsNAC1 Is a Key Transcription Factor in Abiotic Stress Resistance and Growth. *Plant Physiol.* **2018**, *176*, 742–756. [CrossRef] [PubMed]

4. Mizoi, J.; Shinozaki, K.; Yamaguchi-Shinozaki, K. AP2/ERF family transcription factors in plant abiotic stress responses. *Biochim. Biophys. Acta* **2012**, *1819*, 86–96. [CrossRef] [PubMed]

5. Du, C.; Hu, K.; Xian, S.; Liu, C.; Fan, J.; Tu, J.; Fu, T. Dynamic transcriptome analysis reveals AP2/ERF transcription factors responsible for cold stress in rapeseed (*Brassica napus* L.). *Mol. Genet. Genom.* **2016**, *291*, 1053–1067. [CrossRef] [PubMed]

6. Xiong, H.; Li, J.; Liu, P.; Duan, J.; Zhao, Y.; Guo, X.; Li, Y.; Zhang, H.; Ali, J.; Li, Z. Overexpression of *OsMYB48-1*, a novel MYB-related transcription factor, enhances drought and salinity tolerance in rice. *PLoS ONE* **2014**, *9*, e92913. [CrossRef] [PubMed]

7. Wei, Q.; Luo, Q.; Wang, R.; Zhang, F.; He, Y.; Zhang, Y.; Qiu, D.; Li, K.; Chang, J.; Yang, G.; et al. A Wheat R2R3-type MYB Transcription Factor TaODORANT1 Positively Regulates Drought and Salt Stress Responses in Transgenic Tobacco Plants. *Front. Plant Sci.* **2017**, *8*, 1374. [CrossRef] [PubMed]

8. Castilhos, G.; Lazzarotto, F.; Spagnolo-Fonini, L.; Bodanese-Zanettini, M.H.; Margis-Pinheiro, M. Possible roles of basic helix-loop-helix transcription factors in adaptation to drought. *Plant Sci.* **2014**, *223*, 1–7. [CrossRef]

9. Rushton, D.L.; Tripathi, P.; Rabara, R.C.; Lin, J.; Ringler, P.; Boken, A.K.; Langum, T.J.; Smidt, L.; Boomsma, D.D.; Emme, N.J.; et al. WRKY transcription factors: Key components in abscisic acid signalling. *Plant Biotechnol. J.* **2012**, *10*, 2–11. [CrossRef]

10. Chen, L.; Song, Y.; Li, S.; Zhang, L.; Zou, C.; Yu, D. The role of WRKY transcription factors in plant abiotic stresses. *Biochim. Biophys. Acta* **2012**, *1819*, 120–128. [CrossRef]

11. Huang, J.; Sun, S.J.; Xu, D.Q.; Yang, X.; Bao, Y.M.; Wang, Z.F.; Tang, H.J.; Zhang, H. Increased tolerance of rice to cold, drought and oxidative stresses mediated by the overexpression of a gene that encodes the zinc finger protein ZFP245. *Biochem. Biophys. Res. Commun.* **2009**, *389*, 556–561. [CrossRef] [PubMed]

12. Zang, D.; Wang, L.; Zhang, Y.; Zhao, H.; Wang, Y. ThDof1.4 and ThZFP1 constitute a transcriptional regulatory cascade involved in salt or osmotic stress in *Tamarix hispida*. *Plant Mol. Biol.* **2017**, *94*, 495–507. [CrossRef] [PubMed]

13. Xu, Z.; Gongbuzhaxi; Wang, C.; Xue, F.; Zhang, H.; Ji, W. Wheat NAC transcription factor TaNAC29 is involved in response to salt stress. *Plant Physiol. Biochem.* **2015**, *96*, 356–363. [CrossRef]

14. Mao, H.; Yu, L.; Han, R.; Li, Z.; Liu, H. ZmNAC55, a maize stress-responsive NAC transcription factor, confers drought resistance in transgenic Arabidopsis. *Plant Physiol. Biochem.* **2016**, *105*, 55–66. [CrossRef] [PubMed]

15. Liu, X.; Wang, T.; Bartholomew, E.; Black, K.; Dong, M.; Zhang, Y.; Yang, S.; Cai, Y.; Xue, S.; Weng, Y.; et al. Comprehensive analysis of NAC transcription factors and their expression during fruit spine development in cucumber (*Cucumis sativus* L.). *Hortic. Res.* **2018**, *5*, 31. [CrossRef] [PubMed]

16. Zhang, H.; Kang, H.; Su, C.; Qi, Y.; Liu, X.; Pu, J. Genome-wide identification and expression profile analysis of the NAC transcription factor family during abiotic and biotic stress in woodland strawberry. *PLoS ONE* **2018**, *13*, e0197892. [CrossRef] [PubMed]

17. Olsen, A.N.; Ernst, H.A.; Leggio, L.L.; Skriver, K. NAC transcription factors: Structurally distinct, functionally diverse. *Trends Plant Sci.* **2005**, *10*, 79–87. [CrossRef] [PubMed]

18. Yamaguchi, M.; Ohtani, M.; Mitsuda, N.; Kubo, M.; Ohme-Takagi, M.; Fukuda, H.; Demura, T. VND-INTERACTING2, a NAC domain transcription factor, negatively regulates xylem vessel formation in Arabidopsis. *Plant Cell* **2010**, *22*, 1249–1263. [CrossRef] [PubMed]

19. Liu, H.; Zhou, Y.; Li, H.; Wang, T.; Zhang, J.; Ouyang, B.; Ye, Z. Molecular and functional characterization of ShNAC1, an NAC transcription factor from Solanum habrochaites. *Plant Sci.* **2018**, *271*, 9–19. [CrossRef] [PubMed]

20. Mao, X.; Chen, S.; Li, A.; Zhai, C.; Jing, R. Novel NAC transcription factor TaNAC67 confers enhanced multi-abiotic stress tolerances in Arabidopsis. *PLoS ONE* **2014**, *9*, e84359. [CrossRef]

21. Zhang, L.; Zhang, L.; Xia, C.; Zhao, G.; Jia, J.; Kong, X. The Novel Wheat Transcription Factor TaNAC47 Enhances Multiple Abiotic Stress Tolerances in Transgenic Plants. *Front. Plant Sci.* **2015**, *6*, 1174. [CrossRef] [PubMed]

22. Tran, L.S.; Nakashima, K.; Sakuma, Y.; Simpson, S.D.; Fujita, Y.; Maruyama, K.; Fujita, M.; Seki, M.; Shinozaki, K.; Yamaguchi-Shinozaki, K. Isolation and functional analysis of Arabidopsis stress-inducible NAC transcription factors that bind to a drought-responsive cis-element in the early responsive to dehydration stress 1 promoter. *Plant Cell* **2004**, *16*, 2481–2498. [CrossRef] [PubMed]

23. Wu, Y.; Deng, Z.; Lai, J.; Zhang, Y.; Yang, C.; Yin, B.; Zhao, Q.; Zhang, L.; Li, Y.; Yang, C.; et al. Dual function of Arabidopsis ATAF1 in abiotic and biotic stress responses. *Cell Res* **2009**, *19*, 1279–1290. [CrossRef] [PubMed]

24. Cao, H.; Wang, L.; Nawaz, M.A.; Niu, M.; Sun, J.; Xie, J.; Kong, Q.; Huang, Y.; Cheng, F.; Bie, Z. Ectopic Expression of Pumpkin NAC Transcription Factor *CmNAC1* Improves Multiple Abiotic Stress Tolerance in Arabidopsis. *Front. Plant Sci.* **2017**, *8*, 2052. [CrossRef] [PubMed]

25. Gunapati, S.; Naresh, R.; Ranjan, S.; Nigam, D.; Hans, A.; Verma, P.C.; Gadre, R.; Pathre, U.V.; Sane, A.P.; Sane, V.A. Expression of *GhNAC2* from G. herbaceum, improves root growth and imparts tolerance to drought in transgenic cotton and Arabidopsis. *Sci. Rep.* **2016**, *6*, 24978. [CrossRef] [PubMed]

26. Jiang, X.; Zhang, C.; Lü, P.; Jiang, G.; Liu, X.; Dai, F.; Gao, J. RhNAC3, a stress-associated NAC transcription factor, has a role in dehydration tolerance through regulating osmotic stress-related genes in rose petals. *Plant Biotechnol. J.* **2014**, *12*, 38–48. [CrossRef] [PubMed]

27. Sakuraba, Y.; Kim, Y.S.; Han, S.H.; Lee, B.D.; Paek, N.C. The Arabidopsis transcription factor NAC016 promotes drought stress responses by repressing *AREB1* transcription through a trifurcate feed-forward regulatory loop involving NAP. *Plant Cell* **2015**, *27*, 1771–1787. [CrossRef]

28. Lee, S.; Seo, P.J.; Lee, H.J.; Park, C.M. A NAC transcription factor NTL4 promotes reactive oxygen species production during drought-induced leaf senescence in Arabidopsis. *Plant J.* **2012**, *70*, 831–844. [CrossRef]

29. Liu, B.; Ouyang, Z.; Zhang, Y.; Li, X.; Hong, Y.; Huang, L.; Liu, S.; Zhang, H.; Li, D.; Song, F. Tomato NAC transcription factor SlSRN1 positively regulates defense response against biotic stress but negatively regulates abiotic stress response. *PLoS ONE* **2014**, *9*, e102067. [CrossRef]

30. Jin, C.; Li, K.Q.; Xu, X.Y.; Zhang, H.P.; Chen, H.X.; Chen, Y.H.; Hao, J.; Wang, Y.; Huang, X.S.; Zhang, S.L. A Novel NAC transcription factor, PbeNAC1, of *Pyrus betulifolia* confers cold and drought tolerance via interacting with PbeDREBs and activating the expression of stress-responsive genes. *Front. Plant Sci.* **2017**, *8*, 1049. [CrossRef]

31. Wang, K.; Yin, X.R.; Zhang, B.; Grierson, D.; Xu, C.J.; Chen, K.S. Transcriptomic and metabolic analyses provide new insights into chilling injury in peach fruit. *Plantcell. Environ.* **2017**, *40*, 1531–1551. [CrossRef] [PubMed]

32. Zhu, J.; Dong, C.H.; Zhu, J.K. Interplay between cold-responsive gene regulation, metabolism and RNA processing during plant cold acclimation. *Curr. Opin. Plant Biol.* **2007**, *10*, 290–295. [CrossRef] [PubMed]

33. Shan, W.; Kuang, J.-F.; Lu, W.J.; Chen, J.Y. Banana fruit NAC transcription factor MaNAC1 is a direct target of MaICE1 and involved in cold stress through interacting with MaCBF1. *Plantcell. Environ.* **2014**, *37*. [CrossRef] [PubMed]

34. Hao, Y.J.; Wei, W.; Song, Q.X.; Chen, H.W.; Zhang, Y.Q.; Wang, F.; Zou, H.F.; Lei, G.; Tian, A.G.; Zhang, W.K.; et al. Soybean NAC transcription factors promote abiotic stress tolerance and lateral root formation in transgenic plants. *Plant J.* **2011**, *68*, 302–313. [CrossRef] [PubMed]

35. Wang, J.; Wang, Q.; Yang, Y.; Liu, X.; Gu, J.; Li, W.; Ma, S.; Lu, Y. De novo assembly and characterization of stress transcriptome and regulatory networks under temperature, salt and hormone stresses in *Lilium lancifolium. Mol. Biol. Rep.* **2014**, *41*, 8231–8245. [CrossRef] [PubMed]

36. Yong, Y.B.; Li, W.Q.; Wang, J.M.; Zhang, Y.; Lu, Y.M. Identification of gene co-expression networks involved in cold resistance of *Lilium lancifolium. Biol. Plant.* **2018**, *62*, 287–298. [CrossRef]

37. Wang, J.; Yang, Y.; Liu, X.; Huang, J.; Wang, Q.; Gu, J.; Lu, Y. Transcriptome profiling of the cold response and signaling pathways in *Lilium lancifolium. BMC Genom.* **2014**, *15*, 203. [CrossRef]

38. Yang, S.D.; Seo, P.J.; Yoon, H.K.; Park, C.M. The Arabidopsis NAC transcription factor VNI2 integrates abscisic acid signals into leaf senescence via the *COR/RD* Genes. *Plant Cell* **2011**, *23*, 2155–2168. [CrossRef]

39. Pino, M.T.; Skinner, J.S.; Park, E.J.; Jeknić, Z.; Hayes, P.M.; Thomashow, M.F.; Chen, T.H. Use of a stress inducible promoter to drive ectopic *AtCBF* expression improves potato freezing tolerance while minimizing negative effects on tuber yield. *Plant Biotechnol. J.* **2007**, *5*, 591–604. [CrossRef]

40. Huang, L.; Hong, Y.; Zhang, H.; Li, D.; Song, F. Rice NAC transcription factor ONAC095 plays opposite roles in drought and cold stress tolerance. *BMC Plant Biol.* **2016**, *16*, 203. [CrossRef]

41. Wang, L.; Li, Z.; Lu, M.; Wang, Y. ThNAC13, a NAC transcription factor from *Tamarix hispida*, confers salt and osmotic stress tolerance to transgenic Tamarix and Arabidopsis. *Front. Plant Sci.* **2017**, *8*, 635. [CrossRef] [PubMed]

42. Gao, F.; Xiong, A.; Peng, R.; Jin, X.; Xu, J.; Zhu, B.; Chen, J.; Yao, Q. OsNAC52, a rice NAC transcription factor, potentially responds to ABA and confers drought tolerance in transgenic plants. *Plant Celltissue Organ Cult.* **2010**, *100*, 255–262. [CrossRef]

43. Liu, X.; Hong, L.; Li, X.-Y.; Yao, Y.; Hu, B.; Li, L. Improved drought and salt tolerance in transgenic Arabidopsis overexpressing a NAC Transcriptional Factor from *Arachis hypogaea. Biosci. Biotechnol. Biochem.* **2011**, *75*, 443–450. [CrossRef]

44. Jiang, G.; Jiang, X.; Lü, P.; Liu, J.; Gao, J.; Zhang, C. The Rose (*Rosa hybrida*) NAC transcription factor 3 Gene, *RhNAC3*, involved in ABA signaling pathway both in rose and Arabidopsis. *PLoS ONE* **2014**, *9*, e109415. [CrossRef] [PubMed]

45. Yang, X.; Wang, X.; Ji, L.; Yi, Z.; Fu, C.; Ran, J.; Hu, R.; Zhou, G. Overexpression of a *Miscanthus lutarioriparius* NAC gene *MlNAC5* confers enhanced drought and cold tolerance in Arabidopsis. *Plant Cell Rep.* **2015**, *34*, 943–958. [CrossRef] [PubMed]

46. Yoshida, T.; Mogami, J.; Yamaguchi-Shinozaki, K. ABA-dependent and ABA-independent signaling in response to osmotic stress in plants. *Curr. Opin. Plant Biol.* **2014**, *21*, 133–139. [CrossRef]

47. Kim, S.Y.; Chung, H.J.; Thomas, T.L. Isolation of a novel class of bZIP transcription factors that interact with ABA-responsive and embryo-specification elements in the *Dc3* promoter using a modified yeast one-hybrid system. *Plant J. Cell Mol. Biol.* **1997**, *11*, 1237–1251. [CrossRef]

48. Lopez-Molina, L.; Chua, N.H. A null mutation in a bZIP factor confers ABA-insensitivity in Arabidopsis thaliana. *Plant Cell Physiol.* **2000**, *41*, 541–547. [CrossRef]

49. Puranik, S.; Sahu, P.P.; Srivastava, P.S.; Prasad, M. NAC proteins: Regulation and role in stress tolerance. *Trends Plant Sci.* **2012**, *17*, 369–381. [CrossRef]

50. Nakashima, K.; Ito, Y.; Yamaguchi-Shinozaki, K. Transcriptional regulatory networks in response to abiotic stresses in Arabidopsis and grasses. *Plant Physiol.* **2009**, *149*, 88–95. [CrossRef]

51. Jiang, C.; Iu, B.; Singh, J. Requirement of a CCGAC cis-acting element for cold induction of the *BN115* gene from winter *Brassica napus*. *Plant Mol. Biol.* **1996**, *30*, 679–684. [CrossRef] [PubMed]

52. Xue, G. Characterisation of the DNA-binding profile of barley *HvCBF1* using an enzymatic method for rapid, quantitative and high-throughput analysis of the DNA-binding activity. *Nucleic Acids Res.* **2002**, *30*, e77. [CrossRef] [PubMed]

53. Agarwal, M.; Hao, Y.; Kapoor, A.; Dong, C.H.; Fujii, H.; Zheng, X.; Zhu, J.K. A R2R3 type MYB transcription factor is involved in the cold regulation of *CBF* genes and in acquired freezing tolerance. *J. Biol. Chem* **2006**, *281*, 37636–37645. [CrossRef] [PubMed]

54. Svensson, J.T.; Crosatti, C.; Campoli, C.; Bassi, R.; Stanca, A.M.; Close, T.J.; Cattivelli, L. Transcriptome analysis of cold acclimation in barley *albina* and *xantha* mutants. *Plant Physiol.* **2006**, *141*, 257–270. [CrossRef] [PubMed]

55. Tran, L.S.; Nakashima, K.; Sakuma, Y.; Osakabe, Y.; Qin, F.; Simpson, S.D.; Maruyama, K.; Fujita, Y.; Shinozaki, K.; Yamaguchi-Shinozaki, K. Co-expression of the stress-inducible zinc finger homeodomain ZFHD1 and NAC transcription factors enhances expression of the *ERD1* gene in Arabidopsis. *Plant J.* **2007**, *49*, 46–63. [CrossRef] [PubMed]

56. Waadt, R.; Schmidt, L.K.; Lohse, M.; Hashimoto, K.; Bock, R.; Kudla, J. Multicolor bimolecular fluorescence complementation reveals simultaneous formation of alternative CBL/CIPK complexes in planta. *Plant J.* **2008**, *56*, 505–516. [CrossRef] [PubMed]

57. Bent, A. Arabidopsis thaliana floral dip transformation method. *Methods Mol. Biol.* **2006**, *343*, 87–103. [CrossRef] [PubMed]

58. Jefferson, R.A.; Kavanagh, T.A.; Bevan, M.W. GUS fusions: Beta-glucuronidase as a sensitive and versatile gene fusion marker in higher plants. *EMBO J.* **1987**, *6*, 3901–3907. [CrossRef] [PubMed]

59. Cao, W.H.; Liu, J.; He, X.J.; Mu, R.L.; Zhou, H.L.; Chen, S.Y.; Zhang, J.S. Modulation of ethylene responses affects plant salt-stress responses. *Plant. Physiol.* **2007**, *143*, 707–719. [CrossRef]

International Journal of
Molecular Sciences

Article

A MYB-Related Transcription Factor from *Lilium lancifolium* L. (LlMYB3) Is Involved in Anthocyanin Biosynthesis Pathway and Enhances Multiple Abiotic Stress Tolerance in *Arabidopsis thaliana*

Yubing Yong, Yue Zhang and Yingmin Lyu *

Beijing Key Laboratory of Ornamental Germplasm Innovation and Molecular Breeding, National Engineering Research Center for Floriculture, College of Landscape Architecture, Beijing Forestry University, Beijing 100083, China
* Correspondence: luyingmin@bjfu.edu.cn

Received: 13 May 2019; Accepted: 27 June 2019; Published: 29 June 2019

Abstract: Most commercial cultivars of lily are sensitive to abiotic stresses. However, tiger lily (*Lilium lancifolium* L.), one of the most widely distributed wild lilies in Asia, has strong abiotic stresses resistance. Thus, it is indispensable to identify stress-responsive candidate genes in tiger lily for the stress resistance improvement of plants. In this study, a MYB related homolog (LlMYB3) from tiger lily was functionally characterized as a positive regulator in plant stress tolerance. LlMYB3 is a nuclear protein with transcriptional activation activity at C-terminus. The expression of *LlMYB3* gene was induced by multiple stress treatments. Several stress-related cis-acting regulatory elements (MYBRS, MYCRS, LTRE and DRE/CRT) were located within the promoter of *LlMYB3*; however, the promoter activity was not induced sufficiently by various stresses treatments. Overexpressing *LlMYB3* in *Arabidopsis thaliana* L. transgenic plants showed ABA hypersensitivity and enhanced tolerance to cold, drought, and salt stresses. Furthermore, we found *LlMYB3* highly co-expressed with *LlCHS2* gene under cold treatment; yeast one-hybrid (Y1H) assays demonstrated LlMYB3 was able to bind to the promoter of *LlCHS2*. These findings suggest that the stress-responsive LlMYB3 may be involved in anthocyanin biosynthesis pathway to regulate stress tolerance of tiger lily.

Keywords: MYB; CHS; anthocyanin biosynthesis; abiotic stresses; lily

1. Introduction

Environmental stresses, such as drought, high salinity, and extreme temperatures, adversely affect the growth, development, and productivity of plants. To adapt to these environmental stressors, plants have evolved complex signaling cascades to regulate the expression of stress-related genes that can improve stress tolerance either directly or indirectly [1]. Many proteins and genes in the complex signaling networks are regulated by multiple transcription factor (TF) families. As one of the largest TF groups in plants, the MYB family has been proven to be essential for responding to abiotic stresses [2,3].

MYB proteins are characterized by their highly conserved DNA-binding domains [2,4]. According to the number of imperfect repeats of the SANT (for SWI3, ADA2, N-CoR, and TFIIIB) domain (50–53 amino acids) in MYB DNA-binding domains, plant MYB proteins can be mainly sub divided into three subfamilies: MYB-related (one single SANT domain), the R2R3-MYB (two SANT domains) and R1R2R3-MYB (three SANT domains) [5]. Accordingly, research on *MYB* genes has mainly focused on the R2R3-MYB gene family, which has been shown to play important roles in many plant-specific processes including the response to abiotic stress in past decades [6]. In *Arabidopsis thaliana* L., AtMYB14 and AtMYB15 enhance freezing tolerance by regulating *CBF* and its downstream target genes [7,8]; AtMYB20 and AtMYB44 confer salt and drought resistance respectively by downregulating the expression of *PP2Cs* [9,10];

AtMYB2, AtMYB15 and AtMYB96 function in the ABA-mediated drought and salt stress response [11–13]. In rice (*Oryza sativa* L.), OsMYB4 was reported to be a positive regulator in transgenic Arabidopsis, tomato (*Solanum lycopersicum* L. cv. Tondino), and apple (*Malus pumila* Mill.) to cold and drought tolerance [14–16]; Ectopic expression of *OsMYB2* facilitated salt, cold, dehydration tolerance in rice [17]. In wheat (*Triticum aestivum* L.), overexpression of *TaMYBsm1*, *TaMYB33*, *TaMYB2A* and *TaMYB30-B* have been shown to improve the drought tolerance in Arabidopsis [18–21]; TaMYB73 can improve salinity stress tolerance in Arabidopsis [22]; TaMYB19 has been found to participate in responses to abiotic stress in transgenic Arabidopsis [23]. However, compared to 2R-MYB genes, there are few reports of functional studies of other MYB subfamilies in abiotic stress response in plants [5].

MYB TFs have also been identified to be the major determinant regulators in anthocyanin biosynthesis [24]. Interestingly, a cross-talk exists between abiotic stresses responses and anthocyanin biosynthesis. For instance, low temperature induced and high temperature suppressed anthocyanin biosynthesis in Arabidopsis, which involved the altered regulation of *AtMYB3*, *AtMYB6* and *AtMYBL2* [25,26]. Overexpression of *AtPAP1* or *AtMYB12*, two flavonol synthesis regulators, enhances oxidative and drought tolerance in Arabidopsis [27]. *MaMYB10* in apple, *PcMYB10* in pear (*Pyrus communis* L.) and *BrMYB2-2* in *Brassica rapa* are responsible for the temperature affected anthocyanin accumulation [28–30]. Thus, it is pertinent to propose that some MYB proteins may be involved in the correlation between anthocyanin accumulation and abiotic stress tolerance.

As a wild stress-resistant plant, tiger lily (*Lilium lancifolium* L.) has been shown to have capacity for resisting multiple abiotic stresses [31,32], which could be an ideal model to study stress tolerance mechanisms and signaling regulation of a stress-resistant plant. Based on our RNA-seq data published previously [32], we isolated a MYB-related type gene, *LlMYB3*, from *L. lancifolium*, to further study the role of LlMYB3 in plant stress response. Our work showed that *LlMYB3* was induced by cold, drought, salt and exogenous ABA treatments. Ectopic expression of *LlMYB3* could improve cold, drought and salt tolerance by upregulating the transcription of several stress-related genes in Arabidopsis. Moreover, LlMYB3 TF might be involved in anthocyanin biosynthesis, which can bind to the promoter of *LlCHS2*. These results provide valuable insights into the role of the LlMYB3 in regulating plant stress response.

2. Results

2.1. Isolation of LlMYB3 and Sequence Analysis

LlMYB3 gene comprises 1205 bp nucleotides with 462 bp open reading frame, 341 bp 5′ UTR, 402 bp 3′ UTR. It encodes a putative protein of 153 amino acids with a calculated molecular mass of 17.71kD and a pI of 9.54. Amino acid analysis revealed that the LlMYB3 is a MYB-related type protein with one single conserved SANT domain at the N-terminus between 39 and 86 amino acids (Figure 1a). A phylogenetic tree based on the amino acid sequences of some well-studied MYB proteins was constructed [33], which revealed that LlMYB was clustered closely to OsMYB4, OsMYB2 and AtMYB41 (Figure 1b). Furthermore, we BLAST the DNA sequence of *LlMYB3* to the whole-genome sequence of Arabidopsis thaliana using The Arabidopsis Information Resource (TAIR) to locate the chromosomal location. The BLAST result showed that the highest score (bits) significant alignment of *LlMYB3* was *AT5G62320.1* which was located in No.5 chromosome (Figure S1).

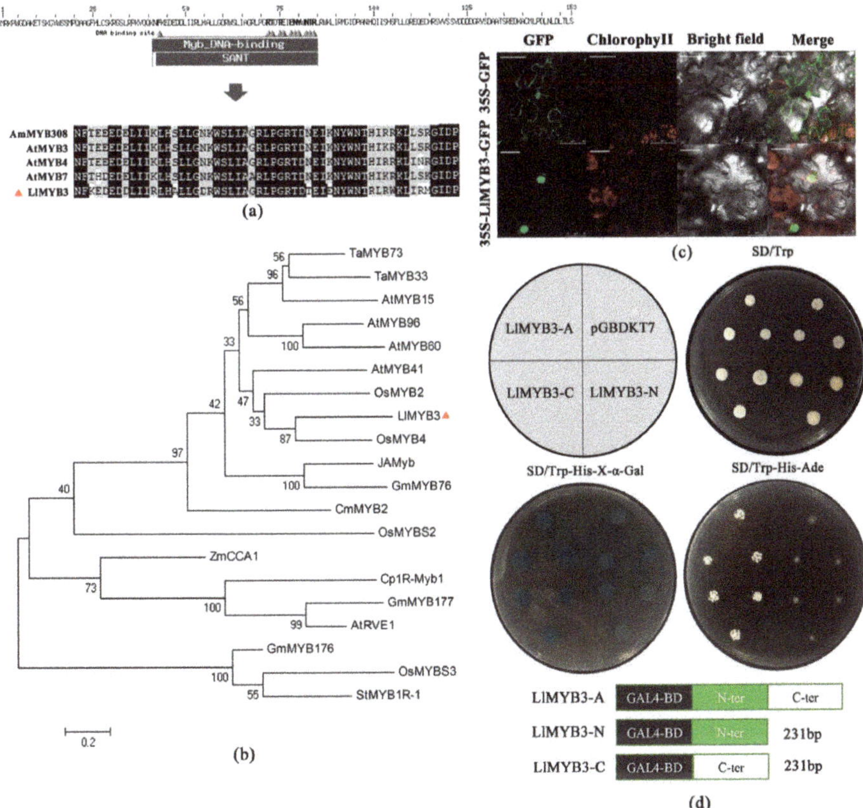

Figure 1. Characterization of tiger lily LlMYB3 protein. (**a**) Alignment of LlMYB3 with *Antirrhinum majus* AmMYB308, *Arabidopsis* AtMYB3, AtMYB4 and AtMYB5. The conserved MYB domain is marked and identical amino acids are shaded in black. (**b**) Phylogenetic tree analysis of LlMYB3 with other known stress-responsive MYB proteins. The LlMYB3 is marked with red triangle. (**c**) GFP and GFP-LlMYB3 fusion proteins were transiently expressed in tobacco leaves under control of the CaMV35S promoter and observed under a laser scanning confocal microscope. Scale bars for 35S-GFP, 50 μm; for 35S-LlMYB3-GFP, 25 μm. (**d**) Full-length protein (LlMYB3-A), N-terminal fragment (LlMYB3-N) and C-terminal fragment (LlMYB3-C) were fused with GAL4 DNA binding domain and expressed in yeast strain Y2HGold. The pGBDKT7 vector was used as a negative control. Transformed yeasts were dripped on the SD/-Trp, SD/-Trp-His-x-gal and SD/-Trp-His-Ade, after being cultured for 3 days in the growth chamber.

2.2. Subcellular Localization and Transactivation Assay of LlMYB3

The GFP-LlMYB3 fusion construct and the GFP control in pBI121-GFP vector driven by CaMV35S promoter were transiently expressed in tobacco epidermal cells and visualized under a laser scanning confocal microscope to determine the subcellular localization of LlMYB3. Results showed that the fluorescence signals from GFP alone were widely distributed throughout the cells, whereas, the GFP-LlMYB3 fusion protein fluorescence signal was mainly detected in the nucleus (Figure 1c). Thus, these results demonstrated that LlMYB3 is a nuclear protein.

To investigate the transcriptional activity of LlMYB3 protein, the entire coding region, N-terminal and C-terminal domain coding sequence were inserted into the pGBDKT7 vector, which contains the GAL4 DNA-binding domain. The transactivation results showed that all transformed yeast cells grew well on SD/-Trp medium (Figure 1d). The yeast strain containing the full-length LlMYB3 (LlMYB3-A) and the C-terminus of LlMYB3 (LlMYB3-C) could grow well on the selection medium SD/-Trp/-His/-Ade, while the cells with the N-terminus of LlMYB3 (LlMYB3-N) and pGBDKT7 empty vector could not grow normally (Figure 1d). Furthermore, the yeast cells that grew well on the SD/-Trp/-His-x-α-gal medium appeared blue in the presence of α-galactosidase, indicating the activation of the reporter gene *Mel1* (Figure 1d). These results indicated that LlMYB3 is a transcriptional activator, and its transactivation domain locates in the C-terminal region.

2.3. Expression Patterns of LlMYB3 under Multiple Stresses and ABA

To explore the possible involvement of *LlMYB3* in abiotic stress response, we analyzed the expression patterns of *LlMYB3* in tiger lily plants after treatment with abiotic stresses and ABA. The qRT-PCR analyses revealed that the *LlMYB3* gene has relatively high expression levels in bulb and flower petal, while its expression was low in the leaf and stem (Figure 2a). The expression of *LlMYB3* was significantly and rapidly induced within 1 h after cold, salt and drought treatment, leading to sevenfold to ninefold increase, but only in the salt treatment, the transcript level of *LlMYB3* increased again during 6–24 h (Figure 2c–e). Treatment of tiger lily plants with ABA induced the expression of *LlMYB3*, showing fivefold to sixfold increases at 24 h after treatment (Figure 2b). These data indicate that *LlMYB3* is a stress-responsive MYB-related gene in tiger lily, and its expression is sensitive to cold, drought and salt signaling molecules.

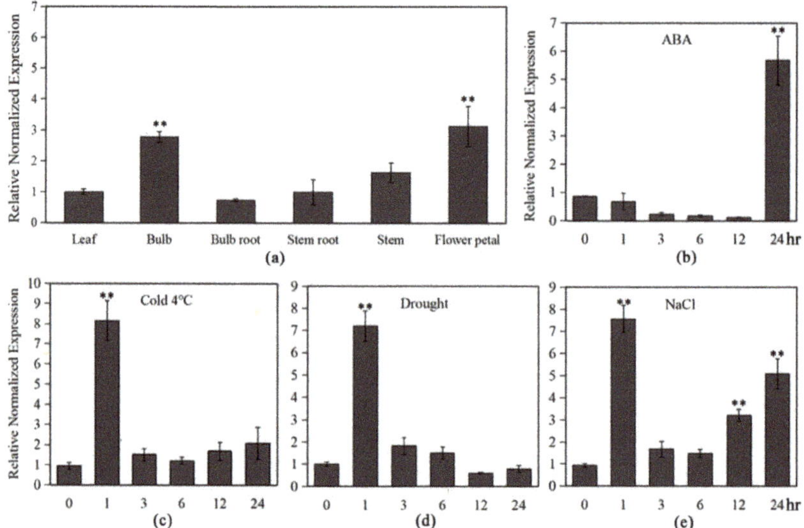

Figure 2. Expression patterns of *LlMYB3* in tiger lily seedlings under different stress treatments. Expression patterns of *LlMYB3* in leaves, bulbs, roots, stems, and flower petal (**a**), and after ABA (**b**), cold (**c**), drought (**d**) and NaCl (**e**) treatments in leaves by qRT-PCR analysis. Transcript levels were normalized to *LlTIP1*. Values are means ± SD of three replicates. Three independent experiments were performed. Asterisks indicate a significant difference (** $p < 0.01$) compared with the corresponding controls.

2.4. Promoter Analysis of LlMYB3 under Multiple Stresses and ABA

To clarify the mechanism underlying the stress-inducible expressions of *LlMYB3*, the 1374 bp upstream of ATG start codon *LlMYB3* promoter sequence was cloned and used to drive the *GUS* expression in Arabidopsis. Sequences of various putative stress-related cis-acting regulatory elements were identified, including MYB, MYC recognition sites, LTRE, DRE/CRT, TGA and ERE elements (Table 1).

Table 1. Stress-related cis-acting regulatory elements identified in the promoter region of *LlMYB3*.

Site Name	(Strand) Position	Sequence	Function
ARE	(+)1277	TGGTTT	cis-acting regulatory element essential for the anaerobic induction
CRT/DRE	(+)1323; (−)44	GTCGAC	Core CRT/DRE motif
LTRE	(+)54	ACCGACA	Putative low temperature responsive element
Box I	(+)696	TTTCAAA	light responsive element
MNF1	(−)32	GTGCCCTATA	light responsive element
MBRS	(+)85,039; (−)91	CAACGG (T/A)AACCA	MYB binding site involved in drought-inducibility
MYC	(+)8,438,501,089; (−)5,251,051	CAA(T/C/A)TG CAT(T/G)TG	MYC recognition site involved in cold and drought-inducibility
CGTCA-motif	(+)855; (−)1009	CGTCA	cis-acting regulatory element involved in the MeJA-responsiveness
TGACG-motif	(+)360; (−)484	TGACG	cis-acting regulatory element involved in the MeJA-responsiveness
ERE	(+)695	ATTTCAAA	ethylene-responsive element
TATC-BOX	(+)1274	TATCCCA	cis-acting element involved in gibberellin-responsiveness
TGA-element	(−)490	AACGAC	auxin-responsive element

By histochemical GUS staining, we only observed prominent GUS staining in leaf veins of over 2-week old transgenic *LlMYB3* promoter plants (Figure 3a). There was no obvious difference in GUS staining between stress-treated and non-treated transgenic plants. Thus, the stress inducible activity of the *LlMYB3* promoter was revealed by measuring *GUS* gene expression levels in the transgenic Arabidopsis through qRT-PCR analyses. The result showed that *GUS* gene transcript level could be induced by cold, ABA, salt and drought treatments with a maximal level at 2 and 12 h, respectively (Figure 3b). By GUS enzyme activity assay, however, only an extremely weak fluorescence signal was detected. These results indicated that the activity of *LlMYB3* promoter can be induced by multiple stress treatments, while it is not strong enough to mediate the GUS enzyme activity.

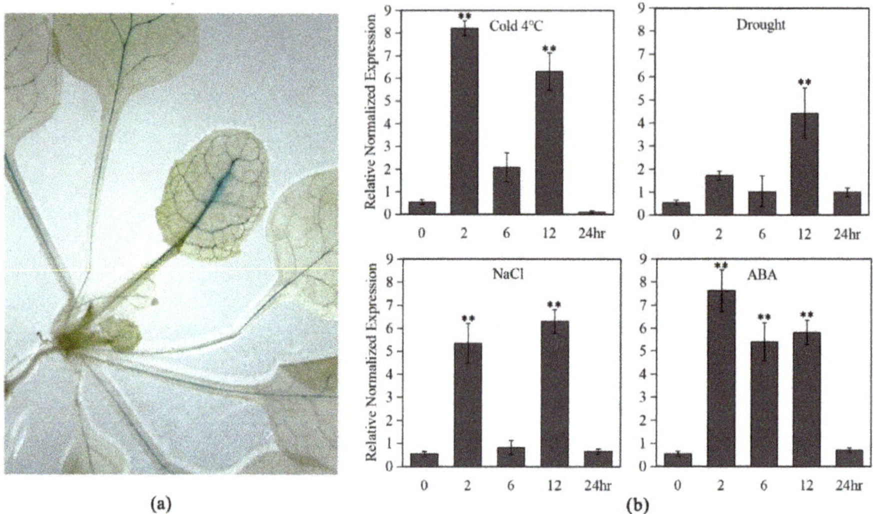

Figure 3. GUS activity of the *LlMYB3* promoter in transgenic Arabidopsis plants. (**a**) Beta-glucuronidase (GUS) expression in untreated transgenic Arabidopsis plants. (**b**) The *GUS* transcript levels in the leaves of the transgenic Arabidopsis under cold (4 °C), drought, salt and ABA treatments. The untreated transformants served as controls. There were 12 transgenic lines acquired. Values are means ±SD of three replicates. Three independent experiments were performed. Asterisks indicate a significant difference (** $p < 0.01$) compared with the corresponding controls.

2.5. Overexpression of LlMYB3 in Arabidopsis Improves Tolerance to Cold and Drought Stresses

To explore the function of LlMYB3 in providing tolerance to abiotic stress in plants, transgenic Arabidopsis plants overexpressing *LlMYB3* driven by the CaMV35S promoter were generated. Two independent homozygous lines LlMYB3-5 (L5) and LlMYB3-8 (L8) with relatively high expression levels (Figure S2) were selected for the analysis.

To study the effect of *LlMYB3* overexpression on cold stress, *LlMYB3* transgenic lines and wild-type (WT) plants were grown in equal amounts of potting soil for 4 weeks under normal conditions, and cold stress was applied by being exposed to −4, −6, −8 °C for 12 h. The results showed that all plants grew well under −4 °C treatment as the same as under normal temperature 22 °C (Figure 4a). When the temperature decreased to −6 °C, most of WT plants were dead with a survival rate at approximately 20%, but over half of transgenic plants survived (Figure 4a,b). Furthermore, all WT plants were dead, whereas the survival rate for transgenic plants was observed at 30–35% under −8 °C treatment (Figure 4a,b). In a further experiment, 4-week-old plants were treated at 4 °C for 3 h, and the relative electrolyte leakage and soluble sugars were measured after treatment. As a result, the electrolyte leakage was lower in transgenic plants relative to WT plants (Figure 4c); and transgenic plants produced remarkably higher levels of soluble sugars under a chilling condition compared to WT plants (Figure 4d).

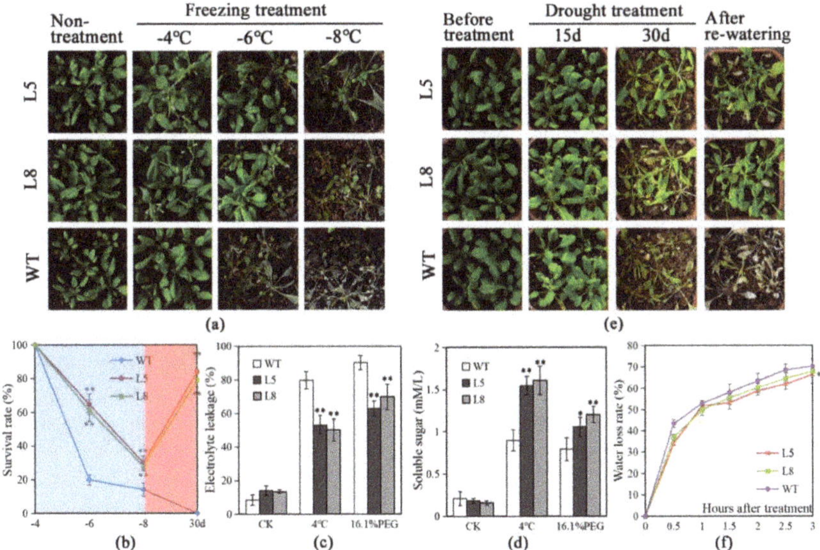

Figure 4. Overexpression of *LlMYB3* in Arabidopsis improves the freezing and drought tolerance. Performance of wild-type (WT) and *LlMYB3* transgenic lines after freezing (**a**) and drought (**e**) treatments. (**b**) Survival rate of plants in (**a**) under freezing temperatures (blue region) and in (**e**) after drought treatment for 30 days (red region). Each data point is the mean of four experiments, and each experiment comprises 30 plants. Relative electrolyte leakage (**c**) and soluble sugar content (**d**) in WT and *LlMYB3* transgenic lines after 4 °C and 16.1% PEG6000 treatments for 3 h. (**f**) Water loss rate of leaves from WT and transgenic Arabidopsis. The data represent the means from three experiments. The bars show the SD. Significant differences between the transgenic and WT lines are indicated as * $0.01 < p < 0.05$ and ** $p < 0.01$.

Similarly, to study drought stress tolerance, after withholding water for 30 days, WT plants showed visible symptoms of drought-induced damage, such as drying, wilting, and even death while some transgenic plants remained green with expanded leaves (Figure 4e). Further analyses showed that after re-watering, few WT plants survived, whereas about 78–82% of transgenic plants continued to grow (Figure 4e,f). Additionally, after being treated with 16.1% PEG6000 (−0.5 MPa) for 3 h, transgenic plants showed lower electrolyte leakage and higher levels of soluble sugars compared to WT plants (Figure 4c,d). The water-loss rates were also slightly lower in transgenic plants (L5) than in WT plants after 3 h treatment (Figure 4f).

2.6. Overexpression of LlMYB3 in Arabidopsis Increases Seed Sensitivity to ABA and Tolerance to NaCl

The salt tolerance and ABA sensitivity of *LlMYB3* transgenic plants was assessed. NaCl significantly inhibited Arabidopsis germination when the seeds were cultivated on MS medium supplemented with 50 mM NaCl (Figure 5a). Only about 30% of the WT seeds germinated in MS medium containing 50 mM NaCl while about 60–65% of transgenic plants seeds were able to germinate (Figure 5a,b). In contrast, except that WT seeds germinated slower in MS medium containing 2 μM ABA, no obvious difference was observed between the germination of WT seeds cultivated on MS medium supplemented with 0 and 2 μM ABA. However, the germination ratio of transgenic plants seeds was remarkably lower than that of WT seeds in MS medium containing 2 μM ABA (Figure 5a,b). Therefore, we suggested that *LlMYB3* transgenic plants are more tolerant to salt stresses and more hypersensitive to ABA than WT plants.

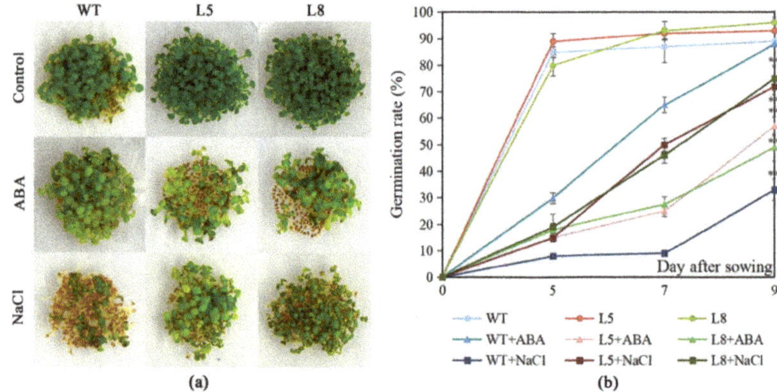

Figure 5. Hypersensitivity and enhancing tolerance of *LlMYB3* transgenic lines to ABA and NaCl. Germination of WT seeds of Col-0 and 35S:LlMYB3 on MS supplemented with 50 mM NaCl and 2 μM ABA after 9 days of incubation at 22 °C (**a**). Germination rate of seeds counted after 5, 7 and 9 days after sowing. The data represent the means from three experiments. The bars show the SD. Significant differences between the transgenic and WT lines are indicated as ** $p < 0.01$.

2.7. Altered Expression of Stress-Responsive Genes in LlMYB3 Transgenic Plants

The *LlMYB3* transgenic plants exhibited an improved tolerance to freezing, drought and salt stresses. We then measured the expression levels of genes involving stress response in the transgenic plants under normal conditions. Except for *AtCOR47*, transcripts of *AtRD29A*, *AtRD29B*, *AtRD20*, *AtABI5*, *AtGolS1*, *AtLEA14* and *AtAPX2* genes (NCBI accession numbers are shown in Table S1) accumulated in *LlMYB3* transgenic plants compared to WT plants (Figure 6). The enhanced expression of these genes in transgenic plants might contribute to the stronger stress tolerance, which also implied that LlMYB3 TF may confer stress tolerance through regulating various stress-responsive genes.

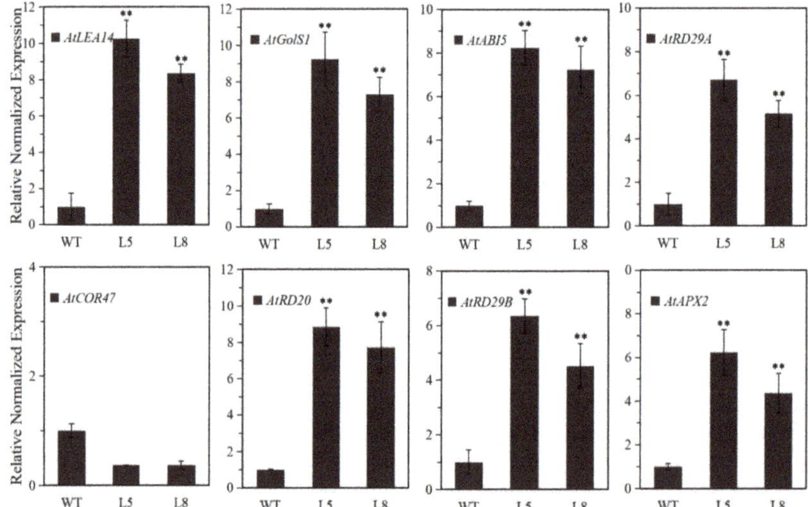

Figure 6. Expression levels of the stress-associated genes under normal condition in WT and *LlMYB3* transgenic plants. Gene-specific primers were used to detect the relative transcript levels of the stress-related genes. Values are means ± SD of three replicates. Three independent experiments were performed. Asterisks indicate a significant difference (** $p < 0.01$) compared with the corresponding controls.

2.8. LlMYB3 Can Bind to the Promoter of LlCHS2

In our previous study, through the analysis of gene co-expression networks involved in cold resistance of tiger lily, we found that the *LlMYB3* was highly co-expressed with genes involved in anthocyanin biosynthesis pathway, including phenylalanine ammonia-lyase (*PAL*), cinnamic acid 4-hydroxylase (*C4H*), 4-hydroxycinnamoyl-CoA ligase (*4CL*), chalcone synthases (*CHS*) and flavonol synthase (*FLS*) [32]. In this study, the results of qRT-PCR and Pearson's correlation coefficient (*r*) confirmed that the expression pattern of *LlMYB3* was significantly similar to *LlCHS2*'s (chalcone synthase2) (*r* > 0.8) under continuous cold treatment (Figure 7a and Table S4). The *LlCHS2* gene information is shown in Figure S3. Thus, we performed the Y1H assay to explore whether there is an interaction between LlMYB3 protein and *LlCHS2* promoter. The 932 bp upstream of ATG start codon *LlCHS2* promoter sequence was cloned, and the fragment (−820 to −553) of the *LlCHS2* promoter containing four MYB binding sites was isolated (Figure S3 and Table S3). The minimal inhibitory concentration of Aureobasidin A (AbA) for bait yeast strains was found to be 200 ng·mL^{-1} (Figure S4). Yeast cells transformed with pGADT7-LlMYB3 and pAbAi-proLlCHS2 grew well on SD/Leu plates with 200 and 250 ng·mL^{-1} AbA (Figure 7b). This showed that LlMYB3 could bind to the promoter of *LlCHS2*; suggesting LlMYB3 is involved in the regulatory pathway of *LlCHS2*.

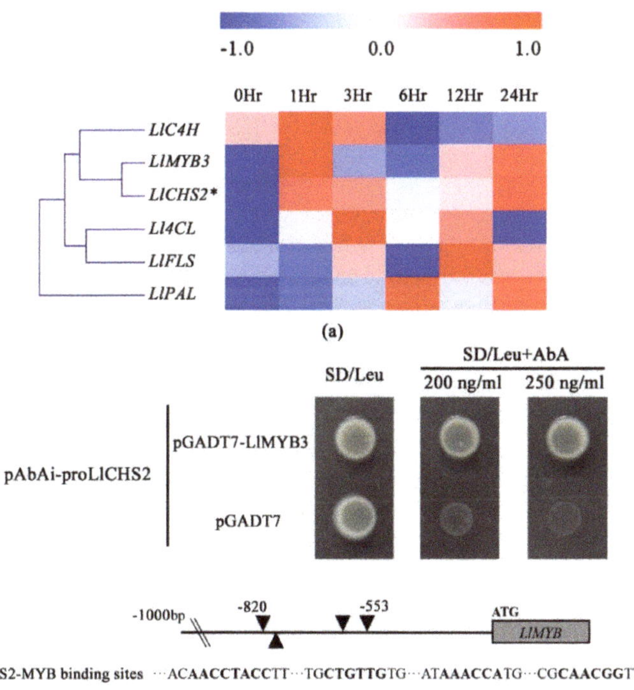

Figure 7. Yeast one-hybrid analysis of LlMYB3 binding to *LlCHS2* promoter. (**a**) Correlation analysis of expression patterns of *LlMYB3* and some structural genes involved in anthocyanin biosynthesis pathway under continuous cold treatment. (**b**) Yeast one-hybrid analysis and schematic representation of MYB binding sites in the *LlCHS2* promoter. Yeast strain Y1HGold was co-transformed with bait (pAbAi-proLlCHS2) and a prey (pGADT7 or pGADT7-LlMYB3) construct. Interaction between bait and prey was determined by cell growth on SD medium lacking Leu in the presence of 200 and 250 ng·mL^{-1} AbA.

3. Discussion

Considerable studies indicate that plant MYB family members play critical roles in response to abiotic stresses. However, the evidence of improved stress resistance for most *MYB* genes is mainly from model species as Arabidopsis and rice. In this study, a cold stress-responsive gene *LlMYB3* was cloned and characterized from tiger lily (*Lilium lancifolium* L.), to investigate the role of this *MYB* gene response to various abiotic stresses. Sequence analysis shows that LlMYB3 is a MYB-related type protein with one single conserved SANT domain and displays high identity with reported stress responsive MYB member OsMYB4, OsMYB2 and AtMYB41. As predicted that LlMYB3 is a TF, LlMYB3 protein is localized in the nucleus and both the C-terminal and full-length *LlMYB3* have high transactivation ability in yeast. Our previous transcriptome data analysis identified a unigene contig 10,499 coding for LlMYB3, which showed significant changes in expression in tiger lily under cold stress [31,32]. Here, we further confirmed that *LlMYB3* was also up-regulated by drought, salt and ABA treatments.

Several stress-related cis-elements are present in the promoter of *LlMYB3*. The *LlMYB3* promoter activity is shown to be induced by multiple stress treatments; however, it is not strong enough to mediate the GUS activity significantly. It indicates that the *LlMYB3* promoter is not an effective stress-inducible promoter, and the expression of *LlMYB3* responding to abiotic stresses might mainly be regulated by the upstream regulatory factors. Furthermore, the promoter of *LlMYB3* shows vascular vein specific expression in transgenic Arabidopsis leaves. Usually, an expression pattern of a gene by promoter analysis could reflect its function [34]. The gene expression in leaf veins is usually regulated by the alteration of environments and physiological metabolic signals of this tissue during leaf development and growth [34]. Thus, it is assumed that *LlMYB3* gene might function in vascular tissues in leaves during stress response.

Compared to R2R3-MYB genes, few studies of the MYB-related genes in abiotic stress response have been reported in plants [23]. For instance, AtMYBC1, a 1R-MYB protein, was reported to be a repressor of freezing tolerance in a CBF independent pathway in Arabidopsis [35]. In rice, MYBS3 was shown to be essential for cold stress tolerance [36]; and overexpression of *OsMYB48-1* enhanced drought and salinity tolerance in rice [23]. In potato, the single MYB domain TF StMYB1R-1 has been shown to involve in drought tolerance by activation of drought-related genes [37]. Overexpression of a single-repeat MYB TF *AmMYB1* from grey mangrove conferred salt tolerance in transgenic tobacco [38]. In the present study, we generated transgenic Arabidopsis plants overexpressing *LlMYB3* gene under the control of the constitutive CaMV35S promoter. Both morphological and physiological evidence strongly demonstrated that transgenic lines had more pronounced tolerances to cold, drought and salt than WT. At gene transcription level, qRT-PCR analysis showed that the expression level of 7 of picked 8 stress-responsive genes, including *AtRD29A, AtRD29B, AtRD20, AtGSTF6, AtGolS1, AtLEA14* and *AtAPX2* genes were higher in the transgenic plants compared with those of WT. It suggests that these genes might be transcriptionally regulated directly or indirectly by LlMYB3, and overexpressed *LlMYB3* gene may enhance stress tolerance by regulating downstream stress-responsive genes in Arabidopsis.

Moreover, we showed overexpression of *LlMYB3* resulted in enhanced ABA sensitivity, which was also observed on many MYB TFs from Arabidopsis, such as AtMYB2, AtMYB15, AtMYB96 and AtDIV2 [10–12,39]. Given that the expression level and promoter activity of *LlMYB3* can be induced by ABA treatment, *LlMYB3* might be involved in ABA signaling pathway to response to stresses. However, promoter analysis showed that there are no known ABA responsive related cis-acting elements located in the promoter of *LlMYB3*. We thus assume that there are two possible reasons. The first is that novel ABA responsive related cis-elements might exist in the promoter of *LlMYB3*; the second is that the transport of ABA signaling molecules to *LlMYB3* is through a complex signaling network rather than by directly recognizing ABA responsive related cis-elements in *LlMYB3* promoter.

On the other hand, MYB proteins play essential roles by regulating the expression of a large number of anthocyanin biosynthesis genes. For example, in Arabidopsis, the anthocyanin regulators MYB75/PAP1, MYB90/PAP2, MYB113, and MYB114 control the expression of the late biosynthetic genes

DFR and *LDOX/ANS* [2,40,41]; the flavonol regulators MYB12/PFG1, MYB11/PFG2, and MYB111/PFG3 regulate expression of the four early biosynthetic genes *CHS*, *CHI*, *F3H*, *F3'H* and *FLS* [42,43]. More importantly, some anthocyanin biosynthetic genes are even the direct targets of MYB proteins in response to abiotic stresses [3]. MYB1 in carrot can bind to the box-L-like sequences of phenylalanine ammonia-lyase 1 (*PAL1*) promoter specifically and activates *PAL1* under UV-B irradiation [44]. MYB134 in poplar, which is essential for wound and UV-B tolerance, regulates stress-responsive proanthocyanidin biosynthesis by binding to the promoter of proanthocyanidin biosynthetic genes, such as *ANR2* [45]. In rice, OsC1-MYB protein is shown to bind to the MYB responsive elements in the promoters of stress-induced flavonoid pathway genes *OsDFR* and *OsANS* [46]; overexpression of *OsMYB4* in transgenic Arabidopsis increases chilling and freezing tolerance by transactivating *PAL2* and other cold inducible genes [15]. Here, we found LlMYB3 can bind to the MYB binding sites in the promoter of cold-responsive *LlCHS2* gene, suggesting LlMYB3 protein may also function in the correlation between anthocyanin accumulation and cold stress tolerance.

In conclusion, LlMYB3 is a nucleus-localized transcriptional activator which is regulated by cold, drought and salt stresses and sensitive to ABA. Overexpressing *LlMYB3* in Arabidopsis showed ABA hypersensitivity and enhancing tolerance of transgenic plants to freezing, dehydration and salt conditions by up-regulate many stress-responsive genes. Furthermore, cold-responsive *LlCHS2* is a direct target of LlMYB3 TF in response to abiotic stresses. Therefore, our findings provide a novel MYB-related gene, which plays a positive role in plant stress resistance and might be involved in anthocyanin biosynthesis pathway in response to cold stress. Our future efforts will be focused on investigating the role of upstream regulatory factors in regulating expression and modulating the function of LlMYB3 under various stress conditions.

4. Materials and Methods

4.1. Plant Materials

The tiger lily seedlings preparation method is described in our previous study [32]. The bulbs of tiger lily were cleaned, disinfected, and then stored at 4 °C; in March, the bulbs were box-cultivated in a greenhouse (116.3° E, 40.0° N) under controlled conditions. The model plant *Arabidopsis thaliana* L. Columbia-0 (Col-0) was selected for the transgenic study of LlMYB3. Arabidopsis plants were grown in 8 cm × 8 cm plastic pots containing a 1:1 mixture of sterile peat soil and vermiculite under controlled conditions (22/16 °C, 16 h light/8 h dark, 65% relative humidity, and 1000lx light intensity). Seeds of *Nicotiana benthamiana* L. were planted and cultured under the same conditions.

4.2. Cloning and Sequence Analysis of LlMYB3

The complete sequence cDNA of *LlMYB3* gene was obtained from the transcriptome data of cold-treated tiger lily leave in our laboratory. Primer pairs (Table S1) were designed to amplify the coding sequence (CDS). The PCR products were cloned into pEASYT1-Blunt vector (TransGen Biotech, Beijing, China). After confirmation by sequencing, plasmid pEASYT1-LlMYB3 was used as a template for all experiments. The homolog genes of *LlMYB3* were searched through BLAST (http://www.ncbi. nlm.nih.gov/BLAST/) database [47]. Multiple sequence alignments were performed using DNAMAN (version 7). Phylogenetic tree analysis was performed using neighbor-joining method in MEGA5 software with 1000 replications [48]. The NCBI accession numbers of genes used in multiple sequence alignments and phylogenetic tree analysis are shown in Table S2. The theoretical molecular weight and isoelectric point were calculated using ExPASy (http://expasy.org/tools/protparam.html) [49].

Genomic DNA was extracted from tiger lily leaves using the DNeasy Plant Mini Kit (Qiagen, Valencia, CA, USA). The promoter of *LlMYB3* gene was isolated using a Genome Walker Kit (Clontech, Mountain View, CA, USA) with nest PCR according to the manufacturer's instructions. Conserved cis-element motifs of *LlMYB3* promoter were predicted using PLACE (http://www.dna.affrc.go.jp/PLACE/signalscan.html) databases [50].

4.3. Abiotic Stresses Treatment and Quantitative Real-Time PCR Analysis

For expression analysis of *LlMYB3* in response to abiotic stress and ABA treatment, 8-week-old tiger lily seedlings were treated with 4 °C, 16.1% PEG6000 (−0.5 MPa), 100 mM NaCl and 100 μM exogenous ABA for 0, 1, 3, 6, 12 and 24 h, respectively. Leaf samples were collected and immediately frozen with liquid nitrogen and stored at −80 °C for RNA isolation.

Total RNA was isolated using an RNAisomate RNA Easyspin isolation system (Aidlab Biotech, Beijing, China). First-strand cDNA synthesis was performed using Prime Script II 1st strand cDNA Synthesis Kit (Takara, Shiga, Japan). The qRT-PCR was performed using a Bio-Rad/CFX Connect™ Real-Time PCR Detection System (Bio-Rad, San Diego, CA, USA) with SYBR® qPCR mix (Takara, Shiga, Japan). Relative mRNA content was calculated using the $2^{-\Delta\Delta Ct}$ method [51] against the internal reference gene encoding tiger lily tonoplast intrinsic protein 1 (LlTIP1) [31] and Arabidopsis *Atactin* gene (NCBI accession No. NM_112764). The primers used in this study were designed with Primer Premier 5 and are listed in Table S1. All reactions were performed in three biological replicates. Student's t-test was performed for all statistical analysis in this study. The heat-map was generated using MeV4.9 and clustered by hierarchical clustering (HCL) with default parameters [52]. Pearson's correlation coefficient (r) to define similarity of expression levels between *LlMYB3* and structural genes involved in anthocyanin biosynthesis pathway.

4.4. Subcellular Localization and Transactivation Assay

To determine its subcellular localization, the whole *LlMYB3* coding region without stop codon was amplified and cloned into pBI121-GFP at *Xho*I and *Sal*I by using ClonExpressII One Step Cloning Kits (Vazyme, Piscataway, NJ, USA) to produce LlMYB3-GFP fusion construct driven by CaMV35S promoter. The recombinant constructs and empty GFP vector were transformed into *Agrobacterium tumefaciens* GV3101 and infiltrated separately into tobacco (*N. benthamiana*) epidermal cells. After agro-infiltration for 32–48 h, GFP fluorescence signals were excited at 488nm and detected under Leica TCS SP8 Confocal Laser Scanning Platform (Leica SP8, Leica, Wetzlar, Germany) using a 500–530 nm emission filter.

The transactivation experiment was carried out according to the manual of Yeast Protocols Handbook (Clontech). The full-length coding region and truncated fragments N-terminus (1–231 bp) and C-terminus (232–462) of *LlMYB3* generated by PCR amplification were fused in frame to the GAL4 DNA binding domain in the vector of pGBKT7 (Invitrogen, Carlsbad, CA, USA). These constructs and negative control pGBKT7 were transformed into yeast strain Y2HGold by using Quick Easy Yeast Transformation Mix (Clontech). The transformed yeast strains were screened on the selective medium plates SD/-Trp, SD/-Trp/-His-Ade and SD/-Trp-His-x-α-gal plates. The transactivation activity was detected according to their growth status and α-galactosidase activity.

4.5. Yeast One-Hybrid (Y1H) Assays

Y1H assay was carried out using the Matchmaker™ Gold Yeast One-Hybrid System (Clontech). The *LlCHS2* promoter was amplified by genome walking nested PCR method described previously for *LlMYB3* promoter, and the fragment (−820 to −553) of *LlCHS2* promoter containing four MYB binding sites was isolated and cloned into the pAbAi (bait) vector (shown in Figure 7b and Figure S3). Full-length *LlMYB3* was inserted into pGADT7 (prey) vector yielding plasmid pGADT7-LlMYB3. The bait plasmids were linearized and transformed into the yeast strain Y1HGold. Positive yeast cells were then transformed with pGADT7-LlMYB3 plasmid. The DNA-protein interaction was determined based on the growth ability of the co-transformants on SD/-Leu medium with Aureobasidin A (AbA) according to the manual.

4.6. Generation of Transgenic Arabidopsis

The *LlMYB3* open read frame (ORF) was cloned into pBI121 vector under control of a CaMV35S promoter; the *LlMYB3* promoter region was inserted into CaMV35S-GUS vector by replacing the CaMV35S promoter. The recombinant vectors and empty GUS vector were transformed into 5-week-old Arabidopsis ecotype Col-0 plants by using *Agrobacterium tumefaciens* GV3101 and the floral-dip method [53]. Transformed seeds were selected on MS medium containing 50 mg/L kanamycin. T3-generation homozygous lines were selected for gene functional analysis.

4.7. Histochemical Staining and Fluorometric GUS Assay

Histochemical staining and fluorometric GUS assay analysis for GUS activity was carried out as described before [54]. To understand the effects of different stresses on GUS expression mediated by the *LlMYB3* promoter, transgenic *LlMYB3* Arabidopsis plants were treated with 4 °C, 16.1% PEG6000 (−0.5 MPa), 100 mM NaCl and 100 µM exogenous ABA for different durations before sampling. The leaves of stress-treated transgenic *LlMYB3* Arabidopsis were incubated in GUS reaction buffer (3 mg/mL X-gluc, 40 mM sodium phosphate pH 7, 10 mM EDTA, 0.1% Triton X-100, 0.5 mM ferricyanatum kalium, 0.5 mM ferrocyanatum kalium and 20% methanol). After overnight incubation at 37 °C, the stained samples were bleached with 70% (*v/v*) ethanol to remove chlorophyll. Photos of those stained samples were obtained by a Leica TL3000 Ergo microscope under white light. Leaves of stress-treated transgenic Arabidopsis were also used to exam *GUS* gene expression level by qRT-PCR, and determine GUS enzyme activity and measuring the fluorescence of 4-methylumbelliferone produced by GUS cleavage of 4-methylumbelliferyl-β-D-glucuronide (4-MUG). Protein amount was determined using a Protein Assay kit (Bio-Rad, Hercules, CA, USA) using bovine serum albumin as a standard.

4.8. Evaluation of Transgenic Plants Abiotic Stress Tolerance and ABA Sensitivity

The seeds of *LlMYB3* T3-generation homozygous lines and the wild type (WT) were sown on vermiculite soil in pots for freezing and drought treatment. There were 3-week-old seedlings at 4 °C for 3 h, then at −4, −6 or −8 °C, respectively, for 12 h. After that, the plants were kept at 4 °C for 3 h before transferring to a normal condition at 22 °C. For the drought treatment, the water intake of 3-week-old potted Arabidopsis plants in water-saturated substrate was withheld for 30 days, followed by rehydrating the seedlings for 7 days. The survival rates of transgenic and WT seedlings were statistical analyzed.

For determining the salt tolerance and ABA sensitivity in transgenic plants, Arabidopsis seeds were cultivated on MS medium supplemented with 0 and 2 µM ABA or 50 mM NaCl, respectively, under continuous light at 22 °C in a growth chamber. The germination rate was scored on the 9th day after planting on the plates.

4.9. Measurements of Relative Electrolyte Leakage, Soluble Sugar, and Water Loss Rate

The relative electrolyte leakage, soluble sugar content and water loss rate were evaluated following the method described previously [55,56]. The relative electrolyte leakage was evaluated by determining the relative conductivity of fresh leaves (100 mg) in solution using a conductivity detector. The anthrone-sulfuric acid colorimetry was used for determining the soluble sugar. The water loss rate was calculated related to the initial fresh weight of the leaf samples; the samples were placed on the lab bench (20–22 °C, humidity 45–60%) and weighed at designated time points. All the measurements were performed with ten plants in triplicate.

Supplementary Materials: Supplementary materials can be found at http://www.mdpi.com/1422-0067/20/13/3195/s1. Table S1 Primers used in this study. Table S2 NCBI accession numbers of genes used in multiple sequence alignments and phylogenetic tree analysis. Table S3 MYB binding sites identified in the promoter region of *LlCHS2*. Table S4 Expression pattern correlation between *LlMYB3* and anthocyanin biosynthesis structural genes under continuous cold stress. Asterisks indicate a significant difference ($0.01 < * p < 0.05$). Figure S1 Chromosomal location of *AT5G62320.1*. The highest score (bits) significant alignment of *LlMYB3* in the whole-genome sequence of *Arabidopsis thaliana* was *AT5G62320.1* which located in No.5 chromosome. Figure S2 Overexpression of *LlMYB3* confirmed by qRT-PCR. 12 independent homozygous were selected for the analysis. Wild type Arabidopsis is served as negative control. Values are means ± SD of three replicates. Three independent experiments were performed. Asterisks indicate a significant difference (** $p < 0.01$) compared with the corresponding controls. The lines 5 and 8, which showed relative high expression levels of *LlMYB3* transcripts, were selected for further study. Figure S3 Sequence information about chalcone synthase 2 gene *LlCHS2* from tiger lily. The fragment (−820 to −553) of the *LlCHS2* promoter containing four MYB binding sites used in Y1H assay is underlined. The MYB binding sites described in Table S3 are highlighted in yellow. The partial ORF sequence of *LlCHS2* is shaded in grey. Figure S4 Minimal inhibitory concentration of Aureobasidin A (AbA) selected for bait yeast strains. 200 ng·mL^{-1} of Aureobasidin A (AbA) was shown to be the minimal inhibitory concentration for bait (pAbAi-proLlCHS2) yeast strains.

Author Contributions: Y.Y. designed of the study and performed the statistical analysis and molecular experiments. Y.Z. conceived of the study, and helped to draft the manuscript. All authors read and approved the final manuscript.

Funding: This research was funded by China National Key Research & Development Project, grant number 2018YFD1000402 and China National Natural Science Foundation (grant no. 31672190, 31872138, 31071815 and No. 31272204).

Conflicts of Interest: The authors declare no conflict of interest. The funders had no role in the design of the study; in the collection, analyses, or interpretation of data; in the writing of the manuscript, or in the decision to publish the results.

Abbreviations

ABA	abscisic acid
HCL	hierarchical clustering HCL
GFP	Green fluorescent protein
AbA	Aureobasidin A
GUS	β-glucuronidase
CHS	chalcone synthase
CHI	chalcone isomerase
F3H	flavanone 3-hydroxylase
F3′H	flavonoid 3′-hydroxylase
FLS	flavonol synthase
DFR	late biosynthetic genes dihydroflavonol reductase
LDOX/ANS	leucoanthocyanidin dioxygenase/anthocyanidin synthase
PAL1	phenylalanine ammonia-lyase 1
ANR2	anthocyanidin reductase2

References

1. Huang, G.T.; Ma, S.L.; Bai, L.P.; Zhang, L.; Ma, H.; Jia, P.; Liu, J.; Zhong, M.; Guo, Z.F. Signal transduction during cold, salt, and drought stresses in plants. *Mol. Biol. Rep.* **2012**, *39*, 969–987. [CrossRef] [PubMed]

2. Dubos, C.; Stracke, R.; Grotewold, E.; Weisshaar, B.; Martin, C.; Lepiniec, L. MYB transcription factors in Arabidopsis. *Trends Plant Sci.* **2010**, *15*, 573–581. [CrossRef] [PubMed]

3. Li, C.; Ng, C.K.Y.; Fan, L.M. MYB transcription factors, active players in abiotic stress signaling. *Environ. Exp. Bot.* **2015**, *114*, 80–91. [CrossRef]

4. Yanhui, C.; Xiaoyuan, Y.; Kun, H.; Meihua, L.; Jigang, L.; Zhaofeng, G.; Zhiqiang, L.; Yunfei, Z.; Xiaoxiao, W.; Xiaoming, Q.; et al. The MYB transcription factor superfamily of Arabidopsis: Expression analysis and phylogenetic comparison with the rice MYB family. *Plant Mol. Biol.* **2006**, *60*, 107–124. [CrossRef] [PubMed]

5. Xiong, H.; Li, J.; Liu, P.; Duan, J.; Zhao, Y.; Guo, X.; Li, Y.; Zhang, H.; Ali, J.; Li, Z. Overexpression of *OsMYB48-1*, a novel MYB-related transcription factor, enhances drought and salinity tolerance in rice. *PLoS ONE* **2014**, *9*, e92913. [CrossRef] [PubMed]

6. Du, H.; Wang, Y.B.; Xie, Y.; Liang, Z.; Jiang, S.J.; Zhang, S.S.; Huang, Y.B.; Tang, Y.X. Genome-wide identification and evolutionary and expression analyses of MYB-related genes in land plants. *DNA Res.* **2013**, *20*, 437–448. [CrossRef] [PubMed]

7. Chen, Y.; Zhangliang, C.; Kang, J.; Kang, D.; Gu, H.; Qin, G. AtMYB14 regulates cold tolerance in Arabidopsis. *Plant Mol. Biol. Rep.* **2013**, *31*, 87–97. [CrossRef]

8. Agarwal, M.; Hao, Y.; Kapoor, A.; Dong, C.-H.; Fujii, H.; Zheng, X.; Zhu, J.K. A R2R3 type MYB transcription factor is involved in the cold regulation of *CBF* genes and in acquired freezing tolerance. *J. Biol. Chem.* **2007**, *281*, 37636–37645. [CrossRef]

9. Cui, M.H.; Yoo, K.S.; Hyoung, S.; Nguyen, H.T.K.; Kim, Y.Y.; Kim, H.J.; Ok, S.H.; Yoo, S.D.; Shin, J.S. An Arabidopsis R2R3-MYB transcription factor, AtMYB20, negatively regulates type 2C serine/threonine protein phosphatases to enhance salt tolerance. *FEBS Lett.* **2013**, *587*, 1773–1778. [CrossRef]

10. Jung, C.; Seo, J.S.; Won Han, S.; Koo, Y.; Ho Kim, C.; Ik Song, S.; Hie Nahm, B.; Do Choi, Y.; Cheong, J.J. Overexpression of *AtMYB44* Enhances stomatal closure to confer abiotic stress tolerance in transgenic Arabidopsis. *Plant Physiol.* **2008**, *146*, 623–635. [CrossRef]

11. Abe, H. Arabidopsis AtMYC2 (bHLH) and AtMYB2 (MYB) function as transcriptional activators in abscisic acid signaling. *Plant Cell* **2003**, *15*, 63–78. [CrossRef] [PubMed]

12. Seo, P.J.; Xiang, F.; Qiao, M.; Park, J.Y.; Na Lee, Y.; Kim, S.G.; Lee, Y.H.; Park, W.J.; Park, C.M. The MYB96 transcription factor mediates abscisic acid signaling during drought stress response in Arabidopsis. *Plant Physiol.* **2009**, *151*, 275–289. [CrossRef] [PubMed]

13. Ding, Z.; Li, S.; An, X.; Liu, X.; Huanju, Q.; Wang, D. Transgenic expression of *MYB15* confers enhanced sensitivity to abscisic acid and improved drought tolerance in *Arabidopsis thaliana*. *J. Genet. Genom.* **2009**, *36*, 17–29. [CrossRef]

14. Pasquali, G.; Biricolti, S.; Locatelli, F.; Baldoni, E.; Mattana, M. *OsMYB4* expression improves adaptive responses to drought and cold stress in transgenic apples. *Plant Cell Rep.* **2008**, *27*, 1677–1686. [CrossRef] [PubMed]

15. Vannini, C.; Campa, M.; Iriti, M.; Genga, A.; Faoro, F.; Carravieri, S.; Rotino, G.; Rossoni, M.; Spinardi, A.; Bracale, M. Evaluation of transgenic tomato plants ectopically expressing the rice *Osmyb4* gene. *Plant Sci.* **2007**, *173*, 231–239. [CrossRef]

16. Vannini, C.; Locatelli, F.; Bracale, M.; Magnani, E.; Marsoni, M.; Osnato, M.; Mattana, M.; Baldoni, E.; Coraggio, I. Overexpression of the rice *OsMYB4* gene increases chilling and freezing tolerance of *Arabidopsis thaliana* plants. *Plant J.* **2004**, *37*, 115–127. [CrossRef]

17. Yang, A.; Dai, X.; Zhang, W.H. A R2R3-type MYB gene, *OsMYB2*, is involved in salt, cold, and dehydration tolerance in rice. *J. Exp. Bot.* **2012**, *63*, 2541–2556. [CrossRef]

18. Mao, X.; Jia, D.; Li, A.; Zhang, H.; Tian, S.; Zhang, X.; Jia, J.; Jing, R. Transgenic expression of *TaMYB2A* confers enhanced tolerance to multiple abiotic stresses in Arabidopsis. *Funct. Integr. Genom.* **2011**, *11*, 445. [CrossRef]

19. Qin, Y.; Wang, M.; Tian, Y.; He, W.; Han, L.; Xia, G. Over-expression of *TaMYB33* encoding a novel wheat MYB transcription factor increases salt and drought tolerance in Arabidopsis. *Mol. Biol. Rep.* **2012**, *39*, 7183–7192. [CrossRef]

20. Li, M.J.; Qiao, Y.; Li, Y.Q.; Shi, Z.L.; Zhang, N.; Bi, C.; Guo, J.K. A R2R3-MYB transcription factor gene in common wheat (namely *TaMYBsm1*) involved in enhancement of drought tolerance in transgenic Arabidopsis. *J. Plant Res.* **2016**, *129*. [CrossRef]

21. Zhang, L.; Zhao, G.; Xia, C.; Jia, J.; Liu, X.; Kong, X. A wheat R2R3-MYB gene, *TaMYB30-B*, improves drought stress tolerance in transgenic Arabidopsis. *J. Exp. Bot.* **2012**, *63*, 5873–5885. [CrossRef] [PubMed]

22. He, Y.; Li, W.; Lv, J.; Jia, Y.; Wang, M.; Xia, G. Ectopic expression of a wheat MYB transcription factor gene, *TaMYB73*, improves salinity stress tolerance in *Arabidopsis thaliana*. *J. Exp. Bot.* **2011**, *63*, 1511–1522. [CrossRef] [PubMed]

23. Zhang, L.; Liu, G.; Zhao, G.; Xia, C.; Jia, J.; Liu, X.; Kong, X. Characterization of a wheat R2R3-MYB transcription factor gene, *TaMYB19*, involved in enhanced abiotic stresses in Arabidopsis. *Plant Cell Physiol.* **2014**, *55*. [CrossRef] [PubMed]

24. Chen, L.; Hu, B.; Qin, Y.; Hu, G.; Zhao, J. Advance of the negative regulation of anthocyanin biosynthesis by MYB transcription factors. *Plant Physiol. Biochem.* **2019**, *136*, 178–187. [CrossRef] [PubMed]

25. Leyva, A.; Jarillo, J.A.; Salinas, J.; Martinez-Zapater, J.M. Low Temperature induces the accumulation of phenylalanine ammonia-lyase and chalcone synthase mRNAs of *Arabidopsis thaliana* in a light-dependent manner. *Plant Physiol.* **1995**, *108*, 39–46. [CrossRef] [PubMed]

26. Rowan, D.; Cao, M.; Lin-Wang, K.; Cooney, J.; Jensen, D.; Austin, P.; B Hunt, M.; Norling, C.; Hellens, R.; J Schaffer, R.; et al. Environmental regulation of leaf colour in red 35S:PAP1 *Arabidopsis thaliana*. *New Phytol.* **2009**, *182*, 102–115. [CrossRef] [PubMed]

27. Nakabayashi, R.; Yonekura-Sakakibara, K.; Urano, K.; Suzuki, M.; Yamada, Y.; Nishizawa, T.; Matsuda, F.; Kojima, M.; Sakakibara, H.; Shinozaki, K.; et al. Enhancement of oxidative and drought tolerance in Arabidopsis by over accumulation of antioxidant flavonoids. *Plant J.* **2013**, *77*. [CrossRef]

28. Lin-Wang, K.; Micheletti, D.; Palmer, J.; Volz, R.; Lozano, L.; Espley, R.; Hellens, R.; Chagne, D.; Rowan, D.; Troggio, M.; et al. High temperature reduces apple fruit colour via modulation of the anthocyanin regulatory complex. *Plant Cell Environ.* **2011**, *34*, 1176–1190. [CrossRef]

29. Li, L.; Ban, Z.; Li, X.H.; Wu, M.Y.; Wang, A.L.; Jiang, Y.; Jiang, Y.H. Differential expression of anthocyanin biosynthetic genes and transcription factor PcMYB10 in pears (*Pyrus communis* L.). *PLoS ONE* **2012**, *7*, e46070. [CrossRef]

30. Ahmed, N.U.; Park, J.I.; Jung, H.J.; Hur, Y.; Nou, I.S. Anthocyanin biosynthesis for cold and freezing stress tolerance and desirable color in *Brassica rapa*. *Funct. Integr. Genom.* **2015**, *15*, 383–394. [CrossRef]

31. Wang, J.; Wang, Q.; Yang, Y.; Liu, X.; Gu, J.; Li, W.; Ma, S.; Lu, Y. De novo assembly and characterization of stress transcriptome and regulatory networks under temperature, salt and hormone stresses in *Lilium lancifolium*. *Mol. Biol. Rep.* **2014**, *41*, 8231–8245. [CrossRef] [PubMed]

32. Yong, Y.B.; Li, W.Q.; Wang, J.M.; Zhang, Y.; Lu, Y.M. Identification of gene co-expression networks involved in cold resistance of *Lilium lancifolium*. *Biol. Plant* **2018**, *62*, 287–298. [CrossRef]

33. Huang, P.; Chen, H.; Mu, R.; Yuan, X.; Zhang, H.S.; Huang, J. *OsMYB511* encodes a MYB domain transcription activator early regulated by abiotic stress in rice. *Genet. Mol. Res.* **2015**, *14*, 9506–9517. [CrossRef] [PubMed]

34. Wu, X.; Huang, R.; Liu, Z.; Zhang, G. Functional characterization of cis-elements conferring vascular vein expression of *At4g34880* amidase family protein gene in Arabidopsis. *PLoS ONE* **2013**, *8*, e67562. [CrossRef] [PubMed]

35. Zhai, H.; Bai, X.; Zhu, Y.; Li, Y.; Cai, H.; Ji, W.; Ji, Z.; Liu, X.; Liu, X.; Li, J. A single-repeat R3-MYB transcription factor MYBC1 negatively regulates freezing tolerance in Arabidopsis. *Biochem. Biophys. Res. Commun.* **2010**, *394*, 1018–1023. [CrossRef] [PubMed]

36. Su, C.F.; Wang, Y.C.; Hsieh, T.H.; Lu, C.A.; Tseng, T.H.; Yu, S.M. A novel MYBS3-dependent pathway confers cold tolerance in rice. *Plant Physiol.* **2010**, *153*, 145–158. [CrossRef]

37. Shin, D.; Moon, S.J.; Han, S.; Kim, B.G.; Park, S.R.; Lee, S.K.; Yoon, H.J.; Lee, H.E.; Kwon, H.B.; Baek, D.; et al. Expression of *StMYB1R-1*, a novel potato single MYB-like domain transcription factor, increases drought tolerance. *Plant Physiol.* **2011**, *155*, 421–432. [CrossRef]

38. Ganesan, G.; Sankararamasubramanian, H.M.; Harikrishnan, M.; Parida, A.; Ashwin, G. A MYB transcription factor from the grey mangrove is induced by stress and confers NaCl tolerance in tobacco. *J. Exp. Bot.* **2012**, *63*, 4549–4561. [CrossRef]

39. Fang, Q.; Wang, Q.; Mao, H.; Xu, J.; Wang, Y.; Hu, H.; He, S.; Tu, J.; Cheng, C.; Tian, G.; et al. AtDIV2, an R-R-type MYB transcription factor of Arabidopsis, negatively regulates salt stress by modulating ABA signaling. *Plant Cell Rep.* **2018**, *37*, 1499–1511. [CrossRef]

40. Borevitz, J.O.; Xia, Y.; Blount, J.; Dixon, R.; Lamb, C. Activation Tagging identifies a conserved MYB regulator of phenylpropanoid biosynthesis. *Plant Cell* **2001**, *12*, 2383–2394. [CrossRef]

41. Gonzalez, A.; Zhao, M.; M Leavitt, J.; Lloyd, A. Regulation of the anthocyanin biosynthetic pathway by the TTG1/BHLH/MYB transcriptional complex in Arabidopsis seedlings. *Plant J.* **2008**, *53*, 814–827. [CrossRef]

42. Stracke, R.; Ishihara, H.; Huep, G.; Barsch, A.; Mehrtens, F.; Niehaus, K.; Weisshaar, B. Differential regulation of closely related R2R3-MYB transcription factors controls flavonol accumulation in different parts of the *Arabidopsis thaliana* seedling. *Plant J.* **2007**, *50*, 660–677. [CrossRef] [PubMed]

43. Stracke, R.; Favory, J.J.; Gruber, H.; Bartelniewoehner, L.; Bartels, S.; Binkert, M.; Funk, M.; Weisshaar, B.; Ulm, R. The Arabidopsis bZIP transcription factor HY5 regulates expression of the *PFG1/MYB12* gene in response to light and ultraviolet-B radiation. *Plant Cell Environ.* **2009**, *33*, 88–103. [CrossRef] [PubMed]

44. Maeda, K.; Kimura, S.; Demura, T.; Takeda, J.; Ozeki, Y. DcMYB1 acts as a transcriptional activator of the carrot phenylalanine ammonia-lyase Gene (*DcPAL1*) in response to elicitor treatment, UV-B irradiation and the dilution effect. *Plant Mol. Biol.* **2005**, *59*, 739–752. [CrossRef] [PubMed]

45. Mellway, R.D.; T Tran, L.; Prouse, M.B.; Campbell, M.M.; Peter Constabel, C. The wound-, pathogen-, and ultraviolet B-responsive *MYB134* gene encodes an R2R3 MYB transcription factor that regulates proanthocyanidin synthesis in poplar. *Plant Physiol.* **2009**, *150*, 924–941. [CrossRef] [PubMed]

46. Ithal, N.; Reddy, A.R. Rice flavonoid pathway genes, *OsDfr* and *OsAns*, are induced by dehydration, high salt and ABA, and contain stress responsive promoter elements that interact with the transcription activator, OsC1-MYB. *Plant Sci.* **2004**, *166*, 1505–1513. [CrossRef]

47. Altschul, S.; Gish, W.; Miller, W.; Myers, E.W.; Lipman, D.J. Basic Local Aligment Search Tool. *J. Mol. Biol.* **1990**, *215*, 403–410. [CrossRef]

48. Hall, B.G. Building Phylogenetic Trees from Molecular Data with MEGA. *Mol. Biol. Evol.* **2013**, *30*, 1229–1235. [CrossRef]

49. Artimo, P.; Jonnalagedda, M.; Arnold, K.; Baratin, D.; Csardi, G.; de Castro, E.; Duvaud, S.; Flegel, V.; Fortier, A.; Gasteiger, E.; et al. ExPASy: SIB bioinformatics resource portal. *Nucleic Acids Res.* **2012**, *40*, W597–W603. [CrossRef]

50. Higo, K.; Ugawa, Y.; Iwamoto, M.; Korenaga, T. Plant cis-acting regulatory DNA elements (PLACE) database: 1999. *Nucleic Acids Res.* **1999**, *27*, 297–300. [CrossRef]

51. Livak, K.J.; Schmittgen, T.D. Analysis of Relative Gene Expression Data Using Real-Time Quantitative PCR and the $2^{-\Delta\Delta CT}$ Method. *Methods* **2001**, *25*, 402–408. [CrossRef] [PubMed]

52. Saeed, A.I.; Sharov, V.; White, J.; Li, J.; Liang, W.; Bhagabati, N.; Braisted, J.; Klapa, M.; Currier, T.; Thiagarajan, M.; et al. TM4: A Free, Open-Source System for Microarray Data Management and Analysis. *Biotechniques* **2003**, *34*, 374–378. [CrossRef] [PubMed]

53. Bent, A. *Arabidopsis thaliana* floral dip transformation method. *Methods Mol. Biol.* **2006**, *343*, 87–103. [CrossRef] [PubMed]

54. Jefferson, R.A.; Kavanagh, T.A.; Bevan, M.W. GUS fusions: Beta-glucuronidase as a sensitive and versatile gene fusion marker in higher plants. *EMBO J.* **1987**, *6*, 3901–3907. [CrossRef] [PubMed]

55. Cao, W.H.; Liu, J.; He, X.J.; Mu, R.-L.; Zhou, H.L.; Chen, S.Y.; Zhang, J.S. Modulation of ethylene responses affects plant salt-stress responses. *Plant Physiol.* **2007**, *143*, 707–719. [CrossRef] [PubMed]

56. Zhang, L.; Zhang, L.; Xia, C.; Zhao, G.; Jia, J.; Kong, X. The novel wheat transcription factor *TaNAC47* enhances multiple abiotic stress tolerances in transgenic plants. *Front. Plant Sci.* **2015**, *6*, 1174. [CrossRef] [PubMed]

International Journal of
Molecular Sciences

Article

QTL Analysis of Resistance to High-Intensity UV-B Irradiation in Soybean (*Glycine max* [L.] Merr.)

Min Young Yoon [1,†], Moon Young Kim [1,2], Jungmin Ha [1,2], Taeyoung Lee [1], Kyung Do Kim [3] and Suk-Ha Lee [1,2,*]

[1] Department of Plant Science and Research Institute of Agriculture and Life Sciences, Seoul National University, Seoul 08826, Korea
[2] Plant Genomics and Breeding Institute, Seoul National University, Seoul 08826, Korea
[3] Corporate R&D, LG Chem, Seoul 07795, Korea
* Correspondence: sukhalee@snu.ac.kr; Tel.: +82-2880-4545; Fax: +82-2877-4550
† Present address: Professional Fine Chemical Business Team, Life Science, LG Chem, Seoul 07795, Korea.

Received: 4 June 2019; Accepted: 3 July 2019; Published: 4 July 2019

Abstract: High-intensity ultraviolet-B (UV-B) irradiation is a complex abiotic stressor resulting in excessive light exposure, heat, and dehydration, thereby affecting crop yields. In the present study, we identified quantitative trait loci (QTLs) for resistance to high-intensity UV-B irradiation in soybean (*Glycine max* [L.]). We used a genotyping-by-sequencing approach using an F6 recombinant inbred line (RIL) population derived from a cross between Cheongja 3 (UV-B sensitive) and Buseok (UV-B resistant). We evaluated the degree of leaf damage by high-intensity UV-B radiation in the RIL population and identified four QTLs, *UVBR12-1*, *6-1*, *10-1*, and *14-1*, for UV-B stress resistance, together explaining 20% of the observed phenotypic variation. The genomic regions containing *UVBR12-1* and *UVBR6-1* and their syntenic blocks included other known biotic and abiotic stress-related QTLs. The QTL with the highest logarithm of odds (LOD) score of 3.76 was *UVBR12-1* on Chromosome 12, containing two genes encoding spectrin beta chain, brain (SPTBN, Glyma.12g088600) and bZIP transcription factor21/TGACG motif-binding 9 (bZIP TF21/TGA9, Glyma.12g088700). Their amino acid sequences did not differ between the mapping parents, but both genes were significantly upregulated by UV-B stress in Buseok but not in Cheongja 3. Among five genes in *UVBR6-1* on Chromosome 6, Glyma.06g319700 (encoding a leucine-rich repeat family protein) had two nonsynonymous single nucleotide polymorphisms differentiating the parental lines. Our findings offer powerful genetic resources for efficient and precise breeding programs aimed at developing resistant soybean cultivars to multiple stresses. Furthermore, functional validation of the candidate genes will improve our understanding of UV-B stress defense mechanisms.

Keywords: soybean; UV-B stress resistance; quantitative trait loci; spectrin beta chain, brain; bZIP transcription factor21/TGA9; leucine-rich repeat family protein; stress defense signaling

---

1. Introduction

Increased solar ultraviolet-B radiation (UV-B, 280–315 nm) since the late 1980s is considered a serious environmental issue owing to the lengthy expected time of the recovery of the destroyed stratospheric ozone layer to pre-1980 levels [1,2]. High-intensity UV-B radiation beyond the level of positive stimulus to sessile plants exerts multiple stresses, such as strong light, high temperatures, and dehydration, causing physiological and morphological damages including reduced photosynthetic capacity, leaf discoloration, and reduced biomass and seed yields [3,4]. According to the United Nations Environment Programme (UNEP) annual report, increased UV-B radiation in terrestrial areas reduces plant productivity by about 6% [5]. To prevent such yield losses in crop plants, genetic studies of resistance to high-intensity UV-B radiation as a complex abiotic stress and the identification of genetic elements involved in the UV-B defense response are needed for major crops across the world.

Soybean (*Glycine max* [L.] Merr.) is one of the most important crops for food, feed, energy production, and industrial resources worldwide. Under supplemental UV-B radiation, soybean cultivars show differences in physiological damage and morphological changes; UV-B-sensitive cultivars show significant yield reductions [6–8]. Quantitative trait loci (QTLs) associated with resistance to supplementary UV-B treatment have been identified on chromosomes (Chrs) 3, 6, 7, and 19 using a recombinant inbred line (RIL) population derived from a cross between Keunol (UV-B sensitive) and Iksan10 (UV-B resistant) [9]. Using the same population, the QTL *qUVBT1* on Chr 7 was identified with the 180K AXIOM SoyaSNP array and a gene encoding a UV excision repair protein was identified as a candidate gene involved in UV-B tolerance [10]. In our previous study, we identified the most resistant and most sensitive genotypes to elevated UV-B, Buseok and Cheongja 3, among 140 soybean genotypes, including 94 *G. max* and 46 *G. soja* accessions [11]. Transcriptome profiling of these two genotypes differing in UV-B resistance has revealed differentially expressed genes involved in stress defense signaling, immune responses, and reactive oxygen species metabolism [12].

In the present study, we identified QTLs associated with resistance to high-intensity UV-B irradiation using an F_6 RIL population of Cheongja 3 (UV-B sensitive) × Buseok (UV-B resistant). Furthermore, we investigated nucleotide variation in genes located in the QTLs in the mapping parents and their expression levels in response to UV-B treatment.

2. Results

2.1. Phenotypic Evaluation of UV-B Stress Resistance in a RIL Population of Cheongja 3 × Buseok

UV-B-resistant Buseok and UV-B-sensitive Cheongja 3 showed leaf damage of 26.8% and 62.4%, respectively, in response to high-intensity UV-B irradiation (Figure 1). The difference in UV-B resistance between the two parents was consistent with the results of previous studies [11,12]. The RIL population of Cheongja 3 × Buseok showed high phenotypic variation in leaf damage, ranging from 10% to 100%, with a mean value of 50.3% (Figure 1). The normal distribution of the degree of UV-B leaf damage in the RIL population (Shapiro-Wilk: W = 0.993, P = 0.665), with transgressive segregation, indicates that UV-B resistance is quantitatively regulated by multiple genes.

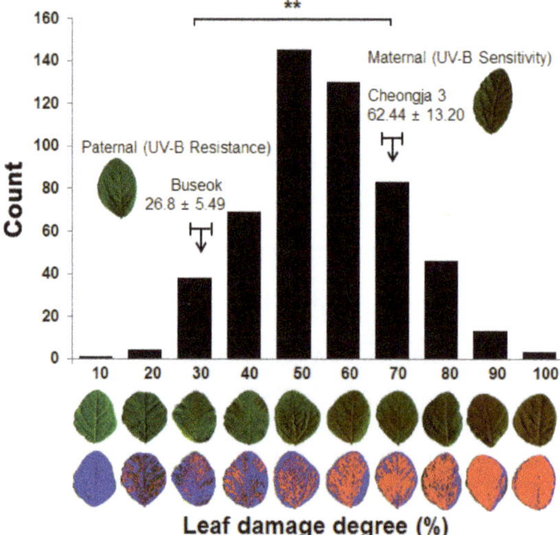

Figure 1. Distribution of the degree of leaf damage under high-intensity ultraviolet-B (UV-B) irradiation in the recombinant inbred line (RIL) population of Cheongja 3 × Buseok. X-axis indicates the degree of

leaf damage from 10% to 100%. For each degree of damage, RGB (red-green-blue) and two-color-transformed images of a representative injured leaf are shown. Blue and red in the two-color-transformed image indicate intact and damaged parts of the leaf exposed to UV-B irradiation, respectively. ** indicates a significant difference between Cheongja 3 and Buseok at $p < 0.01$ based on a Student's *t*-test.

2.2. Genotyping-by-Sequencing Analysis and Genetic Linkage Map Construction

Out of about 562 million genotyping-by-sequencing (GBS) reads generated from two libraries of the two parents and 174 RILs, 441,142,257 (78.4%) high-quality clean reads were obtained (Table S1). The number of reads per line ranged from 1,147,211 to 10,657,712 with an average of 3,197,053 reads. From these reads, we identified 271,254 unfiltered single nucleotide polymorphisms (SNPs) differentiating Cheongja 3 and Buseok. We used 4604 high-quality single nucleotide polymorphisms after filtering to construct a genetic linkage map of the RIL population. The final linkage map contained 3136 SNP markers evenly distributed on 20 chromosomes according to their physical locations on the soybean reference genome (Figure S1 and Table S2). The map spanned 3607.3 cM with an average interval of 1.15 cM between adjacent markers (Table S2). On average, each chromosome contained 156 markers spanning an average length of 180.4 cM. The chromosomes ranged from 65.8 cM (Chr 16) to 269.2 cM (Chr 15). The shortest chromosome, Chr 16, was most saturated, containing 95 SNP markers with an average marker density of 0.7 cM. Chr 4 had the largest intervals of 2.3 cM between adjacent markers. The longest chromosome, Chr 15, harbored 319 markers, covering 269.2 cM with an average marker interval of only 0.8 cM.

2.3. Identification of QTLs for UV-B Resistance

Based on the constructed genetic map, we identified four QTLs underlying UV-B resistance in soybean on Chr 12, 14, 10, and 6 (in order of logarithm of odds [LOD] score), together explaining 20.1% of phenotypic variation (Table 1). The QTL *UVBR12-1* was located at Chr12:7,261,406 ... 7,273,570 and had the highest LOD score of 3.8, explaining 9.5% of the phenotypic variation. The QTLs *UVBR14-1* and *UVBR10-1* on Chr 14 and 16 had LOD scores of 2.2 and 1.1 and explained 5.3% and 2.7% of phenotypic variation, respectively. On Chr 6, a minor QTL for UV-B resistance between markers Chr06:50,843,417 and Chr06:50,873,593 had the lowest R^2 value (2.6%) among all QTLs detected (Table 1).

We searched for genes located within the four QTLs *UVBR6-1, 10-1, 12-1,* and *14-1* according to the physical locations of SNP markers associated with the QTLs (Table 1 and Figure 2). Two genes encoding spectrin beta chain, brain (*SPTBN*, Glyma.12g088600) and bZIP transcription factor21 (*bZIP TF21/TGA9*, Glyma.12g088700) were located within *UVBR12-1* of the 121.6 kb region flanked by the marker positions Chr12:7273570 and Chr12:72761406 (Table 1 and Figure 2). *UVBR6-1* harbored five protein-coding genes anchored in the 30.2 kb region between Chr06:50,843,417 and Chr06:50,873,593 (Table 1 and Figure 2), including two genes (Glyma.06g319600 and Glyma.06g319700) encoding leucine-rich repeat (LRR) family proteins, one gene (Glyma.06g319800) for alfin-like 1, and two genes (Glyma.06g319900 and Glyma.06g320000) encoding family with sequence similarity 136, member A (FAM136A)-like protein (Table 1 and Figure 2). The 1.33 Mb genomic region of *UVBR10-1* (Chr10:41,185,273 ... Chr10:42,517,624) included 140 genes, and 60 genes were located in the 647.6 kb genomic region corresponding to *UVBR14-1* (Chr14:47,368,499 ... Chr14:46,720,930) (Table 1 and Table S3).

Table 1. Quantitative trait loci (QTLs) for UV-B stress resistance identified by inclusive composite interval mapping in 176 RILs derived from Cheongja 3 × Buseok.

Locus	Left Marker	Right Marker	Position [a] (cM)	LOD [b]	Add [c]	R^2 [d] (%)	No. of Genes [e]	Known Stress-Related QTLs [f]
UVBR12-1	Chr12:7261406	Chr12:7273570	74	3.8	-0.4	9.5	2	*Drought tolerance 6-4* *SCN 39-4*
UVBR14-1	Chr14:46720930	Chr14:47368499	189	2.2	0.3	5.3	60	*Fe-effect 3-2, 9-2, 10-3* *Flood tolerance 4-7*
UVBR10-1	Chr10:41185273	Chr10:42517624	160	1.1	-0.2	2.7	142	*Drought tolerance 6-3* *Sclero 2-23, 4-10, 3-18* *Phytoph 5-3*
UVBR6-1	Chr06:50843417	Chr06:50873593	38	1.1	0.2	2.6	5	*Plant damage, UV-B induced 1-2*

[a] Genetic position of a QTL peak in the linkage map constructed in the present study. [b] Maximum-likelihood logarithm of odds (LOD) score for the individual QTL. [c] Allelic effect. [d] Percent of phenotypic variance explained by the QTL. [e] Number of protein-coding genes within marker intervals on the basis of G. max gene models ver. 1.1. [f] Known stress-related QTLs within 2 Mb surrounding the QTLs identified in this study.

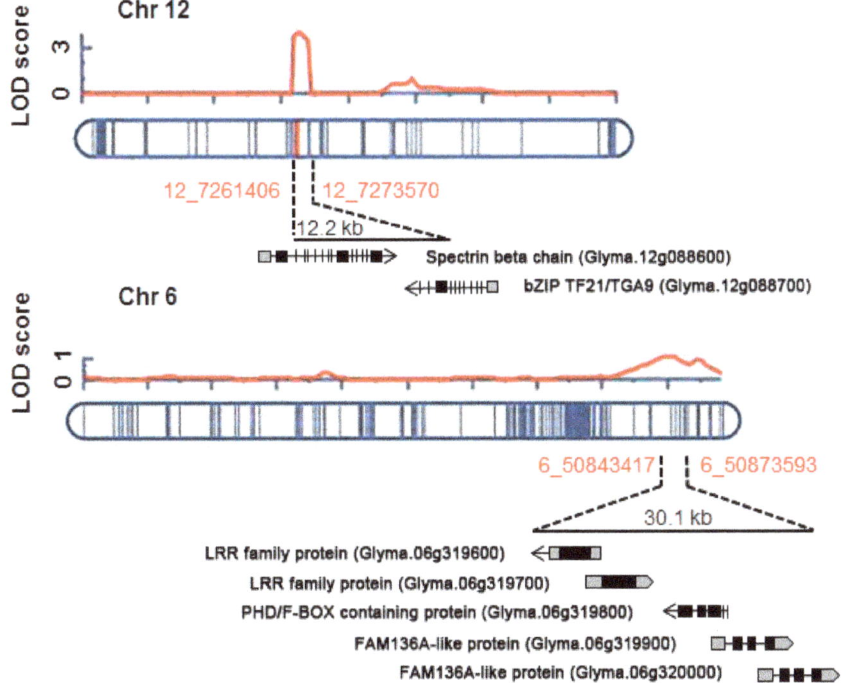

Figure 2. LOD peaks and chromosomal locations of two QTLs *UVBR12-1* and *UVBR6-1* for resistance to high-intensity UV-B irradiation on chromosome (Chr) 12 and Chr 6. Red curves present LOD score distribution of detected QTLs on Chr 12 and Chr 6. These loci contain two and five protein-coding genes, respectively.

2.4. SNPs and Gene Expression Differences in the QTLs in the Mapping Parents

To identify candidate genes in the QTLs associated with UV-B resistance, we investigated SNPs in two and five genes within the two UV-B resistance QTLs *UVBR12-1* and *UVBR6-1*, respectively, by a sequence analysis of the parental genotypes Cheongja 3 and Buseok (Table). In *UVBR12-1*, the two genes *SPTBN* (Glyma.12G088600) and *bZIP TF21/TGA9* (Glyma.12G088700) had nine and 25 SNPs between Cheongja 3 and Buseok, respectively (Table S4). In *SPTBN* (Glyma.12G088600), we detected one and five SNPs in the up- and downstream regions, respectively. We found three additional SNPs in the genic regions of *SPTBN*, one of which was a synonymous SNP in an exon (Figure 3). Among 25 total SNPs in *bZIP TF21/TGA9* (Glyma.12G088700), we found 15 SNPs in the genic region, consisting of three in 3′UTR, 10 in introns and two in exons (Figure 3 and Table S5). Each of the 2 kb up- and downstream regions of *bZIP TF21/TGA9* possessed five SNPs. Two SNPs in the coding sequence of *bZIP TF21/TGA9* were synonymous (Figure 3).

Among five genes within *UVBR6-1*, three genes had nucleotide differences between Cheongja 3 and Buseok (Tables S4 and S5). In Glyma.06g319700, encoding a LRR family protein, we detected two nonsynonymous missense mutations in exons and two SNPs in the 2 kb upstream region (Figure 3). In particular, we discovered amino acid changes from Phe to Leu and Arg to Gly at positions 247 and 254 of Glyma.06g319700, respectively. The two genes Glyma.06g319900 and Glyma.06g320000 (encoding FAM136A-like proteins) had eight and five SNPs in non-coding regions, respectively (Table S5).

Among five genes within *UVBR6-1*, three genes had nucleotide differences between Cheongja 3 and Buseok (Tables S4 and S5). In Glyma.06g319700, encoding a LRR family protein, we detected two nonsynonymous missense mutations in exons and two SNPs in the 2 kb upstream region (Figure 3). In particular, we discovered amino acid changes from Phe to Leu and Arg to Gly at positions 247 and 254 of Glyma.06g319700, respectively. The two genes Glyma.06g319900 and Glyma.06g320000 (encoding FAM136A-like proteins) had eight and five SNPs in non-coding regions, respectively (Table S5).

We further investigated transcriptional differences of *SPTBN* (Glyma.12G088600) and *bZIP TF21/TGA9* (Glyma.12G08870) on *UVBR12-1* between Cheongja 3 and Buseok by qRT-PCR (Figure 4) owing to the lack of protein sequence differences between the mapping parents (Figure 3). Both of the genes were upregulated in Buseok subjected to 6 h of UV-B treatment compared with levels in the control. Relative to UV-B-sensitive Cheongja 3, Buseok showed significantly higher expression levels of the two genes in response to 6 h of UV-B exposure.

2.5. Comparisons of UV-B Stress Resistance QTLs with Known Stress-Related QTLs Based on Synteny

The QTL with the highest LOD, *UVBR12-1*, on Chr 12 was located near *SCN39-4*, a QTL for the reaction to the soybean cyst nematode (SCN; *Heterodera glycines*) linked to Sctt009 (Table 1). The genomic region (Chr12:6,657,383 ... 11,713,748) harboring *UVBR12-1* and *SCN39-4* had a duplicated block on the same chromosome (Chr12:34,009,158 ... 36,919,294), carrying QTLs for drought and flood tolerance (*Drought tolerance 6-4* and *Flood tolerance 7-1*, respectively) linked to Sat 175 (Figure 5). Three homeologous blocks that show syntenic relationships with the two duplicated regions on Chr 12 were also detected on Chr 6, 11, and 13. The homeologous region (Chr11:24,411,218 ... 26,359,492) on Chr 11 with a median Ks value of 0.135 probably resulted from the recent whole genome duplication (WGD) event 13 million years ago [13] and only had collinearity with the beginning of the genomic region (including only *UVBR12-1*) on Chr 12. This syntenic block contained six QTLs for SCN resistance, *SCN 17-1, 18-2, 20-1, 23-1, 24-1,* and *32-2,* associated with Satt583. On Chr 13, the marker Satt554, associated with the QTL *Asian soybean rust 2-3* for resistance to Asian soybean rust, resided in another duplicated region (Chr13:39,150,251 ... 41,460,183) resulting from the recent WGD (median Ks, 0.133). In *UVBR12-1, bZIP TF21/TGA9* was retained in all three duplicated regions but *SPTBN* was only retained in the homeologous block on Chr 11 and was lost in the other blocks (Figure 5). On Chr 6, the remaining duplicated block (Chr06:47,897,433 ... 50,898,694) was mainly conserved with the middle part of the genomic region of *UVBR12-1*. This block with the UV-B resistance QTL *UVBR6-1* included another UV-B-related QTL (*Plant damage UV-B induced 1-2*) as well as biotic stress-related QTLs, including *SCN 17-3* and *20-2*, and *SDS 7-5* for sudden death syndrome resistance (reaction to *Fusarium solani* f. sp. *glycines*), which were associated with Satt371.

The UV-B resistance QTL *UVR14-1* on Chr 14 was mapped near three QTLs for iron efficiency, *Fe-effect 3-2, 9-2,* and *10-3* (Table 1). On Chr 10, *UVR10-1* was co-localized with several QTLs related to biotic as well as abiotic stresses, including *Flood tolerance 4-7, Drought tolerance 6-3, Sclero (Reaction to Sclerotinia sclerotiorum infection) 2-23, 4-10,* and *3-18,* and *Phytoph (Reaction to Phytophthora sojae infection) 5-3*.

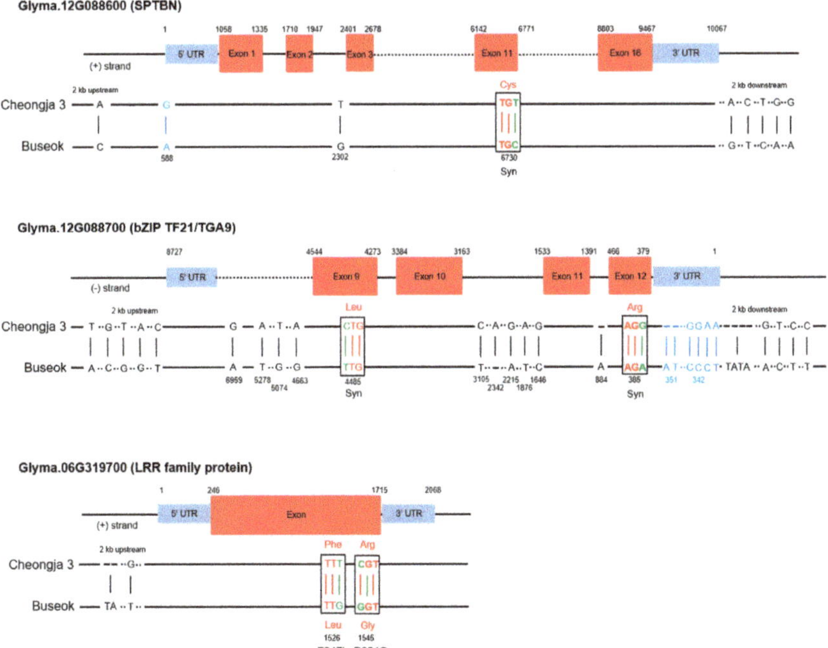

Figure 3. Structures of three genes encoding SPTBN (Glyma.12G088600), bZIP TF21/TGA9 (Glyma.12G088700), and LRR family protein (Glyma.06G319700), and single nucleotide polymorphisms (SNPs) in the mapping parents Cheongja and Buseok. Exons and untranslated regions (UTRs) are indicated by red- and blue-filled boxes, respectively. SNP positions, nucleotide replacements, and amino acid substitutions between Cheongja 3 and Buseok are presented. One and two SNPs in exons of *SPTBN* (Glyma.12G088600) and *bZIP TF21/TGA9* (Glyma.12G088700), respectively, were synonymous. The SNPs T/G and C/G in exons of Glyma.06G319700 (LRR family protein) caused amino acid replacements of Phe to Leu and Arg to Gly at the 247th and 254th residues, respectively.

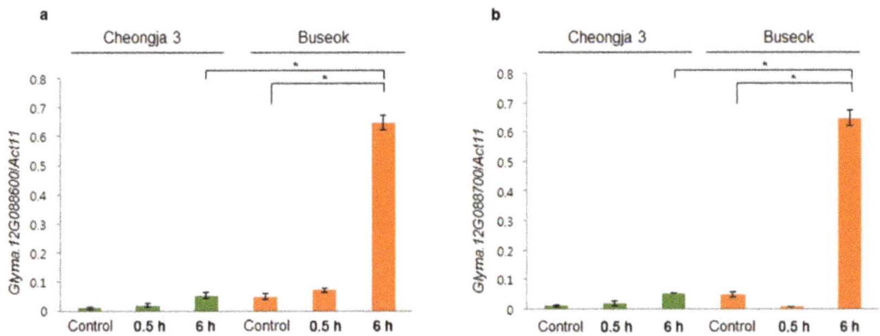

Figure 4. Expression levels of *SPTBN* (Glyma.12G088600) (**a**) and *bZIP TF21/TGA9* (Glyma.12G088700) (**b**) in Cheongja 3 (green) and Buseok (orange). Y-axis represents the relative transcript abundance determined by qRT-PCR. Control, 0.5, and 6 h on the X-axis refer to 0, 0.5, and 6 h UV-B irradiation, respectively. Error bars represent the standard error for three independent replicates. *, significant at $p < 0.05$.

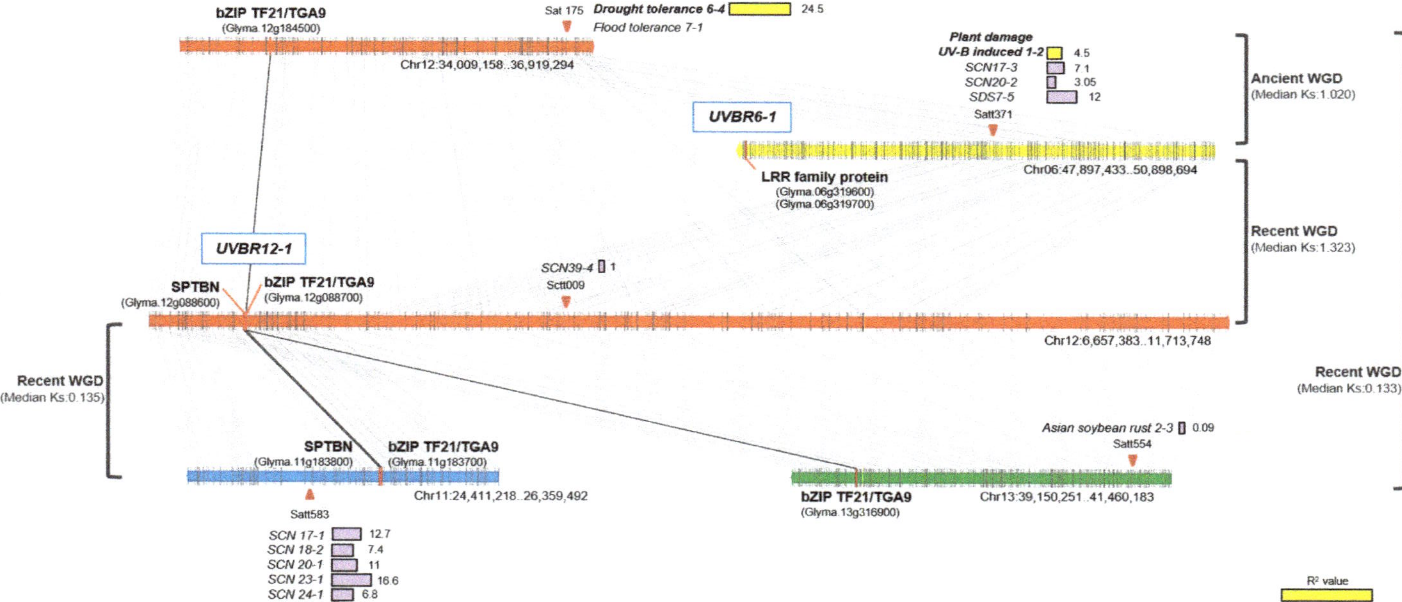

Figure 5. Syntenic conservation of two genomic regions of *UVBR12-1* and *UVBR6-1* and comparisons with other known stress-related QTL regions. The genomic region with *UVBR12-1* (Chr12:6,657,383 … 11,713,748) had a duplicated block on the same Chr 12 (Chr12:34,009,158...36,919,294) and three additional homeologous regions on Chr 6, 11, and 13. Whole genome duplication (WGD) events producing each pair of syntenic blocks are defined as recent or ancient based on median Ks values. Among two genes in *UVBR12-1*, *bZIP TF21/TGA9* was retained in all duplicated regions but *SPTBN* was only retained in the homeologous block on Chr 11 and was lost in the other blocks. Yellow and purple bars indicate phenotypic R2 values of known QTLs related to abiotic and biotic stress, respectively. SPTBN, spectra beta chain, brain; bZIP TF21/TGA9, bZIP transcription factor 21/TGACG motif-binding 9; SCN, Reaction to *Heterodera glycines*; SDS, Reaction to *Fusarium solani* f. sp. *glycines* infection.

3. Discussion

Crop plants under abiotic stresses, such as drought, salinity, and extreme temperatures, induce common cellular signaling mechanisms associated with osmotic stress, which disrupts homeostasis and alters the ion balance in the cell [14–16]. Similarly, high-intensity UV-B radiation above ambient levels induces cellular responses to nonspecific (genotoxic) damage, similar to various stress defense mechanisms but distinct from photomorphogenesis to low-dose UV-B, indicating that it is a complex environmental stressor [17–22]. We previously found that, compared with gene expression in UV-B-sensitive Cheongja 3, UV-B-resistant Buseok exhibits the upregulation of genes involved in phosphatidic acid-dependent signaling pathways as well as several downstream pathways, such as ABA signaling, mitogen-activated protein kinase cascades, and reactive oxygen species overproduction [12]. Therefore, high-density UV-B irradiation is a useful stress treatment for molecular biological and genetic studies of genes involved in extensive resistance to multiple abiotic stresses caused by climate change.

To identify the genetic elements responsible for UV-B resistance by QTL mapping, we developed a RIL population from a cross between UV-B-sensitive Cheongja 3 and UV-B-resistant Buseok. High-intensity UV-B irradiation turns most leaves yellow with red spots in UV-B-sensitive Cheongja 3, resulting in a dramatic loss in aerial dry weight, whereas Buseok retains more healthy leaves [11,12]. Our RIL population showed a wide distribution in the degree of leaf damage caused by UV-B stress (Figure 1). We identified four QTLs associated with UV-B resistance on Chr 12, 14, 10, and 6 (in order of LOD score) in this population, and all LOD scores were less than four (Table 1). Using a different UV-B-resistant soybean genotype, Iksan 10, previous QTL studies based on SSR genotyping and a SoyaSNP assay identified two major loci for UV-B resistance on Chr 19 and 7, respectively [9,10]. These previously reported QTLs do not overlap with our UV-B resistance QTLs. We identified a novel QTL, *UVBR12-1*, with a LOD score of 3.76 on Chr 12, explaining about 10% of phenotypic variation (Table 1). Thus, the genetic determinants of UV-B resistance in the two soybean genotypes Buseok and Iksan 10 are presumed to be different. However, a minor QTL, *UVBR6-1*, on Chr 6 was mapped in the vicinity of a known minor QTL *UV-B induced plant damage 1-2* associated with Satt371 [9] (Table 1 and Figure 5), which was only detected by multiple regression in the previous study [9].

The newly identified UV-B resistance QTLs were co-localized with other known QTLs for resistance to biotic as well as abiotic stresses (Table 1 and Figure 5). These biotic stresses include infections with *H. glycines*, *F. solani* f. sp. *glycines*, *S. sclerotiorum*, and *P. sojae*, and the abiotic stresses include drought, flood, and iron deficiency. Plants subjected to different combinations of multiple stresses show extensive overlap and crosstalk between stress-response signaling pathways, together with specific responses to individual stresses, indicating a high degree of complexity in plant molecular responses to external stresses [23]. Our results indicate that some genetic elements mediating resistance to UV-B stress may be shared with mechanisms underlying responses to other stresses, consistent with previous findings [12,24].

We identified candidate genes controlling resistance to high-intensity UV-B irradiation among two genes on *UVBR12-1* and six genes on *UVBR6-1* (Figures 3 and 4). The gene Glyma.12G088600 on *UVBR12-1* is a homolog of *Arabidopsis SPTBN* (AT5G22450), likely encoding a β-spectrin (βI to βV), which are actin-binding proteins in mammals [25]. Though plants, including *Arabidopsis*, lack spectrins or spectrin-like proteins [26,27], sequences with spectrin repeats and N-terminal calponin homology domains for actin binding are present in the *Arabidopsis* genome [28]. Additionally, spectrins or spectrin-like proteins are localized in plant cellular organelles, such as the Golgi apparatus [29], endoplasmic reticulum [30], and nucleus [31], and in the border plasma membrane of elongating plant cells [32,33]. In mammalian cells, actins bound to β-spectrins and other actin-binding proteins, such as Protein 4.1, Adducin, and Dematin, are connected to the junctional complex at the intracellular side of the plasma membrane [34–36]; these are implicated in signal targeting as well as the maintenance of cell shape and structure [25,36]. β-Spectrins interact with membrane phosphoinositides (PtdIns) via Pleckstrin homology (PH) domains present in a number of proteins involved in cellular signaling [37–39]. Direct evidence for a relationship between the spectrin-based membrane cytoskeleton and plant stress signaling is very limited; however, levels of PtdIns, such as PtdIns4P and PtdIns(4,5)P2, and other phospholipid molecules in plant cells are increased by osmotic stress from salinity and dehydration [40]. Several genes encoding phosphatidylinositol phosphate kinases and phosphatidylinositol-specific phospholipase C involved in phospholipid signaling are upregulated in Buseok after exposure to high-intensity UV-B [12]. In this study, despite a lack of amino acid differences between the mapping parents (Figure 4 and Table S6), elevated *SPTBN* expression in response to UV-B stress was observed in Buseok but not in Cheongja 3 (Figure 4). Further investigations are necessary to characterize the cellular function of SPTBN in soybean using knockout and overexpression mutants.

HY5 (elongated hypocotyl 5) and HYH (HY5 homolog) are bZIP TFs induced by UV-B. They are key components of low-level UV-B photomorphogenic signaling mediated by β-propeller protein and E3 ubiquitin ligase, encoded by *UVR8* (*UV Resistance Locus 8*) and *COP1* (*Constitutively Photomorphogenic 1*), respectively [19,41,42]. The bZIP TF gene Glyma.12g088700 on *UVBR12-1* is an ortholog of *Arabidopsis* bZIP21/TGA9 (AT1G08320), which belongs to Clade IV of the TGACG motif-binding (TGA) protein family and is essential for anther development [43,44]. In a phylogenetic tree of bZIP genes from *G. max* and *A. thaliana*, the bZIP21/TGA9 orthologs Glyma.12g088700 and AT1G08320 belong to a different cluster from that including HY5 (AT5g11260) (Figure S2), corresponding to previous classifications in *A. thaliana* [43]. The expression of Glyma.12g088700 on *UVR12-1* was significantly increased by exposure to high-intensity UV-B in Buseok (Figure 4), but direct evidence for the role of this type of bZIP gene in resistance to abiotic stresses is lacking. bZIP21/TGA9 and TGA10 are only known to be involved in plant pathogen-associated molecular pattern-triggered immunity stimulated by the immunogenic peptide of bacterial flagellin (flg22) [45,46].

In *UVBR6-1*, Glyma.06g319700 showed amino acid differences between the mapping parents; the gene is orthologous to the *Arabidopsis* gene AT1G33590, which encodes a LRR family protein. The LRR motif plays a central role in recognizing different pathogen-associated molecules in the innate host defense of plants and animals [47]. Recent studies have shown that several genes encoding LRR-containing proteins are involved in the abiotic stress response [48], including *LP2* (*LEAF PANICLE 2*), *TaPRK2697* (*Triticum aestivum PROTEIN OF RECEPTOR KINASES 26697*), *AtPXL1* (*Arabidopsis thaliana PHLOEM INTERCALATED WITH XYLEM-LIKE 1*), and *LRR-RLK-VIII* (*LEUCINE-RICH REPEAT RECEPTOR-LIKE KINASE VIII*), which are upregulated by drought, salt, cold, and toxic metals, respectively, in diverse plant species [49–52].

Three candidate genes responsible for UV-B resistance in soybean were identified from two QTLs, *UVBR12-1* and *UVBR6-1*. These genes encode SPTBN, bZIP TF21/TGA9, and a LRR family protein and are likely involved in stress defense signaling; they should be experimentally verified using overexpression or knockout mutants. Further functional studies will improve our understanding of UV-B stress defense mechanisms. Furthermore, our results provide powerful genetic tools for more efficient and precise breeding programs aimed at the development of highly adaptable soybean cultivars under various abiotic stresses caused by global climate change.

4. Materials and Methods

4.1. Plant Materials and DNA Extraction

Two soybean genotypes were used as parents to develop a RIL population for genetic map construction and the QTL analysis. Buseok, a paternal line, is a UV-B-resistant genotype, and Cheongja 3, a maternal line, is a UV-B-sensitive genotype [11,12]. Artificial crossing was performed in summer 2012, and 176 F_6 RILs were generated from F_2 seeds using the single seed descent method from winter 2012 to spring 2016. Healthy young leaves from two parental genotypes and RILs were collected, and high-quality genomic DNA was extracted using the GeneAll® Exgene Plant SV Kit (GeneAll Biotechnology, Seoul, Korea). DNA quality was assessed by the 260/280 nm ratio using a Nanodrop 3000 spectrometer (Thermo Scientific, Wilmington, DE, USA). DNA was quantified using the Invitrogen Quant-iT PicoGreen® dsDNA Assay Kit (Life Technologies, Burlington, ON, Canada) and adjusted to 20 ng/μL.

4.2. UV-B Treatment and Phenotypic Evaluation

Soybean seeds of the mapping parents and their RILs were planted (three seeds per pot) in 3 L pots containing a 1:1 mixture of desalinated sand and commercial potting soil (Baroker, Seoul Bio Co., Seoul, Korea) in a greenhouse at the experimental farm of Seoul National University, Suwon, Korea in August 2016. Supplemental UV-B irradiation was used to treat soybean plants at the V3 growth stage two to three weeks after emergence according to a previous study [11]. Supplemental UV-B radiation was applied using G40T10E UV-B lamps (Sankyo Denki, Nagano, Japan) with a mean UV-B intensity of 5.68 ± 0.4 Wm^{-2} at the plant level under the lamps. The plants were exposed to high-intensity UV-B stress for 1 h every day at 11:00 am from 17 August to 20 August. The intensity of 1 h of UV-B irradiation was equivalent to two times 11.5 kJ/m^2, the daily soybean UV-B biological effective dose [12,53]. Three sets of first and second trifoliate leaves above unifoliate leaves from three plants after UV-B treatment were collected with three replications per line. To investigate leaf color changes caused by high-intensity UV-B exposure, the collected leaf samples were scanned with an EPSON Perfection V33 scanner (Epson Inc., Long Beach, CA, USA) and scanned images were analyzed using the WinDIAS 3 Leaf Image Analysis System (DELTA-T DEVICES LTD., Cambridge, UK). The degree of damage was scored on a scale of 1–10 (where 1 = 0%–10%, 2 = 11%–20%, and up to 10 = 91%–100%; damaged area/healthy area × 100 (%)) and these scores was used in QTL mapping for UV-B resistance. Phenotypic distribution was tested for deviation from normality with the Shapiro–Wilk test.

4.3. Genotyping-by-Sequencing

A GBS technique was used to detect SNPs to genotype the RIL population. A GBS library was constructed following the protocol described by [54]. DNAs from the mapping parents and 176 RILs were digested individually with ApeKI, which recognizes a degenerate 5 bp sequence (GCWGC, where W is either A or T). Barcoded adapters were ligated with digested DNA fragments (Table S6), and ligated DNA was amplified with appropriate primers. Then, two separate libraries were constructed by pooling amplified DNA samples of 88 RILs for each library. Single-end sequencing of the GBS libraries was performed on two lanes of an Illumina HiSeq2000 instrument (Illumina Inc., San Diego, CA, USA).

4.4. Sequence Analysis, Genetic Map Construction, and QTL Analysis

Raw GBS reads were mapped against the *G. max* reference genome (Wm82.a2) downloaded from Phytozome (https://phytozome.jgi.doe.gov/pz/portal.html) using Bowtie v2.1 after sequence quality control, such as the removal of barcode, adapter, and ApeKI overhang sequences as well as reads with Phred scores of <15 for at least 80% of bases [55,56]. SNPs were called using in-house python scripts with the following filtering criteria: read depth ≥3, Q value ≥30, and missing error rate ≤10%. All SNPs were deposited in The European Variation Archive (https://www.ebi.ac.uk/eva,PRJEB32685). Using identified SNP markers, a genetic map was constructed with JoinMap 4.1 (https://www.kyazma.nl/index.php/mc.JoinMap). SNP markers were grouped using the Kosambi mapping function, and segregation distortion of individual markers was calculated using the χ^2 test in JoinMap 4.1. The SNP genotyping data and degrees of leaf damage by UV-B treatment for 176 RILs of Cheongja 3 × Buseok were evaluated using QTL IciMapping v.4.1.0.0, and QTLs for resistance to high-intensity UV-B were identified by inclusive composite interval mapping (https://www.integratedbreeding.net/386/breeding-services/more-software-tools/icimapping). To determine the statistically significant threshold for the LOD score, 1000 permutation test was performed at the 5% significance level.

4.5. SNP Survey and qRT-PCR Analysis of Genes in the QTLs for UV-B Stress Resistance

To investigate SNPs in genes located in the QTLs for UV-B resistance in Cheongja 3 and Buseok, whole genome sequences of these genotypes reported in a previous study were used [11]; the sequences are available from the website of the Crop Genomics Laboratory at Seoul National University (http://plantgenomics.snu.ac.kr/). To detect high-confidence SNPs, the following cut-off values were used: minimum read depth of 5, maximum read depth of 20, and Phred-scaled probability score above 30. A maximum read depth of 20 was applied to remove false positive SNP calls that may result from duplicated or repetitive sequences. The SNP positions were determined based on the Glyma 2.0 gene models (Wm82.a2.v1). To compare gene expression levels within the QTL *UVBR12-1* for UV-B resistance between Cheongja 3 and Buseok, gene-specific primers were designed for qRT-PCR using Primer3 (http://bioinfo.ut.ee/primer3-0.4.0/) (Table S7). Total RNA from each sample for control, 0.5 h, and 6 h UV-B treatments in Cheongja 3 and Buseok was used to synthesize cDNA using a Bio-Rad iScript™ cDNA Synthesis Kit (Hercules, CA, USA). UV-B treatments for qRT-PCR analysis followed as described in the previous RNA-seq study [12]: 0.5 h exposure (equivalent to 11.5 KJ/m^2 soybean UV-B$_{BE}$) induced photomorphogenic (non-damaging) cellular response to low-dose UV-B radiation, while 6 h exposure (12-times higher than daily UV-B$_{BE}$) resulted in nonspecific (genotoxic) damage induced by abiotic stress. qRT-PCR was performed using the synthesized cDNA as a template and a Bio-Rad iQ™ SYBR Green Supermix Kit on a LightCycler® 480 (Roche Diagnostics, Laval, QC, Canada). The qRT-PCR mixture in a total volume of 20 µL contained 100 ng of cDNA, each primer at 300 µM, 8 µL of sterile water, and 10 µL of Bio-Rad iQ™ SYBR Green Supermix. The amplification conditions were as follows: 5 min of denaturation at 95 °C followed by 40 cycles of 95 °C for 10 s and 60 °C for 1 min. The samples were analyzed in triplicate, and actin was used as a reference gene for the normalization of target gene expression in soybean. Data were analyzed based on the stable expression level of the reference gene according to the method of Livak and Schmittgen [57].

4.6. Co-Localization of Stress-Related QTLs with UV-B Resistance QTLs

Abiotic and biotic stress-related QTLs for soybean were obtained from the SoyBase website (http://soybase.org/). The chromosomal positions of QTLs on soybean chromosomes were determined using marker information from version 4.0 of the soybean map from SoyBase [58]. Syntenic blocks in the soybean reference genome sequence were analyzed using MCScanX with default parameters [59], and syntenic blocks overlapping with genomic regions of approximately 2 to 5 Mbp encompassing SNP markers associated with UV-B resistance were identified.

Supplementary Materials: Supplementary materials can be found at http://www.mdpi.com/1422-0067/20/13/3287/s1.

Author Contributions: Conceptualization, M.Y.Y. and M.Y.K.; methodology, M.Y.Y. and M.Y.K.; software, M.Y.Y. and T.L.; validation, M.Y.Y.; formal analysis, M.Y.Y.; investigation, M.Y.Y.; resources, M.Y.Y.; data curation, J.H.; writing—original draft preparation, Y.M.Y. and M.Y.K.; writing—review and editing, M.Y.K., J.H., K.D.K and S.-H.L.; visualization, M.Y.Y., T.L. and M.Y.K.; supervision, S.-H.L.; project administration, S.-H.L.; funding acquisition, S.-H.L.

Funding: This work was supported by the Next Generation BioGreen 21 Program (Code No. PJ01322401), Rural Development Administration, Republic of Korea

Conflicts of Interest: The authors declare no conflict of interest.

Abbreviations

UV-B	Ultraviolet-B
QTL	Quantitative trait locus
RIL	Recombinant inbred line
LOD	Logarithm of odds
GBS	Genotyping-by-sequencing
SNP	Single nucleotide polymorphism
WGD	Whole genome duplication

SCN	Soybean cyst nematode
SPTBN	Spectrin beta chain, brain
bZIP TF21/TGA9	bZIP transcription factor21/TGACC motif-binding 9
LRR	Leucine-rich repeat

References

1. World Meteorological Organization, 2008. WMO Statement on the Status of the Global Climate in 2007. WMO-No. 1031. Available online: https://library.wmo.int/doc_num.php?explnum_id=3457 (accessed on 20 February 2019).
2. Petrescu, R.V.; Aversa, R.; Apicella, A.; Petrescu, F.I. NASA sees first in 2018 the direct proof of ozone hole recovery. *J. Aircr. Spacecr. Technolog.* **2018**, *2*, 53–64. [CrossRef]
3. Rozema, J.; van de Staaij, J.; Björn, L.O.; Caldwell, M. UV-B as an environmental factor in plant life: Stress and regulation. *Trends Ecol. Evolut.* **1997**, *12*, 22–28. [CrossRef]
4. Frohnmeyer, H.; Staiger, D. Ultraviolet-B radiation-mediated responses in plants. *Balancing damage and protection. Plant Physiol.* **2003**, *133*, 1420–1428. [CrossRef] [PubMed]
5. UNEP. 2010 Annual Report. 2011. Available online: https://wedocs.unep.org/handle/20.500.11822/7915 (accessed on 20 February 2019).
6. Murali, N.; Teramura, A.H.; Randall, S.K. Response differences between two soybean cultivars with contrasting UV-B radiation sensitivities. *Photochem. Photobiol.* **1988**, *48*, 653–657. [CrossRef]
7. Teramura, A.H.; Sullivan, J.H.; Lydon, J. Effects of UV-B radiation on soybean yield and seed quality: A 6-year field study. *Physiol. Plantarum.* **1990**, *80*, 5–11. [CrossRef]
8. Baroniya, S.S.; Kataria, S.; Pandey, G.P.; Guruprasad, K.N. Intraspecific variation in sensitivity to ambient ultraviolet-B radiation in growth and yield characteristics of eight soybean cultivars grown under field conditions. *Braz. J. Plant Physiol.* **2011**, *23*, 197–202. [CrossRef]
9. Shim, H.C.; Ha, B.K.; Yoo, M.; Kang, S.T. Detection of quantitative trait loci controlling UV-B resistance in soybean. *Euphytica* **2015**, *202*, 109–118. [CrossRef]
10. Lee, J.S.; Kim, S.; Ha, B.K.; Kang, S.T. Positional mapping and identification of novel quantitative trait locus responsible for UV-B radiation tolerance in soybean (*Glycine max* [L.] Merr.). *Mol. Breed.* **2016**, *36*, 1–10. [CrossRef]
11. Kim, K.D.; Yun, M.Y.; Shin, J.H.; Kang, Y.J.; Kim, M.Y.; Lee, S.H. Underlying genetic variation in the response of cultivated and wild soybean to enhanced ultraviolet-B radiation. *Euphytica* **2015**, *202*, 207–217. [CrossRef]
12. Yoon, M.Y.; Kim, M.Y.; Shim, S.; Kim, K.D.; Ha, J.; Shin, J.H.; Lee, S.H. Transcriptomic profiling of soybean in response to high-intensity UV-B irradiation reveals stress defense signaling. *Front. Plant Sci.* **2016**, *7*, 1917. [CrossRef]
13. Schmutz, J.; Cannon, S.B.; Schlueter, J.; Ma, J.; Mitros, T.; Nelson, W.; Hyten, D.L.; Song, Q.; Thelen, J.J.; Cheng, J.; et al. Genome sequence of the palaeopolyploid soybean. *Nature* **2010**, *463*, 178–183. [CrossRef] [PubMed]
14. Wang, W.; Vinocur, B.; Altman, A. Plant responses to drought, salinity and extreme temperatures: Towards genetic engineering for stress tolerance. *Planta* **2003**, *218*, 1–14. [CrossRef] [PubMed]
15. Shinozaki, K.; Yamaguchi-Shinozaki, K. Molecular responses to dehydration and low temperature: Differences and cross-talk between two stress signaling pathways. *Curr. Opin. Plant Biol.* **2000**, *3*, 217–223. [CrossRef]
16. Knight, H.; Knight, M.R. Abiotic stress signaling pathways: Specificity and cross-talk. *Trends Plant Sci.* **2001**, *6*, 262–267. [CrossRef]
17. Jordan, B.R. Review: Molecular response of plant cells to UV-B stress. *Funct. Plant Biol.* **2002**, *29*, 909–916. [CrossRef]
18. Paul, N.D.; Gwynn-Jones, D. Ecological roles of solar UV radiation: Towards an integrated approach. *Trends Ecol. Evolut.* **2003**, *18*, 48–55. [CrossRef]
19. Jenkins, G.I. Signal transduction in responses to UV-B radiation. *Annu. Rev. Plant Biol.* **2009**, *60*, 407–431. [CrossRef]
20. Heijde, M.; Ulm, R. Reversion of the Arabidopsis UV-B photoreceptor UVR8 to the homodimeric ground state. *Proc. Natl. Acad. Sci. USA* **2013**, *110*, 1113–1118. [CrossRef]

21. Müller-Xing, R.; Xing, Q.; Goodrich, J. Footprints of the sun: Memory of UV and light stress in plants. *Front. Plant Sci.* **2014**, *5*, 474. [CrossRef]

22. Parihar, P.; Singh, S.; Singh, R.; Singh, V.P.; Prasad, S.M. Changing scenario in plant UV-B research: UV-B from a generic stressor to a specific regulator. *J. Photochem. Photobiol. B Biol.* **2015**, *153*, 334–343. [CrossRef]

23. Suzuki, N.; Rivero, R.M.; Shulaev, V.; Blumwald, E.; Mittler, R. Abiotic and biotic stress combinations. *New Phytol.* **2014**, *203*, 32–43. [CrossRef] [PubMed]

24. Yoon, M.Y.; Kim, M.Y.; Lee, J.; Lee, T.; Kim, K.H.; Ha, J.; Kim, Y.H.; Lee, S.H. Transcriptomic profiling of soybean in response to UV-B and Xanthomonas axonopodis treatment reveals shared gene components in stress defense pathways. *Genes Genom.* **2017**, *39*, 225–236. [CrossRef]

25. Machnicka, B.; Czogalla, A.; Hryniewicz-Jankowska, A.; Bogusławska, D.M.; Grochowalska, R.; Heger, E.; Sikorski, A.F. Spectrins: A structural platform for stabilization and activation of membrane channels, receptors and transporters. *Biochim. Biophys. Acta Biomembr* **2014**, *1838*, 620–634. [CrossRef] [PubMed]

26. Bennett, V.; Baines, A.J. Spectrin and ankyrin-based pathways: Metazoan inventions for integrating cells into tissues. *Physiol. Rev.* **2001**, *81*, 1353–1392. [CrossRef] [PubMed]

27. Drøbak, B.K.; Franklin-Tong, V.E.; Staiger, C.J. The role of the actin cytoskeleton in plant cell signaling. *New Phytol.* **2004**, *163*, 13–30. [CrossRef]

28. Young, K.G.; Kothary, R. Spectrin repeat proteins in the nucleus. *BioEssays* **2005**, *2 7*, 144–152. [CrossRef]

29. Holzinger, A.; De Ruijter, N.; Emons, A.M.; LüTz-Meindl, U. Spectrin-like proteins in green algae (Desmidiaceae). *Cell Biol. Int.* **1999**, *23*, 335–344. [CrossRef]

30. Braun, M. Association of spectrin-like proteins with the actin-organized aggregate of endoplasmic reticulum in the spitzenkörper of gravitropically tip-growing plant cells. *Plant Physiol.* **2001**, *125*, 1611–1619. [CrossRef]

31. Pérez-Munive, C.; Moreno Diaz de la Espina, S. Nuclear spectrin-like proteins are structural actin-binding proteins in plants. *Biol. Cell* **2011**, *103*, 145–157. [CrossRef]

32. Michaud, D.; Guillet, G.; Roger, P.A.; Charest, P.M. Identification of a 220-kDa membrane-associated plant cell protein immunologically related to human β-spectrin. *FEBS Lett.* **1991**, *294*, 77–80. [CrossRef]

33. De Ruijter, N.C.A.; Emons, A.M.C. Immunodetection of spectrin in plant cells. *Cell Biol. Int.* **1993**, *17*, 169–182. [CrossRef]

34. Khan, A.A.; Hanada, T.; Mohseni, M.; Jeong, J.J.; Zeng, L.; Gaetani, M.; Li, D.; Reed, B.C.; Speicher, D.W.; Chishti, A.H. Dematin and adducing provide a novel link between the spectrin cytoskeleton and human erythrocyte membrane by directly interacting with glucose transporter-1. *J. Biol. Chem.* **2008**, *283*, 14600–14609. [CrossRef] [PubMed]

35. Salomao, M.; Zhang, X.; Yang, Y.; Lee, S.; Hartwig, J.H.; Chasis, J.A.; Mohandas, N.; An, X. Protein 4.1R-dependent multiprotein complex: New insights into the structural organization of the red blood cell membrane. *Proc. Natl. Acad. Sci. USA* **2008**, *105*, 8026–8031. [CrossRef] [PubMed]

36. Machnicka, B.; Grochowalska, R.; Bogusławska, D.M.; Sikorski, A.F.; Lecomte, M.C. Spectrin-based skeleton as an actor in cell signaling. *Cell Mol. Life Sci.* **2012**, *69*, 191–201. [CrossRef] [PubMed]

37. Haslam, R.J.; Koide, H.B.; Hemmings, B.A. Pleckstrin domain homology. *Nature* **1993**, *363*, 309–310. [CrossRef] [PubMed]

38. Musacchio, A.; Gibson, T.; Rice, P.; Thompson, J.; Saraste, M. The PH domain: A common piece in the structural patchwork of signaling proteins. *Trends Biochem. Sci.* **1993**, *18*, 343–348. [CrossRef]

39. Macias, M.J.; Musacchio, A.; Ponstingl, H.; Nilges, M.; Saraste, M.; Oschkinat, H. Structure of the pleckstrin homology domain from beta-spectrin. *Nature* **1994**, *369*, 675–677. [CrossRef] [PubMed]

40. DeWald, D.B.; Torabinejad, J.; Jones, C.A.; Shope, J.C.; Cangelosi, A.R.; Thompson, J.E.; Prestwich, G.D.; Hama, H. Rapid accumulation of phosphatidylinositol 4,5-bisphosphate and inositol 1,4,5-trisphosphate correlates with calcium mobilization in salt-stressed Arabidopsis. *Plant Physiol.* **2001**, *126*, 759–769. [CrossRef]

41. Ulm, R.; Baumann, A.; Oravecz, A.; Máté, Z.; Ádám, É.; Oakeley, E.J.; Nagy, F. Genome-wide analysis of gene expression reveals function of the bZIP transcription factor HY5 in the UV-B response of Arabidopsis. *Proc. Natl. Acad. Sci. USA* **2004**, *101*, 1397–1402. [CrossRef]

42. Favory, J.J.; Stec, A.; Gruber, H.; Rizzini, L.; Oravecz, A.; Funk, M.; Seidlitz, H.K. Interaction of COP1 and UVR8 regulates UV-B-induced photomorphogenesis and stress acclimation in Arabidopsis. *EMBO J.* **2009**, *28*, 591–601. [CrossRef]

43. Jakoby, M.; Weisshaar, B.; Dröge-Laser, W.; Vicente-Carbajosa, J.; Tiedemann, J.; Kroj, T.; Parcy, F. bZIP transcription factors in Arabidopsis. *Trends Plant Sci.* **2002**, *7*, 106–111. [CrossRef]

44. Murmu, J.; Bush, M.J.; DeLong, C.; Li, S.; Xu, M.; Khan, M.; Hepworth, S.R. Arabidopsis basic leucine-zipper transcription factors TGA9 and TGA10 interact with floral glutaredoxins ROXY1 and ROXY2 and are redundantly required for anther development. *Plant Physiol.* **2010**, *154*, 1492–1504. [CrossRef] [PubMed]

45. Shigeoka, S.; Maruta, T. Cellular redox regulation, signaling, and stress response in plants. *Biosci. Biotechnol. Biochem.* **2014**, *78*, 1457–1470. [CrossRef] [PubMed]

46. Noshi, M.; Mori, D.; Tanabe, N.; Maruta, T.; Shigeoka, S. Arabidopsis clade IV TGA transcription factors, TGA10 and TGA9, are involved in ROS-mediated responses to bacterial PAMP flg22. *Plant Sci.* **2016**, *252*, 12–21. [CrossRef] [PubMed]

47. Gunawardena, H.P.; Huang, Y.; Kenjale, R.; Wang, H.; Xie, L.; Chen, X. Unambiguous characterization of site-specific phosphorylation of leucine-rich repeat Fli-I-interacting protein 2 (LRRFIP2) in toll-like receptor 4 (TLR4)-mediated signaling. *J. Biol. Chem.* **2011**, *286*, 10897–10910. [CrossRef] [PubMed]

48. Ye, Y.; Ding, Y.; Jiang, Q.; Wang, F.; Sun, J.; Zhu, C. The role of receptor-like protein kinases (RLKs) in abiotic stress response in plants. *Plant Cell Rep.* **2017**, *36*, 235–242. [CrossRef]

49. Fu, S.F.; Chen, P.Y.; Nguyen, Q.T.T.; Huang, L.Y.; Zeng, G.R.; Huang, T.L.; Lin, C.Y.; Huang, H.J. Transcriptome profiling of genes and pathways associated with arsenic toxicity and tolerance in Arabidopsis. *BMC Plant Biol.* **2014**, *14*, 94. [CrossRef]

50. Chang, G.J.; Hwang, S.G.; Yong, C.P.; Hyeon, M.P.; Dong, S.K.; Duck, H.P.; Cheol, S.J. Molecular characterization of the cold- and heat-induced Arabidopsis PXL1 gene and its potential role in transduction pathways under temperature fluctuations. *J. Plant Physiol.* **2015**, *176*, 138–146. [CrossRef]

51. Ma, X.L.; Cui, W.N.; Zhao, Q.; Zhao, J.; Hou, X.N.; Li, D.Y.; Chen, Z.L.; Shen, Y.Z.; Huang, Z.J. Functional study of a salt-inducible TaSR gene in Triticum aestivum. *Physiol. Plant.* **2015**, *156*, 40–53. [CrossRef]

52. Wu, F.; Sheng, P.; Tan, J.; Chen, X.L.; Lu, G.W.; Ma, W.W.; Heng, Y.Q.; Lin, Q.B.; Zhu, S.S.; Wang, J.L.; et al. Plasma membrane receptor-like kinase leaf panicle 2 acts downstream of the drought and salt tolerance transcription factor to regulate drought sensitivity in rice. *J. Exp. Bot.* **2015**, *66*, 271–281. [CrossRef]

53. Caldwell, M.M. Solar UV irradiation and the growth and development of higher plants. *Photophysiology* **1971**, *6*, 131–177. [CrossRef]

54. Elshire, R.J.; Glaubitz, J.C.; Sun, Q.; Poland, J.A.; Kawamoto, K.; Buckler, E.S.; Mitchellet, S.E. A robust, simple genotyping-by-sequencing (GBS) approach for high diversity species. *PLoS ONE* **2011**, *6*, e19379. [CrossRef] [PubMed]

55. Langmead, B.; Salzberg, S.L. Fast gapped-read alignment with Bowtie 2. *Nat. Methods* **2012**, *9*, 357–359. [CrossRef] [PubMed]

56. Glaubitz, J.C.; Casstevens, T.M.; Lu, F.; Harriman, J.; Elshire, R.J.; Sun, Q.; Buckler, E.S. TASSEL-GBS: A high capacity genotyping by sequencing analysis pipeline. *PLoS ONE* **2014**, *9*, e90346. [CrossRef] [PubMed]

57. Livak, K.J.; Schmittgen, T.D. Analysis of relative gene expression data using real-time quantitative PCR and the $2^{-\Delta\Delta C_T}$ method. *Methods* **2001**, *25*, 402–408. [CrossRef] [PubMed]

58. Grant, D.; Nelson, R.T.; Cannon, S.B.; Shoemaker, R.C. SoyBase, the USDA-ARS soybean genetics and genomics database. *Nucleic Acids Res.* **2010**, *38*, D843–D846. [CrossRef] [PubMed]

59. Wang, Y.; Tang, H.; DeBarry, J.D.; Tan, X.; Li, J.; Wang, X.; Lee, T.; Jin, H.; Marler, B.; Hui, G.; et al. MCScanX: A toolkit for detection and evolutionary analysis of gene synteny and collinearity. *Nucleic. Acids Res.* **2012**, *40*, e49. [CrossRef]

International Journal of
Molecular Sciences

Article

Identification and Expression Analysis of GRAS Transcription Factors to Elucidate Candidate Genes Related to Stolons, Fruit Ripening and Abiotic Stresses in Woodland Strawberry (*Fragaria vesca*)

Hong Chen [1], Huihui Li [2], Xiaoqing Lu [1], Longzheng Chen [3], Jing Liu [3,*] and Han Wu [4,*]

[1] Jiangsu Key Laboratory for the Research and Utilization of Plant Resources, Institute of Botany, Jiangsu Province and Chinese Academy of Sciences, Nanjing 210014, China
[2] Fuyang Academy of Agricultural Sciences, Fuyang 236065, China
[3] Jiangsu Key Laboratory for Horticultural Crop Genetic Improvement, Institute of Vegetable Crops, Jiangsu Academy of Agricultural Sciences, Nanjing 210014, China
[4] State Key Laboratory of Crop Genetics and Germplasm Enhancement, College of Horticulture, Nanjing Agricultural University, Nanjing 210095, China
* Correspondence: wuhan@njau.edu.cn (H.W.); liujing200809@163.com (J.L.)

Received: 25 August 2019; Accepted: 14 September 2019; Published: 17 September 2019

Abstract: The cultivated strawberry (*Fragaria × ananassa*), an allo-octoploid with non-climacteric fleshy fruits, is a popular Rosaceae horticultural crop worldwide that is mainly propagated via stolons during cultivation. Woodland strawberry (*Fragaria vesca*), one of the four diploid progenitor species of cultivated strawberry, is widely used as a model plant in the study of Rosaceae fruit trees, non-climacteric fruits and stolons. One GRAS transcription factor has been shown to regulate stolon formation; the other GRAS proteins in woodland strawberry remain unknown. In this study, we identified 54 FveGRAS proteins in woodland strawberry, and divided them into 14 subfamilies. Conserved motif analysis revealed that the motif composition of FveGRAS proteins was conserved within each subfamily, but diverged widely among subfamilies. We found 56 orthologous pairs of GRAS proteins between woodland strawberry and *Arabidopsis thaliana*, 47 orthologous pairs between woodland strawberry and rice and 92 paralogous pairs within woodland strawberry. The expression patterns of *FveGRAS* genes in various organs and tissues, and changes therein under cold, heat and GA3 treatments, were characterized using transcriptomic analysis. The results showed that 34 *FveGRAS* genes were expressed with different degrees in at least four organs, including stolons; only a few genes displayed organ-specific expression. The expression levels of 16 genes decreased, while that of four genes increased during fruit ripening; *FveGRAS54* showed the largest increase in expression. Under cold, heat and GA3 treatments, around half of the *FveGRAS* genes displayed increased or decreased expression to some extent, suggesting differing functions of these *FveGRAS* genes in the responses to cold, heat and GAs. This study provides insight into the potential functions of *FveGRAS* genes in woodland strawberry. A few *FveGRAS* genes were identified as candidate genes for further study, in terms of their functions in stolon formation, fruit ripening and abiotic stresses.

Keywords: genome-wide identification; expression analysis; GRAS transcription factors; stolons; non-climacteric fruit ripening; gibberellins; abiotic stresses; woodland strawberry

---

1. Introduction

Cultivated strawberry (*Fragaria × ananassa*) is a popular Rosaceae horticultural crop worldwide, the fleshy fruit of cultivated strawberry is a non-climacteric fruit [1]. Cultivated strawberry mainly propagates via stolons (also called runners) in agricultural production. However, the complexity of the cultivated strawberry genome (about 805 Mb) has made molecular, genetic and functional studies difficult. Cultivated strawberry is derived from the hybridization of two wild octoploid species (*F.virginiana* and *F.chiloensis*), both of which descended from the merger of four diploid progenitor species (*F. vesca*, *F.iinumae*, *F.viridis* and *F.nipponica*) [2]. The genome of *F. vesca* (woodland strawberry) has been published; its small genome (240 Mb) and ease of genetic transformation make it useful as a model plant for the study of Rosaceae fruit trees, non-climacteric fruits and stolons [3]. To date, only a few genes have been demonstrated to regulate stolon formation in woodland strawberry, such as *CONSTANS(CO)* [4], *SUPPRESSOR OF OVEREXPRESSION OF CONSTANS1(SOC1)* [5], *Gibberellin 20-oxidase 4(FveGA20ox4)* [6] and *DELLA* [7,8]. *DELLA* is the only functionally analyzed *GRAS* gene in woodland strawberry [7,8]. Other *GRAS* genes in woodland strawberry remain unidentified and uncharacterized.

The GRAS family was initially named for the first three functionally identified members in *Arabidopsis thaliana*, GIBBERELLIN-INSENSITIVE (GAI), REPRESSOR OF GA1-3 (RGA) and SCARECROW (SCR) [9]. Most GRAS proteins contain a conserved GRAS domain at the C-terminus and a variable region at the N-terminus, but a small number have their GRAS domain at the N-terminus, or have two GRAS domains. In addition, some GRAS proteins, including DELLA, contain a conserved GRAS domain in the C-terminal region and other conserved domains in the N-terminal region [10]. Generally, the conserved GRAS domain contains five ordered conserved motifs: LHR I (leucine heptad repeat I), VHIID, LHR II (leucine heptad repeat II), PFYRE, SAW. LRI, VHIID and LRII, individually and in combination as LRI-VHIID-LRII, are putative DNA-binding sites or protein–protein interaction regions of GRAS proteins, while the functions of the PFYRE and SAW motifs remain unclear [10–12].

Evolutionary analyses have suggested that GRAS proteins could have originated in bacteria and been transferred into land plants by lateral transfer, and then undergone differentiation and expansion in higher plants [10,13]. Using genome-wide analysis, GRAS family proteins have been identified in numerous plants. According to functional analysis of *GRAS* genes and phylogenetic trees constructed for the model plants *Arabidopsis thaliana* and rice, GRAS proteins can be initially divided into eight subfamilies: LlSCL, PAT1, SCL3, DELLA, SCR, SHR, LS and HAM [14]. However, recent studies have divided GRAS proteins into 10–17 subfamilies based on more detailed data, and about 33–184 GRAS proteins have been identified in various plant species [10,15].

GRAS proteins play important roles in regulating a wide range of developmental and signal transduction processes in higher plants, such as the development of root, shoot, leaf, shoot apical meristem, axillary meristem, flower, embryo and seed and the signaling of gibberellins (GAs), light and stresses [9,16,17]. A large number of *GRAS* genes have been identified and functionally analyzed in plants, especially in *Arabidopsis thaliana* and rice. For example, proteins in the DELLA subfamily are well-known GRAS proteins, and some members have been demonstrated to negatively regulate GA signal transduction [18–20]. DELLA proteins also participate in the signaling pathways of auxin, brassinosteroids (BRs), cytokinins (CKs), abscisic acid (ABA), jasmonate (JA) strigolactones (SLs) and ethylene. In fact, DELLA proteins act as a major hub in multiple hormone signaling networks, thereby regulating a variety of developmental processes in plants related to these hormones [20,21]. In addition, AtSCR (belonging to the SCR subfamily) and AtSHR (belonging to the SHR subfamily) in *Arabidopsis thaliana* are involved in various root and shoot development stages, as well as in high salinity and osmotic stress [16]; SCL3 subfamily protein AtSCL3 in *Arabidopsis thaliana* plays a positive role in integrating and maintaining the GA pathway by attenuating DELLA repressors in the root [22,23]; HAM subfamily proteins AtSCL6, AtSCL22 and AtSCL27 can regulate root and leaf development and shoot branching in *Arabidopsis thaliana* [24–26]; LAS subfamily proteins MOC1 of rice, AtLAS of *Arabidopsis thaliana* and Ls of tomato are involved in axillary meristem formation [27–31]; DLT subfamily protein GS6 in rice negatively regulates grain size [32] and PAT1 subfamily proteins AtPAT1, AtSCL13 and AtSCL21 act as positive regulators in phytochrome signaling pathways [33–35].

The stolon of strawberries is above the soil that develops from axillary buds in the aboveground crown, axillary buds can also develop to new rosette stems called branch crowns. An axillary bud will develop into a stolon or a branch crown, depending on environmental conditions such as photoperiod and temperature [36,37]. Potato (*Solanum tuberosum*) also has stolons, but potato stolons are below the soil, which develop only from axillary buds in the belowground shoots under darkness and moist atmosphere conditions, and axillary buds in the aerial part of the shoot mainly develop into leafy shoots [38]. Therefore, strawberry stolons are very different from potato stolons. In woodland strawberry, although the *GRAS* gene *DELLA* has been shown to control stolon formation during asexual reproduction [7,8], knowledge of the other *GRAS* genes remains limited. In this study, we aimed to identify all *GRAS* genes in woodland strawberry and analyze their protein motif compositions, vegetative and reproductive organ expression patterns and responses to cold, heat and GA treatments. According to the findings described above, we selected several *FveGRAS* genes for further functional analysis with respect to stolon formation, fruit ripening and abiotic stress responses in woodland strawberry. This research may provide useful information about the functions of *GRAS* genes in woodland strawberry.

2. Results

2.1. Identification and Phylogenetic Analysis of GRAS Proteins in Woodland Strawberry

To identify all possible GRAS proteins in woodland strawberry, the protein sequence of the conserved GRAS domain (PF03514.13) was used as a query to identify similar proteins in *Arabidopsis thaliana*, rice and woodland strawberry. After removing redundant results, 34, 60 and 54 GRAS proteins were identified in *Arabidopsis thaliana*, rice and woodland strawberry, respectively. The *GRAS* genes of woodland strawberry were distributed among all seven chromosomes (Chr) as follows: Chr1 (6), Chr2 (5), Chr3 (21), Chr4 (2), Chr5 (6), Chr6 (8) and Chr7 (6). Four *GRAS* gene clusters were located on Chr3, Chr5, Chr6 and Chr7. The largest cluster was on Chr3, where 16 *GRAS* genes (*FveGRAS16–FveGRAS31*) were distributed within an interval of 194 kb. The other three gene clusters each contain two *GRAS* genes: *FveGRAS39* and *FveGRAS40* on Chr5, *FveGRAS42* and *FveGRAS43* on Chr6 and *FveGRAS50* and *FveGRAS51* on Chr7 (Table S1). In addition, the open reading frames (ORFs) of *FveGRAS* genes ranged from 1320 to 4904 bp, and the length of their encoded proteins ranged from 375 to 836 amino acids (aa). Interestingly, the GRAS domains of 53 proteins ranged from 302–416 aa in length, while FveGRAS16 had a GRAS domain of only 83 aa. The molecular weights (kDa) and isoelectric points (pI) of these FveGRAS proteins ranged from 41.46 to 94.21 kDa and 4.64 to 9.20, respectively (Table S1).

We identified 34 and 60 GRAS proteins from *Arabidopsis thaliana* and rice respectively, consistent with previous studies [39,40]. However, one *Arabidopsis thaliana* protein and ten rice proteins were considered to be putative pseudogenes because these members contained partial GRAS domains with missing motifs [39,40]. Therefore, the other 33 and 50 GRAS proteins from *Arabidopsis thaliana* and rice were selected to perform phylogenetic analysis with 54 GRAS proteins from woodland strawberry. On the basis of previous GRAS family studies, we grouped the GRAS proteins of woodland strawberry into 14 subfamilies: SHR, PAT1, SCL3, DELLA, Os43, Os4, SCR, DLT, LAS, ASCL4/7, HAM, Os19, Fve39 and LlSCL (Figure 1). Fifty GRAS proteins identified in woodland strawberry have extensive similarities with proteins from *Arabidopsis thaliana* or rice, and could be divided into 13 subfamilies with high bootstrap values, and the distribution of these GRAS proteins of woodland strawberry among these 13 subfamilies was as follows: SHR (4), PAT1 (6), SCL3 (1), DELLA (2), Os43 (1), Os4 (3), SCR (3), DLT (1), LAS (1), ASCL4/7 (1), HAM (7), Os19 (1) and LlSCL (19). The other four woodland strawberry GRAS proteins (FveGRAS39, 40, 50 and 51) were grouped into a single subfamily, which lacked homologs in either *Arabidopsis thaliana* or rice; we named this new subfamily Fve39. The LlSCL subfamily had the largest number of GRAS proteins in woodland strawberry, and 17 of 19 FveGRAS proteins in the LlSCL subfamily were clustered together into three small branches that lacked homologous proteins in *Arabidopsis thaliana* and rice (Figure 1), suggesting that these genes may be derived from gene duplication events in woodland strawberry.

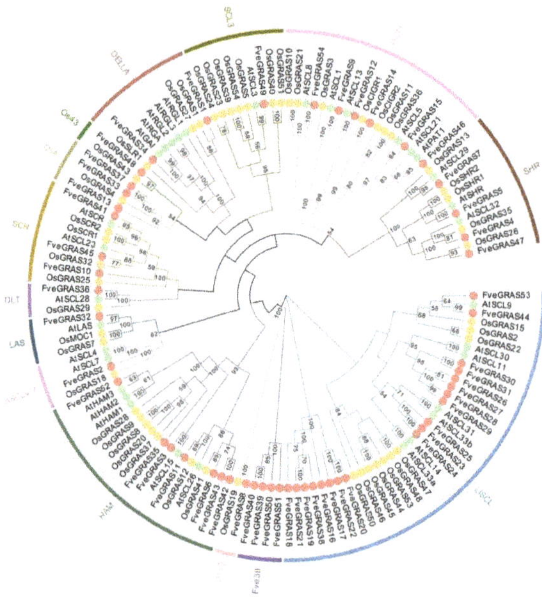

Figure 1. Phylogenetic relationship of GRAS proteins in *Arabidopsis thaliana*, rice and woodland strawberry. 33, 50 and 54 GRAS proteins from *Arabidopsis thaliana*, rice and woodland strawberry were selected to construct phylogenetic tree using MEGA6.0 by the NJ (neighbor-joining) method with the bootstrap test replicated 1000 times. GRAS proteins from *Arabidopsis thaliana*, rice and woodland strawberry are represented by green, yellow and red dots.

2.2. Conserved Domain and Motif Compositions of GRAS Proteins in Woodland Strawberry

The domain and motif composition of transcription factors are critical to their DNA-binding ability and function. To further elucidate functional conservation and divergence among GRAS subfamilies, we analyzed the conserved domain and motif composition of GRAS proteins in *Arabidopsis thaliana* and woodland strawberry (Figure 2). The results showed that aside from FveGRAS16, all FveGRAS proteins contained a typical GRAS domain (a minimum length of about 350 aa) in the C-terminus, while FveGRAS16 contained only a partial GRAS domain (83 aa), suggesting that it is a pseudogene. In addition, another conserved domain, the DELLA domain, is present in the DELLA subfamily; two FveGRAS proteins (FveGRAS1 and FveGRAS34) and five AtGRAS proteins (AtGAI, AtRGA, AtRGL1, AtRGL2 and AtRGL3) were clustered to this group with a bootstrap value of 100 (Figure 2A,B), suggesting that FveGRAS1 and FveGRAS34 have the same function as DELLA proteins in *Arabidopsis thaliana*.

We used MEME5.0.5 to analyze the motif composition of GRAS proteins in woodland strawberry and *Arabidopsis thaliana*, and found that the C-terminus of GRAS proteins contained many more motifs than the N-terminus, and that GRAS proteins from the same subfamily shared similar motif compositions (Figure 2C); this suggested that GRAS proteins in the same subfamily have similar biological functions. Motifs 1, 3, 5, 6, 7 and 8 were found in all 14 GRAS subfamilies, and are located in the C-terminal region of GRAS proteins, suggesting important roles in the conserved function of the *GRAS* gene family. In addition, motif 2 was completely absent from the HAM, SCR, Os19 and SCL4/7 subfamilies, motif 9 was lost in SCL4/7 and some HAM proteins, motif 10 was missing from the SCR, Os43 and Os19 subfamilies, and some HAM members, and motif 14 was absent from LlSCL, Os19, SCL4/7 and most PAT1 proteins. Meanwhile, motif 14 was found only in LlSCL, PAT1 and SCL4/7, motif 18 was present in most members of LlSCL and PAT1; motifs 11, 15, 16, 23, 24 and 29 were found only in members of LlSCL, and motifs 20 and 28 were only found in DELLA proteins (Figure 2C; Table S2). In conclusion, the diversity in motif number and composition among GRAS subfamilies reveals that the functions of GRAS proteins likely diverged during evolution.

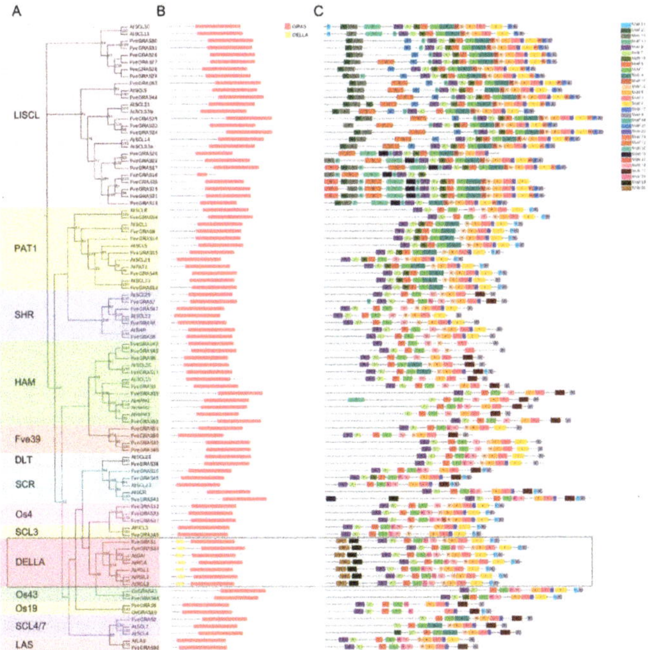

Figure 2. Domain and motif compositions of GRAS proteins from *Arabidopsis thaliana* and woodland strawberry. (**A**) Phylogenetic tree of GRAS proteins in *Arabidopsis thaliana* and woodland strawberry. Different subfamilies are marked with different color backgrounds. (**B**) Domain compositions of GRAS proteins from *Arabidopsis thaliana* and woodland strawberry. Red boxes represent GRAS domains, and yellow boxes represent DELLA domains. Domains were identified by Pfam website. (**C**) Motif compositions of GRAS proteins from *Arabidopsis thaliana* and woodland strawberry. Motifs were identified by MEME software, up to 30 motifs were permitted and other parameters were default settings. Thirty motifs are indicated by different color boxes. The distribution of conserved motifs is presented in Supplementary Table S2. The black rectangle represents DELLA subfamily.

2.3. Identification of Orthologous and Paralogous GRAS Genes in Woodland Strawberry, Arabidopsis thaliana and Rice

To compare the genetic relationships of *GRAS* genes in woodland strawberry to those of *Arabidopsis thaliana* and rice, we used OrthoMCL to identify orthologous and co-orthologous genes among the three plant genomes, as well as paralogous genes within woodland strawberry (Figure 3). The results showed 56 orthologous and 75 co-orthologous gene pairs of GRAS proteins between woodland strawberry and *Arabidopsis thaliana* (Figure 3A), and 47 orthologous and 118 co-orthologous gene pairs between woodland strawberry and rice (Figure 3B). Furthermore, 23 *FveGRAS* genes had orthologous genes in both *Arabidopsis thaliana* and rice, nine *FveGRAS* genes had orthologous genes in *Arabidopsis thaliana* but not in rice, eight *FveGRAS* genes had orthologous genes in rice but not in *Arabidopsis thaliana* and 14 *FveGRAS* genes did not have orthologous genes in either *Arabidopsis thaliana* or rice (Table S3). Additionally, 92 paralogous gene pairs were identified in the woodland strawberry, of which 77 pairs were in the LlSCL subfamily (Figure 3C; Table S4), suggesting that *LlSCL* family genes in woodland strawberry have undergone gene duplication events during the process of evolution.

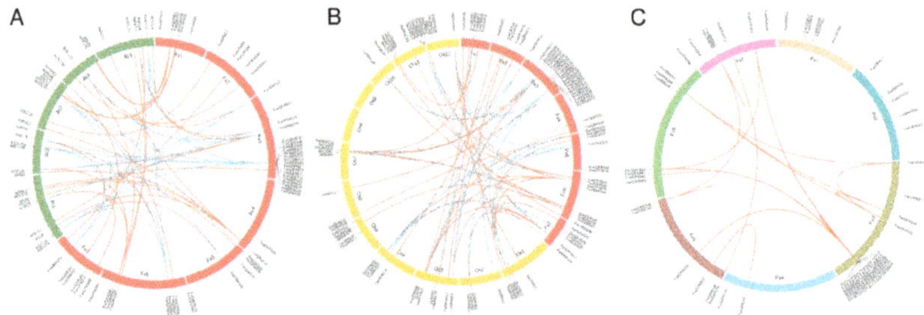

Figure 3. Orthologs, co-orthologs and paralogs of *GRAS* genes from woodland strawberry, *Arabidopsis thaliana* and rice. (**A**) Orthologs and co-orthologs of *GRAS* genes between *Arabidopsis thaliana* and woodland strawberry. (**B**) Orthologs and co-orthologs of *GRAS* genes between rice and woodland strawberry. (**C**) Paralogs of *GRAS* genes in woodland strawberry. OrthoMCL was used to identify orthologs, co-orthologs and paralogs. Red, green and yellow boxes in (**A**) and (**B**) represent chromosomes of woodland strawberry, *Arabidopsis thaliana* and rice. Red and blue lines in (**A**) and (**B**) represent orthologous and co-orthologous gene pairs. Different color boxes in (**C**) represent the seven chromosomes of woodland strawberry, and red lines in (**C**) represent paralogous gene pairs. The distribution of orthologs, co-orthologs and paralogs of *GRAS* genes are presented in Supplementary Tables S3 and S4.

2.4. Expression Profile Analysis of GRAS Genes in Various Organs of Woodland Strawberry

To investigate the expression profiles of *FveGRAS* genes in various organs of woodland strawberry, we collected roots, crowns, stolons, stolon tips, leaves, petioles, open flowers and immature fruits of the woodland strawberry variety "Hawaii 4" for transcriptomic analysis. *FveGRAS* genes enumerated in normalized fragments per kilobase per million (FPKM) were selected for display in a heat map of expression profiles. An FPKM value greater than 1 indicates that the gene is expressed [41,42]. Our results showed that six *FveGRAS* genes (*FveGRAS6, 50, 32, 10, 48* and *16*) have FPKM values lower than 1 in all eight organs, while the other 48 *FveGRAS* genes have FPKM values greater than 1 in at least one organ (Table S5). The detailed results were as follows: 34 *FveGRAS* genes were expressed to varying degrees in at least four organs, including nine *FveGRAS* genes (*FveGRAS35, 52, 2, 1, 34, 54, 12, 15* and *46*) with FPKM greater than 10 in all eight organs; this suggests that any of these *GRAS* genes may play roles in the growth and development of several organs in woodland strawberry. We found that only a few genes showed organ-specific expression, with seven genes (*FveGRAS42, 43, 11, 39, 8, 7* and *26*) having the highest expression in roots and nearly no expression in other seven organs; this indicates that these genes may participate in root growth and development. *FveGRAS13* and *FveGRAS29* were the only genes showing elevated expression solely in open flowers, and *FveGRAS33* had high expression in leaves. *FveGRAS4* had its highest expression level in roots, and moderate expression in crowns and stolon tips. *FveGRAS37* was expressed mainly in petioles, at levels about three-fold greater than in roots and stolons. *FveGRAS51* was most highly expressed in leaves, and showed moderate expression in roots and open flowers (Figure 4A). In conclusion, the expression profiles of *FveGRAS* genes in various organs suggest that numerous *FveGRAS* genes play roles in organ growth and development in woodland strawberry.

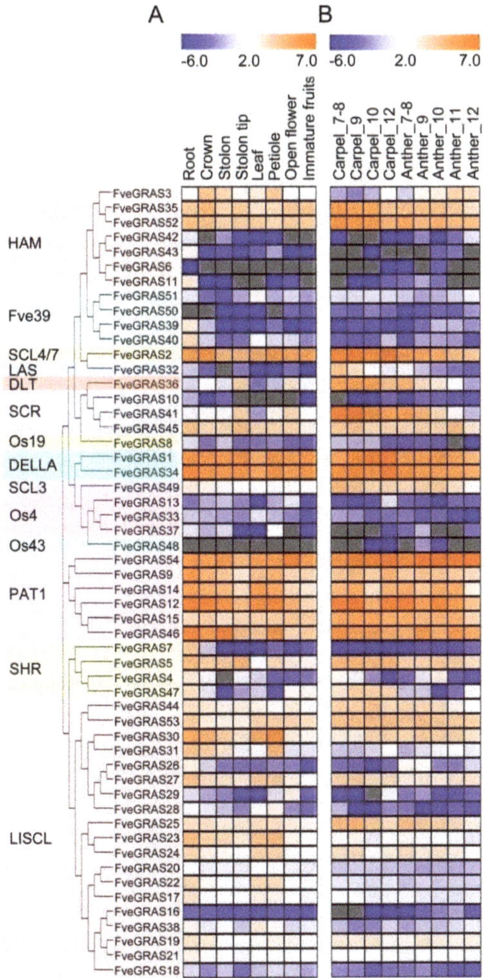

Figure 4. Expression pattern of *GRAS* genes in various organs of woodland strawberry. (**A**) Expression pattern of *GRAS* genes in the root, crown, stolon, stolon tip, leaf, petiole, open flower and immature fruit of woodland strawberry. Samples were collected in our lab, and then RNA-seq was performed. (**B**) Expression pattern of *GRAS* genes in the developing carpel and anther of woodland strawberry. The transcriptome data was downloaded from Li et al. (2019) [43]. Fragments per kilobase per million (FPKM) values in (**A**) and reads per kilobase per million (RPKM) values in (**B**) were in Supplementary Tables S5 and S6, and the heat map showed log2 level. Numbers after carpel and anther represent the developmental stages. The details of the stages were on the website (http://bioinformatics.towson.edu/strawberry/newpage/Tissue_Description.aspx. Access date: 25 May 2019).

The expression levels of *FveGRAS* genes in developing carpels and anthers were investigated using published RNA-sequencing (RNA-Seq) data [43]. The results showed that 32 *FveGRAS* genes had higher expression levels than the other 22 genes (Table S6). Specifically, the expression levels of four genes (*FveGRAS3, 54, 47* and *31*) were elevated, while those of 15 genes (*FveGRAS35, 52, 32, 36, 41, 45, 49, 12, 15, 4, 44, 25, 23, 24* and *17*) were reduced in carpel-12 compared to carpel-7; the same pattern was observed in the anther. Six gene expression levels (*FveGRAS2, 1, 34, 9, 14* and *30*) were higher in carpel-12 than carpel-7, but decreased in the anther. Two genes (*FveGRAS46* and *27*) showed reduced expression in carpel-12 compared to carpel-7, but increased expression in the anther; three other genes (*FveGRAS5, 53* and *26*) had similar expression levels in the developing carpel, but slightly decreased expression in the developing anther. Furthermore, some fluctuations in *FveGRAS* gene expression were noted during carpel and anther development. For example, the expression of *FveGRAS2* at carpel-10 was lower than that at carpel-7, but the level at carpel-12 was higher than at carpel-7. This gene expression pattern of "first decrease and then increase" also occurred in *FveGRAS34, 9, 14, 12, 23* and *17* in the carpel. A greater number of genes displayed a "first increase and then decrease" pattern during the development of carpel and anther, such as *FveGRAS3, 45, 1, 49, 54, 9, 46, 5, 4, 30* and *31* in anther and *FveGRAS41* in carpel (Figure 4B). Therefore, different *GRAS* genes play different roles in the development of the woodland strawberry carpel and anther.

2.5. Expression Analysis of GRAS Genes in Developing and Ripening Fruits of Woodland Strawberry

The period from anthesis to green fruit (including achene and receptacle) is divided into five stages: Stage 1 (pre-fertilization), stage 2 (2–4 days post-anthesis (DPA)), stage 3 (6–9 DPA), stage 4 (8–10 DPA) and stage 5 (10–13 DPA). The achene can be divided into the wall and seeds, the latter of which are further divided into embryo and ghost (seeds without an embryo) [44]. The heat map of *FveGRAS* gene expression based on published RNA-Seq data [43] showed that from stage 1 to stage 2, the expression levels of seven genes (*FveGRAS3, 35, 52, 51, 40, 34* and *5*) increased markedly, by more than 30%, while the expression of 16 genes (*FveGRAS36, 41, 45, 33, 9, 14, 12, 15, 30, 31, 27, 25, 23, 24, 22* and *17*) decreased by more than 20% in seed 2 (vs. ovule 1; Figure 5A; Table S7), suggesting that the expression of these *FveGRAS* genes was affected by fertilization. Then, from stage 3 to stage 5, the expression of 14 *FveGRAS* genes (*FveGRAS35, 52, 2, 36, 41, 1, 34, 49, 9, 5, 44, 53, 25* and *24*) first increased at embryo 4 and then decreased at embryo 5; four genes (*FveGRAS32, 45, 54* and *4*) continuously decreased, and seven genes (*FveGRAS14, 12, 15, 46, 30, 27* and *17*) continuously increased in the embryo from stage 3 to stage 5. However, the same genes showed distinct expression patterns in ghost, where in five gene expressions (*FveGRAS3, 52, 54, 44* and *30*) increased, in 15 gene expressions (*FveGRAS36, 41, 45, 1, 34, 49, 14, 12, 53, 31, 27, 25, 23, 24* and *19*) decreased, in three gene expressions (*FveGRAS35, 2* and *5*) first decreased then increased, and in two gene expressions (*FveGRAS15* and *46*) first increased and then decreased (from stage 3 to stage 5). Furthermore, in the achene wall from stage 1 to stage 5, the expression levels of four genes (*FveGRAS3, 34, 54* and *53*) increased, while those of 13 genes (*FveGRAS36, 41, 33, 15, 44, 30, 31, 27, 25, 23, 24, 22* and *17*) decreased; furthermore, eight genes (*FveGRAS35, 52, 40, 1, 49, 46, 5* and *47*) showed the "first increase and then decrease" expression pattern, and five genes (*FveGRAS2, 45, 9, 14* and *12*) showed the "first decrease and then increase" expression pattern. Notably, the greatest increase or decrease in expression for most *FveGRAS* genes occurred from wall 1 to wall 2 (Figure 5A; Table S7), suggesting that fertilization of the ovule also affects *FveGRAS* gene expression levels in the achene wall.

The receptacle of woodland strawberry fruits was divided into cortex and pith [44], and the result of published RNA-Seq data [43] showed that most *FveGRAS* genes had similar expression patterns in the cortex and pith among stages 1 to 5 (Figure 5A; Table S7). In both cortex and pith, 18 genes (*FveGRAS35, 52, 2, 36, 41, 45, 1, 49, 12, 15, 5, 47, 44, 27, 29, 25, 23* and *24*) showed decreased expression, four genes (*FveGRAS34, 53, 30* and *31*) showed the "first increase and then decrease" expression pattern, and two genes (*FveGRAS54* and *46*) showed the "first decrease and then increase" expression pattern. The expression of *FveGRAS3* decreased in the cortex, but increased in the pith, while *FveGRAS9* and *FveGRAS14* expression decreased markedly in the cortex, but first increased and then decreased in pith (Figure 5A; Table S7). These results suggest that these *FveGRAS* genes play roles in the early development of fruits in woodland strawberry.

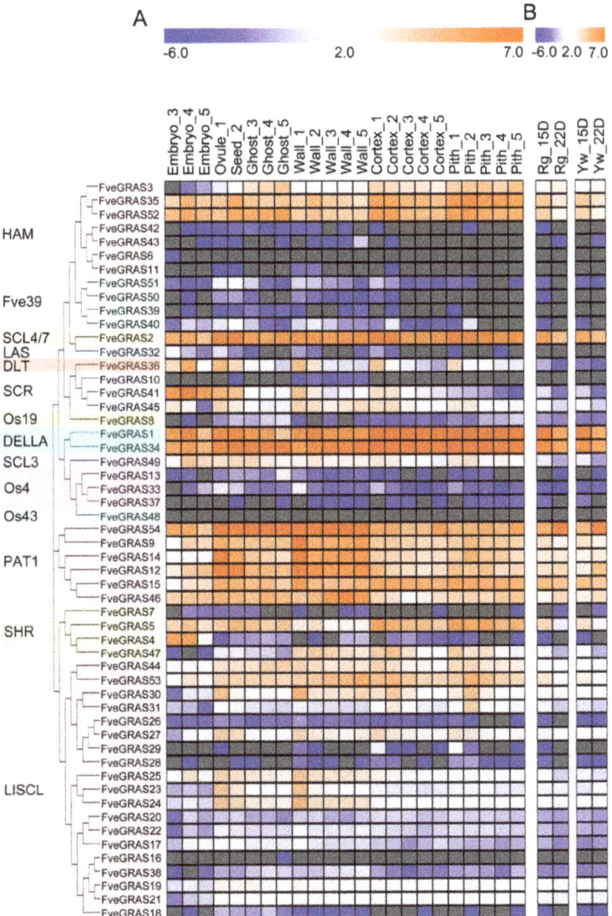

Figure 5. Expression pattern of *GRAS* genes in developing and ripening fruit tissues of woodland strawberry. (**A**) Expression pattern of *GRAS* genes in developing fruit tissues (including embryo, ovule, ghost, wall, cortex and pith). (**B**) Expression change of *GRAS* genes from immature to ripening fruits in two varieties "Ruegen" and "Yellow Wonder". RPKM values in (**A**,**B**) downloaded from Li et al. (2019) [43] were in Supplementary Tables S7 and S8, and the heat map showed log2 level. Numbers after tissues in (**A**) represent different developmental stages: Stage 1 (pre-fertilization), stage 2 (2–4 days post-anthesis (DPA)), stage 3 (6–9 DPA), stage 4 (8–10 DPA) and stage 5 (10–13 DPA). Rg and Yw in (**B**) represent "Ruegen" and "Yellow Wonder", and 15D and 22D represent 15 and 22 DPA.

We also analyzed changes in the expression of *FveGRAS* genes between immature and ripening fruits in two woodland strawberry varieties, "Ruegen" (red fruit) and "Yellow Wonder" (yellow fruit) using published RNA-seq data [43]. The results showed that 26 *FveGRAS* genes had relatively high expression levels, among which 16 genes showed decreased, and only four genes (*FveGRAS3, 54, 12* and *46*) showed increased expression in both "Ruegen" and "Yellow Wonder" between immature and ripening fruits (Figure 5B; Table S8). Notably, the expression of *FveGRAS54* increased by more than 10-fold in ripening fruits compared to immature fruits, suggesting that *FveGRAS54* may play an important role in fruit ripening. The expression of another gene, *FveGRAS27*, decreased in "Ruegen", but showed a slight increase in "Yellow Wonder" (Figure 5B; Table S8), suggesting that this gene may have different roles in the ripening processes of red versus yellow fruits.

2.6. Expression Analysis of GRAS Genes of Woodland Strawberry under Cold and Heat Stresses

To assess the functions of *FveGRAS* genes in plant defenses against abiotic stresses, we performed a transcriptomic analysis using "Hawaii 4" seedlings to analyze changes in the expression of *SlGRAS* genes under cold and heat stresses. In cold-treated seedlings, the FPKM values of 36 *FveGRAS* genes were greater than 1. Among these genes, the expression levels of 24 genes (*FveGRAS3, 52, 51, 40, 2, 45, 49, 54, 9, 14, 12, 15, 7, 5, 47, 44, 53, 30, 27, 25, 23, 24, 22* and *17*) were elevated, of which 12 gene expressions (*FveGRAS3, 51, 2, 45, 14, 12, 15, 7, 5, 30, 23* and *24*) increased by more than two-fold. Especially, *FveGRAS45* expression increased by 5.6-fold at 24 h, and *FveGRAS14* expression increased by 5.6-fold at 48 h. In addition, ten genes (*FveGRAS36, 41, 1, 34, 46, 31,26, 38, 19* and *21*) showed reduced expression, of which seven gene expressions (*FveGRAS36, 41, 1, 46,31, 26* and *21*) decreased by more than two-fold (Figure 6A; Table S9). Under heat stress, the FPKM values of 37 *FveGRAS* genes were greater than 1. The expression levels of 18 genes (*FveGRAS51, 2, 49, 54, 14, 12, 47, 44, 30, 27, 25, 23, 24, 22, 17, 19, 21* and *18*) were elevated compared with the control, of which 12 gene expressions (*FveGRAS51, 2, 54, 14, 12, 47, 27, 25, 23, 24, 19* and *18*) increased by more than two-fold. Especially, *FveGRAS14* and *FveGRAS23* expressions increased by about 19- and 16-fold at 48 h, respectively (Figure 6B; Table S9). In addition, the expressions of 18 genes (*FveGRAS3, 35, 52, 40, 36, 41, 45, 1, 34, 9, 15, 46, 7, 5, 53, 31, 26* and *38*) were reduced compared with the control, of which 11 genes (*FveGRAS3, 35, 52, 40, 36, 41, 1, 46, 7, 5* and *31*) were reduced at least five-fold at a certain hour post treatment (Figure 6B; Table S9).

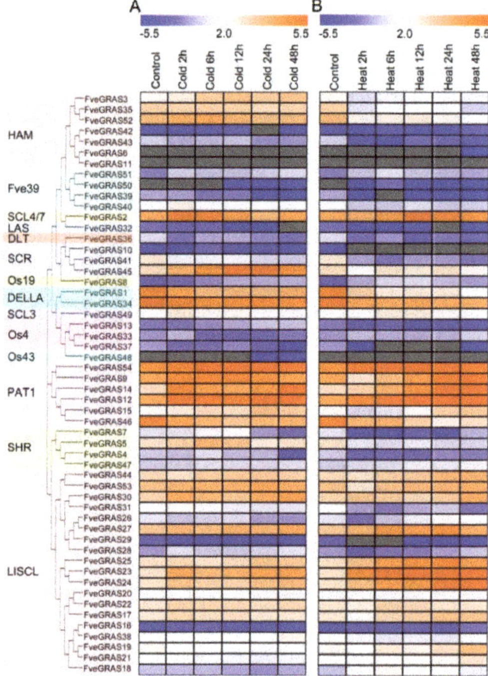

Figure 6. Expression changes of *GRAS* genes of woodland strawberry under cold and heat stresses. (**A**) Expression changes of *GRAS* genes of woodland strawberry under cold stress. (**B**) Expression changes of *GRAS* genes of woodland strawberry under heat stress. Woodland strawberry variety "Hawaii 4" seedlings were used for cold (4 °C) and heat (40 °C) treatments. FPKM values were in Supplementary Table S9, and the heat map showed log2 level.

Notably, 15 gene expressions (*FveGRAS51, 2, 49, 54, 14, 12, 47, 44, 30, 27, 25, 23, 24, 22* and *17*) were increased, and eight gene expressions (*FveGRAS36, 41, 1, 34, 46, 31, 26* and *38*) were decreased both in cold and heat treatment seedlings, which distributed among six chromosomes: Chr1 (2), Chr3 (11), Chr4 (1), Chr5 (2), Chr6 (4) and Chr7 (3). In addition, nine gene expressions (*FveGRAS3, 52, 40, 45, 9, 15, 7, 5* and *53*) were increased by cold treatment, but decreased by heat treatment, and three gene expressions (*FveGRAS19, 21* and *18*) were decreased by cold treatment, but increased by heat treatment, these 12 genes distributed among six chromosomes: Chr1 (2), Chr2 (2), Chr3 (4), Chr5 (1), Chr6 (1) and Chr7 (2) (Figure 7). The result suggests that *GRAS* genes have multiple roles in responses to cold and heat stresses.

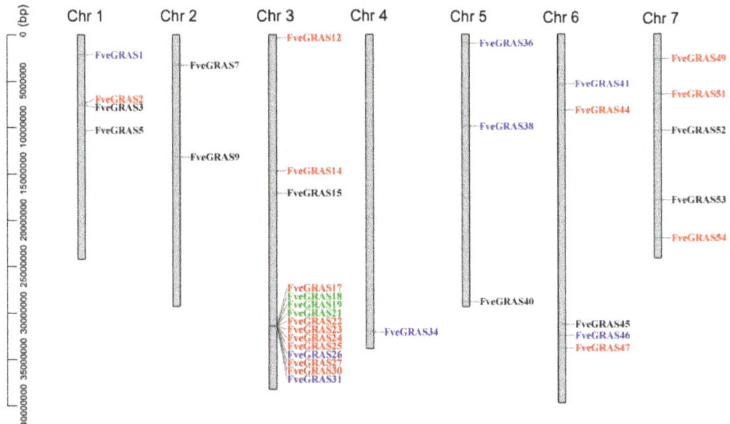

Figure 7. Chromosomal positions of woodland strawberry *GRAS* genes that respond to cold and/or heat treatments. Fifteen up-regulated genes and eight down-regulated genes both in cold and heat treatments are indicated by red and blue fonts, respectively. Nine genes with increased expression in cold stress, but decreased expression in heat stress are indicated by black fonts, and three genes with decreased expression in cold stress, but increased expression in heat stress are indicated by green fonts.

2.7. Analysis of GRAS Gene Expression during Woodland Strawberry Responses to GA Phytohormone.

The differentiation and elongation of stolons in strawberry are regulated by the phytohormone group known as GAs, and many *GRAS* genes are involved in the response to GAs. Therefore, we used woodland strawberry seedlings treated with exogenous GA_3 to analyze the response of *FveGRAS* genes to GAs. The results showed that after GA_3 treatment, the expression levels of nine genes (*FveGRAS52, 40, 1, 49, 54, 46, 47, 44* and *53*) first decreased, and then increased, including three genes (*FveGRAS40, 46* and *44*) that were reduced by more than two-fold at 2 h, and began to increase thereafter(Figure 8; Table S10). In addition, the expression of 17 genes (*FveGRAS2, 45, 34, 9, 14, 12, 15, 5, 30, 31, 27, 25, 23, 24, 22, 17* and *38*) first increased and then decreased, including ten genes (*FveGRAS2, 45, 9, 14, 12, 15, 30, 27, 23* and *24*) with a more than two-fold increase in expression level at 2 h, which later began to decrease. Especially, *FveGRAS14* and *FveGRAS30* had the highest increase by about 14- and 7-fold at 2 h, respectively (Figure 8; Table S10). This result showed that GA-regulated *FveGRAS* genes might be involved in GA-related biological processes in woodland strawberry.

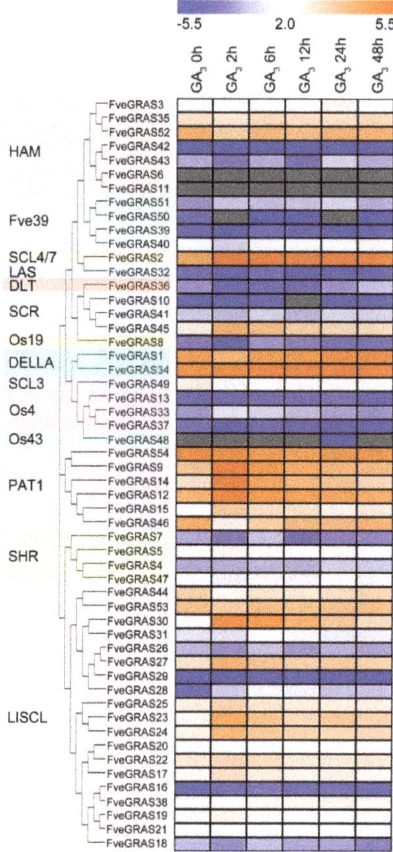

Figure 8. Expression changes of *GRAS* genes of woodland strawberry under GA₃ treatment. Woodland strawberry variety "Hawaii 4" seedlings were used for GA₃ treatment. FPKM values were in Supplementary Table S10, and the heat map showed log2 level.

3. Discussion

3.1. Evolution and Expansion of GRAS Genes

To date, GRAS proteins have been identified in many higher plants, as well as in lower plants such as the lycophyte *Selaginella moellendorffii* and the bryophyte *Physcomitrella patens* [45,46], as well as in the aquatic alga *Spirogyra pratensis*, which belongs to the charophytes, the ancestral taxon of all land plants [46]. Interestingly, GRAS proteins have also been found in some bacteria, but not in any fungi, Metazoa or other algae, suggesting that horizontal gene transfer (HGT) of the GRAS domain from a bacterial source to the common ancestor of land plants is possible and may explain the origin of plant *GRAS* genes [47]. Only one or two GRAS homologs appear to be present per bacterial genome, but at least 30 *GRAS* genes have been identified in all higher plant species investigated. Moreover, bacterial GRAS proteins clearly cluster into a separate clade, rather than the clades of plants, suggesting that *GRAS* genes have undergone differentiation and expansion in higher plants, which may have aided their adaptation to the land environment [10,13,47].

Over the last two decades, several *GRAS* genes have been isolated and functionally analyzed, especially in the model plants *Arabidopsis thaliana* and rice. *GRAS* genes are continuously being found in various plants using genome-wide identification methods. The number of *GRAS* genes in plants of various species ranges widely, 33–184 GRAS proteins have been identified in various plant species, and these genes are clustered into 8–17 subfamilies, depending on the study [10,15]. In woodland strawberry, we identified 54 *GRAS* genes, and divided them into 14 subfamilies based on similar motif composition (Figures 1 and 2); of these subfamilies, four (Os4, Os19, Os43 and Fve39) were absent in *Arabidopsis thaliana*. The GRAS proteins Os4, Os19 and Os43 can be found in other plants, such as rice [14], *Populus* [39], tomato [48], castor beans [49], *Prunus mume* [50] and tea plant [15], and we therefore suggest that these three gene families existed before the divergence of dicotyledons and monocotyledons, but were lost in *Arabidopsis thaliana*. Interestingly, the Fve39 family appears to be a woodland strawberry-specific family (Figure 1), because no orthologous genes were found in *Arabidopsis thaliana* or rice (Figure 3). Some other plant species have similar specific families, such as the Pt20 family in *Populus* and tomato [39,48], G_GRAS in cotton [51], and Rc_GRAS in castor beans [49]; other plant species do not have specific families, including maize [52], tea [15] and Chinese cabbage [53]. Therefore, we speculated that these species-specific families originated from ancestors that occurred after the divergence of dicotyledons and monocotyledons, or arose from ancestors before the divergence of dicotyledons and monocotyledons but were entirely lost in some species during their evolutionary process, such as *Arabidopsis thaliana*, rice and so on. It is tempting to speculate that the genes in the Fve39 subfamily play special roles in woodland strawberry.

Another notable finding is that the LlSCL subfamily contains the largest number of *GRAS* genes. Expansion of the *LlSCL* subfamily genes has been suggested to occur independently in multiple plant species, as *LlSCL* genes within a given plant cluster into an independent group [54], such as those in rice [14], tea [15], tomato [48] and *Populus* [39], as well as in woodland strawberry, as observed in this study. In total, 35% (19 of 54 genes) of the *GRAS* genes in woodland strawberry belong to the LlSCL subfamily, which is a much higher proportion than in other plants, such as *Arabidopsis thaliana* (21%, 7/33) [14], rice (20%, 10/50) [14], cotton (13%, 20/150) [51], tea plant (19%, 10/52) [15], castor beans (15%, 7/46) [49] and *Populus* (11%, 12/106) [39], indicating that a marked expansion of *GRAS* genes may have occurred in the LlSCL subfamily during the evolution of the woodland strawberry. Consistent with this, 16 *GRAS* genes (*FveGRAS16–FveGRAS31*) formed a large gene cluster on chromosome 3 (Table S1), showing that the expansion of *GRAS* genes in the LlSCL subfamily may have been due to tandem gene duplication events. Duplications have long been considered a primary source of genetic redundancy, as well as functional novelty, leading to new gene functions and expression patterns [55]. Accordingly, we found differing *LlSCL* subfamily gene expression patterns between woodland strawberry and *Arabidopsis thaliana*. All *LlSCL* subfamily genes (*AtSCL9, 11, 14, 30, 31* and *33*) in *Arabidopsis thaliana* were expressed in all tested organs (roots, leaves, flowers and seedlings) [56]. In woodland strawberry, 13 genes in the *LlSCL* subfamily (*FveGRAS44, 53, 30, 27, 25, 23, 24, 20, 22, 17, 38, 19* and *21*) that were orthologous with *Arabidopsis thaliana* genes were expressed to varying degrees in all tested organs, while the other six *LlSCL* subfamily genes, which were not orthologous with *Arabidopsis thaliana* genes, showed organ-specific expression or negligible expression (Figure 4A). Therefore, duplication of *GRAS* genes in the *LlSCL* family may have produced new functions related to the growth and development of woodland strawberry compared to orthologous genes in *Arabidopsis thaliana*.

3.2. Possible Roles of GRAS Genes in the Vegetative Organs of Woodland Strawberry

Several *GRAS* genes have been functionally characterized, particularly in *Arabidopsis thaliana*, demonstrating that *GRAS* family genes play important roles in a wide variety of biological process. However, few *GRAS* genes have been studied in woodland strawberry, so the expression patterns of *GRAS* genes identified herein could help in the assessment of their possible functions in woodland strawberry. The *AtSHR* and *AtSCR* genes in *Arabidopsis thaliana* have received the most research attention in the *GRAS* family; these genes can interact with each other, or with other proteins, to regulate root, shoot and leaf development, and play similar roles in rice [16]. In addition, other subfamilies of genes in the *GRAS* family have been found to participate in root meristem arrest, lateral root development, root modification and development of the leaf and shoot apical meristem, including the *HAM* subfamily genes *AtSCL6, 22* and *27* (in *Arabidopsis thaliana*), *MtNSP2* (in *Medicago truncatula*), *PhHAM* (in petunia), the SCL3 subfamily gene *AtSCL3* (in *Arabidopsis thaliana*) and the *PAT1* subfamily gene *SIN1* (in *Phaseolus vulgaris*) [16,17]. This study showed that all of these subfamilies contain some genes expressed in the root, stem, leaf and petiole of woodland strawberry (Figure 4A), suggesting conserved functions with homologous genes in other plant species, whereas other genes (*FveGRAS42, 43, 6, 11, 10* and *7*) in these subfamilies were specifically expressed in roots or were undetectable (Figure 4A), showing that these genes may be involved in root development or other biological processes.

The cultivated strawberry plant propagates sexually through seeds and vegetatively through stolons. Stolon propagation is the main method used in cultivated strawberry production. The model plants *Arabidopsis thaliana*, rice and tomato do not produce stolons, so diploid woodland strawberry is an ideal model for studying the mechanism of stolon formation. CO, SOC1, FveGA20ox4 and DELLA have been demonstrated to regulate stolon formation [4–8]. DELLA proteins belong to the GRAS family, and act as repressors of GA signaling, thereby regulating numerous processes during growth and development [19]. Five FveGRAS proteins (gene06210, gene30958, gene01356, gene22702 and gene06947) were identified as probable DELLA subfamily proteins [7,44], but only gene06210 (FveGRAS34 in this study) and gene30958 (FveGRAS1 in this study) belonged to the DELLA subfamily in our study; gene01356, gene22702 and gene06947 were grouped into the Os43 and Os4 subfamilies. Furthermore, only gene06210 (FveGRAS34) contains a full DELLA motif, and mutation of gene06210 in a runnerless variety (Yellow Wonder) rescues the ability to develop runners [7], in addition, silencing the expression of *FveRGA1* gene (*DELLA, gene06210*) in the naturally non-runnering woodland strawberry cultivars "Ruegen" and "Yellow Wonder" produced many runners [8], demonstrating that this DELLA protein controls runner formation during asexual reproduction in woodland strawberry [7,8]. We found that *gene06210 (FveGRAS34)* was highly expressed in all tested organs (Figure 4A), and is the only gene containing a full DELLA motif in woodland strawberry [7]; this indicates that *gene06210 (FveGRAS34)* participates in biological processes related to GAs in woodland strawberry. Lateral shoots of *Arabidopsis thaliana* and tillers of rice also develop from axillary buds; in *Arabidopsis thaliana* mutants of three *HAM* genes (*AtSCL6, AtSCL22*and *AtSCL27*) or one *LAS* gene (*AtLAS*), phenotypes with reduced lateral shoots could be observed during vegetative development [24–26,31], and mutation of the *LAS* subfamily gene *MOC1* in rice caused defective tiller formation [27–29]. Three *HAM* genes (*FveGRAS3, 35* and *52*) were found to be expressed in the crown, stolon and stolon tip (Figure 4A), indicating that these three genes may regulate stolon or branch crown development. However, the only one *LAS* gene, *FveGRAS32*, showed very low expression in all vegetative organs. A possible explanation for this result is that *FveGRAS32* does not participate in stolon or branch crown development, while an alternative possibility is that *FveGRAS32* was expressed in tissues or cells not tested in this study. In addition, we did not observe stolon-specific expression of *FveGRAS* genes, but many *FveGRAS* genes belonging to various subfamilies showed high expression in the crown, stolon, stolon tip, leaf and petiole (Figure 4A). Since photoperiod and temperature regulate axillary bud development into a stolon or branch crown, and as leaves are the main organ used to sense photoperiod and temperature, a signaling pathway from leaf to crown through the petiole may exist to regulate stolon or branch crown development [5,36,37]; thus, *FveGRAS* genes expressed in the crown, stolon, stolon tip, leaf or petiole may regulate stolon and branch crown initiation or elongation in woodland strawberry, although this requires further investigation.

3.3. Possible Roles of GRAS Genes in the Reproductive Organs of Woodland Strawberry

A large number of studies of *GRAS* genes have focused on their roles in the development of vegetative organs, but not much is known about the functions of GRAS protein during the reproductive stages from flower development to seed formation. *LlSCL* family genes were first functionally characterized in lily, where they were predominantly expressed during the premeiotic phase within the anther and specifically enhanced the activity of a meiosis-associated promoter during microsporogenesis [57]. However, we did not observe anther-specific expression of *LlSCL* genes in woodland strawberry; five *LlSCL* genes (*FveGRAS25, 23, 24, 19* and *21*) were highly expressed in both the carpel and anther (Figure 4B), suggesting that these genes may regulate carpel and anther development. All other subfamilies had some *GRAS* genes that were expressed in the carpel or anther or both (Figure 4B), indicating that many *FveGRAS* genes may participate in the development of those structures. Evidence supporting this finding is still lacking for *Arabidopsis thaliana* and rice, so further research is needed.

In addition, few studies have reported GRAS functions during seed development. For example, the expression of *AtSCR* and *AtSHR* in *Arabidopsis thaliana* can be detected at the heart stage of embryo development and may be responsible for meristem development during this stage [58,59]. In woodland strawberry, two *SCR* genes (*FveGRAS41* and *45*) and two *SHR* genes (*FveGRAS5* and *4*) showed expression in the embryo (Figure 5A), suggesting that these genes may have similar functions as *AtSCR* and *AtSHR* during embryo development. The *HAM* subfamily gene *AtSCL15* can be expressed in the seed coat, and its encoded protein can recruit HDA19 to prevent the transition from seed maturation to vegetative growth [60]; the orthologous gene *FveGRAS3* was not expressed in the embryo, but was expressed in ghost and the seed wall (Figure 5A), suggesting similar functions to *AtSCL15*. In rice, the *DLT* subfamily gene *GS6* negatively regulates grain size [32], while the *DLT* gene in woodland strawberry, *FveGRAS36*, was mainly expressed during the early stages of seed and receptacle (Figure 5A), revealing a role in early stage fruit development in woodland strawberry. Other genes in various *GRAS* subfamilies have not been studied in terms of their roles in seed development, including *HAM, SCL4/7, SCL3, PAT1* and *LlSCL* subfamily genes, but showed similar or obviously differing expression levels among the embryo, ghost and wall in woodland strawberry (Figure 5A); this suggests that many *FveGRAS* genes in various subfamilies may participate in seed development in woodland strawberry.

In contrast to the dry fruits of *Arabidopsis thaliana* and rice, woodland strawberry fruit is fleshy and develops from a receptacle with embedded seeds. ABA is the major phytohormone regulating strawberry fruit ripening [1], whereas tomato fruit ripening is regulated by ethylene [61]; thus, woodland strawberry is used as a model organism in studies of non-climacteric fruits. Therefore, we focused on the analysis of candidate *FveGRAS* genes, which may regulate woodland strawberry fruit ripening. At present, little is known about the GRAS function during fruit development and ripening, even in tomato (the model plant of climacteric fruits), although *GRAS* family genes have been identified in genome-wide analysis of tomato [48]. Knockdown of the *PAT1* subfamily gene *SlGRAS2* significantly reduced tomato fruit weight [62], silencing of *DELLA* resulted in facultative parthenocarpy of tomato fruits [63], while overexpression of the *HAM* subfamily gene *SlGRAS24* and *SlGRAS24* resulted in reduced fruit set, smaller fruit size with fewer seeds [64,65], the homologs of these four genes in woodland strawberry *FveGRAS14, FvGRAS34* and *FveGRAS52* showed high expressions in all tested organs, including carpel, anther, seeds and immature fruits (Figure 5A). In addition, higher expression levels of tomato *GRAS* genes are generally seen in immature versus mature fruits [48], in accordance with results for woodland strawberry in the present study (Figure 5A); this suggests that most *GRAS* genes are involved in early fruit development. Five *SlGRAS* genes in tomato [48], and four *FveGRAS* genes (*FveGRAS3, 54, 12* and *46)* in woodland strawberry, showed dramatic increases in expression from the immature stage to the ripening stage (Figure 5B), indicating that these genes may play roles during the onset of ripening in tomato and woodland strawberry. These five tomato *GRAS* genes belong to the *HAM, PAT1, SCL3* and *SHR* subfamilies [48], and the four woodland strawberry *GRAS* genes belong to the *HAM* and *PAT1* subfamilies. In particular, the *SHR* family gene *SlGRAS38* and *SCL3* family gene *SlGRAS18* showed relatively strong and specific expression during fruit ripening, and have been reported as target genes of tomato-ripening key transcriptional regulator RIN [66,67]. In woodland strawberry, the expression of all *SHR* and *SCL3* family genes was negligible or decreased from the immature stage to the ripening stage (Figure 5B), indicating that the *SHR* and *SCL3* subfamily genes are likely not involved in woodland strawberry fruit ripening, which does not require ethylene.

Furthermore, the *PAT1* family gene *FveGRAS54* showed more than a 10-fold increased expression, which is the largest increase during fruit ripening among *FveGRAS* genes (Figure 5B), indicating that *FveGRAS54* may play an important role in the fruit ripening of woodland strawberry; however, this requires validation.

3.4. Response of GRAS Genes of Woodland Strawberry to Environmental Factors

Cultivated strawberry is a perennial rosette plant found throughout the Northern Hemisphere. During the long and warm days of summer, axillary buds in the crown of cultivated strawberry generally develop into stolons. When the days become shorter and temperatures decrease at the end of summer, stolon production ceases and axillary buds develop into branch crowns. During the short days of autumn, inflorescences are produced at the terminal apices of the main and branch crowns until growth and development cease in the winter. During spring, growth resumes and the inflorescences complete their development, followed by flowering and the development of fruits [36,37].

Light is a very important environmental factor for woodland strawberry growth and development; it not only regulates the differentiation of stolons, branch crowns and flowers, but also affects flowering and fruit growth, ripening and quality [36,37]. It has been demonstrated that members of the PAT1 subgroup of the GRAS family are downstream members of the phytochrome signal transduction pathway. AtPAT1 and AtSCL21 in *Arabidopsis thaliana* are closely related and can interact with each other; both are positive regulators of phytochrome A (phyA) signal transduction, because their mutants develop an elongated hypocotyl specifically under far-red light, which is a phyA-dependent trait [33,35]. Another PAT1 family protein, AtSCL13, is downstream of phytochrome B (phyB) and acts as a positive regulator of red-light signals. AtSCL13 can also modulate phyA signaling in a phyB-independent manner [34]. In this study, six PAT1 subfamily proteins were identified in woodland strawberry, which all had high expression levels in every organ and tissue tested (Figures 4 and 5), suggesting that *PAT1* subfamily genes may participate in the regulation of light signals for many biological processes, including stolon and branch crown differentiation and fruit growth. More importantly, during fruit ripening, the expression levels of *FveGRAS46* and *FveGRAS12*, the most homologous genes of *AtPAT1*, *AtSCL21* and *AtSCL13*, increased; moreover, the expression of another *PAT1* subfamily gene, *FveGRAS54*, increased more than 10-fold (Figure 5B), indicating all three of these genes may play positive roles in light regulation on fruit ripening or the quality of woodland strawberry fruits.

Hot temperatures in summer and cold temperatures in winter can seriously affect cultivated strawberry vegetative and reproductive growth. To identify possible *FveGRAS* genes related to cold and heat tolerance, we assessed the responses of *FveGRAS* genes to heat and cold treatments. It has been reported that *GRAS* family genes can be regulated by abiotic stresses, including heat, salt and drought, and that overexpression of some *GRAS* genes can enhance tolerance to these stresses. For example, overexpression of the poplar *GRAS* gene *PeSCL7* enhanced salt and drought tolerance in *Arabidopsis thaliana* [68], overexpression of *BnLAS* from *Brassica napus* resulted in enhanced drought tolerance in *Arabidopsis thaliana* [69] and overexpression of *VaPAT1* from *Vitis amurensis* conferred cold, drought and high-salinity tolerance in *Arabidopsis thaliana* [70]. *FveGRAS2*, *FveGRAS32* and *FveGRAS14* in woodland strawberry are homologous genes of *PeSCL7*, *BnLAS* and *VaPAT1*, and we found that expression of *FveGRAS2* and *FveGRAS14* increased significantly with cold and heat treatments; in contrast, the change in *FveGRAS32* was negligible (Figure 6), suggesting that *FveGRAS2* and *FveGRAS14* may participate in cold and heat tolerance, while *FveGRAS32* does not. In addition, the *SCR* and *SHR* genes are involved in low-phosphate stress [71], *DELLA* genes are involved in many abiotic stresses including low temperature, phosphate starvation, osmotic stress, and high NO concentration [18,21], and the *LISCL* family gene *OsGRAS23* is involved in drought stress [72]. However, the functions of other *GRAS* genes under abiotic stress conditions, especially heat and cold stresses, have not been studied. In our study, the expression levels of some *FveGRAS* genes showed opposite trends under heat and cold stresses (Figure 6), suggesting that these genes may regulate different pathways under heat and cold stresses. Meanwhile, some *FveGRAS* gene expression levels uniformly increased or decreased under both heat and cold stresses (Figure 6), suggesting that these genes may regulate the same response pathway to heat and cold stresses, and could possibly be used to enhance woodland strawberry tolerance to multiple environmental stresses.

3.5. Response of GRAS Genes in Woodland Strawberry to GAs

As noted above, stolon propagation is the main reproductive method used during cultivated strawberry cultivation, and the stolon is an organ specific to cultivated strawberry not found in the model plants *Arabidopsis thaliana* and rice. Phytohormones GAs can regulate stolon initiation and elongation. Exogenous GA_3 treatment promotes stolon formation, which is suppressed by GA biosynthetic inhibitor treatment [55,56]. The runnerless trait of the diploid woodland strawberry variety "Alpine" is due to mutation of the GA biosynthetic gene *GA20ox4*, which is expressed mainly in the axillary meristem dome and primordia, and in developing stolons [6]; this suggests that GA signaling genes, GA-regulated genes and regulators of GA biosynthesis may be involved in stolon initiation and elongation. GRAS proteins are important components of the GA signaling pathway, and the best studied GRAS proteins in GA signaling are DELLAs, which are master repressors that can directly interact with GA receptors to inhibit GA signaling [18]. In woodland strawberry, one GRAS protein with a full DELLA motif can regulate the development of internodes, flowering shoots, leaves and stolons [6], and expression of this gene (*FveGRAS34*) was induced by GA_3 treatment in this study. Some other *GRAS* genes have also been functionally characterized in terms of their roles in GA signaling in *Arabidopsis thaliana* and rice, but not in woodland strawberry. For example, AtSCL3 promotes GA signaling by antagonizing the master growth repressor DELLA during seed germination and seedling growth in *Arabidopsis thaliana*, while *AtSCL3* expression is induced by DELLAs and repressed by GA [73]. As an orthologous gene of *AtSCL3*, *FveGRAS49* expression was also repressed by GA_3 treatment in this study (Figure 8), suggesting conserved function of *FveGRAS49* with *AtSCL3* in GA signal transduction. In addition, overexpression of the *SCL4/7* subfamily gene *Ha-GRASL* from sunflower reduced the metabolic flow of GAs and increased axillary meristem outgrowth in *Arabidopsis thaliana* [74], while the *SCL4/7* subfamily gene *FveGRAS2* in woodland strawberry was induced by GA_3 treatment (Figure 8). In addition to GAs, *GRAS* family genes are also involved in other hormones. DELLA proteins also mediate auxin, BR, JA and ethylene signaling, thereby acting as a major hub in hormonal signaling [21]. Overexpression of the tomato *HAM* subfamily gene *SlGRAS24* and *SlGRAS40*, and *PAT1* subfamily gene *SlGRAS7* all affected multiple agronomic traits through regulation of GA and auxin signaling, and the expression of *SlGRAS40* and *SlGRAS7* were down-regulated, but *SlGRAS24* was up-regulated by auxin and GA treatments [64,65,75], while in woodland strawberry, *SlGRAS7* homologous gene *FveGRAS46*, and *SlGRAS24* and *SlGRAS40* homologous gene *FveGRAS52* were all down-regulated after GA_3 treatment (Figure 8). The rice *DLT* subfamily gene *GS6* not only regulates GA biosynthesis, but also plays a role in BR signaling [76], and the rice *SCL4/7* subfamily gene OsGRAS19 (OsGRAS18 in this study) acts as a positive regulator in BR signaling [77], suggesting that *GRAS* genes are involved in crosstalk among hormonal signaling pathways. In this study, the expression levels of 26 *GRAS* genes in woodland strawberry either increased or decreased after GA_3 treatment (Figure 8), indicating that these genes may play roles in biological processes related to GA biosynthesis, metabolism or signaling, such as stolon initiation and elongation. In particular, the expression of *FveGRAS14* showed the largest increase, of more than 10-fold, after GA_3 treatment (Figure 8), This result, together with the *FveGRAS14* orthologous gene *AtPAT1* in *Arabidopsis thaliana* being a positive regulator of phyA signal transduction, and the photoperiod regulating stolon formation in woodland strawberry, led us to speculate that *FveGRAS14* may play a role in the regulation of stolon formation by photoperiod. However, the effects of *FveGRAS* genes on GAs and other hormones, and their roles in the growth and development of woodland strawberry, require further investigation.

4. Materials and Methods

4.1. Identification and Phylogenetic Analysis of GRAS Proteins

The protein databases and annotation information of *Arabidopsis thaliana* and rice (*Oryza sativa*) were downloaded from Phytozome (https://phytozome.jgi.doe.gov/pz/portal.html. Access date: 16 May 2019), and the protein database and corresponding annotation of woodland strawberry (*F.vesca*) was downloaded from GDR (https://www.rosaceae.org/species/fragaria/fragaria_vesca. Access date: 16 May 2019). The full-alignment sequences of the GRAS domain (PF03514.13) was downloaded from the Pfam database, HMMER software was used to identify similar proteins in *Arabidopsis thaliana*, rice and woodland strawberry with an E-value cut-off of $1 \times e^{-4}$ using PF03514.13 as query. The longest proteins were selected when there was alternative splicing, and the GRAS domain of identified proteins was confirmed using the Pfam (http://pfam.xfam.org/search. Access date: 3 June 2019) and SMART (http://smart.embl-heidelberg.de/ Access date: 3 June 2019). The 54 putative *FveGRAS* genes were renamed as *FveGRAS1* to *FveGRAS54* according to their chromosomal locations.

34 and 60 GRAS proteins were identified from *Arabidopsis thaliana* and rice respectively, but one *Arabidopsis thaliana* protein and ten rice proteins containing partial GRAS domains were considered as pseudogenes. Therefore, the protein sequences of other 33 and 50 GRAS proteins from *Arabidopsis thaliana* and rice, together with 54 GRAS proteins from woodland strawberry were aligned using the ClustalX2.0 program with the default settings. A phylogenetic tree based on the alignment was constructed using MEGA6.0 by the NJ (neighbor-joining) method with the bootstrap test replicated 1000 times.

4.2. Conserved Domain and Motif Analysis of GRAS Proteins

All complete protein sequences of GRASs were used to analyze conserved domains and motifs. The conserved domains were identified by Pfam website, and the conserved motifs were identified by MEME website (Version 5.0.5, http://meme-suite.org/tools/meme. Access date: 4 June 2019) [78], with the maximum number of motifs was 30, and other parameters were default settings. The illustration containing conserved domains and motifs was constructed using TBtools software (Version No.0.66763, South China Agricultural University, Guangzhou, China) [79].

4.3. Identification of Orthologs, Coorthologs and Paralogs of GRAS Genes

Orthologous, coorthologous and paralogous gene pairs were identified by submitting all complete protein sequences of GRAS from *Arabidopsis thaliana*, rice and woodland strawberry to OrthoMCL software (Version 2.0, University of Pennsylvania, Philadelphia, USA) [80]. Illustrations of orthologous, coorthologous and paralogous gene pairs among the three species were constructed using Circos software (Canada's Michael Smith Genome Sciences Center, Vancouver, Canada) [81].

4.4. Plant Materials and Cold, Heat and GA_3 Treatments

The genome-sequenced diploid woodland strawberry "Hawaii 4" (*F.vesca*) was used as plant material. "Hawaii 4" seeds were sterilized and sowed as previously studied [82], and the seedlings were grown in a climate chamber under a 16 h light (22 °C) with 15,000 Lux irradiance and 8 h of dark (20 °C) for about two months. Then the seedlings were transferred into pots containing a mixture of perlite, vermiculite and sphagnum (ratio, 1:2:3) in January 2019, and grown in a glass-enclosed greenhouse of Nanjing Agricultural University under native photoperiod. On 23th April 2019, we collected immature fruits with seeds about 8–15 DPA, stolon tips and fully open flowersfrom about 100 plants because of limited organs. Then the root was sufficiently washed by running water to remove soils, the clean roots, crowns, leaves, petioles and stolons were sampled from 24 plants. All of the samples were rapidly frozen in liquid nitrogen, and then stored at −80 °C for RNA extraction. Roots, crowns, leaves, petioles, stolons and immature fruits had three biological replicates, and stolon tips and fully open flowers had two biological replicates because of limited samples.

For cold, heat and GA$_3$ treatments, "Hawaii 4" seeds were sterilized and sowed as previously studied [82], and the seedlings were grown in a climate chamber under 16 h of light (22 °C) with 15000 Lux irradiance and 8 h of dark (20 °C) for about two months. To avoid a photoperiod effect on gene expression during treatments, the growth conditions were reset as 24 h light (22 °C), and the seedlings were re-adapted for three days before treatments. Afterwards, seedlings were transferred to another chamber maintained at 40 °C for heat treatment and at 4 °C for cold treatment. For the GA$_3$ treatment, the seedlings were sprayed with 50mg/L GA$_3$ solution [5,6]. At 0, 2, 6, 12, 24 and 48 h after treatments, the whole seedlings with roots were collected, rapidly frozen in liquid nitrogen, and then stored at −80 °C for RNA extraction. Three biological replicates were performed.

4.5. RNA Extraction, Transcriptome Sequencing and Data Analysis

RNA extraction and transcriptome sequencing were performed according to Gu et al. (2019) [38]. Afterward, Raw data from transcriptome sequencing were firstly processed through in-house perl scripts. Clean data were obtained by removing reads with adapter, ploy-N and low quality from raw data. Q20, Q30 and GC content of the clean data were calculated at the same time. Reference genome of woodland strawberry *F.vesca* v4.0.a1 and gene annotation files *F.vesca* v4.0.a2 were downloaded from website (https://www.rosaceae.org/species/fragaria/fragaria_vesca. Access date: 16 May 2019). The index of the reference genome was built by Hisat2 v2.0.4 and paired-end clean reads were aligned to the reference genome by Hisat2 v2.0.4. The reads numbers mapped to each gene was counted by HTSeq v0.9.1, then FPKM value of each gene was calculated based on the length of the gene and reads count mapped to this gene.

4.6. Heat Map Construction of FveGRAS Gene Expressions

The reads per kilobase per million (RPKM) values of transcriptome data in Figures 4B and 5 of woodland strawberry were downloaded from Li et al. (2019) [43]. The FPKM values of *FveGRAS* genes in Figures 4A, 6 and 7 were selected from our own transcriptome data. Then RPKM and FPKM values were transformed in log2 level, and heat maps were constructed by MeV4.8 software.

5. Conclusions

In this study, we identified 54 FveGRAS proteins in woodland strawberry, and divided them into 14 subfamilies. Phylogenetic analysis, motif composition, orthologous and paralogous analysis were performed to study the genetic relationship among woodland strawberry, *Arabidopsis thaliana* and rice. The RNA-seq analysis showed that *FveGRAS* genes were expressed with different degrees in different organs and tissues. Sixteen genes showed decreased expression, while four genes showed increased expression during fruit ripening. In addition, around half of the *FveGRAS* genes displayed increased or decreased expression to some extent under cold, heat and GA$_3$ treatments. A few *FveGRAS* genes were predicted as candidate genes to study their functions in stolon formation, fruit ripening and abiotic stresses.

Supplementary Materials: Supplementary materials can be found at http://www.mdpi.com/1422-0067/20/18/4593/s1. Supplementary Table S1: Identification and characterization of GRAS proteins in woodland strawberry; Supplementary Table S2: Motif composition of GRAS family in woodland strawberry; Supplementary Table S3: Orthologs and co-orthologs of GRAS proteins among woodland strawberry, Arabidopsis and rice; Supplementary Table S4: Paralogs of GRAS proteins in woodland strawberry; Supplementary Table S5: The FPKM value of GRAS genes in different organs of woodland strawberry; Supplementary Table S6: The RPKM value of GRAS genes in the flowers of woodland strawberry; Supplementary Table S7: The RPKM value of GRAS genes in the flowers, early-stage fruits and ripening fruits of woodland strawberry; Supplementary Table S8: The RPKM value of GRAS genes in the ripening fruits of woodland strawberry; Supplementary Table S9: The FPKM value of GRAS genes uder cold and heat treatments of woodland strawberry seedlings; Supplementary Table S10: The FPKM value of GRAS genes uder GA$_3$ treatment of woodland strawberry seedlings.

Author Contributions: Investigation, H.C., H.L. and X.L.; Supervision, J.L. and H.W.; Validation, J.L. and H.W.; Visualization, H.C., H.L. and X.L.; Writing—original draft, H.C. and H.W.; Writing—review & editing, L.C. and J.L.

Funding: This research was funded by the National Natural Science Foundation of China (NSFC 31601736).

Acknowledgments: There is no content in this part.

Conflicts of Interest: The authors declare no conflict of interest.

Abbreviations

CO	CONSTANS
SOC1	SUPPRESSOR OF OVEREXPRESSION OF CONSTANS1
GA20ox4	Gibberellin 20-oxidase 4
ABA	Abscisic Acid
GAI	GIBBERELLIN-INSENSITIVE
RGA	REPRESSOR OF GA1-3
SCR	SCARECROW
LHR	Leucine Heptad Repeat
GAs	Gibberellins
BRs	Brassinosteroids
CKs	Cytokinins
JA	Jasmonate
SLs	Strigolactones
Chr	Chromosome
FPKM	Fragments Per Kilobase Per Million
HGT	Horizontal Gene Transfer
PhyA	Phytochrome A
PhyB	Phytochrome B

References

1. Li, C.; Jia, H.; Chai, Y.; Shen, Y. Abscisic acid perception and signaling transduction in strawberry: A model for non-climacteric fruit ripening. *Plant Signal. Behav.* **2011**, *6*, 1950–1953. [CrossRef] [PubMed]
2. Edger, P.P.; Poorten, T.J.; VanBuren, R.; Hardigan, M.A.; Colle, M.; McKain, M.R.; Smith, R.D.; Teresi, S.J.; Nelson, A.; Wai, C.M.; et al. Origin and evolution of the octoploid strawberry genome. *Nat. Genet.* **2019**, *51*, 541–547. [CrossRef] [PubMed]
3. Shulaev, V.; Sargent, D.J.; Crowhurst, R.N.; Mockler, T.C.; Folkerts, O.; Delcher, A.L.; Jaiswal, P.; Mockaitis, K.; Liston, A.; Mane, S.P.; et al. The genome of woodland strawberry (*Fragaria vesca*). *Nat. Genet.* **2011**, *43*, 109–116. [CrossRef] [PubMed]
4. Kurokura, T.; Samad, S.; Koskela, E.; Mouhu, K.; Hytonen, T. *Fragaria vesca* CONSTANS controls photoperiodic flowering and vegetative development. *J. Exp. Bot.* **2017**, *68*, 4839–4850. [CrossRef] [PubMed]
5. Mouhu, K.; Kurokura, T.; Koskela, E.A.; Albert, V.A.; Elomaa, P.; Hytonen, T. The *Fragaria vesca* homolog of SUPPRESSOR OF OVEREXPRESSION OF CONSTANS1 represses flowering and promotes vegetative growth. *Plant Cell* **2013**, *25*, 3296–3310. [CrossRef] [PubMed]
6. Tenreira, T.; Lange, M.; Lange, T.; Bres, C.; Labadie, M.; Monfort, A.; Hernould, M.; Rothan, C.; Denoyes, B. A specific gibberellin 20-oxidase dictates the flowering-runnering decision in diploid strawberry. *Plant Cell* **2017**, *29*, 2168–2182. [CrossRef] [PubMed]
7. Caruana, J.C.; Sittmann, J.W.; Wang, W.; Liu, Z. *Suppressor of runnerless* encodes a DELLA protein that controls runner formation for asexual reproduction in strawberry. *Mol. Plant* **2018**, *11*, 230–233. [CrossRef] [PubMed]
8. Li, W.; Zhang, J.; Sun, H.; Wang, S.; Chen, K.; Liu, Y.; Li, H.; Ma, Y.; Zhang, Z. *FveRGA1*, encoding a DELLA protein, negatively regulates runner production in *Fragaria vesca*. *Planta* **2018**, *247*, 941–951. [CrossRef] [PubMed]
9. Bolle, C. The role of GRAS proteins in plant signal transduction and development. *Planta* **2004**, *218*, 683–692. [CrossRef]
10. Bolle, C. Chapter 10—Structure and evolution of plant GRAS family proteins. In *Plant Transcription Factors*; Gonzalez, D.H., Ed.; Academic Press: Boston, MA, USA, 2016; pp. 153–161.

11. Pysh, L.D.; Wysocka-Diller, J.W.; Camilleri, C.; Bouchez, D.; Benfey, P.N. The GRAS gene family in Arabidopsis: Sequence characterization and basic expression analysis of the *SCARECROW-LIKE* genes. *Plant J.* **1999**, *18*, 111–119. [CrossRef]

12. Sun, X.; Xue, B.; Jones, W.T.; Rikkerink, E.; Dunker, A.K.; Uversky, V.N. A functionally required unfoldome from the plant kingdom: Intrinsically disordered N-terminal domains of GRAS proteins are involved in molecular recognition during plant development. *Plant Mol. Biol.* **2011**, *77*, 205–223. [CrossRef] [PubMed]

13. Cenci, A.; Rouard, M. Evolutionary analyses of GRAS transcription factors in angiosperms. *Front. Plant Sci.* **2017**, *8*, 273. [CrossRef] [PubMed]

14. Tian, C.; Wan, P.; Sun, S.; Li, J.; Chen, M. Genome-wide analysis of the GRAS gene family in rice and *Arabidopsis*. *Plant Mol. Biol.* **2004**, *54*, 519–532. [CrossRef] [PubMed]

15. Wang, Y.X.; Liu, Z.W.; Wu, Z.J.; Li, H.; Wang, W.L.; Cui, X.; Zhuang, J. Genome-wide identification and expression analysis of GRAS family transcription factors in tea plant (*Camellia sinensis*). *Sci. Rep.* **2018**, *8*, 3949. [CrossRef] [PubMed]

16. Bolle, C. Chapter 19—Functional aspects of GRAS family proteins. In *Plant Transcription Factors*; Gonzalez, D.H., Ed.; Academic Press: Boston, MA, USA, 2016; pp. 295–311.

17. Hirsch, S.; Oldroyd, G.E. GRAS-domain transcription factors that regulate plant development. *Plant Signal. Behav.* **2009**, *4*, 698–700. [CrossRef] [PubMed]

18. Sun, T.P. Gibberellin-GID1-DELLA: A pivotal regulatory module for plant growth and development. *Plant Physiol.* **2010**, *154*, 567–570. [CrossRef] [PubMed]

19. Hauvermale, A.L.; Ariizumi, T.; Steber, C.M. Gibberellin signaling: A theme and variations on DELLA repression. *Plant Physiol.* **2012**, *160*, 83–92. [CrossRef]

20. Van De Velde, K.; Ruelens, P.; Geuten, K.; Rohde, A.; Van Der Straeten, D. Exploiting DELLA signaling in cereals. *Trends Plant Sci.* **2017**, *22*, 880–893. [CrossRef]

21. Daviere, J.M.; Achard, P. A pivotal role of DELLAs in regulating multiple hormone signals. *Mol. Plant* **2016**, *9*, 10–20. [CrossRef] [PubMed]

22. Zhang, Z.L.; Ogawa, M.; Fleet, C.M.; Zentella, R.; Hu, J.; Heo, J.O.; Lim, J.; Kamiya, Y.; Yamaguchi, S.; Sun, T.P. Scarecrow-like 3 promotes gibberellin signaling by antagonizing master growth repressor DELLA in Arabidopsis. *Proc. Natl. Acad. Sci. USA* **2011**, *108*, 2160–2165. [CrossRef] [PubMed]

23. Heo, J.O.; Chang, K.S.; Kim, I.A.; Lee, M.H.; Lee, S.A.; Song, S.K.; Lee, M.M.; Lim, J. Funneling of gibberellin signaling by the GRAS transcription regulator scarecrow-like 3 in the *Arabidopsis* root. *Proc. Natl. Acad. Sci. USA* **2011**, *108*, 2166–2171. [CrossRef] [PubMed]

24. Engstrom, E.M.; Andersen, C.M.; Gumulak-Smith, J.; Hu, J.; Orlova, E.; Sozzani, R.; Bowman, J.L. Arabidopsis homologs of the petunia *HAIRY MERISTEM* gene are required for maintenance of shoot and root indeterminacy. *Plant Physiol.* **2011**, *155*, 735–750. [CrossRef]

25. Schulze, S.; Schafer, B.N.; Parizotto, E.A.; Voinnet, O.; Theres, K. *LOST MERISTEMS* genes regulate cell differentiation of central zone descendants in Arabidopsis shoot meristems. *Plant J.* **2010**, *64*, 668–678. [CrossRef] [PubMed]

26. Ma, Z.; Hu, X.; Cai, W.; Huang, W.; Zhou, X.; Luo, Q.; Yang, H.; Wang, J.; Huang, J. *Arabidopsis* miR171-targeted scarecrow-like proteins bind to GT cis-elements and mediate gibberellin-regulated chlorophyll biosynthesis under light conditions. *PLoS Genet.* **2014**, *10*, e1004519. [CrossRef]

27. Li, X.; Qian, Q.; Fu, Z.; Wang, Y.; Xiong, G.; Zeng, D.; Wang, X.; Liu, X.; Teng, S.; Hiroshi, F.; et al. Control of tillering in rice. *Nature* **2003**, *422*, 618–621. [CrossRef] [PubMed]

28. Xu, C.; Wang, Y.; Yu, Y.; Duan, J.; Liao, Z.; Xiong, G.; Meng, X.; Liu, G.; Qian, Q.; Li, J. Degradation of MONOCULM 1 by APC/C(TAD1) regulates rice tillering. *Nat. Commun.* **2012**, *3*, 750. [CrossRef]

29. Lin, Q.; Wang, D.; Dong, H.; Gu, S.; Cheng, Z.; Gong, J.; Qin, R.; Jiang, L.; Li, G.; Wang, J.L.; et al. Rice APC/C(TE) controls tillering by mediating the degradation of MONOCULM 1. *Nat. Commun.* **2012**, *3*, 752. [CrossRef] [PubMed]

30. Schumacher, K.; Schmitt, T.; Rossberg, M.; Schmitz, G.; Theres, K. The *Lateral suppressor* (*Ls*) gene of tomato encodes a new member of the VHIID protein family. *Proc. Natl. Acad. Sci. USA* **1999**, *96*, 290–295. [CrossRef]

31. Greb, T.; Clarenz, O.; Schafer, E.; Muller, D.; Herrero, R.; Schmitz, G.; Theres, K. Molecular analysis of the *LATERAL SUPPRESSOR* gene in *Arabidopsis* reveals a conserved control mechanism for axillary meristem formation. *Genes Dev.* **2003**, *17*, 1175–1187. [CrossRef]

32. Sun, L.; Li, X.; Fu, Y.; Zhu, Z.; Tan, L.; Liu, F.; Sun, X.; Sun, X.; Sun, C. *GS6*, a member of the GRAS gene family, negatively regulates grain size in rice. *J. Integr. Plant Biol.* **2013**, *55*, 938–949. [CrossRef]

33. Bolle, C.; Koncz, C.; Chua, N.H. PAT1, a new member of the GRAS family, is involved in phytochrome a signal transduction. *Genes Dev.* **2000**, *14*, 1269–1278. [PubMed]

34. Torres-Galea, P.; Huang, L.F.; Chua, N.H.; Bolle, C. The GRAS protein SCL13 is a positive regulator of phytochrome-dependent red light signaling, but can also modulate phytochrome A responses. *Mol. Genet.Genom.* **2006**, *276*, 13–30. [CrossRef] [PubMed]

35. Torres-Galea, P.; Hirtreiter, B.; Bolle, C. Two GRAS proteins, SCARECROW-LIKE21 and PHYTOCHROME A SIGNAL TRANSDUCTION1, function cooperatively in phytochrome A signal transduction. *Plant Physiol.* **2013**, *161*, 291–304. [CrossRef]

36. Koskela, E. *Genetic and Environmental Control of Flowering in Wild and Cultivated Strawberries*; University of Helsinki: Helsinki, Finland, 2016.

37. Rantanen, M.M.M. *Light and Temperature as Developmental Signals in Woodland Strawberry and Red Raspberry*; University of Helsinki: Helsinki, Finland, 2017.

38. Kumar, D.; Wareing, P.F. Factors controlling stolon development in the potato plant. *New Phytol.* **1972**, *71*, 639–648. [CrossRef]

39. Liu, X.; Widmer, A. Genome-wide comparative analysis of the GRAS gene family in *Populus*, *Arabidopsis* and rice. *Plant Mol. Biol. Rep.* **2014**, *32*, 1129–1145. [CrossRef]

40. Wang, Y.; Shi, S.; Zhou, Y.; Zhou, Y.; Yang, J.; Tang, X. Genome-wide identification and characterization of GRAS transcription factors in sacred lotus (*Nelumbo nucifera*). *PeerJ* **2016**, *4*, e2388. [CrossRef] [PubMed]

41. Trapnell, C.; Williams, B.A.; Pertea, G.; Mortazavi, A.; Kwan, G.; van Baren, M.J.; Salzberg, S.L.; Wold, B.J.; Pachter, L. Transcript assembly and quantification by RNA-Seq reveals unannotated transcripts and isoform switching during cell differentiation. *Nat. Biotechnol.* **2010**, *28*, 511–515. [CrossRef]

42. Gu, T.; Jia, S.; Huang, X.; Wang, L.; Fu, W.; Huo, G.; Gan, L.; Ding, J.; Li, Y. Transcriptome and hormone analyses provide insights into hormonal regulation in strawberry ripening. *Planta* **2019**, *250*, 145–162. [CrossRef]

43. Li, Y.; Pi, M.; Gao, Q.; Liu, Z.; Kang, C. Updated annotation of the wild strawberry *Fragaria vesca* V4 genome. *Hortic. Res.* **2019**, *6*, 61. [CrossRef]

44. Kang, C.; Darwish, O.; Geretz, A.; Shahan, R.; Alkharouf, N.; Liu, Z. Genome-scale transcriptomic insights into early-stage fruit development in woodland strawberry *Fragaria vesca*. *Plant Cell* **2013**, *25*, 1960–1978. [CrossRef] [PubMed]

45. Yasumura, Y.; Crumpton-Taylor, M.; Fuentes, S.; Harberd, N.P. Step-by-step acquisition of the gibberellin-DELLA growth-regulatory mechanism during land-plant evolution. *Curr. Biol.* **2007**, *17*, 1225–1230. [CrossRef] [PubMed]

46. Engstrom, E.M. Phylogenetic analysis of GRAS proteins from moss, lycophyte and vascular plant lineages reveals that GRAS genes arose and underwent substantial diversification in the ancestral lineage common to bryophytes and vascular plants. *Plant Signal. Behav.* **2011**, *6*, 850–854. [CrossRef] [PubMed]

47. Zhang, D.; Iyer, L.M.; Aravind, L. Bacterial GRAS domain proteins throw new light on gibberellic acid response mechanisms. *Bioinformatics* **2012**, *28*, 2407–2411. [CrossRef] [PubMed]

48. Huang, W.; Xian, Z.; Kang, X.; Tang, N.; Li, Z. Genome-wide identification, phylogeny and expression analysis of GRAS gene family in tomato. *BMC Plant Biol.* **2015**, *15*, 209. [CrossRef] [PubMed]

49. Xu, W.; Chen, Z.; Ahmed, N.; Han, B.; Cui, Q.; Liu, A. Genome-Wide identification, evolutionary analysis, and stress responses of the GRAS gene family in castor beans. *Int. J. Mol. Sci.* **2016**, *17*, 1004. [CrossRef]

50. Lu, J.; Wang, T.; Xu, Z.; Sun, L.; Zhang, Q. Genome-wide analysis of the GRAS gene family in *Prunus mume*. *Mol Genet. Genom.* **2015**, *290*, 303–317. [CrossRef] [PubMed]

51. Zhang, B.; Liu, J.; Yang, Z.E.; Chen, E.Y.; Zhang, C.J.; Zhang, X.Y.; Li, F.G. Genome-wide analysis of GRAS transcription factor gene family in *Gossypium hirsutum* L. *BMC Genom.* **2018**, *19*, 348. [CrossRef] [PubMed]

52. Guo, Y.; Wu, H.; Li, X.; Li, Q.; Zhao, X.; Duan, X.; An, Y.; Lv, W.; An, H. Identification and expression of GRAS family genes in maize (*Zea mays* L.). *PLoS ONE* **2017**, *12*, e185418. [CrossRef]

53. Song, X.M.; Liu, T.K.; Duan, W.K.; Ma, Q.H.; Ren, J.; Wang, Z.; Li, Y.; Hou, X.L. Genome-wide analysis of the GRAS gene family in Chinese cabbage (*Brassica rapa* ssp. Pekinensis). *Genomics* **2014**, *103*, 135–146. [CrossRef]

54. Wu, Z.Y.; Wu, P.Z.; Chen, Y.P.; Li, M.R.; Wu, G.J.; Jiang, H.W. Genome-wide analysis of the GRAS gene family in physic nut (*Jatropha curcas* L.). *Genet. Mol. Res.* **2015**, *14*, 19211–19224. [CrossRef]

55. Flagel, L.E.; Wendel, J.F. Gene duplication and evolutionary novelty in plants. *New Phytol.* **2009**, *183*, 557–564. [CrossRef]

56. Lee, M.H.; Kim, B.; Song, S.K.; Heo, J.O.; Yu, N.I.; Lee, S.A.; Kim, M.; Kim, D.G.; Sohn, S.O.; Lim, C.E.; et al. Large-scale analysis of the GRAS gene family in *Arabidopsis thaliana*. *Plant Mol. Biol.* **2008**, *67*, 659–670. [CrossRef] [PubMed]

57. Morohashi, K.; Minami, M.; Takase, H.; Hotta, Y.; Hiratsuka, K. Isolation and characterization of a novel GRAS gene that regulates meiosis-associated gene expression. *J. Biol. Chem.* **2003**, *278*, 20865–20873. [CrossRef] [PubMed]

58. Helariutta, Y.; Fukaki, H.; Wysocka-Diller, J.; Nakajima, K.; Jung, J.; Sena, G.; Hauser, M.T.; Benfey, P.N. The *SHORT-ROOT* gene controls radial patterning of the *Arabidopsis* root through radial signaling. *Cell* **2000**, *101*, 555–567. [CrossRef]

59. Wysocka-Diller, J.W.; Helariutta, Y.; Fukaki, H.; Malamy, J.E.; Benfey, P.N. Molecular analysis of SCARECROW function reveals a radial patterning mechanism common to root and shoot. *Development* **2000**, *127*, 595–603.

60. Gao, M.J.; Li, X.; Huang, J.; Gropp, G.M.; Gjetvaj, B.; Lindsay, D.L.; Wei, S.; Coutu, C.; Chen, Z.; Wan, X.C.; et al. SCARECROW-LIKE15 interacts with HISTONE DEACETYLASE19 and is essential for repressing the seed maturation programme. *Nat. Commun.* **2015**, *6*, 7243. [CrossRef] [PubMed]

61. Liu, M.; Pirrello, J.; Chervin, C.; Roustan, J.P.; Bouzayen, M. Ethylene control of fruit ripening: Revisiting the complex network of transcriptional regulation. *Plant Physiol.* **2015**, *169*, 2380–2390. [CrossRef]

62. Li, M.; Wang, X.; Li, C.; Li, H.; Zhang, J.; Ye, Z. Silencing *GRAS2* reduces fruit weight in tomato. *J. Integr. Plant Biol.* **2018**, *60*, 498–513. [CrossRef]

63. Marti, C.; Orzaez, D.; Ellul, P.; Moreno, V.; Carbonell, J.; Granell, A. Silencing of *DELLA* induces facultative parthenocarpy in tomato fruits. *Plant J.* **2007**, *52*, 865–876. [CrossRef] [PubMed]

64. Huang, W.; Peng, S.; Xian, Z.; Lin, D.; Hu, G.; Yang, L.; Ren, M.; Li, Z. Overexpression of a tomato miR171 target gene *SlGRAS24* impacts multiple agronomical traits via regulating gibberellin and auxin homeostasis. *Plant Biotechnol. J.* **2017**, *15*, 472–488. [CrossRef]

65. Liu, Y.; Huang, W.; Xian, Z.; Hu, N.; Lin, D.; Ren, H.; Chen, J.; Su, D.; Li, Z. Overexpression of *SlGRAS40* in tomato enhances tolerance to abiotic stresses and influences auxin and gibberellin signaling. *Front. Plant Sci.* **2017**, *8*, 1659. [CrossRef] [PubMed]

66. Fujisawa, M.; Shima, Y.; Higuchi, N.; Nakano, T.; Koyama, Y.; Kasumi, T.; Ito, Y. Direct targets of the tomato-ripening regulator RIN identified by transcriptome and chromatin immunoprecipitation analyses. *Planta* **2012**, *235*, 1107–1122. [CrossRef] [PubMed]

67. Fujisawa, M.; Nakano, T.; Shima, Y.; Ito, Y. A large-scale identification of direct targets of the tomato MADS box transcription factor RIPENING INHIBITOR reveals the regulation of fruit ripening. *Plant Cell* **2013**, *25*, 371–386. [CrossRef] [PubMed]

68. Ma, H.S.; Liang, D.; Shuai, P.; Xia, X.L.; Yin, W.L. The salt- and drought-inducible poplar GRAS protein SCL7 confers salt and drought tolerance in *Arabidopsis thaliana*. *J. Exp. Bot.* **2010**, *61*, 4011–4019. [CrossRef] [PubMed]

69. Yang, M.; Yang, Q.; Fu, T.; Zhou, Y. Overexpression of the *Brassica napus BnLAS* gene in *Arabidopsis* affects plant development and increases drought tolerance. *Plant Cell Rep.* **2011**, *30*, 373–388. [CrossRef]

70. Yuan, Y.; Fang, L.; Karungo, S.K.; Zhang, L.; Gao, Y.; Li, S.; Xin, H. Overexpression of *VaPAT1*, a GRAS transcription factor from *Vitis amurensis*, confers abiotic stress tolerance in Arabidopsis. *Plant Cell Rep.* **2016**, *35*, 655–666. [CrossRef] [PubMed]

71. Sato, A.; Miura, K. Root architecture remodeling induced by phosphate starvation. *Plant Signal Behav.* **2011**, *6*, 1122–1126. [CrossRef]

72. Xu, K.; Chen, S.; Li, T.; Ma, X.; Liang, X.; Ding, X.; Liu, H.; Luo, L. *OsGRAS23*, a rice GRAS transcription factor gene, is involved in drought stress response through regulating expression of stress-responsive genes. *BMC Plant Biol.* **2015**, *15*, 141. [CrossRef]

73. Yoshida, H.; Ueguchi-Tanaka, M. DELLA and SCL3 balance gibberellin feedback regulation by utilizing INDETERMINATE DOMAIN proteins as transcriptional scaffolds. *Plant Signal. Behav.* **2014**, *9*, e29726. [CrossRef]

74. Fambrini, M.; Mariotti, L.; Parlanti, S.; Salvini, M.; Pugliesi, C. A *GRAS*-like gene of sunflower (*Helianthus annuus* L.) alters the gibberellin content and axillary meristem outgrowth in transgenic *Arabidopsis* plants. *Plant Biol. (Stuttg)* **2015**, *17*, 1123–1134. [CrossRef]
75. Habib, S.; Waseem, M.; Li, N.; Yang, L.; Li, Z. Overexpression of *SlGRAS7* affects multiple behaviors leading to confer abiotic stresses tolerance and impacts gibberellin and auxin signaling in tomato. *Int. J.Genom.* **2019**, *2019*, 4051981. [CrossRef]
76. Tong, H.; Chu, C. Roles of DLT in fine modulation on brassinosteroid response in rice. *Plant Signal. Behav.* **2009**, *4*, 438–439. [CrossRef]
77. Chen, L.; Xiong, G.; Cui, X.; Yan, M.; Xu, T.; Qian, Q.; Xue, Y.; Li, J.; Wang, Y. OsGRAS19 may be a novel component involved in the brassinosteroid signaling pathway in rice. *Mol. Plant* **2013**, *6*, 988–991. [CrossRef]
78. Bailey, T.L.; Boden, M.; Buske, F.A.; Frith, M.; Grant, C.E.; Clementi, L.; Ren, J.; Li, W.W.; Noble, W.S. MEME SUITE: tools for motif discovery and searching. *Nucleic Acids Res.* **2009**, *37*, W202–W208. [CrossRef]
79. Chen, C.; Xia, R.; Chen, H.; He, Y. TBtools, a Toolkit for Biologists integrating various HTS-data handling tools with a user-friendly interface. *BioRxiv* **2018**. [CrossRef]
80. Fischer, S.; Brunk, B.P.; Chen, F.; Gao, X.; Harb, O.S.; Iodice, J.B.; Shanmugam, D.; Roos, D.S.; Stoeckert, C.J. Using OrthoMCL to assign proteins to OrthoMCL-DB groups or to cluster proteomes into new ortholog groups. *Curr. Protoc. Bioinform.* **2011**, *35*, 6.12.1–6.12.19.
81. Krzywinski, M.; Schein, J.; Birol, I.; Connors, J.; Gascoyne, R.; Horsman, D.; Jones, S.J.; Marra, M.A. Circos: An information aesthetic for comparative genomics. *Genome Res.* **2009**, *19*, 1639–1645. [CrossRef]
82. Wu, H.; Li, H.; Chen, H.; Qi, Q.; Ding, Q.; Xue, J.; Ding, J.; Jiang, X.; Hou, X.; Li, Y. Identification and expression analysis of strigolactone biosynthetic and signaling genes reveal strigolactones are involved in fruit development of the woodland strawberry (*Fragaria vesca*). *BMC Plant Biol.* **2019**, *19*, 73. [CrossRef]

 International Journal of
Molecular Sciences

Article

Arabidopsis NDL-AGB1 modules Play Role in Abiotic Stress and Hormonal Responses Along with Their Specific Functions

Arpana Katiyar and Yashwanti Mudgil *

Department of Botany, University of Delhi, New Delhi 110007, India; katiyar.arpana@gmail.com
* Correspondence: ymudgil@botany.du.ac.in

Received: 18 August 2019; Accepted: 19 September 2019; Published: 24 September 2019

Abstract: *Arabidopsis* N-MYC Downregulated Like Proteins (NDLs) are interacting partners of G-Protein core components. Animal homologs of the gene family N-myc downstream regulated gene (NDRG) has been found to be induced during hypoxia, DNA damage, in presence of reducing agent, increased intracellular calcium level and in response to metal ions like nickel and cobalt, which indicates the involvement of the gene family during stress responses. *Arabidopsis NDL* gene family contains three homologs *NDL1*, *NDL2* and *NDL3* which share up to 75% identity at protein level. Previous studies on NDL proteins involved detailed characterization of the role of NDL1; roles of other two members were also established in root and shoot development using miRNA knockdown approach. Role of entire family in development has been established but specific functions of NDL2 and NDL3 if any are still unknown. Our *in-silico* analysis of *NDLs* promoters reveled that all three members share some common and some specific transcription factors (TFs) binding sites, hinting towards their common as well as specific functions. Based on promoter elements characteristics, present study was designed to carry out comparative analysis of the *Arabidopsis NDL* family during different stages of plant development, under various abiotic stresses and plant hormonal responses, in order to find out their specific and combined roles in plant growth and development. Developmental analysis using GUS fusion revealed specific localization/expression during different stages of development for all three family members. Stress analysis after treatment with various hormonal and abiotic stresses showed stress and tissue-specific differential expression patterns for all three *NDL* members. All three *NDL* members were collectively showed role in dehydration stress along with specific responses to various treatments. Their specific expression patterns were affected by presence of interacting partner the *Arabidopsis* heterotrimeric G-protein β subunit 1 (AGB1). The present study will improve our understanding of the possible molecular mechanisms of action of the independent *NDL–AGB1* modules during stress and hormonal responses. These findings also suggest potential use of this knowledge for crop improvement.

Keywords: *Arabidopsis*; N-Myc downregulated like; abiotic stress; mannitol; PEG; promoter elements; GUS staining; plant hormones

---

1. Introduction

G-proteins are well characterized for signal transduction in response to many environmental stresses like drought, heat, salinity and light intensity. Internalization of AtRGS1 (Regulator of G protein signaling protein) in response to NaCl indicates role of G-protein signaling in stress responses [1]. Expression levels of *AGB1* were significantly upregulated during salt treatment, whereas showed down regulation during heat and cold stress [2]. G-protein interactome has been published using Yeast two hybrid (Y2H) and in planta interaction studies, NDL interactome has also been solved as part of same initiative and hints towards role of NDLs in various stress responses [3]. Animal homolog of plant NDLs consist of four members NDRG1-4. NDRG1 is a ubiquitously expressed intracellular protein that is induced under number of stresses and pathological responses [4]. NDRG1 upregulates under various stress conditions like hypoxia, in the presence of a reducing agent, DNA damage and in response to increased intracellular calcium concentration [5–9]. Various metal ions like nickel, cobalt, and iron are also reported for inducing expression of NDRG1 [10].

Plant NDR proteins were first reported in sunflower (SF21) as pollen, stigma and transmitting tissue cell-specific proteins with putative role, as a signaling molecule in pollen-pistil interaction [11]. Detailed analysis of expression profiles of gene family members in sunflower revealed multiple alternative and organ-specific transcripts indicating spatial and developmental regulation and roles [12,13].

In *Arabidopsis* NDL1 was re-discovered as AGB1/AGG dimmer interacting protein, three homologs had been identified, showing presence of a conserved NDR domain and an alpha/beta hydrolase fold. All NDLs interact with AGB1/AGG (βγ) dimer and C4 domain of RGS1 [14]. Detailed characterization of *Arabidopsis* NDL1 and widespread presence of NDLs from bryophytes to angiosperm hints towards their important functions, though their specific molecular and cellular mechanism of action which still needs to be discovered.

A recent study involving correlation of the gene expression and morpho-physiological traits under severe water-deficient conditions reports expression of *NDL1* positively correlated with rate of transpiration and projected rosette area [15]. NDLs interactome predicts AtNDL1 interaction with ANNEXIN (drought-responsive gene) Sodium and Lithium-Tolerant 1 (SLT1-salt stress-responsive gene) O-Acetylserine (Thiol) Lyase (OAS-TL) Isoform A1 (characterized for its role during cadmium tolerance) *Arabidopsis* Ribosomal Protein (ARS27A involved in genotoxic stress) hinting us to speculate that AtNDL proteins playing important role in abiotic stress signaling through these NDLs stress related interactors (NSRI), stress signaling downstream components (Figure 1A showing location of all NSRI members on *Arabidopsis* genome) [3].

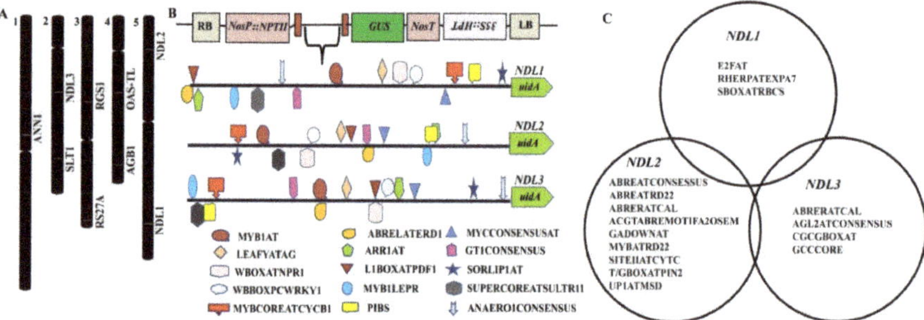

Figure 1. Mapping of the *Arabidopsis* N-MYC Downregulated Like Proteins (*AtNDL*) gene family and putative abiotic stress interacting partners, NDLs stress related interactors (NSRI) on *Arabidopsis* genome along with the stress related elements in the promoter region of *AtNDL* gene family. (**A**) Physical Map of AGI at TAIR using Chromosome Map tool to indicate Chromosomal location of *NDLs* (*NDL1*, *NDL2*, and *NDL3*) and their stress related interactors in *Arabidopsis*. The names of the genes are shown to the right of each chromosome and location of genes is indicating their absolute location on respective chromosome. *RGS1*-Regulator of G protein signaling, *ANN1*-ANNEXIN, *SLTI*-Sodium and Lithium-Tolerant 1, *RS27A*- *Arabidopsis* Ribosomal Protein, *OAS-TL*-O-Acetylserine (Thiol) Lyase. (**B**) In silico analysis of the promoter region of *AtNDL* gene family promoter regions analyzed to get information about common *cis*-regulatory elements present on each of the member. (**C**) Specific *cis*-regulatory elements present on each *AtNDL* member. Different colors are used to indicate for each *cis*-regulatory element.

Previous *NDL* characterization study used miRNA knockdown approach of the entire *NDL* gene family, as well as the detailed overexpression and localization of NDL1, revealed their important role in regulating auxin transport and distribution. Detailed expression/localization profiling of the *NDL1* homologs *NDL2* and *NDL3* had not been done, though all three proteins share 75% identity at amino acid level [14]. Current study aims to dissect out independent functional significance of all three *NDL* gene family members and their role in plant stress management, comparative transcription profiling of all three *NDL* gene family members is performed in silico and *in vivo*. Previously it has been reported that all *NDL* members play role in root as well as shoot development, AGB1 presence is not essential for transcriptional stability of *NDL1* but needed for the stability of NDL1 protein *in planta* [14,16] in order to find-out effect of AGB1 availability on expression profile of *NDL2* and *NDL3* their expression is also studied in *agb1-2* background. The main objectives of the current study are:

1. In silico comparative account of the regulatory elements and expression profiles of *NDL1, NDL2,* and *NDL3* to ascertain their characteristics, similarities, and differences.
2. *In planta* comparative analysis of the all three members of the *NDL* gene family during different stages of plant growth and development
3. Comparative expression profile of *NDL* members in response to various abiotic stresses and hormonal treatments in presence and absence of AGB1.

2. Results

2.1. In Silico Analysis of the Upstream Regulatory Regions of Arabidopsis NDL1, NDL2 and NDL3

Comparative analysis of the *cis*-elements present in the promoter sequences of all three *NDL* members was done using the PLANTPAN 2.0 software. In silico comparative analysis revealed presence of many common TFs binding sites. All three members showed presence of AtMYB2 and AtMYC2 binding sites which are already established characteristic of *NDL* gene family in animals and reason for their nomenclature [17,18]. Abiotic stress-responsive TFs that are involved in dehydration (MYB1AT and ABRELATERD1), sulfur-responsive (SURECOREATSULTR11), phosphate-responsive (PIBS) and SA-induced (WBOXATNPR1), cytokinin-responsive element (ARR1) and defense related TFs like MYB1LEPR binding motif are also present in all the three members of the family. Presence of all these common elements in all three *NDL* genes hints towards the common role of *NDL* family during abiotic and biotic responses. Light-induced TFs like GT1CONSENSUS and SORLIP1AT and developmental and cell cycle-responsive TFs like LEAFYATG and MYBCOREAYCYCB1 are also present in all three members indicating diverse role of *NDL* family members during various stages of plant growth and development (Figure 1B and Table 1).

In addition to these common regulatory elements unique TFs binding sites specific to the promoter region of *NDL1, NDL 2* and *NDL3* are also observed (Figure 1C). *NDL1* promoter showed presence of E2F binding sites indicating its cell cycle-specific functions. Other noted specific elements of *NDL1* promoter region are SBOXATRBCS, important for sugar and abscisic acid (ABA) responses, and RHERPATEXPA7, a root hair specific cis element (Table 2).

NDL2, promoter harbors three ABA-responsive elements, ABREATCONSENSUS, ABREATRD22, and ACGTABREMOTIFA2OSEM. It is well established that drought stress triggers production of ABA which in turn induces various other drought inducible genes [19]. Thereby presence of ABA-responsive elements in the promoter sequence of *NDL2* along with MYBATRD22 dehydration-responsive element suggests regulation and involvement of *NDL2* during dehydration/drought stress (Table 2). Along with ABA-responsive elements other specific sites present in the promoter region of *NDL2* are Gibberellic acid (GA) (GADOWNAT) and Jasmonic acid (JA) (T/GBOXATPIN2)-responsive elements. *NDL3* promoter sequence shows presence of one pathogen-responsive element (GCCCORE), and two Ca^{2+}/Calmodulin binding sites (CGCGBOXAT) along with AGAMOUS-LIKE2 and HD-ZIP transcription binding sites.

Table 1. List of common transcription factor (TF) binding sites in *NDL1*, *NDL2*, and *NDL3* promoter.

S.NO.	TFs	Function
1	MYB1AT	Dehydration-responsive elements
2	ARR1AT	Cytokinin response regulators
3	GT1CONSENSUS	Salicylic acid-responsive elements, Light-responsive elements
4	MYCCONSENSUSAT	dehydration-responsive
5	SURECOREATSULTR11	SURE contains auxin response factor (ARF) binding sequence
6	L1BOXATPDF1	MYB binding motif
7	WBOXATNPR1	Salicyclic acid-responsive elements
8	MYBCOREATCYCB1	Cyclin B1-responsive elements
9	ABRELATERD1	ABA-responsive elements
10	ANAERO1CONSENSUS	Anaerobic-responsive elements
11	WBBOXPCWRKY1	Pathogenesis-related elements
12	LEAFYATAG	Target sequence of LEAFY
13	P1BS	Phosphate starvation-responsive elements
14	SORLIP1AT	Light-responsive elements
15	MYB1LEPR	Defence-related elements

Table 2. List of specific TFs binding sites in promoter region of *NDL1*, *NDL2*, and *NDL3*.

S.NO.	TFs	Function
NDL1	E2FAT	Cell cycle-responsive elements
	RHERPATEXPA7	Root hair specific-cis elements
	SBOXATRBCS	Sugar and ABA-responsive elements
NDL2	ABREATCONSENSUS	ABA-responsive elements
	ABREATRD22	ABA-responsive elements
	ABRERATCAL	Ca^{2+}-responsive elements
	ACGTABREMOTIFA2OSEM	ABA-responsive elements
	GADOWNAT	GA-responsive elements
	MYBATRD22	Dehydration-responsive elements
	SITEIIATCYTC	Responsible for oxidative phosphorylation
	T/GBOXATPIN2	JA-responsive elements
	UP1ATMSD	Upregulation after main stem decapitation
NDL3	ABRERATCAL	Ca^{2+}-responsive elements
	AGL2ATCONSENSUS	AGAMOUS-LIKE 2
	CGCGBOXAT	Ca^{2+}/Calmodulin response elements
	GCCCORE	Pathogen-responsive elements

Presence of common TFs binding sites in the upstream promoter regions of all three *NDL* members postulates their involvement in common regulatory mechanisms, while the presence of specific TFs binding sites might attribute unique functions to each *NDL* family member.

2.2. Comparative In Silico Expression Analysis of NDL Members

Comparative expression analysis of *NDL1*, *NDL2*, and *NDL3* has been performed using genevestigator microarray data. All three *NDL* members showed expression across all developmental stages, *NDL2* showed expression little higher than *NDL1* and *NDL3* (Figure 2A). Microarray analysis in different plant tissues revealed that *NDL* genes exhibit differential expression pattern across various organs. *NDL1* and *NDL2* showed highest expression in pollen while *NDL3* showed highest expression in carpel and shoot apex (Figure 2B). Since, the role of *NDLs* has been already established in auxin transport during root development [14]. We closely looked at the expression levels of the *NDLs* members in different parts of roots and found that expression level of all the members were comparable, in primary roots *NDL1* and *NDL2* is comparatively little higher than *NDL3*. Overall expression pattern revealed ubiquitous expression of the gene family which indicates important function of the family members during plant growth and developmental responses (Figure 2B).

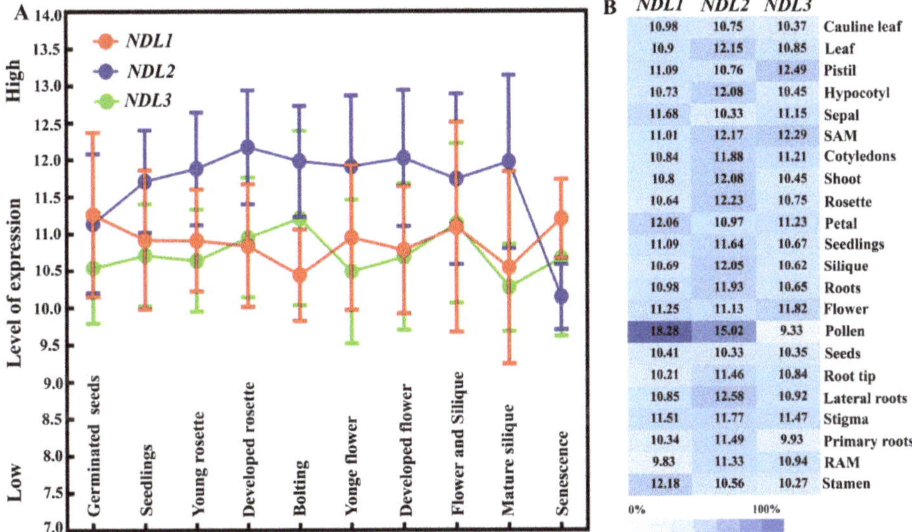

B	NDL1	NDL2	NDL3	
	10.98	10.75	10.37	Cauline leaf
	10.9	12.15	10.85	Leaf
	11.09	10.76	12.49	Pistil
	10.73	12.08	10.45	Hypocotyl
	11.68	10.33	11.15	Sepal
	11.01	12.17	12.29	SAM
	10.84	11.88	11.21	Cotyledons
	10.8	12.08	10.45	Shoot
	10.64	12.23	10.75	Rosette
	12.06	10.97	11.23	Petal
	11.09	11.64	10.67	Seedlings
	10.69	12.05	10.62	Silique
	10.98	11.93	10.65	Roots
	11.25	11.13	11.82	Flower
	18.28	15.02	9.33	Pollen
	10.41	10.33	10.35	Seeds
	10.21	11.46	10.84	Root tip
	10.85	12.58	10.92	Lateral roots
	11.51	11.77	11.47	Stigma
	10.34	11.49	9.93	Primary roots
	9.83	11.33	10.94	RAM
	12.18	10.56	10.27	Stamen

Figure 2. In silico expression analysis of *NDL* gene family. (**A**) GENEVESTIGATOR microarray expression analysis of *NDL1*, *NDL2* and *NDL3* during different developmental stages of plant growth. (**B**) In silico microarray analysis showing tissue-specific expression of *NDL1*, *NDL2*, and *NDL3* using GENEVESTIGATOR.

2.3. Comparative In Vivo Expression Analysis of NDLs during Early Stages of the Plant Growth

In-vivo expression analysis of all three *NDL* members showed that the localization pattern of GUS is quite unique for each member. During the initial stages of plant growth (4–12 day-old seedlings) *pNDL1-NDL1-GUS* showed localization in cotyledonary leaves, true leaves, various parts of primary and lateral roots including elongation zone and root apical meristem (RAM) (Figure 3A–C and Table 3). *pNDL2-GUS* showed GUS expression in cotyledonary leaves, true leaves, hypocotyl and maturation zone of primary root, but strikingly missing from root tip and in lateral roots (Figure 3G–I and Table 3). *pNDL3-GUS* expression was missing in cotyledonary leaves, present in newly emerging true leaves, present in the RAM of the primary root and in lateral roots (Figure 3M–O and Table 3).

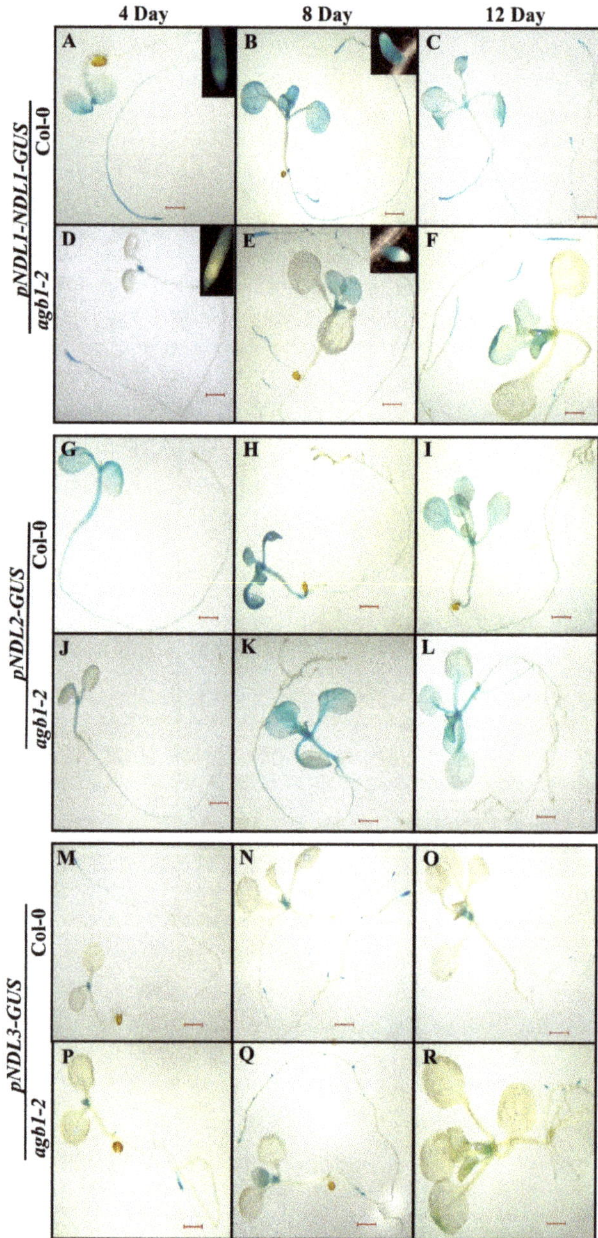

Figure 3. *In planta* expression analysis of *NDL* gene family during different developmental stagesin presence and absence of *AGB1*: (**A–C**, **G–I** and **M–O**). Histochemical GUS staining of *pNDL1-NDL1-GUS*, *pNDL2-GUS* and *pNDL3-GUS* seedlings in wild type Col-0 background. (**D–F**, **J–L** and **P–R**) Histochemical GUS staining of *pNDL1-NDL1-GUS*, *pNDL2-GUS* and *pNDL3-GUS* seedlings in *agb1-2* mutant background. Histochemical GUS analysis in transgenic *Arabidopsis* plants during different stages of development was carried out 4 day, 8 day, and 12 day old seedlings were analyzed. GUS expression is under the control of *NDL2* and *NDL3* promoter and translational fusion in case of *NDL1*, Scale Bar = 0.2 μM.

Table 3. *AtNDLs* localization during various stages of development and after abiotic and hormonal treatments.

Stage	Organ	Tissue	Col-0			agb1-2		
			AtNDL1	*AtNDL2*	*AtNDL3*	*AtNDL1*	*AtNDL2*	*AtNDL3*
8-day old seedlings	PR	RT	++	–	++	–	–	++
		CDZ	++	–	++	++	–	++
		EZ	++	–	–	++	–	–
		MZ	++	++	–	–	++	–
	LR		++	–	++	++	–	++
	Hypocotyl		–	++	–	–	++	–
	Cotyledons		++	++	–	–	++	–
	Leaves		++	++	++	++	++	++
6-day old seedlings (Cold)	PR	RT	++	–	++	–	–	++
		CDZ	++	–	++	++	–	++
		EZ	++	–	–	++	–	–
		MZ	+	+	–	+	++	–
	Hypocotyl		++	++	++	–	++	–
	Cotyledons		+++	+++	++	–	++	–
6-day old seedlings (Heat)	PR	RT	–	–	++	–	–	++
		CDZ	–	–	++	–	–	++
		EZ	–	–	–	–	–	–
		MZ	–	++	–	–	++	–
	Hypocotyl		+	++	–	–	++	–
	Cotyledons		–	++	+	–	++	–

Table 3. *Cont.*

Stage	Organ	Tissue	Col-0			agb1-2		
			AtNDL1	AtNDL2	AtNDL3	AtNDL1	AtNDL2	AtNDL3
6-day old seedlings (Mannitol)	PR	RT	++	−	++	−	−	++
		CDZ	+	−	+	++	−	++
		EZ	+	−	−	−	−	−
		MZ	−	++	−	−	++	−
	Hypocotyl		++	++	−	−	++	−
	Cotyledons		++	++	−	+	++	+
6-day old seedlings (PEG)	PR	RT	+++	−	−	−	−	+++
		CDZ	+++	−	−	+++	−	++
		EZ	++	−	−	+	−	−
		MZ	−	++	−	−	++	−
	Hypocotyl		−	++	+	−	++	−
	Cotyledons		++	++	+	−	++	−
6-day old seedlings (Salt)	PR	RT	++	−	++	−	−	++
		CDZ	++	−	++	++	−	++
		EZ	+	−	−	++	−	−
		MZ	+	++	−	−	++	−
	Hypocotyl		++	++	+	−	++	+
	Cotyledons		++	++	+	−	++	−
6-day old seedlings (ABA)	PR	RT	++	−	++	−	−	++
		CDZ	++	−	++	++	−	++
		EZ	++	−	−	++	−	−
		MZ	−	++	−	++	++	−
	Hypocotyl		−	++	−	−	++	−
	Cotyledons		++	+++	++	−	++	−

Table 3. *Cont.*

			Col-0			agb1-2		
Stage	**Organ**	**Tissue**	*AtNDL1*	*AtNDL2*	*AtNDL3*	*AtNDL1*	*AtNDL2*	*AtNDL3*
6-day old seedlings (GA)	PR	RT	++	−	++	−	−	++
		CDZ	++	−	++	++	−	++
		EZ	++	−	−	+++	−	−
		MZ	+	+	−	++	++	−
	Hypocotyl		+	++	+	−	++	−
	Cotyledons		++	++	+	−	++	−
6-day old seedlings (IAA)	PR	RT	++	−	++	−	−	++
		CDZ	+++	−	++	++	−	++
		EZ	+++	−	−	++	−	−
		MZ	++	+	−	+	++	−
	Hypocotyl		++	++	+	−	++	−
	Cotyledons		++	++	++	−	++	−
6-day old seedlings (JA)	PR	RT	++	−	++	+++	−	++
		CDZ	++	−	++	+++	−	++
		EZ	++	−	−	+++	−	−
		MZ	++	+	−	++	++	−
	Hypocotyl		++	++	−	−	++	−
	Cotyledons		++	+++	−	−	++	−
6-day old seedlings (SA)	PR	RT	++	−	++	−	−	++
		CDZ	++	−	++	+	−	++
		EZ	+	−	−	+	−	−
		MZ	−	++	−	−	++	−
	Hypocotyl		+	++	+	−	++	−
	Cotyledons		++	++	+	−	++	+

Table 3. *Cont.*

Stage	Organ	Tissue	Col-0			agb1-2		
			AtNDL1	*AtNDL2*	*AtNDL3*	*AtNDL1*	*AtNDL2*	*AtNDL3*
6-day old seedlings (BAP)	PR	RT	++	−	++	−	−	++
		CDZ	++	−	++	++	−	++
		EZ	++	−	−	++	−	−
		MZ	+	++	−	++	++	−
	Hypocotyl		−	++	−	−	++	−
	Cotyledons		+++	++	+	−	++	+
6-day old seedlings (Control)	PR	RT	++	−	++	−	−	++
		CDZ	++	−	++	++	−	++
		EZ	++	−	−	++	−	−
		MZ	−	++	−	++	++	−
	Hypocotyl		−	++	++	−	++	−
	Cotyledons		++	++	++	−	++	−

PR = Primary Root, LR = Lateral Root, CDZ = Cell division Zone, EZ = Elongation Zone, MZ = Maturation Zone, Root tip = RT. (-) No GUS staining, (++) basal staining level in each genotype without any treatment, (+++) increased level of staining, (+) decreased levels of staining compared to basal genotypic levels.

2.4. Comparative In Vivo Expression Analysis of NDLs in Absence of AGB1

Previously, we have found that presence of AGB1 doesn't affect transcriptional levels of *NDL1* but necessary to maintain the post translational steady state level of NDL1 [14]. Since all three NDL members had shown physical interaction with AGB1 in Y2H we hypothesized that NDL2 and NDL3 might require its partner, AGB1, either to regulate expression levels or for post-translational stability similar to NDL1. In order to test this speculation, expression of *pNDL2-GUS* and *pNDL3-GUS* was studied in *agb1-2* background.

In *agb1-2* background *pNDL1-NDL1-GUS* localization was absent from cotyledonary leaves (Figure 3D–F and Table 3) but present in true leaves. In case of roots the localization was absent from both primary and lateral roots RAM (Figure 3D,E inset) whereas present in other parts of roots similar to Col-0 background. In case of *pNDL2-GUS* and *pNDL3-GUS* localization pattern of the GUS was same in *agb1-2* mutant background as compared to wild type Col-0. (Figure 3J–L for *NDL2*, Figure 3P–R for *NDL3* and Table 3). These finding indicate that expression levels of both *NDL2* and *NDL3* are also unaffected by the absence of *AGB1* similar to previously reported for *NDL1*.

2.5. In Silico Expression Analysis of NDLs under Various Stress and Hormonal Treatments

Previous study has confirmed that *NDLs* play role in auxin transport during lateral root formation [14]. Correlation of gene expression and responses of morpho-physiological traits under severe water-deficient conditions indicate substantial alteration in expression levels of *AtNDL1* during drought stress responses using tanscriptomic analysis [15]. In order to find out if independent NDL members play independent specific roles or have combined redundant functions during abiotic stress exposure and hormonal treatments, we have analyzed their comparative expression profiles from publically available databases. (Genevestigator and eFP browser).

Genevestigator Microarray data revealed that after auxin (IAA) treatment, expression levels of *NDL1* decreased to nearly 1.77 fold, expression of *NDL2* and *NDL3* showed a similar trend, decrease of around 1.3 fold. (Figure 4A). In presence of ABA, *NDL2* showed a decrease (2.21 fold) in expression level, as compared to *NDL1* and *NDL3* which showed an upregulation up to ~1.12 fold and down regulation of nearly 1.25 fold respectively. Unlike *NDL2* and *NDL3*, *NDL1* showed slight decrease (1.1 fold) in its expression in response to GA (Figure 4A).

All *NDL* members also showed a response toward various abiotic stresses (heat, cold and drought). In case of heat treatment in green tissue *NDL1* showed increase (2.08 fold) as compared to *NDL2* and *NDL3* which showed a decrease in expression (1.44 and 1.52 fold respectively) in case of roots also similar trend was there. Even for cold treatment *NDL1* showed positive regulation (1.06 fold up) while down regulation was observed for *NDL2* and *NDL3* (3.59 and 1.43 fold respectively). Under drought stress all members have positive response, *NDL2* showed upregulation in expression (3.59 fold) followed by *NDL3* and *NDL1* (increase of 2.11 and 1.60 fold respectively, Figure 4A). The observed high expression of *NDL2* gene in drought stress thereby could possibly be linked to the fact that multiple ABA-responsive factor binding elements are present in promoter sequence of *NDL2*.

To find out optimal time point for the in vivo expression analysis we also performed comparative in silico expression analysis of *NDLs* at different time points during various abiotic stress treatments using eFP browser. For *NDL1* maximum transcript levels compared to no-treatment control were detected at 24 h. In case of *NDL2* expression, levels fluctuated during different stress treatments and variation was found at different time points but overall maximum levels were observed at 24 h. A trend similar to *NDL2* was observed for *NDL3* expression during different treatments and time points (Figure 4B)

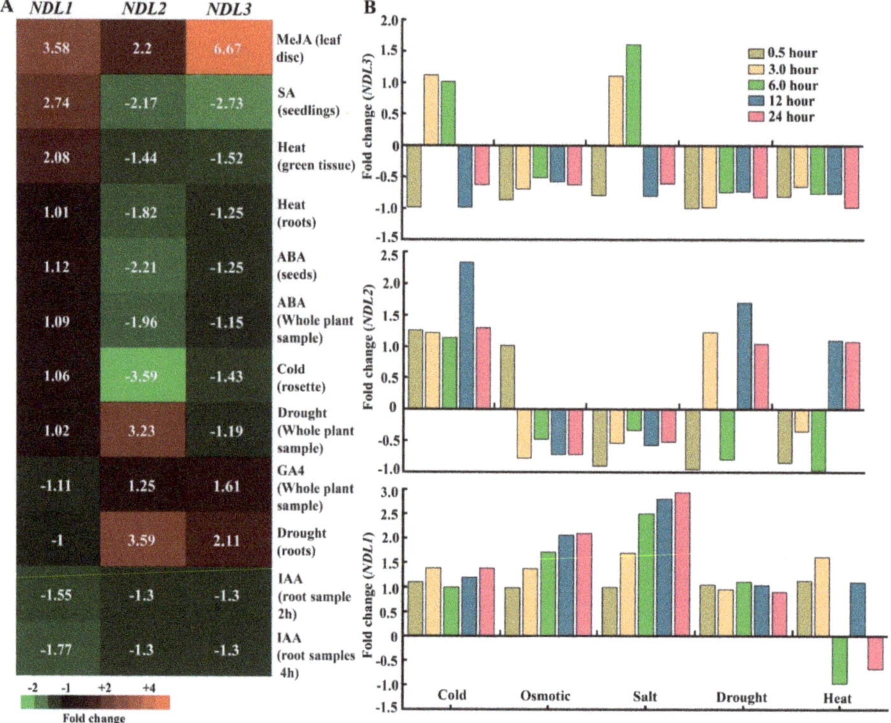

Figure 4. In silico expression analysis of *NDL* gene family during abiotic stress and hormonal treatment using genevestigator and eFP browser: (**A**). Comparative *in-silico* expression analysis of *NDL1*, *NDL2*, and *NDL3* using GENEVESTIGATOR microarray data, under different abiotic stress and hormonal treatments (**B**). In silico analysis of *NDL1*, *NDL2*, and *NDL3* during abiotic stress treatments at different time points (0.5, 3, 6, 12, and 24 h) comparative expression is shown as fold change compared to no stress control.

2.6. In Vivo Expression Analysis under Abiotic Stresses and Hormonal Treatments

In order to find out maximum expression levels for all three *NDLs* promoter GUS/β-glucuronidase activity was tested using 4-Methylumbelliferyl-β-D-glucuronide (MUG) assay for two time points (6 h and 24 h) in Col-0 background, no significant difference was found between expression levels of *NDLs* and two time points. (Supplementary Figure S1). Based on the in vivo GUS/β-glucuronidase activity levels and in silico analysis of the expression profiles of all three *NDL* members, treatment with various hormones and abiotic stresses were designed. Six-day-old seedlings were subjected to 24 h treatment with various hormones (ABA, GA, IAA, JA, SA, and BAP) and abiotic stresses (cold, heat, Mannitol, PEG and salt) followed by measurement of GUS activity either histochemically or by fluorometry.

pNDL1-NDL1-GUS, showed increase in the GUS staining intensity when treated with cold, IAA (auxin), JA, Salicyclic acid (SA), BAP (cytokinin), and Poly ethylene glycol (PEG) and almost no staining observed with heat treatment (Figure 5). GUS staining remained unaffected after treatment with ABA (Supplementary Figure S2), GA (Supplementary Figure S3) and mannitol (Supplementary Figure S4). In case of salt treatment, no difference was found in root but hypocotyle showed increase in the GUS staining intensity (Supplementary Figure S5) compared to no stress treatment control.

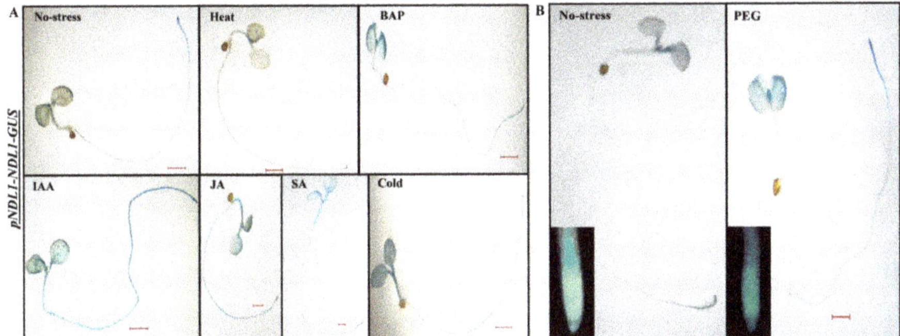

Figure 5. In vivo histochemical GUS activity for *NDL1* in wild type background in response to different abiotic stresses and hormones: (**A**). Histochemical analysis of GUS activity for *NDL1* in wild type background in response to different abiotic stress and hormones. Changes in intensity of GUS staining were detected with Heat, Cold, IA, JA, SA and BAP. GUS staining was done in seedlings that were six days old and 24 h of stress treatment followed. (**B**) Histochemical GUS staining of 6-D-old *pNDL1-NDL1-GUS* seedlings in wild type Col-0 background. Increased levels of GUS staining intensity detected with PEG at root tip (inset). Result shown is representative of three independent biological replicates ($n \geq 10$ in each experiment). Scale Bar = 0.2 μM for seedlings and 0.02 μM for inset RAM.

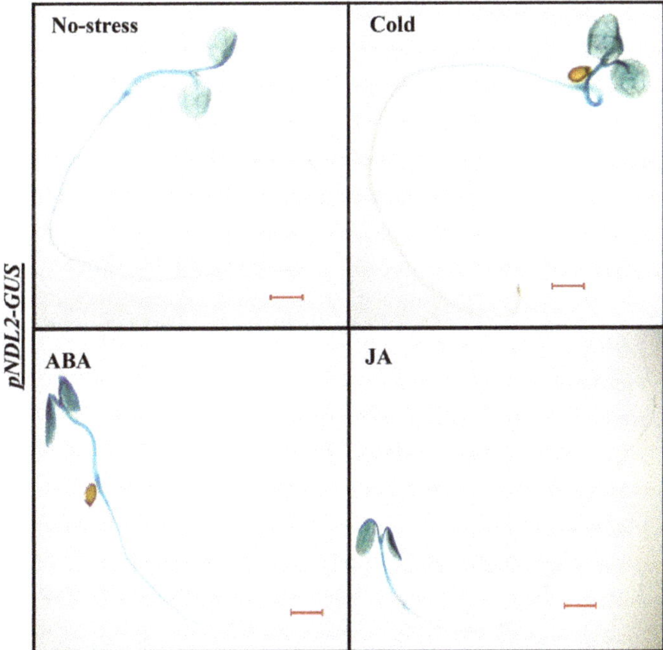

Figure 6. In vivo histochemical analysis of GUS activity for *NDL2* in wild type Col-0 background in response to different abiotic stresses and hormones: Histochemical GUS staining of 6-D-old transgenic *Arabidopsis* seedlings (*pNDL2-GUS*) in wild type Col-0 background. Increased intensity levels of GUS staining detected with ABA, cold, and JA. Result shown is representative of three independent biological replicates ($n \geq 10$ in each experiment). Scale Bar = 0.2 μM.

pNDL2-GUS showed increased levels of GUS staining in the cotyledons and hypocotyl after treatment with ABA, cold and JA (Figure 6), primary root showed no expression similar to no treatment control. BAP (Supplementary Figure S2), IAA, GA, heat (Supplementary Figure S3), Mannitol and PEG (Supplementary Figure S4) and SA (Supplementary Figure S5) and other treatments had no effect on expression level as no difference was observed compared to no stress treatment control.

In case of *pNDL3-GUS,* no difference in the GUS staining was detected when seedlings were treated with ABA, BAP, cold (Supplementary Figure S2), IAA, GA, JA, heat (Supplementary Figure S3), salt and SA (Supplementary Figure S5). Mannitol and PEG treatment resulted in decreased level of *pNDL3-GUS* expression in RAM and hypocotyl in comparison to no treatment control which showed diminished staining at both the places. GUS staining was absent in the rest of the seedlings, same as in case of no treatment control of six-day old seedling (Figure 7 and Supplementary Figure S4).

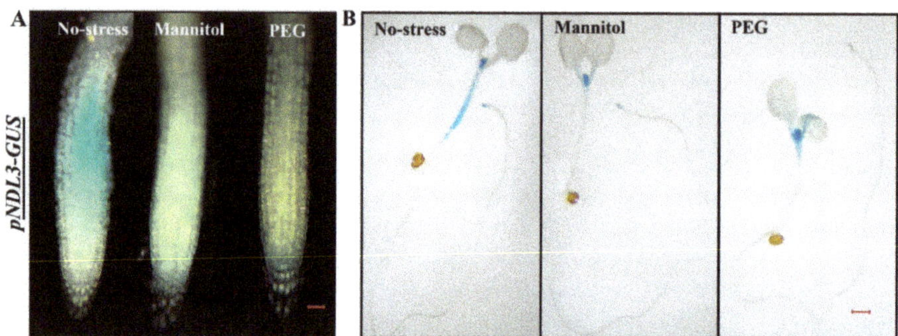

Figure 7. In vivo histochemical analysis of GUS activity for *NDL3* in wild type Col-0 background in response to different abiotic stresses and hormones: (**A**). Histochemical GUS staining of 6-D-old transgenic *Arabidopsis* seedlings (*pNDL3-GUS*) in wild type Col-0 background. Changes in intensity of GUS staining detected with PEG and Mannitol in RAM. (**B**) and in hypocotyl part of the seedlings. Result shown is representative of three independent biological replicates (*n* ≥ 10 in each experiment). Scale Bar = 0.2μM for seedlings and 0.02 μM for RAM.

Results indicate that all the three family members' show independent response when encountering various stress/hormonal treatments. Both *NDL1* and *NDL3* after treatment with Mannitol and PEG showed difference in GUS intensity indicates that both the members are working when plants encounter with drought and osmotic stress. Although both the members have altered expression in same stress mechanism of functioning seems opposite as *NDL1* showed up regulation after PEG treatment while *NDL3* showed down regulation when treated with PEG (compare Figures 5B and 7), while *NDL2* expression remains unaffected by both treatments in tested plant stage (Supplementary Figure S4).

2.7. In Vivo Expression Analysis of NDLs in Absence of AGB1 under Abiotic Stress and Hormonal Treatments

Previous study has reported that expression levels of *NDL1* were not affected but protein steady state levels were affected by AGB1 [14]. In order to find out role of AGB1 on expression levels of *NDL2* and *NDL3* their expression patterns were studied in *agb1-2* mutant. We found that expression of *NDL2* and *NDL3* were not affected in *agb1-2* mutant and stays same as in wild type Col-0 background (Figure 3). In order to find-out whether AGB1 has any role on expression during abiotic stress or hormonal treatments expression pattern was observed for all the three members in absence of AGB1 after abiotic (cold, heat, Mannitol, salt and PEG), and hormonal (ABA, GA, IAA, JA, SA and BAP) treatments.

pNDL1-NDL1-GUS showed increased GUS intensity in primary root of the seedlings compared to no treatment control for GA, Mannitol, PEG and JA while no staining was detected in the primary root after heat treatment (Figure 8A,B). MUG assay was also performed and significantly increased GUS activity was observed only for PEG and Mannitol treatment in comparison to no-treatment control seedlings (Figure 8C). GUS staining intensity remains unaffected for ABA, BAP, cold (Supplementary Figure S2), IAA (Supplementary Figure S3), salt and SA (Supplementary Figure S5) treatments. Response of ABA and salt for *pNDL1-NDL1-GUS* remained similar in both the background (in wild type Col-0 as well as in *agb1-2* mutant) meaning presence of AGB1 don't make any difference on expression, while for cold, IAA, SA and BAP upregulation was detected in the cotyledons and in primary root in comparison to no treatment control seedlings in case of wild type

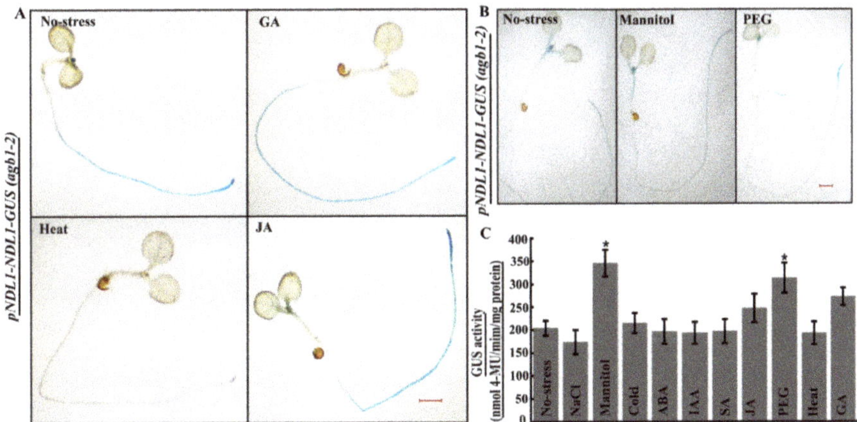

Figure 8. In vivo histochemical GUS staining of 6-day old transgenic *Arabidopsis* seedlings (*pNDL1-NDL1-GUS*) in *agb1-2* mutant background: Histochemical GUS staining was done in transgenic *Arabidopsis* seedlings that were six days old treatment was given for 24 h. (**A**) Changes in GUS intensity levels were detected with GA, heat and JA. (**B**). Increased GUS staining detected for Mannitol and PEG at the cell division and elongation part of the primary root. (**C**) Abiotic stress treatments followed by fluorometric MUG assay was done for quantitative estimation of stress response. Significantly higher GUS activity was obtained for Mannitol and PEG treatment compared to No-stress control indicative of NDL1 involvement during osmotic and drought responses (Students *t*-test = * *p* value < 0.05, Error bars represent SD). Result shown is representative of three independent biological replicates (*n* ≥ 10 in each experiment). Scale Bar = 0.2 µM.

Col-0 background while remains unaffected in *agb1-2* mutant background meaning presence of AGB1 is vital for NDL1 stability during cold, IAA, SA and BAP treatment, indicating stress specific role of NDL1-AGB1 module.

The expression analysis for *pNDL2-GUS* in terms of GUS staining was also observed in absence of AGB1. Seedlings were treated with ABA, cold, IAA, GA, heat, JA, Mannitol, PEG, salt, SA and BAP. No difference in GUS staining was detected in *agb1-2* mutant background compared to no treatment control (Supplementary Figures S2–S5) suggesting no role/requirement of AGB1 for *NDL2* expression during treatment in *agb1-2* background. In case of cold, ABA and JA treatment upregulation of *pNDL2-GUS* was detected in the cotyledons in wild type Col-0 background (Figure 6) but in *agb1-2* background this effect was not present (Supplementary Figures S2–S5) meaning AGB1 presence is required during cold, ABA and JA treatments for *NDL2* expression upregulation.

In case of *pNDL3-GUS*, no difference in expression was detected in for ABA, BAP, Cold (Supplementary Figure S2), IAA, GA, JA, heat (Supplementary Figure S3), salt, SA (Supplementary Figure S5) in comparison to no treatment control. For all these treatments results were similar in both the background in wild type as well as in *agb1-2* mutant background. Very interestingly after treatment with PEG and Mannitol in *agb1-2* mutant background intensity of GUS staining showed increase in the RAM which is opposite effect compared to the Col-0 background where a decline in hypocotyl and almost absence of expression was observed in RAM (Figure 7). This means differential and opposite role of AGB1 is operating for regulating *NDL3* expression under normal and dehydration stress condition (Figure 9). Presence of AGB1 is needed under stress condition to downregulate *NDL3* expression, which might need to be downregulated under dehydration stress.

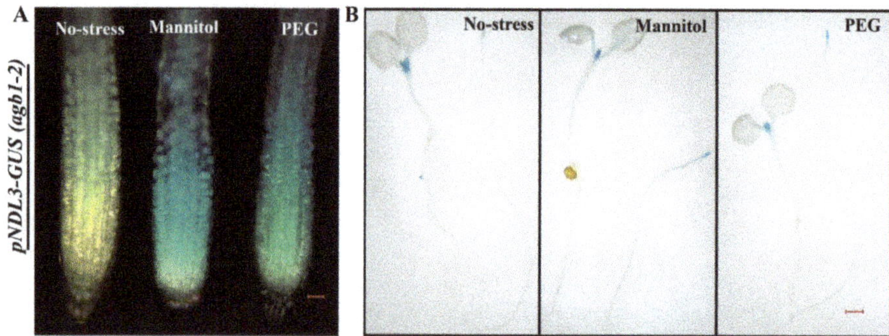

Figure 9. In vivo histochemical GUS staining of 6-day old transgenic *Arabidopsis* seedlings (*pNDL3-GUS*) in *agb1-2* mutant background: (**A**). Increased levels of GUS staining intensity were detected in RAM compare to almost no staining in no treatment control with Mannitol and PEG treatment. (**B**) Histochemical GUS staining was analysed in whole seedling also. ne Six days old *Arabidopsis* transgenic seedlings were subjected to treatment for 24 h. Result shown is representative of three independent biological replicates (*n* ≥ 10 in each experiment). Scale Bar = 0.2 μM for seedlings and 0.02 μM for RAM.

3. Discussion

Comparative analysis for all three members of *NDL* gene family revealed their role in many common and unique functions. *In-silico* analysis revealed that all three members share many common transcription factor binding sites (Table 1). Presence of MYC binding factors postulate evolutionary conserved regulatory mechanisms of downregulation similar to animal systems for *AtNDL* family. Regulatory regions of all the three members contain motifs for stresses (biotic and abiotic), development, and for hormonal responses. MYB recognition sites like MYB1AT and MYB1LEPR are present in all the *NDL* members. MYB1AT is a MYB recognition site which is also present in the promoter of *RD22* which is ABA-induced and drought-responsive gene in *Arabidopsis* [18]. Another dehydration-responsive motif ABRELATERD1 is also present in the *NDL* family members, this motif has been reported in the up-stream region of the *ERD1* (early response to dehydration 1). Promoters with *ERD1* motif are involved in response to ABA and also show significant upregulation under water stress [20]. Both AtMYC and AtMYB are known to function as transcriptional activators in abscisic acid signaling presence of these motifs in combination with dehydration-responsive motifs in *NDL* family members indicates their role in ABA-dependent early response to dehydration.

In the present study, we have only analysed early stage of plant growth but Rymaszewski et al., 2017 has analysed plant growth starting from early stages till reproductive stage. They found positive co-relation of morpho-physiological traits like projected rosette area, and increased transpiration rate under low water-deficient conditions and expression levels for *NDL1*. They also speculate that these acclimations may be driven by largely stress hormone ABA. In another study Du et al., 2018 [21] has also found that during low water treatment conditions plants shifts for drought escape mechanism and leads to early flowering to complete their life cycle. Activation of the flowering genes are both ABA-dependent and ABA-independent [21]. All these finding indicates the complex network between stress, hormones and developmental responses.

Interestingly both AGB1 and GPA1 both are interacting partners of NDL members [14] are also interacting partners of one of the MYB (AT3G24120) in G protein interactome using in Y2H [3].

ANAERO1CONSENSUS motif is indicative of involvement of the gene family during anaerobic conditions. MYCCONSENSUSAT regulates transcription of *CBF/DREB1* genes during cold. Presence MYC recognition sites had been postulated to regulate the gene transcription by a basic bHLH transcription activator during cold stress [22]. MYCCONSENSUSAT sites are present mainly at the genes which have more expression in seed and the response to cold is ABA-mediated [23]. MYCCONSENSUSAT motif was also detected in the promoter of the gene which showed response for JA. WBOXATNPR1 is a disease related motif that is established as a site for binding SA-induced WRKY TFs [23]. Presence of a P1BS and SURECOREATSULTR11 motif indicative of the involvement of the family members for nutrient availability. P1BS motifs are associated with phosphate starvation response, while SURECOREATSULTR11 are found in the genes involved during sulfur deficiency responses (*SULTR1; 1*) [24,25].

All the three *NDL* members contain various hormone-responsive transcription factors binding sites. Responses of MYCCONSENSUSAT and ABRELATERD1 are ABA-dependent during desiccation and cold responses. MYCCONSENSUSAT motif also present in the genes which are involved in JA-mediated defense responses. Cytokinin response regulator ARR1AT and SA-responsive WBOXATNPR1 is also present commonly in all the three family members. Light-responsive motifs GT1CONSENSUS and SORLIP1AT are also common in all the members. GT1CONSENSUS a light-responsive motif, also showed tissue-specific localization, it is present in the up-stream region of the right regulated genes which are highly expressed in leaf [23]. Other than these common shared regulatory elements unique TFs binding sites to *NDL1*, *NDL2* and *NDL3* were also observed (Table 2). *NDL1* showed presence of exclusive binding sites for E2F, which targets cell cycle-regulated expression. E2F controls cell cycle by the regulation of the transcription of the genes that are involved during cell cycle and DNA replication [26,27] *Arabidopsis* E2F are classified as classical E2F proteins (E2Fa-c) and atypical E2F proteins (E2Fd-f) [28]. In G-protein interactome database NDL1 shows interaction with CKS2 (cell cycle regulatory subunit) and cyclin-dependent kinase G1 hinting specific function during cell cycle. Previously it is established that sucrose and D-glucose enhances the stability of NDL1 protein [14] which goes in accordance with presence of SBOXATRBCS, important for sugar and ABA responses. Role of NDL1 during primary and lateral root growth and development is already established presence of RHERPATEXPA7, a root hair specific cis-element needs further probing to study role of NDL1 in root hair development.

Likewise, *NDL2* strikingly shows presence of various kinds of ABA-responsive elements like ABREATCONSENSUS, which play role during ABA signaling and responsible for abiotic stress tolerance [29], ABREATRD22 and ACGTABREMOTIFA2OSEM. ACGTABREMOTIFA2OSEM motif present in the promoter of rice *OsEm* gene, which is regulated by seed specific transcription factor and ABA-responsive [30]. ACGTABREMOTIFA2OSEM motif also found in maturing seeds and acts as a binding site for bZIP TFs ABI5 [31]. Those genes which are highly expressed in mature seeds of *Arabidopsis* are found to be enriched with ABRE motif [32]. A single copy of ABRE is not sufficient for ABA-mediated responses. Multiple copy number of ABRE along with a coupling element forms the active complex for ABA-mediated responses [33]. Along with ABA-responsive binding sites several other sites like, GA and JA-responsive sites GADOWNAT and T/GBOXATPIN2 are also present in the promoter region of *NDL2*. GADOWNAT is common sequence found in genes which downregulates after GA treatments. GADOWNAT is shown to be identical as ABRE [34], also hints the regulation of gene under ABA responses. T/GBOXATPIN2 is a wounding response motif, found in the promoter of JA-responsive genes [35]. As per the *in-silico* analysis presence of different kinds of ABA response factors and seed specific transcription factors which are again ABA-dependent for their responses. Since our in vivo expression analysis proved absence of the *NDL2* expression in the roots. *NDL2* members might be specifically involved during seed germination and growth responses.

The *NDL3* promoter showed presence of -ABRERATCAL, an early stress-induced motif related to ABRE and found in the up-stream region of Ca^{2+} ion-responsive genes, CGCGBOXAT is a Ca^{2+}-dependent Calmodulin binding motif is also present both indicative of involvement of *NDL3* during calcium and ABA-mediated stress adaptations [36–38]. GCCCORE has been found in the promoter region of the several pathogen-responsive genes showed JA-dependent defense responses [39]. HDZIPIIIAT involved in vascular differentiation and patterning [40] and required for polar auxin transport in shoot [41], *NDLs* combined role is already established for polar auxin transport in roots [14] Increased HDZIPIIIAT activity leads to formation of extra cotyledons [42], a phenotype somewhat similar to tricot phenotype of *ndl* microRNA downregulated lines [16]. While downregulation of HDZIPIIIAT showed the loss of cotyledons formation [42,43]. All these findings indicate that *NDL3* might be a Ca^{2+}-responsive gene during stress signaling. *NDL3* also involved in similar functions like *NDL1* as both share similar in vivo localization during early stages of plant growth

The combinatorial effect of unique and common transcription factors present in *NDL* gene family essentially suggests that *NDL1*, *NDL2*, and *NDL3* do share a basic common regulatory mechanism of downregulation by MYC/MYB. Our in vivo analysis also shows that during different stages of life cycle, and when plant encounter specific stress conditions differential regulation of all *NDL* members is possible and this could be attributed to the unique transcription factor binding sites present in each gene respectively.

Initial developmental study for all three members of the family showed that they are very specific about their expression pattern. In wild type background, *NDL1* was expressed in cotyledonary leaves, true leaves, primary and lateral roots. Functional characterization of *NDL1* has been done in detail and it has been found that *NDL1* is required for primary root and shoot meristem initiation growth and lateral root/shoot branching [14,16]. For *NDL2* the expression was detected in cotyledonary leaves, true leaves and in maturation zone of primary root, no expression was detected in RAM and in lateral roots (Figure 3) indicating function in different organs compared to *NDL1* and *NDL3*, which share overlapping expression zones. Presence of GA and cold-responsive elements, along with different kind of ABA-responsive factors in *NDL2* promoter is suggestive of its involvement during seed germination and early growth as during low temperature ABA biosynthesis and GA catabolism is up regulated that lead to seed dormancy [44].

In silico and in vivo expression analysis in case of *NDL3* was detected in newly emerging true leaves and at primary and lateral root tips (Figure 3). *NDL3* show response during drought stress and expression pattern is again quite similar with *NDL1* response (Figure 7; Figure 9) meaning, both the members share overlapping expression patterns in vivo and are playing role in similar physiological processes.

Previously it has been found that in *agb1* mutant *NDL1* expression levels are of wild type level meaning *NDL1* transcript is unaffected by the AGB1 (Figure 5D) [14] Similar to *NDL1* study we found that in vivo expression levels/patterns of *NDL2* and *NDL3* remains similar to Col-0 levels even in *agb1-2* mutant, meaning AGB1 does not affect the transcript levels of *NDLs*.

NDL1 protein stability in young primary root needs presence of AGB1, as NDL1 undergoes proteosomal degradation in absence of AGB1 [14], effect of AGB1 on stability and localization of NDL2 and NDL3 is still pending and needs to be characterized.

The effect of different hormones and abiotic stress responses also shows that along with the common response of *NDL1* and *NDL3* during osmotic/drought stress the different family members show specific responses for various treatments. NDL1 localization is upregulated during cold, IAA, JA, SA and cytokinin treatment and downregulated after heat stress (Figure 5A), meaning presence of AGB1 is essential for NDL1 stability during cold, IAA, JA, SA and BAP treatment, indicating stress specific role of NDL1-AGB1 module.

Although *NDL1* and *NDL3* share similar expression domains, *NDL3* expression remains unaffected by cold, IAA, JA, SA, cytokinin and heat treatments (Supplementary Figures S2–S5) meaning differential specificity in their functions.

Various ABA binding sites are present in the promoter region of *NDL2* and in accordance after ABA treatment (also for cold and JA) the *NDL2* expression shows upregulation in Col-0 background but did not showed any difference in *agb1-2* mutant background indicating presence of AGB1 is essential for *NDL2* expression upregulation during these treatments.

In case of *NDL3* expression upon treatment with Mannitol and dehydration stress it was found that AGB1 is negatively regulating *NDL3* expression under normal and dehydration stress condition (Figure 9). Presence of AGB1 is needed under stress condition to downregulate *NDL3* expression, which may be downregulated under dehydration stress.

Although various common TFs have been found in the promoter analysis but all the three members did not show the response for each of them. Every members of the family behaves differentially even though they share the common regulatory motifs. This happens because the family members also showed differential expression pattern and response to various stresses and hormones could be specific for particular developmental stage, our analysis is limited for initial growth stage of the plant. Previously it was found that AGB1 protein is needed for NDL1 stability and here we also found that the protein stability was affected in mutant background. In addition, *AGB1* was also confirmed to be involved in drought stress [45]. *NDL1* is also predicted to be a stress marker for drought [15]. Our *in-vivo* study proves that during the combined function out of three members two of them (*NDL1* and *NDL3*) are involved during drought stress response. *NDL1* is acting as a general abiotic stress responder during heat, cold, IAA, JA, SA and in cytokinin responses as protein steady state levels show upregulation after all these treatments, while *NDL3* expression remains unaffected by all these treatments, meaning limited and specific role. *NDL2* doesn't show any alteration in expression after Mannitol and PEG treatments but showed significant upregulation for ABA, cold, and JA treatment.

In summary dehydration stress (Mannitol and PEG treatments) caused increase in steady state protein levels of NDL1 in both wild type Col-0 as well as in *agb1-2* mutant background. Whereas *NDL3* showed downregulation of expression in wild type Col-0 background while upregulation in *agb1-2* mutant background in RAM. Expression levels of *pNDL2-GUS* were affected in AGB1-dependent manner during cold, ABA and JA treatment. These results strongly support the direct involvement of *NDL1* and *NDL3* during osmotic/drought stress responses and *NDL2* might be playing ABA-dependent indirect role. AGB1 is also playing role in differential regulation of expression of these *NDLs* members during different treatments by affecting protein stability (seen in case of NDL1) or transcript levels (as seen in case of *NDL3* expression) in stress specific manner.

4. Material and Methods

4.1. Plant Material and Growth Conditions

Arabidopsis Col-0 ecotype was used in the present study. *Agrobacterium* (strain GV3101)-mediated floral dip method [46] was used to generate transcriptional fusion transgenic (*pNDL2, 3-GUS*) in wild type Col-0 as well as in the *agb1-2* mutant background. *pNDL1-NDL1-GUS* translational fusion lines were taken from previous study [14]. Transformed seeds were selected on $\frac{1}{2}$ MS medium (Himedia, Mumbai, India) containing 25 mg/L Hygromycin (Duchefa, Amsterdam, Netherland), resistant plants were moved to the soil and grown to maturity in a growth room with a photoperiod 16h light/8h dark at 22 °C, and the light intensity of 100 μmolm^{-2}s^{-1}. Three independent single insertion T3 homozygous lines were obtained and used for developmental and stress treatments. For developmental study, seeds were grown vertically for, 4-day, 8-day and 12-day respectively followed by in vivo GUS assay. For stress treatment and fluorometric analysis six day old seedlings were used.

4.2. Isolation and Cloning of NDL2 and NDL3 Promoters

Genomic DNA was isolated with minor modifications from the *Arabidopsis* thaliana Col-0 using Doyle method [47]. Primers listed in Supplementary Table S1 were used to amplify the promoter region of *NDL2* and *NDL3*. Amplified fragments were then cloned into a pENTR/D-TOPO entry vector (Invitrogen, Massachusetts, United States) gateway, followed by pGWB3 destination vectors) with C-terminus GUS reporter. For NDL1 previously published lines were used [14].

4.3. In-Silico Analysis

Sequence information and location of *NDL1*, *NDL2* and *NDL3* was retrieved from https://www.arabidopsis.org/. In silico analysis was done using online programs PLANTPAN 2.0 [48]. For the expression pattern of *NDL* genes in different stages of development in different tissues, and time points data from the free version of the GENEVESTIGATOR online portal (https://www.genevestigator.com/gv/plant.jsp) and eFP browser (http://bar.utoronto.ca/efp/cgi-bin/efpWeb.cgi) was used for analysis.

4.4. GUS Staining Assay

Three independent lines for *pNDL1-NDL1-GUS* and published *pNDL1-NDL1-GUS* lines were germinated and grown on $\frac{1}{2}$ MS media for 4-day, 8-day and 12-days. Gus staining on the seedlings was done using Jefferson method [49].

4.5. Fluorometric GUS Assay

For quantitatively GUS activity. 0.5 g of seedlings (6-day old) were harvested and frozen into liquid N_2 in 1.5 mL micro centrifuge tube, followed by grinding and protein extraction in extraction buffer (50 mM sodium phosphate buffer, pH 7.0, 10 mM EDTA, 0.1% Triton X-100, and 10 mM β-mercaptoethanol). The homogenate was centrifuged at 10,000 g for 20 min at 4 °C. Supernatant was collected and assayed for protein concentration using Bradford method [50]. Five microgram of protein was added to GUS assay buffer ($1X = 50$ mM sodium phosphate buffer, pH 7.0, 10 mM EDTA, 0.1% Triton X-100, and 10 mM β-mercaptoethanol containing 2 mM MUG (Himedia, Mumbai, India) total volume was made up to 500 µL using ddH$_2$O. Samples were incubated at 37 °C for one hour followed termination of the reaction using 400 µL of 0.2 M Na$_2$CO$_3$. For quantitative GUS analysis duplicate samples were assayed. The reaction product 4-methylumbelliferon (MU) was detected fluorometrically at excitation and emission of 365 nm and 455 nm respectively [51] using Tecan Spark multimode micro plate reader. GUS activity was expressed in nmol MU min^{-1} µg^{-1} protein.

4.6. Hormone and Abiotic Stress Treatments

Seeds were stratified and grown vertically on $\frac{1}{2}$ MS media. Six day old seedlings were subjected to 24 h treatment of sodium chloride (NaCl; 150 mM), Mannitol (300 mM), cold (4 °C), heat (37 °C), abscisic acid (ABA; 20 µM), Indole-3-acetic acid (IAA; 10 µM), salicylic acid (SA; 10 µM), methyl jasmonate (JA; 10 µM), polyethylene glycol (PEG 6000, 20%) and Gibberellic acid (GA; 20 µM) in liquid MS. NaCl, Mannitol and PEG were obtained from Himedia (Himedia, Mumbai, India) all the other hormones and fine chemicals were obtained from Sigma (Sigma, Missouri, United States).

4.7. Accession Numbers

Sequence data from the article can be found from https://www.arabidopsis.org/using accession number: AT5G56750 (*NDL1*), AT5G11790 (*NDL2*), AT2G19620 (*NDL3*) and AT4G34460 (*AGB1*). Chromosomal locations of the respective genes are as follows: NDL1: Chromosome 5 (22957629-22960916), NDL2: Chromosome 5 (3799408-3803216), NDL3: Chromosome 2 (8485991-8488963), AGB1: Chromosome 4 (16477031-16479620), Regulator of G protein signaling protein-RGS1: Chromosome 3 (9532613-9535629), ANNEXIN-ANN1: Chromosome 1 (13225168-13227239), Sodium and Lithium-Tolerant 1-SLT1: Chromosome 2 (15761295-15763451), *Arabidopsis* Ribosomal Protein -RS27A: Chromosome 3 (22611521-22612905), O-Acetylserine (Thiol) Lyase -OAS-TL: Chromosome 4 (8517960-8520596).

5. Conclusions

Activation of specific AGB1-NDL module interaction might be stress specific and hormone regulated, specificity of action is further added in different stages of growth and development due to differential expression patterns of *NDL* members. NDL1 protein stability in young primary root needs presence of AGB1, as NDL1 undergoes proteosomal degradation in absence of AGB1 [14]. Though AGB1 physically interacts with all NDLs it might differentially regulate their stability and hence their function during different stresses, effect of AGB1 on protein stability of NDL2 and NDL3 is still pending and needs to be characterized. Interestingly, Yeast 2 Hybrid (Y2H) confirmed NDL1 interactors includes-Annexin 1 (ANNAT1-has role in drought stress), Sodium and Lithium-Tolerant 1 (SLT1 involved in salt stress), Lesion Stimulating Disease 1 (LSD1 regulates cold stress), O-Acetylserine (Thiol) Lyase (OAS-TL) Isoform A1 (OASA1 role in cadmium tolerance) and *Arabidopsis* Ribosomal Protein S27 (ARS27A involved in genotoxic stress) [3]. These interactions speculate that NDL proteins might plays role in stress signaling in the form of multimeric complexes with these stress effectors and other G protein core components (RGS1 and AGB1) in stress specific manner.

All of these findings together including regulatory elements in silico and in vivo expression profiling indicate that *NDL* family members along with *AGB1* play–key differential roles in different organs during different stages of plant growth and developmental. Collectively, our data suggest that *NDL-AGB1* modules are abiotic stress and hormone treatment-responsive and could be used as stress markers. In long run they could be potential candidates for crop improvement strategies (Figure 10).

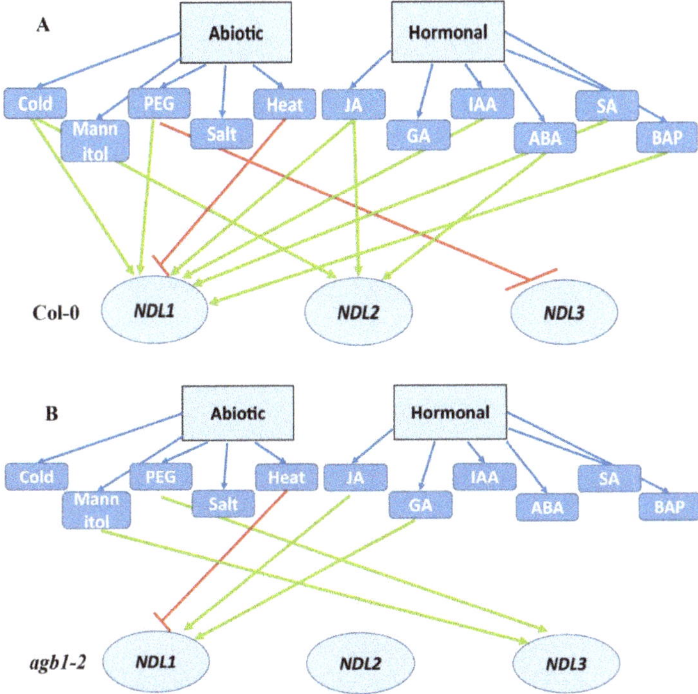

Figure 10. *AtNDLs* responses to various abiotic and hormonal treatments in presence and absence of AGB1: (**A**) Diagrammatic representation of *AtNDLs* in wild type background in response to various abiotic and hormonal treatments. (**B**) *AtNDLs* responses to various abiotic and hormonal treatments in *agb1-2* mutant background. The green lines indicates that the gene was up regulated while the red lines indicates the downregulation of gene after the respective treatments.

Supplementary Materials: Can be found at http://www.mdpi.com/1422-0067/20/19/4736/s1.

Author Contributions: Conceptualization, Y.M.; methodology, Y.M. and A.K.; validation, A.K.; formal analysis, A.K.; investigation, A.K.; resources, Y.M.; data curation, A.K.; writing—original draft preparation, A.K.; writing—review and editing, Y.M.; visualization, A.K.; supervision, Y.M.; project administration, Y.M.; funding acquisition, Y.M.

Funding: Work in the Mudgil's Lab is supported by grants from the DST-SERB (EMR/2016/002780), DBT (BT/PR20657/BPA/118/206/2016) and Delhi University R&D and DU-DST PURSE grant. A.K is supported by JRF/SRF fellowship from CSIR.

Acknowledgments: We thank Swati Singh (Delhi University, Botany Department) for cloning of NDL2 and NDL3 promoter in Entry Vector and Diwakar Bhardwaj (Delhi University, Botany Department) for his help in generating T3 transgenic lines used in the study. We would also like to extend our thanks to Aniket (Delhi University, Ambedkar Centre for Biomedical Research) for his help with fluorometer.

Conflicts of Interest: The authors declare no conflict of interest.

References

1. Colaneri, A.C.; Tunc-Ozdemir, M.; Huang, J.P.; Jones, A.M. Growth attenuation under saline stress is mediated by the heterotrimeric G protein complex. *BMC Plant Biol.* **2014**, *14*, 129. [CrossRef] [PubMed]

2. Ming, C.H.; Xu, D.B.; Fang, G.N.; Wang, E.H.; Gao, S.Q.; Xu, Z.S.; Li, L.C.; Zhang, X.H.; Miin, D.H. G-protein β subunit AGB1 positively regulates salt stress tolerance in *Arabidopsis*. *J. Integr. Agric.* **2015**, *14*, 314–325.

3. Klopffleisch, K.; Phan, N.; Augustin, K.; Bayne, R.S.; Booker, K.S.; Botella, J.R.; Carpita, N.C.; Carr, T.; Chen, J.G.; Cooke, T.R.; et al. *Arabidopsis* G-protein interactome reveals connections to cell wall carbohydrates and morphogenesis. *Mol. Syst. Biol.* **2011**, *7*, 532. [CrossRef] [PubMed]

4. Kalaydjieva, L.; Gresham, D.; Gooding, R.; Heather, L.; Baas, F.; De Jonge, R.; Blechschmidt, K.; Angelicheva, D.; Chandler, D.; Worsley, P.; et al. N-myc downstream-regulated gene 1 is mutated in hereditary motor and sensory neuropathy–Lom. *Am. J. Hum. Genet.* **2000**, *67*, 47–58. [CrossRef] [PubMed]

5. Yan, S.F.; Lu, J.; Zou, Y.S.; Soh-Won, J.; Cohen, D.M.; Buttrick, P.M.; Cooper, D.R.; Steinberg, S.F.; Mackman, N.; Pinsky, D.J.; et al. Hypoxia-associated induction of early growth response-1 gene expression. *J. Biol. Chem.* **1999**, *274*, 15030–15040. [CrossRef] [PubMed]

6. Salnikow, K.; Su, W.; Blagosklonny, M.V.; Costa, M. Carcinogenic metals induce hypoxia-inducible factor-stimulated transcription by reactive oxygen species-independent mechanism. *Cancer Res.* **2000**, *60*, 3375–3378. [PubMed]

7. Lachat, P.; Shaw, P.; Gebhard, S.; Van Belzen, N.; Chaubert, P.; Bosman, F.T. Expression of NDRG1, a differentiation-related gene, in human tissues. *Histochem. Cell Biol.* **2002**, *118*, 399–408. [CrossRef] [PubMed]

8. Salnikow, K.; Kluz, T.; Costa, M.; Piquemal, D.; Demidenko, Z.N.; Xie, K.; Blagosklonny, M.V. The regulation of hypoxic genes by calcium involves c-Jun/AP-1, which cooperates with hypoxia-inducible factor 1 in response to hypoxia. *Mol. Cell. Boil.* **2002**, *22*, 1734–1741. [CrossRef] [PubMed]

9. Wu, D.; Zhau, H.E.; Huang, W.C.; Iqbal, S.; Habib, F.K.; Sartor, O.; Cvitanovic, L.; Marshall, F.F.; Xu, Z.; Chung, L.W.K. cAMP-responsive element-binding protein regulates vascular endothelial growth factor expression: Implication in human prostate cancer bone metastasis. *Oncogene* **2007**, *26*, 5070–5077. [CrossRef] [PubMed]

10. Zhou, D.; Salnikow, K.; Costa, M. Cap43, a novel gene specifically induced by Ni^{2+} compounds. *Cancer Res.* **1998**, *58*, 2182–2189. [PubMed]

11. Krauter-Canham, R.; Bronner, R.; Evrard, J.L.; Hahne, G.; Friedt, W.; Steinmetz, A. A transmitting tissue-and pollen-expressed protein from sunflower with sequence similarity to the human RTP protein. *Plant Sci.* **1997**, *129*, 191–202. [CrossRef]

12. Lazarescu, E.; Friedt, W.; Horn, R.; Steinmetz, A. Expression analysis of the sunflower SF21 gene family reveals multiple alternative and organ-specific splicing of transcripts. *Gene* **2006**, *374*, 77–86. [CrossRef] [PubMed]

13. Lazarescu, E.; Friedt, W.; Steinmetz, A. Organ-specific alternatively spliced transcript isoforms of the sunflower *SF21C* gene. *Plant Cell Rep.* **2010**, *29*, 673–683. [CrossRef] [PubMed]

14. Mudgil, Y.; Uhrig, J.F.; Zhou, J.; Temple, B.; Jiang, K.; Jones, A.M. *Arabidopsis N-MYC DOWNREGULATED-LIKE1*, a positive regulator of auxin transport in a G protein–mediated pathway. *Plant Cell.* **2009**, *21*, 3591–3609. [CrossRef] [PubMed]

15. Rymaszewski, W.; Vile, D.; Bediee, A.; Dauzat, M.; Luchaire, N.; Kamrowska, D.; Granier, C.; Hennig, J. Stress-related gene expression reflects morphophysiological responses to water deficit. *Plant Physiol.* **2017**, *174*, 1913–1930. [CrossRef] [PubMed]

16. Mudgil, Y.; Ghawana, S.; Jones, A.M. N-MYC down-regulated-like proteins regulate meristem initiation by modulating auxin transport and MAX2 expression. *PLoS ONE* **2013**, *8*, e77863. [CrossRef] [PubMed]

17. Luscher, B.; Eisenman, R.N. New light on Myc and Myb. Part II. Myb. *Genes Dev.* **1990**, *4*, 2235–2241. [CrossRef] [PubMed]

18. Abe, H.; Urao, T.; Ito, T.; Seki, M.; Shinozaki, K.; Yamaguchi-Shinozaki, K. *Arabidopsis AtMYC2* (bHLH) and *AtMYB2* (MYB) function as transcriptional activators in abscisic acid signaling. *Plant Cell.* **2003**, *15*, 63–78. [CrossRef]

19. Yamaguchi-Shinozaki, K.; Shinozaki, K. Organization of *cis*-acting regulatory elements in osmotic-and cold-stress-responsive promoters. *Trends Plant Sci.* **2005**, *10*, 88–94. [CrossRef]

20. Simpson, S.D.; Nakashima, K.; Narusaka, Y.; Seki, M.; Shinozaki, K.; Yamaguchi-Shinozaki, K. Two different novel *cis*-acting elements of *erd1*, a *clpA* homologous *Arabidopsis* gene function in induction by dehydration stress and dark-induced senescence. *Plant J.* **2003**, *33*, 259–270. [CrossRef]

21. Du, H.; Huang, F.; Wu, N.; Li, X.; Hu, H.; Xiong, L. Integrative regulation of drought escape through ABA-dependent and-independent pathways in rice. *Mol. Plant.* **2018**, *4*, 584–597. [CrossRef]

22. Chinnusamy, V.; Ohta, M.; Kanrar, S.; Lee, B.H.; Hong, X.; Agarwal, M.; Zhu, J.K. ICE1: A regulator of cold-induced transcriptome and freezing tolerance in *Arabidopsis*. *Genes Dev.* **2003**, *17*, 1043–1054. [CrossRef]

23. Wang, C.; Wang, Y.; Pan, Q.; Chen, S.; Feng, C.; Hai, J.; Li, H. Comparison of Trihelix transcription factors between wheat and *Brachypodiumdistachyon* at genome-wide. *BMC Genom.* **2019**, *20*, 142. [CrossRef]

24. Rouached, H.; Secco, D.; Arpat, B.; Poirier, Y. The transcription factor PHR1 plays a key role in the regulation of sulfate shoot-to-root flux upon phosphate starvation in *Arabidopsis*. *BMC Plant Biol.* **2011**, *11*, 19. [CrossRef]

25. Sobkowiak, L.; Bielewicz, D.; Małecka, E.; Jakobsen, I.; Albrechtsen, M.; Szweykowska-Kulinska, Z.; Pacak, A.M. The role of the P1BS element containing promoter-driven genes in Pi transport and homeostasis in plants. *Front. Plant Sci.* **2012**, *3*, 58. [CrossRef]

26. Helin, K. Regulation of cell proliferation by the E2F transcription factors. *Curr. Opin. Genet. Dev.* **1998**, *8*, 28–35. [CrossRef]

27. Lammens, T.; Li, J.; Leone, G.; De Veylder, L. Atypical E2Fs: New players in the E2F transcription factor family. *Trends Cell Biol.* **2009**, *19*, 111–118. [CrossRef]

28. De Veylder, L.; Beeckman, T.; Inze, D. The ins and outs of the plant cell cycle. *Nat. Rev. Mol. Cell Biol.* **2007**, *8*, 655. [CrossRef]

29. Narusaka, Y.; Nakashima, K.; Shinwari, Z.K.; Sakuma, Y.; Furihata, T.; Abe, H.; Narusaka, M.; Shinozaki, K.; Yamaguchi-Shinozaki, K. Interaction between two cis-acting elements, ABRE and DRE, in ABA-dependent expression of *Arabidopsis* rd29A gene in response to dehydration and high-salinity stresses. *Plant J.* **2003**, *34*, 137–148. [CrossRef]

30. Hattori, T.; Totsuka, M.; Hobo, T.; Kagaya, Y.; Yamamoto-Toyoda, A. Experimentally determined sequence requirement of ACGT-containing abscisic acid response element. *Plant Cell Physiol.* **2002**, *43*, 136–140. [CrossRef]

31. Yang, X.; Yang, Y.N.; Xue, L.J.; Zou, M.J.; Liu, J.Y.; Chen, F.; Xue, H.W. Rice ABI5-Like1 regulates abscisic acid and auxin responses by affecting the expression of ABRE-containing genes. *Plant Physiol.* **2011**, *156*, 1397–1409. [CrossRef]

32. Nakabayashi, K.; Okamoto, M.; Koshiba, T.; Kamiya, Y.; Nambara, E. Genome-wide profiling of stored mRNA in *Arabidopsis thaliana* seed germination: Epigenetic and genetic regulation of transcription in seed. *Plant J.* **2005**, *41*, 697–709. [CrossRef]

33. Skriver, K.; Olsen, F.L.; Rogers, J.C.; Mundy, J. *Cis*-acting DNA elements responsive to gibberellin and its antagonist abscisic acid. *Proc. Natl. Acad. Sci. USA* **1991**, *88*, 7266–7270. [CrossRef]

34. Ogawa, M.; Hanada, A.; Yamauchi, Y.; Kuwahara, A.; Kamiya, Y.; Yamaguchi, S. Gibberellin biosynthesis and response during *Arabidopsis* seed germination. *Plant Cell.* **2003**, *15*, 1591–1604. [CrossRef]

35. Boter, M.; Ruiz-Rivero, O.; Abdeen, A.; Prat, S. Conserved MYC transcription factors play a key role in jasmonate signaling both in tomato and *Arabidopsis*. *Genes Dev.* **2004**, *18*, 1577–1591. [CrossRef]

36. Kaplan, B.; Davydov, O.; Knight, H.; Galon, Y.; Knight, M.R.; Fluhr, R.; Fromm, H. Rapid transcriptome changes induced by cytosolic Ca^{2+} transients reveal ABRE-related sequences as Ca^{2+}-responsive *cis*-elements in *Arabidopsis*. *Plant Cell.* **2006**, *18*, 2733–2748. [CrossRef]

37. Yang, T.; Poovaiah, B.W. A calmodulin-binding/CGCG box DNA-binding protein family involved in multiple signaling pathways in plants. *J. Biol. Chem.* **2002**, *277*, 45049–45058. [CrossRef]

38. Du, L.; Ali, G.S.; Simons, K.A.; Hou, J.; Yang, T.; Reddy, A.S.; Poovaiah, B.W. Ca^{2+}/calmodulin regulates salicylic-acid-mediated plant immunity. *Nature* **2009**, *457*, 1154–1158. [CrossRef]

39. Brown, R.L.; Kazan, K.; McGrath, K.C.; Maclean, D.J.; Manners, J.M.A. Role for the GCC-box in jasmonate-mediated activation of the *PDF1.2* gene of *Arabidopsis*. *Plant Physiol.* **2003**, *132*, 1020–1032. [CrossRef]

40. Yang, J.H.; Wang, H. Molecular mechanisms for vascular development and secondary cell wall formation. *Front. Plant Sci.* **2016**, *7*, 356. [CrossRef]

41. Huang, T.; Harrar, Y.; Lin, C.; Reinhart, B.; Newell, N.R.; Talavera-Rauh, F.; Hokin, S.A.; Barton, M.K.; Kerstetter, R.A. *Arabidopsis* KANADI1 acts as a transcriptional repressor by interacting with a specific *cis*-element and regulates auxin biosynthesis, transport, and signaling in opposition to HD-ZIPIII factors. *Plant Cell.* **2014**, *26*, 246–262. [CrossRef]

42. Emery, J.F.; Floyd, S.K.; Alvarez, J.; Eshed, Y.; Hawker, N.P.; Izhaki, A.; Baum, S.F.; Bowman, J.L. Radial patterning of *Arabidopsis* shoots by class III HD-ZIP and KANADI genes. *Curr. Biol.* **2003**, *13*, 1768–1774. [CrossRef]

43. Izhaki, A.; Bowman, J.L. KANADI and class III HD-Zip gene families regulate embryo patterning and modulate auxin flow during embryogenesis in *Arabidopsis*. *Plant Cell.* **2007**, *19*, 495–508. [CrossRef]

44. Footitt, S.; Douterelo-Soler, I.; Clay, H.; Finch-Savage, W.E. Dormancy cycling in *Arabidopsis* seeds is controlled by seasonally distinct hormone-signaling pathways. *Proc. Natl. Acad. Sci. USA* **2011**, *108*, 20236–20241. [CrossRef]

45. Xu, D.B.; Chen, M.; Ma, Y.N.; Xu, Z.S.; Li, L.C.; Chen, Y.F.; Ma, Y.Z.A. G-protein β subunit, AGB1, negatively regulates the ABA response and drought tolerance by down-regulating AtMPK6-related pathway in *Arabidopsis*. *PLoS ONE* **2015**, *10*, e0116385. [CrossRef]

46. Clough, S.J.; Bent, A.F. Floral dip: A simplified method for *Agrobacterium*-mediated transformation of *Arabidopsis thaliana*. *Plant J.* **1998**, *16*, 735–743. [CrossRef]

47. Doyle, J.J.; Doyle, J.L. Isolation of plant DNA from fresh tissue. *Focus* **1990**, *12*, 39–40.

48. Chang, W.C.; Lee, T.Y.; Huang, H.D.; Huang, H.Y.; Pan, R.L. PlantPAN: Plant promoter analysis navigator, for identifying combinatorial *cis*-regulatory elements with distance constraint in plant gene groups. *BMC Genom.* **2008**, *9*, 561. [CrossRef]

49. Jefferson, R.A.; Kavanagh, T.A.; Bevan, M.W. GUS fusions: Beta-glucuronidase as a sensitive and versatile gene fusion marker in higher plants. *EMBO J.* **1987**, *6*, 3901–3907. [CrossRef]

50. Bradford, M.M. ARapid and sensitive method for the quantitation of microgram quantities of protein utilizing the principle of protein-dye binding. *Anal. Biochem.* **1976**, *72*, 248–254. [CrossRef]

51. Ren, Y.; Zhao, J. Functional analysis of the rice metallothionein gene OsMT2b promoter in transgenic *Arabidopsis* plants and rice germinated embryos. *Plant Sci.* **2009**, *176*, 528–538. [CrossRef]

International Journal of
Molecular Sciences

Article

Overexpression of *GmCAMTA12* Enhanced Drought Tolerance in Arabidopsis and Soybean

Muhammad Noman, Aysha Jameel, Wei-Dong Qiang, Naveed Ahmad, Wei-Can Liu, Fa-Wei Wang * and Hai-Yan Li *

College of Life Sciences, Engineering Research Center of the Chinese Ministry of Education for Bioreactor and Pharmaceutical Development, Jilin Agricultural University, Changchun 130118, Jilin, China
* Correspondence: fw-1980@163.com (F.-W.W.); hyli99@163.com (H.-Y.L.)

Received: 15 August 2019; Accepted: 26 September 2019; Published: 29 September 2019

Abstract: Fifteen transcription factors in the CAMTA (calmodulin binding transcription activator) family of soybean were reported to differentially regulate in multiple stresses; however, their functional analyses had not yet been attempted. To characterize their role in stresses, we first comprehensively analyzed the *GmCAMTA* family in silico and thereafter determined their expression pattern under drought. The bioinformatics analysis revealed multiple stress-related *cis*-regulatory elements including *ABRE, SARE, G-box* and *W-box*, 10 unique miRNA (microRNA) targets in *GmCAMTA* transcripts and 48 proteins in GmCAMTAs' interaction network. We then cloned the 2769 bp CDS (coding sequence) of *GmCAMTA12* in an expression vector and overexpressed in soybean and Arabidopsis through *Agrobacterium*-mediated transformation. The T3 (Transgenic generation 3) stably transformed homozygous lines of Arabidopsis exhibited enhanced tolerance to drought in soil as well as on MS (Murashige and Skoog) media containing mannitol. In their drought assay, the average survival rate of transgenic Arabidopsis lines OE5 and OE12 (Overexpression Line 5 and Line 12) was 83.66% and 87.87%, respectively, which was ~30% higher than that of wild type. In addition, the germination and root length assays as well as physiological indexes such as proline and malondialdehyde contents, catalase activity and leakage of electrolytes affirmed the better performance of OE lines. Similarly, *GmCAMTA12* overexpression in soybean promoted drought-efficient hairy roots in OE chimeric plants as compare to that of VC (Vector control). In parallel, the improved growth performance of OE in Hoagland-PEG (polyethylene glycol) and on MS-mannitol was revealed by their phenotypic, physiological and molecular measures. Furthermore, with the overexpression of *GmCAMTA12*, the downstream genes including *AtAnnexin5, AtCaMHSP, At2G433110* and *AtWRKY14* were upregulated in Arabidopsis. Likewise, in soybean hairy roots, *GmELO, GmNAB* and *GmPLA1-IId* were significantly upregulated as a result of *GmCAMTA12* overexpression and majority of these upregulated genes in both plants possess CAMTA binding *CGCG/CGTG* motif in their promoters. Taken together, we report that *GmCAMTA12* plays substantial role in tolerance of soybean against drought stress and could prove to be a novel candidate for engineering soybean and other plants against drought stress. Some research gaps were also identified for future studies to extend our comprehension of *Ca-CaM-CAMTA*-mediated stress regulatory mechanisms.

Keywords: arabidopsis; CaM (Calmodulin); calmodulin-binding transcription activators (CAMTA); *cis*-elements; drought; qPCR; soybean hairy roots

---

1. Introduction

A successful sustainable agriculture should ensure food security, must be eco-friendly and safe to humans [1]. With the existing population growth rate, the current food production rate needs to be increased at least up to 70% by 2050 [2,3]. Despite advanced farming practices, the abiotic stresses due to drought, salinity, water and temperature fluctuations are causing 50–80% losses in crop yield, and therefore, should be as effectively managed as possible [4]. Soon, the warmer earth will cause a more humid atmosphere but less humid soil, leading to more frequent drought that would negatively affect the rate of photosynthesis, uptake of CO_2, accumulation of biomass and yield [5,6]. Thus, developing stress-resistant crops with stable yields under adverse conditions is an important strategy to ensure future food security [2,7]. Deep insights into the mechanisms underlying signaling crosstalk mediating stress tolerance would aid plant researchers to upgrade the plant's indigenous natural machinery through biotechnology and genetic engineering.

In plants, the ubiquitous secondary messenger calcium is the key to maintain the harmonious and homeostatic conditions via signaling [8,9]. Calcium ions perceive and encode the environmental, developmental or hormonal signals into a definite frequency that is decoded and relayed by the protein molecules next to them including calmodulin (CaM). By interacting with calcium, calmodulin (calcium-modulated) proteins sense and convey the signals to calmodulin-binding proteins [10]. Here, CaM through signal frequency readjustment channelizes them before transducing to CAMTA TFs (transcription factors), thus CaM acts like a prism (as prism dissects white light into its components). A transcription factor family reported and initially referred to as Ethylene-induced calmodulin-binding protein [11] or signal responsive (SR) protein, while now known as Calmodulin-binding transcription activator (CAMTA), is present in almost all eukaryotes. CAMTAs were first detected in *Nicotiana tabaccum* while studying calmodulin-binding proteins [11–13]. After their emergence, all multicellular eukaryotes studied to date have been reported to be equipped with variable number of *CAMTA* genes such as *Arabidopsis thaliana* (6) [13], *Lycopersicum esculantum* (7) [14], *Medicago truncatula* (7) [15], Citrus (9) [16], *Populus trichocarpa* (7) [17], *Nicotiana tabacum* (13) [18], *Musa acuminata* (5) [19] and *Phaseolus vulgaris* (8) [20]. Glycine max possesses 15 *CAMTA* genes and all of them differentially express under various stress conditions [21].

CAMTA TFs are an integral element of Calcium-mediated biotic/abiotic stress and hormonal signaling pathway [22–25]. Stress signals are conveyed and modulated through *Ca-CaM-CAMTA* pathway and a rapid and calculated response is observed by carrying out the transcription of stimulus-specific genes [26]. Calmodulin Binding Transcription Activators through their CG-1 domain recognize and specifically bind the ((A/C)CGCG(C/G/T), (A/C)CGTGT)) sequence of the target genes thereby directly interacting and regulating their transcription [12]. The CG-1 motif is a pivotal member of a rapid stress response element (RSRE) found in the promoters of many genes that are rapidly activated in response to stress [27]. Previous experiments report the negative role of Arabidopsis *CAMTA3* in regulating plant immunity as demonstrated in the *AtMCAMTA3* (loss-of-function) mutant Arabidopsis [28–30]. Similarly, *AtCAMTA1* and *AtCAMTA3* have been shown to function in drought and regulate auxin [22,31]. Moreover, *AtCAMTA1* and *AtCAMTA3* regulated cold response by inducing CBF (C-Repeat/DRE-Binding Factor) pathway genes as the double *ATCAMTA1* and *ATCAMTA2* mutant exhibited impaired freezing tolerance [32,33]. The role of *AtCAMTA3* in regulating SA (salicylic acid) pathway genes working in freezing tolerance was only recently determined [34]. Two recent studies highlighted the role of *AtCAMTA3* [35] and *AtCAMTA6* [36] in salt tolerance. *TaCAMTA4* in wheat was demonstrated to negatively regulate defense response against *Puccinia triticina* [25].

Soybean is an important crop cultivated globally for food, feed [37], pharmaceutical and soil nitrogen improving purposes [38]. However, environmental adversities including drought affect soybean growth and yield. Earlier, the in silico analysis of *GmCAMTAs* was conducted, yet the potential targets of miRNA in *GmCAMTA* transcripts and the protein-protein interaction network were not reported [21]. In addition, the previous study also did not analyze the expression pattern of *GmCAMTAs* in soybean leaves in response to drought [21]. In an extension to that study and in order to decipher the role of soybean CAMTA family in drought, we first comprehensively analyzed (in silico) the *GmCAMTA* family including their physicochemical properties, chromosomal distribution, *cis*-motifs, miRNA targets and protein-protein interaction network. Secondly, we determined the spatiotemporal expression pattern of GmCAMTAs in roots and leaves of soybean under PEG stress and selected an efficient member of the GmCAMTA family to functionally characterize. We constructed the overexpression construct by cloning the 2769 bp CDS (coding sequence) of *GmCAMTA12* and transformed into Arabidopsis and soybean hairy roots. Through various drought assays, we demonstrated that the transgenic Arabidopsis and chimeric soybean (OE-Overexpressing *GmCAMTA12*) plants exhibited enhanced tolerance and performed better under drought stress than their non-transgenic counterpart at phenotypic, physiological and molecular level. qPCR (quantitative PCR) of the downstream genes in Arabidopsis and soybean also displayed altered expression as a result of *GmCAMTA12* overexpression. From our analyses, we report that *GmCAMTA12* as a transcription factor plays role in drought stress by regulating the downstream genes involved in drought tolerance and could be exploited in developing drought-tolerant crops.

2. Results

2.1. Physico-Chemical Properties of GmCAMTA Proteins

Various physico-chemical properties including the number of amino acids, protein molecular weight (MW), pI (isoelectric point), number of atoms, instability and aliphatic indexes and GRAVY (grand average of hydropathy) determined online with the ProtParam tool are given in Table S1 and File S1. GmCAMTA11 and GmCAMTA9 are the shortest polypeptides comprising of 910 and 911 aa (amino acid), while GmCAMTA1 is the longest possessing 1122 aa. Overall, their average length is ~1004 aa with a range of ~200 aa mutual difference. The molecular weight of GmCAMTAs ranges between 126,989.43 and 102,394.56 kDa with an average MW of 112848.3 and the number of atoms is proportional to the molecular weight of each protein. Similarly, GmCAMTA8 has the highest pI value of 7.64 showing that GmCAMTAs have relatively lower pI. Moreover, they are also hydrophilic in nature as the GRAVY ranges between –0.625 (GmCAMTA1) and –0.394 (GmCAMTA12). Almost all of the GmCAMTAs are thermally stable as their aliphatic indexes match with that of the other globular proteins (highest in GmCAMTA12). While none of them are stable in a test tube, as the instability index of all GmCAMTAs is higher than 40. Except GmCAMTA8, the number of Asp+Glu (negatively charged aa) is higher than Arg+Lys (positively charged aa).

2.2. Phylogenetics and Structure of GmCAMTAs

As mentioned earlier, CAMTAs are multiple-stress responsive transcription factors. Enrichment of *cis*-motifs involved in signal response in the promoters of Medicago *CAMTA* genes hints that they are likely to respond variedly to various signals like other *CAMTAs* [15]. We dissected the regulatory region of *GmCAMTAs* (~2 kb upstream) online with PLANTCare, which detected stimulus-specific *cis*-motifs in their promoters. Overall, there are light (*G-box*, *MRE* and *AE-box*), drought (*MBS*), salt (*MYB*), pathogen (*TC-rich repeats*), wound (*WUN*, *WRE*), low temperature (*LTR*), gibberellin (*GARE*, *P-box*), auxin (*AUXRE*) and abscisic acid (*ABRE*) responsive *cis*-elements as shown in figure 1A. The presence of multiple *cis*-motifs in *GmCAMTA* genes represents their responsivity to multiple stimuli. Moreover, the *cis*-element (*CG-box*) is the binding site of CAMTA TFs; thus, the presence of *G-box* within *GmCAMTA* promoters indicates the interaction of one GmCAMTA TF with another. *GmCAMTA12* possesses *MYC*, *MYB*, *MBS* and *G-box*, which shows its potential role in drought stress.

Using the online GSDS tool, the gene structure of all the *GmCAMTAs* was visualized in order to mutually compare their structural diversity. The length of GmCAMTA genes lie in the range between 3196 bp (*GmCAMTA14*) to 3947 bp (*GmCAMTA6*) with an average length of 3607 bp. The exons (yellow), introns (black lines) as well as 5′ and 3′ UTR (untranslated) regions (blue) of each gene are shown in figure 1B. A close observation of the number of exon-introns reveals a similar pattern, namely, three genes (*GmCAMTA7*, *GCAMTA10* and *GmCAMTA15*) that are comprised of 12 exons; the rest have 13 exons and 12 introns of variable length, indicating their close mutual evolutionary relationship. This fixed numbering of intron and exon is a conserved characteristic of *CAMTA*, which is descended from ancestors and is also demonstrated in the *CAMTA* family of other species [23]. Similarly, except the last one/two exons, an ascending order in the exon size from 5′ to 3′ UTR can be observed across all the genes.

The four domains (CG-1, ANK (ankyrin repeat), IQ and CaMBD (CaM binding domain)) is the common conserved characteristic of all CAMTA TFs [23]. Scanning the amino acid sequences of GmCAMTAs with online protein domains illustrator tool showed the same four domains in all 15 members (figure 1C). GmCAMTA TFs through IQ (calcium-independent)/CaMBD (calcium-dependent) interact with Calmodulin, while through the CG-1 domain, they bind the DNA in a sequence-specific manner (*CGCG/CGTG*) at their promoter region. "ANK repeats" mediate protein-protein interactions. All these conserved domains, along with other properties, make GmCAMTA proteins, the "transcription factors". The high sequence specificity is common in the Calmodulin binding domain of Arabidopsis and soybean CAMTAs as shown in figure 1D.

An ML (Maximum Likelihood) tree was constructed which traced the evolutionary relationship among the CAMTA families of soybean, Arabidopsis, maize and tomato (Figure 1E). Using MEGAX (molecular evolutionary genetic analysis), the evolutionary tree was constructed from the full length aligned CAMTA protein sequences of the four species. A total of 37 CAMTAs including six from Arabidopsis, 15 from soybean, seven from tomato and nine from maize clustered into four distinct groups, I, II, III and IV. GmCAMTA1–6, AtCAMTA1–3, SlCAMTA1 and 2 and ZmCAMTA3, 6, 7a and 7b might have co-evolved and thus clustered together in group I, representing the largest clade. Similarly, GmCAMTA10, 11, 14, 15, AtCAMTA4, SlCAMTA3, 4 and ZmCAMTA1 clustered in group III showing their mutually high homology. GmCAMTA8, 9, 12 and 13 grouped with AtCAMTA5, 6, SlCAMTA5, 6 and ZmCAMTA5 making clade IV. Two orthologs (GmCAMTA7 and SlCAMTA7) comprised group II, which is the smallest clade. Clustering in the phylogenetic reconstruction indicates more mutual similarity and probably weak homology to the members of the other three bigger clusters. It is noticeable that except II, in the rest of clades, CAMTAs of the same species are at the tips of the same branches and vice versa retaining their intraspecific homology.

2.3. miRNA Targets in GmCAMTA Transcripts

miRNA target prediction is important in finding their role in plant growth development in normal as well as stress conditions [18]. Keeping the Expectation score ≤5, the 639 miRNAs were scanned and miRNA, where the minimum E. value was selected using the online psRNATarget tool [39]. A total of 10 unique miRNAs were predicted which potentially target the *GmCAMTA* transcripts by inhibiting translation or through cleavage (Table S2 and File S1). Their length ranges between 19 bp (gma-miR1533) to 24 bp (gma-miR343b). The accessibility of target site (UPE), which is associated with identification of target site and energy required to cleave the transcript, varies from 11.8 (gma-miR1533) to 21.6 (gma-miR5780c). The translation of *GmCAMTA1* and *GmCAMTA2* transcripts is potentially inhibited by the common gma-miR5780c, while gma-miR6299 cleaves four *GmCAMTA8, GmCAMTA9, GmCAMTA12* and *GmCAMTA13*). Similarly, *GmCAMTA5* and *GmCAMTA 6* have potential targets for gma-miR1533 while *GmCAMTA11* and *GmCAMTA14* are predicted to be cleaved by gma-miR2111b. gma-miR9726 is predicted to cleave *GmCAMTA3* and gma-miR1522 *GmCAMTA15*. These in silico predictions require experimental validation, which would extend our understanding the mechanisms of *Ca-CaM-CAMTA*-mediated stress tolerance in plants.

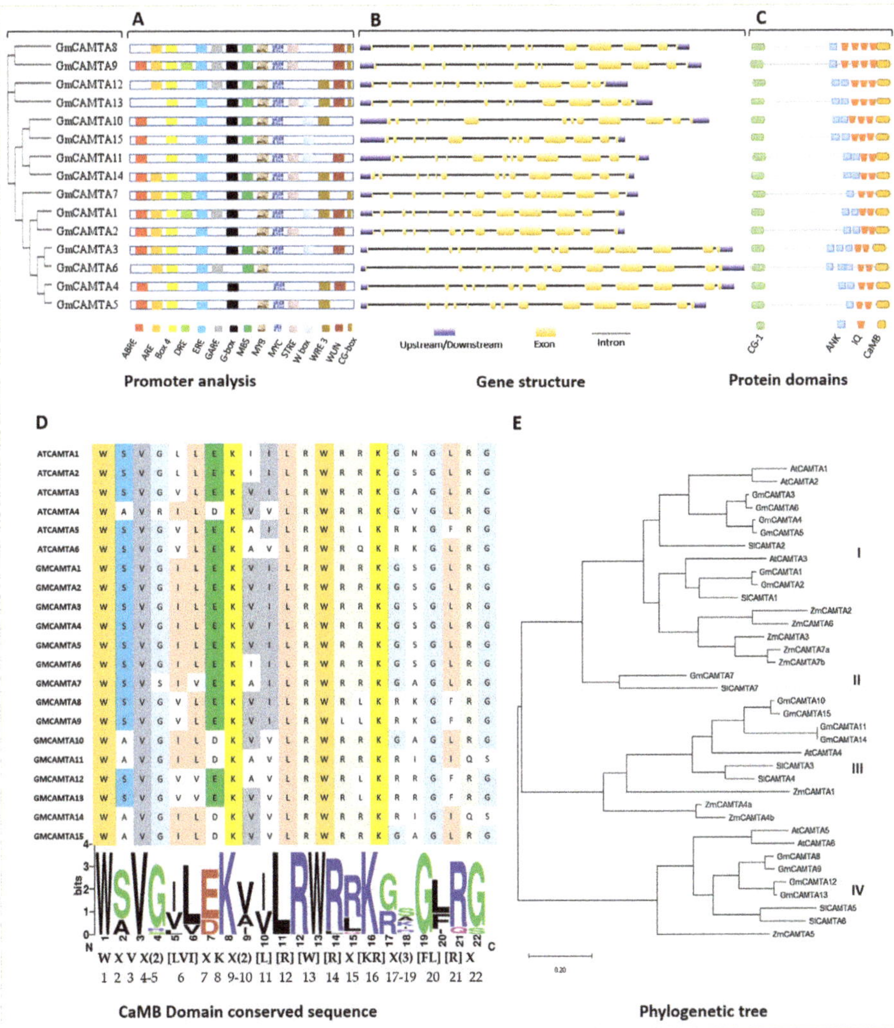

Figure 1. Comprehensive in silico analysis of the *GmCAMTA* family. (**A**) *Cis*-elements in the *GmCAMTA* promoter region. Each type of *cis*-motif present in *GmCAMTA* promoters is shown in unique color/color pattern. (**B**) Exon-intron assembly of *GmCAMTA* genes. (**C**) Domain organization of GmCAMTA proteins. Four different domains are represented in different colored shapes. (**D**) Alignment showing the conserved motif sequence of the Calmodulin Binding Domain of Arabidopsis and Soybean CAMTA TFs. Each conserved residue at the definite position along the row (throughout the orthologues) is shaded in unique color. The functional residues in CaMB domain of these CAMTAs are indicated in the motif below the alignment. In the square brackets "[ ]" are the amino acids allowed in this position of the motif; "X" represents any amino acid and the round brackets "( )" indicate the number of amino acids. (**E**) ML (maximum likelihood) phylogenetic tree.

2.4. Chromosomal Distribution and Regulatory Network of GmCAMTAs

The genome browser tool in NCBI (National Center for Biotechnology Information) mapped *GmCAMTA* genes to their respective chromosomes. The 15 *GmCAMTA* genes are unequally distributed over eight out of 20 chromosomes of soybean as shown in Figure 2. The figure depicts the complete size of each chromosome with the exact position of genes. Chromosome 8 has the highest number of *GmCAMTA* genes, i.e., four, while chromosome 7, 9, 11 and 18 have got only one in each. In prokaryotes, due to the absence of nucleus, transcription and translation occur simultaneously in coupling phase. On the other hand, translation of mRNA is always executed outside the nucleus in the eukaryotic cytoplasm and the proteins that work only in nucleus have a nuclear localization sequence/signal (NLS). Transcription factors also work in the nucleus; thus, after their translation in cytoplasm, they are directed to the nucleus which is mediated through their NLS. To find NLS, protein sequences of each GmCAMTA were submitted to the online cNLS mapper tool at http://nls-mapper.iab.keio.ac.jp/cgi-bin/NLS_Mapper_form.cgi. Keeping a cut off score of 5, at least one NLS in all GmCAMTAs and even more than one in some proteins were detected. Likewise, all of the six Arabidopsis CAMTAs possess only one NLS in the CG-1 domain of each protein [12]. In contrast, rice CAMTAs have one NLS in the C-terminal and another in N of CG-1 domain [40]. Experimental evidence shows that these domains perform diverse functions in the regulation of gene expression [41]. Table S3 (File S1) shows the nuclear localization sequence in each of the 15 GmCAMTAs and predicts that all these transcription factors are localized in the nucleus.

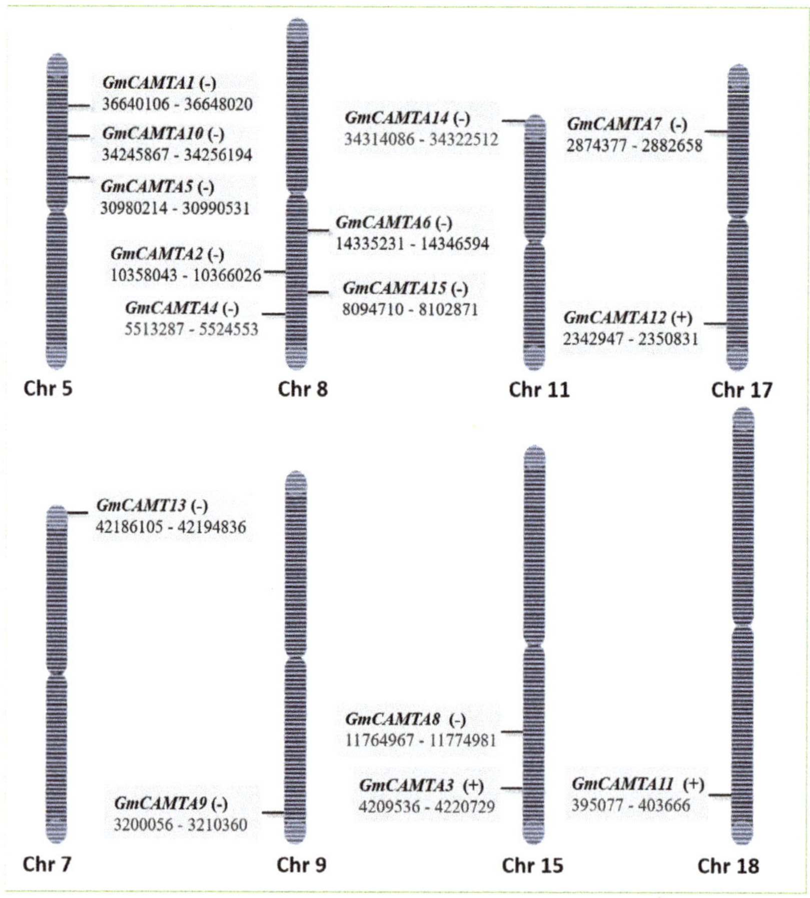

Figure 2. Chromosomal distribution of *GmCAMTA* genes. The 15 *GmCAMTA* genes are located on chromosome 5, 7, 8, 9, 15, 17 and 18.

In order to find the interaction network of GmCAMTA proteins to relate them with other pathways, the protein sequences were individually put in the STRING (Search tool for the retrieval of interacting gene/proteins) database, which predicted a number of interactors [42]. Thus, they can aid in linking proteins of interest to other pathways and could lead to the discovery of novel pathways as well. The STRING database displayed a network of ~10 interactors for each GmCAMTA protein among which some sets were redundant. Thus, a total of 48 unique proteins were predicted for 15 GmCAMTAs as shown in Figure 3 (Table S4 and File S1). This vast interaction network (experimentally determined and software predicted) indicates the complex regulatory upstream/downstream pathways of CAMTAs. However, these *in silico* predicted interactions require experimental validation. Besides these predicted interactions, their orthologues in other species such as Arabidopsis or other legumes could also be exploited to search other potential interactors in soybean.

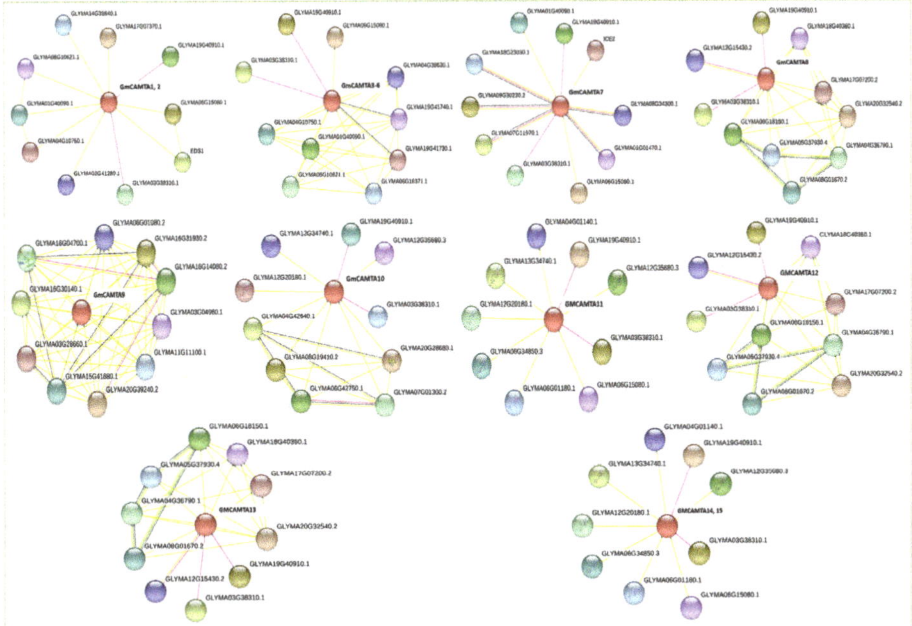

Figure 3. Protein-protein interaction network of GmCAMTAs detected in silico using the STRING database (https://string-db.org/). The experimentally determined interactions are denoted by pink strings, interactions on the basis of textmining are indicated by yellow strings while interactions on the basis of gene neighborhood shown by green strings.

2.5. GmCAMTAs as Early Drought Stress-Responsive TFs

The spatiotemporal expression in roots and leaves under 0, 1, 3, 6, 9 and 12 h of simulated drought stress is shown graphically in Figure 4. In roots, *GmCAMTA2* was highly expressed during 3 h of drought followed by *GmCAMTA7* and *GmCAMTA10*. In contrast, *GmCAMTA14* was downregulated during all the five time points followed by *GmCAMTA8*, *GmCAMTA9* and *GmCAMTA11*. Overall, these transcription factors upregulated abruptly during 1 and 3 h of stress and downregulated afterwards (Figure 4A). The expression profile of *GmCAMTAs* in leaves at different stress durations is different as compared to roots (Figure 4B). In leaves, they look uniform, except *GmCAMTA4*, which is the highest upregulated gene followed by *GmCAMTA5*, *GmCAMTA11* and *GmCAMTA12*. Interestingly, like roots, majority of these 15 genes retained the 3 h trend in the leaves as well. It is obvious that the expression was relatively high at 3 h, after which it decreased until 12 h.

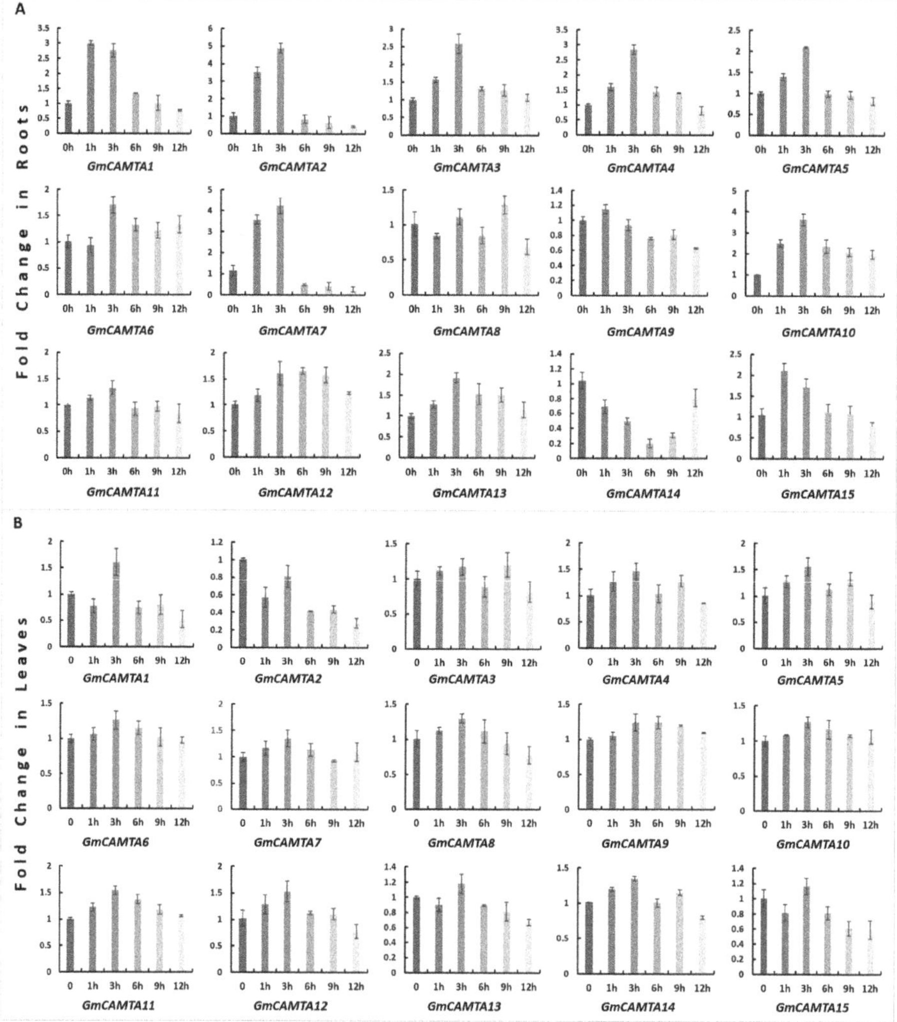

Figure 4. Spatiotemporal expression analysis of GmCAMTA in drought. (**A**) Relative fold expression of *GmCAMTAs* in Roots. Soybean plants were treated with 6% PEG6000 in Hoagland's solution for five different durations (1, 3, 6, 9 and 12 h) along with a control (0 h). (**B**) Relative fold expression of *GmCAMTAs* in Leaves.

The differential expression of *GmCAMTA* family is the result of their tightly regulated transcription. We speculate that in stress conditions, although the calcium ions continuously convey the stress signals through calcium signatures to the cytoplasm as well as the nucleus; however, the intensity/amount of these signals is weighed and adjusted by the next signal relaying molecules, such as CaM, before conveying to *CAMTA* transcription factors. In case, this signal transduction to *CAMTA* is continuing with the same intensity, yet after certain period (3 h in our case), *CAMTAs'* response is not the same throughout the course and seems to be unconcerned and even downregulated as the stress period continues. From the control samples (0 h) in leaves and roots, it is also obvious that the expression of all *CAMTAs* is active at all times. In brief, the spatiotemporal expression pattern revealed that

GmCAMTAs are upregulated in the early phase of drought thus are early drought stress-responsive transcription factors.

2.6. Arabidopsis Overexpressing GmCAMTA12 Exhibited Enhanced Drought Tolerance

To evaluate the contribution of GmCAMTA12 protein in drought stress, we engineered Arabidopsis plants to constitutively express *GmCAMTA12* gene under 35 s promoter. Prior to Arabidopsis transformation, the *Agrobacterium tumefacians* strain EHA105 transformants were verified (through PCR) to harbor the overexpression cassette. (Figure S1 and File S1). After floral dip, several overexpressing lines were obtained of which we selected two independent stable homozygous lines (OE5 and OE12) for functional analysis. The expression of *GmCAMTA12* in two transgenic Arabidopsis lines was validated through qPCR.

For drought assay, the two lines of OE *GmCAMTA12* (OE5 and OE12) and the wild type (WT) plants were subjected to drought stress by withholding water for two weeks and then re-watered as shown in Figure 5A. Initially, all the plants were growing normally until water was withheld. However, upon encountering drought, nearly all the wild type and transgenic Arabidopsis stopped growth, wilted and started turning yellow afterwards. After 14 days of continuous drought treatment, most of the wild type plants were completely dried as obvious from their phenotype (dried leaves). Unlike WT, most of the OE lines retained life processes, which was evident from chlorophyll they retained in their leaves. After re-watering, majority of transgenic Arabidopsis rejuvenated but most of the wild type did not. The plants were then allowed to grow under normal conditions until seed harvest. As expected, the seed yield of WT and transgenic lines was unequal and OE lines developed more seeds than wild type Arabidopsis. Under well-watered conditions, the survival rate in soil under drought was ~100%; however withholding water for two weeks and then re-watering, less than 60% WT plants survived while OE5 and OE12 showed about 83% and 87% survival rate as shown graphically in Figure 5D. Obviously, the constitutive overexpression of *GmCAMTA12* had enhanced the drought survival efficiency of transgenic Arabidopsis leading to better growth and development.

In their root length assay on $\frac{1}{2}$ MS-mannitol medium as shown in Figure 5B, WT plants grew longer roots than OE plants at 0 mM concentration of mannitol; however, on mannitol, OE plants, specifically OE12, developed longer roots than WT at all three concentrations (50, 100 and 200 mM). Interestingly, the roots of OE12 plants were the longest at 200 mM mannitol. Root length (cm) is shown graphically in Figure 5E. As flaunted through root length assay, the comparatively longer roots of OE than WT were the result of *GmCAMTA12* overexpression in T3 seeds.

Between the two overexpression lines, OE12 performed better than OE5 in both drought assays, thus, we subsequently inoculated seeds of WT and OE12 on $\frac{1}{2}$ MS with various concentrations of mannitol (50, 100, 150 and 200 mM) to evaluate the germination rate (Figure 5C). As expected, we observed higher germination rate of transgenic line OE12 than that of WT at all four concentrations of mannitol. The germination rate of OE12 was ~30% higher than WT at 50 mM mannitol. At 100 mM, the germination rate decreased in both types at a similar pace (~75% in OE and ~45% in WT). As the mannitol concentration increased to 150 and 200 mM, we saw a dramatic decline in the germination efficiency of WT (30% and 25%) as compared to OE (>65% and >60%) (Figure 5F). We can say that the constitutive expression on GmCAMTA12 has enhanced the germination efficiency of transgenic seeds under drought.

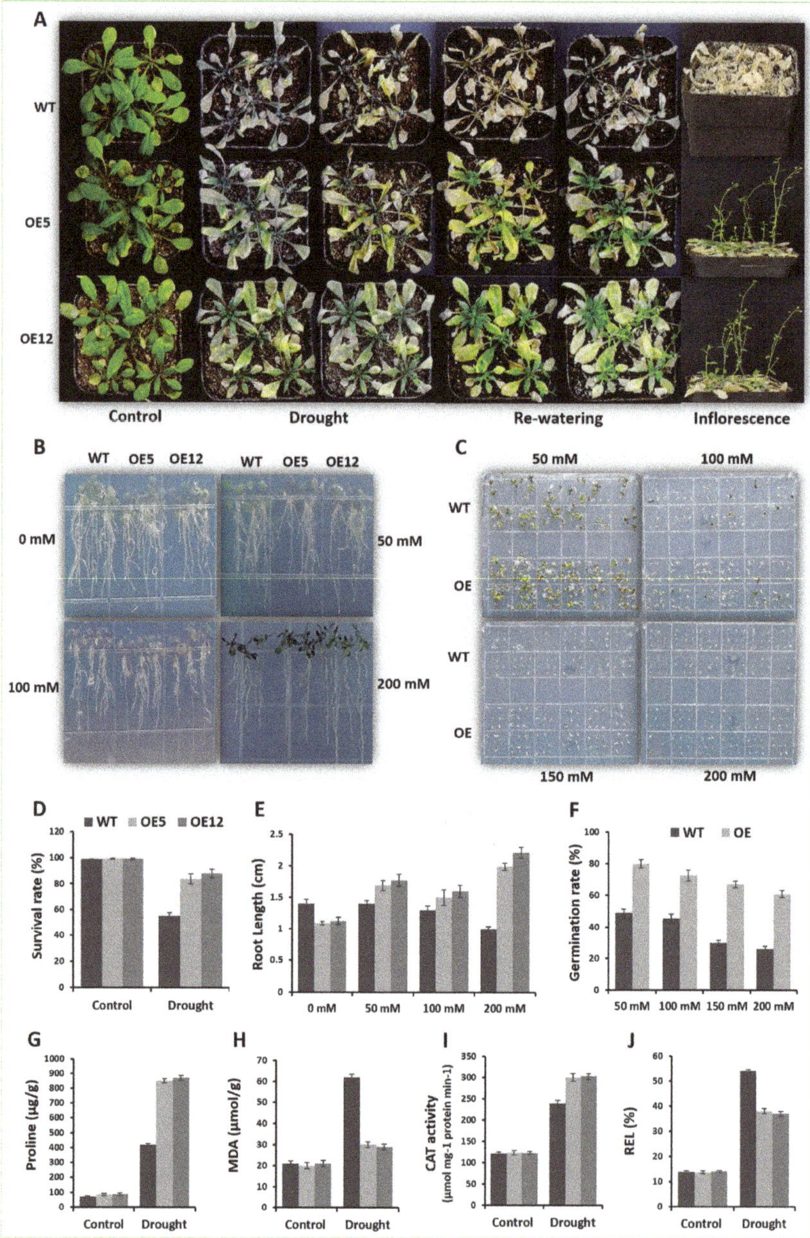

Figure 5. Phenotypic and physiological assays of wild type (WT) and OE under drought stress. (**A**) Drought assay of wild type (WT) and transgenic (OE5 and OE12) Arabidopsis grown on soilrite subjected to 14 days of drought stress. (**B**) Root length assay on MS-mannitol. (**C**) Germination assay on MS-mannitol. (**D**) Column chart showing the survival rate (%) of WT and OE in soil. (**E**) Root length (cm) of WT, OE5 and OE12 lines. (**F**) Germination rate (%) of WT and OE plants. (**G**) Proline contents, (**H**) malonaldehyde (MDA) contents, (**I**) CAT activity and (**J**) relative electrolyte leakage (REL %) of WT and OE plants in control and drought treatment.

In order to check the performance of WT and OE lines at physiological level, the physiological indexes, such as proline and MDA contents, CAT activity and relative electrolyte leakage were determined in all plants subjected to stress. Under well-watered conditions, we observed no significant difference in the level of proline contents, which was quite low in WT and OE lines. In contrast, proline contents calculated in drought-stressed plants had significantly elevated. In WT, the average value of proline was ~400 µg/g, while in OE, it was recorded in the range of 850 to 900 µg/g (Figure 5G). Malonaldehyde (MDA) is a well-known biomarker for sorting out stress-induced membrane damage due to oxidative stress. MDA contents in WT and OE plants during normal conditions matched mutually but its level was doubled in the drought-stressed wt plants compared with drought treated OE lines (Figure 5H). Catalase (CAT) is a major antioxidant enzyme, which is accumulated during abiotic stresses and scavenges H_2O_2. The CAT activity in WT and OE plants under usual conditions was nearly equal (Figure 5I); however, drought treatment enhanced CAT activity in transgenic plants as compare to WT. During stress, electrolytes, specifically K^+ ions leak out of the cells through various channels and thus damage due to stress could be monitored by comparing the electrolyte leakage in WT and transgenic lines. We determined relative electrolyte leakage (REL) in WT and OE lines during normal as well as drought conditions. REL % was nearly leveled under well-watered conditions, but was more pronounced in WT as compared to OE lines during drought (Figure 5J). Noticeably, the amino acid sequence analysis revealed high level of sequence similarity between GmCAMTA12 and AtCAMTA5. We can speculate that GmCAMTA12 being a transcription factor interacted with the downstream target genes (including AtCAMTA5 interactors) and modulated their expression in transgenic Arabidopsis, which contributed to their better performance under drought (determined at molecular level in later section).

2.7. GmCAMTA12 Overexpression Regenerated More Developed and Drought-Efficient Hairy Roots in Soybean

For further functional validation of *GmCAMTA12* in response to drought, the hairy roots system was exploited to overexpress the target gene in soybean. Prior to generating transgenic hairy roots, the *Agrobacterium rhizogenes* strain K599 cells harboring vector control (VC) and overexpressing *GmCAMTA12* construct (OE) were verified through gene-specific PCR (Figure S2 and File S1). After 1–2 weeks of infection of soybean seedlings with K599 (VC and OE), hairy roots had started regenerating with various frequency. Prior to subsequent experiments, we made sure that the hairy roots were transgenic by using gene-specific PCR from the genomic DNA of a small piece of hairy roots. Using qPCR, we validated the overexpression of the target gene in OE roots which was over three times higher than the VC hairy roots. After hairy roots had become long and strong enough by growing for two more weeks, the primary roots were removed and the chimeric plants (with transgenic roots and non-transgenic shoots) were shifted to fresh vermiculite.

When the transgenic roots were ~10 cm long, the VC and OE chimeric soybean plants were extirpated from vermiculite and transferred to Hoagland solution as shown in Figure 6A. After 3 days of acclimation, both VC and OE chimeric plants were treated with 6% PEG6000. The plants started to wilt with leaves curling and shoot apexes drooping on encountering drought. However, the wilting was more in VC than in OE chimeric plants as VC leaves were more wilted. *GmCAMTA12* overexpression induced profuse hairy roots (Figure 6B) due to which the aerial non transgenic part of the OE chimeric soybean plant was also more developed than the VC chimeric plants. The roots were analyzed using the root scanning system (Figure 6C). The OE hairy roots showed higher values with total root length (Figure 6J), surface area (Figure 5K), root volume (Figure 6L), number of branches (Figure 6M) and projected area (Figure 6N).

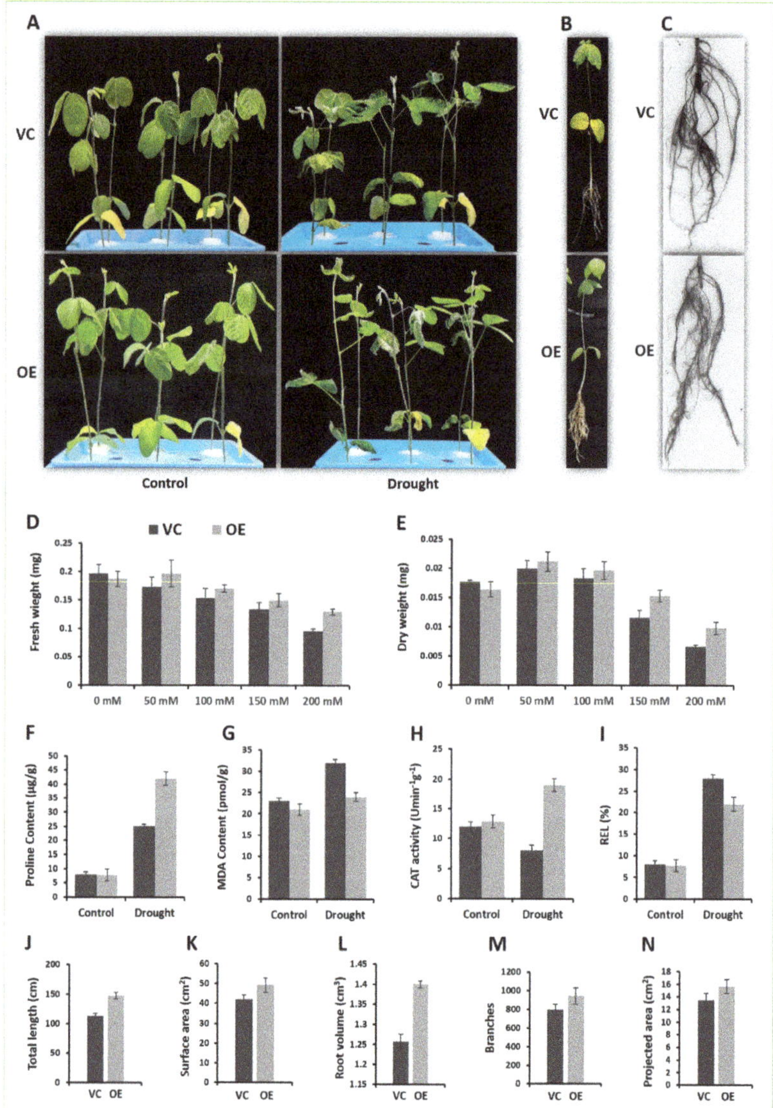

Figure 6. Comparative phenotype and physiology of chimeric sobean plants. (**A**) Chimeric soybean plants bearing VC (Vector control) roots and OE (overexpressing GmCAMTA12) roots in control and drought conditions. (**B**) Comparison of chimeric soybean plants having VC and OE roots. (**C**) VC and OE hairy roots observed with a root scanner. (**D**) Fresh and (**E**) dry weight of VC and OE hairy roots after culturing on MS-mannitol for 10 days. Comparison of VC and OE hairy roots in terms of (**F**) Proline contents, (**G**) MDA contents, (**H**) CAT activity and (**I**) relative electrolyte leakage between VC and OE hairy roots. (**J**) Analysis of the total length, (**K**) surface area, (**L**) root volume, (**M**) number of branches and (**N**) projected area.

The proline and MDA contents, CAT activity and relative electrolyte leakage were determined to check the impact of *GmCAMTA12* overexpression at physiological level. In control samples (Hoagland), proline contents had nearly equal amount in VC and OE hairy roots; however, under drought, proline content level was recorded to be significantly higher in OE hairy roots (Figure 6F). MDA shows the level of membrane damage as it is the final product of lipid peroxidation. In the absence of stress, MDA contents in OE type were slightly less than VC hairy roots; however, with PEG treatment, VC had substantial increase in MDA contents as compare to OE roots (Figure 6G). In contrast, CAT activity was significantly higher in OE hairy roots than VC under drought stress (Figure 6H). For REL, VC hairy roots had higher electrolyte leakage (%) than OE in response to drought (Figure 6I). Comparative physiology of OE and VC hairy roots shows that the overexpression of GmCAMTA improved the drought tolerance of OE by altering physiology.

To analyze their growth efficiency under drought stress, 0.1 g of hairy root from VC and OE hairy roots was weighed under sterile conditions and inoculated on GM media containing four various mannitol concentrations (50, 100, 150 and 200 mM) including a control (0 mM mannitol) with three replicates of each type (Figure S3 and File S1). All the plates were kept in dark in growth room with 28 °C for 10 days of culturing after which the fresh and dry weights were determined. The control samples of both types (VC and OE) hairy roots had nearly same fresh and dry weights; however, on stress media, OE hairy roots showed better performance as the weight of OE roots was more than VC roots at all four concentrations of mannitol (Figure 6D,E). It further flaunted that the overexpression of GmCAMTA12 has improved the drought survival efficiency of OE roots.

2.8. Expression Analysis of GmCAMTA12 Orthologues' Regulatory Network in Arabidopsis

In Arabidopsis, *AtCAMTA5* is the orthologue of *GmCAMTA12*; thus, the regulatory network of *AtCAMTA5* was predicted with STRING database. To test our hypothesis, whether the overexpression of GmCAMTA12 TF in transgenic Arabidopsis would interact with genes in the regulatory network of *AtCAMTA5*, we analyzed the expression of 10 interactors predicted with STRING database (Figure 7A). Total RNA from drought treated WT and OE plants (Figure 7B) was isolated and reverse transcribed to cDNA. Using cDNA and gene-specific primers (Table S6 and File S1), we conducted a qPCR which deciphered the differential expression pattern of the 10 genes in wt and OE plants (Figure 7C). Among them, AtNIP30 (NEFA-interacting nuclear protein—*AT3G62140*), which in humans is involved in negative regulation of proteasomal protein catabolic process, is slightly upregulated in response to drought. AtWRKY14 (*AT1G30650*) encoding WRKY transcription factor 14 possesses a DNA binding domain and specifically interacts with *W box* (a common elicitor-responsive *cis*-element). WRKY14 is also nuclear localized transcription factor and regulates many important processes through gene regulation. *GmCAMTA12* overexpression upregulated *AtWRKY14*, which indicates that along with other transcription factors including WRKY TFs, the spectrum of CAMTA TFs regulatory networks is much wider than we think. Thus, the mutual interaction of CAMTA and WRKY should be further investigated. Peptdyl-prolyl *cis-trans* isomerase AtCYP59 (*AT1G53720*) mediates posttranslational modifications specifically protein folding. With *GmCAMTA12* overexpression, the upregulation of AtCYP59 is almost equal to that of WRKY14. AtANN5 (*AT1G68090*) is a calcium binding protein, plays role in pollen development and is induced by cold, heat, drought and salt stresses. As expected, its expression is relatively the highest among the 10 interactors. Serine racemase (AtSR—*AT4G11640*) is involved in serine biosynthetic pathway. Its expression also seems to elevate with *GmCAMTA12* overexpression in OE Arabidopsis. Elongator complex 3 (AtELO3—*AT5G50320*) is a part of Elongator multiprotein complex and regulates initiation and elongation of transcription. No apparent change in the expression level of AtELO3 in WT and OE was observed. Similar to ANN5, CaMHSP (Calmodulin Binding Heat Shock Protein—*AT3G49050*), also called Alpha/beta-Hydrolases superfamily protein, exhibited a higher transcript level. It is involved in the lipid metabolic pathway. B120 (G-type lectin S-receptor-like serine/threonine-protein kinase—*AT4G21390*) is involved in protein kinase activity and recognition of pollen. Its expression level was downregulated in transgenic Arabidopsis. *AT2G43110* (U3 containing 90S pre-ribosomal complex subunit) was also upregulated with the overexpression of *GmCAMTA12*. *AT3G19850* (BTB (for BR-C, ttk and bab) or POZ (Pox virus and Zinc finger)) domain-containing) mediates protein degradation by facilitating ubiquitination. Its expression level was elevated with *GmCAMTA12* upregulation. All of these genes possess a *CGCG/CGTG* motif in their promoter sequences, which also validates the sequence-specific binding of CAMTA TFs. Most of these interactions are based on text-mining and should be determined experimentally.

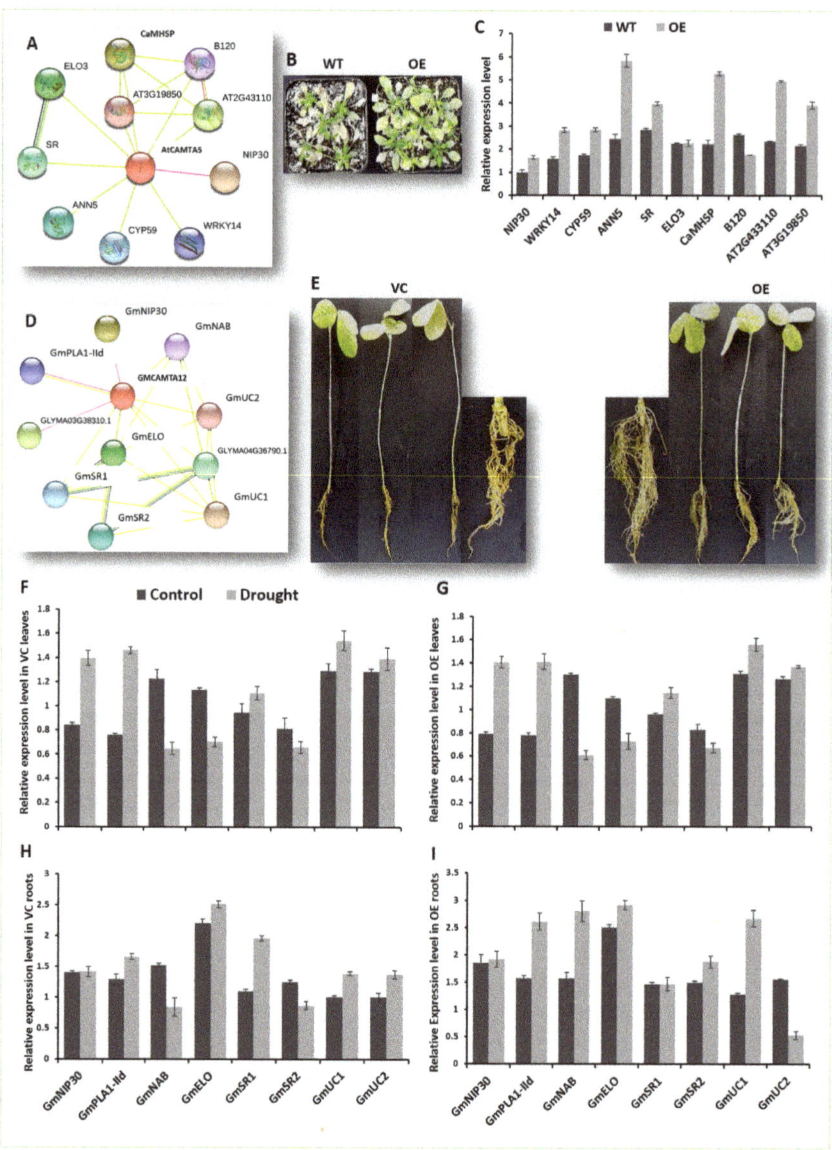

Figure 7. Expression analysis of genes in the regulatory network of *GmCAMTA12* and *AtCAMTA5*. (**A**) STRING-predicted regulatory network of *AtCAMTA5*. (**B**) Drought treated WT and OE Arabidopsis. (**C**) Expression profile of the 10 interactors of *AtCAMTA5* in WT and OE Arabidopsis. (**D**) STRING-predicted regulatory network of GmCAMTA12 in soybean. (**E**) Chimeric soybean plants with VC and OE hairy roots. (**F**) Expression analysis of the eight interactors of *GmCAMTA12* in VC and (**G**) OE leaves. (**H**) Expression analysis of the eight interactors in VC and (**I**) OE roots.

2.9. GmCAMTA12 Overexpression Orchestrated Downstream Genes in Transgenic Hairy Roots

In order to find whether the constitutive overexpression of *GmCAMTA12* modulates the genes with which GmCAMTA12 TF interacts, we selected eight genes in the *GmCAMTA12* regulatory network in soybean predicted with STRING (Figure 7D). In chimeric soybean plants possessing VC and OE hairy roots (Figure 7E) treated with 6% PEG6000, we analyzed the expression pattern of the eight of the predicted interactors of *GmCAMTA12* to know their *GmCAMTA12*-mediated regulation. Using gene-specific primers (Table S7 and File S1), qPCR of the selected interacting proteins in transgenic hairy roots as well as non-transgenic leaves was carried out. The genes displayed a differential expression profile in the control and drought treated chimeric soybean plants. In VC roots (Figure 7H), GmNIP30 (NEFA interacting protein—*GLYMA19G40910*) was slightly upregulated in response to drought and its expression was indifferent to *GmCAMTA12* overexpression in OE roots (Figure 7I). GmPLA1-IId (Phospholipase A1-IId - *GLYMA12G15430*), involved in the lipid metabolic process, was upregulated in VC roots under drought stress, while in OE roots, its upregulation was two-fold higher than that in VC. *GmNAB* (Nucleic Acids Binding—*GLYMA18G40360*) was downregulated in VC roots while upregulated with *GmCAMTA12* overexpression in OE roots in response to drought. The expression of *GmELO* (Catalytic histone acetyltransferase subunit of the RNA polymerase II elongator complex—*GLYMA06G18150*) was the highest and equally expressed gene in both VC and OE roots. GmSR1 (Serine racemase-1 involved in D-Serine biosynthetic process—*GLYMA05G37930*) was upregulated in VC roots; however, with *GmCAMTA12* overexpression, GmSRI was downregulated in OE roots under drought. Similarly, GmSR2 (Serine racemase-2—*GLYMA08G01670*) was repressed in the absence of GmCAMTA12 overexpression; however, in OE roots under drought, it was slightly upregulated. Simultaneously, GmUC1 (uncharacterized, but possessing a *Myb*-DNA binding domain—*GLYMA20G32540*) transcript was slightly upregulated in VC but was significantly induced in OE hairy roots in response to drought. In contrast, GmUC2 (Uncharacterized2—*GLYMA17G07200*) was positively regulated in VC and negatively regulated in OE roots. Unlike the roots, the expression profile of these eight genes was nearly similar in chimeric soybean leaves possessing VC (Figure 7F) and OE (Figure 7G) hairy roots in response to drought. Five out of eight of these genes possess *CGCG/CGTTG* motif.

3. Discussion

Calcium as a ubiquitous secondary messenger orchestrates almost every cellular process in response to environmental stimulus. Plants employ the divalent cation, calcium (Ca^{2+}) in relaying these endogenous (developmental) and exogenous (environmental) signals to appropriate cellular responses. Calcium alone specifically encodes a myriad of distinct signals by using spatial and temporal Ca^{2+} spikes as well as the frequency and amplitude of Ca^{2+} oscillations [10]. Next to the secondary messenger lie the signal relaying molecules including Calmodulin, which further tune the calcium signatures and pass them to *CAMTAs*. Calmodulin Binding Transcription Activators have a short history of two decades, which after first being reported have been genome-wide identified in numerous plant species [23]. *CAMTAs* are important in the sense that they are the transcription factors and an intermediate in the calcium-mediated stress signaling (*Ca-CaM-CAMTA*).

Earlier, the identification and expression analysis of Soybean CAMTAs was conducted [21]; however, their functional characterization remained unexplored. Thus, in order to better comprehend the role of *GmCAMTAs* in drought, and take a holistic snapshot, we first attempted to fill the research gaps through comprehensive in silico analysis of *GmCAMTA* family. Soybean possesses 15 *GmCAMTAs* [21], the second highest number after *Brassica napus*, which possesses 18 genes [24]. Interestingly, such a large stress responsive transcription factor family must have substantial contribution to the drought tolerance of soybean. *CAMTA* has important role in an array of biotic/abiotic stresses as reported in earlier studies in Arabidopsis [22,31–34,43], tomato [14,28], tobacco [18] and *Brassica napus* [24], Arabidopsis [36] and wheat [25]. Dissecting *GmCAMTAs* with various bioinformatics tools, their gene structure depicted 13 exon pattern, which is consistent with its orthologues in Arabidopsis, maize, tomato and others [23]. Promoter enrichment analysis revealed the *cis*-elements including *ABRE, DRE, G-box, W-box, WRKY, ARE* and *MYB*, all of which respond to various stresses [21]. *SlCAMTAs* are differentially expressed during the development and ripening of tomato and the presence of *ERE cis*-element in all *GmCAMTAs* provides the basis for *CAMTA* role in fruit development and ripening [28]. Light responsive elements (*G-box*) are common in all *GmCAMTAs* and we suggest that their role in light stress should be investigated [44]. Additionally, their specific stress-responsive *cis*-motifs should be tested individually or collectively in designing stress-inducible synthetic promoters.

Protein structure is important to know for understanding the action mechanism of protein. The major basic domains, CG-1, ANK, IQ and CaMB are common in all CAMTAs and a close look into the motif sequence of CaMB domain shows residues which are highly conserved across the species. Phylogeny of CAMTAs of four species traced the evolutionary relationship among the homologues as well as orthologues, which is consistent with the previous results [15,17–19,21,23,24]. Some homologues show more similarity and might have co-evolved (Figure 1). Later, the same set of proteins was found to interact with these homologues. A protein interaction network analysis found a few experimentally determined as well as predicted interactions together of 48 unique proteins of various pathways (Figure 3). Thus, we recommend to experimentally determine those interacting proteins to unveil GmCAMTA TFs' link with the pathways of the interacting proteins. In Arabidopsis, 32 proteins were predicted with STRING in the interaction network of *AtCAMTAs* [23]. An experiment-based detailed map of all the *GmCAMTA* interactors would deepen our understanding of intricate mechanisms of *GmCAMTA*-mediated development and immunity against biotic/abiotic factors. Similarly, miRNAs are important transcriptional regulators by directly cleaving their target transcripts; thus, their potential target sites in *GmCAMTAs* were important to determine to understand *CAMTA*-mediated regulation. In silico analysis of *PtCAMTAs* revealed potential targets for four distinct miRNAs [18]. A set of 10 unique miRNAs was detected bringing more complexity in *CAMTA*-mediated stress response mechanisms in soybean (Table S2 and File S1). We recommend to experimentally determine the miRNAs targeting *GmCAMTAs*, which might also lead to the ability to engineer soybean and other crops against drought more accurately.

The expression analysis of *GmCAMTAs* in soybean leaf and root reveals that although all of them express constitutively, but *GmCAMTA2, GmCAMTA4, GmCAMTA5, GmCAMTA11* and *GmCAMTA12* are highly upregulated against drought. Secondly, almost all *GmCAMTAs* expressed and peaked in 3 h after which they were repressed indicating their early responsivity to drought stress (Figure 4). Based on the previous report [21], as well as our qPCR results, GmCAMTA12 was the common drought-efficient transcription factor, and thus, was selected to functionally characterize. As expected, overexpressing *GmCAMTA12* in Arabidopsis and hairy roots and drought assays thereof, validated the previous and current qPCR-based results. Interestingly, *GmCAMTA12* enhanced the drought survivability and growth performance of transgenic Arabidopsis (Figure 5), as well as hairy roots (Figure 6), which was validated at phenotypic, physiological and molecular level. Similar to *GmCAMTA12*, another transcription factor from soybean *GmNAC85* is also drought stress-responsive and its constitutive overexpression significantly enhanced the drought tolerance in transgenic Arabidopsis [45]. Similarly, RNAseq of soybean overexpressing *GmWRKY54* revealed that, *GmWRKY54* conferred drought tolerance by enhancing ABA(Abscisic acid)/Ca^{2+} signaling to close stomata as well as regulating numerous stress responsive transcription factors [6]. Similar results were recently reported about GmWRKY12 transcription factor [46]. In contrast, *AtCAMTA1*-mutant Arabidopsis exhibited enhanced drought sensitivity while also affecting the expression of other drought responsive genes [31]. A recent global transcriptome analysis using RNAseq revealed that a large number of genes involved in diverse stress responses are regulated either directly or indirectly by *AtCAMTA3* as about 3000 genes were misregulated in the *AtCAMTA3*-mutant Arabidopsis [34]. We recommend to experimentally verify the downstream targets of GmCAMTAs through protein-protein interaction experiments such as two-hybrid/pull down assays, which might link/lead to novel pathways. *GmCAMTA12* overexpression altered the expression of the genes in the regulatory network of *AtCAMTA5*, which is the orthologue of

GmCAMTA12 in Arabidopsis. All of the 10 interacting genes possess *CGCG/CGTG* motif in their 2000 bp upstream region validating the sequence-specific interaction of *GmCAMTAs* [12]. In conclusion, the better performance of OE Arabidopsis and chimeric soybean plants on MS-mannitol and Hoagland-PEG was validated at physiologic and molecular level. The comparison of Proline and MDA contents, CAT activity and relative electrolyte leakage between VC and OE hairy roots shown graphically, flaunts the improved development of chimeric plants with hairy roots overexpressing *GmCAMTA12*. This was further verified at a molecular level by determining the transcript level of the genes in the regulatory network of *GmCAMTA12*. The *GmCAMTA* overexpression should also be analyzed in other tissues including soybean flower and seeds as well as in various developmental stages to find its integrated role in processes besides stress tolerance. However, GmCAMTA12 nuclear localization, promoter-GUS analysis, CRISPR/cas9-mediated gene-knockout and global transcriptomics of stably transformed transgenic soybean and Arabidopsis are under investigation. Taken together, a more systematic approach should be adopted to decipher the integrated role of GmCAMTA TFs with the Calcium-Calmodulin signaling crosstalk in drought.

4. Materials and Methods

4.1. In Silico Analysis

By keyword search (Gene ID/locus/accession no.) in three databases (NCBI https://www.ncbi.nlm.nih.gov/, Phytozome https://phytozome.jgi.doe.gov/pz/portal.html and Plantgrnnoble http://plantgrn.noble.org/), we retrieved the genomic, transcriptomic and proteomic sequences of the 15 members of *Glycine max CAMTA* gene family (File S1). Similarly, protein sequences of the corresponding orthologues in *Arabidopsis thaliana*, *Zea mays* and *Solanum lycopersicum* were also retrieved and a dataset was created for bioinformatics analyses (File S1).

The physico-chemical properties including molecular weights, theoretical isoelectric points, Aspartate+Glutamate, Arginine+Lysine, number of atoms, instability index, aliphatic index and GRAVY (Grand average of hydrophathicity) of all *GmCAMTA* proteins were calculated using the ProtParam tool in the ExPASy (https://web.expasy.org/protparam/).

To trace their evolutionary relationship, an ML tree was constructed with MEGAX using default parameters after multiply aligning the protein sequences of *Glycine max*, *Arabidopsis thaliana*, *Zea mays* and *Solanum lycopersicum*.

The exons and introns along the length of *GmCAMTA* genomic sequences were monitored using the online tool GSDS (Gene Structure Display Server (http://gsds.cbi.pku.edu.cn/index.php) [13]. To check the *cis*-motifs within the regulatory regions of *GmCAMTAs*, (2kb upstream) 5′ UTR of each gene was resolved at online database PLACE (https://sogo.dna.affrc.go.jp/). The promoter sequences are in File S1. All the 15 *GmCAMTA* protein sequences were aligned with ClustalX and the conserved motifs were identified using the online SMART tool (Simple Modular Architecture Research Tool) (http://smart.embl-heidelberg.de/). The CaMBD in Arabidopsis and Soybean CAMTA family proteins was particularly analyzed for conserved sequences. The protein alignment is shown in File S1.

The potential miRNA targets along the length of *GmCAMTA* transcripts were predicted online at psRNATarget server (http://plantgrn.noble.org/psRNATarget/). To predict the proteins interacting with *GmCAMTAs*, each protein was separately submitted to STRING database (https://string-db.org/). The online prediction software displayed the proteins experimentally proved/hypothetical proteins interacting with *GmCAMTAs*.

The loci of *CAMTA* family over 20 chromosomes were determined using NCBI Genome Data Viewer at https://www.ncbi.nlm.nih.gov/genome/gdv/. Similarly, the subcellular localization of *GmCAMTA* transcription factors was determined by screening NLS in the protein sequence with online tool cNLS mapper at http://nls-mapper.iab.keio.ac.jp/cgi-bin/NLS_Mapper_form.cgi. NLS (nuclear localization signal/sequence) is an amino acid sequence, which, if present in a protein, indicates its nuclear localization.

Primers for qPCR analysis were designed with Primer premier 5 software as well as NCBI online primer tool https://www.ncbi.nlm.nih.gov/tools/primer-blast/. The multiple alignments of *GmCAMTAs* CDS showed frequent conserved sequences; thus, to minimize ambiguity, the primers were designed with special care against the unique sites in each *GmCAMTA* CDS. Moreover, it was ensured that the primers amplify all alternative transcripts of respective gene. We attempted to design the pair targeting two exons amplifying a stretch of 200–300 bp cDNA. Those primers which failed to give amplification, created more than one peak or did not appear on agarose gel under UV were redesigned until corrected. Moreover, the primers specificity was determined with online nBLAST tool in NCBI https://blast.ncbi.nlm.nih.gov/Blast.cgi?PROGRAM=blastn&PAGE_TYPE=BlastSearch&LINK_LOC=blasthome. Suzhou GENEWIZ Biological Technology Services Company, China, synthesized all primers for this experiment listed in Table S5 and File S1.

4.2. Expression Analysis

The seeds of soybean variety "Willliams 82" were washed with 75% ethanol and then sterile water. They were hydroponically cultured in Hoagland nutrient solution in a growth room with ~28 °C room temperature, 16–8 h (light-dark) photoperiod and a relative humidity of 50%. When the plants have opened their unifoliate leaf pair completely, they were subjected to 6% (PEG 6000) stress simulated drought. The plants were treated with drought for 1, 3, 6, 9 and 12 h in triplicate along with control (Hoagland). At the exact time points, we collected leaf and root samples, froze in liquid nitrogen and stored at –80 C for further processing.

Total RNA from the root and leaf samples (ground in liquid nitrogen) of soybean plants (treated with PEG for various time periods) was extracted using RNAiso Plus (Takara, Japan) according to the manufacturer's protocol. RNA concentration was determined with NanoDrop2000Spectrophotometer (Thermo Fisher Scientific, Waltham, MA, USA). The quality of RNA samples was checked over 1% agarose gel using 0.5X TBE. (Tris/Borate/EDTA) buffer. Bright and clear bands of 25S/18S RNA showed the intact RNA. The samples that showed degraded RNA or relatively unclear bands were re-extracted until intact RNA was detected. The concentration of all good quality RNA samples was adjusted to ~500 ng/µL and it was reverse transcribed into cDNA using the PrimeScript RT reagent kit with gDNA Eraser (Takara, Japan) according to the company's protocol. After the removal of genomic DNA with gDNA eraser, 1 µg total RNA was used as template to synthesize cDNA and stored at –20 °C.

After the reverse transcription of RNA into cDNA and qPCR primer synthesis, qPCR was run according to the experimental design. Prior to the qPCR of all the samples, a standard curve was created with a reaction using a serially diluted cDNA in duplicate. Actin 11 was used as reference with leaf cDNA while Elongation factorα with that of root. Before determining the expression pattern of soybean *CAMTA* gene family in leaves and roots, the personal error was minimized by the standardization of experiment. An initial reaction was run on Stratagene Mx3000 P thermocycler and primers validity was determined first by confirming the single peak (amplification plot), the C_t value for each gene, and then by running the reaction product on agarose gel. Each product gave a single band parallel to right marker band. A 20 µL reaction mixture contained 10 µL SYBR Premix Ex Taq, 0.4 µL ROX Reference Dye II, 0.4 µL of each gene-specific Primer, 2 µL of cDNA and 6.8 µL ddH$_2$O. High PCR efficiency is indispensable for robust and more precise RT-qPCR. The formula $E = 10(-1/Slope)^{-1}$ was used to calculate amplification efficiency and the slope of the calibration curve. The primers having ~100% PCR amplification efficiencies were used. The RT-qPCR profile for our samples consisted of an initial denaturation of 95 °C for 2 min followed by 40 cycles of 94 °C for 5 s and 62 °C for 20 s. The fold change in relative expression level was evaluated using $2^{-\Delta\Delta Ct}$ formula.

4.3. Gene Transformation and Drought Assays

From the bioinformatics as well as qPCR analyses, we selected the gene *GmCAMTA12* (*Glyma.17 G031900*) to clone and transform for functional validation. The complete CDS (2769 bp) of *GmCAMTA12* was PCR-amplified using cDNA as template, forward primer ACTAGTATGGCGAATAACTTAGCGG and reverse primer CCCGGGCTATGTCTTCAGTTGCATGTCAA. For amplification, LA Taq kit (Takara) was used according to the manufacturer's protocol. The PCR profile was; an initial denaturation at 94 °C for 1 min followed 35 cycles of 94 °C for 30 s, 55 °C for 30 s and 72 °C for 150 s, and a final extension at 72 °C for 10 min. Using CT101 cloning kit (Transgene, Beijing, China), the gene was cloned into pEASY®-T1 cloning vector and transformed into Trans1-T1 chemically competent cells (Transgene) following heat shock method according to the manufacturer's instructions. The positive clones were obtained on a selective LB-Kan⁺ agar plate and screened through vector and gene specific PCR as well as restriction. After double check, three separate clones were sequenced using M13 forward and reverse primers. When the sequences of all three clones were analyzed without any point mutation, the gene was restricted from pEASY®-T1 and sub-cloned by restricting from cloning vector using *SpeI* and *SmaI* (Takara) and ligated into the corresponding sites of expression vector pBASTA (with Kanamycin and Glyphosate resistance genes) using T4 Ligase (Promega, Madison, WI, USA) constituting the recombinant overexpression construct, pBASTA-*GmCAMTA12*. The pBASTA-*GmCAMTA12* construct was transformed into DH5α strain of *E.coli* (Transgene) and positive clones were obtained on a selective LB-Kan⁺ agar plate. After a double check (PCR and restriction) of single colonies harboring the overexpression construct, the colonies were preserved in sterile 80% glycerol. The recombinant vector was transformed into chemically competent *Agrobacterium tumefacians* (EHA105) and *Agrobacterium rhizogenes* (K599) by a 5 min liquid nitrogen treatment followed by a heat shock of 5 min at 37 °C. The positive clones of EHA105 were identified on YEP Kan+Rif plates while K599 were selected using YEP Kan+Strep plates. After PCR verification, the cells were preserved at −80 °C for future use.

Arabidopsis Col seeds were kindly provided by Engineering Research Center of the Ministry of Education of Bioreactor and Pharmaceutical Development. To synchronize the germination of seeds, wt seeds were soaked in water and kept at 4 °C for 48 h and then sowed on a mixture of humus + vermiculite (3:7) in a growth room in dark at 23 °C. The germinated seedlings were then allowed to grow in 16 h photoperiod until 40 days with regular watering. We transformed the inflorescence (unopened flowers) of mature healthy Arabidopsis plants through Floral Dip method [47] and harvested T1 (Transgenic generation 1) seeds. The T1 seeds were germinated in excess under the same conditions as for wild type seeds. A primary screening of T1 plants (at six leaf stage) was carried out by spraying with 1% Basta (glyphosate) and the basta-survived plants were grown in separate pots in fresh soil (humus + vermiculite) under the same conditions as for wild type. Each plant represented a separate transformation event. The non-transformants died, while from the green plants, we extracted genomic DNA from leaves by 2 X CTAB and a PCR with gene specific primers using the gDNA as template was performed for secondary screening. T1 plants positive with PCR were grown to harvest T₂ seeds. Similarly, T₃ seeds from homozygous lines were harvested following the primary screening by 1% Basta and secondary with PCR to ensure stable transformation.

The seeds of Arabidopsis were surface sterilized by keeping in 70% ethanol for 1 min followed by 50% bleach for 10 min after which the seeds were washed six times by sterile water. The wild-type and two T₃ transgenic Arabidopsis lines were germinated under sterile conditions on half MS plates added with various mannitol concentration. For germination, we vernalized wt and OE seeds by keeping in dark for 3 days at 4 °C. Each petri plate was marked and divided in to three parts, one of which was allotted to wt while the other two for Line-5 (OE5) and Line-12 (OE12) seeds. The seeded plates were kept in a growth chamber with 16 h photoperiod and 23 °C temperature. The germination of each line was observed and recorded for one week and the germination rate was determined by the number of seeds germinated divided by the total number of seeds. For root length analysis, wt and T3 lines were first germinated on MS media under sterile conditions and one week old seedlings of same length were transferred to square plates containing MS with various concentrations of mannitol. The plates were placed in a rack vertically so that the roots grow downwards to ground. The root growth was observed and the plants were photographed. The seeds of wild-type and two T3 transgenic Arabidopsis lines were germinated and grown for one month in soil at 23 °C, watered regularly and then subjected to drought stress by withholding water for 14 days and photographed before, during and on drought recovery after re-watering.

Various physiological parameters like MDA [48] and proline contents [49] as well as CAT activity [50] and relative electrolyte leakage were determined. Cell membrane integrity is lost as K^+ ions (a chief electrolyte) leaks out of the cell leading to cell death in stress. This measurement is an indicator of the cell damage caused by stress [51]. To determine REL, leaf sample (~2 g) was thoroughly rinsed with deionized water and then subjected to vibration for 2 h at 25 °C in a test tube containing 10 mL deionized H_2O. Using a conductivity meter, we determined the conductivity of solution (C1). A second measurement of conductivity (C2) was taken after boiling the solution for 25 min and then cooled to RT. REL % was calculated using the formula (C1/C2) × 100.

Soybean hairy root is an established platform for the functional analysis of a gene by overexpression. For hairy roots, seeds of soybean variety JU 72 were washed with 75% ethanol and rinsed in sterile water a few times. The seeds were sown in pots containing autoclaved vermiculite and kept in a growth room with 16 h photoperiod and 28 °C temperature and watered regularly. In parallel, K599 from glycerol was streaked on YEP Kan + Strep plates and incubated at 28 °C for 48 h. A single colony was picked and inoculated in YEP containing Kanamycin and Streptomycin. The culture was incubated in a shaking incubator at 200 rpm and 28 °C for 48 h. A 100 µL of inoculum rom the culture was spread on YEP Kan + Strep plates and incubated at 28 °C for 48 h. When the seedlings had just sprouted, the cotyledons unfolded and the first unifoliate leaves had not yet appeared, they were ready to be infected with K599, harboring the overexpression construct. The *Agrobacterium rhizogenes* culture from the plate was picked with a needle and injected into the hypocotyl right at the base of cotyledon [52]. K599 harboring empty pBASTA were used to regeneration of VC (Vector Control) hairy roots. After infection, the seedlings were covered with a transparent plastic to ensure high humidity. Within two weeks, hairy roots started to sprout from the site of infection at variable frequencies with at least one root per seedling. The soybean plants with hairy roots were allowed to grow for two weeks after which excess of autoclaved vermiculite was added to cover the hairy roots and watered regularly. We extracted the genomic DNA from both type of the transgenic roots to confirm VC and OE through gene specific primers. The roots were scanned with a scanner and analyzed with the software.

After the hairy roots were ~10 cm long, the plants were uprooted, primary roots were cut and the chimeric plants (transgenic root and non-transgenic shoot) were transferred to fresh autoclaved vermiculite and watered regularly. It is important to notice that the shoot of each plant was cut after the second trifoliate leaves. Thus, the newly grown leaves were compared between the chimeric plants VC and OE transgenic roots. To check the performance of transgenic hairy roots under drought, plants with 10 cm long transgenic hairy roots were transferred to Hoagland solution and after acclimation, they were treated with 6% PEG6000. Proline and MDA contents, CAT activity and REL were determined in VC and OE hairy roots. Moreover, the 0.1 g of the hairy roots (VC and OE) were weighed in sterile conditions and grown on MS media at various concentrations of mannitol including 0, 50, 100, 150 and 200 mM. After 10 days, fresh and dry weights of the hairy roots were determined. The dry weight was measured keeping hairy roots at 60 °C overnight.

4.4. Expression Analysis of the GmCAMTA12 Orthologue's Regulatory Network in wt and OE Arabidopsis

The regulatory network of the orthologue of *GmCAMTA12* (*AtCAMTA5*) in Arabidopsis was predicted online with STRING database. The genomic and proteomic sequences of these 10 interactors are given in File S1. The primers for qPCR were designed against the 10 genes with Primer-BLAST tool (Table S6 and File S1). In parallel, RNA from wt and OE lines was extracted, quantified and reverse transcribed into cDNA with Takara kit following kit's instructions. To determine the expression pattern of the 10 genes, qPCR was run using the primers of each gene with three biological replicates. *AtActin11* was used an internal reference. The data was analyzed with 2^{-ddCt} method.

4.5. Analysis of GmCAMTA12 Regulatory Network in Chimeric Soybean Plants

Using the STRING database, the interactors of *GmCAMTA12* in soybean were predicted. The primers for qPCR of eight genes designed with Primer-BLAST are given in Table S7 and File S1. The genomic and proteomic sequences of these 10 interactors are given in File S1. To determine the effect of overexpression of *GmCAMTA12* on its interacting proteins, RNA from all samples was extracted using RNAiso plus (Takara) and reverse transcribed into cDNA with Takara rt kit and qPCR was run for all samples in triplicate. Actin 11 was used as internal control. Relative expression level was determined using 2^{-ddCt} method.

Supplementary Materials: Supplementary materials can be found at http://www.mdpi.com/1422-0067/20/19/4849/s1.

Author Contributions: Conceptualization, H.-Y.L.; methodology, W.-C.L. and M.N.; software, M.N.; validation, A.J.; investigation, M.N.; resources, H.-Y.L.; data curation, N.A.; writing—original draft preparation, M.N.; writing—review and editing, M.N. and F.-W.W.; visualization, W.-D.Q.; supervision, F.-W.W.; project administration, H.-Y.L.; funding acquisition, H.-Y.L. All authors read and approved the final manuscript.

Funding: This research was funded by the National Key R&D Program of China, grant number (2016 YFD0101005), the National Natural Science Foundation of China, grant number (31601323), the National Natural Science Foundation of Jilin Province, grant number (20170101015 JC, 20180101028 JC, 20190201259 JC) and Special Program for Research of Transgenic Plants grant number (2016 YFD0101005).

Acknowledgments: We are grateful to Haiyan Li for making all the materials available and critically reviewing the manuscript.

Conflicts of Interest: The authors declare no conflict of interest.

Abbreviations

CAMTA	Calmodulin Binding Transcription Activator
CaM	Calmodulin (Calcium-Modulated Protein)
CML	Calmodulin-Like
CBL	Calcineurin B-like
ML	Maximum Likelihood
OE	Overexpression
PEG	Poly Ethylene Glycol
TF	Transcription Factor
VC	Vector Control
WT	Wild Type

References

1. Foley, J.A.; Ramankutty, N.; Brauman, K.A.; Cassidy, E.S.; Gerber, J.S.; Johnston, M.; Mueller, N.D.; O'Connell, C.; Ray, D.K.; West, P.C.; et al. Solutions for a cultivated planet. *Nature* **2011**, *478*, 337. [CrossRef] [PubMed]

2. Godfray, H.C.J.; Beddington, J.R.; Crute, I.R.; Haddad, L.; Lawrence, D.; Muir, J.F.; Pretty, J.; Robinson, S.; Thomas, S.M. CToulmin, Food Security: The Challenge of Feeding 9 Billion People. *Science* **2010**, *327*, 812–818. [CrossRef] [PubMed]

3. Hickey, L.T.; Hafeez, A.N.; Robinson, H.; Jackson, S.A.; Leal-Bertioli, S.C.M.; Tester, M.; Gao, C.; Godwin, I.D.; Hayes, B.J.; Wulff, B.B.H. Breeding crops to feed 10 billion. *Nat. Biotechnol.* **2019**, *37*, 744–754. [CrossRef] [PubMed]

4. Shinozaki, K.; Uemura, M.; Bailey-Serres, J.; Bray, E. Responses to abiotic stress. In *Biochemistry and Molecular Biology of Plants*, 2nd ed.; Buchanan, B.B., Gruissem, W., Jones, R.L., Eds.; Wiley: Oxford, UK, 2015; pp. 1051–1100.

5. Prudhomme, C.; Giuntoli, I.; Robinson, E.L.; Clark, D.B.; Arnell, N.W.; Dankers, R.; Fekete, B.M.; Franssen, W.; Gerten, D.; Gosling, S.N.; et al. DWisser, Hydrological droughts in the 21 st century, hotspots and uncertainties from a global multimodel ensemble experiment. *Proc. Natl. Acad. Sci. USA* **2014**, *111*, 3262–3267. [CrossRef]

6. Wei, W.; Liang, D.-W.; Bian, X.-H.; Shen, M.; Xiao, J.-H.; Zhang, W.-K.; Ma, B.; Lin, Q.; Lv, J.; Chen, X.; et al. GmWRKY54 improves drought tolerance through activating genes in abscisic acid and Ca2+ signaling pathways in transgenic soybean. *Plant J.* **2019**. [CrossRef] [PubMed]

7. Zhang, H.; Li, Y.; Zhu, J.-K. Developing naturally stress-resistant crops for a sustainable agriculture. *Nat. Plants* **2018**, *4*, 989–996. [CrossRef] [PubMed]

8. Wu, M.; Li, Y.; Chen, D.; Liu, H.; Zhu, D.; Xiang, Y. Genome-wide identification and expression analysis of the IQD gene family in moso bamboo (Phyllostachys edulis). *Sci. Rep.* **2016**, *6*, 24520. [CrossRef] [PubMed]

9. Khan, S.-A.; Li, M.-Z.; Wang, S.-M.; Yin, H.-J. Revisiting the Role of Plant Transcription Factors in the Battle against Abiotic Stress. *Int. J. Mol. Sci.* **2018**, *19*, 1634. [CrossRef] [PubMed]

10. Kim, M.C.; Chung, W.S.; Yun, D.-J.; Cho, M.J. Calcium and calmodulin-mediated regulation of gene expression in plants. *Mol. Plant* **2009**, *2*, 13–21. [CrossRef]

11. Reddy, A.S.N.; Reddy, V.S.; Golovkin, M. A Calmodulin Binding Protein from Arabidopsis Is Induced by Ethylene and Contains a DNA-Binding Motif. *Biochem. Biophys. Res. Commun.* **2000**, *279*, 762–769. [CrossRef] [PubMed]

12. Yang, T.; Poovaiah, B.W. A Calmodulin-binding/CGCG Box DNA-binding Protein Family Involved in Multiple Signaling Pathways in Plants. *J. Biol. Chem.* **2002**, *277*, 45049–45058. [CrossRef] [PubMed]

13. Bouché, N.; Scharlat, A.; Snedden, W.; Bouchez, D.; Fromm, H. A Novel Family of Calmodulin-binding Transcription Activators in Multicellular Organisms. *J. Biol. Chem.* **2002**, *277*, 21851–21861. [CrossRef] [PubMed]

14. Zegzouti, H.; Jones, B.; Frasse, P.; Marty, C.; Maitre, B.; Latché, A.; Pech, J.-C.; Bouzayen, M. Ethylene-regulated gene expression in tomato fruit: Characterization of novel ethylene-responsive and ripening-related genes isolated by differential display. *Plant J.* **1999**, *18*, 589–600. [CrossRef] [PubMed]

15. Yang, Y.; Sun, T.; Xu, L.; Pi, E.; Wang, S.; Wang, H.; Shen, C. Genome-wide identification of CAMTA gene family members in Medicago truncatula and their expression during root nodule symbiosis and hormone treatments. *Front. Plant Sci.* **2015**, *6*, 459. [CrossRef] [PubMed]

16. Zhang, J.; Pan, X.; Ge, T.; Yi, S.; Lv, Q.; Zheng, Y.; Ma, Y.; Liu, X.; Xie, R. Genome-wide identification of citrus CAMTA genes and their expression analysis under stress and hormone treatments. *J. Hortic. Sci. Biotechnol.* **2019**, *94*, 331–340. [CrossRef]

17. Wei, M.; Xu, X.; Li, C. Identification and expression of CAMTA genes in Populus trichocarpa under biotic and abiotic stress. *Sci. Rep.* **2017**, *7*, 17910. [CrossRef]

18. Kakar, K.U.; Nawaz, Z.; Cui, Z.; Cao, P.; Jin, J.; Shu, Q.; Ren, X. Evolutionary and expression analysis of CAMTA gene family in Nicotiana tabacum yielded insights into their origin, expansion and stress responses. *Sci. Rep.* **2018**, *8*, 10322. [CrossRef]

19. Meer, L.; Mumtaz, S.; Labbo, A.M.; Khan, M.J.; Sadiq, I. Genome-wide identification and expression analysis of calmodulin-binding transcription activator genes in banana under drought stress. *Sci. Hortic.* **2019**, *244*, 10–14. [CrossRef]

20. Büyük, İ.; İlhan, E.; Şener, D.; Özsoy, A.U.; Aras, S. Genome-wide identification of CAMTA gene family members in *Phaseolus vulgaris* L. and their expression profiling during salt stress. *Mol. Biol. Rep.* **2019**, *46*, 2721–2732. [CrossRef]

21. Wang, G.; Zeng, H.; Hu, X.; Zhu, Y.; Chen, Y.; Shen, C.; Wang, H.; Poovaiah, B.W.; Du, L. Identification and expression analyses of calmodulin-binding transcription activator genes in soybean. *Plant Soil.* **2015**, *386*, 205–221. [CrossRef]

22. Galon, Y.; Aloni, R.; Nachmias, D.; Snir, O.; Feldmesser, E.; Scrase-Field, S.; Boyce, J.M.; Bouché, N.; Knight, M.R.; Fromm, H. Calmodulin-binding transcription activator 1 mediates auxin signaling and responds to stresses in Arabidopsis. *Planta* **2010**, *232*, 165–178. [CrossRef] [PubMed]

23. Rahman, H.; Yang, J.; Xu, Y.-P.; Munyampundu, J.-P.; Cai, X.-Z. Phylogeny of Plant CAMTAs and Role of AtCAMTAs in Nonhost Resistance to Xanthomonas oryzae pv. oryzae. *Front. Plant Sci.* **2016**, *7*, 177. [CrossRef] [PubMed]

24. Rahman, H.; Xu, Y.-P.; Zhang, X.-R.; Cai, X.-Z. Brassica napus Genome Possesses Extraordinary High Number of CAMTA Genes and CAMTA3 Contributes to PAMP Triggered Immunity and Resistance to Sclerotinia sclerotiorum. *Front. Plant Sci.* **2016**, *7*, 581. [CrossRef] [PubMed]

25. Wang, Y.; Wei, F.; Zhou, H.; Liu, N.; Niu, X.; Yan, C.; Zhang, L.; Han, S.; Hou, C.; Wang, D. TaCAMTA4, a Calmodulin-Interacting Protein, Involved in Defense Response of Wheat to Puccinia triticina. *Sci. Rep.* **2019**, *9*, 641. [CrossRef] [PubMed]

26. Yang, T.; Poovaiah, B.W. An early ethylene up-regulated gene encoding a calmodulin-binding protein involved in plant senescence and death. *J. Biol. Chem.* **2000**, *275*, 38467–38473. [CrossRef] [PubMed]

27. [Benn, G.; Wang, C.-Q.; Hicks, D.R.; Stein, J.; Guthrie, C.; Dehesh, K. A key general stress response motif is regulated non-uniformly by CAMTA transcription factors. *Plant J.* **2014**, *80*, 82–92. [CrossRef] [PubMed]

28. Yang, T.; Peng, H.; Whitaker, B.D.; Jurick, W.M. Differential expression of calcium/calmodulin-regulated SlSRs in response to abiotic and biotic stresses in tomato fruit. *Physiol. Plant.* **2013**, *148*, 445–455. [CrossRef] [PubMed]

29. Du, L.; Ali, G.S.; Simons, K.A.; Hou, J.; Yang, T.; Reddy, A.S.N.; Poovaiah, B.W. Ca2+/calmodulin regulates salicylic-acid-mediated plant immunity. *Nature* **2009**, *457*, 1154. [CrossRef]

30. Nie, H.; Zhao, C.; Wu, G.; Wu, Y.; Chen, Y.; Tang, D. SR1, a calmodulin-binding transcription factor, modulates plant defense and ethylene-induced senescence by directly regulating NDR1 and EIN3. *Plant Physiol.* **2012**, *158*, 1847–1859. [CrossRef]

31. Pandey, N.; Ranjan, A.; Pant, P.; Tripathi, R.K.; Ateek, F.; Pandey, H.P.; Patre, U.V.; Sawant, S.V. CAMTA 1 regulates drought responses in Arabidopsis thaliana. *BMC Genom.* **2013**, *14*, 216. [CrossRef]

32. Doherty, C.J.; van Buskirk, H.A.; Myers, S.J.; Thomashow, M.F. Roles for Arabidopsis CAMTA transcription factors in cold-regulated gene expression and freezing tolerance. *Plant Cell* **2009**, *21*, 972–984. [CrossRef] [PubMed]

33. Kim, Y.; Park, S.; Gilmour, S.J.; Thomashow, M.F. Roles of CAMTA transcription factors and salicylic acid in configuring the low-temperature transcriptome and freezing tolerance of Arabidopsis. *Plant J.* **2013**, *75*, 364–376. [CrossRef]

34. Prasad, K.V.S.K.; Abdel-Hameed, A.A.E.; Xing, D.; Reddy, A.S.N. Global gene expression analysis using RNA-seq uncovered a new role for SR1/CAMTA3 transcription factor in salt stress. *Sci. Rep.* **2016**, *6*, 27021. [CrossRef] [PubMed]

35. Kim, Y.S.; An, C.; Park, S.; Gilmour, S.J.; Wang, L.; Renna, L.; Brandizzi, F.; Grumet, R.; Thomashow, M.F. CAMTA-Mediated Regulation of Salicylic Acid Immunity Pathway Genes in Arabidopsis Exposed to Low Temperature and Pathogen Infection. *Plant Cell* **2017**, *29*, 2465–2477. [CrossRef] [PubMed]

36. Shkolnik, D.; Finkler, A.; Pasmanik-Chor, M.; Fromm, H. CALMODULIN-BINDING TRANSCRIPTION ACTIVATOR 6: A Key Regulator of Na+ Homeostasis during Germination. *Plant Physiol.* **2019**, *180*, 1101–1118. [CrossRef] [PubMed]

37. 2019 Soystats. Available online: http://soystats.com/wp-content/uploads/2019-SoyStats-Web.pdf (accessed on 10 August 2019).

38. Li, Y.; Chen, Q.; Nan, H.; Li, X.; Lu, S.; Zhao, X.; Liu, B.; Guo, C.; Kong, F.; Cao, D. Overexpression of GmFDL19 enhances tolerance to drought and salt stresses in soybean. *PLoS ONE* **2017**, *12*, e0179554. [CrossRef] [PubMed]

39. Dai, X.; Zhao, P.X. psRNATarget: A plant small RNA target analysis server. *Nucleic Acids Res.* **2011**, *39*, W155–W159. [CrossRef] [PubMed]

40. Choi, M.S.; Kim, M.C.; Yoo, J.H.; Moon, B.C.; Koo, S.C.; Park, B.O.; Lee, J.H.; Koo, Y.D.; Han, H.J.; Lee, S.Y.; et al. Isolation of a Calmodulin-binding Transcription Factor from Rice (*Oryza sativa* L.). *J. Biol. Chem.* **2005**, *280*, 40820–40831. [CrossRef] [PubMed]

41. Han, J.; Gong, P.; Reddig, K.; Mitra, M.; Guo, P.; Li, H.-S. The fly CAMTA transcription factor potentiates deactivation of rhodopsin, a G protein-coupled light receptor. *Cell* **2006**, *127*, 847–858. [CrossRef] [PubMed]

42. Szklarczyk, D.; Gable, A.L.; Lyon, D.; Junge, A.; Wyder, S.; Huerta-Cepas, J.; Simonovic, M.; Doncheva, N.T.; Morris, J.H.; Bork, P.; et al. STRING v11: Protein-protein association networks with increased coverage, supporting functional discovery in genome-wide experimental datasets. *Nucleic Acids Res.* **2019**, *47*, D607–D613. [CrossRef]

43. Galon, Y.; Nave, R.; Boyce, J.M.; Nachmias, D.; Knight, M.R.; Fromm, H. Calmodulin-binding transcription activator (CAMTA) 3 mediates biotic defense responses in Arabidopsis. *FEBS Lett.* **2008**, *582*, 943–948. [CrossRef] [PubMed]

44. Finkler, A.; Kaplan, B.; Fromm, H. Ca-Responsive cis-Elements in Plants. *Plant Signal. Behav.* **2007**, *2*, 17–19. [CrossRef] [PubMed]

45. Nguyen, K.H.; Mostofa, M.G.; Li, W.; van Ha, C.; Watanabe, Y.; Le, D.T.; Thao, N.P.; Tran, L.-S.P. The soybean transcription factor GmNAC085 enhances drought tolerance in Arabidopsis. *Environ. Exp. Bot.* **2018**, *151*, 12–20. [CrossRef]

46. Shi, W.-Y.; Du, Y.-T.; Ma, J.; Min, D.-H.; Jin, L.-G.; Chen, J.; Chen, M.; Zhou, Y.-B.; Ma, Y.-Z.; Xu, Z.-S.; et al. The WRKY Transcription Factor GmWRKY12 Confers Drought and Salt Tolerance in Soybean. *Int. J. Mol. Sci.* **2018**, *19*, 4087. [CrossRef] [PubMed]

47. Zhang, X.; Henriques, R.; Lin, S.-S.; Niu, Q.-W.; Chua, N.-H. Agrobacterium-mediated transformation of Arabidopsis thaliana using the floral dip method. *Nat. Protoc.* **2006**, *1*, 641–646. [CrossRef]

48. Xia, Z.; Xu, Z.; Wei, Y.; Wang, M. Overexpression of the Maize Sulfite Oxidase Increases Sulfate and GSH Levels and Enhances Drought Tolerance in Transgenic Tobacco. *Front. Plant Sci.* **2018**, *9*, 298. [CrossRef] [PubMed]

49. Bates, L.S.; Waldren, R.P.; Teare, I.D. Rapid determination of free proline for water-stress studies. *Plant Soil* **1973**, *39*, 205–207. [CrossRef]

50. Tang, Y.; Liu, K.; Zhang, J.; Li, X.; Xu, K.; Zhang, Y.; Qi, J.; Yu, D.; Wang, J.; Li, C. JcDREB2, a Physic Nut AP2/ERF Gene, Alters Plant Growth and Salinity Stress Responses in Transgenic Rice. *Front. Plant Sci.* **2017**, *8*, 306. [CrossRef]

51. Demidchik, V.; Straltsova, D.; Medvedev, S.S.; Pozhvanov, G.A.; Sokolik, A.; Yurin, V. Stress-induced electrolyte leakage: The role of K+-permeable channels and involvement in programmed cell death and metabolic adjustment. *J. Exp. Bot.* **2014**, *65*, 1259–1270. [CrossRef]

52. Kereszt, A.; Li, D.; Indrasumunar, A.; Nguyen, C.D.T.; Nontachaiyapoom, S.; Kinkema, M.; Gresshoff, P.M. Agrobacterium rhizogenes-mediated transformation of soybean to study root biology. *Nat. Protoc.* **2007**, *2*, 948–952. [CrossRef]

International Journal of
Molecular Sciences

Map-Based Functional Analysis of the *GhNLP* Genes Reveals Their Roles in Enhancing Tolerance to N-Deficiency in Cotton

Richard Odongo Magwanga [1,2], Joy Nyangasi Kirungu [1], Pu Lu [1], Xiaoyan Cai [1], Zhongli Zhou [1], Yanchao Xu [1], Yuqing Hou [1], Stephen Gaya Agong [2], Kunbo Wang [1,*] and Fang Liu [1,*]

[1] Research Base in Anyang Institute of Technology, State Key Laboratory of Cotton Biology/Institute of Cotton Research, Chinese Academy of Agricultural Science (ICR, CAAS), Anyang 455000, China; magwangarichard@yahoo.com (R.O.M.); nyangasijoy@yahoo.com (J.N.K.); lupu1992@cricaas.com.cn (P.L.); caixy@cricaas.com.cn (X.C.); zhouzl@cricaas.com.cn (Z.Z.); xuyanchao2016@163.com (Y.X.); houyp@cricaas.com.cn (Y.H.)

[2] School of Biological and Physical sciences (SBPS), Main campus, Jaramogi Oginga Odinga University of Science and Technology (JOOUST), P.O. Box 210-40601, Bondo, Kenya; sgagong@joooust.ac.ke

[*] Correspondence: wkbcri@cricaas.com.cn (K.W.); liufcri@caas.com (F.L.); Tel.: +86-139-4950-7902 (F.L.)

Received: 22 June 2019; Accepted: 1 October 2019; Published: 8 October 2019

Abstract: Nitrogen is a key macronutrient needed by plants to boost their production, but the development of cotton genotypes through conventional approaches has hit a bottleneck due to the narrow genetic base of the elite cotton cultivars, due to intensive selection and inbreeding. Based on our previous research, in which the BC_2F_2 generations developed from two upland cotton genotypes, an abiotic stress-tolerant genotype, *G. tomentosum* (donor parent) and a highly-susceptible, and a highly-susceptible, but very productive, *G. hirsutum* (recurrent parent), were profiled under drought stress conditions. The phenotypic and the genotypic data generated through genotyping by sequencing (GBS) were integrated to map drought-tolerant quantitative trait loci (QTLs). Within the stable QTLs region for the various drought tolerance traits, a nodule-inception-like protein (NLP) gene was identified. We performed a phylogenetic analysis of the NLP proteins, mapped their chromosomal positions, intron-exon structures and conducted ds/dn analysis, which showed that most *NLP* genes underwent negative or purifying selection. Moreover, the functions of one of the highly upregulated genes, *Gh_A05G3286* (*Gh NLP5*), were evaluated using the virus gene silencing (VIGS) mechanism. A total of 226 proteins encoded by the *NLP* genes were identified, with 105, 61, and 60 in *Gossypium hirsutum*, *G. raimondii*, and *G. arboreum*, respectively. Comprehensive Insilico analysis revealed that the proteins encoded by the *NLP* genes had varying molecular weights, protein lengths, isoelectric points (*pI*), and grand hydropathy values (GRAVY). The GRAVY values ranged from a negative one to zero, showing that proteins were hydrophilic. Moreover, various *cis*-regulatory elements that are the binding sites for stress-associated transcription factors were found in the promoters of various *NLP* genes. In addition, many miRNAs were predicted to target *NLP* genes, notably miR167a, miR167b, miR160, and miR167 that were previously shown to target five *NAC* genes, including *NAC1* and *CUC1*, under N-limited conditions. The real-time quantitative polymerase chain reaction (RT-qPCR) analysis, revealed that five genes, *Gh_D02G2018*, *Gh_A12G0439*, *Gh_A03G0493*, *Gh_A03G1178*, and *Gh_A05G3286* were significantly upregulated and perhaps could be the key *NLP* genes regulating plant response under N-limited conditions. Furthermore, the knockdown of the *Gh_A05G3286* (*GhNLP5*) gene by virus-induced silencing (VIGS) significantly reduced the ability of these plants to the knockdown of the *Gh_A05G3286* (GhNLP5) gene by virus-induced gene silencing (VIGS) significantly reduced the ability of the VIGS-plants to tolerate N-limited conditions compared to the wild types (WT). The VIGS-plants registered lower chlorophyll content, fresh shoot biomass, and fresh root biomass, addition to higher levels of malondialdehyde (MDA) and significantly reduced levels of proline, and superoxide dismutase (SOD) compared to the WT under N-limited conditions. Subsequently, the expression levels of the Nitrogen-stress responsive genes, *GhTap46*,

GhRPL18A, and *GhKLU* were shown to be significantly downregulated in VIGS-plants compared to their WT under N-limited conditions. The downregulation of the nitrogen-stress responsive genes provided evidence that the silenced gene had an integral role in enhancing cotton plant tolerance to N-limited conditions.

Keywords: nitrogen fertilizer; nodule-inception-like proteins; cotton plant; miRNAs; oxidant and antioxidant enzymes; virus-induced gene silencing (VIGS) plants

1. Introduction

Nitrogen (N) is an essential macro-element for all forms of life [1]. Plants and fungi are the only eukaryotic organisms, which are able to assimilate inorganic N. Among the higher plants, it is only the leguminous plants that are the most common or the most important group of plants that can form associations with N-fixing bacteria, known as the *Rhizobiae*, which enables them to fix free atmospheric nitrogen [2]. The bacteria colonize the roots, forming nodules, the nodule formation is governed by a protein known as nodule inception-like protein (NLP) [3]. The main source of nitrogen for non-leguminous plants, is through nitrogen fertilizer application, however, boosting soil fertility through nitrogen application is expensive and requires constant soil moisture levels [4]. Therefore, farmers and breeders have employed several strategies to solve the problem of N-deficiency through the initiation of conventional approaches, such as crop rotation and intercropping with leguminous plants, though minimal success has been achieved [5]. The only mechanism to boost cotton production under an N-limited environment is by adopting molecular approaches, by determining the suitable plant transcription factors with a higher role in enhancing the nitrogen metabolism in plants. Some of the genes known to be highly correlated to nitrogen and nitrogen metabolism in plants are the members of the nodule-inception-like protein (NLP) family, they are known to be highly conserved and enhance plant's response to N-deficiency [6]. Plants being sessile, they are constantly exposed to various environmental stresses, including growing in nitrogen (N) deficient conditions. The plants have evolved various survival strategies to tolerate the low N levels in the soil one of which is through the induction of the NLP proteins in order to enhance their survival under such conditions. Plant response to N-deficiency usually begins with limitations in uptake. Nitrogen exists in the soil as nitrate nitrogen, ammonium nitrogen, amino acids, proteins, and other nitrogenous substances, the cotton plant being non-leguminous, can only use inorganic forms of N, either as nitrate (NO_3^-) or ammonium (NH_4^+) [7]. Nitrate is the principal source of N since ammonium is quickly transformed to nitrate in the soil solution, just like other higher plants, cotton plants absorb nitrate through the roots and transport it directly to the leaves through the transpiration stream. Once on the leaf, nitrate is reduced to ammonium and combined with organic acids to form amino acids and proteins [8].

The nodule inception-like proteins (NLPs) are widely distributed across the plant species, even to non-leguminous plants [9]. Even in non-leguminous plants, they function as nodule initiators and regulate the number of nodules that are formed [2]. For instance, in maize, a total of 9 *ZmNLPs* have been identified and found to have an integral role in the primary response to nitrogen [10]. Moreover, overexpression of maize NODULE-INCEPTION-like proteins, *ZmNLP6*, and *ZmNLP8* in the Arabidopsis, restored nitrate assimilation and induction of nitrate-responsive genes in the *NLP7-4* mutant Arabidopsis [11]. The restoration of nitrate assimilation in the mutant Arabidopsis showed that the maize *NLP* genes had a similar regulatory role to the Arabidopsis *NLP7* gene (*AtNLP7*) in nitrate signaling and metabolism [11]. Moreover, studies have shown that plants have developed an advanced mechanism to deal with nitrogen deficiency, through uptake, transport, and assimilation of nitrogen under N-limited environment [11]. Furthermore, the plants only utilize 30% of the applied nitrogenous fertilizers, while the rest is either lost through leaching and or fixed in the soil [12].

The nitrate form of nitrogen is mainly translocated in the plants through two families of Nitrate transporters (NRT/NPF), the nitrate transporter 1/peptide transporter family previously known as the *NRT1/PTR* family and the *NRT2* [13]. In Arabidopsis, there are 7 and 53 members in the *NRT2* and *NPF* families, respectively [14]. The phosphorylation of the *NRT1.1* by two kinds of calcineurin B-like (CBL)-interacting protein kinases, *CIPK8* and *CIPK23* enhance the activity of the NRT1.1 in plants by improving their response to either N-deficient or higher N concentration level conditions [15]. Moreover, *NRT1.1* has dual-affinity of N transportation and or nitrate-sensing functions, acting in the upstream region of the Arabidopsis nitrate regulated-1 (ANR1), in the N signaling pathway [16]. The members of the *NRT2* transporters have high-affinity to nitrate, and the majority of them require *NAR2* (NRT3), to facilitate the transportation of nitrate within the plants [17]. Several studies have shown that the high nitrate affinity transporters do play an integral role in N-uptake efficiency under N-limited environment [18]. Furthermore, analysis of the model plant, *A. thaliana NRT2* genes showed that *AtNRT2.1*, *AtNRT2.2*, *AtNRT2.4*, and *AtNRT2.5* were highly upregulated in the roots compared to other tissues under nitrogen-deficient conditions [19]. Moreover, analysis of the VIGS-plants showed that *AtNRT2.1*, *AtNRT2.2*, *AtNRT2.4*, and *AtNRT2.5*, *NRT2* transporters accounted for over 95% of high-affinity nitrate influx activity under N-limited condition, in which the *AtNRT2.1* was the dominant [20].

Due to cost aspects and environment effects of nitrogen fertilizers, there is increased pressure to develop new crop genotypes with improved performance under N-limited conditions or with minimal N-application [21]. Due to the significance of nitrogen in plants, improving nitrogen uptake efficiency is vital, thus it is important to identify the intrinsic traits plants possess to improve their nitrogen intake efficiency under N-limited environments. Even though a great deal of progress has been made in elucidating the roles of genes and signal pathways that control the N acquisition and plant root phenology in the model plant, but, little progress has been made in understanding the role of N transporters in cotton. in the past, there has been a positive correlation between the quantitative trait loci (QTLs) for nitrogen absorption and plant traits such as the root architecture [22]. Furthermore, other studies have shown that the application of Ammonium chloride and Sodium nitrite increased cell membrane stability (CMS) in wheat plants up-to 46.62% and 84.46%, respectively compared to non-fertilized plants [23]. Moreover, N-deficiency induces cell wall loosening, thereby affecting the cell membrane stability [24].

Cotton is an important industrial crop, the primary source of natural fiber [25]. The commonly grown cotton cultivars are the tetraploid type (4n), which emerged due to polyploidization between two diploid cotton genomes, A and D [26]. The parental line of the tetraploid cotton (AD), are believed to be *Gossypium arboreum* of the A genome and *Gossypium raimondii* of the D genome [27]. Completion of the genome sequences of *G. hirsutum* [28], *G. arboreum* [29], and *G. raimondii* [30] provides valuable resources for exploring the whole cotton genome assembly and their evolution pattern. In the analysis of QTLs in the backcross inbred lines, (BC$_2$F$_2$) population derived from two tetraploid cotton. The drought-tolerant donor parent *Gossypium tomentosum* and drought-sensitive parental line, *G. hirsutum*, under drought condition, Magwanga et al. (in press), found a stable QTL, contributed by the drought-tolerant parental line, *G. tomentosum*, and determining the genes within the QTL flanking region, *Gh_A05G3286*, a member of the NLP5 Protein was found to be the gene responsible. Therefore, in this research work, we carried out genome wide identification of the NLP proteins in cotton, determined their evolution pattern, physiochemical properties, chromosome mapping, their gene structure, expression level under N-limited condition. Moreover, we further characterized the function of the *Gh_A05G3286* gene through virus-induced gene silencing (VIGS) mechanisms in order to determine the possible role of the protein encoded by the *NLP* genes in cotton under N-limited conditions. The results obtained provide the foundation for the role of the NLP proteins in cotton and thus allowing for future exploration of these genes in developing a high nitrogen use efficient cotton genotypes.

2. Results

2.1. Identification and Sequence Analysis of the Cotton NLP Proteins

To identify the *NLP* genes in the three cotton sequenced species of A, D, and AD genomes, the conserved domain of NLP protein and the NLP-related domains, RWP-RK (PF02042) and Phox and Bem1 (PBl), with the Pfam number of PF00564 were downloaded from the protein families (PFAM) database (http://pfam.sanger.ac.uk/). The Hidden Markov models (HMM) profile of the NLP protein was subsequently employed as a query to perform an HMMER search (http://hmmer.janelia.org/) against the *G. hirsutum*, *G. raimondii*, and *G. arboreum*. In the identification of the NLP proteins encoded by the *NLP* genes in cotton, two hundred and 26 NLP proteins were identified, with 105, 61, and 60 distribution in *G. hirsutum*, *G. raimondii*, and *G. arboreum*, respectively. All the two conserved NLP protein domains, RWP-RK (PF02042) and PB1 (PF00564) were identified in all the three cotton species. However, the numbers of NPL proteins of the PB1 (PF00564) domain were higher in proportion compared to those of the RWP-RK domain across the three cotton species. The two domain proportions were in the ratio of one RWP-RK: Three PB1, which indicated that the cotton NLP proteins of the PB1 were the most abundant. The number of NLP proteins identified in cotton was relatively higher compared to the proportions so far identified in other plants such as maize, with nine NLP proteins [10], 56 NLP proteins in the Brassica species [6] and only 9 in Arabidopsis [31]. Moreover, in the evaluation of their physiochemical properties through an online tool, the ExPASy Server (http://www.web.xpasy.org/compute_pi/). The cotton NLP protein characteristics were varied, in which the molecular weights ranged from 7.169 kDa to 151.633 kDa, grand hydropathy values (GRAVY) ranged from −0.904 to a maximum value of 0.399, protein lengths ranged from 63 aa to 1403 aa (Table S1). The results obtained were consistent with previous findings, such as the analysis of the NLP proteins in Brassica species [6].

2.2. Phylogenetic Tree Analysis

To determine the evolutionary pattern of the cotton NLP proteins, the NLP proteins sequences of *G. hirsutum*, *G. raimondii*, and *G. arboreum* were retrieved from cotton functional genome database (https://cottonfgd.org/), while the sequences for *A. thaliana*, *T. cacao*, and *Vitis vinifera* were retrieved phytozome (https://phytozome.jgi.doe.gov/pz/portal.html). All the proteins sequences were aligned, by adopting multiple sequence alignment using ClustalW a component of MEGA 6 tool. The phylogenetic tree was then generated, by adopting neighboring joint (NJ) and Maximum Likelihood Method with P distance Model, with a 1000 bootstrap replications, site cutoff coverage of 100% and missing gaps/missing data treatment with complete deletion [32]. In the analysis of the proteins encoded by the *NPL* genes in cotton and other plants, the NPL proteins were classified into three main groups, with members of group three being the largest (Figure 1 and Supplementary File 1). In the previous studies of the *NPL* genes in various plants, such as Arabidopsis, the NLP proteins were classified into three clades, designated as *NLP1*, *NLP2*, and *NLP3* [31]. This showed that the cotton NLP protein classification was congruent to previous studies, which could perhaps mean that the NLP proteins are highly conserved. In analyzing the possibility of orthologous gene pair formation between the cotton NLPs and other plants used in the phylogenetic tree, only two genes were found to form orthologous gene pairs with the upland cotton *NLP* genes, *Gh_D05G2521*, and *LOC_Os11g30350* (*NLP* gene from *Oryza sativa*), and the other pair was, *Ghsca101252G01* and *GSVIVG01026649001* (*NLP* gene from *Vitis vinifera*). None of the orthologous gene pairs was formed between cotton and the NLP proteins obtained for *T. cacao* the closest relative of the genus *Gossypium* compared to other plant species analyzed. However, the majority of the cotton NLP proteins were members of group 1 and 2, while few were located in group 3 despite being the largest group. In the analysis of the selected type, we evaluated the synonymous (ds) and non-synonymous (dn) rate of substitution. The ds/dn ratio is significant in evaluating the type of selection pressure, which acted on the proteins encoded by the *NLP* genes, the ds/dn value greater than 1 indicates beneficial selection, those that are less than 1 signify negative selection or purifying selection and those of unit value shows that the mutational effect was neutral [33]. The lowest ds/dn value was estimated at 0.389 for two ortholog genes of *G. raimondii*, while the highest ds/dn value was estimated to be 4.7615 for two paralogous gene pairs between *G. arboreum* and *G. hirsutum* (Table S2). However, the majority of the gene pairs had their ds/dn values of less than one, which indicated that a number of the NLP paralog gene pairs underwent negative/purifying selection. The result obtained is in agreement with previous investigations on the stress-responsive genes, the late embryogenesis abundant proteins (LEAs), in which the majority of the paralogous gene pairs had da/ds values of less than one [34].

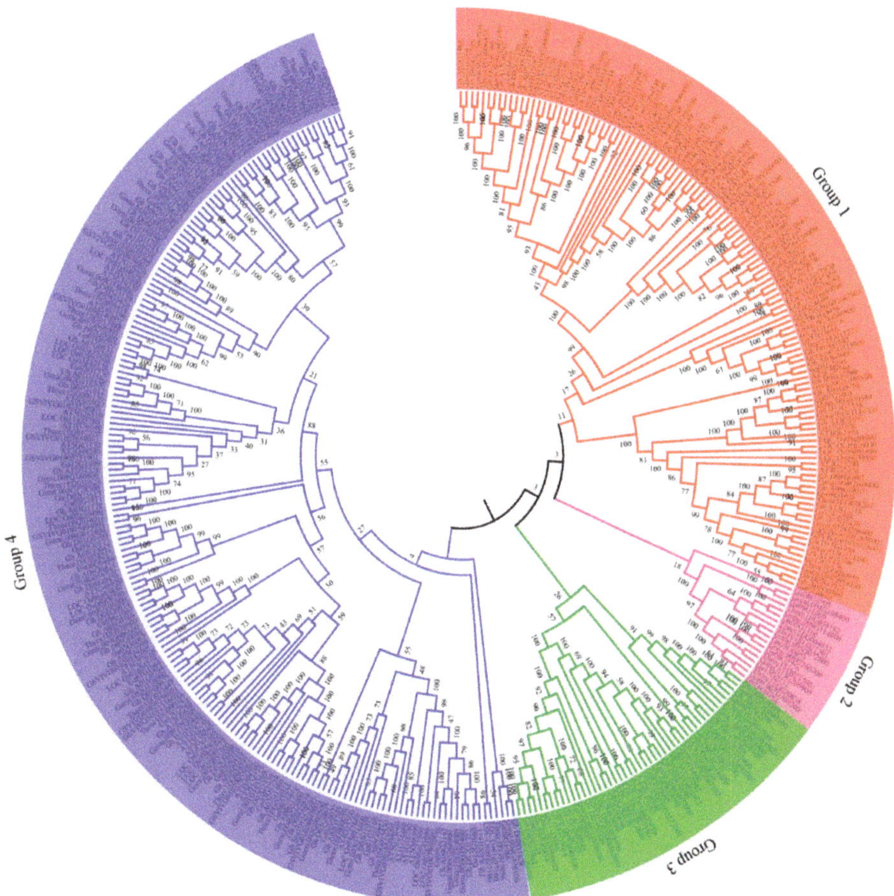

Figure 1. Phylogenetic tree analysis of the proteins encoded by the *NLP* genes in cotton and other plants. Different colours represent different groups, red: Group 1, purple: Group 2, green: Group 3' and blue: Members of group 4. AT: *Arabidopsis thaliana*, LOC: *Oryza sativa*, Thecc: *Theobroma cacao*, GSVIVG: *Vitis vinifera*, Gorai: *Gossypium raimondii*, Ga: *Gossypium arboreum* and Gh: *Gossypium hirsutum*.

2.3. Chromosome Mapping and Subcellular Localization Prediction Analysis of the Cotton Proteins Encoded by the NLP Genes in Cotton

To determine the chromosomal locations of the cotton *NLP* genes based on their positions, data retrieved from the whole cotton genome sequences were used. Chromosome distribution was done using the Basic Local Alignment Search Tool Nucleotide (BLASTN) search against *G. hirsutum* and *G. arboreum* in the cotton genome project (https://jgi.doe.gov/cotton-genome-project-in-al-com/) and *G. raimondii* genome database in Phytozome (http://www.phytozome.net/cotton.php). Out of the 105 proteins found to be encoded by the *NLP* genes in *G. hirsutum*, all were distributed across the 26 chromosomes with only three genes being in the scaffold region. More of the proteins were located in the Dt_sub genome with 53 (50.5%) compared to 49 (46.7%) proteins located in the At_sub genome chromosomes. The highest number of gene loci was noted in chromosome A_h05 and D_h05 with eight and nine genes, respectively. While the lowest gene loci densities were observed in A_h04, A_h10, and D_h10 with a single gene in each (Figure 2A). The gene distribution pattern within the two diploid cotton species was similar, all their 13 chromosomes were found to harbor at least one gene. In A genome, the highest gene loci were noted in chromosome A_208, A_201, A_203, and A_205 with 6, 7, 7 and 10 genes, respectively, while chromosome A_210 harbored only one gene (Figure 2B). In the *G. raimondii* of the D genome, chromosome D_503, D_504, D_505, and D_509 with 6, 6, 8, and 9 genes, respectively (Figure 2C). In the evaluation of the subcellular localization of the cotton proteins encoded by the *NLP* genes, an online tool Wolfsport (https://www.Wolfpsort.hgc.jp/) was employed. The coding sequences (CDS) of all the proteins encoded by the *NLP* genes were retrieved from cotton functional genome database (https://cottonfgd.org/analyze/). Then uploaded to an online tool Wolfsport (https://www.Wolfpsort.hgc.jp/) to predict the subcellular localization of the proteins encoded by the cotton *NLP* genes. The tool provides a different multi-location with varying probability, and the site with the highest probability becomes the main site in which the various proteins are likely to be found. The nucleus was the dominant subcellular structure with 51, 33, and 29 proteins encoded by the *NLP* genes in *G. hirsutum*, *G. raimondii*, and *G. arboreum*, respectively. The number of NLP proteins predicted to be sublocalized within the nucleus accounted for 48.57%, 47.54% and 55% of all the NLP proteins in *G. hirsutum*, *G. arboreum* and *G. raimondii*, respectively the other subcellular structures observed to harbor the proteins encoded by the cotton *NLP* genes were endoplasmic reticulum (E.R), plasma membrane, mitochondrion, and extracellular structures. In *G. hirsutum*, 26 (24.76%) proteins were found to be localized within the E.R, 24 (22.86%) in the plasma membrane, 3 (2.86%) in mitochondria and only one (0.95%) in the extracellular structures. In the two diploid cotton species, *G. raimondii*, 14 (23.33%) proteins were predicted to be located within the E.R, 12 (20%) in the plasma membrane, 1 (1.67%) located in the cytoplasm and the other single protein was harbored in the extracellular structures. There was minimal variation in *G. arboreum*, 15 (24.59%) in E.R, 15 (24.59%) in the plasma membrane and a single protein accounting for 1.64% of all the NLP proteins in *G. arboreum* was predicted to be embedded within the cytoplasm.

2.4. Gene Structure Analysis of the Cotton NLP Proteins

To gain further information into the structural diversity of cotton *NLP* genes, the exon/intron organization in the full-length cDNAs with their corresponding genomic DNA sequences of individual *NLP* genes in cotton were analyzed using an online, the gene structure displayer server (http://gsds.cbi.pku.edu.cn/). In the analysis of the gene structures of the upland cotton, *G. hirsutum* almost all the genes were disrupted by intron except 13 genes, such as *Gh_Sca101252G01* (uncharacterized), *Gh_D01G0769* (tetratricopeptide repeat protein 1), *Gh_A01G0750* (tetratricopeptide repeat protein 1), *Gh_D13G1358* (tetratricopeptide repeat protein 1), *Gh_A13G1093* (tetratricopeptide repeat protein 1), *Gh_D04G1546* (tetratricopeptide repeat protein 1), *Gh_D04G0995* (uncharacterized), *Gh_D02G2018* (serine/threonine-protein phosphatase 4 regulatory subunit 3), *Gh_A03G1567* (uncharacterized), *Gh_A05G2263* (uncharacterized), *Gh_D05G2522* (uncharacterized), *Gh_D05G2083* (uncharacterized), and *Gh_D05G3757* (uncharacterized). The level of intron disruption ranged from a single intron to a maximum of thirteen introns as evident among the following genes *Gh_A03G0454*, *Gh_A07G0445*, *Gh_A11G3016*, *Gh_D03G1084*, *Gh_D07G0509*, *Gh_D09G0055*, *Gh_D11G0397*, and *Gh_D11G0626* (Figure 3A). In relation to the two diploid cotton species, in *G arboreum*, the cotton species of the A genome, only five genes were found to be intronless while the rest were interrupted either one to a maximum of thirteen introns. The highest level of intron disruption was observed in five genes, *Ga01G2121*, *Ga07G0597*, *Ga11G3460*, *Ga09G0065*, and *Ga11G3698* with 13 introns (Figure 3B). Moreover, the same trend was also noted in *G. raimondii*, the cotton species of the D genome, in which the member of the CBS domain-containing protein CBSCBSPB5 was the most interrupted by introns (Figure 3C). Previous investigations have shown that overexpression of the CBS-domain containing protein in rice improved tolerance levels of transgenic tobacco to oxidative stress, salinity and heavy metal toxicity [35]. Moreover, expression of a CBS domain-containing protein, *GmCBS21* a candidate gene for nitrogen use efficiency (NUE) was found to enhance abiotic stress tolerance and improved performance of transgenic *Arabidopsis thaliana* under low nitrogen condition [36]. However, the exact role of the intron requires further investigation, though we preempt that the presence of introns may not be causing any alteration on the gene action.

Figure 2. Chromosome mapping of the cotton *NLP* genes. (**A**) *NLP* genes mapped on *G. hirsutum* chromosomes. (**B**) *NLP* genes for diploid cotton, *G. arboreum* of the A genome. (**C**) *NLP* genes for *G. raimondii* diploid cotton of the D genome.

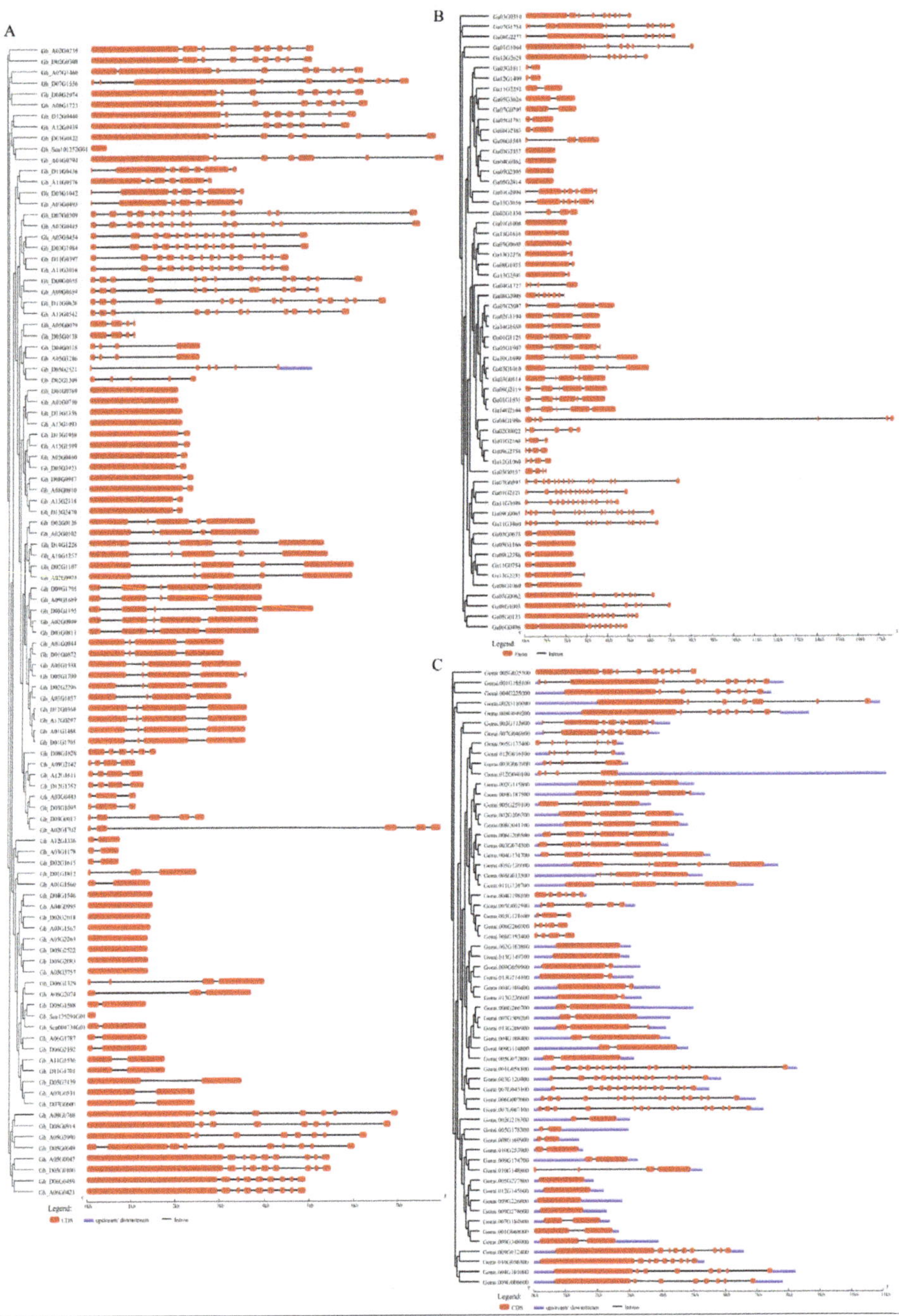

Figure 3. Cotton *NLP* gene structures. (**A**) Gene structures for the upland cotton *G. hirsutum*. (**B**) Gene structures for the *G. arboreum*. (**C**) Gene structural analysis of the *NLP* genes of *G. raimondii*.

2.5. Cis-regulatory Element Analysis and miRNA Target Prediction on the Cotton NLP Genes

To determine the *cis*-regulatory element, a 1 kb up and down stream region of the sequence of each gene was submitted to an online tool the Plant CARE (http://bioinformatics.psb.ugent.be/webtools/plantcare/html/) to obtain the various *cis*-regulatory elements associated with the *NLP* genes. The *cis*-regulatory sequences are linear nucleotide fragments of non-coding DNA with the main role of regulating gene expression and in turn, controls the development and physiology of an organism [37]. We sought to determine any possible interactions of the various cotton *NLP* genes to any of the known *cis*-regulatory elements. We found that a number of the genes were associated with various elements such as MYCCONSENSUSAT (CANNTG), which functions as an elicitor in abiotic stress signaling in plants, MYB2AT (TAACTG) mainly induced by dehydration stress, MYBCORE (CNGTTR) a *cis*-regulatory element that is responsive to water stress among others (Table S3). The detection of these myriads of *cis*-regulatory elements showed that these genes are critical in enhancing the plant's survival under abiotic stress conditions. Similar *cis*-regulatory elements have been found to correlate to various plant transcriptome factors such as the *NAC*, *MYB*, *LEA* genes among others [38]. Moreover, the miRNA targeting the various cotton *NLP* genes were predicted. The CDS of all the cotton NLPs were retrieved from cotton functional genome database (https://cottonfgd.org/analyze/). The retrieved sequences were uploaded onto the online tool the psRNATarget server (http://plantgrn.noble.org/psRNATarget/), non-cotton miRNAs were removed and only cotton miRNAs were analysed A microRNA (miRNA) is a small non-coding RNA molecule with approximately 21 nucleotides in length, found in plants, animals, and some viruses mainly function in transcriptional and post-transcriptional regulation of gene expression [39]. We sought to investigate any possible target by the miRNAs, for the *NLP* genes obtained from the tetraploid cotton 67 genes were found to be targeted by 47 miRNAs. Four (4) sets of miRNAs were found to have the highest level of genes associated with them, such as ghr-miR7484a (14 genes), ghr-miR7484b (14 genes), ghr-miR7495a (12 genes) and ghr-miR7495b (12 genes). previous studies in cotton have shown that miR7484a has a functional role in cotton under drought stress conditions [40]. More interestingly, some of the genes were found to be targeted by very high numbers of miRNAs of 10 and above, *Gh_A07G1460* (10 miRNAs), *Gh_A08G1723* (11 miRNAs), *Gh_D07G1556* (11 miRNAs), and *Gh_D08G2074* (10 miRNAs), while the rested with miRNAs ranging from about one to a maximum of nine miRNAs (Table S4). No miRNA was found to target any of the *NLP* genes obtained from *G. arboreum*, but a huge number of miRNAs were found to target the *NLP* genes obtained from *G. raimondii*, 171 miRNAs were found to target 57 *NLP* genes (Table S5). Previous reports have shown that miRNAs play significant roles in enhancing plants survival under extreme conditions. For instance, miR164 has been found to target the *NAC* gene through cleavage, the miR164 direct cleavage to *NAC* gene family in maize, *zmNAC1* enhance lateral root growth in maize and in turn improves their performance under water limiting condition [41]. Furthermore, miR164 targeted four different genes, *Gh_A03G0443*, *Gh_A12G0439*, *Gh_D03G1095*, and *Gh_D12G0440*. Moreover, a miR169a has been found to play the mediation role in N-limited environments, and its overexpression improved transgenic tomato performance under N-limited condition [42]. In our investigation, miR169a targeted *Gh_A08G1723*, *Gh_A09G0059*, *Gh_D08G2074*, and *Gh_D09G0055*, which demonstrated that the NLP proteins could be playing a significant role in enhancing cotton plant's tolerance to N-limited conditions.

2.6. RT-qPCR Validation of the Selected GhNLP Genes

Based on the phylogenetic tree and gene structures of the upland cotton, *NLP* genes. Fifty three genes were used for determining their expression levels in the tissues of *G. hirsutum* exposed to a nitrogen-limited condition. The genes were classified into three groups. Group 1 members were highly upregulated in all the tissues examined. The group 2 members, exhibited partial upregulation, which implied that they exhibited normal expression range while the group three members, showed differential expression, across the three tissues examined and at different time points (Figure 4). The RT-qPCR analysis was done on the three plant organs, roots, stems, and leaves. The plants were exposed to N-limited conditions, and tissues collected at 0 h, 3 h, 6 h, and 12 h of post-stress exposure.

Cotton *GhActin* was used as the reference gene. Based on the RT-qPCR analysis, five (5 genes) were found to exhibit significantly higher expression levels, *Gh_D02G2018*, *Gh_A12G0439*, *Gh_A03G0493*, *Gh_A03G1178*, and *Gh_A05G3286*, perhaps could be the candidate genes responsible for enhancing cotton plants to tolerate low nitrogen level environments. Moreover, *Gh_A05G3286 (NLP5)* was further evaluated, being it showed significantly higher upregulation under N-limited conditions among all the five genes.

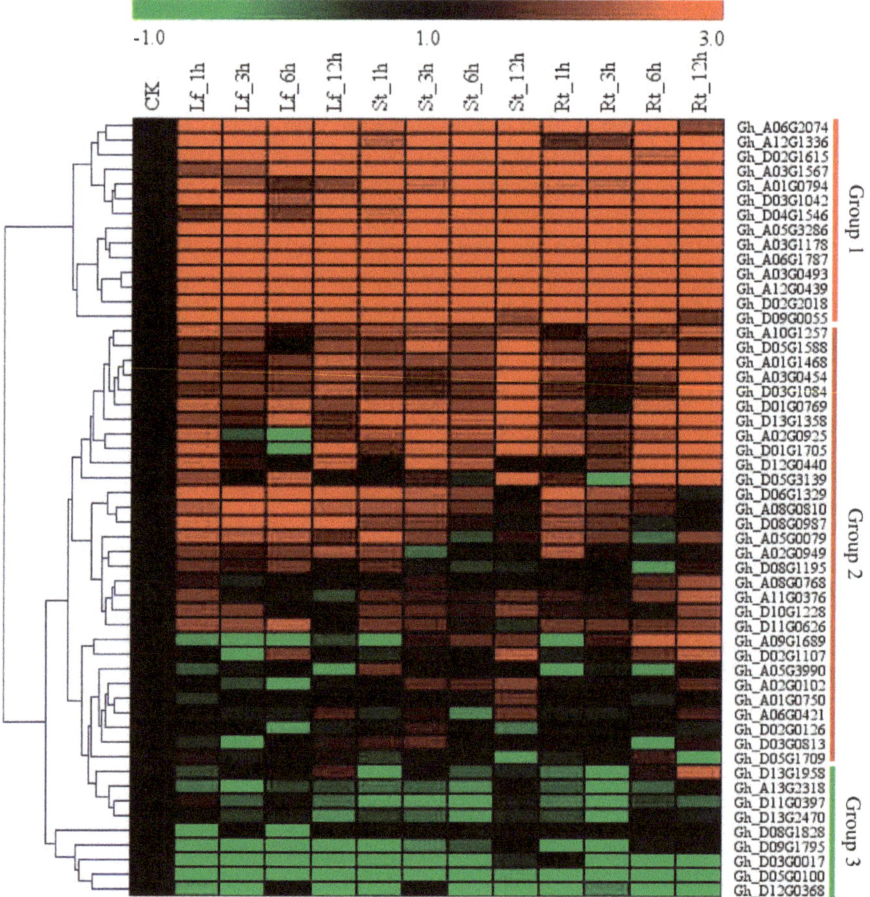

Figure 4. Expression Analysis of the selected *GhNLP* genes. Cotton seedlings were grown in a hydroponic setup with a modified Hoagland nutrient solution, N-limited condition imposed by adjusting the N-content to 0.05 mM NO_3^- concentration, and normal N-condition maintained at 7.5 mM NO_3^- concentration. The leaf, stem and root tissues were harvested at 0 h, 1 h, 3 h, 6 h, and 12 h. RNA extracted and the expression levels of the 53 selected *GhNLP* genes analysed by RT-qPCR with *GhActin* as the internal control. The heat map was visualized using Mev.exe program (Showed by log 2 values). Red: Up-regulated, green: Down-regulated and black-no significant difference in expression levels compared to control (ANOVA, $p < 0.05$).

2.7. Silencing of Gh_A05G3286 (NLP5), Physiological and Morphological Evaluation of the VIGS-Plants and the Non-Cotton Seedlings Under Nitrogen Limited Condition

In order to understand the role of the NLP proteins encoded by the upland cotton NLP genes, a highly upregulated gene as determined by the RNA seq and RT-qPCR validation, *Gh_A05G3286 (NLP5)*. The *NLP* gene, *Gh_A05G3286 (NLP5)* with a sequence length of 819 bp fragment was amplified from cDNA using the VIGS-F (5′ACACGTGCTTGGACTCTGTC3′) and VIGS-R (5′CGAATTTGATGTCAGCGCGT3′) primers. The fragment was amplified using KAPA HiFi HotStart ReadyMix, cloned into the pTRV2 vector to yield pTRV2:Gh_A05G3286 (NLP5) and verified through sequencing. The construction of the pTRV2 silencing vector and method of viral inoculation followed the protocol by Scofield et al. [43]. The phytoene desaturase (PDS) was used in order to determine the effectiveness of the vector [38]. After seven days of post infiltration, the plants infused with the vector containing the TRV:PDS exhibited an albino appearance on their leaves, which progressed to 100% bleaching after fourteen days (Figure 5A). The plants infused with the PDS showed albino-like traits on their leaves. The PDS-infused plants behave like wild-type plants in their capacity to etiolate and produce anthocyanins indicating that the light signal transduction pathway seems to be unaffected, the effect of PDS on the plant is similar to CLA1-1 also referred to cloroplastos alterados or altered chloroplast) [44]. In addition, we analyzed the gene-silenced plants and wild types in order to determine the expression levels and the net effect of the knockdown of the gene. The RNA inference is a highly conserved process in eukaryotes, which do, involves the degradation of the target messenger RNAs (mRNAs) using sequence-specific small RNAs, which do result in a reduction level in expression or knockdown of the target gene. The gene knockdown is achieved via small RNA (sRNA) pathways that use both endogenous and exogenous short interfering RNA (siRNA) and microRNA (miRNA) pathways, the pathways do. interact to form a well-balanced gene regulation system [45]. The expression of the target gene was sharply reduced in the leaf tissues of the VIGS-plants but highly upregulated in the leaf tissues of the wild plants, an indication that the expression level of the gene was significantly suppressed (Figure 5B,C). Moreover, the expression level of the target in the VIGS-plants and the wild types under normal condition was significantly reduced two-fold, however, in the wild type and the TRV: 00 plants, the target gene was significantly upregulated compared to the VIGS-plants. The TRV:00 is the empty-vector control, this was done in order to determine the vector had any effect on the plant, the performance of the plants infused with TRV:00 and the wild types exhibited no significant differences [46]. The albino phenotypic appearance and downregulation as revealed by the RT-qPCR results indicated that the role of the knocked gene was significantly suppressed. The results were in agreement with previous findings, which recorded that the expression of a gene is known to be downregulated through RNAi when the albino appearance is more or equal to 75% [47]. Moreover, N-limitation has a negative impact on chlorophyll content, being N deficiency increases the rate of senescence as a result of the rapid decline in chlorophyll content and soluble proteins [48]. Since, Nitrogen is an essential macro-nutrient element of all the amino acids which are the building blocks of plant proteins, which are vital for the growth and development of important tissues and cells like the cell membranes and chlorophyll content. There was a significant reduction in chlorophyll content on the VIGS-plants compared to the wild types under N-limited environment, the reduction in chlorophyll is an indication that the plant's ability to tolerate low or N-limited conditions was significantly affected thus the decline in chlorophyll content (Figure 5D). Furthermore, we evaluated the fresh shoot, and root biomass of the VIGS-plants and the wild types under N-limited conditions, the VIGS-plants shoot, and root biomasses were significantly reduced compared to their wild types (Figure 5E,F). The results obtained were in agreement with previous findings, which have shown that reduction in N, leads to a reduction in plant growth and development, and in turn, lowers their overall biomass accumulations [49].

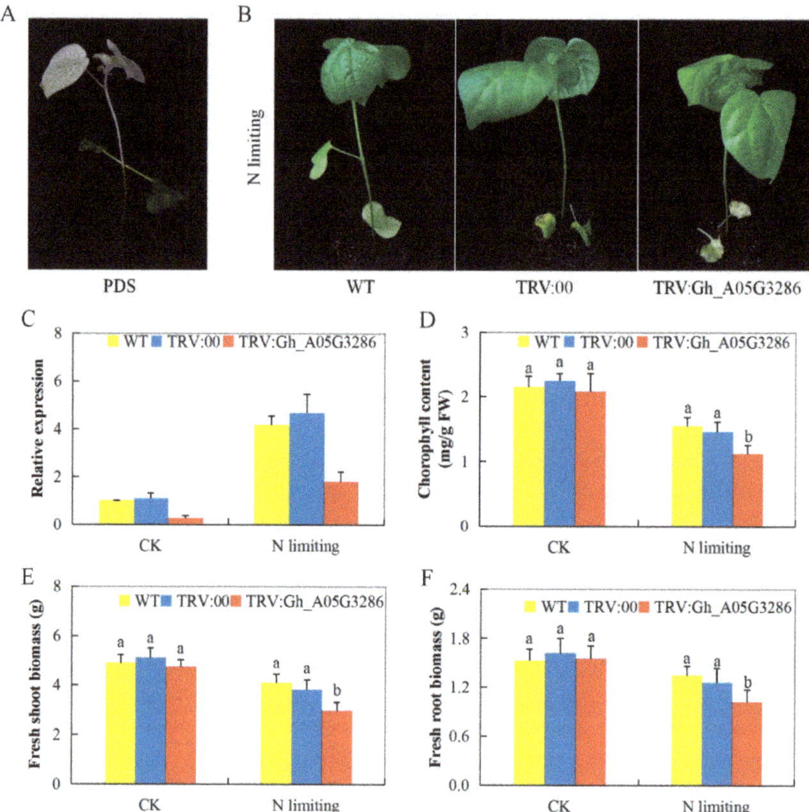

Figure 5. Phenotype observed in the silenced plants with the TRV: 00 empty vector, wild type plants and *Gh_A05G3286 (NLP5)*-silenced plants at 12 days post-inoculation. (**A**) Albino appearance on the leaves of the PDS infused plants. (**B**) Photograph of the plants taken after 15 days of N-deficiency exposure. (**C**) RT-qPCR analysis of the change in the expression level of the *Gh_A05G3286 (NLP5)* gene in cotton plants treated with VIGS. (**D–F**) Evaluation of chlorophyll, fresh shoot biomass and fresh root biomass, "TRV:00" represents the plants carrying control the TRV2 empty vector, "TRV: *Gh_A05G3286 (NLP5)*" represents the *Gh_A05G3286 (NLP5)*-silenced plants. Letters a/b indicate statistically significant differences (two-tailed, $p < 0.05$). The error bars of the *Gh_A05G3286 (NLP5)* gene expression level represent the standard deviation of three biological replicates.

2.8. Transcripts Investigation of Nitrogen Stress-Responsive Genes, analysis of Oxidant, and Antioxidant Content on the Tissues of VIGS and Non-VIGS Cotton Seedling Exposed to Nitrogen Limited Condition

In order to understand the effect of *GhNLP* gene suppression in the VIGS-plants, we carried out RT-qPCR expression analysis of some of the nitrogen deficiency responsive genes such as 60S ribosomal protein 18A (*RPL18A*), PP2A regulatory subunit (*Tap46*) [50], and cytochrome P450 CYP78A5 monooxygenase (*KLU*) [51]. All the three genes expression was downregulated in the VIGS-plants but was upregulated in the wild cotton species (Figure 6A i–iii). Several approaches have manipulated N-metabolism and transporter genes to increase N use efficiency in plants. For instance, overexpression of cytosolic GS increased plant height and dry weight under low-N conditions in tobacco [52]. Transgenic maize constitutively overexpressing *GLN1-3*, which encodes a cytosolic GS, in leaves, showed a 30% increase in kernel number [53]. The SOD and proline, together with malondialdehyde (MDA) were evaluated. The MDA concentration level was higher in the leaf tissues of the VIGS cotton seedlings, whereas proline and SOD levels were significantly reduced (Figure 6B i–iii). The reduced concentration levels of the antioxidant enzyme assayed revealed that VIGS cotton was highly affected compared to their wild types. Hydrogen peroxide is known as a plant stress signaling compound, and its increased level within the plant tissues triggers the release of various antioxidant enzymes [54]. However, when antioxidants are not induced, the H_2O_2 levels becomes lethal, resulting in oxidative damage to the plant cells, and eventually lead to cell death [55]. The excessively produced H_2O_2 is down-regulated by various antioxidant enzymes. However, in this research work, we found that the SOD concentration level was significantly low, an indication that the VIGS-plants suffered an extensive oxidative damage, which was further evidenced by the concentration, levels of MDA. Under stress condition, the level of ROS becomes elevated in the plant tissues as a result of uncontrolled process in the electron transport chain and accumulation of photoreducing power. This excess of electrochemical energy can be dissipated through the Mehler reaction, resulting in ROS production, including hydrogen peroxide [56,57], and damage of membranes, reflected in elevated MDA levels.

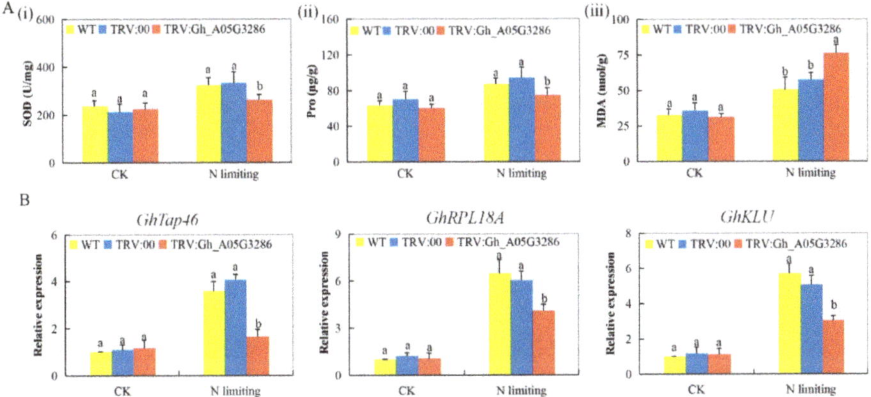

Figure 6. Evaluation of biochemical components, oxidant and antioxidant enzyme concentration levels in *Gh_A05G3286 (NLP5)*. (**A**) (i): Quantitative determination of SOD concentration. (ii): Quantitative determination of proline concentration. (iii): Quantitative determination of MDA concentration (**B**) (i–iii): Stress responsive transcription analysis of the various plants exposed to N-limited conditions for 15 days. Letters a/b indicate statistically significant differences (two-tailed, $p < 0.05$). The *GhActin* gene was used as internal control. The error bars of the *Gh_A05G3286 (NLP5)* gene expression level represent the standard deviation of three biological replicates.

3. Discussion

Enhanced use of nitrogenous fertilizers has led to increased production, change of soil micro and macro-environments, and massive pollution of various water bodies, as a result there is a massive wash off nitrogenous compounds from the agricultural fields leading to eutrophication [58]. For effective application and utilization of nitrogenous fertilizers, the adequate soil moisture content is necessary, thus this requires sufficient rainfall amounts and or maintenance of constant soil moisture through irrigation [59]. The erratic weather pattern, low amount of precipitation and alkalization of arable lands, has affected nitrogen fertilizer application and nitrogen use efficiency, leading to massive losses in agricultural production. The pressure on the arable lands due to human settlement and food demand has led to non-edible crops such as cotton cultivation to be restricted to a much drier, more saline, low nutrient load and heavily polluted lands with heavy metal [60]. Thus in order to maintain cotton production, nitrogen fertilizer application might not be the only solution, being freshwater for agricultural production is also a limiting factor. The plants with the ability to fix free atmospheric nitrogen have been found to have an important protein known as the NODULE-INCEPTION-like proteins (NLPs), this protein has a critical role in plants under N-deficiency condition [9]. In the recent past, the NLPs proteins have also been discovered in non-leguminous plants and found to play an important role in improving the plant's performance under N-limited environments [61]. The purpose of this experiment was to identify *NLP* genes that were significantly up-regulated under N-deficiency and might therefore be candidates for regulating N-deficiency responses.

In the whole genome identification of the cotton NLP proteins, we found a relatively higher number of the NLP proteins in the three cotton species, the upland cotton *G. hirsutum* harbored almost twice the number of proteins contained in either of the two diploid kinds of cotton combined. A total of 105 NLP proteins were identified in *G. hirsutum*, 60 and 61 in *G. arboreum* and *G. raimondii*, respectively. The overall number of the proteins encoded by the *NLP* genes in *G. hirsutum* was lower than the sum total of the two diploid cotton species, the low number of the NLP proteins in the tetraploid cotton could be attributed to either gene loss or chromosome rearrangement as a consequence of whole-genome duplication. Analysis of various functional genes in the three cotton species tend to have a similar pattern, for instance, in the whole genome identification of the calcineurin B-like (CBL) proteins in the three cotton species, 13, 13, and 22 were identified in *G. raimondii*, *G. arboreum*, and the cultivated tetraploid cotton, *G. hirsutum* [62]. Similarly, in the identification of the *LEA* genes in cotton, 242, 136 and 142 proteins encoded by the *LEA* genes were identified in *G. hirsutum*, *G. arboreum*, and *G. raimondii*, respectively [34]. The number of NLP proteins in cotton, seems to be significantly higher than the number obtained for other plants such as Arabidopsis with only nine (9) NLP proteins [63], 31 in *Brassica napus* [6], eight (8) in maize [64], six (6) in rice [65], and only five (5) in sorghum [31]. The high proportion of NLPs in cotton could be a pointer to their significant role in enhancing survival under nitrogen-deficient conditions. The tetraploid cotton genome, underwent whole-genome duplication (WGD), thus explaining the almost double the number of NLP proteins in *G. hirsutum*, a similar observation was also observed in the variation of the number of NLP proteins in the *Brassica* families in which *B. napus* harbored the highest number of NLP proteins. Moreover, the NLP proteins were found to be members of a dual domain, the RWP-RK, and PBI domains, though the PBI members were the dominant group.

The three cotton species NLP proteins were grouped into three as per the phylogenetic tree analysis, the classification of the cotton NLP proteins was coherent to previous studies in which three clades have been the dominant grouping number [6]. We further investigate to determine the evolution pattern of the cotton NLP proteins by determining the non-synonymous (dn) and synonymous (ds) substitution rate, majority of the paralogous gene pairs showed that their ds/dn ratios were less than 1, an indication that they underwent negative or purifying type of selection pressure [66]. The genetic fingerprint of an organism is contained in its DNA, and the processes by which DNA is translated using Ribonucleic acid (RNA) into a protein is called transcription and translation [67]. The transcription and translational process are rather quick and mistake are bound to occur in the process, non-lethal mistakes occur through synonymous mutation while lethal effects occur through non-synonymous mutation, being it results in to complete synthesis of new proteins or stops the process altogether [68]. However, non-synonymous type of mutation results in positive change, favoring the expression of genes, which in turn improves the adaptability of an organism to its environment, the negative/purifying selection could be due to the integral role played by the NLP proteins in cotton. Similar observations were made in the analysis of the late embryogenesis abundant (LEA) proteins, in which the majority of the proteins in all the eight subfamilies underwent a negative/purifying type of mutation [34].

The cotton *NLP* genes were found to be distributed in all the chromosomes, this was evident in both tetraploid and diploid types. The presence of the genes in all the chromosomes could perhaps explain their functions within the plant as an integral component in enhancing the plant's adaptability to the N-limited environment. In addition, four subcellular compartments were predicted to harbor the proteins encoded by the cotton *NLP* genes, the highest proportions of the proteins were predicted to be located within the nucleus. The results were in agreement with previous reports in which the entire NLP proteins in *Brassica napus* except one were located within the nucleus [6], moreover, the nucleus plays an important role in protein formation. Nitrogen metabolism occurs partly in the plastids, the plastids exist in arrange of forms, from the photosynthetic chloroplast to the pigment-storing chromoplast of flowers and fruit, and the starch-accumulating amyloplasts found in storage organs. However, despite their varied forms, the plastids have a unique ability to transform from one form to another. Despite the existence of plastids in various forms, all plastids in a particular organism have common DNA, inherited from the maternal parent, showing that the nuclear genome is vital regulatory role over the plastid morphology and function [69]. Furthermore, the plastids are estimated to contain at least 1000, and possibly as many as 5000 different proteins, the vast majority of these proteins are encoded within the nucleus and their corresponding DNA sequences are believed to have migrated from the plastid genome during the course of evolution [70]. Thus, the high number of the cotton NLP proteins predicted to be sublocalized within the nucleus could possibly be explained by either duplication or transfer events from the plastids to the nucleus. Moreover, nitrogen utilization has been well investigated in yeasts and filamentous fungi and found to be regulated by a process known as the nitrogen catabolite repression (NCR), which modulates the gene expression through a family of GATA binding zinc finger transcription factors [71]. In *Saccharomyces cerevisiae*, when exposed to N-limited conditions, the two forms of NCR, *gln3* and *gat1* move to the nucleus and activate genes containing upstream GATA sites [72], thus the high prediction of the NLP proteins encoded by the *NLP* genes within the nucleus is an indication that these groups of proteins are integral in nitrogen assimilation and metabolism.

Even though several studies have linked the plant miRNAs to environmental stresses such as drought, cold, salt, and heat, new pieces of evidence have shown that miRNAs are important in enhancing plant's adaptability to low nitrogen and phosphorus conditions [73]. Differential expression of a number of plant miRNAs has been observed in both dicotyledonous plants such as *Glycine max* [74] and *Arabidopsis thaliana* [75], similar, observations have been made among the monocotyledonous plants, maize [76] and rice [77]. In this investigation, a number of miRNAs were found to target various *NLP* genes in *G. hirsutum* and *G. raimondii*. Among the *NLP* genes obtained from the tetraploid upland cotton, *G. hirsutum*, ghr-miR167a and ghr-miR167b, were found to target four (4) genes each, which were, *Gh_A08G0810* (Protein unc-45 homolog B), *Gh_A13G2318* (tetratricopeptide repeat protein 1), *Gh_D08G0987* (protein unc-45 homolog B), and *Gh_D13G2470* (tetratricopeptide repeat protein 1). The same miRNA has been found to be targeted by two genes of the Auxin Response Factors (ARF transcription factors), *ARF6* and *ARF8* [78]. The ARF transcription factor, *ARF8* do regulate the development of the lateral roots, and its expression was induced in the pericycle and lateral root cap cells under N-limited conditions [79]. Further pieces of evidence have shown that overexpression of miR160 and miR167, do enhance lateral roots development under N-limited conditions, which is believed to improve or boost the capability of the plants to maximize the uptake of the little available nitrogen. The NAC plant's transcription factors (TFs) are among the top-ranked plant stress-responsive TFs, the NACs consist of three gene families, *NAM* (No Apical Meristem), *ATAF* (Arabidopsis Transcription Activation Factor), *CUC* (CUp shaped Cotyledon) [80]. Five members of the *NAC* genes have been found to be targeted by miR164, including *NAC1* and *CUC1* genes, *NAC1* is integral in auxin signal transduction for the growth and development of lateral roots [81], while *CUC1* is important for the normal embryonic, vegetative, and floral development in plants [82]. The same miRNAs have been found to be upregulated in maize under N-limited conditions. In our investigation, miR164 was found to target four genes such as *Gh_A03G0443* (protein RKD4), *Gh_A12G0439* (serine/threonine-protein kinase *EDR1*), *Gh_D03G1095* (protein *RKD4*), and *Gh_D12G0440* (serine/threonine-protein kinase *EDR1*). The results showed that these genes could possibly be involved in enhancing lateral root development, and improving adaptability of cotton plant to N-limited environments. The miRNAs are an extensive class of ~22-nucleotide noncoding RNAs thought to regulate gene expression in metazoans, moreover, plants are often exposed to various stress factors and the defense mechanism devised by plants to cope with adverse climatic conditions is the reprogramming of gene expression by microRNAs (miRNAs) [83]. The high number of miRNA target on the various genes obtained for *G. raimondii* possibly could explain in part why various vital QTLs are often mapped on the D subgenome chromosomes as opposed to A subgenome chromosomes. Polyploidization contributed significantly to the alteration of the gene functions in the tetraploid cotton, moreover, higher number of transcription factors in tetraploid cotton are contributed by Dt subgenomes compared to the At subgenome, furthermore, previous reports have shown that a number of significant QTLs were mapped in Dt subgenome than At subgenome [84]. For instance, QTL mapping of drought and salt tolerance in an introgressed recombinant inbred line population of Upland cotton, 11 QTL were detected on 8 chromosomes, in which 10 of the QTLs were located on the D subgenome [85]. Moreover, in comprehensive Meta QTL analysis for fiber quality, yield, yield-related and morphological traits, drought tolerance, and disease resistance in tetraploid cotton, A subgenome harbored 536 QTL, while the D subgenome had 687 QTL [86]. Furthermore, a detailed analysis of the RFLP map revealed that a number of resistance genes (R genes) are located on the D subgenome chromosomes compared to A subgenome chromosomes, with five out of six R genes located on the D subgenome chromosomes [87]. Moreover, analysis of the genome structure of tetraploid cotton, *G. hirsutum* showed that 75% (125/166) of the polymorphic loci were tagged on the D-subgenome [88]. The merger of divergent genomes in a common nucleus has been argued to present a shift from genetic flexibility to genetic fixation providing a mechanism for response to selection [89]. These findings provide a unique contribution of the D subgenome to whole-genome evolution and adaptability of tetraploid cotton to its environment, and the high miRNA target could be playing a role in fine-tuning the expression of the D subgenome genes. A number of reasons have been forwarded, explaining the significance of the D subgenome in tetraploid cotton, a higher underlying mutation rate, a higher level of DNA polymorphism, and a non-homologous chromosome rearrangement [90].

Gene expression profiling is a powerful mechanism for studying biological processes, especially tissue/organ-specific ones, at the molecular level, and thus we carried out a detailed analysis of the upland cotton *NLP* genes in root, stem, and leaf tissues at three true leaf stage, under nitrogen-deficient condition. The results showed that the proportions of the *NLP* genes increased with increased exposure to N-limited conditions, more significantly, the novel gene *Gh_A05G3286* (NLP5) showed upregulation with increased exposure to N-limited conditions. The stress-responsive genes have been found to show higher expression with an increase in stress level and duration of exposure [91]. Moreover, gene induction has been found to be governed by time and the organ profiled [92]. The ability of the VIGS-plants to tolerant low nitrogen condition was highly compromised, being there was a significant reduction in root biomass, shoot biomass and leaf chlorophyll content. Furthermore, the antioxidant enzymes, proline, and SOD levels were significantly scaled-down, whereas the MDA level was significantly increased. The increased level of MDA and a significant reduction in SOD and proline indicated that the VIGS-plants were affected under N-limited conditions. The results were in agreement to previous findings in which the accumulation of abscisic acid (ABA) and H_2O_2 in leaves of rice, was found to be induced by N deficiency [93], moreover, overexpression of Arabidopsis *NLP7* gene improved plant growth while the nlp7 mutant plants growth showed constitutive N-deficient phenotypes on both nitrate-rich and limiting media [94]. The significant reduction in the growth of the VIGS-plants both under N-limited and none-limited conditions showed that the cotton *NLP* genes play an integral role in enhancing plant growth and development. When plants are exposed to any form of stress, either biotic or abiotic stress factors, the internal and external cell environments become altered, this is due to increased production of reactive oxygen species (ROS), shut down or slowdown of other cellular and biological processes and or release of reactive nitrogen species (RNS) which modify enzyme activity and gene regulation [95]. The VIGS-plants exhibited a higher level of malondialdehyde (MDA) and a significantly low level of leaf chlorophyll content. Moreover, it has been found that N-deficiency has a negative effect on the leaf chlorophyll content concentration [96]. Moreover, all three N-limited stress-responsive genes were all downregulated. PP2A is a regulatory subunit of the Tap46 an integral component of the target of rapamycin (TOR) signal pathway, the downregulation of the gene indicated that the TOR signal pathway was deactivated leading to repression of nitrogen mobilization. Moreover, the silencing of *Tap46* in tobacco-caused chromatin bridge formation at anaphase, which elucidated the integral role of the *Tap46* in plant growth and development as a component of the TOR signaling pathway [97].

4. Materials and Methods

4.1. Plant Materials and Treatments

Seeds of the upland cotton species, *G. hirsutum* (AD)$_1$, accession number ZM-30014, was obtained from the cotton research institute seed bank. The seeds were germinated in the plant growth chambers with conditions programmed at 16h photoperiod 25/18 °C day/night and with 60% humidity. After three days of germination, the seedlings were transplanted into a hydroponic setup, in the greenhouse of the cotton research institute, Chinese Academy of Agricultural Sciences, CRI-CAAS, China. To analyze the expression patterns of the cotton *NLP* gene family members in different organs, the root (Rt), leaf (leaf), and stem (St) were profiled. To investigate the expression patterns of the cotton *NLP* genes under N-deficiency stress. The cotton seedlings were grown in sand and watered with nutrient solution [98]. Moreover, at three-leaf stages, the seedlings were subjected to N-limited conditions by watering with nutrient solution. The experimental unit was a set of three adjacent plants. Previous studies indicated that 0.05 mM NO_3^- concentration in the nutrient solution provided an N-deficient supply to plants and 7.5 mM NO_3^- provided an N-abundance supply [99]. The treatment nutrient solutions was a modified nutrient solution containing 0.05 mM or 7.5 mM of NO_3^-, with changes in NO_3^- concentration balanced by changes in SO_4^{2-} concentration [99]. The pH value of the nutrient solution was adjusted with KOH and H_2SO_4 and maintained between 5.8 to a maximum value of 6.4. The solution was regularly replaced after every three days. The plants were grown for 15 days. The samples were collected in three replicates after 0 h, 3 d, 6 d, 9 d, 12 d, and 15 d of treatment and stored under −80 °C for RNA extraction.

4.2. Identification and Protein Physiochemical Properties Analysis of the NLP Family Genes in Plants

Genomic, coding sequences (CDS) and protein sequences from three cotton species, *G. hirsutum*, *G. arboreum*, and *G. raimondii* together with other plants, *Arabidopsis thaliana*, *Theobroma cacao*, and *Brassica Rapa* were downloaded. The NLP proteins for *G. hirsutum* were obtained from the cotton research institute genome database (http://mascotton.njau.edu.cn). The *G. arboreum* NLP proteins were downloaded from the Beijing genome institute database (https://www.bgi.com/), and for the D genome *G. raimondii*, *T. cacao*, and *Glycine max* were retrieved from Phytozome 12.0 database (http://www.phytozome.net/), with E-value < 0.01, while for the model plant, *A. thaliana*, the NLP proteins were downloaded from the TAIR (http://www.arabidopsis.org). Since extensive research on Arabidopsis NLP proteins has been done, the protein sequences of Arabidopsis NLP proteins were used to carry out blast search for all the cotton and other plants NLP proteins. The redundant sequences were removed from further analysis. Moreover, PROSITE (http://prosite.expasy.org/scanprosite/) and SMART (http://smart.embl-heidelberg.de/) were used to confirm the presence of the NLP protein domain. SMART and PFAM database were used to verify the presence of the NLP-related domains, RWP-RK (PF02042) and Phox and Bem1 (PB1), Pfam number of PF00564, with an E-value cutoff of 1×10^{-5}. Other protein's physiochemical properties of the three cotton NLP proteins such as the isoelectric charge (*pI*), grand hydropathy values (GRAVY), the charge, molecular weights (MW) and other protein properties were estimated by ExPASy Server tool (http://www.web.xpasy.org/compute_pi/).

4.3. Phylogenetic Analysis of the NLP Family in Cotton Species with Other Plants

The obtained NLP protein sequences from the three cotton species, together with NLP proteins from *A. thaliana*, *T. cacao*, and *Vitis vinifera* were used to carry out a phylogenetic analysis of the NLP proteins in order to understand their evolution pattern. The proteins were aligned, by adopting multiple sequence alignment using ClustalW a component of MEGA 6 tool, the phylogenetic tree was then generated, by adopting neighboring joint (NJ) and Maximum Likelihood (ML) method with P distance Model, with 1000 bootstrap replications, site cutoff coverage of 100% and missing gaps/missing data treatment with complete deletion [32]. The generated tree was visualized by the use of Fig Tree version 1.4.0 (http://tree.bio.ed.ac.uk/software/figtree/). We further analyzed the coding sequences to calculate synonymous (ds) and non-synonymous (dn) rate of mutation by use of dn/ds _calculator version 2.0 and further validated by use of an online tool, Synonymous Non-synonymous Analysis Program (https://www.hiv.lanl.gov/content/sequence/SNAP/SNAP.html).

4.4. Chromosomal Locations, Subcellular Localization Prediction, and Gene Structure Analysis

The chromosomal locations of the three cotton species *NLP* genes were determined based on the results of the reciprocal BLASTP analysis, the cotton NLP homolog genes were determined with a threshold of > 80% in similarity and at least in 80% alignment ratio to their protein total lengths. MapChart tool (https://mapchart.net/) was employed to plot the *NLP* genes in the various cotton chromosomes by use of their physical position in base pairs (bp). The subcellular predictions of the proteins encoded by the three cotton *NLP* genes were predicted by the use of an online tool Wolfsport (https://www.Wolfpsort.hgc.jp/). The results were further validated by the use of TargetP1.1 (http://www.cbs.dtu.dk/services/TargetP/) server and Protein Prowler Subcellular Localisation Predictor version 1.2 (http://www.bioinf.scmb.uq.edu.au/pprowler_webapp_1-2/). Finally, the gene structures were analyzed by the use of a gene structure displayer server (http://gsds.cbi.pku.edu.cn).

4.5. Cis-regulatory Element Analysis and the miRNA Target Prediction of the Cotton NLP Proteins

The promoter sequences of 1 kb up and downstream of the coding sequence regions were analyzed to obtain the possible *cis*-regulatory elements by employing an online tool the Plant CARE (http://bioinformatics.psb.ugent.be/webtools/plantcare/html/). The results were further validated by the use of the PLACE database (http://www.dna.affrc.go.jp/PLACE/signalscan.html). Moreover, we employed the use of the miRNA database in order to determine the possible miRNA targeting the various cotton NLP proteins, the CDS for various cotton species were uploaded onto the online tool the psRNATarget server with default parameters (http://plantgrn.noble.org/psRNATarget/).

4.6. RNA Isolation and Quantitative Reverse-Transcription PCR

Total RNA was extracted from the plant organs using an RNA Extraction Kit obtained from Aid Lab, The RNA sample quality was evaluated by use of the gel electrophoresis and a NanoDrop 2000 spectrophotometer, only RNAs with specification criterion of 260/280 ratio of 1.8–2.1, 260/230 ratio ≥ 2.0 were retained and used for further analysis. The cDNA was synthesized from 1 μg of total RNA using M-MLV transcriptase (TaKaRa Biotechnology, Dalian, China) as described within the manufacturer's user instructions manual. To determine the organ-specific expression profiles of *GhNLP* genes, publicly available *G. hirsutum* RNA-seq data available at cotton functional genome database (https://cottonfgd.org/analyze/) were used to quantify the expression levels of the *G. hirsutum* genes in different organs as fragments per kilobase (Kb) of exon per million reads mapped (FPKM) values using Cufflinks with default parameters. Two independent biological replicates were analyzed per sample. Fifty-three upland cotton *NLP* genes were selected based on the phylogenetic tree and gene structure analysis, for the RT-qPCR validation, the same method had been applied by Magwanga et al. [34], in the selection of the *LEA* genes for RT-qPCR analysis. Reverse transcription-quantitative polymerase chain reaction (RT-qPCR) was performed on a Bio-Rad CFX96 Real-Time System (USA) according to a previously described method [10]. The BatchPrimer3 Primer software (http://batchprimer3.bioinformatics.ucdavis.edu/) used to synthesize the *GhNLP* specific primers (Table S6). Each reaction was carried out in three biological replicates in a reaction volume of 20 μl containing 1.6 μL of gene-specific primers (1.0 μM), 2.0 μL of cDNA, 10 μL of SYBR green, 0.2 μl of ROX Reference Dye II, and 6 μl of sterile distilled water. The PCR program was as follows: 95 °C for 3 min, 45 cycles of 10 s at 95 °C and 30 s at 60 °C, and then melt curve 65 °C to 95 °C, increment 0.5 °C for 5 s. Melting curves were generated to estimate the specificity of these reactions. Relative expression levels were calculated using the $2^{-\Delta\Delta Ct}$ method, with *GhActin* used as internal controls [100].

4.7. Generation of Transiently Transformed G. hirsutum Plants with Repression of Gh_A05G3286 (NLP5)

The *NLP* gene, *Gh_A05G3286 (NLP5)* with a sequence length of 819 bp fragment was amplified from cDNA using the VIGS-F (5′ACACGTGCTTGGACTCTGTC3′) and VIGS-R (5′CGAATTTGAT GTCAGCGCGT3′) primers. The fragment was amplified using KAPA HiFi HotStart ReadyMix, cloned into the pTRV2 vector to yield pTRV2:*Gh_A05G3286 (NLP5)* and verified through sequencing. The construction of the pTRV2 silencing vector and method of viral inoculation followed the protocol by Scofield et al. [43]. Viral controls included pTRV2:00, which is derived from the empty pTRV2 vector, and pTRV2: PDS that included a transcript that targets the phytoene desaturase (*PDS*) gene and acts as a visual marker of correct viral reconstitution. *G. hirsutum* seedlings were treated with the virus constructs to determine the effect of silencing *Gh_A05G3286 (NLP5)*. VIGS-infiltrated seedlings were left to grow for 14 days, upon the emergence of two true leaves, the leaves were harvested and stored at −80 °C for RNA extraction to confirm the VIGS efficiency. Seedlings were then simultaneously subjected to N-deficiency, stress treatment by growing in the sand (nutrient-deficient rooting material), and were watered with two different nutrient solutions with NO_3^- concentrations of 0.05 mM and 7.5 mM to provide N-deficient and N- abundant supply, respectively. The experimental unit was a set of three adjacent plants. Previous studies indicated that 0.05 mM NO_3^- concentration in the nutrient solution provided an N-deficient supply to plants [101] and 7.5 mM NO_3^- provided an N-abundant supply [99]. The treatment nutrient solutions were modified nutrient solution containing 0.05 or 7.5 mM NO_3^-, with changes in NO_3^- concentration balanced by changes in SO_4^{2-} concentration [99]. The pH value of the nutrient solution was adjusted with KOH and H_2SO_4 and maintained between 5.8 to a maximum value of 6.4. The plants were irrigated with a nutrient solution every three days. The plants were grown for 15 days.

4.8. Evaluation of Physiological, Morphological, Biochemical Traits and Expression Analysis of the N-Limited Responsive Genes

We evaluated various physiological and morphological parameters such as chlorophyll content, shoot biomass and root biomass was measured. The chlorophyll content was evaluated through the non-destructive method [102], while shoot and root biomass were measured by the use of a weighing balance in grams. Three biological replicates and three technical replicates were performed for all the measurements. Moreover, we evaluated the concentration of the antioxidant and oxidant enzymes in the leaves of the VIGS and the non-VIGS cotton plants grown under N-limited conditions. The superoxide dismutase (SOD) activity was evaluated by monitoring the inhibition of the photochemical reduction of nitro blue tetrazolium (NBT), as described by Giannopolitis and Ries though with slight modifications [103] and finally proline was evaluated by the method described by Van Assche [104]. Malondialdehyde (MDA) was evaluated between the two plants, wild types and the VIGS-plants under the three stress levels. MDA was determined as a measure of lipid peroxidation [105]. Moreover, three N-limited stress-responsive genes were profiled in order to evaluate their expression levels on the VIGS-plants and the wild types under N-limited conditions. The cotton 60S ribosomal protein L18A (*GhRPL18A*), with forward sequence "TTTCCGGTTTCACCAGTACC", and the reverse sequence "AACAAGTTGGGTTTCGATGC", cotton PP2A regulatory subunit (*GhTap46*), with forward sequence "GGCTGCAGAGGCAAAA CTTA), and reverse sequence "CTTCTCTTGGGCAGCATCA T" [50], cotton cytochrome P450 CYP78A5 monooxygenase (*KLU*), with forward sequence "GCTGATAGGCCAGTGAAGGA" and the reverse sequence "TCTTGAGCCTTTGCTTGGAT" [51], were profiled on the leaf tissues of *Gh_A05G3286 (NLP5)*-silenced plants, wild type (WT) and the positive controlled plants (TRV2:00), grown under N-limited conditions. All evaluations were carried after 15 days of growth under N-limited conditions, grown in sand and watered with a modified nutrient solution. The cotton *GhActin* was used as the internal controls.

5. Conclusions

The loss due to excessive use of nitrogenous fertilizer is estimated at 2–20% through evaporation, 15–25% get fixed by reacting with the organic compounds in the clay soil and the remaining 2–10% is lost through runoff and groundwater which eventually end up into the water bodies [106]. Based on the effects and the price of the nitrogenous fertilizers, a different approach is needed by plant breeders not only to develop nitrogen-use-efficient cotton germplasm but also to develop more genotypes that are resilient with higher adaptability to N-limited conditions. In this research work, we found 226 *NLP* genes in the three cotton species, with 105, 61, and 60 genes in *G. hirsutum*, *G. raimondii*, and *G. arboreum*, respectively. The proteins encoded by the *NLP* genes were classified into three clades and were found to possess a wider genome distribution, covering all the cotton chromosomes. Deep Insilico analysis of the proteins encoded by the cotton *NLP* genes showed that over 50% of the proteins were predicted to be sublocalized in the nucleus. The nucleus regulates all the cellular activities, and the presence of the NLP proteins in the nucleus could perhaps explain their critical role in enhancing plants ability to tolerate low or nitrogen deficient conditions. Moreover, the GRAVY values of most the proteins encoded by the *NLP* genes was less than one, an indicated that these proteins are hydrophilic in nature. Hydrophilicity is a property associated with a number of proteins encoded by various stress-responsive genes such as the *LEA* genes [34]. Moreover, the functional characterization of the novel gene through VIGS approached, revealed that the VIGS-plant's ability to tolerate N-limited conditions was significantly reduced compared to their wild types. Furthermore, the VIGS-plants registered a significant reduction in chlorophyll content, fresh shoot biomass, and fresh root biomass compared to the WT under N-limited condition. Moreover, evaluation of oxidant and antioxidant enzymes showed that the VIGS-plants had a higher concentration level of oxidant enzyme, MDA but significant reduction in antioxidant enzymes, proline, and SOD compared to the wild types under N-limited conditions. Finally, profiling of the N-limited responsive genes, *GhTap46*, *GhRPL18A*, and *GhKLU* exhibited significant reduction on the tissues of the VIGS-plants, but were upregulated on the tissues of the wild types under N-limited condition. These results showed that the cotton NLP proteins encoded by the *NLP* genes could be playing a significant role in enhancing the tolerance level of cotton to N-limited conditions.

Supplementary Materials: Supplementary materials can be found at http://www.mdpi.com/1422-0067/20/19/4953/s1.

Author Contributions: R.O.M., F.L., and K.W. Conceived and designed the study. R.O.M, J.N.K., and P.L. Performed the experiments. X.C., Y.X., Z.Z., S.G.A., and Y.H. Supervised experiments and contributed to drafting the manuscript. All authors contributed to the analysis and interpretation of the data K.W. and F.L. Supervised the experiments. R.O.M. Wrote and prepared the final version of the manuscript. All authors approved the final manuscript.

Funding: This research program was financially sponsored by the National key research and development plan (2016YFD0100306), and the National Natural Science Foundation of China (31671745, 31530053).

Acknowledgments: The authors would like to deeply appreciate the great support of the research team; your cooperation and advice were the pillars towards this research work.

Conflicts of Interest: The authors declare no conflict of interest.

Abbreviations

NLP	NODULE-INCEPTION-like proteins
VIGS	virus-induced gene silencing
GRAVY	Grand hydropathy values
SOD	Superoxide dismutase
MDA	Malondialdehyde
NIN	Nodule inception protein

References

1. Ohyama, T. Nitrogen as a major essential element of plants. *Nitrogen Assim. Plants* **2010**, *37*, 1–17.
2. Qureshi, M.I.; Muneer, S.; Bashir, H.; Ahmad, J.; Iqbal, M. Nodule Physiology and Proteomics of Stressed Legumes. *Adv. Bot. Res.* **2010**, *56*, 1–48. [CrossRef]
3. Kawaharada, Y.; James, E.K.; Kelly, S.; Sandal, N.; Stougaard, J. The Ethylene Responsive Factor Required for Nodulation 1 (ERN1) Transcription Factor Is Required for Infection-Thread Formation in Lotus japonicus. *Mol. Plant-Microbe Interact.* **2017**, *30*, 194–204. [CrossRef] [PubMed]
4. Eyhorn, F.; Ramakrishnan, M.; Mäder, P. The viability of cotton-based organic farming systems in India. *Int. J. Agric. Sustain.* **2007**, *5*, 25–38. [CrossRef]
5. Wang, X.X.; Wang, X.; Sun, Y.; Cheng, Y.; Liu, S.; Chen, X.; Feng, G.; Kuyper, T.W. Arbuscular mycorrhizal fungi negatively affect nitrogen acquisition and grain yield of maize in a N deficient soil. *Front. Microbiol.* **2018**, *9*, 418. [CrossRef] [PubMed]
6. Liu, M.; Chang, W.; Fan, Y.; Sun, W.; Qu, C.; Zhang, K.; Liu, L.; Xu, X.; Tang, Z.; Li, J.; et al. Genome-Wide Identification and Characterization of NODULE-INCEPTION-Like Protein (NLP) Family Genes in Brassica napus. *Int. J. Mol. Sci.* **2018**, *19*, 2270. [CrossRef]
7. Olk, D.C. Organic Forms of Soil Nitrogen. *Nitrogen Agric. Syst.* **2008**, 57–100. [CrossRef]
8. Huppe, H.C.; Turpin, D.H. Integration of Carbon and Nitrogen Metabolism in Plant and Algal Cells. *Annu. Rev. Plant Physiol. Plant Mol. Biol.* **1994**, *45*, 577–607. [CrossRef]
9. Lin, J.S.; Li, X.; Luo, Z.; Mysore, K.S.; Wen, J.; Xie, F. NIN interacts with NLPs to mediate nitrate inhibition of nodulation in *Medicago truncatula*. *Nat. Plants* **2018**, *4*, 942–952. [CrossRef]
10. Ge, M.; Liu, Y.; Jiang, L.; Wang, Y.; Lv, Y.; Zhou, L.; Liang, S.; Bao, H.; Zhao, H. Genome-wide analysis of maize NLP transcription factor family revealed the roles in nitrogen response. *Plant Growth Regul.* **2018**, *84*, 95–105. [CrossRef]
11. Cao, H.; Qi, S.; Sun, M.; Li, Z.; Yang, Y.; Crawford, N.M.; Wang, Y. Overexpression of the Maize ZmNLP6 and ZmNLP8 Can Complement the Arabidopsis Nitrate Regulatory Mutant nlp7 by Restoring Nitrate Signaling and Assimilation. *Front. Plant Sci.* **2017**, *8*, 1703. [CrossRef] [PubMed]
12. Reddy, M.M.; Ulaganathan, K. Nitrogen Nutrition, Its Regulation and Biotechnological Approaches to Improve Crop Productivity. *Am. J. Plant Sci.* **2015**, *6*, 2745–2798. [CrossRef]
13. Krapp, A.; David, L.C.; Chardin, C.; Girin, T.; Marmagne, A.; Leprince, A.S.; Chaillou, S.; Ferrario-Méry, S.; Meyer, C.; Daniel-Vedele, F. Nitrate transport and signalling in Arabidopsis. *J. Exp. Bot.* **2014**, *65*, 789–798. [CrossRef] [PubMed]
14. Wang, R.; Xing, X.; Wang, Y.; Tran, A.; Crawford, N.M. A Genetic Screen for Nitrate Regulatory Mutants Captures the Nitrate Transporter Gene NRT1.1. *Plant Physiol.* **2009**, *151*, 472–478. [CrossRef] [PubMed]
15. Hu, H.C.; Wang, Y.Y.; Tsay, Y.F. AtCIPK8, a CBL-interacting protein kinase, regulates the low-affinity phase of the primary nitrate response. *Plant J.* **2009**, *57*, 264–278. [CrossRef] [PubMed]
16. Mounier, E.; Pervent, M.; Ljung, K.; Gojon, A.; Nacry, P. Auxin-mediated nitrate signalling by NRT1.1 participates in the adaptive response of Arabidopsis root architecture to the spatial heterogeneity of nitrate availability. *Plant Cell Environ.* **2014**, *37*, 162–174. [CrossRef]
17. Liu, X.; Huang, D.; Tao, J.; Miller, A.J.; Fan, X.; Xu, G. Identification and functional assay of the interaction motifs in the partner protein OsNAR2.1 of the two-component system for high-affinity nitrate transport. *New Phytol.* **2014**, *204*, 74–80. [CrossRef]
18. Gu, R.; Duan, F.; An, X.; Zhang, F.; Von Wirén, N.; Yuan, L. Characterization of AMT-mediated high-affinity ammonium uptake in roots of maize (*Zea mays* L.). *Plant Cell Physiol.* **2013**, *54*, 1515–1524. [CrossRef]
19. Okamoto, M.; Vidmar, J.J.; Glass, A.D.M. Regulation of *NRT1* and *NRT2* gene families of *Arabidopsis thaliana*: Responses to nitrate provision. *Plant Cell Physiol.* **2003**, *44*, 304–317. [CrossRef]
20. Lezhneva, L.; Kiba, T.; Feria-Bourrellier, A.B.; Lafouge, F.; Boutet-Mercey, S.; Zoufan, P.; Sakakibara, H.; Daniel-Vedele, F.; Krapp, A. The Arabidopsis nitrate transporter NRT2.5 plays a role in nitrate acquisition and remobilization in nitrogen-starved plants. *Plant J.* **2014**, *80*, 230–241. [CrossRef]
21. Mueller, N.D.; Gerber, J.S.; Johnston, M.; Ray, D.K.; Ramankutty, N.; Foley, J.A. Closing yield gaps through nutrient and water management. *Nature* **2012**, *490*, 254–257. [CrossRef] [PubMed]
22. Garnett, T.; Conn, V.; Kaiser, B.N. Root based approaches to improving nitrogen use efficiency in plants. *Plant Cell Environ.* **2009**, *32*, 1272–1283. [CrossRef] [PubMed]

23. Bharali, B.; Haloi, B.; Chutia, J.; Chack, S.; Hazarika, K. Susceptibility of Some Wheat (*Triticum aestivum* L.) Varieties to Aerosols of Oxidised and Reduced Nitrogen. *Adv. Crop Sci. Technol.* **2015**, *3*, 4. [CrossRef]

24. Fernandes, J.C.; Goulao, L.F.; Amâncio, S. Regulation of cell wall remodeling in grapevine (*Vitis vinifera* L.) callus under individual mineral stress deficiency. *J. Plant Physiol.* **2016**, *190*, 95–105. [CrossRef] [PubMed]

25. Azhar, M.T.; Rehman, A. Overview on Effects of Water Stress on Cotton Plants and Productivity. In *Biochemical, Physiological and Molecular Avenues for Combating Abiotic Stress Tolerance in Plants*; Academic Press: London, UK, 2018; pp. 297–316.

26. Paterson, A.H.; May, O.L.; Desai, A.; Rong, J.; Chee, P.W. Chromosome structural changes in diploid and tetraploid A genomes of Gossypium. *Genome* **2006**, *49*, 336–345. [CrossRef]

27. Meshram, L.D.; Tayyab, M.A. Cytomorphology of two interspecific hybrids of Gossypium. *PKV Res. J.* **1994**, *18*, 63–67.

28. Wang, M.; Tu, L.; Yuan, D.; Zhu, D.; Shen, C.; Li, J.; Liu, F.; Pei, L.; Wang, P.; Zhao, G.; et al. Reference genome sequences of two cultivated allotetraploid cottons, *Gossypium hirsutum* and *Gossypium barbadense*. *Nat. Genet.* **2019**, *51*, 224–229. [CrossRef]

29. Li, F.; Fan, G.; Wang, K.; Sun, F.; Yuan, Y.; Song, G.; Li, Q.; Ma, Z.; Lu, C.; Zou, C.; et al. Genome sequence of the cultivated cotton *Gossypium arboreum*. *Nat. Genet.* **2014**, *46*, 567–572. [CrossRef]

30. Wang, K.; Wang, Z.; Li, F.; Ye, W.; Wang, J.; Song, G.; Yue, Z.; Cong, L.; Shang, H.; Zhu, S.; et al. The draft genome of a diploid cotton *Gossypium raimondii*. *Nat. Genet.* **2012**, *44*, 1098–1103. [CrossRef]

31. Schauser, L.; Wieloch, W.; Stougaard, J. Evolution of NIN-like proteins in Arabidopsis, rice, and *Lotus japonicus*. *J. Mol. Evol.* **2005**, *60*, 229–237. [CrossRef]

32. Tamura, K.; Stecher, G.; Peterson, D.; Filipski, A.; Kumar, S. MEGA6: Molecular evolutionary genetics analysis version 6.0. *Mol. Biol. Evol.* **2013**, *30*, 2725–2729. [CrossRef] [PubMed]

33. Wang, Y.; Pan, F.; Chen, D.; Chu, W.; Liu, H.; Xiang, Y. Genome-wide identification and analysis of the *Populus trichocarpa TIFY* gene family. *Plant Physiol. Biochem.* **2017**, *115*, 360–371. [CrossRef] [PubMed]

34. Magwanga, R.O.; Lu, P.; Kirungu, J.N.; Lu, H.; Wang, X.; Cai, X.; Zhou, Z.; Zhang, Z.; Salih, H.; Wang, K.; et al. Characterization of the late embryogenesis abundant (LEA) proteins family and their role in drought stress tolerance in upland cotton. *BMC Genet.* **2018**, *19*, 6. [CrossRef] [PubMed]

35. Singh, A.K.; Kumar, R.; Pareek, A.; Sopory, S.K.; Singla-Pareek, S.L. Overexpression of rice CBS domain containing protein improves salinity, oxidative, and heavy metal tolerance in transgenic tobacco. *Mol. Biotechnol.* **2012**, *52*, 205–216. [CrossRef] [PubMed]

36. Hao, Q.N.; Shang, W.; Zhang, C.; Chen, H.; Chen, L.; Yuan, S.; Chen, S.; Zhang, X.; Zhou, X. Identification and comparative analysis of CBS domain-containing proteins in soybean (*Glycine max*) and the primary function of GmCBS21 in enhanced tolerance to low nitrogen stress. *Int. J. Mol. Sci.* **2016**, *17*, 620. [CrossRef] [PubMed]

37. Wittkopp, P.J.; Kalay, G. Cis-regulatory elements: Molecular mechanisms and evolutionary processes underlying divergence. *Nat. Rev. Genet.* **2011**, *13*, 59–69. [CrossRef] [PubMed]

38. Magwanga, R.O.; Kirungu, J.N.; Lu, P.; Yang, X.; Dong, Q.; Cai, X.; Xu, Y.; Wang, X.; Zhou, Z.; Hou, Y.; et al. Genome wide identification of the trihelix transcription factors and overexpression of Gh_A05G2067 (GT-2), a novel gene contributing to increased drought and salt stresses tolerance in cotton. *Physiol. Plant.* **2019**. [CrossRef] [PubMed]

39. Chauhan, S.; Yogindran, S.; Rajam, M.V. Role of miRNAs in biotic stress reactions in plants. *Indian J. Plant Physiol.* **2017**, *22*, 514–529. [CrossRef]

40. Dong, Z.; Zhang, J.; Zhu, Q.; Zhao, L.; Sui, S.; Li, Z.; Zhang, Y.; Wang, H.; Tian, D.; Zhao, Y. Identification of microRNAs involved in drought stress responses in early-maturing cotton by high-throughput sequencing. *Genes Genom.* **2018**, *40*, 305–314. [CrossRef]

41. Li, J.; Guo, G.; Guo, W.; Guo, G.; Tong, D.; Ni, Z.; Sun, Q.; Yao, Y. miRNA164-directed cleavage of ZmNAC1 confers lateral root development in maize (*Zea mays* L.). *BMC Plant Biol.* **2012**, *12*, 220. [CrossRef]

42. Zhao, M.; Ding, H.; Zhu, J.K.; Zhang, F.; Li, W.X. Involvement of miR169 in the nitrogen-starvation responses in Arabidopsis. *New Phytol.* **2011**, *190*, 906–915. [CrossRef] [PubMed]

43. Scofield, S.R.; Huang, L.; Brandt, A.S.; Gill, B.S. Development of a Virus-Induced Gene-Silencing System for Hexaploid Wheat and Its Use in Functional Analysis of the Lr21-Mediated Leaf Rust Resistance Pathway. *Plant Physiol.* **2005**, *138*, 2165–2173. [CrossRef] [PubMed]

44. Estévez, J.M.; Cantero, A.; Romero, C.; Kawaide, H.; Jiménez, L.F.; Kuzuyama, T.; Seto, H.; Kamiya, Y.; León, P. Analysis of the Expression of CLA1, a Gene That Encodes the 1-Deoxyxylulose 5-Phosphate Synthase of the 2-C-Methyl-D-Erythritol-4-Phosphate Pathway in Arabidopsis. *Plant Physiol.* **2002**, *124*, 95–104. [CrossRef] [PubMed]

45. Almeida, M.V.; Andrade-Navarro, M.A.; Ketting, R.F. Function and Evolution of Nematode RNAi Pathways. *Non-Coding RNA* **2019**, *5*, 8. [CrossRef] [PubMed]

46. Magwanga, R.O.; Lu, P.; Kirungu, J.N.; Dong, Q.; Cai, X.; Zhou, Z.; Wang, X.; Hou, Y.; Xu, Y.; Peng, R.; et al. Knockdown of Cytochrome P450 Genes Gh_D07G1197 and Gh_A13G2057 on Chromosomes D07 and A13 Reveals Their Putative Role in Enhancing Drought and Salt Stress Tolerance in *Gossypium hirsutum*. *Genes* **2019**, *10*, 226. [CrossRef]

47. Bachan, S.; Dinesh-Kumar, S.P. Tobacco rattle virus (TRV)-based virus-induced gene silencing. *Methods Mol. Biol.* **2012**, *894*, 83–92. [CrossRef]

48. Ding, L.; Wang, K.J.; Jiang, G.M.; Biswas, D.K.; Xu, H.; Li, L.F.; Li, Y.H. Effects of nitrogen deficiency on photosynthetic traits of maize hybrids released in different years. *Ann. Bot.* **2005**, *96*, 925–930. [CrossRef]

49. Nguyen, N.T.; Mohapatra, P.K.; Fujita, K.; Nakabayashi, K.; Thompson, J. Effect of nitrogen deficiency on biomass production, photosynthesis, carbon partitioning, and nitrogen nutrition status of Melaleuca and Eucalyptus species. *Soil Sci. Plant Nutr.* **2003**, *49*, 99–109. [CrossRef]

50. Løvdal, T.; Lillo, C. Reference gene selection for quantitative real-time PCR normalization in tomato subjected to nitrogen, cold, and light stress. *Anal. Biochem.* **2009**, *387*, 238–242. [CrossRef]

51. Warzybok, A.; Migocka, M. Reliable Reference Genes for Normalization of Gene Expression in Cucumber Grown under Different Nitrogen Nutrition. *PLoS ONE* **2013**, *8*, e72887. [CrossRef]

52. Fuentes, S.I.; Allen, D.J.; Ortiz-Lopez, A.; Hernández, G. Over-expression of cytosolic glutamine synthetase increases photosynthesis and growth at low nitrogen concentrations. *J. Exp. Bot.* **2001**, *52*, 1071–1081. [CrossRef] [PubMed]

53. Martin, A.; Lee, J.; Kichey, T.; Gerentes, D.; Zivy, M.; Tatout, C.; Dubois, F.; Balliau, T.; Valot, B.; Davanture, M.; et al. Two Cytosolic Glutamine Synthetase Isoforms of Maize Are Specifically Involved in the Control of Grain Production. *Plant Cell* **2006**, *18*, 3252–3274. [CrossRef] [PubMed]

54. Liheng, H.; Zhiqiang, G.; Runzhi, L. Pretreatment of seed with H_2O_2 enhances drought tolerance of wheat (*Triticum aestivum* L.) seedlings. *Afr. J. Biotechnol.* **2009**, *8*, 6151–6157. [CrossRef]

55. Smirnoff, N.; Arnaud, D. Hydrogen peroxide metabolism and functions in plants. *New Phytol.* **2019**, *221*, 1197–1214. [CrossRef]

56. Matsuki, S.; Ogawa, K.; Tanaka, A.; Hara, T. Morphological and photosynthetic responses of Quercus crispula seedlings to high-light conditions. *Tree Physiol.* **2003**, *23*, 769–775. [CrossRef] [PubMed]

57. Losciale, P.; Zibordi, M.; Manfrini, L.; Morandi, B.; Corelli-Grappadelli, L. Effect of moderate light reduction on absorbed energy management, water use, photoprotection and photo-damage in peach. *Acta Hortic.* **2011**, *907*, 169–174. [CrossRef]

58. Yevenes, M.A.; Mannaerts, C.M. Seasonal and land use impacts on the nitrate budget and export of a mesoscale catchment in Southern Portugal. *Agric. Water Manag.* **2011**, *102*, 54–65. [CrossRef]

59. Bushong, J.T.; Arnall, D.B.; Raun, W.R. Effect of Preplant Irrigation, Nitrogen Fertilizer Application Timing, and Phosphorus and Potassium Fertilization on Winter Wheat Grain Yield and Water Use Efficiency. *Int. J. Agron.* **2014**, *2014*. [CrossRef]

60. Chao, Z.; Zhang, P. Assessing the impact of urbanization on vegetation change and arable land resources change in Shandong province. In Proceedings of the 3rd International Conference on Agro-Geoinformatics, Beijing, China, 11–14 August 2014.

61. Suzuki, W.; Konishi, M.; Yanagisawa, S. The evolutionary events necessary for the emergence of symbiotic nitrogen fixation in legumes may involve a loss of nitrate responsiveness of the NIN transcription factor. *Plant Signal. Behav.* **2013**, *8*, e25975. [CrossRef]

62. Lu, T.; Zhang, G.; Sun, L.; Wang, J.; Hao, F. Genome-wide identification of CBL family and expression analysis of CBLs in response to potassium deficiency in cotton. *PeerJ* **2017**, *5*, e3653. [CrossRef]

63. Hachiya, T.; Mizokami, Y.; Miyata, K.; Tholen, D.; Watanabe, C.K.; Noguchi, K. Evidence for a nitrate-independent function of the nitrate sensor NRT1.1 in *Arabidopsis thaliana*. *J. Plant Res.* **2011**, *124*, 425–430. [CrossRef]

64. Bush, M.; Dixon, R. The Role of Bacterial Enhancer Binding Proteins as Specialized Activators of 54-Dependent Transcription. *Microbiol. Mol. Biol. Rev.* **2012**, *76*, 497–529. [CrossRef] [PubMed]

65. Chardin, C.; Girin, T.; Roudier, F.; Meyer, C.; Krapp, A. The plant RWP-RK transcription factors: Key regulators of nitrogen responses and of gametophyte development. *J. Exp. Bot.* **2014**, *65*, 5577–5587. [CrossRef] [PubMed]

66. Lu, P.; Magwanga, R.O.; Guo, X.; Kirungu, J.N.; Lu, H.; Cai, X.; Zhou, Z.; Wei, Y.; Wang, X.; Zhang, Z.; et al. Genome-Wide Analysis of Multidrug and Toxic Compound Extrusion (MATE) Family in Diploid Cotton, *Gossypium raimondii* and *Gossypium arboreum* and Its Expression Analysis Under Salt, Cadmium and Drought Stress. *G3* **2018**, *8*, 2483–2500. [CrossRef] [PubMed]

67. Morselli, M.; Pastor, W.A.; Montanini, B.; Nee, K.; Ferrari, R.; Fu, K.; Bonora, G.; Rubbi, L.; Clark, A.T.; Ottonello, S.; et al. In vivo targeting of de novo DNA methylation by histone modifications in yeast and mouse. *Elife* **2015**, *4*, e06205. [CrossRef] [PubMed]

68. Sloan, D.B.; Taylor, D.R. Testing for selection on synonymous sites in plant mitochondrial DNA: The role of codon bias and RNA editing. *J. Mol. Evol.* **2010**, *70*, 479–491. [CrossRef] [PubMed]

69. Greiner, S.; Rauwolf, U.; Meurer, J.; Herrmann, R.G. The role of plastids in plant speciation. *Mol. Ecol.* **2011**, *20*, 671–691. [CrossRef] [PubMed]

70. Ondřej, V.; Lukášová, E.; Krejčí, J.; Kozubek, S. Intranuclear trafficking of plasmid DNA is mediated by nuclear polymeric proteins lamins and actin. *Acta Biochim. Pol.* **2008**, *55*, 307–315. [PubMed]

71. Stanbrough, M.; Rowen, D.W.; Magasanik, B. Role of the GATA factors Gln3p and Nil1p of Saccharomyces cerevisiae in the expression of nitrogen-regulated genes. *Proc. Natl. Acad. Sci. USA* **1995**, *92*, 9450–9454. [CrossRef]

72. Beck, T.; Hall, M.N. The TOR signalling pathway controls nuclear localization of nutrient-regulated transcription factors. *Nature* **1999**, *402*, 689–692. [CrossRef]

73. Chen, M.; Bao, H.; Wu, Q.; Wang, Y. Transcriptome-wide identification of miRNA targets under nitrogen deficiency in *Populus tomentosa* using degradome sequencing. *Int. J. Mol. Sci.* **2015**, *16*, 13937–13958. [CrossRef]

74. Wang, Y.; Zhang, C.; Hao, Q.; Sha, A.; Zhou, R.; Zhou, X.; Yuan, L. Elucidation of miRNAs-Mediated Responses to Low Nitrogen Stress by Deep Sequencing of Two Soybean Genotypes. *PLoS ONE* **2013**, *8*, e67423. [CrossRef] [PubMed]

75. Liang, G.; He, H.; Yu, D. Identification of Nitrogen Starvation-Responsive MicroRNAs in Arabidopsis thaliana. *PLoS ONE* **2012**, *7*, e48951. [CrossRef] [PubMed]

76. Zhao, Y.; Xu, Z.; Mo, Q.; Zou, C.; Li, W.; Xu, Y.; Xie, C. Combined small RNA and degradome sequencing reveals novel miRNAs and their targets in response to low nitrate availability in maize. *Ann. Bot.* **2013**, *112*, 633–642. [CrossRef] [PubMed]

77. Yan, Y.; Wang, H.; Hamera, S.; Chen, X.; Fang, R. MiR444a has multiple functions in the rice nitrate-signaling pathway. *Plant J.* **2014**, *78*, 44–55. [CrossRef] [PubMed]

78. Wu, M.-F.; Tian, Q.; Reed, J.W. Arabidopsis microRNA167 controls patterns of ARF6 and ARF8 expression, and regulates both female and male reproduction. *Development* **2006**, *133*, 4211–4218. [CrossRef]

79. Gifford, M.L.; Dean, A.; Gutierrez, R.A.; Coruzzi, G.M.; Birnbaum, K.D. Cell-specific nitrogen responses mediate developmental plasticity. *Proc. Natl. Acad. Sci. USA* **2008**, *105*, 803–808. [CrossRef]

80. Olsen, A.N.; Ernst, H.A.; Leggio, L.L.; Skriver, K. NAC transcription factors: Structurally distinct, functionally diverse. *Trends Plant Sci.* **2005**, *10*, 79–87. [CrossRef]

81. Guo, H.-S. MicroRNA Directs mRNA Cleavage of the Transcription Factor NAC1 to Downregulate Auxin Signals for Arabidopsis Lateral Root Development. *Plant Cell* **2005**, *17*, 1376–1386. [CrossRef]

82. Mallory, A.C.; Dugas, D.V.; Bartel, D.P.; Bartel, B. MicroRNA regulation of NAC-domain targets is required for proper formation and separation of adjacent embryonic, vegetative, and floral organs. *Curr. Biol.* **2004**, *14*, 1035–1046. [CrossRef]

83. Mishra, A.K.; Duraisamy, G.S.; Matoušek, J. Discovering MicroRNAs and Their Targets in Plants. *CRC Crit. Rev. Plant Sci.* **2015**, *34*, 553–571. [CrossRef]

84. Xu, Z.; Yu, J.Z.; Cho, J.; Yu, J.; Kohel, R.J.; Percy, R.G. Polyploidization Altered Gene Functions in Cotton (*Gossypium* spp.). *PLoS ONE* **2010**, *5*, e14351. [CrossRef] [PubMed]

85. Abdelraheem, A.; Fang, D.D.; Zhang, J. Quantitative trait locus mapping of drought and salt tolerance in an introgressed recombinant inbred line population of Upland cotton under the greenhouse and field conditions. *Euphytica* **2018**, *214*, 8. [CrossRef]

86. Said, J.I.; Lin, Z.; Zhang, X.; Song, M.; Zhang, J. A comprehensive meta QTL analysis for fiber quality, yield, yield related and morphological traits, drought tolerance, and disease resistance in tetraploid cotton. *BMC Genom.* **2013**, *14*, 776. [CrossRef] [PubMed]

87. Wright, R.J.; Thaxton, P.M.; El-Zik, K.M.; Paterson, A.H. D-subgenome bias of Xcm resistance genes in tetraploid *Gossypium* (cotton) suggests that polyploid formation has created novel avenues for evolution. *Genetics* **1998**, *149*, 1987–1996. [PubMed]

88. Guo, W.; Cai, C.; Wang, C.; Zhao, L.; Wang, L.; Zhang, T. A preliminary analysis of genome structure and composition in *Gossypium hirsutum*. *BMC Genom.* **2008**, *9*, 314. [CrossRef] [PubMed]

89. Key, J. MAC Significance of mating systems for chromosomes and gametes in polyploids. *Hereditas* **1970**, *66*, 165–176. [CrossRef] [PubMed]

90. Reinisch, A.J.; Dong, J.M.; Brubaker, C.L.; Stelly, D.M.; Wendel, J.F.; Paterson, A.H. A detailed RFLP map of cotton, *Gossypium hirsutum* x *Gossypium barbadense*: Chromosome organization and evolution in a disomic polyploid genome. *Genetics* **1994**, *138*, 829–847.

91. Basu, S.; Roychoudhury, A. Expression profiling of abiotic stress-inducible genes in response to multiple stresses in rice (*Oryza sativa* L.) varieties with contrasting level of stress tolerance. *BioMed Res. Int.* **2014**, *2014*, 706890. [CrossRef]

92. Swindell, W.R. The association among gene expression responses to nine abiotic stress treatments in *Arabidopsis thaliana*. *Genetics* **2006**, *174*, 1811–1824. [CrossRef]

93. Lin, Y.L.; Chao, Y.Y.; Huang, W.D.; Kao, C.H. Effect of nitrogen deficiency on antioxidant status and Cd toxicity in rice seedlings. *Plant Growth Regul.* **2011**, *64*, 263–273. [CrossRef]

94. Lin, W.; Hagen, E.; Fulcher, A.; Hren, M.T.; Cheng, Z.M. Overexpressing the ZmDof1 gene in Populus does not improve growth and nitrogen assimilation under low-nitrogen conditions. *Plant Cell Tissue Organ Cult.* **2013**, *113*, 51–61. [CrossRef]

95. Mittler, R.; Vanderauwera, S.; Suzuki, N.; Miller, G.; Tognetti, V.B.; Vandepoele, K.; Gollery, M.; Shulaev, V.; Van Breusegem, F. ROS signaling: The new wave? *Trends Plant Sci.* **2011**, *16*, 300–309. [CrossRef] [PubMed]

96. Prsa, I.; Stampar, F.; Vodnik, D.; Veberic, R. Influence of nitrogen on leaf chlorophyll content and photosynthesis of "Golden Delicious" apple. *Acta Agric. Scand. Sect. B Soil Plant Sci.* **2007**, *57*, 283–289. [CrossRef]

97. Ahn, C.S.; Ahn, H.K.; Pai, H.S. Overexpression of the PP2A regulatory subunit Tap46 leads to enhanced plant growth through stimulation of the TOR signalling pathway. *J. Exp. Bot.* **2015**, *66*, 827–840. [CrossRef] [PubMed]

98. Alam, S.M. Effect of nutrient solution pH and N-Sources (NH_4/NO_3) on the growth and elemental content of rice plants. *Agronomie* **1984**, *4*, 361–365. [CrossRef]

99. Li, X.Q.; Sveshnikov, D.; Zebarth, B.J.; Tai, H.; de Koeyer, D.; Millard, P.; Haroon, M.; Singh, M. Detection of nitrogen sufficiency in potato plants using gene expression markers. *Am. J. Potato Res.* **2010**, *87*, 50–59. [CrossRef]

100. Sun, Z.; Wang, X.; Liu, Z.; Gu, Q.; Zhang, Y.; Li, Z.; Ke, H.; Yang, J.; Wu, J.; Wu, L.; et al. Genome-wide association study discovered genetic variation and candidate genes of fibre quality traits in *Gossypium hirsutum* L. *Plant Biotechnol. J.* **2017**, *15*, 982–996. [CrossRef]

101. Gálvez, J.H.; Tai, H.H.; Lagüe, M.; Zebarth, B.J.; Strömvik, M.V. The nitrogen responsive transcriptome in potato (*Solanum tuberosum* L.) reveals significant gene regulatory motifs. *Sci. Rep.* **2016**, *6*, 26090. [CrossRef]

102. do, A.C.; Bisognin, D.; Steffens, C. Non-destructive quantification of chlorophylls in leaves by means of a colorimetric method. *Hortic. Bras.* **2008**, *26*, 471–475. [CrossRef]

103. Giannopolitis, C.N.; Ries, S.K. Superoxide Dismutases: II. Purification and Quantitative Relationship with Water-soluble Protein in Seedlings. *Plant Physiol.* **1977**, *59*, 315–318. [CrossRef] [PubMed]

104. Van Assche, F.; Cardinaels, C.; Clijsters, H. Induction of enzyme capacity in plants as a result of heavy metal toxicity: Dose-response relations in *Phaseolus vulgaris* L., treated with zinc and cadmium. *Environ. Pollut.* **1988**, *52*, 103–115. [CrossRef]

105. Cakmak, I.; Horst, W.J. Effect of aluminium on lipid peroxidation, siiperoxide dismutase, catalase, and peroxidase activities in root tips of soybean (*Glycine max*). *Physiol. Plant.* **1991**, *83*, 463–468. [CrossRef]
106. Chen, X.; Kou, M.; Tang, Z.; Zhang, A.; Li, H. The use of humic acid urea fertilizer for increasing yield and utilization of nitrogen in sweet potato. *Plant Soil Environ.* **2017**, *63*, 201–206. [CrossRef]

International Journal of
Molecular Sciences

MDPI

Article

Genome-Wide Identification and Characterization of Cucumber BPC Transcription Factors and Their Responses to Abiotic Stresses and Exogenous Phytohormones

Shuzhen Li [1,2], Li Miao [1], Bin Huang [1], Lihong Gao [2], Chaoxing He [1], Yan Yan [1], Jun Wang [1], Xianchang Yu [1,*] and Yansu Li [1,*]

[1] Institute of Vegetables and Flowers, Chinese Academy of Agricultural Sciences, Beijing 100081, China; limengfbj@126.com (S.L.); happymml@163.com (L.M.); 82101175049@caas.cn (B.H.); hechaoxing@caas.cn (C.H.); yanyan@caas.cn (Y.Y.); wangjun01@caas.cn (J.W.)

[2] Beijing Key Laboratory of Growth and Developmental Regulation for Protected Vegetable Crops, College of Horticulture, China Agricultural University, Beijing 100193, China; gaolh@cau.edu.cn

* Correspondence: yuxianchang@caas.cn (X.Y.); liyansu@caas.cn (Y.L.)

Received: 20 August 2019; Accepted: 2 October 2019; Published: 11 October 2019

Abstract: BASIC PENTACYSTEINE (BPC) is a small transcription factor family that functions in diverse growth and development processes in plants. However, the roles of BPCs in plants, especially cucumber (*Cucumis sativus* L.), in response to abiotic stress and exogenous phytohormones are still unclear. Here, we identified four BPC genes in the cucumber genome, and classified them into two groups according to phylogenetic analysis. We also investigated the gene structures and detected five conserved motifs in these *CsBPCs*. Tissue expression pattern analysis revealed that the four *CsBPCs* were expressed ubiquitously in both vegetative and reproductive organs. Additionally, the transcriptional levels of the four *CsBPCs* were induced by various abiotic stress and hormone treatments. Overexpression of *CsBPC2* in tobacco (*Nicotiana tabacum*) inhibited seed germination under saline, polyethylene glycol, and abscisic acid (ABA) conditions. The results suggest that the CsBPC genes may play crucial roles in cucumber growth and development, as well as responses to abiotic stresses and plant hormones. *CsBPC2* overexpression in tobacco negatively affected seed germination under hyperosmotic conditions. Additionally, *CsBPC2* functioned in ABA-inhibited seed germination and hypersensitivity to ABA-mediated responses. Our results provide fundamental information for further research on the biological functions of BPCs in development and abiotic stress responses in cucumber and other plant species.

Keywords: BASIC PENTACYSTEINE (BPC); cucumber; expression analysis; abiotic stress; plant hormones

---

1. Introduction

Cucumber (*Cucumis sativus* L.) is a major vegetable crop worldwide and has served as a model system for studies on sex determination [1] and plant vascular biology [2]. However, its growth and development are frequently affected by various stresses, such as low temperature [3], high salinity [4,5], water deficit [6], and pathogen attack [7], which severely reduce production and quality. Therefore, functional studies of cucumber stress responses and identification of stress-related genes are required to elucidate the molecular mechanisms of cucumber stress tolerance and protect them from detrimental surroundings.

BASIC PENTACYSTEINE (BPC)/BARLEY B RECOMBINANT (BBR), a plant-specific transcription factor family, is characterized by the ability to bind gene promoter sequences at the GAGA motif; the soybean GAGA-binding protein (GBP) binds to a (GA)$_9$ repeat sequence of the *Glutamate 1-Semialdehyde Aminotransferase* (*Gsa1*) promoter [8], the barley *BARLEY B RECOMBINANT* (*BBR*) factor binds specifically to the (GA)$_8$ repeat in vitro [9], and *Arabidopsis* BPC proteins recognize (GA)$_6$ and (GA)$_9$ sequences [10,11]. According this characteristic, BPC transcription factors are also named GAGA-binding transcriptional activators. To date, based on gene sequence similarity and protein domain structures, seven *Arabidopsis* BPC genes have been identified and classified into three classes: class I (BPC1–3), class II (BPC4–6), and class III (BPC7). All of the BPC proteins contain a highly conserved DNA-binding domain with five conserved cysteine residues at their C terminus [10]. Among these genes, BPC5 is thought to be a pseudogene that is unable to produce an active protein [10,12].

As newly identified transcription factors, BPCs have been proposed to function in diverse plant growth and development responses. For example, transcripts of the barley BBR gene were detected in all tissues, including roots, stems, leaves, inflorescences, and embryos, among which the highest level was observed in embryos and the lowest in leaves. Overexpression of the gene in tobacco resulted in pronounced leaf and flower shapes or structural modifications [9]. In *Arabidopsis*, all seven BPC genes except for *BPC5* were expressed ubiquitously in both vegetative and reproductive organs. Multiple *BPC* allele mutants displayed pleiotropic developmental defects, including dwarfism, small rosettes, early flowering, aberrant ovules, unopened floral buds, and even high sterility [12–15]. The reason for these morphological changes may be that BPCs bind to and regulate the activities and expression levels of target genes associated with development. BPC genes have been reported to be regulators of *INNER NO OUTER* (*INO*), a gene involved in ovule development [10]. BPC genes were also found to regulate the expression of the homeotic MADS box gene *SEEDSTICK* (*STK*), to control the ovule identity [11,13,16,17]. The *LEAFY COTYLEDON 2* (*LEC2*) gene is only expressed in embryos and acts as a master regulator of seed development, and is regulated by BPCs in *Arabidopsis* [18–20]. Additionally, BPCs downregulated the expression of the genes *SHOOTMERISTEMLESS* (*STM*) and *BREVIPEDICELLUS/KNAT1* (*BP*), which resulted in floral organ malformation [14]. Additionally, class I BPCs also act as direct regulators of several HOMEOBOX genes, such as *KNOX*, *WUS*, and *BELL* family members [14]. In our previous research, we found that cucumber BPCs are involved in seed germination as regulators of *ABSCISIC ACID INSENSITIVE3* (*ABI3*) [21]. In rice, the GAGA-binding transcription factors *OsGBP1* and *OsGBP3* displayed functional divergence in the regulation of grain size and plant growth [22]. *OsGBP1* also delayed flowering time by directly binding to the promoter of *OsLFL1*, a LEC2/FUS3-like gene, which constitutively inhibits the expression of a flowering activator, *Early heading date 1* (*Ehd1*) [22–24]. Moreover, increasing evidence has also demonstrated a role for BPCs in the regulation of plant responses to hormones, such as ethylene [12] and cytokinins [14,25]. While BPCs are known to participate in numerous developmental processes, the involvement of BPCs in stress responses is not clear.

The completion of genome sequencing for more species has enabled many genes to be identified and characterized. However, comprehensive analyses of BPC genes in cucumber and other plant species are still limited. Thus, in the present study, we identified four *BPC* family members in the cucumber genome, named *CsBPC1* to *CsBPC4*. They were classified into two groups based on phylogenetic analysis. Their predicted gene structures and conserved motifs were subsequently analyzed. Furthermore, we investigated their expression patterns in various tissues and in response to different stresses and plant hormones by qRT-PCR. To further verify the cellular functions of the *CsBPCs*, transgenic tobacco plants constitutively overexpressing *CsBPC2* were generated and seed germination experiments with different concentrations of salt, polyethylene glycol (PEG), and abscisic acid (ABA) were conducted. This work provides a basis for exploring the potential functions of BPC genes, especially in stress resistance. This will enrich the stress tolerance theory of plants and lay a theoretical foundation to alleviate the detrimental effects on cucumber growth caused by various abiotic stresses.

2. Results

2.1. Identification and Characterization of Cucumber BPC Genes

To identify BPC family genes in cucumber, we performed BLASTP searches against the Cucumber Genome Database using seven *Arabidopsis* BPC proteins as query sequences, and confirmed the candidate sequences using the Pfam and SMART databases. We detected four BPC family proteins containing the GAGA binding-like domain (Table 1), which was consistent with our previous identification [21]. However, here we found that *Csa5G092920.2* and *Csa7G007860.1* had two and three transcripts, respectively, and we selected genes with longest encoding protein sequences for subsequent analysis. Two of the four members were distributed on chromosome five; the other two were distributed on chromosomes two and seven, respectively (Figure 1). All proteins shared similar parameters. The amino acid lengths of these four genes ranged from 279 to 338 aa, with molecular weights ranging from 31.1 to 37.8 kDa. The isoelectric points of all four BPC proteins were relatively high (pI > 9), indicating that they are rich in alkaline amino acids. Subcellular location prediction showed that all four members were localized to the nucleus.

Table 1. Characteristics of the BASIC PENTACYSTEINE (BPC) family in cucumber. Gene ID with black bold means more than one transcripts.

Gene ID	Length (aa)	Molecular Weight (KD)	Chromosome	Location	pI	Strand Direction	Subcellular Location
Csa2G365700.1	338	37.8	2	17669793–17673023	9.52	+	Nuclear
Csa5G092910.1	279	31.1	5	2693816–2695298	9.62	-	Nuclear
Csa5G092920.2	284	31.7	5	2696954–2700269	9.84	-	Nuclear
Csa7G007860.1	313	35	7	388479–391899	9.6	+	Nuclear

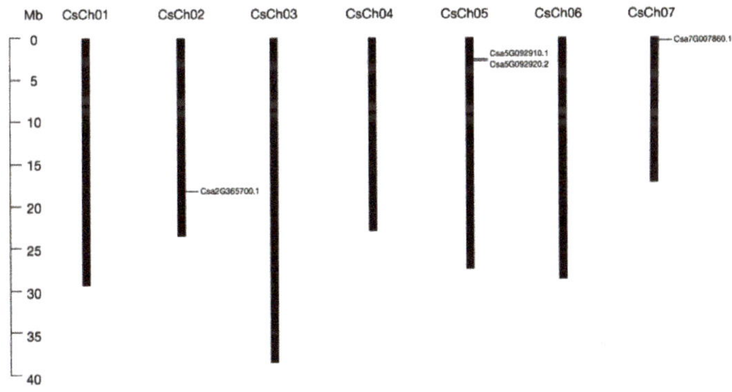

Figure 1. Localization of four cucumber BPC genes on chromosomes. There are seven chromosomes in cucumber, and the chromosome number is indicated at the top of each chromosome. The scale bar on the left side indicates the size of chromosome. The relative positions of four BPC genes are marked on the chromosomes.

2.2. Phylogenetic Gene Structure and Motif Analysis

To gain insight into the evolution of the *BPCs* in cucumber, a phylogenetic tree was constructed using BPC proteins from different plant species, including cucumber, watermelon, melon, cucurbita pepo, grape, western balsam poplar, common sunflower, *Arabidopsis*, rice and tomato. As shown in Figure 2, the 52 BPC proteins were classified into three groups, which was consistent with previous classification in *Arabidopsis* [10]. *Csa5G092910.1* and *Csa5G092920.2* belonged to group I; here, we named them *CsBPC1* and *CsBPC2*, respectively. *Csa7G007860.1* and *Csa2G365700.1* belonged to group II; here, we named them *CsBPC3* and *CsBPC4*, respectively. The numbers of BPC genes in group I, group II, and group III were 24, 26, and 2, respectively. BPC genes in *Arabidopsis* and western balsam poplar were clustered in each group, whereas, those in other eight species were just clustered in group I and group II. This indicated that BPC genes in group I and group II were relatively more conserved than those in group III, and that BPC genes in group III might be lost more easily during the evolutionary process. Besides, since cucumber, watermelon, melon, and cucurbita pepo belong to the Cucurbitaceae family, their evolutionary processes were more similar with each other than that with other species. Moreover, each cucumber BPC gene was closest with that in melon. Figure 3A illustrates the exon–intron organizations of the CsBPC genes. All the CsBPC genes possessed one intron, with the exception of *CsBPC1*, which lacked an intron. Additionally, the introns of *CsBPC3* and *CsBPC4* were phase zero introns, while *CsBPC2* had a phase two intron. The conserved domains of the CsBPC proteins were then analyzed and five conserved motifs were identified (Figure 3B). In general, members in the same group shared similar motif compositions, whereas members in different groups had differing motif compositions. Both group I members possessed four motifs, while both group II members had three motifs. Motifs one and two were found in all four members. Motifs three and five were only found in the group I members, and motif four only in the group II members. Furthermore, motif one contained five conserved cysteine residues (Figure 3C), which is the unique hallmark of the BPC transcription factor family [10,11].

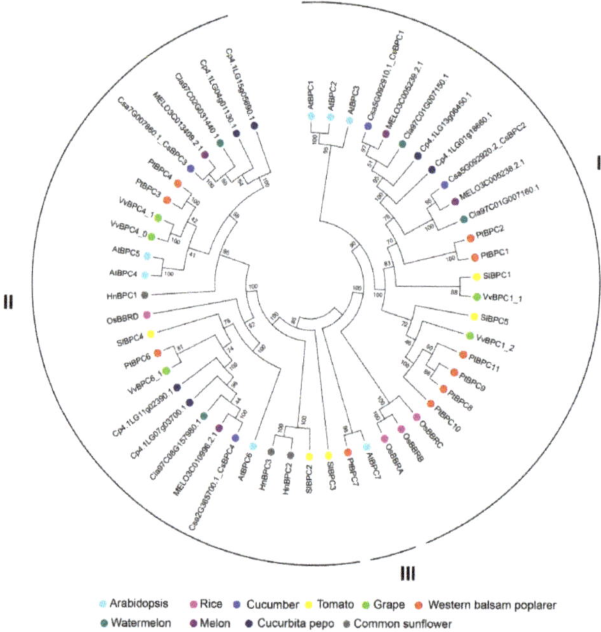

Figure 2. Phylogenetic analysis of BPC proteins from cucumber and other species. BPC polypeptide sequences generated from four cucumber, seven *Arabidopsis*, four rice, four watermelon, four melon, six cucurbita pepo, five grape, ten western balsam poplar, three common sunflower, and five tomato (SlBPCs) plants were used to construct an unrooted neighbor-joining phylogenetic tree by MEGA7 software with 100 bootstrap replicates. The different colored dots indicate different species.

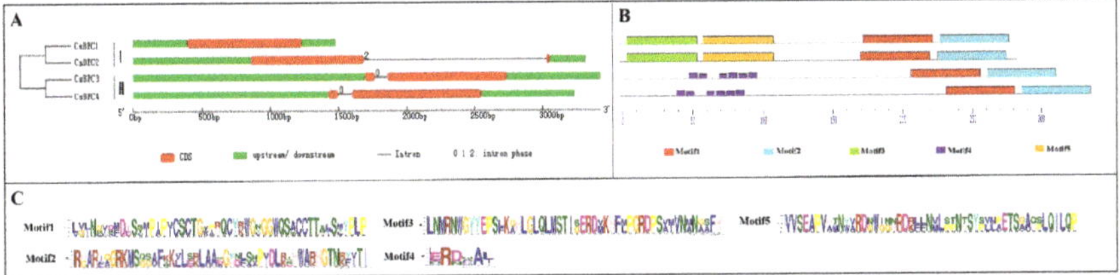

Figure 3. Gene structure and conserved motif analyses of cucumber BPC members. (**A**) Exon–intron organization of CsBPC genes. Red boxes represent exons and the black lines represent introns. The numbers 0, 1, and 2 represent the intron phases. The untranslated 5′- and 3′- regions (UTR) are shown with green boxes. (**B**) Conserved motif in cucumber BPC proteins. Five motifs with E-value < 0.05 are presented with different colored boxes. (**C**) The sequences of five identified motifs in cucumber BPC proteins.

2.3. Expression Profiles in Different Tissues

To confirm the potential functions of the *CsBPCs* in cucumber growth and development, the expression patterns of the four *CsBPC* genes were analyzed by qRT-PCR in 12 different tissues. As shown in Figure 4, all four *CsBPC* genes were expressed to varying degrees in the tissues tested, and they shared the highest expression levels in seeds and the lowest levels in tendrils and stems. The expression levels of the *CsBPCs* in different floral organs, including male flowers (MF), female flowers with ovaries removed (FF), and ovaries (O), showed that *CsBPC1*, *CsBPC2*, and *CsBPC3* were highly expressed in ovaries compared with the other two organs, whereas *CsBPC4* was expressed at almost the same level in all three organs. Additionally, when comparing the expression levels in leaves at different developmental stages, we found that *CsBPC1* and *CsBPC4* had relatively higher transcription levels in young leaves than in mature and old leaves, which indicated that they may play an important role in the early stages of leaf growth and development. Collectively, the results indicated the *CsBPC* genes may play vital roles in many cellular processes in cucumber growth and development, and that the different members have overlapping but distinct functions. However, it is necessary to do plenty of complex experiments to investigate the functions of *CsBPCs* in regulating cucumber growth and development.

2.4. Expression Patterns of the CsBPC Genes under Different Abiotic Stress and Phytohormone Treatments

Increasing evidence suggests that BPC transcription factors play important roles in regulating plant growth and development. However, the involvement of *BPCs* in responses to abiotic stresses and phytohormones is not clear. To confirm the involvement of the *CsBPCs* in abiotic stress and hormone responses, their transcript abundances were investigated under NaCl, PEG, cold (5 °C), heat (38 °C), ABA, SA, JA, ETH, 2,4-D, and GA treatments. Overall, the expression of all four CsBPC genes was induced by all the treatments tested (Figures 5 and 6). For instance, under the NaCl, PEG, and cold treatments, the expression levels of *CsBPC2* in roots or leaves increased by a maximum of 6.1-, 6.1-, and 1.7-fold, respectively, the most among the four genes. Under heat stress, the transcript level of *CsBPC4* was the highest at 12 h in roots (increased by 15.6-fold), followed by *CsBPC2* in roots (increased by 14.0-fold). In response to the application of exogenous hormones, including ABA, GA, JA, and SA, the transcript levels of all four CsBPC genes, especially *CsBPC2*, in roots were dramatically induced with prolonged treatment, whereas the transcript levels in leaves were increased slightly. Additionally, the application of ABA more significantly induced gene expression in roots compared with the other three hormones. Notably, the expression level of *CsBPC2* was increased by 112.1-fold after 24 h of ABA treatment. However, the expression patterns under ETH treatment were just the opposite, with higher gene expression induction in leaves than in roots. Lastly, application of 2,4-D highly induced the expression of all four genes in both roots and leaves, with maximal transcript levels at 12 h in leaves and 24 h in roots. Thus, the *CsBPCs* are involved in, and positively induced by, various abiotic stresses and phytohormones.

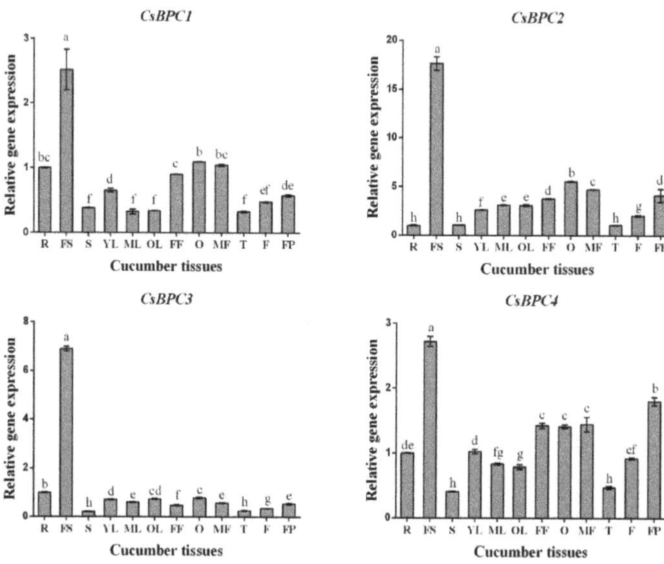

Figure 4. Expression analysis of CsBPC genes in different tissues. Transcript levels of *CsBPCs* were analyzed by qRT-PCR in roots (R), stems (S), top young leaves (YL), middle mature leaves (ML), basal old leaves (OL), blooming male flowers (MF), blooming female flowers with ovaries removed (FF), ovaries (O), tendrils (T), fruits (F), mature fruits' seeds (FS), and pulps (FP). The expression levels of roots were defined as 1. Values are means ± SD ($n = 3$). Different letters indicate significant differences ($p < 0.05$).

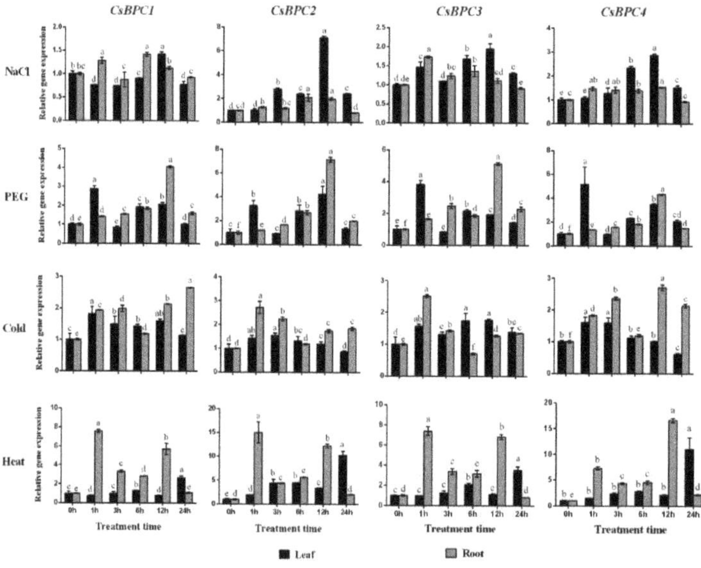

Figure 5. Expression patterns of CsBPC genes under different abiotic stress treatments. Cucumber seedlings at three-leaf-stage were exposed to 100 mM NaCl, 10% PEG 6000, cold (5 °C), and heat (38 °C) conditions, and qRT-PCR was performed to examine the transcript levels of *CsBPCs* in response to different abiotic stresses. The expression levels in non-stressed leaves and roots were set as 1. Values are means ± SD ($n = 3$). Different letters indicate significant differences ($p < 0.05$).

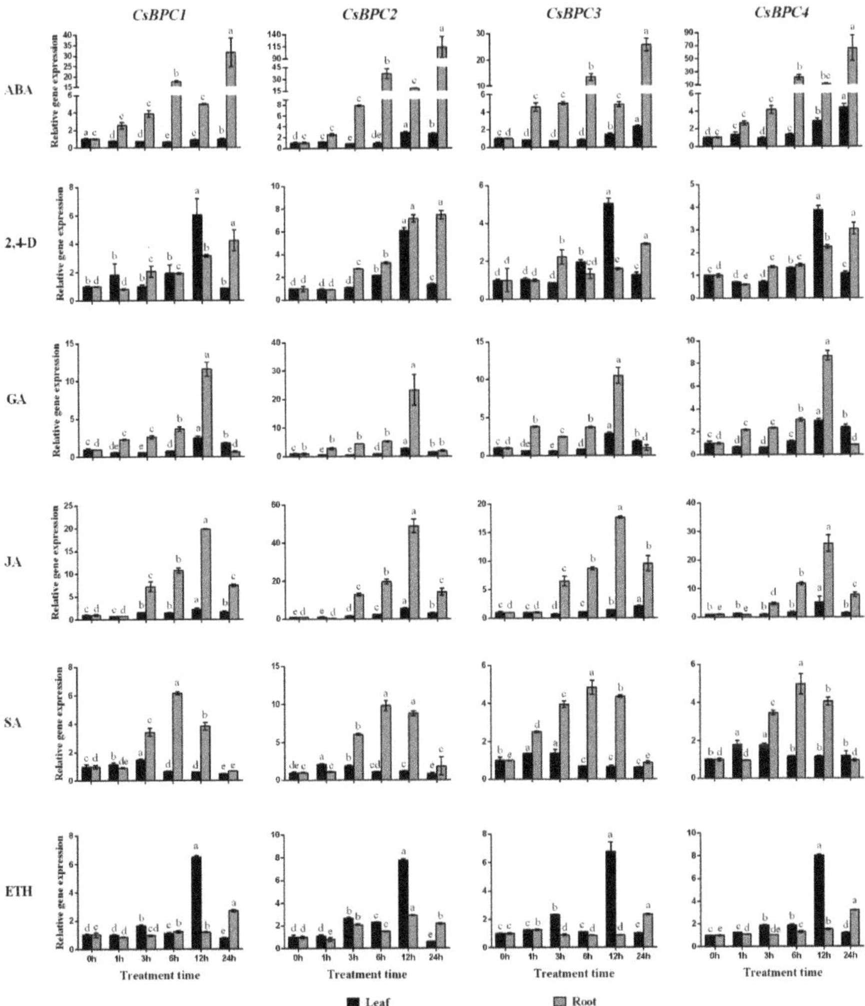

Figure 6. Expression patterns of CsBPC genes under different phytohormone treatments. Cucumber seedlings at three-leaf-stage were exposed to ABA, SA, JA, ETH, 2,4-D, and GA conditions, and qRT-PCR was performed to examine the transcript levels of *CsBPCs* in response to different phytohormones. The expression levels in non-stressed leaves and roots were set as 1. Values are means ± SD ($n = 3$). Different letters indicate significant differences ($p < 0.05$).

2.5. Germination Assays under Stress and ABA Treatments

Due to the fact that the above results showed that *CsBPC* expression was induced by various abiotic stresses and phytohormones, especially *CsBPC2*, whose expression level increased the most under the majority of treatments, a vector for constitutive overexpression of *CsBPC2* was constructed and transferred into tobacco plants to further study the cellular functions mediated by *CsBPC2* in response to environmental stresses and plant hormones. Genomic DNA of 10 independent T0 progeny transgenic tobacco lines (1, 2, 3, etc.) was extracted for PCR confirmation (Additional file 3). The results showed that *CsBPC2* was successfully integrated into the tobacco genome. Additionally, qRT-PCR indicated that *CsBPC2* was highly expressed in the transgenic lines, but no expression was observed in wild-type (WT) plants (Additional file 3). Thus, we chose two highly expressing lines, 5 and 6, for further experiments.

To further verify the effects mediated by *CsBPC2* in abiotic stresses and plant hormone responses, seeds of WT and *CsBPC2*-overexpression plants (OE) were sown on MS solid medium containing different concentrations of NaCl, PEG, and ABA. As shown in Figure 7A,D, the germination times and rates of the WT and OE seeds displayed no difference when sown on MS medium. However, in the presence of 100 mM NaCl, the germination of OE was slightly inhibited compared with the WT (Figure 7B,E). Under 200 mM NaCl, the germination rates and times of all genotypes were significantly inhibited, but the transgenic lines were inhibited much more than the WT (Figure 7C,F). The seeds of T1-5 and T1-6 began to germinate at day 14, six days later than the WT, and their germination rates were just 2.67% and 4.67%, respectively, while that of the WT was 70%. On the last day of the treatment, the germination rates of T1-5 and T1-6 were 48.0% and 56.67%, respectively, which were significantly lower than that of the WT (91.33%).

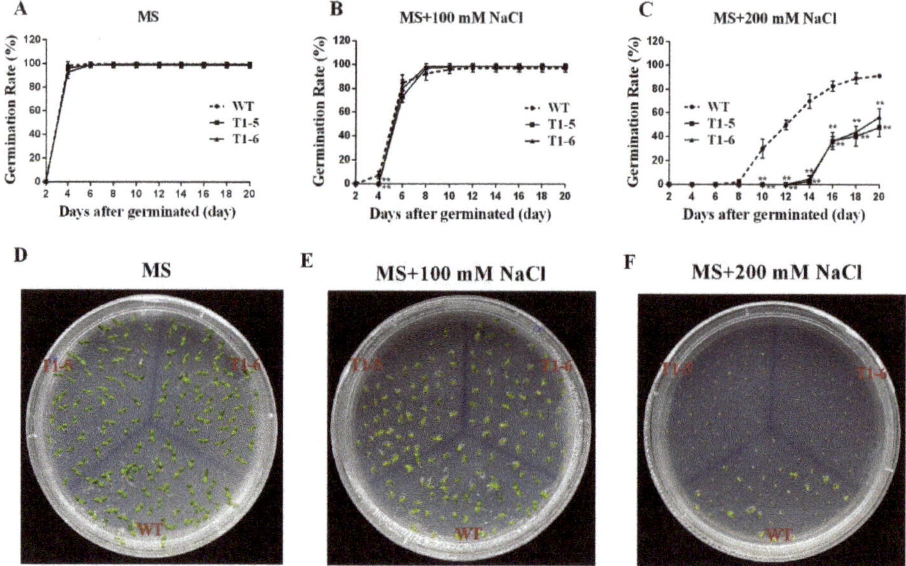

Figure 7. Effects of NaCl application on germination rates between wild-type (WT) and transgenic plants. Fifty seeds of WT and T1 transgenic plants were sown on sterile MS solid medium supplemented with different concentrations of NaCl (**A–C**). Values are means ± SD (n = 3). The photographs (**D,E**) were taken after being sown for 9 d, photograph **F** was taken after being sown for 13 d. * and ** Significant at $p < 0.05$ and $p < 0.01$ compared with WT, respectively.

In the presence of 10% PEG, although there was no difference in the final germination rates of the WT and OE seeds, the T1-5 and T1-6 seeds started to germinate at day eight, four days later than the WT, and their germination rates were 45.33% and 60.67%, respectively, while that of the WT was 97.33% (Figure 8A,C). Under 20% PEG, the WT seeds started to germinate at day six, and the germination rate was 18.0%, while T1-5 and T1-6 seeds started to germinate at day 10 and day eight, respectively, and their germination rates were 40.0% and 2.67%, respectively; however, the final germination rates of all seeds displayed no difference (Figure 8B,D).

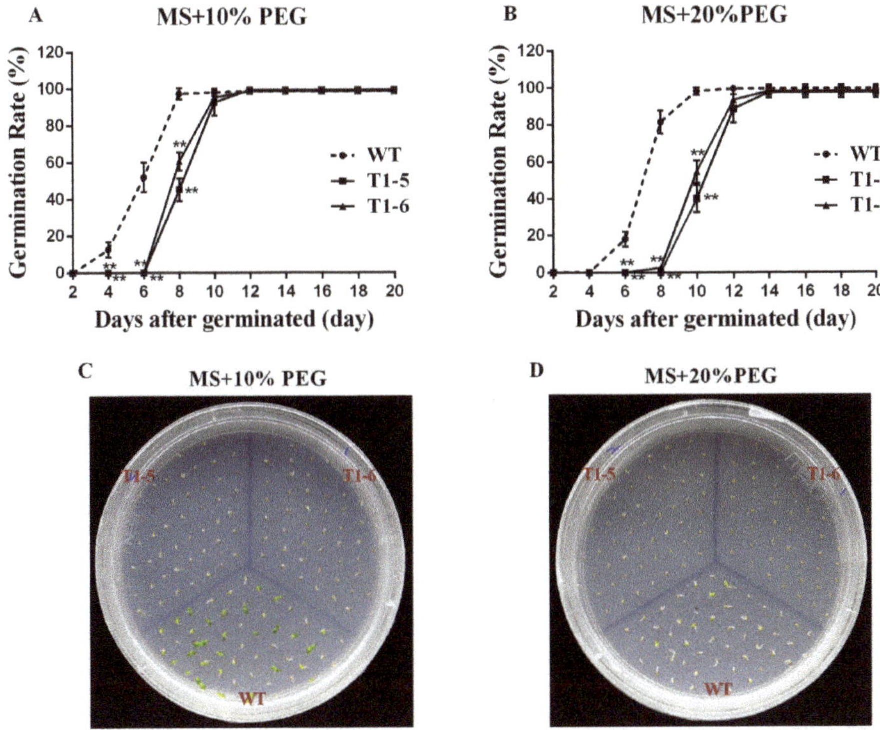

Figure 8. Effects of PEG application on germination rates between wild-type (WT) and transgenic plants. Fifty seeds of WT and T1 transgenic plants were sown on sterile MS solid medium supplemented with different concentrations of PEG (**A**,**B**). Values are means ± SD ($n = 3$). The photographs (**C**,**D**) were taken after being sown for 9 days. * and ** Significant at $p < 0.05$ and $p < 0.01$ compared with WT, respectively.

In the presence of 1 μM ABA, the OE plants displayed great inhibition of germination compared with the WT (Figure 9A,D). When the ABA concentration was increased to 2 μM, the inhibition was even greater (Figure 9B,E). The T1-5 and T1-6 seeds started to germinate at day 10, four days later than the WT, and their germination rates were 3.33% and 1.33%, respectively, while that of the WT was 90.67%. At the end of the treatment, the germination rates of T1-5 and T1-6 were 79.33% and 78.0%, respectively, which were significantly lower than that of the WT (98.67%). When the ABA concentration was increased to 4 μM, the final germination rates of T1-5 and T1-6 were just 30.0% and 16.0%, respectively, which were dramatically lower than that of WT (95.33%) (Figure 9C,F).

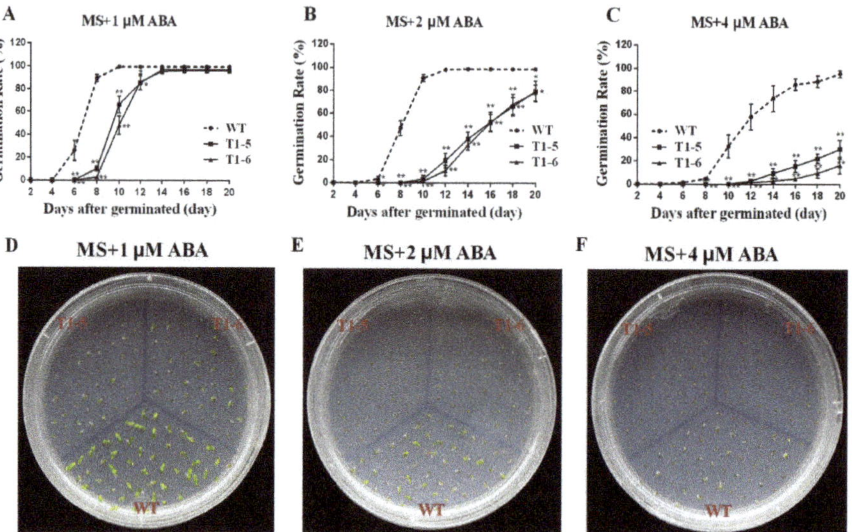

Figure 9. Effects of ABA application on germination rates between wild type (WT) and transgenic plants. Per 50 seeds of WT and T1 transgenic plants were sown on sterile MS solid medium supplemented with different concentrations of ABA (**A–C**). Values are means ± SD ($n = 3$). The photographs (**B–D**) were taken after being sown for 9 d. * and ** Significant at $p < 0.05$ and $p < 0.01$ compared with WT, respectively.

3. Discussion

Transcription factors play essential roles in regulating plant growth and development as well as responses to diverse abiotic stresses by activating or repressing related downstream genes [26]. With the completion of genome sequencing for more species, many major transcription factor families with large numbers of members have been identified and characterized in numerous plants, such as *bHLH* [27], *MYB* [28], and *WRKY* [29]. The functions of these transcription factors in growth, development, and stress responses have been studied extensively. However, no extensive studies have been done on the BPC family in recent years. Studies of BPC transcription factors have mainly focused on the regulation of plant growth and development [12,14,15,25], and almost no research has been reported on their regulation of stress resistance. Additionally, functional studies of the BPC family have primarily been done in the model plants *Arabidopsis* [12,14,15,25] and rice [22]; there is little information available about the BPC genes in other plant species.

In the present study, we identified and characterized four predicted BPC proteins in cucumber at the whole-genome level, and compared them with four watermelon, four melon, six cucurbita pepo, five grape, ten western balsam poplar, three common sunflower, seven *Arabidopsis*, four rice, and five tomato BPC proteins. Phylogenetic analysis classified these 52 BPC proteins into three groups (Figure 2), the same as in *Arabidopsis* [10]. Additionally, group I and group II contained similar members, whereas group III just contained two members, indicating that the BPC genes had diversified before various species diverged and that members in group III might be lost during the evolutionary process, however, the functions of these genes remain unclear. Here, the four CsBPC genes were classified into group I and group II, with each group containing two members. Gene structure (Figure 3A) and conserved motif (Figure 3B) analyses indicated that the CsBPC genes in the same group shared similar exon–intron and motif organizations, whereas members in different groups had differing exon–intron and motif compositions, which may suggest a closer evolutionary relationship between members of the same group and functional diversification among the members of different groups. Moreover, the two BPC groups showed a distinct difference in their N-terminal structural domains, which have been predicted to form zipper-like coiled-coil structures and may function as dimerization domains, protein–protein interactions domains, or nuclear and nucleolar localization signals [9,10,30]. Conversely, all four CsBPC members shared a highly conserved domain at their C-terminus that is important for DNA binding. The hallmark of this domain is the existence of five cysteine residues with defined positions and spacing [9,10,30]. These cysteines were previously believed to form a zinc finger-like structure for direct recognition of GAGA motifs [10]. However, the results of Theune et al.'s [31] research contradicted the suggestion of a zinc finger-like DNA-binding mechanism. Instead, they proposed that the conserved cysteines form inter- and intramolecular disulfide bonds to stabilize a parallel conformation of monomers, and that this conformation is required for neighboring GAGA motif recognition and binding.

Tissue-specific expression analysis is usually performed to predict the biological functions of genes in plant growth and development. Therefore, we investigated the expression of the four CsBPC genes by qRT-PCR in 12 different tissues. The results revealed that all four cucumber BPC genes were ubiquitously expressed in all the tissues tested, and that they shared the highest expression levels in seeds and the lowest levels in tendrils and stems (Figure 4), which may suggest a vital role for the *CsBPCs* in cucumber growth and development processes and, particularly, in seed development. Taking these results together, BPC transcription factors are essential for maintaining a wide range of normal growth and developmental processes, but further studies are needed to determine the functions of *CsBPCs* in regulating cucumber growth and development.

As there were few studies on *BPCs* in response to abiotic stresses and phytohormones, we did the expression analysis of *CsBPCs* under various stresses and hormones treatments. qRT-PCR showed that the transcriptional levels of all four CsBPC genes were induced by various abiotic stress treatments (NaCl, PEG, cold, and heat) (Figure 5) and hormone treatments (ABA, SA, JA, ETH, 2,4-D, and GA) (Figure 6), indicating that abiotic stresses and hormones might be activators of the *CsBPCs*.

To further research the cellular functions of the *CsBPCs*, especially in plant responses to hyperosmotic stress, transgenic tobacco plants constitutively overexpressing *CsBPC2*, whose expression level increased the most under the majority of abiotic stress and hormone treatments, were generated. No apparent differences in seed germination rates or seedling size were observed when seeds of the WT and transgenic lines were germinated under normal growth conditions (Figure 7). However, when seeds of these plants were germinated on MS solid medium supplemented with different concentrations of NaCl (Figure 7) and PEG (Figure 8), the germination rates and times of the transgenic lines were significantly inhibited compared with those of the WT. Additionally, when seeds were germinated on MS solid medium supplemented with different concentrations of ABA, the inhibition was even more severe (Figure 9). Collectively, these results indicated that *CsBPC2* plays a negative role in modulating seed germination under hyperosmotic conditions. *CsBPC2* may also function in ABA-inhibited seed germination and hypersensitivity to ABA-mediated responses. The altered stress responses exhibited by *CsBPC2* transgenic seeds may be caused by changes in phytohormones levels. Numerous studies have demonstrated that plant hormones are critical for regulation of seed dormancy and germination [32,33]. For example, the plant hormones gibberellin [34,35], cytokinin [36], brassinosteroids [37], and ethylene [38] are positive regulators of seed germination, whereas abscisic acid is a negative regulator of seed germination [39,40]. A previous study confirmed a role for *BPCs* in the regulation of the cytokinin content in the meristem [14]. In *Arabidopsis bpc1-1 bpc2 bpc3* triple mutants, the expression levels of both *ISOPENTENYLTRANSFERASE 7* (*IPT7*) and *ARABIDOPSIS RESPONSE REGULATOR 7* (*ARR7*), two genes involved in cytokinin biosynthesis and responsiveness, respectively [41–43], were upregulated, resulting in an increase in the cytokinin concentration in inflorescence meristems. Moreover, *Arabidopsis bpc1-1 bpc2 bpc4 bpc6* quadruple mutants showed reduced sensitivity to ethylene responses, the lack of an apical hook, and delayed senescence [12]. The hormone-mediated regulatory mechanisms may be complex or there may be other signal perception or transduction pathways involved in seed germination regulation under hyperosmotic conditions;

we need to research these issues in future studies. Additionally, further studies are needed to test the function of *CsBPC2* at the seedling stage in responses to abiotic stresses as genes can participate in different regulatory mechanisms at different growth stages and, thus, exhibit different functions. Understanding these matters will aid in understanding the *BPC*-mediated signaling cascades in plant resistance to stresses.

4. Materials and Methods

4.1. Identification of BPC Gene Family Members in Cucumber

Seven *Arabidopsis* BPC proteins sequences were downloaded from TAIR (http://www.arabidopsis.org) and the obtained sequences were used as queries to identify all GAGA-binding transcriptional activators in cucumber by searching against the Cucumber Genome Database (http://cucurbitgenomics.org/organism/2) with the BLASTP program and default E-value. All predicted cucumber BPC proteins were confirmed using the Pfam (http://pfam.xfam.org) [44] and SMART (http://smart.embl-heidelberg.de/) databases [45]. The amino acid lengths, chromosome locations, and strand directions of the BPC genes were also obtained from the databases. Physicochemical parameters, including the isoelectric point (pI) and molecular weight (kDa), were calculated using the compute pI/Mw tool in ExPASy (http://web.expasy.org/compute_pi/). Subcellular location prediction was performed using the software CELLO 2.5 (http://cello.life.nctu.edu.tw/) [46]. The information on chromosome locations of the four BPC genes were drawn using MapInspect tool.

4.2. Phylogenetic Gene Structure and Conserved Motif Analyses

The sequences of the four cucumber, four watermelon, four melon, six cucurbita pepo, and seven *Arabidopsis* BPC genes were downloaded from the Cucumber (Chinese Long) v2 Genome Database (http://cucurbitgenomics.org/organism/2), Watermelon (97103) v2 Genome Database (http://cucurbitgenomics.org/organism/21), Melon (DHL92) v3.6.1 Genome Database (http://cucurbitgenomics.org/organism/18), Cucurbita pepo (Zucchini) Genome Database (http://cucurbitgenomics.org/organism/14), and TAIR (https://www.arabidopsis.org/), respectively. The sequences of four rice, five grape, ten western balsam poplar, three common sunflower, and five tomato BPC proteins were obtained from UniProt (https://www.uniprot.org/). All full-length protein sequences of cucumber, *Arabidopsis*, tomato, watermelon, melon, cucurbita pepo, grape, western balsam poplar, common sunflower, and rice were aligned using the Clustal W program with default parameters, and then an unrooted neighbor-joining phylogenetic tree was constructed using the MEGA7 software with 100 bootstrap replicates [47]. The exon–intron compositions of the cucumber BPC genes were predicted by comparing the CDS sequences with their corresponding genomic sequences using the online program GSDS 2.0 (http://gsds.cbi.pku.edu.cn/) [48]. The conserved motifs of the cucumber BPC proteins were identified using the online tool MEME (http://meme-suite.org/tools/meme) [49] with the following parameters: any number of repetitions site distribution, 5 motifs found, and minimum and maximum motif widths—6 and 50, respectively.

4.3. Plant Material, Growth Conditions, and Treatments

Cucumber (*Cucumis sativus* L. cv. 'changchunmici') seeds (saved by our laboratory) were germinated on moist gauze in an incubator at 28 °C under dark conditions. The germinated seeds were sown in plastic plugs (32 holes) filled with nursery substrate (peat: vermiculite: perlite = 2:1:1) in a climate chamber at The Institute of Vegetables and Flowers, Chinese Academy of Agricultural Sciences. When the seedlings were at the two-leaf-stage, they were transplanted into a greenhouse. When their fruits became ripe, samples of various tissues were collected, including fresh roots (R) and stems (S), top young leaves (YL), middle mature leaves (ML), basal old leaves (OL), blooming male flowers (MF), blooming female flowers with ovaries removed (FF), ovaries (O), tendrils (T), around 10-day-old fruits (F), and seeds (FS) and pulp (FP) gathered from the mature fruits. All samples were immediately frozen in liquid nitrogen and stored at −80 °C for RNA extraction and tissue expression analysis.

For different stress and phytohormone treatments, batches of five seedlings at the one-leaf-stage were transferred to a 5 L (33 cm × 25 cm × 11 cm) plastic tank filled with an aerated complete nutrient solution (pH 6.0–6.5) containing: $Ca(NO_3)_2·4H_2O$ 4 mM, KNO_3 6 mM, $MgSO_4·7H_2O$ 2 mM, $NH_4H_2PO_4$ 1 mM, EDTA-FeNa 80 µM, H_3BO_3 46.3 µM, $MnSO_4·H_2O$ 9.5 µM, $ZnSO_4·7H_2O$ 0.8 µM, $CuSO_4·5H_2O$ 0.3 µM, and $(NH_4)_6Mo_7O_2·4H_2O$ 0.02 µM. The experiment was carried out under normal conditions (28 °C/14 h light, 18 °C/10 h dark) in a climate chamber. When the cucumber seedlings were at the three-leaf-stage, 100 mM NaCl, 10% polyethylene glycol (PEG) 6000, 100 µmol ABA, 100 µmol salicylic acid (SA), 100 µmol jasmonic acid (JA), 50 µmol ethephon (ETH), 50 µmol 2,4-dichlorophenoxyacetic acid (2,4-D), and 50 µmol gibberellin (GA) were added to the nutrient solution. For cold and heat treatments, cucumber seedlings were cultivated at 5 °C and 38 °C, respectively. Samples of cucumber seedlings were taken after 0, 1, 3, 6, 12, and 24 h of the different treatments for gene expression analysis.

4.4. Vector Construction and Tobacco Transformation

Two primers were designed to amplify the *CsBPC2* gene based on its CDS sequence (Additional file 2. S2). Then, the *CsBPC2* CDS was inserted into the *Bam*HI/*Pml*I restriction sites downstream of the 35S promoter of the vector 1305 (35S:*CsBPC2*) using In-Fusion technology. The construct was introduced into the *Agrobacterium tumefaciens* strain EHA105 for tobacco (*Nicotiana tabacum* cv. NC89) transformation using the leaf disc method [50]. Tobacco genomic DNA from young leaves was extracted for PCR confirmation using primers for *Hyg* (Additional file 2. S3) and *CsBPC2* CDS amplification to determine whether *CsBPC2* was integrated into the tobacco genome. Additionally, qRT-PCR was performed to further test the expression of *CsBPC2* in the transgenic tobacco plants using primers for *CsBPC2*, which is unique to cucumber (Additional file 2. S4).

4.5. RNA Extraction and qRT-PCR Analysis

Total RNA was extracted using an RNA prep pure Plant Kit (TANGEN) and first-strand cDNA was synthesized using a PrimeScript™ RT reagent Kit with gDNA Eraser (Perfect Real Time) (TaKaRa) according to the manufacturer's instructions. Quantitative RT-PCR was performed following the instructions of the SYBR® Premix Ex Taq™ Kit (TaKaRa) on a Mx3000P real-time PCR instrument (Agilent). The experiments were performed with three biological replicates, each with three plants. Relative gene expression was calculated using the $2^{-\triangle\triangle Ct}$ method [51]. The *CsBPC* specific primers used for expression pattern analysis in different tissues and under abiotic stress and hormone treatments are shown in Additional file 2. S1.

4.6. Seed Germination of Transgenic Tobacco Plants under Hyperosmotic Stress Conditions

Seeds of the wild-type (WT) and T1 progeny obtained from transgenic tobacco plants were germinated on sterile MS solid medium containing 30 g/L sucrose. To test the effects of hyperosmotic stress or phytohormones on germination, 100 and 200 mM NaCl, 10% and 20% PEG-6000, and 1, 2, and 4 µM ABA were added to the medium. Germination rate assays were carried out on three replicates of 50 seeds. The seeds were geminated in an incubator (26 °C/14 h light, 10 h dark) for 20 days.

4.7. Statistical Analyses

Values presented are means ± standard deviation (SD) of three replicates. Statistical analyses were carried out by an analysis of variance (ANOVA) using SAS (SAS Institute, Cary, NC, USA) software.

5. Conclusions

The ubiquitous expression of the four *CsBPCs* in various tissues perhaps to some extent implied they have crucial roles in regulating cucumber growth and development. Moreover, analysis of *CsBPC* expression under various abiotic stress (NaCl, PEG, cold, and heat) and hormone (ABA, SA, JA, ETH, 2,4-D, and GA) treatments revealed that the *CsBPCs* respond to phytohormones and abiotic stresses. Additionally, tobacco plants overexpressing *CsBPC2* showed an inhibited germination phenotype when treated with NaCl, PEG, and ABA, compared with the wild type. This further confirmed the regulatory function of the *CsBPCs* in stress and hormone responses. The present results reveal the potential roles of *CsBPCs* in cucumber development and abiotic stresses responses, and provide a basis for further functional studies of the BPC genes in cucumber and other plant species.

Supplementary Materials: Supplementary materials can be found at http://www.mdpi.com/1422-0067/20/20/5048/s1. Additional file 1: Sequences of CsBPC genes. A: The coding sequences of *CsBPCs*. B: The protein sequences of *CsBPCs*. C: The 2 kb genomic DNA sequences upstream of the initiation codon. Additional file 2: Sequences of the primers used in this study. Additional file 3: Confirmation of transgenic tobacco plants by PCR and qRT-PCR methods.

Author Contributions: X.Y., Y.L., and S.L. conceived and designed the research. S.L. performed the research. L.G. and C.H. made important comments on design of the trial, the article writing, and the revisions. L.M., B.H., Y.Y., and J.W. analyzed the data. S.L. wrote the first draft of the manuscript. X.Y. and Y.L. improved the first draft of the manuscript. All of the authors read and approved the final manuscript.

Funding: This work was financially supported by funds from the National Natural Sciences Foundations of China (No.31972480), the National Key Research and Development Plan of China (2018YFD0201207), the Earmarked fund for Modern Agro-industry Technology Research System in China (CARS-25-C-01), the Science and Technology Innovation Program of the Chinese Academy of Agricultural Sciences (CAASASTIP-IVFCAAS), and the Key Laboratory of Horticultural Crop Biology and Germplasm Innovation, Ministry of Agriculture, China. The funders played roles in the design of the study, the collection and analysis of data, and in writing the manuscript.

Conflicts of Interest: The authors declare no conflicts of interest.

Abbreviations

wt	wild-type
qRT-PCR	quantitative real-time polymerase chain reaction
PEG	polyethylene glycol
MS	Murashige and Skoog
ABA	abscisic acid

References

1. Tanurdzic, M.; Banks, J.A. Sex-determining mechanisms in land plants. *Plant. Cell* **2004**, *16*, S61–S71. [CrossRef]

2. Zhang, B.C.; Tolstikov, V.; Turnbull, C.; Hicks, L.M.; Fiehn, O. Divergent metabolome and proteome suggest functional independence of dual phloem transport systems in cucurbits. *Proc. Natl. Acad. Sci. USA* **2010**, *107*, 13532–13537. [CrossRef]

3. Lee, S.H.; Singh, A.P.; Chung, G.C. Rapid accumulation of hydrogen peroxide in cucumber roots due to exposure to low temperature appears to mediate decreases in water transport. *J. Exp. Bot* **2004**, *55*, 1733–1741. [CrossRef]

4. Klobus, G.; Janicka-Russak, M. Modulation by cytosolic components of proton pump activities in plasma membrane and tonoplast from *Cucumis sativus* roots during salt stress. *Physiol Plant.* **2004**, *121*, 84–92. [CrossRef]

5. Zhong, M.; Yuan, Y.H.; Shu, S.; Sun, J.; Guo, S.R.; Yuan, R.N.; Tang, Y.Y. Effects of exogenous putrescine on glycolysis and Krebs cycle metabolism in cucumber leaves subjected to salt stress. *Plant. Growth Regul* **2016**, *79*, 319–330. [CrossRef]

6. Janoudi, A.K.; Widders, I.E.; Flore, J.A. Water deficits and environmental factors affect photosynthesis in leaves of cucumber (*Cucumis sativus*). *J. Am. Soc. Hortic Sci.* **1993**, *118*, 366–370. [CrossRef]

7. Klein, E.; Katan, J.; Gamliel, A. Soil suppressiveness by organic amendment to Fusarium disease in cucumber: effect on pathogen and host. *Phytoparasitica* **2016**, *44*, 239–249. [CrossRef]

8. Sangwan, I.; O'Brian, M.R. Identification of a soybean protein that interacts with GAGA element dinucleotide repeat DNA. *Plant. Physiol.* **2002**, *129*, 1788–1794. [CrossRef]

9. Santi, L.; Wang, Y.M.; Stile, M.R.; Berendzen, K.; Wanke, D.; Roig, C.; Pozzi, C.; Muller, K.; Muller, J.; Rohde, W.; et al. The GA octodinucleotide repeat binding factor BBR participates in the transcriptional regulation of the homeobox gene Bkn3. *Plant. J.* **2003**, *34*, 813–826. [CrossRef]

10. Meister, R.J.; Williams, L.A.; Monfared, M.M.; Gallagher, T.L.; Kraft, E.A.; Nelson, C.G.; Gasser, C.S. Definition and interactions of a positive regulatory element of the Arabidopsis *INNER NOOUTER* promoter. *Plant. J.* **2004**, *37*, 426–438. [CrossRef]

11. Kooiker, M.; Airoldi, C.A.; Losa, A.; Manzotti, P.S.; Finzi, L.; Kater, M.M.; Colombo, L. BASIC PENTACYSTEINE1, a GA binding protein that induces conformational changes in the regulatory region of the homeotic Arabidopsis gene SEEDSTICK. *Plant. Cell* **2005**, *17*, 722–729. [CrossRef]

12. Monfared, M.M.; Simon, M.K.; Meister, R.J.; Roig-Villanova, I.; Kooiker, M.; Colombo, L.; Fletcher, J.C.; Gasser, C.S. Overlapping and antagonistic activities of BASIC PENTACYSTEINE genes affect a range of developmental processes in Arabidopsis. *Plant. J.* **2011**, *66*, 1020–1031. [CrossRef]

13. Simonini, S.; Roig-Villanova, I.; Gregis, V.; Colombo, B.; Colombo, L.; Kater, M.M. Basic pentacysteine proteins mediate MADS domain complex binding to the DNA for tissue-specific expression of target genes in Arabidopsis. *Plant. Cell* **2012**, *24*, 4163–4172. [CrossRef]

14. Simonini, S.; Kater, M.M. Class I BASIC PENTACYSTEINE factors regulate HOMEOBOX genes involved in meristem size maintenance. *J. Exp. Bot.* **2014**, *65*, 1455–1465. [CrossRef]

15. Hecker, A.; Brand, L.H.; Peter, S.; Simoncello, N.; Kilian, J.; Harter, K.; Gaudin, V.; Wanke, D. The Arabidopsis GAGA-Binding Factor BASIC PENTACYSTEINE6 Recruits the POLYCOMB-REPRESSIVE COMPLEX1 Component LIKE HETEROCHROMATIN PROTEIN1 to GAGA DNA Motifs. *Plant. Physiol* **2015**, *168*, 1013–1024. [CrossRef]

16. Anusak, P.; Ditta, G.S.; Beth, S.; Liljegren, S.J.; Elvira, B.; Ellen, W.; Yanofsky, M.F. Assessing the redundancy of MADS-box genes during carpel and ovule development. *Nature* **2003**, *424*, 85–88.

17. Rebecca, F.; Anusak, P.; Raffaella, B.; Maarten, K.; Lorenzo, B.; Gary, D.; Yanofsky, M.F.; Kater, M.M.; Lucia, C. MADS-box protein complexes control carpel and ovule development in Arabidopsis. *Plant. Cell* **2003**, *15*, 2603–2611.

18. Berger, N.; Dubreucq, B.; Roudier, F.; Dubos, C.; Lepiniec, L. Transcriptional regulation of Arabidopsis LEAFY COTYLEDON2 involves RLE, a cis-element that regulates trimethylation of histone H3 at lysine-27. *Plant. Cell* **2011**, *23*, 4065–4078. [CrossRef]

19. Stone, S.L.; Kwong, L.W.; Yee, K.M.; Pelletier, J.; Lepiniec, L.; Fischer, R.L.; Goldberg, R.B.; Harada, J.J. LEAFY COTYLEDON2 encodes a B3 domain transcription factor that induces embryo development. *Proc. Natl. Acad. Sci. USA* **2001**, *98*, 11806–11811. [CrossRef]

20. Kroj, T.; Savino, G.; Valon, C.; Giraudat, J.; Parcy, F. Regulation of storage protein gene expression in Arabidopsis. *Development* **2003**, *130*, 6065–6073. [CrossRef]

21. Mu, Y.; Liu, Y.M.; Bai, L.Q.; Li, S.Z.; He, C.X.; Yan, Y.; Yu, X.C.; Li, Y.S. Cucumber *CsBPCs* regulate the expression of *CsABI3* during seed germination. *Front. Plant. Sci.* **2017**, *8*, 459. [CrossRef]

22. Gong, R.; Cao, H.; Zhang, J.; Xie, K.; Wang, D.; Yu, S. Divergent functions of the GAGA-binding transcription factor family in rice. *Plant. J.* **2018**, *94*, 32–47. [CrossRef]

23. Peng, L.T.; Shi, Z.Y.; Li, L.; Shen, G.Z.; Zhang, J.L. Overexpression of transcription factor *OsLFL1* delays flowering time in *Oryza sativa*. *J. Plant. Physiol* **2008**, *165*, 876–885. [CrossRef]

24. Peng, L.T.; Shi, Z.Y.; Li, L.; Shen, G.Z.; Zhang, J.L. Ectopic expression of *OsLFL1* in rice represses Ehd1 by binding on its promoter. *Biochem. Biophys. Res. Commun.* **2007**, *360*, 251–256. [CrossRef]

25. Shanks, C.M.; Hecker, A.; Cheng, C.Y.; Brand, L.; Collani, S.; Schmid, M.; Schaller, G.E.; Wanke, D.; Harter, K.; Kieber, J.J. Role of BASIC PENTACYSTEINE transcription factors in a subset of cytokinin signaling responses. *Plant. J.* **2018**, *95*, 458–473. [CrossRef]

26. Wu, S.; Gallagher, K.L. Transcription factors on the move. *Curr. Opin. Plant. Biol.* **2012**, *15*, 645–651. [CrossRef]

27. Mao, K.; Dong, Q.; Li, C.; Liu, C.; Ma, F. Genome wide identification and characterization of apple bHLH transcription factors and expression analysis in response to drought and salt stress. *Front. Plant. Sci.* **2017**, *8*, 480. [CrossRef]

28. Wang, Y.; Zhan, D.F.; Li, H.L.; Guo, D.; Zhu, J.H.; Peng, S.Q. Transcriptome-wide identification and characterization of MYB transcription factor genes in the laticifer cells of *Hevea brasiliensis*. *Front. Plant. Sci.* **2017**, *8*, 1974. [CrossRef]

29. Xie, T.; Chen, C.; Li, C.; Liu, J.; Liu, C.; He, Y. Genome-wide investigation of WRKY gene family in pineapple: evolution and expression profiles during development and stress. *BMC Genomics* **2018**, *19*, 490. [CrossRef]

30. Wanke, D.; Hohenstatt, M.L.; Dynowski, M.; Bloss, U.; Hecker, A.; Elgass, K.; Hummel, S.; Hahn, A.; Caesar, K.; Schleifenbaum, F.; et al. Alanine zipper-like coiled-coil domains are necessary for homotypic dimerization of plant GAGA-factors in the nucleus and nucleolus. *PLoS ONE* **2011**, *6*, e16070. [CrossRef]

31. Theune, M.L.; Hummel, S.; Jaspert, N.; Lafos, M.; Wanke, D. Dimerization of the BASIC PENTACYSTEINE domain in plant GAGA-factors is mediated by disulfide bonds and required for DNA-binding. *J. Adv. Plant. Biol.* **2017**, *1*, 26–39. [CrossRef]

32. Kucera, B.; Cohn, M.A.; Leubner-Metzger, G. Plant hormone interactions during seed dormancy release and germination. *Seed Sci. Res.* **2005**, *15*, 281–307. [CrossRef]

33. Warpeha, K.M.; Montgomery, B.L. Light and hormone interactions in the seed-to-seedling transition. *Environ. Exp. Bot.* **2016**, *121*, 56–65. [CrossRef]

34. Gupta, R.; Chakrabarty, S.K. Gibberellic acid in plant: still a mystery unresolved. *Plant. Signal. Behav.* **2013**, *8*. [CrossRef]

35. Hedden, P.; Thomas, S.G. Gibberellin biosynthesis and its regulation. *Biochem. J.* **2012**, *444*, 11–25. [CrossRef]

36. Perilli, S.; Moubayidin, L.; Sabatini, S. The molecular basis of cytokinin function. *Curr. Opin. Plant. Biol.* **2010**, *13*, 21–26. [CrossRef]

37. Clouse, S.D. Brassinosteroids. *The Arabidopsis Book* **2011**, *9*, e0151. [CrossRef]

38. Corbineau, F.; Xia, Q.; Bailly, C.; El-Maarouf-Bouteau, H. Ethylene, a key factor in the regulation of seed dormancy. *Front. Plant. Sci.* **2014**, *5*, 539. [CrossRef]

39. Sakata, Y.; Komatsu, K.; Takezawa, D. ABA as a universal plant hormone. *Progress Botany* **2014**, *75*, 57–96.

40. Cutler, S.R.; Rodriguez, P.L.; Finkelstein, R.R.; Abrams, S.R. Abscisic acid: Emergence of a core signaling network. *Annu. Rev. Plant. Biol.* **2010**, *61*, 651–679. [CrossRef]

41. Kakimoto, T. Identification of plant cytokinin biosynthetic enzymes as dimethylallyl diphosphate: ATP/ADP isopentenyltransferases. *Plant. Cell Physiol.* **2001**, *42*, 677–685. [CrossRef]

42. Buechel, S.; Leibfried, A.; To, J.P.C.; Zhao, Z.; Andersen, S.U.; Kieber, J.J.; Lohmann, J.U. Role of A-type ARABIDOPSIS RESPONSE REGULATORS in meristem maintenance and regeneration. *Eur. J. Cell Biol.* **2010**, *89*, 279–284. [CrossRef]

43. Zhong, Z.; Andersen, S.U.; Karin, L.; Karel, D.; Andrej, M.; Schultheiss, S.J.; Lohmann, J.U. Hormonal control of the shoot stem-cell niche. *Nature* **2010**, *465*, 1089–1092.

44. Bateman, A.; Birney, E.; Durbin, R.; Eddy, S.R.; Howe, K.L.; Sonnhammer, E.L. The Pfam protein families database. *Nucleic Acids Res.* **2000**, *28*, 263–266. [CrossRef]

45. Letunic, I.; Bork, P. 20 years of the SMART protein domain annotation resource. *Nucleic Acids Res.* **2018**, *46*, D493–D496. [CrossRef]

46. Yu, C.S.; Chen, Y.C.; Lu, C.H.; Hwang, J.K. Prediction of protein subcellular localization. *Proteins* **2006**, *64*, 643–651. [CrossRef]

47. Wang, P.; Wang, S.; Chen, Y.; Xu, X.; Guang, X.; Zhang, Y. Genome-wide Analysis of the MADS-Box Gene Family in Watermelon. *Comput. Biol. Chem.* **2019**, *80*, 341–350. [CrossRef]

48. Hu, B.; Jin, J.; Guo, A.Y.; Zhang, H.; Luo, J.; Gao, G. GSDS 2.0: an upgraded gene feature visualization server. *Bioinformatics* **2015**, *31*, 1296–1297. [CrossRef]

49. Bailey, T.L.; Boden, M.; Buske, F.A.; Frith, M.; Grant, C.E.; Clementi, L.; Ren, J.; Li, W.W.; Noble, W.S. MEME SUITE: tools for motif discovery and searching. *Nucleic Acids Res.* **2009**, *37*, W202–W208. [CrossRef]

50. Horsch, R.B.; Fry, J.E.; Hoffmann, N.L.; Eichholtz, D.; Rogers, S.G.; Fraley, R.T. A simple and general method for transferring genes into plants. *Science* **1985**, *227*, 1229–1231.

51. Livak, K.J.; Schmittgen, T.D. Analysis of relative gene expression data using real-time quantitative PCR and the $2^{-\Delta\Delta CT}$ Method. *Methods* **2001**, *25*, 402–408. [CrossRef]

 International Journal of
Molecular Sciences

Article

Genome-Wide Identification, Classification, and Expression Analysis of the *Hsf* Gene Family in Carnation (*Dianthus caryophyllus*)

Wei Li [1,†], Xue-Li Wan [1,†], Jia-Yu Yu [1], Kui-Ling Wang [1,*] and Jin Zhang [2,3,*]

[1]   College of Landscape Architecture and Forestry, Qingdao Agricultural University, Qingdao 266000, China; lwcsu_caf@163.com (W.L.); wanxueli86@163.com (X.-L.W.); yjy13583204761@163.com (J.-Y.Y.)
[2]   State Key Laboratory of Tree Genetics and Breeding, Key Laboratory of Tree Breeding and Cultivation of the National Forestry and Grassland Administration, Research Institute of Forestry, Chinese Academy of Forestry, Beijing 100091, China
[3]   Biosciences Division, Oak Ridge National Laboratory, Oak Ridge, TN 37831, USA
*   Correspondence: wkl6310@163.com (K.-L.W.); zhangj1@ornl.gov (J.Z.)
†   These authors contributed equally to this work.

Received: 21 September 2019; Accepted: 18 October 2019; Published: 22 October 2019

Abstract: Heat shock transcription factors (Hsfs) are a class of important transcription factors (TFs) which play crucial roles in the protection of plants from damages caused by various abiotic stresses. The present study aimed to characterize the *Hsf* genes in carnation (*Dianthus caryophyllus*), which is one of the four largest cut flowers worldwide. In this study, a total of 17 non-redundant *Hsf* genes were identified from the *D. caryophyllus* genome. Specifically, the gene structure and motifs of each *DcaHsf* were comprehensively analyzed. Phylogenetic analysis of the *DcaHsf* family distinctly separated nine class A, seven class B, and one class C *Hsf* genes. Additionally, promoter analysis indicated that the *DcaHsf* promoters included various *cis*-acting elements that were related to stress, hormones, as well as development processes. In addition, *cis*-elements, such as STRE, MYB, and ABRE binding sites, were identified in the promoters of most *DcaHsf* genes. According to qRT-PCR data, the expression of *DcaHsfs* varied in eight tissues and six flowering stages and among different *DcaHsfs*, even in the same class. Moreover, *DcaHsf-A1, A2a, A9a, B2a, B3a* revealed their putative involvement in the early flowering stages. The time-course expression profile of *DcaHsf* during stress responses illustrated that all the *DcaHsfs* were heat- and drought-responsive, and almost all *DcaHsfs* were down-regulated by cold, salt, and abscisic acid (ABA) stress. Meanwhile, *DcaHsf-A3, A7, A9a, A9b, B3a* were primarily up-regulated at an early stage in response to salicylic acid (SA). This study provides an overview of the *Hsf* gene family in *D. caryophyllus* and a basis for the breeding of stress-resistant carnation.

Keywords: heat shock factor; *Dianthus caryophyllus*; abiotic stresses; gene expression

1. Introduction

Plant growth and production are affected by abiotic stresses such as heat, cold, drought, and salinity [1–3]. Unlike animals, plants are sessile organisms. Consequently, to cope with environmental stresses, plants have evolved a series of defense or signaling mechanisms. Furthermore, each process involves different types of transcription factors (TFs). These include heat shock transcription factors (Hsfs), such as WRKY, MYB, AP2/ERF, and NAC, which regulate the expression of thousands of genes under various stress conditions [4–6]. In plants, the Hsf family is one of the most important TF families in plants involved in resistance to heat [7] and other abiotic stresses or chemical stressors, such as abscisic acid (ABA) and salicylic acid (SA) [8,9]. Hsfs regulate the expression of *Heat shock proteins*

(*Hsps*) as well as other stress-responsive proteins, such as reactive oxygen species (ROS)-scavenging enzymes [ascorbate peroxidase (APX) and catalase (CAT)] [4]. Besides their roles in stress responses, Hsfs are also involved in plant growth and development [10–12].

Similar to many other TFs, *Hsfs* are a part of an evolutionarily conserved gene family. *Hsf* genes are composed of several structurally and functionally conserved domains, including DNA-binding domains (DBD), N-terminal adjacent bipartite oligomerization domains (HR-A/B), nuclear localization signals (NLS), nuclear export signals (NES), C-terminal activator peptide proteins (AHA), and repressor domains (RD) [13]. Among these conserved domains, DBD is characterized by a central helix–turn–helix motif and is responsible for binding to the heat shock elements (HSEs) of the target genes [7]. Notably, the HSEs are palindromic binding motifs (5′-AGAAnnTTCT-3′) conserved in the promoters of heat stress-inducible genes [7,14]. According to the flexible linker of variable lengths (about 15–80 amino acids) and HR-A/B regions, plant *Hsfs* can be divided into at least three types, i.e., class A (subclasses A1, A2, A3, A4, A5, A6, A7, A8, and A9), class B (subclasses B1, B2, B3, and B4), and class C (subclasses C1 and C2) [15–17].

The size of the *Hsf* gene family varies significantly in different plant species. For instance, there are 22 *Hsf* members in the model plant *Arabidopsis thaliana* [18], 25 members in *Oryza sativa* [18], 16 members in *Medicago truncatula* [19], 26 members in *Glycine max* [20], 25 members in *Zea mays* [21], 25 members in *Malus domestica* [22], 21 members in *Cucumis sativus* [23], 28 members in *Populus trichocarpa* [24], 40 members in *Gossypium hirsutum* [25], and 56 members in members in *Triticum aestivum* [26]. To date, the largest *Hsf* gene family has been identified in *Brassica napus*, with 64 *Hsf*s [27].

Carnation (*Dianthus caryophyllus* L.) is a major floricultural crop and one of the four largest cut flowers [28,29]. Until now, more than 300 *Dianthus* species have been identified worldwide. Carnations are cultivated widely for their attractive characteristics such as flower color, flower size, fragrance, and flower longevity. However, the vegetative and reproductive growth of carnations are severely impaired in heat stress conditions, resulting in flower wilting and quality decline [30]. The completion of the draft genome sequence of *D. caryophyllus* L. has greatly facilitated the identification of *Hsfs* at the whole-genome level and it is extremely important to study the heat-resistant mechanism of carnation [31]. To our knowledge, there are no reports on the identification and functional analysis of carnation *Hsfs* to date. In this study, we aimed to comprehensively study the structural and expression profiles of the *Hsf* gene family in *D. caryophyllus*. A total of 17 putative genes were identified and characterized as members of the *Hsf* gene family from *D. caryophyllus*. Additionally, we performed bioinformatic analyses of phylogenetic relationships, conserved domains, motifs, and other. Furthermore, the expression level of these genes in various tissues and in response to abiotic stresses were compared. Our results will be a reference and provide valuable information for the functional analysis of the *Hsf* genes in *D. caryophyllus*.

2. Results

2.1. Identification of DcaHsfs in Carnation

The amino acid sequences of putative Hsf proteins were examined using the conserved Hsf domain (PF00447) from the carnation database (DB, http://carnation.kazusa.or.jp). Additionally, searches using the BLASTP program resulted in the identification of putative *Hsf* gene candidates. In total, 17 proteins were retrieved as DcaHsfs in *D. caryophyllus*.

The physical and chemical properties of the 17 DcaHsfs were analyzed (Table 1). The DcaHsfs ranged from 133 amino acids (aa; DcaHsf-C1, incomplete) to 495 aa (DcaHsf-A5) in length. The predicted isoelectric points (pI) varied from 4.74 (DcaHsf-A1) to 8.88 (DcaHsf-B3a and DcaHsf-B3b), and the molecular weight (MW) varied from 15.89 kDa (DcaHsf-C1) to 54.52 kDa (DcaHsf-A5). The instability index, i.e., the stability of the protein in a test tube, indicated that all DcaHsfs were unstable, except for DcaHsf-A9b and DcaHsf-B1. The GRAVY value reflects the hydropathicity of a protein; the low GRAVY values (<0) of DcaHsfs suggest that all DcaHsfs are hydrophilic. The total number of negatively charged residues (Asp + Glu, n.c.r.) and the total number of positively charged residues (p.c.r.) of class A were all greater than those of class B and C. These differences might be caused by differences in the amino acid composition of the non-conserved region. We determined the scaffold locations of *DcaHsfs* on the basis of the information from the Carnation genomic database. We mapped 17 *DcaHsfs* to 17 scaffolds, and these genes were distributed evenly in the Carnation genome (Figure 1).

Table 1. Summary information of *Dianthus caryophyllus* heat shock transcription factors (DcaHsfs) in carnation. Notes: I.I., instability index; Stability, (U: unstable protein, S: stable protein); A.I., aliphatic index; n.c.r., total number of negatively charged residues (Asp + Glu); p.c.r., total number of positively charged residues (Arg + Lys); GRAVY, grand average of hydropathicity; pI, isoelectric point; MW, molecular weight.

Protein Name	Gene ID	Subfamily	Size	I.I.	Stability	A.I.	n.c.r. (%)	p.c.r.(%)	GRAVY	pI	MW (kDa)
DcaHsf-A1	Dca57201.1	A1	488	55.97	U	66.72	69	47	−0.690	4.74	54.49
DcaHsf-A2a	Dca14360.1	A2	380	59.24	U	83.55	59	45	−0.503	5.12	42.96
DcaHsf-A2b	Dca52568.1	A2	359	58.97	U	80.03	57	43	−0.548	5.05	40.55
DcaHsf-A3	Dca41810.1	A3	244	64.58	U	67.33	67	50	−0.566	4.98	51.75
DcaHsf-A4	Dca23163.1	A4	390	47.75	U	71.95	59	47	−0.859	5.7	44.99
DcaHsf-A5	Dca19769.1	A5	489	48.53	U	72.17	67	52	−0.745	5.45	54.52
DcaHsf-A7	Dca4574.1	A7	425	48.11	U	67.36	63	61	−0.771	6.72	48.91
DcaHsf-A9a	Dca9629.1	A9	401	47.92	U	69.73	67	46	−0.691	5	46.12
DcaHsf-A9b	Dca41703.1	A9	331	36.63	S	69.43	45	44	−0.805	6.23	38.21
DcaHsf-B1	Dca60410.1	B1	276	34.47	S	68.77	38	39	−0.804	7.61	31.0
DcaHsf-B2a	Dca22545.1	B2	337	54.72	U	61.64	35	29	−0.721	6	36.48
DcaHsf-B2b	Dca48996.1	B2	337	60.89	U	60.98	39	32	−0.709	6	36.48
DcaHsf-B2c	Dca54105.1	B2	318	61.35	U	68.57	33	32	−0.592	5.91	33.80
DcaHsf-B3a	Dca44175.1	B3	244	48.80	U	70.33	31	37	−0.763	8.88	28.38
DcaHsf-B3b	Dca48010.1	B3	457	48.40	U	68.96	31	37	−0.784	8.88	28.33
DcaHsf-B4	Dca39623.1	B4	287	54.07	U	72.30	30	32	−0.874	8.52	33.72
DcaHsf-C1 (incomplete)	Dca24054.1	C1	133	58.77	U	76.24	21	22	−0.881	8.01	15.89

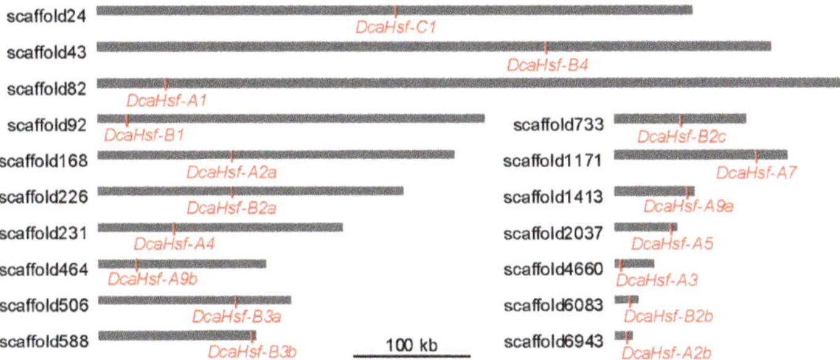

Figure 1. Scaffold locations of *DcaHsfs*. Bars represent the scaffolds, *DcaHsfs* are marked by redlines.

2.2. Phylogenetic and Sequence Conservation Analysis of DcaHsfs

To explore the phylogenetic relationship of *Hsfs* in *D. caryophyllus* and other species, the amino acid sequences of Hsfs from *A. thaliana*, *O. sativa*, and *P. trichocarpa* were used, together with those of DcaHsfs, as a means to construct a phylogenetic tree. In this study, 21 Hsf proteins from *A. thaliana* [18], 25 from *O. sativa* [18], 31 from *P. trichocarpa* [32], and 17 from *D. caryophyllus* were utilized for the phylogenetic analysis. A total of 94 Hsf proteins from the four species were clearly divided into three classes (class A, B, and C) with well-supported bootstrap values (Figure 2).

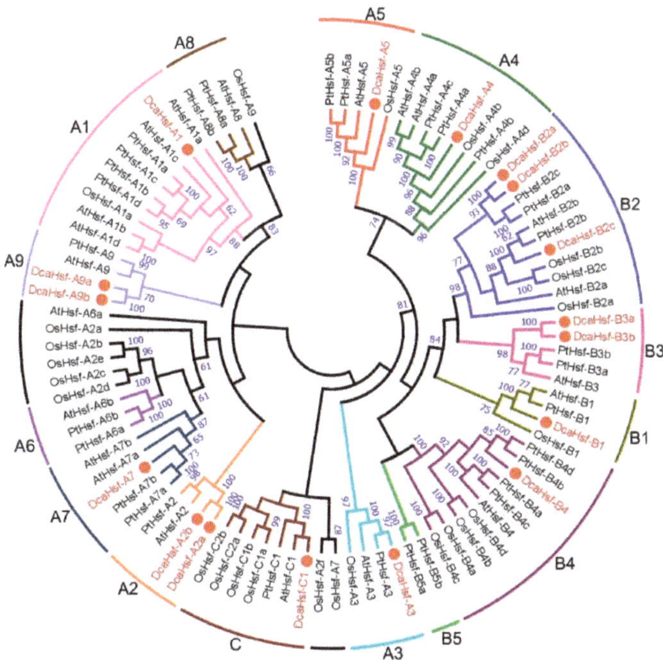

Figure 2. The phylogenetic tree of Hsf proteins. The phylogenetic tree of Hsf proteins in carnation and other plant species was generated by MEGA 7 using the neighbor-joining method. Dca, *D. caryophyllus*; At, *Arabidopsis thaliana*; Os, *Oryza sativa* and Pt, *Populus trichocarpa*.

In *D. caryophyllus*, 9 DcaHsfs out of 17 proteins belonged to class A, making it the largest subclass, followed by 7 DcaHsfs belonging to class B. The number of class B Hsfs in *D. caryophyllus* (7) was greater than that in *Arabidopsis* (5). The class C Hsf was present as a single copy in *D. caryophyllus*, *Arabidopsis*, and *P. trichocarpa*, whereas four copies of class C Hsfs were discovered in *O. sativa*. However, none of the DcaHsfs belonged to subclasses A6 and A8. Sequence conservation among DcaHsfs was also supported by their identity at the amino acid level. Detailed information on the identity of AtHsfs, OsHsfs, and PtHsfs amino acid sequences is illustrated in Table S1.

2.3. Structural and Motif Analysis of DcaHsfs

The structural diversity of the DcaHsf family was analyzed in terms of the exon/intron arrangement of the coding sequences. The number of introns in *DcaHsfs* ranged from one to three. The detailed gene structure of *DcaHsfs* is pictured in Figure 3a. Three introns were identified in *DcaHsfs-A7*, whereas all the other *DcaHsfs* had only one intron. Most closely related *DcaHsfs* in the same class or subfamily shared a similar gene structure in terms of intron number and intron and exon length (Figure 3a).

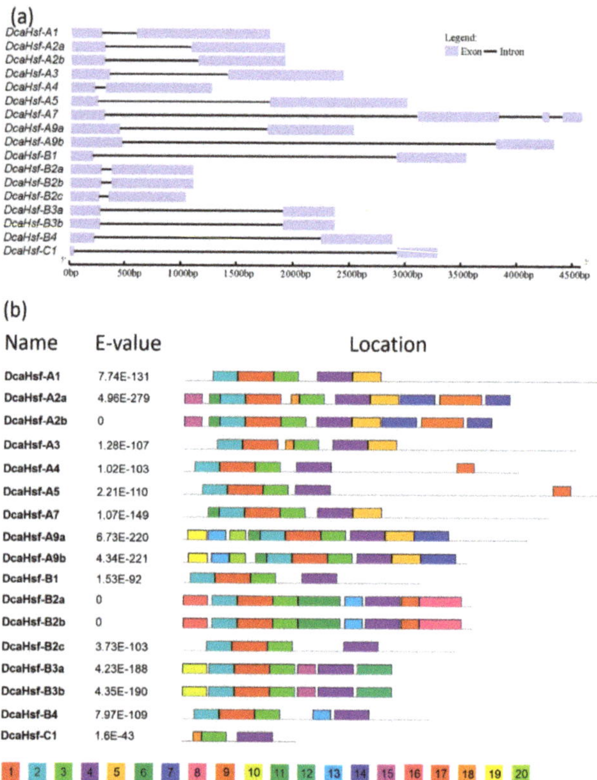

Figure 3. Intron and exon structure (**a**) and amino acid motifs (**b**) of members of the DcaHsf family. (**a**) Boxes and lines represent exons and introns, respectively. (**b**) A total of 20 conserved motifs were identified using Multiple Em for Motif Elicitation (MEME).

To investigate the protein sequence features of DcaHsfs, 20 different motifs were identified in DcaHsfs, with lengths ranging from 10 to 50 aa. (Figure 3b, Table 2). All members showed similar motif composition, but small differences between different groups were also found (Figure 3b). The conserved motifs in *Hsf* genes indicated that all DcaHsfs contained motif 1, motif 2, motif 3, and motif 4, except for DcaHsf-C1. Additionally, some motifs were only discovered in a certain subfamily of DcaHsfs. For instance, motif 6, motif 8, motif 9, and motif 14 were present in the B2 subfamily, whereas motif 10 and motif 12 were present in the B3 subfamily. Specifically, the phylogenetic analysis showed that the same clusters of DcaHsfs shared similar conserved domain composition. This indicates that *Hsf* genes are evolutionarily well conserved or possess similar regulatory functions in *D. caryophyllus* (Figure 3b). Additionally, motifs 1–3 represent the Hsf DBD domains (~100 aa). The DBD domain contains three α-helices and a four-stranded antiparallel β-sheet (α1-β1-β2-α2-α3-β3-β4) (Figure S1).

The 20 motifs consist of six different domains, including DBD, HR-A/B, NLS, NES, RD, and AHA domains. Among these, the highly structured DBD domain is the most conserved section in the DcaHsf family (Table 3). In addition to DBD, HR-A/B is critical for Hsf–Hsf interactions in the formation of a trimer [7], HR-A/B is also present in all DcaHsfs, and class A Hsfs have longer HR-A/B regions compared with class B and class C Hsfs (Figure S1, Table 3) Meanwhile, the other four conserved domains were only identified in specific DcaHsf members. The majority of class A DcaHsfs contained an NLS sequence rich in basic amino acid residues (K/R), except DcaHsf-A1/A3, whereas two or three NLS domains were located in seven DcaHsfs (A2a, A2b, A7, A9a, A9b, B3a, and B3b). NLS domains were not identified in DcaHsf-C1 and in some class B proteins (DcaHsf-B1, B2a, B2b, B2c). NES motifs were found in nine DcaHsfs. Also, five Class B Hsfs, except DcaHsf-B2a and DcaHsf-B2b, contained an RD in the C-terminus, characterized by the tetrapeptide LFGV. Transcription activator AHA motifs were located in class A DcaHsfs (Figure S1, Table 3). Sequence conservation among DcaHsfs was also supported by their identity at the amino acid level (0.023–0.83, Table S2). Four pairs of DcaHsfs (A2a–A2b, A9a–A9b, B2a–B2b, and B3a–B3b) exhibited high sequence identity (Table S2).

Table 2. Analysis and distribution of conserved motifs in carnation DcaHsfs.

Motif	E-Value	Width	Best Possible Match
Motif1	2.70×10^{-426}	41	VWDPAEFARDLLPRYFKHNNFSSFVRQLNTYGFRKVV PDRW
Motif2	5.20×10^{-225}	29	PFLTKTYDMVDDPSTDDIVSWSEDGTSFV
Motif3	1.10×10^{-187}	29	EFANEGFLRGQKHLLKNIKRRKTTTAHSQ
Motif4	4.20×10^{-115}	41	GLEGENERLRRENEVLMSELVKLKQQQQNTFSLLQA MESRL
Motif5	4.80×10^{-54}	34	QSTEWKQKQMMTFLAKAMQNPTFVQQLVQKKDER
Motif6	1.60×10^{-22}	49	PPQQQPSTAAPTNSSDEQVISNSNSPPPLAIPSVIMHR HHHHHHLYHNNN
Motif7	1.20×10^{-19}	41	KSVKAIRVSMKRRLTSTLSAPNLNDVVEPELVRSMAV SSDN
Motif8	2.60×10^{-16}	50	CGGGGGGGSPMIFGVSIGGKRGREGGDDGGGEVVGGGEGLG ATEVHDDHMH
Motif9	2.60×10^{-13}	50	ATLDNGTDGDIKEQKVDDSMPPEIDTNVGDVSQTSW EELLWAEDEEGFRQ
Motif10	7.70×10^{-10}	29	MRELSIKGLFDDHDDDDECGIIMRRKMTK
Motif11	1.50×10^{-9}	13	PKPMEGLNEMNPP
Motif12	1.80×10^{-5}	41	DSDGDDGNNKNRPKLFGVRLDLQDESE RKRRKKLALDYTRT
Motif13	5.20×10^{-8}	21	NNNNNNNVVITRKNNENEMNN
Motif14	1.40×10^{-4}	29	PVENVVPESGNWGEDVEDLIEQLGFLGPM
Motif15	1.80×10^{-4}	21	MTAVLVTVSDLVSSSTTSSSS
Motif16	2.10×10^{-4}	29	MSPPPSPPAEEKPEKLTAVVVGGGGGETQ
Motif17	1.70×10^{-3}	21	AAPSRVNDAVWTQLLTLPRGS
Motif18	2.90×10^{-3}	10	GFRKVDPDKW
Motif19	1.30×10^{-2}	23	TSTTCTCTPLSTESPQLGLQLSP
Motif20	1.00×10^{-2}	19	YWYDFDGEDEVELEERVPC

Table 3. Functional domains of DcaHsfs.

Gene Name	DBD	HR-A/B	NLS	NES	RD	AHA
DcaHsf-A1	31–124	162–182/201–212	N.D.	(403) L	N.D.	N.D.
DcaHsf-A2a	40–154	183–201/222–233	(147–156) KTIKRRRNVT (258–267) AGMKRRLTST	N.D.	N.D.	(329–338) QTSWEELLWA
DcaHsf-A2b	40–133	162–180/201–212	(126–137) LLKTIKRRRNVT (237–246) AGMKRRLTST	(232–237) LDITHL	N.D.	(308–317) QTSWEELLWA
DcaHsf-A3	37–148	181–199/220–230	N.D.	(319) L	N.D.	N.D.
DcaHsf-A4	11–104	139–157/178–189	(204–213) HDRKRRFSRP	N.D.	N.D.	(325–334) DVFWEQFLTE
DcaHsf-A5	20–113	138–156/177–187	(206–217) LSAYNKKRRLPP	N.D.	ND	(438–447) DLFWEQFLTE
DcaHsf-A7	40–133	164–182/203–213	(126–140) LLKNIKRRKNPSQTF (237–246) LSKKRRRPIE	N.D.	ND	(322–331) DDFWEDLLNE
DcaHsf-A9a	87–180	202–220/241–251	(173–184) LLKSIKRKRHGS (275–285) RVSKKRRLAST	(204) LDQEALKVEI	ND	N.D.
DcaHsf-A9b	95–188	210–228/249–259	(181–192) LLKSIKRKRHGS (283–293) RVSKKRRLAST	(217–221) LKVEI	ND	N.D.
DcaHsf-B1	6–99	125–131	N.D.	(155–157) LEL	(220-226) KLFGVWL	N.D.
DcaHsf-B2a	32–125	151–169/190–200	N.D.	N.D.	ND	N.D.
DcaHsf-B2b	32–125	151–169/190–200	N.D.	N.D.	ND	N.D.
DcaHsf-B2c	26–119	145–153/172–178	N.D.	(197–202) KENMSL	(284–290) KLFGVSI	N.D.
DcaHsf-B3a	29–122	147–152	(5–30) SIKGLFDDHDDDDECGIIMRRKMTKP (178–187) NAMKRKCQEL (207–235) KNRPKLFGVRLDLQDESERKRRKKLALDY	(222–238) LDLQDESERKRRKKLAL	(216–222) KLFGVRL	N.D.
DcaHsf-B3b	29–122	147–152	(5–30) SIKGLFDDHDDDDECGIIMRRKMTKP (178–187) NAMKRKCQEL (207–235) KNRPKLFGVRLDLQDESERKRRKKLALDY	N.D.	(216–222) KLFGVRL	N.D.
DcaHsf-B4	12–105	131–149/163–166	(275–283) HSKKRLHLA	N.D.	(268–274) RLFGVPL	N.D.
DcaHsf-C1 (not full)	1–44	73–89/98–108	N.D.	(66) L	N.D.	N.D.

N.D., not detected.

2.4. Cis-Acting Element Analysis in the Promoters of DcaHsfs

To predict the biological function of *DcaHsfs*, 1500-bp upstream sequences from the translation start sites of *DcaHsfs* were analyzed through the PlantCARE database. The promoter of each *DcaHsf* consists of several *cis*-acting elements, such as phytohormone-, abiotic stress-, and developmental process-related elements. As illustrated in Figure 4, the MYB element, ARE element (essential for anaerobic induction), and STRE element (activated by heat shock, osmotic stress, low pH, and nutrient starvation) were identified in the promoters of 15, 12, and 12 *DcaHsf* genes, respectively. The promoters of 11 *DcaHsfs* contained the ABA-responsive element (ABRE), methyl jasmonate (MeJA)-responsive element (CGTCA-motif), and TGACG motif involved in MeJA responsiveness (Figure 4). Also, the ethylene-responsive element (ERE), *cis*-acting element involved in salicylic acid responsiveness (TCA-element), stress-inducible element (TCA), wounding and pathogen responsiveness elements (W-box) were all found in 10, 10, 8, and 8 *DcaHsfs*, respectively (Figure 4). In total, 17 *DcaHsf* promoters contained 30 MYB, 30 STRE, 23 ARE, 21 ABRE, and 17 CGTCA-motif elements (Figure 4). These findings demonstrate that *DcaHsfs* might be associated with various transcriptional regulations involving development, hormones, and stress responses.

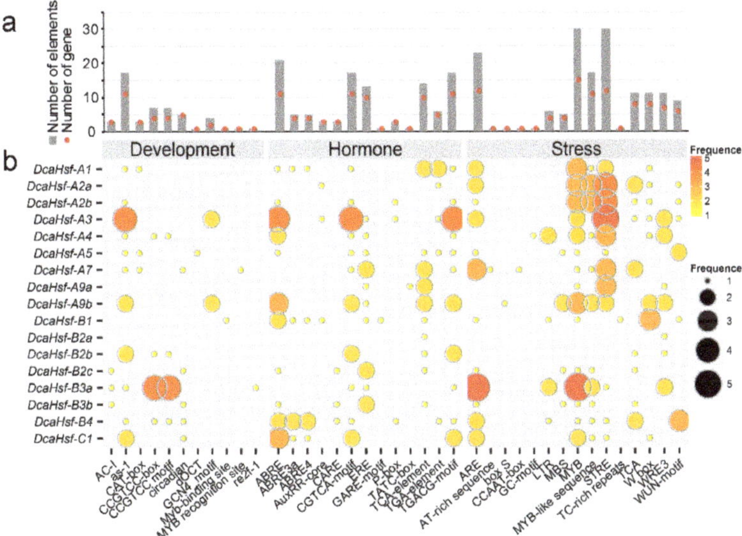

Figure 4. *Cis*-regulatory elements in the promoter region of *DcaHsfs*. (**a**) Number of each *cis*-acting element in the promoter region (1.5 kb upstream of the translation initiation site) of *DcaHsfs*. Statistics of the total number of *DcaHsfs* including the corresponding *cis*-acting elements (red dot) and the total number of *cis*-acting elements in the *DcaHsf* gene family (gray box). (**b**) Frequency of the *cis*-acting elements in each gene. Based on the functional annotation, the *cis*-acting elements were classified into three major classes: stress-, hormone-, and development-related *cis*-acting elements.

2.5. The Expression Pattern of DcaHsfs in Different Tissues and Flower Development

To elucidate the tissue-specific expression patterns of *DcaHsfs*, qRT-PCR was utilized to determine the expression levels of 17 *DcaHsfs* in 8 carnation tissues [root (R), stem (S), calyx (CA), young leaf (YL), mature leaf (ML), stigma (ST), ovary (OV), and flower (F)] and at 6 flowering stages (FS1, FS2, FS3, FS4, FS5, and FS6) (Figure 5, Table S4). Interestingly, the expression levels differed in different tissues and flowering stages, and the expression patterns of different members of DcaHsfs also differed, even for the same class. Among the different tissues, *DcaHsf-A1, A2a, 2b, A7, A9b, B2a, 2c, B4* were up-regulated in S, CA, YL, and ML. Meanwhile, all *DcaHsf*s were down-regulated in ST, OV, and F

(Figure 5). Twelve *DcaHsfs* were more highly expressed in CA, and 10 out of 17 *DcaHsfs* had higher expression levels in ML (Figure 5). Some genes demonstrated tissue-specific expression patterns. For instance, *DcaHsf-A9a, B1, B3a, B3b, C1* were up-regulated in R, and *DcaHsf-A3, A4, A5* were expressed at high levels in CA.

During the six flowering stages of carnations, all *DcaHsfs* showed relatively high expression levels at FS1. Additionally, *DcaHsfA1, A2a, A9a, B2a, B3a* were up-regulated at FS2 (Figure 5), implying that these genes may be involved in the early development of carnation flowers. In contrast, *DcaHsf-A5* and *DcaHsf-B2b* exhibited a high expression level at FS6 (Figure 5).

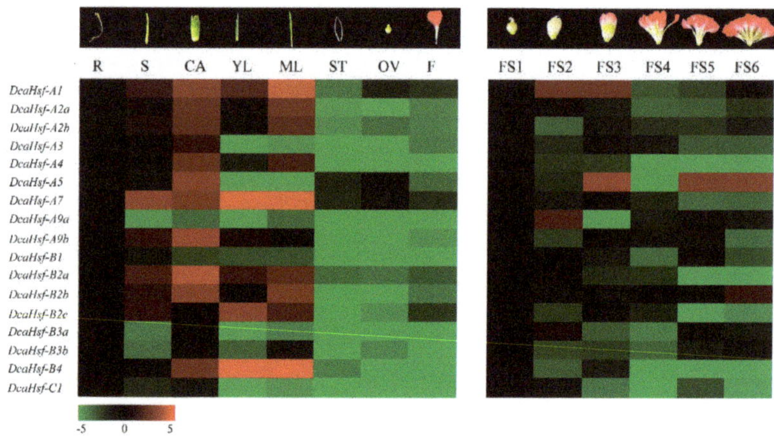

Figure 5. The expression levels of *DcaHsfs* in different tissues and flowering stages. The different colors correspond to log$_2$-transformed fold change, green indicates down-regulation, and red represents up-regulation.

2.6. DcaHsfs Response to Various Stresses

To determine the potential roles of the *DcaHsfs* in plant responses to various environmental stresses, qRT-PCR was conducted on the 17 *DcaHsfs* using the leaves of carnations exposed to heat, cold, drought, salt, ABA, and SA. The results illustrated that almost all *DcaHsfs* revealed three types of expression patterns under different stress conditions: (1) the expression of all genes was up-regulated; (2) the expression of all genes was down-regulated; and (3) some genes were expressed at higher levels in the early stage of stress, while others were up-regulated in the later stage of stress (Figure 6). Regarding the first category (1), all *DcaHsfs* were up-regulated after leaf exposure to heat and polyethylene glycol (PEG) treatments (Figure 6). For the second category of genes (2), almost all *DcaHsfs* displayed a decrease in their expression levels under cold, salt, or ABA stresses (except for individual *DcaHsfs*) (Figure 6). For example, four *DcaHsfs* (*DcaHsf-A2a, A5, B2b, C1*) were slightly induced at different time points at 4 °C., while the transcription levels of the remaining 13 genes were down-regulated at the tested time points (Figure 6). *DcaHsf-A5* demonstrated higher transcript accumulation compared to the other genes at 12 h under 200 mM NaCl treatment. Meanwhile, *DcaHsf-A3* was slightly up-regulated at 12 h under ABA treatment (Figure 6). Finally, the genes in category (3) and the expression of *DcaHsf-A3, A7, A9a, A9b, B3a* were primarily up-regulated at the earlier stage of SA treatment, whereas other *DcaHsfs* were strongly up-regulated after 12 h of SA treatment (Figure 6). These findings indicate that *DcaHsf* genes might play crucial roles in different stress response pathways.

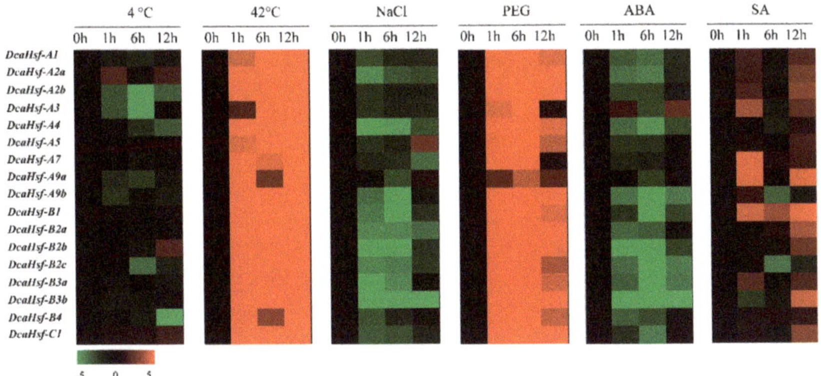

Figure 6. Expression levels of *DcaHsfs* under various abiotic stresses, determined by qRT-PCR. The different colors correspond to log$_2$-transformed fold change, green indicates down-regulation, and red represents up-regulation.

3. Discussion

3.1. Characterization of the Carnation Hsf Genes Family

Hsfs exist extensively in all plant species and act as the key regulatory components involved in various abiotic stresses to protect the plant cellular machinery under stress conditions [4,13,26]. In this study, a comprehensive genome-wide analysis of the *DcaHsf* family in carnations was carried out for the first time. A total of 17 *DcaHsf* genes were identified from the Carnation genome database [31]. The size of the carnation *Hsf* gene family is smaller compared with that of three other plant species, i.e., *A. thaliana*, *O. sativa*, and *P. trichocarpa*. Meanwhile, all four species have a similar subfamily distribution, which indicates that parallel evolutionary events of *Hsf* genes occurred in dicots and monocots. Additionally, the subclasses A6 and A8 are absent in carnation, and the diversification of *Hsf* members could provide some clues about the biological function of the corresponding *Hsf* counterparts in carnation. This suggests that gene loss and gene duplication events occurred at different stages of the evolutionary process, resulting in *Hsf* diversity [18] (Table S1).

3.2. Cis-Element Analysis in the Promoters of DcaHsfs

The number and form of *cis*-elements in promoter regions might play an essential function in the regulation of gene expression related to metabolic pathways [33]. The results illustrate that abiotic stress-related *cis*-elements, including MYB, STRE, ARE, ABRE, CGTCA-motif element, ERE, TCA-element, and W-box, are major regulatory elements in *DcaHsfs* promoters activated by heat shock or other abiotic stress. The presence of these stress-related elements is related to the expression response of *DcaHsfs* to heat, drought, ABA, and SA treatments. STRE is a marker element for plant Hsf proteins, which has been located in the promoters of the 17 *DcaHsfs* (Figure 4). In our study, a large number of STRE elements were identified in the promoter of 12 *DcaHsf* genes, which coincides with their expression (Figure 6). These findings suggest that STRE plays a vital role in transcriptional regulation under heat conditions in carnation. *DcaHsf* subclass A promoters contained MYB binding sites which participate in drought, low temperature, salt, ABA, and gibberellic acid (GA) stress responses [34]. We found that 15 *DcaHsfs* included 30 MYB binding sites in their promoter regions. However, the presence of MYB elements seems to be correlated with the positive regulation of *DcaHsfs* during drought and the negative regulation of *DcaHsfs* in response to salt and ABA treatments (Figure 6, Table S3). Other abiotic stress-related *cis*-elements, including the CGTCA-motif, TGACG-motif, ERE element, and TCA-element, were also major regulatory elements identified in *DcaHsfs*. Furthermore, the presence of these stress-related elements appears to be correlated with MeJA,

SA, and stress responsiveness, suggesting their potential roles in the response to pathogen infections. Consequently, *DcaHsfs* could be taken as candidate genes to understand the responses to drought and other biotic stresses.

3.3. Structural Analysis of DcaHsfs

The detailed knowledge of *A. thaliana*, *O. sativa*, and *P. trichocarpa Hsf* functional domains enabled us to analyze similar domains in *D. caryophyllus Hsf* gene family. It has been reported that the number of introns both regulate gene expression and participate in gene evolution [35]. Analysis of *Hsf* gene structure revealed that 16 of 17 *DcaHsfs* have one intron in their DBD domain (Figure 3a), which is an evolutionarily conserved intron [36]. However, *DcaHsf-A7* contains three introns (Figure 3a), which might affect its expression under stress conditions. All 17 DcaHsfs proteins contain the necessary DBD domain and specific protein domains (HR-A/B, NLS, NES, RD, and AHA) (Table 3, Figure 3b), which provide the structural basis for their conserved function [22]. The Hsf DBD domain of approximately 100 amino acid residues is highly conserved in different organisms, from plants to animals [7]. However, the DBD of DcaHsf-C1 contains only 44 aa and is shorter than the other DcaHsfs, lacking the full α1-helix, β1-sheet, β1-sheet, and α2-helix. Notably, this might be caused by the current genome assembly. It is interesting that AHA, an essential domain for the activator function in the HsfA class [7], was not found in several members of class A DcaHsfs (A1, A3, A9a, and A9b) (Table 3). The members of Hsfs lacking AHA domains might contribute differently to the activator function or bind to other HsfAs to form hetero-oligomers [18].

3.4. DcaHsfs Involvement in Carnation Development Processes

The expression patterns of *DcaHsfs* in seven different organs or tissues uncovered that *DcaHsfs* have different expression profiles in carnation. This suggests that they may participate in various developmental processes or regulatory pathways. In this study, nearly all the *DcaHsfs* were found to display high transcription levels in ML and at FS1 (Figure 5). Within the potato *HsfA1* group, *StHsf002* is highly expressed in flowers, petals, and sepals, whereas *StHsf003* is highly expressed in roots, flowers, carpels, and sepals [37]. In our study, *DcaHsf-A1* demonstrated up-regulation in S, CA, YL, and ML (Figure 5). *Phyllostachys edulis PheHsfA2a-2* is predicted to play an important role in flower and shoot development [38] and *Cicer arietinum CarHsfA2* is up-regulated in shoot, root, and flower [12]. Their orthologs in carnations, *DcaHsf-A2a, A2b*, were constitutively expressed in S, CA, YL, and ML at relatively high levels (Figure 5). These findings indicate that the members of the *Hsf-A1* and *A2* sub-families are conserved and involved in the development of vegetative organs.

HsfA5 has been reported to play a vital role in stress tolerance during anther/pollen development as well as in other stages of plant reproduction in tomato and *Arabidopsis* [22,39]. In this study, *DcaHsf-A5* was highly expressed in CA and OV (Figure 5). This implies that the function of *DcaHsf-A5* might be conserved for regulating reproductive organ development and the growth of carnation. *Salix suchowensis SsuHsf-A9* is specifically expressed in the female catkin [32]. In *Populus* female catkin development, *PtHsf-A9* displays relatively high transcription levels [40]. Our results indicate that carnation *DcaHsf-A9b* was up-regulated in S and CA and is possibly widely involved in the development of both vegetative and reproductive tissues.

For Class B *Hsfs*, Chickpea *CarHsfB2c* is highly expressed in the late flowering stages, while *CarHsfB2a* is expressed in root, flower, pod wall, and grain. *CarHsfB4b* is specifically expressed in flower and grain [12]. In carnations, *DcaHsfs-B2a, B2c, B4* are highly expressed in S, CA, YL, and ML. However, *DcaHsf-B2a/B3a* and *DcaHsf B2b* are highly expressed in FS2 and FS6, respectively (Figure 5). This indicates that members of the Class B *DcaHsfs* might be widely involved in the development of both vegetative and reproductive organs and tissues. The expression patterns of *Hsf-C1* genes were diverse in different tissues. For example, the transcripts of *Vitis pseudoreticulata VpHsfC1a* remain at relatively lower levels (even undetectable) in roots, stems, leaves, and tendrils [41]. Similarly, *SaHsfC1a* is expressed at low levels in all the tested tissues in *Sedum alfredii* [42]. Carnation *DcaHsf-C1* was down-regulated in almost all tested tissues (except for R) (Figure 5), which is consistent with the expression pattern of *SaHsfC1b* [42]. Specifically, it may be attributed to the fact that *DcaHsf-C1* acts as a negative regulator in the development of organs.

3.5. DcaHsfs are Involved in Carnation Stress Response

The genome-wide expression profile analyses indicated that the majority of the *Hsf* genes are involved in heat, cold, drought, and salt stress responses [22,26]. Under heat or other stress conditions, plant *Hsfs* display diversity in patterns of expression [14]. In our study, all 17 *DcaHsfs* were found to be induced by a high temperature of 42 °C (Figure 6), which is in agreement with a previous study [43]. All *DcaHsfs* accumulated during drought treatment (Figure 6), and a previous study revealed that approximately 90% of sesame *Hsfs* are drought-responsive [44]. *DcaHsf-A2a, A2b* revealed to be strongly induced under heat stress conditions. This indicates that HsfA2 is a dominant regulator during the heat stress response in carnation, which is consistent with the studies of *Arabidopsis*, tomato, apple, *Populus euphratica*, and *Phyllostachys edulis* [11,22,38,39,45]. *HsfA3* has been identified as an important player in the responses to heat, high salinity, and drought stresses in *Solanum lycopersicum* [46], whereas a similar function for *HsfA3* is not detected in tomato [14]. In this study, *DcaHsfA3* was also up-regulated in response to four analyzed abiotic stresses (heat, drought, ABA, and SA) (Figure 6). Group A4 Hsfs are involved in controlling reactive oxygen species homeostasis in plants, and group A5 Hsfs act as specific repressors of HsfA4 [47,48]. *Fragaria vesca FvHsfA4a, A5a* were both distinctly up-regulated in response to abiotic stresses such as cold, drought, and salt and hormone treatments (ABA, Eth, MeJA, and SA) [49]. Our data are highly similar, indicating *DcaHsf-A4* accumulation during heat, drought, and SA treatment, as well as *DcaHsf-A5* upregulation in response to cold, heat, drought, and salt and SA treatments (Figure 6). *Arabidopsis Hsf-A9a* is associated with ABA-mediated stress signaling and drought resistance [50]. Similarly, *DcaHsfA9a* was also induced in response to salt, ABA, and SA (Figure 6). Compared to Class A *Hsfs*, the members in Class B and C still have not been well studied. *Arabidopsis AtHsfB1a* and *F. vesca FvHsfB1a* were highly induced and accumulated in response to SA treatment [45,49,51]. Additionally, additional evidence demonstrated that *AtHsfB1a, B2b* are crucial components in primed defense gene activation and pathogen-induced acquired immune response [51]. Similarly, in this study, *DcaHsf-B1, B2a, B2b* accumulated at high levels at the later stage of SA treatment (Figure 6) Therefore, it is reasonable to speculate that *DcaHsf-B1, B2a, B2b* play a crucial role in the acquired immune response to pathogens. Additionally, *DcaHsf-C1* acts as a positive regulator of heat shock proteins under heat stress conditions or PEG stress. The expression of rice *OsHsfC1b* was induced by salt, mannitol, and ABA [52]. In *V. pseudoreticulata*, *VpHsfC1a* was up-regulated in response to ABA treatment but significantly down-regulated during both MeJA and Eth treatments [41]. However, the expression of *DcaHsf-C1* was not up-regulated by cold, ABA, or salt stress (Figure 6). We can speculate that *DcaHsf-C1* might be involved in ABA-independent pathways in carnations. However, gene expression is a complex biological process, and more thorough studies are required to decipher the regulatory mechanisms.

4. Materials and Methods

4.1. Identification and Characterization of Hsf Genes in D. caryophyllus

The protein and nucleotide sequences of *D. caryophyllus* were downloaded from the carnation DB (http://carnation.kazusa.or.jp). The conserved domain of Hsf DBD (Pfam: PF00447) was submitted as a query in a BLASTP search of the *D. caryophyllus* proteome. The SMART 7 software (http://smart.embl-heidelberg.de/) was used to identify integrated DBD domain and (HR-A/B) domain in the putative Hsfs. Candidate proteins without integrated DBD domain and HR-A/B domain were removed. The ExPaSy-Protparam tool (https://www.expasy.org/tools/ProtParam.html) was used to analyze the physical properties of the predicted Hsf proteins.

4.2. Phylogenetic Analysis

Multiple sequence alignments of full-length Hsf proteins from *D. caryophyllus* and other three model species, i.e., *A. thaliana*, *O. sativa*, and *P. trichocarpa*, were performed using Clustal W2 (https://www.ebi.ac.uk/Tools/msa/clustalw2/, Dublin, Ireland). An unrooted neighbor-joining (NJ) phylogenetic tree was constructed using MEGA7.0 (Philadelphia, PA, U.S.A.) with 1000 bootstrap replicates. Distinctive names for each of the *Hsfs* identified in *D. caryophyllus* were given according to the classification of *Hsfs* in classes A, B, and C, referred to as *DcaHsf* genes.

4.3. Structural and Motif Analyses of DcaHsf Genes

The gene structures including exons and introns were displayed using Gene Structure Display Server (GSDS, http://gsds.cbi.pku.edu.cn/index.php, Beijing, China). The conserved motifs of DcaHsfs were defined by Multiple Em for Motif Elicitation (MEME, http://meme-suite.org/, U.S.A.) using the following parameters: number of repetitions = any, maximum number of motifs = 20, minimum width $\geq$10, maximum width $\leq$200, and only motifs with an *E*-value < 0.01 were retained for further analysis. NLS domains were predicted using cNLS Mapper software (http://nls-mapper.iab.keio.ac.jp/cgi-bin/NLS_Mapper_form.cgi, Tsuruoka, Japan). NES domains in the *DcaHsfs* were predicted with the NetNES 1.1 server software (http://www.cbs.dtudk/services/NetNES/, Lyngby Denmark).

4.4. Cis-acting Element Analysis of DcaHsfs

The 1500-bp sequence upstream from the initiation codon of each *DcaHsf* gene was obtained from the *D. caryophyllus* genome database. These sequences were used to identify *cis*-acting regulatory elements with the online program PlantCARE (http://bioinformatics.psb.ugent.be/webtools/plantcare/html/, Ghent, Belgium).

4.5. Plant Materials, Growth Conditions, and Stress Treatments

Tissue culture seedlings of carnation were grown in a chamber at Qingdao Agriculture University (Qingdao, China) under a 12 h light (300 μmol·m^{-2}·s^{-1})/12 h dark cycle at 23–25 °C ambient temperature and 70% relative humidity. Various tissues, including the root (R), shoot (S), calyx (CA), young leaf (YL), mature leaf (ML), stigma (ST), ovary (OV), and flower petals (F), and six flowering stages (FS1, FS2, FS3, FS4, FS5, and FS6) were collected from the carnation seedlings. For abiotic stress and hormone treatments, the seedlings were treated at 42 °C (for heat stress), with 20% (*w/v*) polyethylene glycol (PEG) 6000 (for drought stress), 200 mM NaCl (for salt stress), 100 μM ABA, or 100 μM SA. The first or second tender leaves of the seedlings were collected at 0, 1, 6, and 12 h, immediately frozen in liquid nitrogen, and then stored at −80 °C for further analysis. Three biological replicates were performed for each sample.

4.6. RNA Isolation and Expression Analysis of DcaHsf Genes

Total RNA from carnation leaves was extracted using the Plant RNA Kit (Omega, Norcross, GA, USA) according to the instructions. Subsequently, 500 ng of total RNA was reverse-transcribed to first-strand cDNA by the PrimeScrip RT reagent Kit with gDNA Eraser (TaKaRa, Dalian, China) according to the manufacturer's protocol, and the cDNA was diluted 10-fold for quantitative real-time PCR (qRT-PCR). qRT-PCR was performed using 2 μL of cDNA in a 20 μL reaction volume with SYBR® Premix Ex Taq™ II (TaKaRa, Dalian, China) on a StepOnePlus Real-Time PCR System (ABI, USA), using the following PCR program: 95 °C for 3 min, followed by 40 cycles at 95 °C for 30 s and at 60 °C for 1 min 30 s. Melting curves were obtained to verify the amplification specificity through a stepwise heating of the amplicon from 60 to 95 °C. Primer pairs were designed by Primer Premier 5.0 (Table S4). The *GAPDH* gene was used as an internal control gene. Three independent biological replicates were performed, and the relative expression levels of the *DcaHsf* genes were calculated with the $2^{-\Delta\Delta Ct}$ method [53].

5. Conclusions

In this study, 17 *DcaHsf* genes were identified in the carnation genome for the first time. Comprehensive analyses of these genes, including phylogeny, genes structure, conserved motifs, and expression profiles in various tissues and under abiotic stresses were performed. Structural characteristics and comparisons with *A. thaliana*, *O. sativa*, and *P. trichocarpa* assisted in classifying these genes into three major classes (A, B, and C), with members of class A being the most abundant. The *DcaHsf* members were expressed in at least one tissue among root, stem, calyx, young leaf, mature leaf, stigma, ovary, and flower. In addition, *DcaHsfA1, A2a, A9a, B2a, B3a* revealed their putative involvement in the early flowering stages. The results of qRT-PCR revealed that all *DcaHsfs* responded to heat and drought, and many *DcaHsfs* were also regulated by cold, salt, and osmotic stress, as well as by the phytohormones ABA and SA. Our research suggests that *DcaHsf A2a, A2b* may be used as candidate genes for the breeding of heat-resistant carnation. Meanwhile, *DcaHsf-B2a, B3a* and *DcaHsfA5, B2b* could be considered as probable candidate genes for promoting early blooming and prolonging florescence in carnations.

Supplementary Materials: Supplementary Materials can be found at http://www.mdpi.com/1422-0067/20/20/5233/s1.

Author Contributions: J.Z. conceived and designed the research. W.L., X.-L.W., and J.-Y.Y. performed the experiments. K.-L.W. contributed with valuable discussions. W.L. wrote the original draft. J.Z. revised the manuscript. All authors read and approved the final manuscript.

Funding: This research received no external funding.

Conflicts of Interest: The authors declare no conflict of interest.

Abbreviations

ABA	Abscisic acid
ABRE	ABA-responsive element
APX	Ascorbate peroxidase
CA	Calyx
CAT	Catalase
DBD	DNA-binding domains
ERE	Ethylene-responsive element
F	Flower
FS	Flowering stages
HSE	Heat shock element
Hsf	Heat shock transcription factor
MEME	Multiple Em for Motif Elicitation
ML	Mature leaf
MW	Molecular weight
NES	Nuclear export signals
NLS	Nuclear localization signals
OV	Ovary
PEG	Polyethylene glycol
qRT-PCR	Quantitative real-time PCR
R	Root
RD	Repressor domains
ROS	Reactive oxygen species
S	Stem
SA	Salicylic acid
ST	Stigma
TFs	Transcription factors
YL	Young leaf

References

1. Hu, H.; Xiong, L. Genetic engineering and breeding of drought-resistant crops. *Annu. Rev. Plant Biol.* **2014**, *65*, 715–741. [CrossRef] [PubMed]
2. Pereira, A. Plant abiotic stress challenges from the changing environment. *Front. Plant Sci.* **2016**, *7*, 1123. [CrossRef] [PubMed]
3. Zhu, J.K. Abiotic stress signaling and responses in plants. *Cell* **2016**, *167*, 313–324. [CrossRef] [PubMed]
4. Ohama, N.; Sato, H.; Shinozaki, K.; Yamaguchi-Shinozaki, K. Transcriptional regulatory network of plant heat stress response. *Trends Plant Sci.* **2017**, *22*, 53–65. [CrossRef]
5. Rushton, P.J.; Somssich, I.E.; Ringler, P.; Shen, Q.J. WRKY transcription factors. *Trends Plant Sci.* **2010**, *15*, 247–258. [CrossRef]
6. Puranik, S.; Sahu, P.P.; Srivastava, P.S.; Prasad, M. NAC proteins: Regulation and role in stress tolerance. *Trends Plant Sci.* **2012**, *17*, 369–381. [CrossRef]
7. Scharf, K.D.; Berberich, T.; Ebersberger, I.; Nover, L. The plant heat stress transcription factor (Hsf) family: Structure, function and evolution. *Biochim. Biophys. Acta Gene Regul. Mech.* **2012**, *1819*, 104–119. [CrossRef]
8. Kotak, S.; Larkindale, J.; Lee, U.; Koskull-Döring, P.V.; Vierling, E.; Scharf, K.D. Complexity of the heat stress response in plants. *Curr. Opin. Plant Biol.* **2007**, *10*, 310–316. [CrossRef]
9. Swindell, W.R.; Huebner, M.; Weber, A.P. Transcriptional profiling of *Arabidopsis* heat shock proteins and transcription factors reveals extensive overlap between heat and non-heat stress response pathways. *BMC Genom.* **2007**, *8*, 125. [CrossRef]
10. Wei, Y.X.; Hu, W.; Xia, F.Y.; Zeng, H.Q.; Li, X.L.; Yan, Y.; He, C.Z.; Shi, H.T. Heat shock transcription factors in banana: Genome-wide characterization and expression profile analysis during development and stress response. *Sci. Rep.* **2016**, *6*, 36864. [CrossRef]
11. Zhang, J.; Jia, H.X.; Li, J.B.; Li, Y.; Lu, M.Z.; Hu, J.J. Molecular evolution and expression divergence of the *Populus euphratica Hsf* genes provide insight into the stress acclimation of desert poplar. *Sci. Rep.* **2016**, *6*, 30050. [CrossRef] [PubMed]
12. Chidambaranathan, P.; Jagannadham, P.T.K.; Satheesh, V.; Kohli, D.; Basavarajappa, S.H.; Chellapilla, B.; Kumar, J.; Jain, P.K.; Srinivasan, R. Genome-wide analysis identifies chickpea (*Cicer arietinum*) heat stress transcription factors (Hsfs) responsive to heat stress at the pod development stage. *J. Plant Res.* **2017**, *131*, 525–542. [CrossRef] [PubMed]
13. Guo, M.; Liu, J.H.; Ma, X.; Luo, D.X.; Gong, Z.H.; Lu, M.H. The plant Heat Stress Transcription Factors (HSFs): Structure, regulation, and gunction in response to abiotic stresses. *Front. Plant Sci.* **2016**, *7*, 114. [CrossRef] [PubMed]
14. Koskull-DöRing, P.V.; Scharf, K.D.; Nover, L. The diversity of plant heat stress transcription factors. *Trends Plant Sci.* **2007**, *12*, 452–457. [CrossRef]
15. Singh, A.; Mittal, D.; Lavania, D.; Agarwal, M.; Mishra, R.C.; Grover, A. OsHsfA2c and OsHsfB4b are involved in the transcriptional regulation of cytoplasmic *OsClpB* (*Hsp100*) gene in rice (*Oryza sativa* L.). *Cell Stress Chaperones* **2012**, *17*, 243–254. [CrossRef]
16. Cheng, Q.; Zhou, Y.; Liu, Z.; Zhang, L.; Song, G.; Guo, Z.; Wang, W.; Qu, X.; Zhu, Y.; Yang, D. An alternatively spliced heat shock transcription factor, *OsHSFA2dI*, functions in the heat stress-induced unfolded protein response in rice. *Plant Biol.* **2015**, *17*, 419–429. [CrossRef]
17. Giesguth, M.; Sahm, A.; Simon, S.; Dietz, K.J. Redox-dependent translocation of the heat shock transcription factor AtHSFA8 from the cytosol to the nucleus in *Arabidopsis thaliana*. *FEBS Lett.* **2015**, *589*, 718–725. [CrossRef]
18. Guo, J.K.; Wu, J.; Ji, Q.; Wang, C.; Luo, L.; Yuan, Y.; Wang, Y.H.; Wang, J. Genome-wide analysis of heat shock transcription factor families in rice and *Arabidopsis*. *J. Genet. Genom.* **2008**, *35*, 105–118. [CrossRef]
19. Wang, F.M.; Dong, Q.; Jiang, H.Y.; Zhu, S.W.; Chen, B.J.; Xiang, Y. Genome-wide analysis of the heat shock transcription factors in *Populus trichocarpa* and *Medicago truncatula*. *Mol. Biol. Rep.* **2012**, *39*, 1877–1886. [CrossRef]
20. Chung, E.; Kim, K.M.; Lee, J.H. Genome-wide analysis and molecular characterization of heat shock transcription factor family in *Glycine max*. *J. Genet. Genom.* **2013**, *40*, 127–135. [CrossRef]

21. Lin, Y.X.; Jiang, H.Y.; Chu, Z.X.; Tang, X.L.; Zhu, S.W.; Cheng, B.J. Genome-wide identification, classification and analysis of heat shock transcription factor family in maize. *BMC Genom.* **2011**, *12*, 76. [CrossRef] [PubMed]

22. Giorno, F.; Guerriero, G.; Baric, S.; Mariani, C. Heat shock transcriptional factors in *Malus domestica*: Identification, classification and expression analysis. *BMC Genom.* **2012**, *13*, 639. [CrossRef] [PubMed]

23. Zhou, S.J.; Zhang, P.; Jing, Z.; Shi, J.L. Genome-wide identification and analysis of heat shock transcription factor family in cucumber (*Cucumis sativus* L.). *Plant Omics* **2013**, *6*, 449–455.

24. Zhang, J.; Liu, B.B.; Li, J.B.; Zhang, L.; Wang, Y.; Zheng, H.Q.; Lu, M.Z.; Chen, J. *Hsf* and *Hsp* gene families in *Populus*: Genome-wide identification, organization and correlated expression during development and in stress responses. *BMC Genom.* **2015**, *16*, 181. [CrossRef]

25. Wang, J.; Sun, N.; Deng, T.; Zhang, L.D.; Zuo, K.J. Genome-wide cloning, identification, classification and functional analysis of cotton heat shock transcription factors in cotton (*Gossypium hirsutum*). *BMC Genom.* **2014**, *15*, 961. [CrossRef]

26. Xue, G.P.; Sadat, S.; Drenth, J.; Mcintyre, C.L. The heat shock factor family from *Triticum aestivum* in response to heat and other major abiotic stresses and their role in regulation of heat shock protein genes. *J. Exp. Bot.* **2014**, *65*, 539–557. [CrossRef]

27. Zhu, X.Y.; Huang, C.Q.; Zhang, L.; Liu, H.F.; Yu, J.H.; Hu, Z.Y.; Hua, W. Systematic analysis of *Hsf* family genes in the *Brassica napus* genome reveals novel responses to heat, drought and high CO_2 stresses. *Front. Plant Sci.* **2017**, *8*, 1174. [CrossRef]

28. Boxriker, M.; Boehm, R.; Krezdorn, N.; Rotter, B.; Piepho, H.P. Comparative transcriptome analysis of vase life and carnation type in *Dianthus caryophyllus* L. *Sci. Hortic.* **2017**, *217*, 61–72. [CrossRef]

29. Onozaki, T.; Ikeda, H.; Yamaguchi, T. Genetic improvement of vase life of carnation flowers by crossing and selection. *Sci. Hortic.* **2001**, *87*, 107–120. [CrossRef]

30. Muneer, S.; Soundararajan, P.; Jeong, B.R. Proteomic and antioxidant analysis elucidates the underlying mechanism of tolerance to hyperhydricity stress in in vitro shoot cultures of *Dianthus caryophyllus*. *J. Plant Growth Regul.* **2016**, *35*, 667–679. [CrossRef]

31. Yagi, M.; Kosugi, S.; Hirakawa, H.; Ohmiya, A.; Tanase, K.; Harada, T.; Kishimoto, K.; Nakayama, M.; Ichimura, K.; Onozaki, T.; et al. Sequence analysis of the genome of carnation (*Dianthus caryophyllus* L.). *DNA Res.* **2013**, *21*, 231–241. [CrossRef] [PubMed]

32. Zhang, J.; Li, Y.; Jia, H.X.; Li, J.B.; Huang, J.; Lu, M.Z.; Hu, J.J. The heat shock factor gene family in *Salix suchowensis*: A genome-wide survey and expression profiling during development and abiotic stresses. *Front. Plant Sci.* **2015**, *6*, 748. [CrossRef] [PubMed]

33. Lescot, M.; Déhais, P.; Thijs, G.; Marchal, K.; Moreau, Y.; de Peer, Y.V.; Rouzé, P.; Rombauts, S. PlantCARE, a database of plant *cis*-acting regulatory elements and a portal to tools for in silico analysis of promoter sequences. *Nucleic Acids Res.* **2002**, *30*, 325–327. [CrossRef] [PubMed]

34. He, Y.; Li, W.; Lv, J.; Jia, Y.B.; Wang, M.C.; Xia, G.M. Ectopic expression of a wheat MYB transcription factor gene, *TaMYB73*, improves salinity stress tolerance in *Arabidopsis thaliana*. *J. Exp. Bot.* **2012**, *63*, 1511–1522. [CrossRef] [PubMed]

35. Rose, A.B. Intron-mediated regulation of gene expression. *Curr. Top. Microbiol. Immunol.* **2008**, *326*, 277–290. [PubMed]

36. Pelham, H.R.; Bienz, M. A synthetic heat-shock promoter element confers heat-inducibility on the herpes simplex virus thymidine kinase gene. *EMBO J.* **1981**, *1*, 1473–1477. [CrossRef]

37. Tang, R.M.; Zhu, W.J.; Song, X.Y.; Lin, X.Y.; Cai, J.H.; Man, W.; Yang, Q. Genome-wide identification and function analyses of heat shock transcription factors in potato. *Front. Plant Sci.* **2016**, *7*, 490. [CrossRef]

38. Xie, L.H.; Li, X.Y.; Hou, D.; Cheng, Z.C.; Liu, J.; Li, J.; Mu, S.H.; Gao, J. Genome-wide analysis and expression profiling of the heat shock factor gene family in *Phyllostachys edulis* during development and in response to abiotic stresses. *Forests* **2019**, *10*, 100. [CrossRef]

39. Fragkostefanakis, S.; Röth, S.; Schleiff, E.; Scharf, K.D. Prospects of engineering thermotolerance in crops through modulation of heat stress transcription factor and heat shock protein networks. *Plant Cell Environ.* **2015**, *38*, 1881–1895. [CrossRef]

40. Liu, B.; Hu, J.; Zhang, J. Evolutionary divergence of duplicated *Hsf* genes in *Populus*. *Cells* **2019**, *8*, 438. [CrossRef]

41. Hu, Y.; Han, Y.T.; Zhang, K.; Zhao, F.L.; Li, Y.J.; Zheng, Y.; Wang, Y.J.; Wen, Y.Q. Identification and expression analysis of heat shock transcription factors in the wild Chinese grapevine (*Vitis pseudoreticulata*). *Plant Physiol. Biochem.* **2016**, *99*, 1–10. [CrossRef] [PubMed]

42. Chen, S.S.; Jiang, J.; Han, X.J.; Zhang, Y.X.; Zhuo, R.Y. Identification, expression analysis of the Hsf family, and characterization of class A4 in *Sedum Alfredii* Hance under cadmium stress. *Int. J. Mol. Sci.* **2018**, *19*, 1216. [CrossRef] [PubMed]

43. Mittal, D.; Chakrabarti, S.; Sarkar, A.; Singh, A.; Grover, A. Heat shock factor gene family in rice: Genomic organization and transcript expression profiling in response to high temperature, low temperature and oxidative stresses. *Plant Physiol. Biochem.* **2009**, *47*, 785–795. [CrossRef] [PubMed]

44. Dossa, K.; Diouf, D.; Cisse, N. Genome-wide investigation of *Hsf* genes in sesame reveals their segmental duplication expansion and their active role in drought stress response. *Front. Plant Sci.* **2016**, *7*, 1522. [CrossRef]

45. Ikeda, M.; Mitsuda, N.; Ohme-Takagi, M. *Arabidopsis* HsfB1 and HsfB2b act as repressors of the expression of heat-inducible Hsfs but positively regulate the acquired thermotolerance. *Plant Physiol.* **2011**, *157*, 1243–1254. [CrossRef]

46. Li, Z.J.; Zhang, L.L.; Wang, A.X.; Xu, X.Y.; Li, J.F. Ectopic overexpression of *SlHsfA3*, a heat stress transcription factor from tomato, confers increased thermotolerance and salt hypersensitivity in germination in transgenic *Arabidopsis*. *PLoS ONE* **2013**, *8*, e54880. [CrossRef]

47. Yamanouchi, U.; Yano, M.; Lin, H.X.; Ashikari, M.; Yamada, K. A rice spotted leaf gene, *Spl7*, encodes a heat stress transcription factor protein. *Proc. Natl. Acad. Sci. USA* **2002**, *99*, 7530–7535. [CrossRef]

48. Baniwal, S.K.; Chan, K.Y.; Scharf, K.D.; Nover, L. Role of heat stress transcription factor HsfA5 as specific repressor of HsfA4. *J. Biol. Chem.* **2007**, *282*, 3605–3613. [CrossRef]

49. Hu, Y.; Han, Y.T.; Wei, W.; Li, Y.J.; Zhang, K.; Gao, Y.R.; Zhao, F.L.; Feng, J.Y. Identification, isolation, and expression analysis of heat shock transcription factors in the diploid woodland strawberry *Fragaria vesca*. *Front. Plant Sci.* **2015**, *6*, 736. [CrossRef]

50. Schramm, F.; Ganguli, A.; Kiehlmann, E.; Englich, G.; Walch, D.; Koskull-Döring, P.V. The heat stress transcription factor HsfA2 serves as a regulatory amplifier of a subset of genes in the heat stress response in *Arabidopsis*. *Plant Mol. Biol.* **2006**, *60*, 759–772. [CrossRef]

51. Pick, T.; Jaskiewicz, M.; Peterhänsel, C.; Conrath, U. Heat shock factor HsfB1 primes gene transcription and systemic acquired resistance in *Arabidopsis*. *Plant Physiol.* **2012**, *159*, 52–55. [CrossRef] [PubMed]

52. Schmidt, R.; Schippers, J.H.M.; Welker, A.; Mieulet, D.; Guiderdoni, E.; Mueller-Roeber, B. Transcription factor OsHsfC1b regulates salt tolerance and development in *Oryza sativa* ssp. japonica. *AoB Plants* **2012**, *2012*, pls011. [CrossRef] [PubMed]

53. Livak, K.J.; Schmittgenb, T.D. Analysis of relative gene expression data using real-time quantitative PCR and the 2$^{-\Delta\Delta CT}$ Method. *Methods* **2001**, *25*, 402–408. [CrossRef] [PubMed]

 International Journal of
Molecular Sciences

Article

Comparative Genome-wide Analysis and Expression Profiling of Histone Acetyltransferase (HAT) Gene Family in Response to Hormonal Applications, Metal and Abiotic Stresses in Cotton

Muhammad Imran [1,2], Sarfraz Shafiq [1,3,]*, Muhammad Ansar Farooq [4],
Muhammad Kashif Naeem [2], Emilie Widemann [5], Ali Bakhsh [6], Kevin B. Jensen [7] and
Richard R.-C. Wang [7,]*

[1] School of Life Sciences, Tsinghua University, Beijing 100084, China; imran_m1303@yahoo.com
[2] State Key Laboratory of Plant Cell and Chromosome Engineering, Institute of Genetics and Developmental Biology, Chinese Academy of Sciences, Beijing 100101, China; kashifuaar102@hotmail.com
[3] Department of Environmental Sciences, COMSATS University Islamabad, Abbottabad campus, Abbottabad 22060, Pakistan
[4] Institute of Soil & Environmental Sciences, University of Agriculture, Faisalabad 38000, Pakistan; ansar_1264@yahoo.com
[5] Department of Biology, University of Western Ontario, 1151 Richmond St, London, ON N6A5B8 Canada; ewidema4@uwo.ca
[6] Department of Plant breeding and Genetics, Ghazi University, Dera Ghazi Khan 32200, Pakistan; abakhsh@gudgk.edu.pk
[7] Forage & Range Research, United States Department of Agriculture, Agricultural Research Service, Logan, UT 84322, USA; kevin.jensen@usda.gov
* Correspondence: sarfraz@mail.tsinghua.edu.cn (S.S.); richard.wang@usda.gov (R.R.-C.W.)

Received: 26 August 2019; Accepted: 24 October 2019; Published: 25 October 2019

Abstract: Post-translational modifications are involved in regulating diverse developmental processes. Histone acetyltransferases (HATs) play vital roles in the regulation of chromation structure and activate the gene transcription implicated in various cellular processes. However, HATs in cotton, as well as their regulation in response to developmental and environmental cues, remain unidentified. In this study, 9 HATs were identified from *Gossypium raimondi* and *Gossypium arboretum*, while 18 HATs were identified from *Gossypium hirsutum*. Based on their amino acid sequences, *Gossypium* HATs were divided into three groups: CPB, GNAT, and TAF$_{II}$250. Almost all the HATs within each subgroup share similar gene structure and conserved motifs. *Gossypium* HATs are unevenly distributed on the chromosomes, and duplication analysis suggests that *Gossypium* HATs are under strong purifying selection. Gene expression analysis showed that *Gossypium* HATs were differentially expressed in various vegetative tissues and at different stages of fiber development. Furthermore, all the HATs were differentially regulated in response to various stresses (salt, drought, cold, heavy metal and DNA damage) and hormones (abscisic acid (ABA) and auxin (NAA)). Finally, co-localization of HAT genes with reported quantitative trait loci (QTL) of fiber development were reported. Altogether, these results highlight the functional diversification of HATs in cotton growth and fiber development, as well as in response to different environmental cues. This study enhances our understanding of function of histone acetylation in cotton growth, fiber development, and stress adaptation, which will eventually lead to the long-term improvement of stress tolerance and fiber quality in cotton.

Keywords: histone acetyltransferases; genome-wide analysis; fiber; abiotic stress expression profiles; cotton

---

1. Introduction

Nucleosomes, the basic unit of chromatin, are composed of 147 bp of DNA wrapped around a histone octamer (two copies of each of H2A, H2B, H3, and H4 histone proteins). Nucleosomes are dynamic in response to developmental and environmental signals, thus altering the DNA accessibility and DNA-template processes to regulate the various processes in plants, including flowering time, root growth, and response to environmental changes [1,2]. Cells use several mechanisms, including post-translational histone modification and DNA methylation, to regulate the gene expression. The N-terminal tails of histone are subjected to various post-translational modifications, including histone acetylation, methylation, etc. [3]. Histone acetylation is carried out by histone acetyltransferases (HATs) in eukaryotes and is associated with transcriptional activation. Histone acetylation can be detected on different lysine residues of histone H3 and H4. For example, in *Arabidopsis*, K9, K14, K18, K23 and K27 of histone H3, and K5, K8, K12, K16, and K20 of H4 are acetylated [4,5]. HATs are well-conserved in yeast, animals, and plants, suggesting the functional conservation of histone acetylation in transcriptional activation. Plant HATs were classified into different subclasses based on their sequence homology to yeast and animal HATs and their mode of action: (1) The Gcn5-related N-acetyltransferase (GNAT)/MYST (Moz, YBF2, Sas2p, Tip) family, (2) CREB-binding Protein (CBP) family, and (3) TBP-associated factor$_{II}$ 250 (TAF$_{II}$250) family [6].

In plants, the genome-wide analysis of HATs has been performed in several species, including *Arabidopsis* [7], rice [8], *Vitis vinifera* [9], and *Citrus sinensis* [10]. HATs have been widely reported to play an important role in various aspects of plant development, including floral development [11–13], root growth [14], and gametophyte development [15]. In addition to developmental functions, HATs are also involved in plant adaptation to various environmental fluctuations, such as salt stress [16], cold stress [17], heat stress [18], light signaling [19], abscisic acid (ABA) [20], and other hormone signaling [21]. Therefore, the understanding of HATs functions in field crops may play an important role in sustainable agriculture and food security.

Cotton, as a major source of natural and renewable textile fiber, holds a significant value in the world economy and in daily human life [22]. *Gossypium hirsutum* is a natural allotetraploid (AADD) that arose from the interspecific hybridization between *Gossypium arboretum* (AA) and *Gossypium raimondii* (DD), which occurred approximately 1–2 million years ago [23]. Because allopolyploid cotton produces a superior quality of fiber with high yield compared with their diploid progenitors [24], *G. hirsutum* is widely grown in many parts of the world and contributes to more than 90% of commercial cotton production [25]. On the one hand, world climate is changing quickly, and the abiotic stresses are severely affecting the cotton yield and fiber quality [26,27]. On the other hand, the increasing world population demands the improvement of the cotton yield to meet the requirements of the textile industry. Therefore, identifying the potential genes conferring resistance to different stresses for the molecular breeding of cotton is of the utmost importance [28]. Although the role of HATs has been investigated in the plant development [29,30], response to environmental signals [16,18,31], and hormone signaling [16] in other plants, little is known regarding the function of HATs in cotton. In this scenario, *G. hirsutum*, *G. raimondii*, and *G. arboretum*, having recently available genomic data [32–35], provide an excellent opportunity to identify the candidate genes involved in fiber development and biotic and abiotic stress tolerance and to expand our understanding of underlying epigenetic mechanisms.

In this study, we identified HAT genes from the whole genome of *G. hirsutum*, *G raimondii*, and *G. arboretum*. The identified HATs were comprehensively analyzed for phylogenetic classifications, gene structures, identification of conserved motifs and domain organization, and the presence of cis-regulatory elements in their promoters. Furthermore, gene expression profiles of *G. hirsutum HATs* were analyzed during the different stages of fiber development and in response to various abiotic stresses. Such a comprehensive analysis of HATs provides a fundamental understanding of their roles in cotton growth and development. Furthermore, this study will be useful for functional genomic studies on the regulations of histone acetylation and will eventually lead to the long-term improvement of stress tolerance in cotton.

2. Results

2.1. Identification of HATs in Cotton

A systematic blast search was performed to identify the HATs in the genomes of *G. hirsutum*, *G. raimondii* and *G. arboretum* with the query sequence of *Arabidopsis*, and candidate HAT were identified in the cotton genomes. Then, Pfam and InterProScan databases were used to further verify the candidate HAT, and a total of 18 *G. hirsutum* (GhHATs), 9 *G. raimondii* (GrHATs), and 9 *G. arboretum* (GaHATs) were identified (Table S1). The properties of the identified *Gossypium* HATs were analyzed by ExPASy and we found an open reading frame (ORF) length ranging from 1407–6876 bp, which encoded the polypeptides ranging from 468–2291 amino acids with a predicted molecular weight of 53–256 KD. In addition, the theoretical isoelectronic point (PI) values ranged from 0.28–8.84 (Table S1).

2.2. Phylogenetic Analysis of the HAT Gene Family

To investigate the evolutionary relationship of HATs among cotton (*G.raimondii*, *G. arboretum*, and *G. hirsutum*) and other species, we constructed a phylogenetic tree using MEGA 6.0. We used the newly identified *Gossypium* HATs and previously identified HATs from *Arabidopsis thaliana*, *Vitis vinifera*, *Oryza sativa*, and *Brachypodium distachyon* to confirm their evolutionary relationship with the un-rooted phylogenetic tree using the neighbor-joining (NJ) method (Figure 1). The phylogenetic tree results showed that similar to *Vitis* and *Arabidopsis*, *Gossypium* HATs could also be grouped into three distinct classes: CPB, GNAT and TAF$_{II}$250. MYST homologs were found absent in the *Gossypium* genomes. To validate the phylogenetic tree constructed using the NJ method, we also used the minimum evolution method to construct a tree and found that HAT genes could be naturally classified into three groups and the members within each clade were stable with little difference between topology, which indicates that the NJ tree method could be used for further analysis.

We found that the number of genes in CPB (30) was greater than that in either GNAT (27) or TAF$_{II}$250 (9). Furthermore, we found that all the three groups comprised mono and dicot species. It is noteworthy that the genes within each group clustered with a dicot- or monocot-specific pattern. The number of HAT from each species was different within each group. Compared to other species, cotton HATs showed a closer relationship with *Vitis* HATs because they always clustered closely to each other in the phylogenetic tree. Nevertheless, their gene numbers were not similar within the group.

2.3. Phylogenetic Tree Construction, Conserved Motif and Gene Structure Analysis of Cotton HATs Genes

To confirm the subgroup classification and to determine the evolutionary relationship between *Gossypium* HATs, we generated another un-rooted phylogenetic tree using the neighbor-joining method (Figure 2A). We observed one subgroup related to CBP (HACs) and TAF$_{II}$250 (HAFs) in *Gossypium*, whereas three subgroups related to GNAT (HAGs) were observed. The CBP and GNAT subgroups contained four members in *G. raimondii* and *G. arboretum*, while TAF$_{II}$250 contained only one member. Furthermore, each diploid progenitor had its own ortholog in allotetraploid.

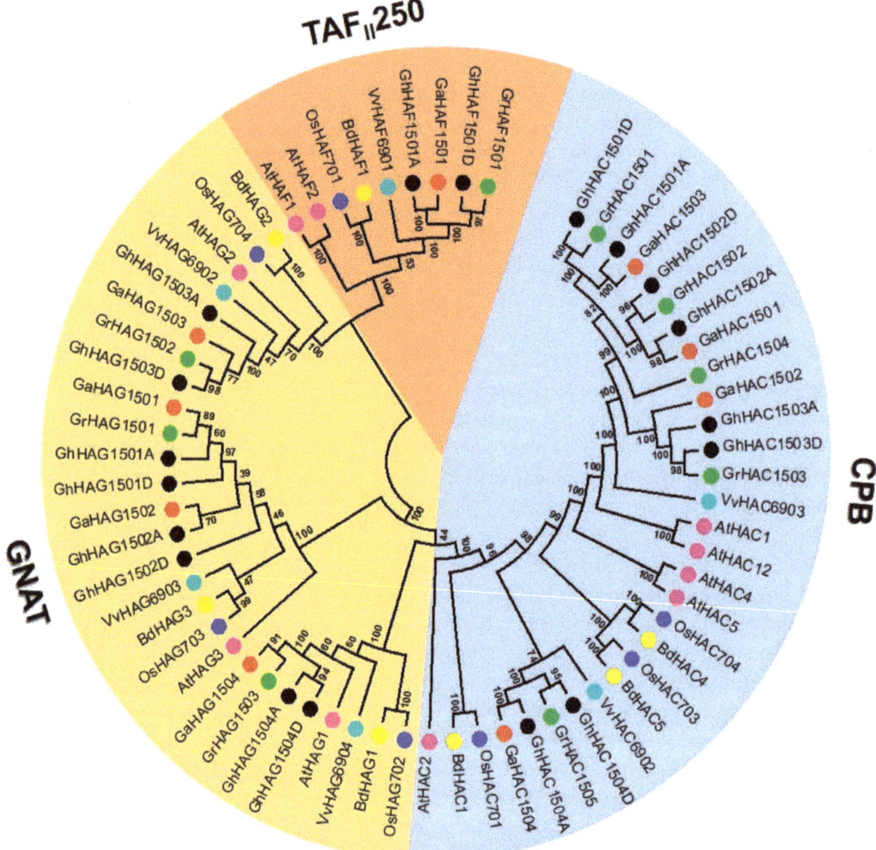

Figure 1. Phylogenetic relationships of histone acetyltransferases (HATs) from *Gossypium hirsutum*, *Gossypium raimondii*, *Gossypium arboretum*, *Arabidopsis thaliana*, *Oryza sativa*, *Brachypodium distachyon* and *Vitis vinifera*. The un-rooted phylogenetic tree was constructed using MEGA 6 by the neighbor-joining (NJ) method, and the bootstrap analysis was performed with 1000 replicates. For the phylogenetic tree, amino acid sequences were used and the classification of CPB, GNAT and TAF$_{II}$250 was performed based on the conserved signature domain of each subgroup. The HATs from *G. hirsutum*, *G. raimondii*, *G. arboretum*, *Arabidopsis*, *V. vinifera*, *Brachypodium distachyon*, and *Oryza sativa* were marked with black, green, red, pink, cyan blue, yellow, and blue dots, respectively.

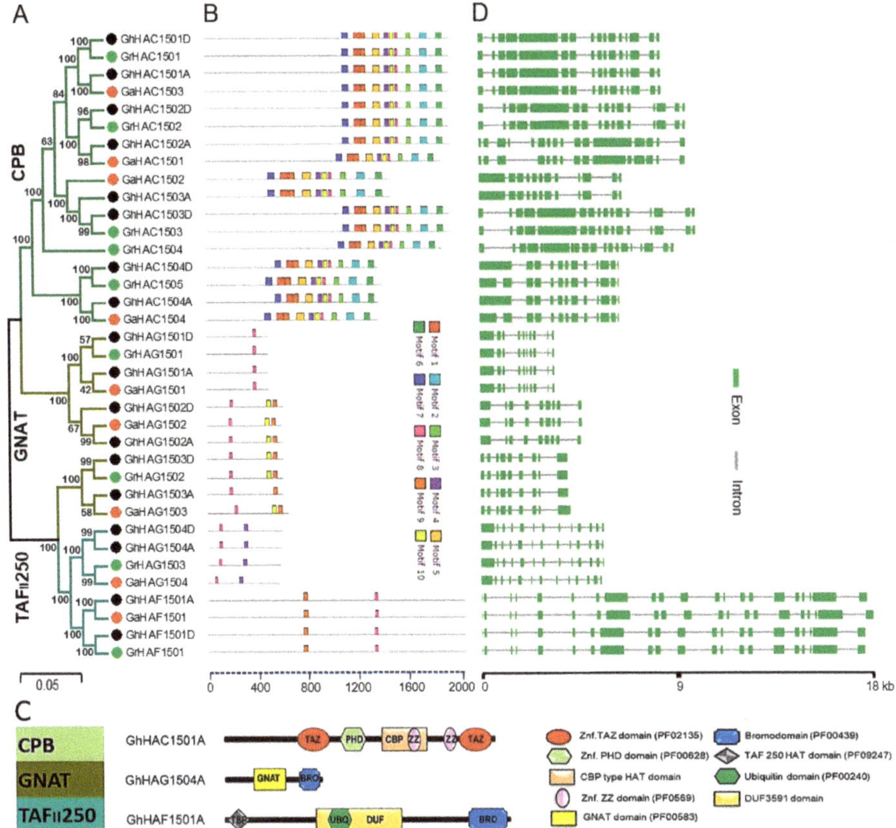

Figure 2. Phylogenetic relationships, conserved motifs, domain organization, and gene structure analysis of HATs in *Gossypium hirsutum*, *Gossypium raimondii* and *Gossypium arboretum*. (**A**) The unrooted phylogenetic tree was constructed using MEGA 6 by the NJ method, and the bootstrap analysis was performed with 1000 replicates. The HAT proteins from *G. hirsutum*, *G. raimondii*, and *G. arboretum* were marked with black, green and red dots, respectively. The branches of each subgroup were indicated in a specific color. (**B**) Motif identification analysis in HAT proteins from *G. hirsutum*, *G. raimondii* and *G. arboretum*. Each motif is shown in unique color, and the regular expression and sequences of the 1–10 motifs are listed in Figure S1. (**C**) Domain organization of HAT proteins from *G. hirsutum*, *G. raimondii* and *G. arboretum*. One representative of each subgroup of HAT from *G. hirsutum* is presented. (**D**) The intron/exon structure of *HAT* genes from *G. hirsutum*, *G. raimondii* and *G. arboretum*. The green boxes and grey lines represent exons and introns, respectively.

To gain insight into the evolutionary origin and putative functional diversification, a Multiple Expectation Maximization for Motif Elicitation (MEME) analysis was performed, which identified a total of 10 conserved motifs in *Gossypium* HATs (Figure 2B and Figure S1). All members of the CPB subfamily contained all the conserved motifs, whereas GNAT and TAF$_{II}$250 only contained a few conserved motifs. However, the motif 8 was found conserved in all the *Gossypium* HATs, suggesting that the potential catalytic site of *Gossypium* HATs was conserved. We further dissected the motif analysis of *Gossypium* HATs by investigating their domain organization (Figure 2C). The CBP subfamily of *Gossypium* HATs contained plant homeodomain (PHD), Znf-TAZ, Znf ZZ and a CBP-type HAT domain, while the GNAT subfamily of *Gossypium* HATs contained a conserved GNAT and bromodomain. Moreover, the TAF$_{II}$250 subfamily of *Gossypium* HATs contained a bromodomain, ubiquitin (UBQ), TATA box–binding protein (TBP), and domain of unknown function (DUF) domain. Furthermore, an alignment information was produced to explore the amino acid conservation of the GhHATs domain sequence (Figure S2). The multiple sequence alignment analysis revealed that all GhHATs proteins shared regions of conserved polypeptide sequences, which could be involved in their molecular functions (Figure S2A–C). In general, *Gossypium* HATs contained a domain organization similar to that of their counterparts in other species.

We then analyzed the gene structure of HATs from *G. hirsutum*, *G raimondii* and *G. arboretum* (Figure 2D). Our results showed that the number and length of intron/exons were different among CPB, GNAT and TAF$_{II}$250 classes. For example, intron/exon numbers and gene length of TAF$_{II}$250 were greater than the CPB and GNAT. Furthermore, the gene structure of *G. raimondii* and *G. arboretum* was extremely similar to their orthologs in the *G. hirsutum*. However, in general, gene structure in terms of intron/exon was greatly similar within the subgroup, which was consistent with the phylogenetic analysis.

2.4. Chromosomal Distribution and Duplication Analysis of HATs

The *G. hirsutum* HATs were mapped to their corresponding chromosomes (Figure 3A), and all the *HATs* were unevenly distributed on the chromosomes of *G. hirsutum*. For example, all the 18 *G. hirsutum* HATs were assigned to 12 chromosomes out of 26 (Figure 3A). The CBP subfamily of *G. hirsutum* HATs was localized on the chromosomes 5, 6, 8 and 10, while the GNAT subfamily of *G. hirsutum* HATs was localized on the chromosomes 6, 7, 10 and 11. We also mapped the HATs from the diploids *G. raimondii* and *G. arboretum* and found that all the HATs from *G. raimondii* and *G. arboretum* were also unevenly distributed on their chromosomes, similar to *G. hirsutum* (Figure 3B). However, compared with *G. raimondii*, the HATs from *G. arboretum* were more evenly distributed on their chromosomes. We also observed that subfamilies of *Gossypium* HATs were localized to different chromosomes in both diploid progenitors (*G. raimondii* and *G. arboretum*), as well as in the allotetraploid *G. hirsutum*. For example, the TAF$_{II}$250 subfamily member HAF1501 was localized on the chromosome 10 in *G. hirsutum*, and was localized on the chromosome 8 and 11 in *G. arboretum* and *G. raimondii*, respectively.

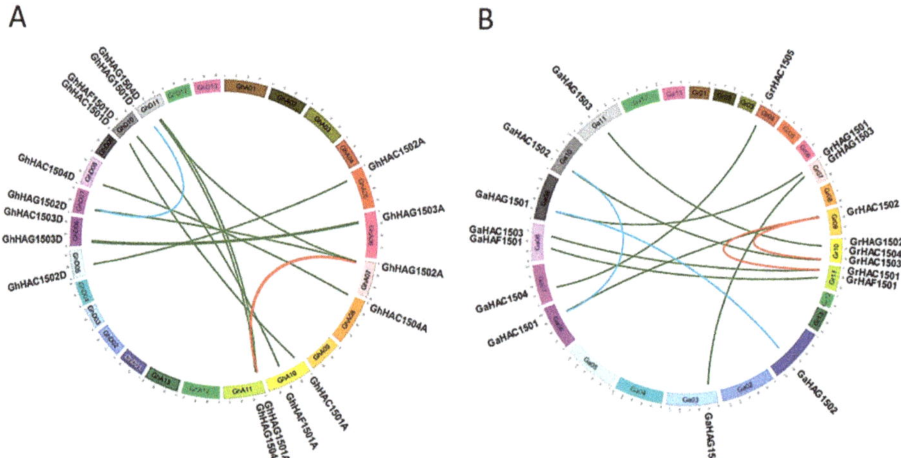

Figure 3. Chromosomal distribution and gene duplication of HATs in *Gossypium hirsutum* (**A**) and in *Gossypium raimondii* and *Gossypium arboretum* (**B**). The chromosome number was indicated in boxes and represented as Gh1-Gh13A/D (**A**), Ga1-Ga13, and Gr1-Gr13 (**B**) for *G. hirsutum*, *G. raimondii*, and *G. arboretum*, respectively. The orthologous HATs are connected by green, while the segment duplication of HATs is represented by light blue and red colors in different genomes.

We also investigated the contribution of gene duplication to the expansion of *Gossypium* HATs (Figure 3A,B). Two segmental duplications were found in *G. hirsutum* (*GhHAC1501A/GhHAC1502A*, *GhHAG1501A/GhHAG1502A*), indicating that duplications occurred in the A genome of *G. hirsutum*. Similarly, two segmental duplications were also found in *G. raimondii* (*GrHAC1501/GrHAC1502*, *GrHAC1502/GrHAC1503*) and in *G. arboretum* (*GaHAC1501/GaHAC1502*, *GaHAG1501/GaHAG1502*). Furthermore, we also checked the Ka/Ks ratio to explore the selective constraints on each pair of duplicated *Gossypium HAT* (Table S2). The Ka/Ks ratio was found to be less than 0.30 in all the duplicated gene pairs of *Gossypium* HATs, suggesting that all the duplicated gene pairs of *Gossypium* HATs experienced strong purifying selection pressure.

2.5. Putative Cis-Elements in the Promoter Regions of GhHATs

To gain more insight into the putative functions of HATs, the putative cis-regulatory elements were scanned in the 1000 bp upstream of the transcription start sites of *G. hirsutum* using the Plant CARE database (Figure S3, Tables S3 and S4). In addition to TATA- and CAAT-box core cis-elements, phytohormone response elements, stress response elements and development response elements were found in the promoters of *G. hirsutum* HATs. Most of the cis-elements were conserved among the CBP, GNAT and TAF$_{II}$250 subfamilies of *G. hirsutum* HATs. However, some cis-elements were absent in some subfamilies. For example, CAT-box (cis-acting regulatory element related to meristem expression), MRE (MYB binding site involved in light responsiveness), P-box (Gibberellin-responsive element), and O2-sites (cis-acting regulatory element involved in zein metabolism regulation) cis-elements were absent in the CBP subfamily, but were found present in GNAT and TAF$_{II}$250 members. Similarly, circadian (cis-acting regulatory element involved in circadian control), skn-1 motifs (cis-regulatory element required for endosperm expression), Box-W1 (fungal elicitor-responsive element), and 5′ UTR Py-rich stretch (a cis-regulatory element conferring high transcription levels) cis-elements were absent in the TAF$_{II}$250 subfamily. Moreover, cis-elements in the promoters of A and D genomes were largely conserved in *G. hirsutum HATs*. However, TGA cis-elements (a cis-acting regulatory element involved in the MeJA responsiveness) were only present in the D genome. We also observed that cis-elements in the promoters of orthologous gene pairs of *G. hirsutum* were largely similar. However, some exceptions were also found where they differed much for the cis-elements, e.g., GhHAC1504-A/GhHAC1504-D and GhHAG1503-A/GhHAG1503-D.

2.6. Gene Expression Analysis

2.6.1. Expression Analysis of *GhHATs* in Different Tissues, Developmental Stages and Multiple Abiotic Stresses by RNA-sequencing

To better understand the potential physiological functions of GhHATs in allotetraploid cotton, we investigated the expression of *GhHAT* genes. RNA-sequencing data were downloaded from the National Center for Biotechnology Information (NCBI) and analyzed. Their analysis revealed that *GhHATs* were widely expressed in the vegetative (root, stem, and leaf) and reproductive (torus, petal, stamen, pistil, calyx, and −3, −1, 0, 1, 3, 5, 10, 20, 25 and 35 days post-anthesis (DPA) ovule) tissues (Figure 4A), highlighting the diverse biological functions of HATs in different tissues. We also noted that some *GhHATs* did not express in vegetative tissues, but had a very weak expression in the reproductive tissue, e.g., *GhHAG1502-D*. We found that A and D genomes showed preferential expression for only some genes in the leaf, root, and stem. For instance, the expression of *GhHAG1502-A* was higher than *GhHAG1502-D* in all the analyzed tissues. However, the opposite correlation was also observed, e.g., the expression of *GhHAG1504-D* and *GhHAG1503-D* was higher in the root than *GhHAG1504-A* and *GhHAG1503-A*. In addition, we also found that *GhHATs* expression was also differently regulated during the course of development (Figure 4B). For example, *GhHAC1501-D* and *GhHAC1502-A/D* expressions were increased with the development of the root, while the expression of *GhHAC1502-D* decreased with seed cotyledon development. We further investigated the gene expression pattern of *GhHATs* in different abiotic stresses. The expression of *GhHAG1501-A/GhHAG1501-D* and *GhHAG1504-A/GhHAG1504-D* was strongly induced by multiple stresses, indicating their potential involvement in stress responses. However, no clear changes in expression levels were observed for more than half of the *GhHATs* under cold, hot, salt, and polyethylene glycol (PEG) 6000 conditions.

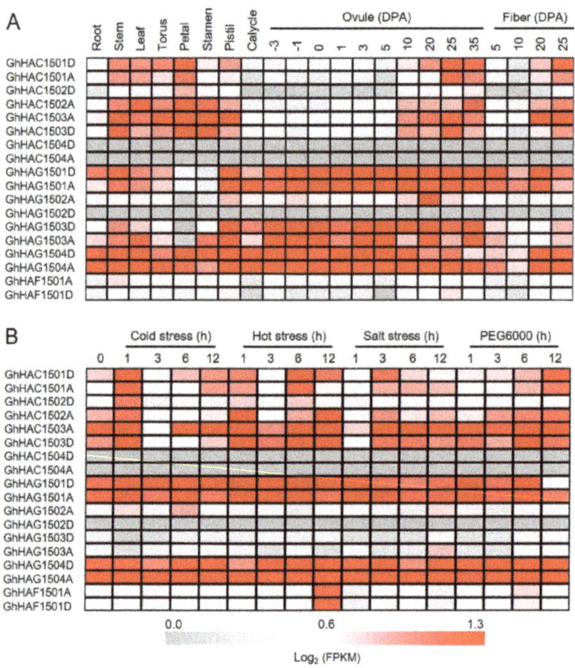

Figure 4. Gene expression pattern of *Gossypium hirsutum HATs* by the analysis of RNA-sequencing in different tissues and at different stages of fiber development (**A**) and in response to cold, hot, salt, and polyethylene glycol (PEG) 6000 stresses (**B**). The illumina reads of RNA-seq data were retrieved from the National Center for Biotechnology Information Sequence Read Archive (NCBI SRA) database. The color scale at the bottom of heat map indicates the fragments per kilobase million (FPKM)-normalized log2 transformed counts.

2.6.2. Expression Pattern of *GhHATs* Genes by qPCR

Tissue Specific Expression Patterns of HATs

We further validated the *GhHATs* gene expression in some vegetative and reproductive tissues by quantitative real-time polymerase chain reaction (qRT-PCR) (Figure 5A). Because the A and D genomes of allotetraploid cotton were extremely similar in mRNA levels, we considered *GhHAT-A* and *GhHAT-D* as one combination (*GhHAT*) and checked the expression levels by qRT-PCR. Among all the nine analyzed *GhHATs* belonging to three different classes, *GhHAG1501*, *GhHAG1502*, *GhHAC1503* and *GhHAG1504* showed the most prominent expression levels in the analyzed tissues, indicating their roles in the development of the leaf, root, stem and flowers. Although *GhHAG1502* and *GhHAC1503* were expressed in the root, stem and leaf, the highest expression was found in leaves. However, only *GhHAG1501, GhHAG1504* and *GhHAG1502* showed the highest expression in flowers, implying the specific function of these *GhHAGs* in flower development. It also indicated the dramatic functional divergence among different classes of GhHATs. Furthermore, *GhHAG1501* and *GhHAG1502* formed a paralogous pair and showed the expression conservation in all the analyzed tissues.

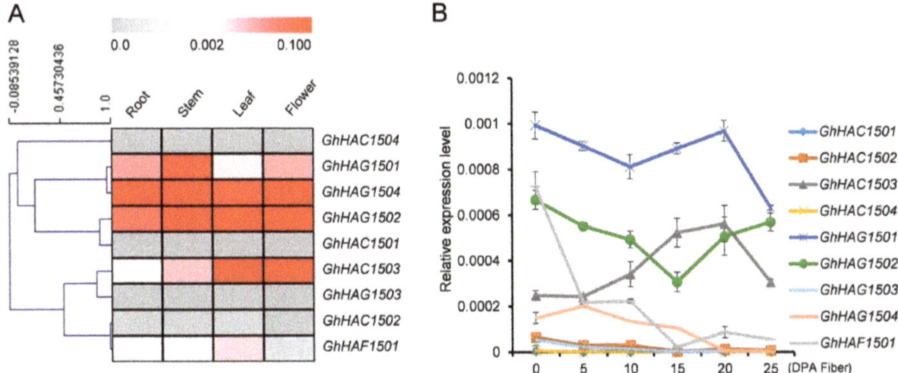

Figure 5. Gene expression validation of *HATs* from *G. hirsutum* by quantitative real-time polymerase chain reaction (qRT-PCR) in different tissues. (**A**) RNA from the root, leaf, stem and flowers was extracted and reverse transcribed. *Ubiquitin 7* (*UBQ7*) was used as an internal control for qRT-PCR. The relative expression is presented in the heat map, and the color scale at the top represents the relative signal intensity. The primer sequences can be found in Table S5. (**B**) Gene expression validation of *HATs* from *G. hirsutum* by qRT-PCR at different stages of fiber development. The data were normalized to *UBQ7*. The data presented are the average of three biological replicates. Bar = standard deviation (SD).

Expression of HATs at Different Fiber Development Stages

To explore the potential role of GhHATs in fiber development, we investigated their expression at different developmental stages of fiber development (0–25 DPA) by qRT-PCR (Figure 5B). Among all the *GhHATs*, only *GhHAG1501*, *GhHAG1502*, *GhHAC1503* and *GhHAF1501* showed the prominent expression levels during the different stages of fiber development (Figure 5B), indicating that these four genes may play dominant roles in the fiber development. However, among these four, the expression of *GhHAG1501* and *GhHAG1502* was the highest among the *GhHAGs* subgroup members at 0 DPA, and then gradually decreased during the fiber development (0–25 DPA). Similar to *GhHAGs* family members, the expression of *GhHAF1501* was also decreased during the fiber development, except at 20 DPA, where only the expression of *GhHAF1501*, *GhHAG1501* and *GhHAG1502* slightly increased and then again decreased at 25 DPA. Contrary to *GhHAF* and *GhHAG* family members, the expression of *GhHAC1503* from *GhHACs* family was gradually increased during the fiber development with a maximum at 20 DPA. Afterward, the expression of *GhHAC1503* was decreased at 25 DPA. We further separated the contribution of A and D genomes for the fiber development of *G. hirsutum* from RNA-seq data analysis and found that the A genome preferentially expressed more than D genome partners in the different fiber developmental stages (Figure 4A).

Expression of HATs in Response to Heavy Metals and Abiotic Stresses

To better understand the role of GhHATs in abiotic stresses, we investigated the gene expression of *GhHATs* in response to metal stress (Cd and Zn), salt stress (NaCl), cold stress and drought stress by qRT-PCR. All the abiotic stresses differentially regulated the expression of *GhHATs* (Figure 6), indicating the functional specificity of GhHATs in response to a particular stress. In response to Cd stress, the expression of *GhHAC1501*, *GhHAC1502* and *GhHAG1501* decreased, while the expression of *GhHAC1503*, *GhHAC1504*, *GhHAG1502* and *GhHAG1503* increased compared with the control. However, the expression of *GhHAG1501* and *GhHAG1504* did not change compared with the control. In response to Zn stress, only the expression levels of *GhHAC1503*, *GhHAC1504*, *GhHAG1501* and *GhHAG1502* increased, while all other studied genes did not show differential expression compared with the control. However, the expression levels of *GhHAC1503*, *GhHAC1504* and *GhHAG1502* were different in response to Cd and Zn stresses despite their increased expression. Furthermore, the expression of *GhHATs* was investigated in response to cold, salt, and drought stresses. The results showed that in response to cold stress, the expression of *GhHAC1502, GhHAC1503, GhHAC1504, GhHAG1501, GhHAG1502, GhHAG1503, GhHAG1504* and *GhHAF1501* decreased compared with the control. In response to salt stress, the expression of *GhHAC1501, GhHAC1502, GhHAC1503, GhHAG1501, GhHAG1504* and *GhHAF1501* increased, while the expression of *GhHAC1504* and *GhHAG1503* decreased compared with the control. In response to PEG treatment, the expression of *GhHAG1501* and *GhHAF1501* increased, while the expression of *GhHAC1503* and *GhHAC1504* decreased compared with the control.

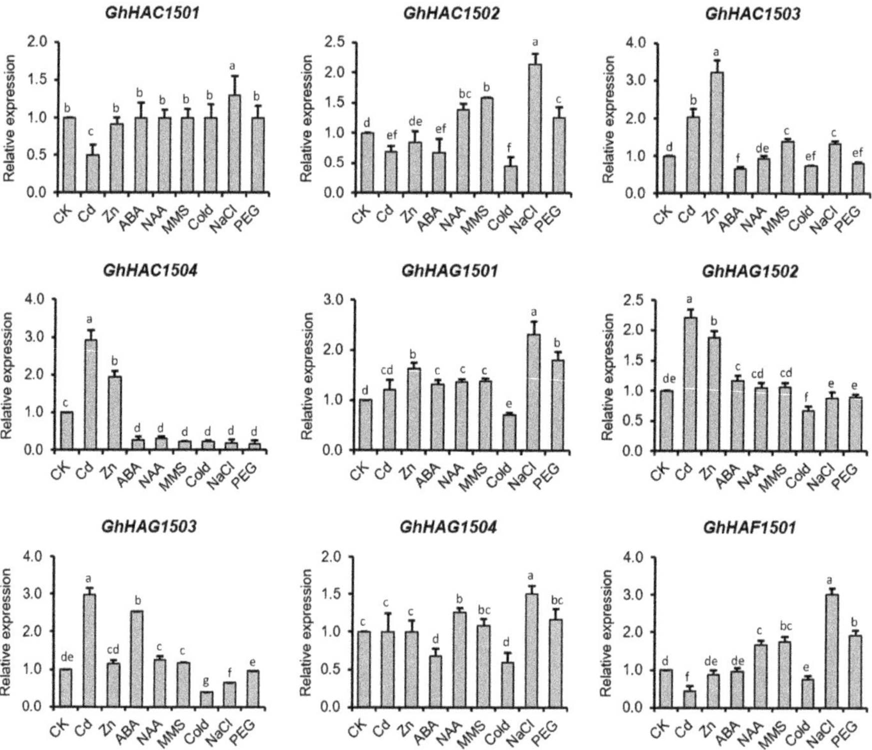

Figure 6. Expression pattern of *Gossypium hirsutum HATs* in response to abiotic stresses and hormones. RNA from roots after 24 h of CK (control), cadmium (Cd), zinc (Zn), abscisic acid (ABA), auxin (NAA), DNA damage (MMS), cold, salt (NaCl), and drought (PEG4000) was extracted and reverse transcribed. *UBQ7* was used as an internal control for qRT-PCR. The data presented are the average of three biological replicates. Different letters indicate significant difference by least significant difference (LSD) test ($p \leq 0.05$). Bar = SD.

Expression of HATs in Response to MMS

We also investigated the *Gossypium HATs* gene expression in response to the DNA damage agent methyl methanesulfonate (MMS) (Figure 6). The results showed that in response to MMS treatment, the expression of *GhHAC1502, GhHAC1503, GhHAG1501* and *GhHAF1501* increased, while the expression of *GhHAC1504* decreased compared with the control, indicating their potential role in DNA damage repair pathways in cotton.

Expression of HATs in Response to Phytohormones

Cis-regulatory elements related to phytohormones were found in the promoter of *GhHATs* (Figure S2, Tables S3 and S4). We therefore investigated the gene expression of *GhHATs* in response to auxin (NAA) and abscisic acid (ABA) by qRT-PCR. Auxin and ABA treatments differentially regulated the expression of *GhHATs* (Figure 6). The results showed that in response to ABA treatment, the expression of *GhHAC1503*, *GhHAC1504* and *GhHAG1504* decreased, while the expression of *GhHAG1501* and *GhHAG1503* increased compared with the control. In response to NAA treatment, the expression of *GhHAC1502*, *GhHAG1501* and *GhHAF1501* increased, while the expression of *GhHAC1504* decreased compared with the control.

2.7. Co-Localization of HATs with QTLs of Fiber Development

To validate the potential function of GhHATs in fiber development, the co-localization of *GhHATs* with reported QTLs/SNPs of fiber development (i.e., fiber length (FL), fiber elongation (FE), fiber micronaire (FM), fiber strength (FS), and fiber uniformity (FU) was analyzed. Nine genes were mapped to eight chromosomes with reported QTLs of FL, FE, FS, FM, and FU (Figure 7). Among nine *GhHATs*, only four genes were co-localized with QTLs of FL, FE, FM, FS, and FU on four chromosomes, i.e., Chr-A05, Chr-A06, Chr-A08 and Chr-A11 of the A subgenome. *GhHAC1502* was located within the qFE-A05-2, *GhHAG1503* gene anchored in qFU-A06-1 and qFE-A06-2, and *GhHAG1504* was mapped in the qFU-A05-2 QTL region, while *GhHAC1504* was 2 Mb from the qFL-A08. Among the D subgenome HATs, *GhHAC1502*, *GhHAC1503* and *GhHAG1503* were anchored in the FL-QTL-9, SNP (i20058Gh, i38606Gh), and qFL-D06, respectively, while *GhHAF1501* and *GhHAG1504* were found in surroundings of reported SNP. Interestingly, some genes were co-localized with multiple QTLs related to different fiber development traits. For example, the *GhHAC1502* gene on chromosome A05 and D05 was anchored in FL and FE QTL, and the *GhHAG1503* gene was localized inside the FL, FE and FU QTL. This reveals that these *GhHATs* had pleiotropic effects on fiber development related traits.

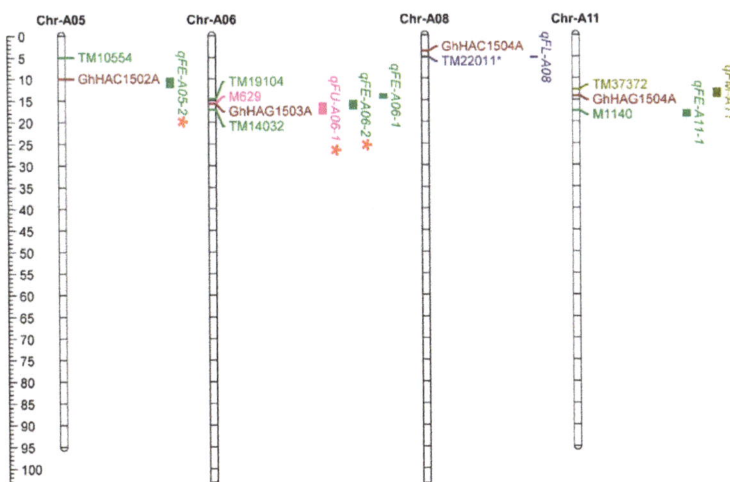

Figure 7. *Cont.*

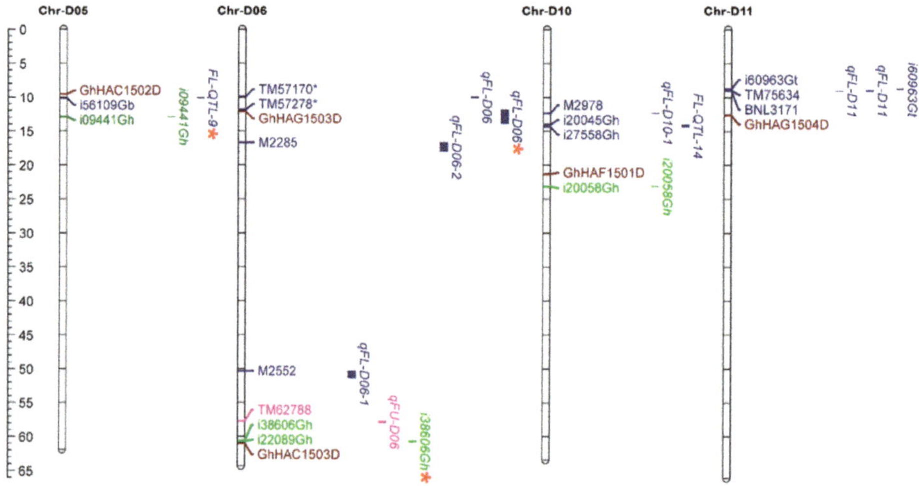

Figure 7. Distribution of co-localized *HAT* genes on chromosomes of A and D subgenomes of *G. hirsutum*. The scale represents the physical position of genes and quantitative trait loci (QTL)-linked markers in megabases (Mb). QTLs/Single Nucleotide Polymorphism (SNPs) related to fiber length (FL), fiber elongation (FE), fiber micronaire (FM), fiber strength (FS), and fiber uniformity (FU) are shown. Asterisks indicate that the *HAT* genes co-localized with QTLs/SNPs related to fiber.

3. Discussion

Histone acetylation is a mark of transcriptional activation and has been reported to play an important role in plant development and response to various biotic and abiotic stresses [12,14–16,18]. Moreover, levels of histone acetylation are tightly linked with gene expression regulation. HATs carry out histone acetylation and have been classified into different distinct groups in *Arabidopsis* [7], rice [8], *Vitis vinifera* [9], and *Citrus sinensis* [10]. In this study, we revealed that *Gossypium* HATs could also be classified into three major subgroups: CBP, GNAT, and TAF$_{II}$250. This suggests that multiple subgroups of *Gossypium* HATs might play specialized roles in the adaptive evolution of cotton. However, similar to rice [8], the MYST domain containing homologs of HATs in all the three genomes of cotton were absent, while *Arabidopsis* [7] and *Vitis* [9] had MYST domain members. This indicates the specific functional divergence of HATs between the different field crop plants. CBP, GNAT, and MYST HATs are also considered as transcriptional co-activators in addition to their HAT activity. For example, TAZ-type, ZZ-type and PHD-type zinc finger domains of CBP have been reported to play an important role in protein recognition and protein-protein interactions [36,37]. The PHD domain has also been reported to interact with histones and other histone-related proteins [38]. Furthermore, bromodomains are known to bind to acetylated lysine residues [39,40]. Along with functional catalytic domains, all the other conserved domains of *Gossypium* GNAT, CBP and TAF$_{II}$250 were largely similar to their counterparts in monocots and dicots (Figure 2C). These observations suggest that all the *Gossypium* HATs may have similar functions as described in other plant species.

Gene duplication plays a significant role in generating new gene subfamilies in the evolution of genome and genetic systems [41]. Tandem duplication, polyploidy, and segmental duplications primarily contribute to the creation of new gene families [41]. Two segmental duplications were found in *G. hirsutum* (*GhHAC1501A/GhHAC1502A, GhHAG1501A/GhHAG1502A*), *G. raimondii* (*GrHAC1501/GrHAC1502, GrHAC1502/GrHAC1503*), and *G. arboretum* (*GaHAC1501/GaHAC1502, GaHAG1501/GaHAG1502*) (Figure 3). Gene duplications occurred in these genes because the identities of the genes flanking both sides of the paralogous *Gossypium HAT* genes were found to be absolutely conserved and located on duplicated segments on two different chromosomes. Moreover, these duplicated genes were not likely diverged much during the evolution based on their Ka/Ks ratios (Table S2), suggesting the functional conservation of duplicated genes. This observation can partially be validated by the overlapping expression patterns of *GhHAG1501* and *GhHAG1502* during the fiber development and in different tissues (Figure 5A,B). However, the expression of *GhHAG1501* did not change in response to Cd, while the expression of *GhHAG1502* increased compared with control (Figure 6), suggesting that these duplicated genes may undergo functional divergence in response to particular stimuli.

Arabidopsis HATs play a crucial role in different aspects of plant development [12,14–16,18]. *G. hirsutum* has a huge contribution in the textile industry. Therefore, the identification of *HATs* genes in *G. hirsutum* and their role in fiber development will provide the fundamental information for future studies. Our results showed that the expression of *GhHAC1503* (CBP subgroup) was increased during the different stages of fiber development, while the expression of *GhHAG1501/GhHAG1502* (GNAT subgroup) and *GhHAF1501* (TAF$_{II}$250 subgroup) was generally decreased (Figure 5B). This suggests that histone acetylation levels are likely to be dynamic during the course of fiber development. However, further studies are required to investigate the function of histone acetylation in cotton fiber development. Phytohormones, including gibberellic acid (GA) [42], jasmonic acid (JA) [43], abscisic acid (ABA) [44], ethylene [45], and auxin (NAA) [46], are also known to regulate the fiber development. Cis-regulatory elements specific for ethylene (ERE), GA (GARE), ABA (ABRE), and JA (CGTCA) were found in the promoter of four highly expressed *GhHATs* (*GhHAC1503*, *GhHAG1501*, *GhHAG1502* and *GhHAF1501*) during the fiber development (Tables S3 and S4, Figure S3). This suggests that the expression of these *GhHATs* might be regulated by phytohormones. Consistent with this, the expression of *GhHAG1501/GhHAG1502* slightly increased in response to ABA treatment, while the expression of *GhHAC1503* decreased (Figure 6). The expression of *GhHAF1501* did not change in response to ABA treatment, but the expression was increased in response to NAA treatment (Figure 6). This suggests that these phytohormones directly or indirectly regulate the expression of these *GhHATs*, which may lead to different acetylation levels. The exogenous application of auxin has been shown to cause higher acetylation levels at the promoter of *SKP2B* [47], while the acetylation levels decrease at the promoter of *KRP7* in *Arabidopsis* [48], indicating the crosstalk between phytohormones and histone acetylation in plants. However, further studies are required to investigate the crosstalk between histone acetylation and phytohormones in cotton. Together, our results suggest that histone acetylation might play an important role in fiber development, as well as cotton growth. However, further studies are required to investigate the function of each GhHATs in phytohormone related pathways.

Histone acetylation has been reported to play a key role in DNA damage repair [49–51]. Histone acetyltransferase 1 (HAT1) is required for the incorporation of H4K5/H4K12 acetylated H3.3 histones at the sites of double strand breaks (DSB), which then facilitate the recruitment of a key DNA repair factor Rad51 in mammals [52]. Furthermore, DSB-inducing agents have been reported to induce the H4K16ac levels in mammals [53], indicating a key role of histone acetylation in DNA damage and repair pathways. In *Arabidopsis*, the histone acetyltransferases HAM1, HAM2 and HAG3 have been reported to participate in UV-B induced DNA damage [54,55]. Furthermore, a treatment with the histone acetyltransferase inhibitor curcumin increased DNA damage in *Arabidopsis* and maize [54], indicating the conserved role of histone acetylation in DNA damage in plants. Our qRT-PCR results showed that the expression of *GhHAC1502*, *GhHAC1503*, *GhHAG1501* and *GhHAF1501* increased, while the expression of *GhHAC1504* decreased compared with the control in response to MMS treatment (Figure 6). In brief, our results suggest the potential implication of GhHATs in DNA damage repair pathways in cotton. However, further studies are required to investigate whether and how GhHATs are involved in DNA damage repair.

Different abiotic stresses, including, cold, drought, salt, and metal stresses, severely affect the cotton growth and yield. Recent studies established the link of histone acetylation and abiotic stresses [32,56–58]. For example, salt treatment caused a global increase of H3K9 and H3K4 acetylations [32], while cold treatment decreased the H3K9, H4K5 and H4K4 acetylations compared with the control in maize [58]. Similarly, in response to dehydration treatment, drought responsive genes such as *RD29A, RD29B, RD20* and *RAP2.4* were found to be differentially acetylated at H3K9, H3K14, H3K23 and H3K27 in *A. thaliana* [56]. We showed that many *GhHATs* were differentially regulated in response to different abiotic stresses, suggesting the dynamic levels of histone acetylation in each abiotic stress adaptation in cotton. Our expression data also suggest the functional diversity and specificity among different subgroups of *Gossypium* HATs (GhHACs, GhHAGs and GhHAFs) in response to stress, as well as hormone treatments (Figure 6). For example, the expression of *GhHAC1502* strongly increased in response to NaCl, while the expression of *GhHAC1503* was strongly induced by Zn. Similarly, in the HAGs subgroup, the expression of *GhHAG1501* was strongly induced by NaCl, while the expression of *GhHAG1502* did not change. These observations suggest that GhHATs likely play an important role in cotton plant adaptation to various abiotic stresses. Furthermore, we also noticed that more than one gene responded to a particular stress. For example, in response to salt stress, the expression of *GhHAC1501, GhHAC1502, GhHAC1503, GhHAG1501, GhHAG1504* and *GhHAF1501* increased, while the expression of *GhHAC1504* and *GhHAG1503* decreased compared to the control. This suggests that different GhHATs may work together in response to particular stimuli and may participate in long-term resistance to different abiotic stresses. However, further molecular and biochemical studies are required to validate GhHATs function and to understand the underlying molecular mechanism.

Four *HAT* genes (*GhHAC1502, GhHAC1503, GhHAG1503* and *GhHAG1504)* on six chromosomes were anchored in fiber development-related QTL regions. These four genes were identified inside multiple QTLs, suggesting the pleiotropic role of these genes in fiber related traits. These results also support the conclusion that both the A and D genomes of *Gossypium hirsutum* jointly participate in different aspects of cotton fiber trait. Interestingly, *GhHAC1502, GhHAC1503, GhHAG1503* and *GhHAG1504* also displayed differential expression at different fiber development stages and in response to abiotic stresses, metal stress, MMS and phytohormones. This suggests that these four *HAT* genes are potentially involved in cotton growth and development, fiber-related traits, and plant response to the environment. However, in general, there were few discrepancies between RNA-seq and qRT-PCR data. RNA-seq expression levels were either from the A or D genome of *G. hirsutum* and it would be difficult to properly separate and count the reads from the homologous regions. On the contrary, we regarded *GhHACxA* and *GhHACxD* as one combination, referred to as *GhHACx*, because of the extremely high similarity between the mRNAs of the *GhHACxA-GhHACxD* gene pairs (for example, *GhHAC1502A/GhHAC1502D* and *GhHAG1501A/GhHAG1501D* have 99 and 99.5% similarity, respectively) and their nearly identical transcript sizes. With such a high similarity, we could not distinguish them using the qRT-PCR, and the results of qRT-PCR are theoretically the average of A and D genome's expression. Thus, potential difficulties of read counts on homologous regions in RNA-seq and inability to separate the A and G genome expressions by qRT-PCR may cause the discrepancy between RNA-seq and qRT-PCR results. In brief, further studies are required to comprehensively elucidate the function of HATs in different aspects of cotton plant.

In this study, we highlighted complex and diverse transcriptional regulations of *GhHATs* which significantly broadened our understanding of the underlying epigenetic mechanisms in allotetraploid cotton and unraveled an extra layer of complexity for better allotetraploid cotton adaptation in response to developmental, environmental, and hormonal cues. Furthermore, our results strongly recommend the comprehensive dissection of the biological and cellular function of GhHATS and argue for the potential implication of histone acetyltransferases in cotton molecular breeding in addition to other existing breeding strategies. This will eventually lead to the long-term improvement of stress tolerance in cotton and will allow high fiber yield and quality.

4. Materials and Methods

4.1. Identification of HATs Gene Family

The data of three cotton species, *G. raimondii* (JGI, version), *G. arboretum* (BJI, version 1.0), and *G. hirsutum* (NAU, version 1.1) were attained from the COTTONGEN (http://www.cottongen.org) [33–35]. The HAT protein sequences from *Arabidopsis*, rice, *Brachypodium distachyon*, and *Vitis Vinfera* were downloaded from Phytozome (http://phytozome.jgi.doe.gov/pz/portal.html) and then used as queries in BLASTP searches [59] against the *G. raimondii*, *G. arboretum* and *G. hirsutum* genomes, respectively. Genes with *E*-values < 1.0 were selected, and redundant sequences were eliminated by following the previously published method [60]. Furthermore, InterProScan (http://www.ebi.ac.uk/interpro/search/sequence-search) was used to confirm the presence of the HAT domain [61]. The physicochemical properties were predicted by ExPASy (http://cn.expasy.org/tools).

4.2. Analysis of Chromosomal Location and Gene Duplication

The loci of *HATs* were deduced from the gff3-files of cotton genome. The localization of *HATs* genes on the chromosomes were visualized using the program Circos [62]. The duplication events of *HATs* and Ka/Ks were calculated using the previously published method [63]. The T = Ks/2λ equation was used to determine the duplication time and deviation of the *HAT* gene pairs, assuming clock-like rates of (λ) 1.5×10^{-8} substitutions per synonymous site per year for cotton [64].

4.3. Sequence Alignment and Phylogenetic Analyses

Multiple sequence alignment was performed for the full-length HAT proteins using Clustal W with standard settings [65]. A neighbor-joining (NJ) phylogenetic tree was constructed using the full length HATs sequences from *G. raimondii*, *G. arboretum* and *G. hirsutum* by MEGA 6.0 [66], with P-distance and pairwise gap deletion parameters engaged. The bootstrap test was used with 1000 replicates to evaluate the statistical consistency of each node. To confirm the grades from the NJ method, the minimal-evolution method of MEGA 6.0 was utilized with 1000 replicates as well.

4.4. Gene Structure, Protein Motif, and Promoter Cis-Element Analysis

The exon/intron structures of the *HAT* genes were acquired from bed-file and displayed using the online tool Gene Structure Display Server (http://gsds.cbi.pku.edu.cn) [67]. The NJ tree was constructed with MEGA 6.0 as explained above. The deduced HAT protein sequences of three cotton species were submitted to the online Multiple Expectation Maximization for Motif Elicitation (MEME) version 4.11.1 (http://meme-suite.org/tools/meme) [68] as described [60]. For cis-element analysis in promoter regions, the 1 kb upstream sequences were analyzed in the PlantCARE database (http://bioinformatics.psb.ugent.be/webtools/plantcare/html/) [69].

4.5. Transcriptome Data Analysis and Gene Expression Heatmap

The raw data of RNA-seq of *G. hirsutum* were downloaded from the NCBI Sequence Read Archive (SRA: PRJNA248163) to calculate the expression level. TopHat (version: 2.0.13) was used for mapping reads, cufflinks (version: 2.2.1) were used to analyze gene expression levels, and fragments per kilobase million values were used to normalize gene expression levels [70]. Finally, these values of the *GhHAT* candidates were extracted from total expression data and the heatmap was generated by MeV 4.0 (http://www.tm4.org/).

4.6. Plant Materials, Stress Treatments, and qRT-PCR

G. hirsutum cultivar 'CRI35' was used for gene expression analysis. All the sampled tissues obtained from cotton plants grown under field condition with standardized cultural practices to determine the expression analysis [60]. For treatments, cotton seeds were surface sterilized and germinated on a moist paper. Young seedlings of same size were selected and exposed to NaCl (200 mM), polyethylene glycol 4000 (PEG4000) (15%), cold (4 °C), methyl methanesulfonate (MMS) (250 ppm), auxin (NAA) (10 µM), abscisic acid (ABA) (10 µM), $ZnSO_4$ (1 mM), and $CdCl_2$ (1 mM) for 24 h. All treatments were performed in three biological replicates. All samples were frozen quickly in liquid nitrogen and kept at −80 °C. The total RNA was extracted from cotton samples using the RNAprep Pure Plant kit (TIANGEN, Beijing, China). A total of 2 µg of RNA was used as the template, and the first-strand cDNAs were synthesized using the SuperScript III (Invitrogen, Waltham, MA, USA). Quantitative real-time PCR (qRT-PCR) analysis was performed as described previously in [71] using the specific primers for each *GhHAT* gene (Table S5). Cotton *UBQ7* (UniProt accession number: AY189972) was used as an internal reference gene for normalization of expression and three biological replicates were performed for each sample. To calculate the relative expression levels, a comparative $2^{-\Delta\Delta Ct}$ method was used [72]. The heat map for the gene expression profiles was generated with Mev 4.0 (http://www.tm4.org/) [73].

4.7. Co-localization of HATs with Fiber Related QTLs

To identify the localization of QTLs and SNPs for fiber development related traits, QTLs and linked molecular markers were retrieved from the Cotton Gen website (https://www.cottongen.org). The sequence of each marker was fetched from Cotton Gen to obtain the physical position information. For this purpose, the sequence of each marker was BLAST against the *G. hirsutum* (AD1) and HAU genome in the CottonGFD database. HAT genes co-localized with QTLs were displayed to show *HAT* gene distribution on chromosomes, along with surrounding loci and QTLs, using mapchart software. Genes identified inside the QTL or ≤500 kb far from SNP were considered as anchored gene in QTL because cotton LD decay was approximately 0.80 Mb [74].

4.8. Statistical Analysis

After performing normal distribution of the data and the homogeneity of variance tests, analysis of variance (ANOVA) was performed, followed by the least significant difference (LSD) test at p value ≤ 0.05 for each parameter. Different letters indicate significant difference by LSD test ($p \leq 0.05$). Statistical analyses were performed using the Statistical Package for Social Sciences (SPSS) software (version 11.5, SPSS Inc., Chicago, IL, USA).

5. Conclusions

In this study, 36 HATs were identified in three genomes of *Gossypium* and clustered into three groups: CPB, GNAT, and $TAF_{II}250$. *Gossypium HATs* are unevenly distributed on the chromosomes, and segmental duplications contributed to the evolution of the *HATs* family. Cis-element analysis discovered several abiotic and biotic stresses and hormonal responsive elements in the promoter region of the *GhHATs*, but each member had peculiar types and numbers. Furthermore, the expression profile of the upland cotton *HAT* gene family exhibited different expression patterns in response to abiotic and hormonal stresses, which disclosed that GhHATs play roles in different aspects of upland cotton abiotic stress tolerance and hormonal signaling. Thus, our study helps to lay the foundation for the functional characterization of the *GhHATs* gene family by overexpression and knockdown/out using RNAi or CRISPR-Cas9 genome editing and provides new insight into the evolution and divergence of *HAT* genes in plants. Furthermore, these results may enhance the understanding of the molecular mechanisms of many agronomic traits of cotton, such as fiber development and other physiological processes.

Supplementary Materials: Supplementary Materials can be found at http://www.mdpi.com/1422-0067/20/21/5311/s1.

Author Contributions: Conceptualization, S.S.; Formal analysis, M.I., S.S., M.A.F., M.K.N., E.W. and A.B.; Funding acquisition, S.S. and R.R.-C.W.; Methodology, M.I.; Writing—original draft, M.I. and S.S.; Writing—review & editing, S.S., K.B.J. and R.R.-C.W.

Funding: This research was supported by the grant PD-IPFP/HRD/HEC/2013/1129 from Higher Education Commission of Pakistan.

Acknowledgments: We would like to thank Yasar Sajjad (Department of Biotechnology, COMSATS University Islamabad, Abbottabad campus, Pakistan) for his valuable suggestions.

Conflicts of Interest: The authors declare no conflict of interest.

References

1. Berr, A.; Shafiq, S.; Shen, W.-H. Histone modifications in transcriptional activation during plant development. *Biochim. Et Biophys. Acta (BBA)-Gene Regul. Mech.* **2011**, *1809*, 567–576. [CrossRef]
2. Ho, L.; Crabtree, G.R. Chromatin remodelling during development. *Nature* **2010**, *463*, 474. [CrossRef] [PubMed]
3. Patel, D.J.; Wang, Z. Readout of epigenetic modifications. *Annu. Rev. Biochem.* **2013**, *82*, 81–118. [CrossRef] [PubMed]
4. Earley, K.W.; Shook, M.S.; Brower-Toland, B.; Hicks, L.; Pikaard, C.S. In vitro specificities of Arabidopsis co-activator histone acetyltransferases: Implications for histone hyperacetylation in gene activation. *Plant J.* **2007**, *52*, 615–626. [CrossRef] [PubMed]
5. Zhang, K.; Sridhar, V.V.; Zhu, J.; Kapoor, A.; Zhu, J.-K. Distinctive core histone post-translational modification patterns in Arabidopsis thaliana. *PLoS ONE* **2007**, *2*, e1210. [CrossRef] [PubMed]
6. Sterner, D.E.; Berger, S.L. Acetylation of histones and transcription-related factors. *Microbiol. Mol. Biol. Rev.* **2000**, *64*, 435–459. [CrossRef]
7. Pandey, R.; MuÈller, A.; Napoli, C.A.; Selinger, D.A.; Pikaard, C.S.; Richards, E.J.; Bender, J.; Mount, D.W.; Jorgensen, R.A. Analysis of histone acetyltransferase and histone deacetylase families of Arabidopsis thaliana suggests functional diversification of chromatin modification among multicellular eukaryotes. *Nucleic Acids Res.* **2002**, *30*, 5036–5055. [CrossRef] [PubMed]
8. Liu, X.; Luo, M.; Zhang, W.; Zhao, J.; Zhang, J.; Wu, K.; Tian, L.; Duan, J. Histone acetyltransferases in rice (Oryza sativa L.): Phylogenetic analysis, subcellular localization and expression. *BMC Plant Biol.* **2012**, *12*, 145. [CrossRef]
9. Aquea, F.; Timmermann, T.; Arce-Johnson, P. Analysis of histone acetyltransferase and deacetylase families of Vitis vinifera. *Plant Physiol. Biochem.* **2010**, *48*, 194–199. [CrossRef]
10. Xu, J.; Xu, H.; Liu, Y.; Wang, X.; Xu, Q.; Deng, X. Genome-wide identification of sweet orange (Citrus sinensis) histone modification gene families and their expression analysis during the fruit development and fruit-blue mold infection process. *Front. Plant Sci.* **2015**, *6*, 607. [CrossRef]
11. Deng, W.; Liu, C.; Pei, Y.; Deng, X.; Niu, L.; Cao, X. Involvement of the Histone Acetyltransferase AtHAC1 in the Regulation of Flowering Time via Repression of *FLOWERING LOCUS C* in Arabidopsis. *Plant Physiol.* **2007**, *143*, 1660–1668. [CrossRef] [PubMed]
12. Bertrand, C.; Bergounioux, C.; Domenichini, S.; Delarue, M.; Zhou, D.-X. Arabidopsis histone acetyltransferase AtGCN5 regulates the floral meristem activity through the WUSCHEL/AGAMOUS pathway. *J. Biol. Chem.* **2003**, *278*, 28246–28251. [CrossRef] [PubMed]
13. Han, S.K.; Song, J.D.; Noh, Y.S.; Noh, B. Role of plant CBP/p300-like genes in the regulation of flowering time. *Plant J.* **2007**, *49*, 103–114. [CrossRef] [PubMed]
14. Kornet, N.; Scheres, B. Members of the GCN5 histone acetyltransferase complex regulate PLETHORA-mediated root stem cell niche maintenance and transit amplifying cell proliferation in Arabidopsis. *Plant Cell* **2009**, *21*, 1070–1079. [CrossRef]
15. Latrasse, D.; Benhamed, M.; Henry, Y.; Domenichini, S.; Kim, W.; Zhou, D.-X.; Delarue, M. The MYST histone acetyltransferases are essential for gametophyte development in Arabidopsis. *BMC Plant Biol.* **2008**, *8*, 121. [CrossRef]
16. Sokol, A.; Kwiatkowska, A.; Jerzmanowski, A.; Prymakowska-Bosak, M. Up-regulation of stress-inducible genes in tobacco and Arabidopsis cells in response to abiotic stresses and ABA treatment correlates with dynamic changes in histone H3 and H4 modifications. *Planta* **2007**, *227*, 245–254. [CrossRef]
17. Stockinger, E.J.; Mao, Y.; Regier, M.K.; Triezenberg, S.J.; Thomashow, M.F. Transcriptional adaptor and histone acetyltransferase proteins in Arabidopsis and their interactions with CBF1, a transcriptional activator involved in cold-regulated gene expression. *Nucleic Acids Res.* **2001**, *29*, 1524–1533. [CrossRef]

18. Bharti, K.; von Koskull-Döring, P.; Bharti, S.; Kumar, P.; Tintschl-Körbitzer, A.; Treuter, E.; Nover, L. Tomato heat stress transcription factor HsfB1 represents a novel type of general transcription coactivator with a histone-like motif interacting with the plant CREB binding protein ortholog HAC1. *Plant Cell* **2004**, *16*, 1521–1535. [CrossRef]

19. Benhamed, M.; Bertrand, C.; Servet, C.; Zhou, D.-X. Arabidopsis GCN5, HD1, and TAF1/HAF2 interact to regulate histone acetylation required for light-responsive gene expression. *Plant Cell* **2006**, *18*, 2893–2903. [CrossRef]

20. Chen, Z.; Zhang, H.; Jablonowski, D.; Zhou, X.; Ren, X.; Hong, X.; Schaffrath, R.; Zhu, J.-K.; Gong, Z. Mutations in ABO1/ELO2, a subunit of holo-Elongator, increase abscisic acid sensitivity and drought tolerance in Arabidopsis thaliana. *Mol. Cell. Biol.* **2006**, *26*, 6902–6912. [CrossRef]

21. Nelissen, H.; De Groeve, S.; Fleury, D.; Neyt, P.; Bruno, L.; Bitonti, M.B.; Vandenbussche, F.; Van Der Straeten, D.; Yamaguchi, T.; Tsukaya, H. Plant Elongator regulates auxin-related genes during RNA polymerase II transcription elongation. *Proc. Natl. Acad. Sci. USA* **2010**, *107*, 1678–1683. [CrossRef] [PubMed]

22. Cao, X. Whole genome sequencing of cotton—A new chapter in cotton genomics. *Sci. China Life Sci.* **2015**, *58*, 515–516. [CrossRef] [PubMed]

23. Chen, Z.J.; Scheffler, B.E.; Dennis, E.; Triplett, B.A.; Zhang, T.; Guo, W.; Chen, X.; Stelly, D.M.; Rabinowicz, P.D.; Town, C.D. Toward sequencing cotton (Gossypium) genomes. *Plant Physiol.* **2007**, *145*, 1303–1310. [CrossRef] [PubMed]

24. Jiang, C.-X.; Wright, R.J.; El-Zik, K.M.; Paterson, A.H. Polyploid formation created unique avenues for response to selection in Gossypium (cotton). *Proc. Natl. Acad. Sci. USA* **1998**, *95*, 4419–4424. [CrossRef]

25. Wendel, J.F. New World tetraploid cottons contain Old World cytoplasm. *Proc. Natl. Acad. Sci. USA* **1989**, *86*, 4132–4136. [CrossRef]

26. Wang, R.; Ji, S.; Zhang, P.; Meng, Y.; Wang, Y.; Chen, B.; Zhou, Z. Drought Effects on Cotton Yield and Fiber Quality on Different Fruiting Branches. *Crop Sci.* **2016**, *56*, 1265–1276. [CrossRef]

27. Imran, M.; Shakeel, A.; Farooq, J.; Saeed, A.; Farooq, A.; Riaz, M. Genetic studies of fiber quality parameter and earliness related traits in upland cotton (Gossypium hirsutum L.). *Adv. Agric. Bot.* **2011**, *3*, 151–159.

28. Imran, M.; Shakeel, A.; Azhar, F.; Farooq, J.; Saleem, M.; Saeed, A.; Nazeer, W.; Riaz, M.; Naeem, M.; Javaid, A. Combining ability analysis for within-boll yield components in upland cotton (Gossypium hirsutum L.). *Genet. Mol. Res.* **2012**, *11*, 2790–2800. [CrossRef]

29. Tian, L.; Fong, M.P.; Wang, J.J.; Wei, N.E.; Jiang, H.; Doerge, R.; Chen, Z.J. Reversible histone acetylation and deacetylation mediate genome-wide, promoter-dependent and locus-specific changes in gene expression during plant development. *Genetics* **2005**, *169*, 337–345. [CrossRef]

30. Chen, Z.J.; Tian, L. Roles of dynamic and reversible histone acetylation in plant development and polyploidy. *Biochim. Et Biophys. Acta (BBA)-Gene Struct. Expr.* **2007**, *1769*, 295–307. [CrossRef]

31. Pavangadkar, K.; Thomashow, M.F.; Triezenberg, S.J. Histone dynamics and roles of histone acetyltransferases during cold-induced gene regulation in Arabidopsis. *Plant Mol. Biol.* **2010**, *74*, 183–200. [CrossRef] [PubMed]

32. Li, H.; Yan, S.; Zhao, L.; Tan, J.; Zhang, Q.; Gao, F.; Wang, P.; Hou, H.; Li, L. Histone acetylation associated up-regulation of the cell wall related genes is involved in salt stress induced maize root swelling. *BMC Plant Biol.* **2014**, *14*, 105. [CrossRef] [PubMed]

33. Zhang, T.; Hu, Y.; Jiang, W.; Fang, L.; Guan, X.; Chen, J.; Zhang, J.; Saski, C.A.; Scheffler, B.E.; Stelly, D.M. Sequencing of allotetraploid cotton (Gossypium hirsutum L. acc. TM-1) provides a resource for fiber improvement. *Nat. Biotechnol.* **2015**, *33*, 531. [CrossRef] [PubMed]

34. Li, F.; Fan, G.; Lu, C.; Xiao, G.; Zou, C.; Kohel, R.J.; Ma, Z.; Shang, H.; Ma, X.; Wu, J. Genome sequence of cultivated Upland cotton (Gossypium hirsutum TM-1) provides insights into genome evolution. *Nat. Biotechnol.* **2015**, *33*, 524. [CrossRef] [PubMed]

35. Wang, K.; Wang, Z.; Li, F.; Ye, W.; Wang, J.; Song, G.; Yue, Z.; Cong, L.; Shang, H.; Zhu, S. The draft genome of a diploid cotton Gossypium raimondii. *Nat. Genet.* **2012**, *44*, 1098. [CrossRef]

36. Gamsjaeger, R.; Liew, C.K.; Loughlin, F.E.; Crossley, M.; Mackay, J.P. Sticky fingers: Zinc-fingers as protein-recognition motifs. *Trends Biochem. Sci.* **2007**, *32*, 63–70. [CrossRef]

37. Bienz, M. The PHD finger, a nuclear protein-interaction domain. *Trends Biochem. Sci.* **2006**, *31*, 35–40. [CrossRef]

38. Lallous, N.; Legrand, P.; McEwen, A.G.; Ramón-Maiques, S.; Samama, J.-P.; Birck, C. The PHD finger of human UHRF1 reveals a new subgroup of unmethylated histone H3 tail readers. *PLoS ONE* **2011**, *6*, e27599. [CrossRef]

39. Servet, C.; e Silva, N.C.; Zhou, D.-X. Histone acetyltransferase AtGCN5/HAG1 is a versatile regulator of developmental and inducible gene expression in Arabidopsis. *Mol. Plant* **2010**, *3*, 670–677. [CrossRef]

40. Marmorstein, R.; Berger, S.L. Structure and function of bromodomains in chromatin-regulating complexes. *Gene* **2001**, *272*, 1–9. [CrossRef]

41. Cannon, S.B.; Mitra, A.; Baumgarten, A.; Young, N.D.; May, G. The roles of segmental and tandem gene duplication in the evolution of large gene families in Arabidopsis thaliana. *BMC Plant Biol.* **2004**, *4*, 10. [CrossRef] [PubMed]

42. Xiao, Y.-H.; Li, D.-M.; Yin, M.-H.; Li, X.-B.; Zhang, M.; Wang, Y.-J.; Dong, J.; Zhao, J.; Luo, M.; Luo, X.-Y. Gibberellin 20-oxidase promotes initiation and elongation of cotton fibers by regulating gibberellin synthesis. *J. Plant Physiol.* **2010**, *167*, 829–837. [CrossRef] [PubMed]

43. Wang, L.; Zhu, Y.; Hu, W.; Zhang, X.; Cai, C.; Guo, W. Comparative transcriptomics reveals jasmonic acid-associated metabolism related to cotton fiber initiation. *PLoS ONE* **2015**, *10*, e0129854. [CrossRef] [PubMed]

44. Samuel Yang, S.; Cheung, F.; Lee, J.J.; Ha, M.; Wei, N.E.; Sze, S.H.; Stelly, D.M.; Thaxton, P.; Triplett, B.; Town, C.D. Accumulation of genome-specific transcripts, transcription factors and phytohormonal regulators during early stages of fiber cell development in allotetraploid cotton. *Plant J.* **2006**, *47*, 761–775. [CrossRef] [PubMed]

45. Shi, Y.-H.; Zhu, S.-W.; Mao, X.-Z.; Feng, J.-X.; Qin, Y.-M.; Zhang, L.; Cheng, J.; Wei, L.-P.; Wang, Z.-Y.; Zhu, Y.-X. Transcriptome profiling, molecular biological, and physiological studies reveal a major role for ethylene in cotton fiber cell elongation. *Plant Cell* **2006**, *18*, 651–664. [CrossRef] [PubMed]

46. Zhang, M.; Zheng, X.; Song, S.; Zeng, Q.; Hou, L.; Li, D.; Zhao, J.; Wei, Y.; Li, X.; Luo, M. Spatiotemporal manipulation of auxin biosynthesis in cotton ovule epidermal cells enhances fiber yield and quality. *Nat. Biotechnol.* **2011**, *29*, 453. [CrossRef] [PubMed]

47. Manzano, C.; Ramirez-Parra, E.; Casimiro, I.; Otero, S.; Desvoyes, B.; De Rybel, B.; Beeckman, T.; Casero, P.; Gutierrez, C.; del Pozo, J.C. Auxin and epigenetic regulation of SKP2B, an F-box that represses lateral root formation. *Plant Physiol.* **2012**, *160*, 749–762. [CrossRef]

48. Anzola, J.M.; Sieberer, T.; Ortbauer, M.; Butt, H.; Korbei, B.; Weinhofer, I.; Mullner, A.E.; Luschnig, C. Putative Arabidopsis transcriptional adaptor protein (PROPORZ1) is required to modulate histone acetylation in response to auxin. *Proc. Natl. Acad. Sci. USA* **2010**, *107*, 10308–10313. [CrossRef]

49. Chailleux, C.; Tyteca, S.; Papin, C.; Boudsocq, F.; Puget, N.; Courilleau, C.; Grigoriev, M.; Canitrot, Y.; Trouche, D. Physical interaction between the histone acetyl transferase Tip60 and the DNA double-strand breaks sensor MRN complex. *Biochem. J.* **2010**, *426*, 365–371. [CrossRef]

50. Dhar, S.; Gursoy-Yuzugullu, O.; Parasuram, R.; Price, B.D. The tale of a tail: Histone H4 acetylation and the repair of DNA breaks. *Philos. Trans. R. Soc. Lond. Ser. B Biol. Sci.* **2017**, *372*. [CrossRef]

51. Murr, R.; Loizou, J.I.; Yang, Y.G.; Cuenin, C.; Li, H.; Wang, Z.Q.; Herceg, Z. Histone acetylation by Trrap-Tip60 modulates loading of repair proteins and repair of DNA double-strand breaks. *Nat. Cell Biol.* **2006**, *8*, 91–99. [CrossRef] [PubMed]

52. Yang, X.; Li, L.; Liang, J.; Shi, L.; Yang, J.; Yi, X.; Zhang, D.; Han, X.; Yu, N.; Shang, Y. Histone acetyltransferase 1 promotes homologous recombination in DNA repair by facilitating histone turnover. *J. Biol. Chem.* **2013**, *288*, 18271–18282. [CrossRef] [PubMed]

53. Li, L.; Wang, Y. Cross-talk between the H3K36me3 and H4K16ac histone epigenetic marks in DNA double-strand break repair. *J. Biol. Chem.* **2017**, *292*, 11951–11959. [CrossRef] [PubMed]

54. Campi, M.; D'Andrea, L.; Emiliani, J.; Casati, P. Participation of chromatin-remodeling proteins in the repair of ultraviolet-B-damaged DNA. *Plant Physiol.* **2012**, *158*, 981–995. [CrossRef] [PubMed]

55. Fina, J.P.; Casati, P. HAG3, a Histone Acetyltransferase, Affects UV-B Responses by Negatively Regulating the Expression of DNA Repair Enzymes and Sunscreen Content in Arabidopsis thaliana. *Plant Cell Physiol.* **2015**, *56*, 1388–1400. [CrossRef]

56. Kim, J.M.; To, T.K.; Ishida, J.; Morosawa, T.; Kawashima, M.; Matsui, A.; Toyoda, T.; Kimura, H.; Shinozaki, K.; Seki, M. Alterations of lysine modifications on the histone H3 N-tail under drought stress conditions in Arabidopsis thaliana. *Plant Cell Physiol.* **2008**, *49*, 1580–1588. [CrossRef]

57. Asensi-Fabado, M.A.; Amtmann, A.; Perrella, G. Plant responses to abiotic stress: The chromatin context of transcriptional regulation. *Biochim. Et Biophys. Acta. Gene Regul. Mech.* **2017**, *1860*, 106–122. [CrossRef]

58. Hu, Y.; Zhang, L.; Zhao, L.; Li, J.; He, S.; Zhou, K.; Yang, F.; Huang, M.; Jiang, L.; Li, L. Trichostatin A selectively suppresses the cold-induced transcription of the ZmDREB1 gene in maize. *PLoS ONE* **2011**, *6*, e22132. [CrossRef]

59. Altschul, S.F.; Madden, T.L.; Schäffer, A.A.; Zhang, J.; Zhang, Z.; Miller, W.; Lipman, D.J. Gapped BLAST and PSI-BLAST: A new generation of protein database search programs. *Nucleic Acids Res.* **1997**, *25*, 3389–3402. [CrossRef]

60. Imran, M.; Tang, K.; Liu, J.-Y. Comparative Genome-Wide Analysis of the Malate Dehydrogenase Gene Families in Cotton. *PLoS ONE* **2016**, *11*, e0166341. [CrossRef]

61. Quevillon, E.; Silventoinen, V.; Pillai, S.; Harte, N.; Mulder, N.; Apweiler, R.; Lopez, R. InterProScan: Protein domains identifier. *Nucleic Acids Res.* **2005**, *33*, W116–W120. [CrossRef] [PubMed]

62. Krzywinski, M.; Schein, J.; Birol, I.; Connors, J.; Gascoyne, R.; Horsman, D.; Jones, S.J.; Marra, M.A. Circos: An information aesthetic for comparative genomics. *Genome Res.* **2009**, *19*, 1639–1645. [CrossRef] [PubMed]

63. Imran, M.; Liu, J.-Y. Genome-wide identification and expression analysis of the malate dehydrogenase gene family in Gossypium arboreum. *Pak. J. Bot.* **2016**, *48*, 1081–1090.

64. Blanc, G.; Wolfe, K.H. Widespread paleopolyploidy in model plant species inferred from age distributions of duplicate genes. *Plant Cell* **2004**, *16*, 1667–1678. [CrossRef] [PubMed]

65. Thompson, J.D.; Higgins, D.G.; Gibson, T.J. CLUSTAL W: Improving the sensitivity of progressive multiple sequence alignment through sequence weighting, position-specific gap penalties and weight matrix choice. *Nucleic Acids Res.* **1994**, *22*, 4673–4680. [CrossRef]

66. Tamura, K.; Stecher, G.; Peterson, D.; Filipski, A.; Kumar, S. MEGA6: Molecular evolutionary genetics analysis version 6.0. *Mol. Biol. Evol.* **2013**, *30*, 2725–2729. [CrossRef]

67. Guo, A.; Zhu, Q.; Chen, X.; Luo, J. GSDS: A gene structure display server. *Yi Chuan = Hered.* **2007**, *29*, 1023–1026. [CrossRef]

68. Bailey, T.L.; Williams, N.; Misleh, C.; Li, W.W. MEME: Discovering and analyzing DNA and protein sequence motifs. *Nucleic Acids Res.* **2006**, *34*, W369–W373. [CrossRef]

69. Lescot, M.; Déhais, P.; Thijs, G.; Marchal, K.; Moreau, Y.; Van de Peer, Y.; Rouzé, P.; Rombauts, S. PlantCARE, a database of plant cis-acting regulatory elements and a portal to tools for in silico analysis of promoter sequences. *Nucleic Acids Res.* **2002**, *30*, 325–327. [CrossRef]

70. Trapnell, C.; Roberts, A.; Goff, L.; Pertea, G.; Kim, D.; Kelley, D.R.; Pimentel, H.; Salzberg, S.L.; Rinn, J.L.; Pachter, L. Differential gene and transcript expression analysis of RNA-seq experiments with TopHat and Cufflinks. *Nat. Protoc.* **2012**, *7*, 562–578. [CrossRef]

71. Shafiq, S.; Chen, C.; Yang, J.; Cheng, L.; Ma, F.; Widemann, E.; Sun, Q. DNA Topoisomerase 1 prevents R-loop accumulation to modulate auxin-regulated root development in rice. *Mol. Plant* **2017**, *10*, 821–833. [CrossRef] [PubMed]

72. Livak, K.J.; Schmittgen, T.D. Analysis of relative gene expression data using real-time quantitative PCR and the 2− ∆∆CT method. *Methods* **2001**, *25*, 402–408. [CrossRef] [PubMed]

73. Saeed, A.; Sharov, V.; White, J.; Li, J.; Liang, W.; Bhagabati, N.; Braisted, J.; Klapa, M.; Currier, T.; Thiagarajan, M. TM4: A free, open-source system for microarray data management and analysis. *Biotechniques* **2003**, *34*, 374–378. [CrossRef] [PubMed]

74. Sun, Z.; Wang, X.; Liu, Z.; Gu, Q.; Zhang, Y.; Li, Z.; Ke, H.; Yang, J.; Wu, J.; Wu, L.; et al. Genome-wide association study discovered genetic variation and candidate genes of fibre quality traits in *Gossypium hirsutum* L. *Plant Biotechnol. J.* **2017**, *15*, 982–996. [CrossRef] [PubMed]

International Journal of
Molecular Sciences

Article

Cloning and Expression Analysis of the *BocMBF1c* Gene Involved in Heat Tolerance in Chinese Kale

Lifang Zou [1,2], Bingwei Yu [1], Xing-Liang Ma [2], Bihao Cao [1], Guoju Chen [1], Changming Chen [1] and Jianjun Lei [1,3,*]

[1] Key Laboratory of Horticultural Crop Biology and Germplasm Innovation in South China, Ministry of Agriculture, College of Horticulture, South China Agricultural University, Guangzhou 510642, China; lifang.zou@gifs.ca (L.Z.); yubingwei@outlook.com (B.Y.); caobh01@scau.edu.cn (B.C.); gjchen@scau.edu.cn (G.C.); cmchen@scau.edu.cn (C.C.)
[2] Seed and Developmental Biology Program, Global Institute for Food Security, University of Saskatchewan, Saskatoon, SK S7N 0W9, Canada; Xingliang.ma@gifs.ca
[3] Henry School of Agricultural Science and Engineering, Shaoguan University, Shaoguan 512005, China
* Correspondence: jjlei@scau.edu.cn; Tel.: +86-133-8005-5679

Received: 3 September 2019; Accepted: 8 November 2019; Published: 11 November 2019

Abstract: Chinese kale (*Brassica oleracea* var. *chinensis* Lei) is an important vegetable crop in South China, valued for its nutritional content and taste. Nonetheless, the thermal tolerance of Chinese kale still needs improvement. Molecular characterization of Chinese kale's heat stress response could provide a timely solution for developing a thermally tolerant Chinese kale variety. Here, we report the cloning of *multi-protein bridging factor (MBF) 1c* from Chinese kale (*BocMBF1c*), an ortholog to the key heat stress responsive gene *MBF1c*. Phylogenetic analysis showed that *BocMBF1c* is highly similar to the stress-response transcriptional coactivator *MBF1c* from *Arabidopsis thaliana* (*AtMBF1c*), and the *BocMBF1c* coding region conserves MBF1 and helix-turn-helix (HTH) domains. Moreover, the promoter region of *BocMBF1c* contains three heat shock elements (HSEs) and, thus, is highly responsive to heat treatment. This was verified in *Nicotiana benthamiana* leaf tissue using a green fluorescent protein (GFP) reporter. In addition, the expression of *BocMBF1c* can be induced by various abiotic stresses in Chinese kale which indicates the involvement of stress responses. The BocMBF1c-eGFP (enhanced green fluorescent protein) chimeric protein quickly translocated into the nucleus under high temperature treatment in *Nicotiana benthamiana* leaf tissue. Overexpression of *BocMBF1c* in *Arabidopsis thaliana* results in a larger size and enhanced thermal tolerance compared with the wild type. Our results provide valuable insight for the role of *BocMBF1c* during heat stress in Chinese kale.

Keywords: Chinese kale; thermal tolerance; *BocMBF1c*; expression; localization

1. Introduction

Chinese kale (*Brassica oleracea* var. *chinensis* Lei Syn:*Brassica alboglabra* Bailey) is a popular vegetable for its health benefits and taste [1]. Chinese kale, belonging to the *Brassica* genus, is native to China and, currently, is predominantly cultivated in South China and Southeast Asia [1]. Recently, the importance of Chinese kale has generated interest in its nutrition and genetics [2]. In particular, Chinese kale contains high levels of glucoraphanin which may possess cancer prevention properties [3].

Unfortunately, Chinese kale is very sensitive to heat stress, which often leads to decreased yield and quality [4]. Consequently, the main growing season and distribution of Chinese kale is limited to the winter season of South China.

High temperature can disturb the membrane, structure of protein, and chromatin architecture of the plant cells which respond with several signaling pathways to address the stress [5]. Many plant hormones are reported in the heat response of plants, and exogenous application of plant hormones can activate expression of heat stress-related genes thus improving thermal tolerance. For example, applying abscisic acid (ABA) can increase the thermal tolerance level by upregulating *ABA responsive transcription factors (ABRFs)*, *heat stress transcription factor (HSF) A2c (FaHSFA2c)*, and *heat shock proteins (HSPs)* in tall fescue [6]; treatment with methyl jasmonate (MeJA) stimulates the expression of *HSP70* [7]; and exogenous salicylic acid (SA) increases the heat stress resistance of wheat by regulating the accumulation of ethylene (ET) and proline, increasing the efficiency of nitrogen usage [8]. Application of ET and jasmonic acid (JA) can activate thermal stress-related genes by upregulating *ethylene response factor (ERF) 1* in *Arabidopsis*.

In plants, enormous progress has recently been made in understanding of the molecular mechanism of thermal tolerance [9]. In addition to HSF- and HSP-responsive pathways [9], the transcriptional coactivator *Multi-protein bridging factor1c (MBF1c)* was identified as a critical regulator for thermal tolerance responses in *Arabidopsis* [10]. In *Arabidopsis*, heat stress can lead to increased expression of *AtMBF1c* and triggers the nuclear localization of AtMBF1c proteins. Nuclear AtMBF1c then function as transcription factors to regulate the downstream SA, trehalose, and ET thermal resistance-related pathways [10]. The *AtMBF1c* is a trans-acting regulatory element which recognizes CTAGA as a potential binding sequence and can regulate 36 genes, including stress-responsive gene *dehydration-responsive element (DRE)-binding protein 2A (DREB2A)* [11]. In addition, overexpression of *AtMBF1c* and its ortholog from wheat can improve the performance of host plants under high temperature [11,12]. Therefore, *MBF1c* is an ideal gene for improving plant thermal tolerance and, thus, increase productivity under heat stress. The expression of the *MBF1c* gene in Antarctic Moss (*PaMBF1c*) can be induced by several abiotic stresses, and overexpression of *PaMBF1c* can enhance heat, cold, and salinity stress tolerance in *Arabidopsis* [12]. However, to date there have been no studies conducted on the *MBF1c* of Chinese kale. In this study, we reported the ortholog of *AtMBF1c* in Chinese kale, termed *BocMBF1c*, and characterized its potential role in regulating heat tolerance in Chinese kale.

2. Results

2.1. Cloning and Analysis of the BocMBF1c Gene

As the genome sequence of Chinese kale is still unavailable, we used the *Brassica rapa* MBF1c gene (LOC103828628) as a reference sequence for primer design and then cloned the coding region of *MBF1c* from Chinese kale (variety "Ai-jiao-xiang-gu") genomic DNA. The resultant amplicon of 707 bp was sequenced and confirmed to be *BocMBF1c* by sequence homology analysis (Figure 1a,b). Open reading fragment (ORF) prediction by ORF finder did not find any intron within the *BocMBF1c* gene, which was then confirmed by agarose gel analysis (Figure S1). Amino acid sequence analysis identified multi-protein bridging factor1 (MBF1) and helix-turn-helix (HTH) domains (Figure 1a). Phylogenetic analysis indicated that BocMBF1c clustered with *Arabidopsis* MBF1c (Figure 1b, Table S1). Next, we obtained the *MBF1c* gene and promoter region of *Brassica oleracea* (Bol013952) by searching the *BocMBF1c* coding sequence (CDS) using BLASTN (http://brassicadb.org/brad/blastPage.php). The 2005 bp DNA fragment containing the promoter of *BocMBF1c* was cloned with primers designed from *BoMBF1c*, from which we used 1409 bp upstream sequence from initiation codon for further analysis. The *BocMBF1c* gene was submitted to GenBank (Accession number: MH685643).

Sequence analysis of the 1409 bp *BocMBF1c* promoter revealed multiple stress and hormonal responsive cis-elements, including heat shock elements (HSEs), ethylene responsive elements (EREs), abscisic acid-responsive element (ABRE), MeJA responsive motifs (CGTCA-motif), drought stress and pathogen-responsive TC-rich repeat, and MYB binding sites (MBS). All of those cis-elements are located upstream of the TATA box (Figure 2).

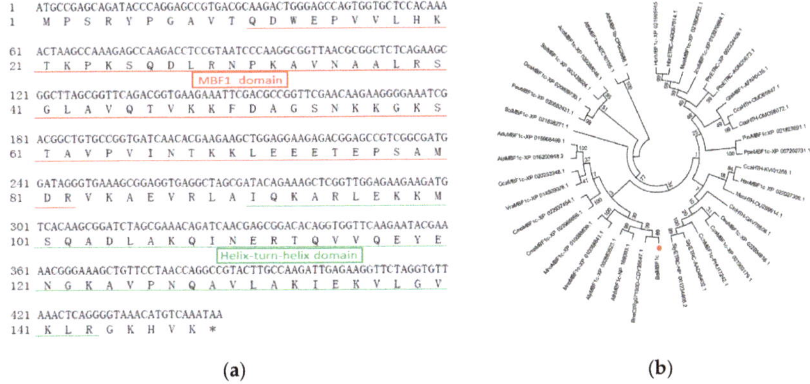

(a)   (b)

Figure 1. Sequence feature of BocMBF1c protein. (**a**) DNA and amino acid sequence of BocMBF1c. Full-length DNA and deduced amino acid sequence of BocMBF1c. Red underline indicates multi-protein bridging factor1 (MBF1) gene family domain, and green underline signatures helix-turn-helix (HTH) domain. (**b**) Phylogenetic tree constructed by the neighbor-joining method based on the MBF1c amino acid sequences.

Figure 2. 1409 bp upstream of the start code are shown as the *BocMBF1c* promoter. DNA elements in the *BocMBF1c* promoter were designated. The start code is designated as bold and the TATA box as grass green. Three heat shock elements (HSEs) are boxed in red, ethylene-responsive elements (EREs) in green boxes, ABA-responsive elements (ABREs) in yellow boxes, methyl jasmonate (MeJA)-responding elements in purple, MYB binding sites (MBS) in the purple red box, and TC-rich repeats as stress-responding elements in the blue box.

We cloned the coding region of *BocMBF1c* from genomic DNA, and the deduced BocMBF1c protein contained the *MBF1* gene family domain and HTH domain (Figure 1a and Figure S2). The amino acid sequence of BocMBF1c shared 99% and 94% similarity with its ortholog from *Brassica napus* (CDY39547.1) and *Arabidopsis*, respectively (NP_189093.1). We queried the BocMBF1 protein sequence using the BLASTP program (https://blast.ncbi.nlm.nih.gov/Blast.cgi), and the resulting sequences all belonged to angiosperm. The phylogenetic tree showed *BocMBF1c* was clustered in one group with *MBF1c* genes from dicots, while all *MBF1c* genes from monocots belonged to another group. In addition, *BocMBF1c* belonged to the *MBF1c* gene clade from the Brassicaceae family including *Brassica napus* and *Arabidopsis*. The *AtMBF1a/b* were clustered in an individual cluster, implying they belonged to different members of the *MBF1* gene family (Figure 1b, Table S1).

2.2. Transient Expression Analysis Showed BocMBF1c Promoter Activity Can Be Induced by Heat Stress

The *BocMBF1c* promoter contains several HSEs, indicating its role in heat stress response. Thus, we analyzed the activity of the *BocMBF1c* promoter under heat treatment. We combined the 1409 bp region upstream of the BocMBF1c start code with the CDS of *green fluorescent protein* (*GFP*), and then transiently expressed in *Nicotiana* (*N.*) *benthamiana* leaf tissue by agroinfiltration under ambient and heat-stressed conditions. As predicted, the activity of the *BocMBF1c* promoter was rapidly stimulated under heat stress in Tabaco leaves (Figure 3a,b), which is consistent with previous reports [13].

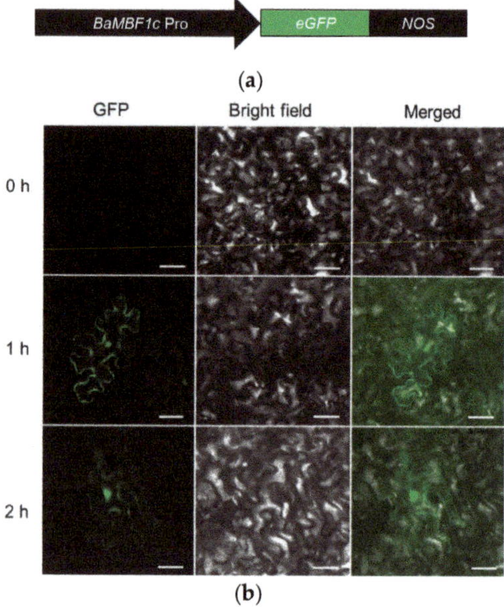

Figure 3. Heat treatment can stimulate *BocMBF1c* promoter activity. (**a**) Legend of pBI101-*pBocMBF1c-GFP* construct; *BocMBF1c* promoter was fused with *green fluorescent protein* (*GFP*) coding sequence (CDS) with the *NOS* terminator. (**b**) Treatment at 37 °C induced increased expression in the *Nicotiana* (*N.*) *benthamiana* leaves after agroinfiltration. Scale bar, 50 μm.

2.3. Expression Pattern of BocMBF1c

Under a standard growth environment, expression of *BocMBF1c* was measured in various tissues of Chinese kale (Figure 4a). The results showed *BocMBF1c* had a comparatively higher expression level in combining sites (CS) and leaf veins (LV) and maintained a comparatively low expression level in other tissues (Figure 4b).

Under heat stress, *BocMBF1c* expression rapidly increased >250 fold within 0.5 h and maintained this high expression level for at least 8 h in the leaves of Chinese kale (Figure 5a). When the Chinese kale was exposed to chilling conditions, *BocMBF1c* transcript abundance increased five-fold within 0.5 h, then slowly returned to unstressed levels by 4 h post-treatment (Figure 5b). As illustrated in Figure 2, the *BocMBF1c* promoter possessed a series of hormone and other related cis-elements. The qPCR analysis confirmed that *BocMBF1c* was moderately responsive to salinity treatment and several plant hormones including MeJA, ABA, SA, and ET (Figure 5c,d). Interestingly, the temporal response patterns of *BocMBF1c* expression were distinct among different treatments. Under heat and cold stress, the increase in the *BocMBF1c* gene reached full extent within 0.5 h. For salinity treatment, the expression level of *BocMBF1c* reached the maximum after 8 h. When MeJA, ABA, SA, and ethephon (CEPA) were applied, the highest *BocMBF1c* transcription activities were detected after 2 h, 8 h, 2 h, and 16 h, respectively. Overall, our results indicated that other abiotic factors could also influence the transcript level of *BocMBF1c* besides responding to high-temperature conditions. These results indicate that *BocMBF1c* in Chinese kale involves resistance to multiple stresses including heat stress.

(**a**)   (**b**)

Figure 4. The expression pattern of *BocMBF1c* in different tissues under standard growth conditions. (**a**) Various tissues of Chinese kale. SL, senescent leaf; ML, mature leaf; YL, young leaf; LV, leaf vein; Pe, petiole; YB, young bolting stem flesh; MB, middle bolting stem flesh; BSS, bolting stem skin; FB, flower buds; CS, combining sites between stem and root; Ro, Roots. (**b**) Bolting stage tissues were harvested for expression specificity analysis. Quantitative PCR (qPCR) was used to measure *BocMBF1c* expression in different tissues, with the *BocTubulin8* gene as the control. Data shown are the mean ± SD of three biological replicates.

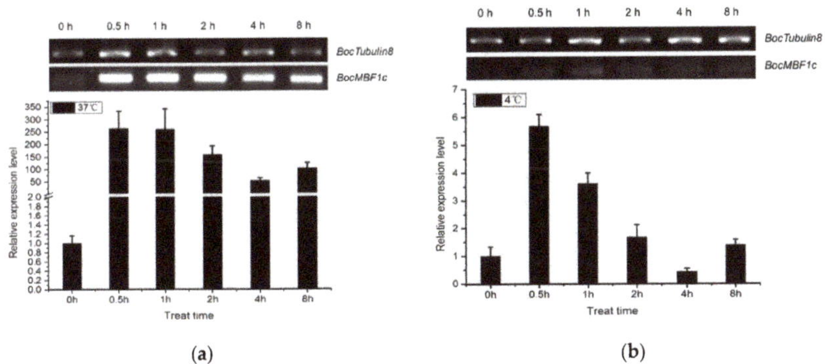

(**a**)   (**b**)

Figure 5. *Cont.*

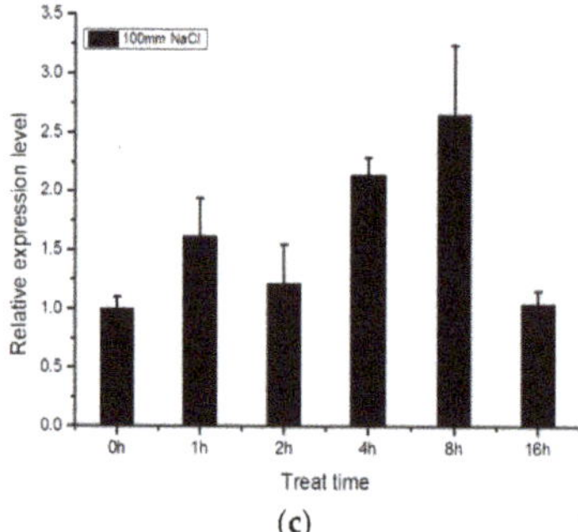
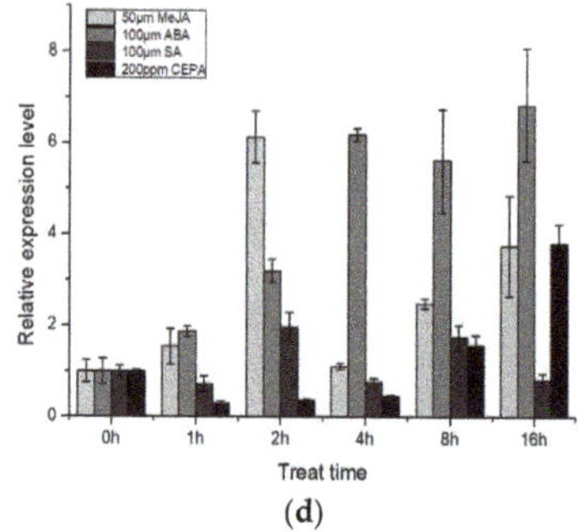

(c)   **(d)**

Figure 5. Responses of *BocMBF1c* to various abiotic treatments. (**a**) semiRT-PCR (reverse transcription PCR) and qPCR were used to determine the expression profile under heat stress (37 °C) in the leaf tissue of Chinese kale. (**b**) Expression of *BocMBF1c* under cold treatment (4 °C). semiRT-PCR and qPCR were used to determine the expression profile under cold condition in the leaf tissue of Chinese Kale. (**c**) Responses of *BocMBF1c* to salinity treatment. (**d**) Expression of *BocMBF1c* after applying hormones. All analyses used *Tubulin8* as an internal reference gene, and all data shown were normalized to mock treatments. Data shown are the mean ± SD of three biological replicates.

2.4. BocMBF1c Protein Localizes to the Nucleus under Heat Stress

Under normal temperature, the GFP-tagged BocMBF1c protein shows no obvious distribution preference when transiently expressed in *N. benthamiana* leaf tissue (Figure 6b). However, the chimeric BocMBF1c-eGFP protein quickly concentrated into the nucleus when exposed to 37 °C for 1 h (Figure 6c). This translocation is critical for the functionality of this heat responsive transcription coactivator.

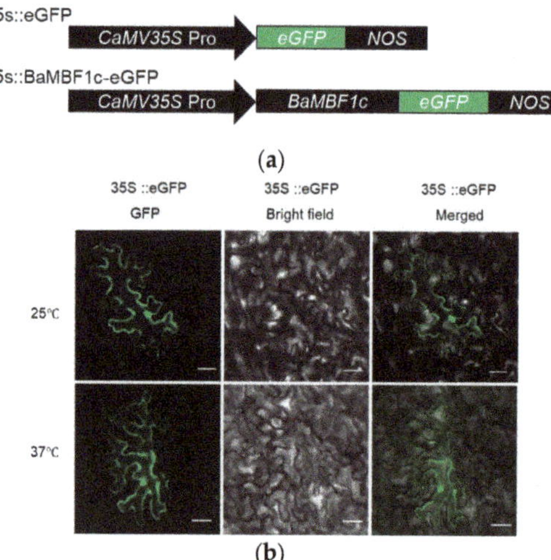

Figure 6. *Cont.*

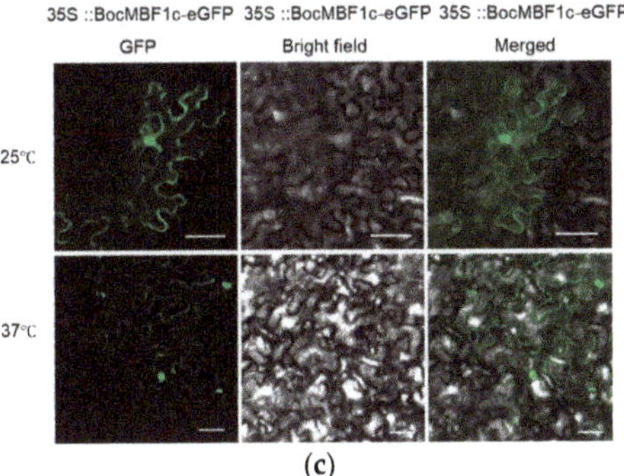

(c)

Figure 6. Heat stress led to the nucleus localization of the BocMBF1c protein. (**a**) Schematics of the constructs for 35s::*eGFP* and 35s::*BocMBF1c-eGFP* (pBE-*BocMBF1c-eGFP*). (**b**) The GFP protein alone has no obvious localization preference either under normal temperature or high temperature in *N. benthamiana* leaf tissue. (**c**) BocMBF1c translocated to the nucleus when exposed to 37 °C for 1 h. Scale bar, 50 μm.

2.5. Phenotype and Thermotolerance Analysis of BocMBF1c Overexpression Lines

We selected three T_4 homozygous transgenic lines (Figure S3) overexpressed with the *BocMBF1c* and β-glucuronidase (*GUS*) fusion gene driven by the 35S *cauliflower mosaic virus* (*CaMV*) promoter (Figure 7a) for further study. We named them *BocMBF1c*–OE1, *BocMBF1c*–OE2, and *BocMBF1c*–OE3. β-glucuronidase activity analyses were performed with 5 day old seedlings to validate the expression of *BocMBF1c* (Figure 7b). We also examined the growth of 2 week old *BocMBF1c*-OE plants. The *BocMBF1c*-OE plants show larger sizes compared to wild type (WT) under normal growth conditions (Figure 7c) which is consistent with a previous report [14]. The five-day-old seedlings grown on MS plates were used for thermal tolerance analysis. After treatment at 46 °C for 2 h, the transgenic lines showed a survival ratio of 21.7% to 30.0%, while only 10.7% of the WT plants survived after heat stress (Figure 7d).

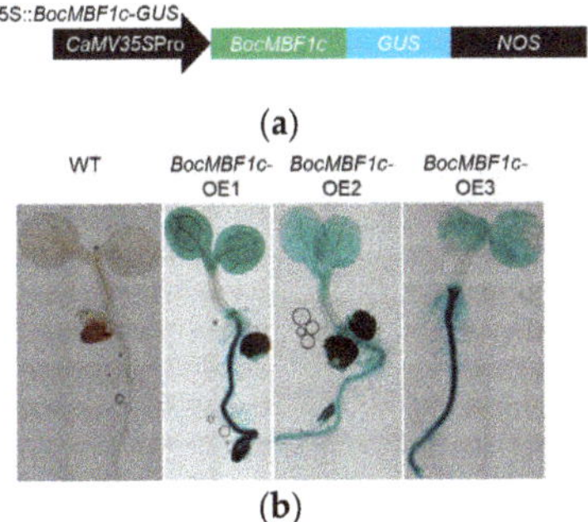

(a)

(b)

Figure 7. *Cont.*

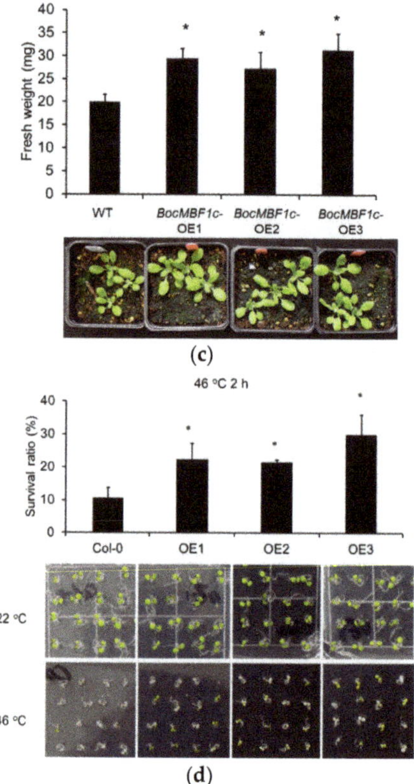

Figure 7. Phenotypic characterization of *BocMBF1c-OE* plants. (**a**) Schematic illustration of the 35s::*BocMBF1c-GUS* construct used for ectopic expression. (**b**) Expression of *BocMBF1c* was validated by β-glucuronidase (GUS) staining ($n \geq 15$). (**c**) Compared with the wild type (WT) seedlings, the fresh weight of the *BocMBF1c*-OE plants indicated better growth, and the plants shown in the pictures were 14 days post-germination. Error bars represent the standard deviation of the means ($n = 20$). (**d**) All three *BocMBF1c*-OE lines showed a higher survival ratio after heat stress than WT ($n = 100$). Error bars are the mean ± SD of three independent experiments. Photos were taken 2 days after heat stress. Asterisk indicates statistical significance by the Student's test ($p < 0.05$).

3. Discussion

3.1. BocMBF1c Can Respond to Multiple Stresses and Hormone Treatment in Addition to Heat Stress in Chinese Kale

We previously measured the effects of heat stress among 10 different Chinese kale varieties with different growth parameters, and the variety "Ai-jiao-xiang-gu" was found to be the most tolerable to heat stress [15]. We further measured the expression level of *BocMBF1c*; "Ai-jiao-xiang-gu" had the highest increase in expression level [15], indicating its important role in heat stress tolerance.

The promoter of *BocMBF1c* contained several HSEs; moreover, in *N. benthamiana* cells, the *BocMBF1c* promoter activity remained steady under ambient temperatures but was activated rapidly by high temperature stress. This is supported by semi-quantitative and qPCR evidence that the transcript abundance of *BocMBF1c* in planta is strongly and rapidly increased in response to heat stress treatment. Together, these data indicate that the expression of the *BocMBF1c* gene responds to heat stress via promoter activity. Our results showed that *BocMBF1c* can respond to cold, salinity, and several hormone treatments thus potentially possessing other functions besides thermal tolerance.

Bioinformatic analysis predicted that the *BocMBF1c* promoter contained ERE- and ABRE-responding cis-elements, implying ERFs and ABFs (ABRE binding factors)/ABEB could regulate the expression of *BocMBF1c*. Accumulation of ABA can initiate the expression of stress-responsive genes and participate in the regulation of the network through ABFs/AREB [16]. Ethylene and JA can stimulate the expression of downstream stress-related genes by activating ERF1 [17]. Previous reports showed that AtMBF1c functions upstream of SA and ET during thermotolerance [10]. From our results, *BocMBF1c* reacted distinctly under different hormone treatments. The expression levels of *BocMBF1c* both increased after exogenous application of MeJA and ABA although with different temporal response feature. For MeJA and ABA treatment, *BocMBF1c* expression peaked after 2 h and 4 h, respectively, which is possibly attributed to the difference in activation times for ERF1 and ABFs/AREB. After SA and ET application, the expression level of *BocMBF1c* both increased after a certain reduction period at the beginning. This indicates that *BocMBF1c* could possibly be negatively regulated by a high concentration of ET and SA, since the applied hormones would slowly degrade after application. With lower concentrations, ET can activate ERF1 thus stimulating the expression of *BocMBF1c*.

Overall, the accumulation of ABA, ET, and JA can activate corresponding transcription factors, which will then interact with the cis-elements located at the *BocMBF1c* promoter. The expressed BocMBF1c protein would then participate in the heat stress regulation network by regulating the downstream SA and ET signaling pathway (Figure 8).

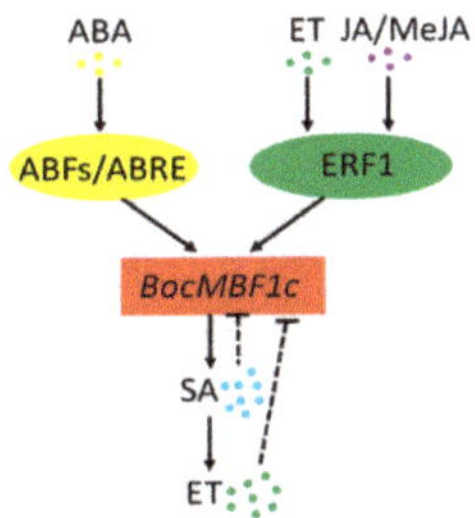

Figure 8. A model for the relation of *BocMBF1c* and other hormones. abscisic acid (ABA) and ethylene (ET) and jasmonic acid (JA)/MeJA activate ABRE binding factors (ABFs)/ABRE and Ethylene response factor1 (ERF1), respectively. ABFs/ABRE and ERF1 then upregulate the *BocMBF1c* gene by interacting with cis-elements in the promoter region. The BocMBF1c protein can regulate the salicylic acid (SA) and ethylene (ET) heat stress responsive pathway. High concentrations of SA and ET will inhibit the expression of *BocMBF1c* through negative feedback.

Through analysis of the *Arabidopsis* public database (http://bar.utoronto.ca/efp_arabidopsis/cgi-bin/efpweb.cgi), we found *AtMBF1c* could be upregulated by heat, salt, and ABA treatment, and that it does not respond to cold and MeJA treatment (Figure S4) (http://bar.utoronto.ca/efp_arabidopsis/cgi-bin/efpWeb.cgi) [18]. The different responsive spectrums between *BocMBF1c* and *AtMBF1c* implies the *BocMBF1c* gene could participate in stress-related tolerance for the host plant.

The BocMBF1c protein quickly relocated to the nucleus when the cells were exposed to high temperature, further suggesting that this is a regulator for the expression of many heat stress-responsive genes. However, the mechanism of this nuclear translocation in a high-temperature environment remains unknown and should be the focus of future study. Likewise, downstream genes activated by this transcription factor remain unknown in Chinese kale, and transcriptome analysis by next-generation sequencing would be useful in the identification of candidate downstream genetic elements. Expression of these elements as well as known targets from other species can be further confirmed by qPCR. Earlier studies found the "CTAGA" element as a potential binding sequence for downstream genes; thus, it may be interesting to analyze the orthologs of the downstream genes in Chinese kale. This may reveal the reason for the relative temperature sensitivity of Chinese kale. Thus, it is highly likely that the *BocMBF1c* gene is necessary for multiple biotic/abiotic stresses in Chinese kale.

3.2. BocMBF1c Can be Applied to Improve the Productivity of Chinese Kale

The high-level conservation of *MBF1c* sequences (Table S1 and Figure S2) implies that this gene is critical for thermal tolerance and climate adaptation. For Chinese kale, bolting stems are the main edible tissue. The genetic mechanism underlying declined yield and quality of bolting stems under heat stress are still elusive in Chinese kale; cloning of *BocMBF1c* genes as a key heat tolerance factor would shed light on improving the heat tolerance of Chinese kale.

Transformation of Chinese kale is readily feasible and has been used for improving drought stress tolerance [19]. Our results indicate that the *BocMBF1c* gene can respond to heat treatment from both transcription and protein localization level. Overexpression *BocMBF1c* in *Arabidopsis* result in larger-sized plants with enhanced heat tolerance, while knockout of the *BocMBF1c* gene will be needed to fully validate its role in the thermal tolerance of Chinese kale, for example, by the CRISPR/Cas9-based gene editing technique [20]. Ectopic expression of *AtMBF1c* can improve tolerance to bacterial infection, heat, and osmotic stress in *Arabidopsis* [14]. Further study showed that overexpression of *AtMBF1c* could improve thermal tolerance without impairing the yield in *Arabidopsis* [10], soybean [11], and rice [21] under controlled conditions. A recent report showed that overexpression of *MBF1c* in Antarctic moss could make the plant bigger and enhance tolerance to salinity stress in *Arabidopsis* [12]. Thus, it will be intriguing to overexpress the *BocMBF1c* gene as a potential solution to improving the biotic/abiotic stress resistance and productivity of Chinese kale.

4. Materials and Methods

4.1. Plant Growth and Abiotic Treatment Conditions

We collected Chinese kale samples for tissue-specific expression analysis from the greenhouse (22–25 °C) at the South China Agricultural University (Guangzhou, China) [2]. For abiotic treatment, Chinese kale (*B. oleracea* var. *chinensis* Lei) variety "Ai-jiao-xiang-gu" were grown in a culture room under controlled conditions at 25 °C, with a 16 h light/8 h dark cycle. Chinese kale seedlings of 4~6 true leaves were transferred to the growth chamber for 24 h with the same conditions, and then the temperature was switched to heat (37 °C) and cold (4 °C) treatments. Leaf tissues were harvested for RNA extraction at 0 h, 0.5 h, 1 h, 2 h, 4 h, and 8 h.

For salinity treatment, each Chinese kale seedling of 4~6 true leaves were sprayed and watered by 100 mL of 100 mM NaCl, and leaf tissues were collected with the timeline of 0 h, 1 h, 2 h, 4 h, 8 h, and 16 h. Similarly, hormone treatments were carried out by applying 50 µM of MeJA solution, 100 µM ABA, 100 µM SA, and 100 ppm CEPA, respectively, and leaf tissues were collected along the same timeline. All samples used in this study were collected from three separated Chinese kale.

Arabidopsis thaliana (ecotype Columbia) were grown in chamber under controlled condition at 22 °C day/18 °C night, 70% relative humidity, and 100 µmol m^{-2} s^{-1} with a 16 h light/8 h dark light cycle. *Arabidopsis* seeds were germinated on MS plates or soil (Sunshine® Mix #8, Sun Gro Horticulture) after being sterilized with 10% bleach for 10 min. All seeds were kept at 4 °C for at least 2 days before transfer to the growing chamber.

4.2. Cloning of the BocMBF1c CDS and Promoter

Leaves of Chinese kale were harvested for genomic DNA extraction [22]. We used the *MBF1c* gene from *Brassica rapa* (LOC103828628) as a reference sequence (Supplementary B1) to design primers for the CDS of *BocMBF1c*. The gDNA fragment containing the *BocMBF1c* CDS was amplified with primer MBF1c-up (5′-TCGAACTCTCCAGAAACTCGT-3′) and MBF1c-dw (5′-CGTTTGCCAGATACGATGT-3′). Next, *BocMBF1c* promoter was amplified with primer MBF1cP-F (5′-AGAATCCCCATTGACAGCCT-3′) and MBF1cP-R (5′-GGCATCGTCGCTGTTGTTTA-3′), designed from the gene sequence of *Brassica oleracea* (Supplementary B2). All PCRs were carried out with HiFiTaq PCR StarMix (GeneStar) according to the manufacturer's protocol.

4.3. Sequence Analysis of the BocMBF1c Gene

The open reading frame (ORF) of *BocMBF1c* was predicted by the ORF finder (http://www.bioinformatics.org/sms2/orf_find.html), and the amino acid sequence was generated by the Expasy-Translate tool (https://web.expasy.org/translate/). Sequence query and alignment were carried out with BLAST (https://blast.ncbi.nlm.nih.gov/Blast.cgi). A phylogenetic tree was constructed with the software Mega 6.05, and the conserved sites were identified by DNAMEN.

The cis-elements of the *BocMBF1c* promoter were predicted by PlantCARE Database (http://bioinformatics.psb.ugent.be/webtools/plantcare/html/).

4.4. Promoter Activity and Subcellular Localization

The putative *BocMBF1c* promoter of the 1409 bp upstream sequence from the initiation codon was cloned into the binary vector pBI101-GFP with In-Fusion cloning, using primers designed following the manufacturer's instructions (http://bioinfo.clontech.com/infusion/convertPcrPrimersInit.do), including pBI101 infusion M-P-F (5′-GCAGGTCGACTCTAGAGGTACCGAATAGCTCTCCCC-3′) and pBI101 infusion M-P-R (5′- CTTTACCCATTCTAGAGGCATCGTCGCTGTTGTTTA-3′). Similarly, CDS of the *BocMBF1c* was inserted into the pBE-eGFP construct by the In-Fusion technique, with primers MBF1-yxb–F (5′-CGCGGGCCCGGGATCCATGCCGAGCAGAT-3′) and MBF1-yxb-R (5′-GGCGACCGGTGGATCCTTGACATGTTT-3′). *Xba* I was used to digest the pBI101-GFP vector, and pBE-eGFP was digested by *Bam*H I, resulting in pBI101-*pBocMBF1c*-GFP and pBE-*BocMBF1c-eGFP* constructions.

Transient expression was carried out in the leaf tissue of *Nicotiana benthamiana* seedlings with 4 to 6 true leaves, and with *Agrobacterium* strain GV3101 harboring the pBI101-*pBocMBF1c*-GFP and pBE-*BocMBF1c-eGFP* constructs. The GFP signal was observed two days after agro-infiltration under fluorescent microscopy [23].

4.5. Gene Expression Patterns of BocMBF1c by qPCR and Semi-Quantitative RT-PCR

The RNA was extracted using HiPure Plant RNA Kits (Magen), and HiScript® II Q Select RT SuperMix (Vazyme) was used for cDNA synthase. Gene expression was assessed by real-time PCR using AceQ® qPCR SYBR® Green Master Mix (Vazyme) and Bio-Rad iQ5 Multicolor Real-Time PCR Detection System. Primers MBF1c-YG1-UP (5′-GTTAACGCGGCTCTCAGAAG-3′) and MBF1c YG1-DW (5′-TCCTCCAGCTTCTTCGTGTT-3′) were used for *BocMBF1c*; *Tubulin8* was used as an internal control with primers Tubulin-F (5′-CTTCTTTCGTGCTCATTTTGCC-3′) and Tubulin-R (5′-CCATTCCCTCGTCTCCACTTCT-3′) [19]. The qPCR conditions: initial denaturation of 95 °C for 5 min, then 40 cycles of 95 °C for 10 s followed by 60 °C for 30 s; melting curves were obtained by gradually increasing the temperature from 60 °C to 95 °C for 15 s at a rate of 0.5 °C/s. Relative expression was calculated by the $\Delta\Delta$Ct threshold method [24]. For heat and cold treatment, semi-quantitative RT-PCR was used to detect the expression pattern of *BocMBF1c* with Recombinant Taq DNA Polymerase TaKaRa Taq™ (Takara) and the same primers of qPCR. Semi-quantitative RT-PCR conditions: initial denaturation of 94 °C for 2 min, then 26 cycles for *Tubulin8* and 30 cycles for *BocMBF1c* of 95 °C for 30 s, 56 °C for 15 s, 72 °C for 10 s. Three biological replicates with triple technical replicates were used for all samples.

4.6. Arabidopsis Transformation and Phenotypic Analysis of Transgenic Plants

The *BocMBF1c* coding sequence was fused with *GUS* and then expressed in *Arabidopsis thaliana* driven by the *35S CaMV* promoter and then integrated into the *pBI121* binary vector. The floral dipping method [25] was used for *Arabidopsis* transformation. Transgenic seeds were selected on a 0.5 MS agar plate with 50 mg/L kanamycin and 300 mg/L timentin and then transplanted to soil for setting the seeds under standard growth conditions. Three homozygous transgenic lines from independent transformation events were used for further study (Figure S3).

Five-day-old *BocMBF1c*-OE and Col-0 *Arabidopsis* seedlings were collected for GUS staining analysis after germination on MS plates [13]. Two-week-old seedlings grown on soil were harvested for measuring fresh weight. For thermal tolerance analysis, five-day-old seedlings grown on MS plates were treated at 46 °C for 2 h, and then these plates were put back into growth chambers under normal condition for another 2 days.

Supplementary Materials: Supplementary materials can be found at http://www.mdpi.com/1422-0067/20/22/5637/s1.

Author Contributions: Conceptualization, C.C. and J.L.; Data curation, L.Z.; Formal analysis, L.Z., X.-L.M. and B.C.; Investigation, L.Z.; Methodology, L.Z., B.Y. and C.C.; Project administration, G.C. and J.L.; Supervision, J.L.; Validation, L.Z. and X.-L.M.; Visualization, L.Z. and X.-L.M.

Funding: This work was supported by the key project of Guangdong Science and Technology Section, grant numbers 2013B051000069 and 2014B020202005, and the key project of the Guangzhou Science Technology and Innovation Commission, grant number 201508030021.

Acknowledgments: We acknowledge Joanne Ernest (Global Institute for Food Security) for providing very valuable comments when preparing this manuscript; we thank Yehua He for providing the pBI-101-eGFP construct and Fengpin Wang for aiding in the agro-infiltration in tobacco leaves.

Conflicts of Interest: The authors declare no conflict of interest.

Abbreviations

MBF	Multi-protein bridging factor
HTH	Helix-turn-helix
HSEs	Heat shock elements
GFP	Green fluorescent protein
eGFP	Enhanced green fluorescent protein
ABA	Abscisic acid
ABRFs	ABA responsive transcription factors
HSF	Heat stress transcription factor
HSPs	Heat shock proteins
MeJA	Methyl jasmonate
SA	Salicylic acid
ET	Ethylene
JA	Jasmonic acid
ERF	Ethylene response factor
DRE	Dehydration-responsive element
ORF	Open reading frame
CDS	Coding sequence
ERE	Ethylene-responsive elements
ABRE	Abscisic acid-responsive element
MBS	MYB binding site
N.	Nicotiana
qPCR	Quantitative PCR
CEPA	Ethephon
RT-PCR	Reverse transcription PCR
GUS	β-Glucuronidase
CaMV	Cauliflower mosaic virus
WT	Wild type
ABFs	ABRE binding factors

References

1. Lei, J.; Chen, G.; Chen, C.; Cao, B. Germplasm Diversity of Chinese Kale in China. *Hortic. Plant J.* **2017**, *3*, 101–104. [CrossRef]
2. Wu, S.; Lei, J.; Chen, G.; Chen, H.; Cao, B.; Chen, C. De novo Transcriptome Assembly of Chinese Kale and Global Expression Analysis of Genes Involved in Glucosinolate Metabolism in Multiple Tissues. *Front. Plant Sci.* **2017**, *8*, 92. [CrossRef] [PubMed]
3. Si, Y.; Chen, G.; Lei, J.; Cao, B.; Feng, E. Analysis on Composition and Content of Glucosinolates in Different Genotypes of Chinese Kale. *China Veg.* **2009**, *6*, 7–13.
4. Yang, X.; Yang, Y. The Effects of Temperature on Flower Bud Differentiation, Yield and Quality Formation in Chinese Kale (Brassica alboglabra Bailey). *J. S. China Agric. Univ.* **2002**, *23*, 5–7.
5. Bita, C.E.; Gerats, T. Plant tolerance to high temperature in a changing environment: Scientific fundamentals and production of heat stress-tolerant crops. *Front. Plant Sci.* **2013**, *4*, 273. [CrossRef] [PubMed]
6. Wang, X.; Zhuang, L.; Shi, Y.; Huang, B. Up-Regulation of HSFA2c and HSPs by ABA Contributing to Improved Heat Tolerance in Tall Fescue and Arabidopsis. *Int. J. Mol. Sci.* **2017**, *18*, 1981. [CrossRef] [PubMed]
7. Duan, Y.-H.; Guo, J.; Ding, K.; Wang, S.-J.; Zhang, H.; Dai, X.-W.; Chen, Y.-Y.; Govers, F.; Huang, L.-L.; Kang, Z.-S. Characterization of a wheat HSP70 gene and its expression in response to stripe rust infection and abiotic stresses. *Mol. Biol. Rep.* **2011**, *38*, 301–307. [CrossRef] [PubMed]
8. Khan, M.I.R.; Iqbal, N.; Masood, A.; Per, T.S.; Khan, N.A. Salicylic acid alleviates adverse effects of heat stress on photosynthesis through changes in proline production and ethylene formation. *Plant Signal. Behav.* **2013**, *8*, e26374. [CrossRef] [PubMed]
9. Qu, A.-L.; Ding, Y.-F.; Jiang, Q.; Zhu, C. Molecular mechanisms of the plant heat stress response. *Biochem. Biophys. Res. Commun.* **2013**, *432*, 203–207. [CrossRef] [PubMed]
10. Suzuki, N.; Bajad, S.; Shuman, J.; Shulaev, V.; Mittler, R. The transcriptional co-activator MBF1c is a key regulator of thermotolerance in Arabidopsis thaliana. *J. Biol. Chem.* **2008**, *283*, 9269–9275. [CrossRef] [PubMed]
11. Suzuki, N.; Sejima, H.; Tam, R.; Schlauch, K.; Mittler, R. Identification of the MBF1 heat-response regulon of Arabidopsis thaliana. *Plant J.* **2011**, *66*, 844–851. [CrossRef] [PubMed]
12. Alavilli, H.; Lee, H.; Park, M.; Lee, B. Antarctic Moss Multiprotein Bridging Factor 1c Overexpression in Arabidopsis Resulted in Enhanced Tolerance to Salt Stress. *Front. Plant Sci.* **2017**, *8*, 1206. [CrossRef] [PubMed]
13. Tsuda, K.; Yamazaki, K.I. Structure and expression analysis of three subtypes of Arabidopsis MBF1 genes. *Biochim. Biophys. Acta—Gene Struct. Expr.* **2004**, *1680*, 1–10. [CrossRef] [PubMed]
14. Suzuki, N. Enhanced Tolerance to Environmental Stress in Transgenic Plants Expressing the Transcriptional Coactivator Multiprotein Bridging Factor 1c. *Plant Physiol.* **2005**, *139*, 1313–1322. [CrossRef] [PubMed]
15. Wang, H.; Chen, G.; Chen, C.; Cao, B.; Zou, L.; Lei, J. Identification of Heat Tolerance in Chinese Kale and the Expression Analysis of Heat Tolerance Transcription Factor MBF1c. *China Veg.* **2017**, *2*, 30–37.
16. Jacob, P.; Hirt, H.; Bendahmane, A. The heat-shock protein/chaperone network and multiple stress resistance. *Plant Biotechnol. J.* **2017**, *15*, 405–414. [CrossRef] [PubMed]
17. Cheng, M.-C.; Liao, P.-M.; Kuo, W.-W.; Lin, T.-P. The Arabidopsis ETHYLENE RESPONSE FACTOR1 Regulates Abiotic Stress-Responsive Gene Expression by Binding to Different cis-Acting Elements in Response to Different Stress Signals. *Plant Physiol.* **2013**, *162*, 1566–1582. [CrossRef] [PubMed]
18. Toufighi, K.; Brady, S.M.; Austin, R.; Ly, E.; Provart, N.J. The botany array resource: E-Northerns, expression angling, and promoter analyses. *Plant J.* **2005**, *43*, 153–163. [CrossRef] [PubMed]
19. Zhu, Z.; Sun, B.; Xu, X.; Chen, H.; Zou, L.; Chen, G.; Cao, B.; Chen, C.; Lei, J. Overexpression of AtEDT1/HDG11 in Chinese Kale (Brassica oleracea var. alboglabra) Enhances Drought and Osmotic Stress Tolerance. *Front. Plant Sci.* **2016**, *7*, 1285. [CrossRef] [PubMed]
20. Ma, X.; Zhang, Q.; Zhu, Q.; Liu, W.; Chen, Y.; Qiu, R.; Wang, B.; Yang, Z.; Li, H.; Lin, Y.; et al. A Robust CRISPR/Cas9 System for Convenient, High-Efficiency Multiplex Genome Editing in Monocot and Dicot Plants. *Mol. Plant* **2015**, *8*, 1274–1284. [CrossRef] [PubMed]
21. Qin, D.; Wang, F.; Geng, X.; Zhang, L.; Yao, Y.; Ni, Z.; Peng, H.; Sun, Q. Overexpression of heat stress-responsive TaMBF1c, a wheat (Triticum aestivum L.) Multiprotein Bridging Factor, confers heat tolerance in both yeast and rice. *Plant Mol. Biol.* **2015**, *87*, 31–45. [CrossRef] [PubMed]

22. Doyle, J.J.; Doyle, J.L. A rapid DNA isolation procedure for small quantities of fresh leaf tissue. *Phytochem. Bull.* **1987**, *19*, 11–15.

23. Goodin, M.M.; Zaitlin, D.; Naidu, R.A.; Lommel, S.A. *Nicotiana benthamiana*: Its History and Future as a Model for Plant–Pathogen Interactions. *Mol. Plant-Microbe Interact.* **2008**, *21*, 1015–1026. [CrossRef] [PubMed]

24. Livak, K.J.; Schmittgen, T.D. Analysis of relative gene expression data using real-time quantitative PCR and. *Methods* **2001**, *25*, 402–408. [CrossRef] [PubMed]

25. Clough, S.J.; Bent, A.F. Floral dip: A simplified method for Agrobacterium-mediated transformation of Arabidopsis thaliana. *Plant J.* **1998**, *16*, 735–743. [CrossRef] [PubMed]

International Journal of
Molecular Sciences

Article

Genome-Wide Analysis of the DYW Subgroup PPR Gene Family and Identification of *GmPPR4* Responses to Drought Stress

Hong-Gang Su [1,2,†], Bo Li [1,†], Xin-Yuan Song [3,†], Jian Ma [4], Jun Chen [1], Yong-Bin Zhou [1], Ming Chen [1], Dong-Hong Min [2], Zhao-Shi Xu [1,*] and You-Zhi Ma [1,*]

[1]   Institute of Crop Science, Chinese Academy of Agricultural Sciences (CAAS)/National Key Facility for Crop Gene Resources and Genetic Improvement, Key Laboratory of Biology and Genetic Improvement of Triticeae Crops, Ministry of Agriculture, Beijing 100081, China; suhonggang924@126.com (H.-G.S.); libo708@126.com (B.L.); chenjun01@caas.cn (J.C.); zhouyongbin@caas.cn (Y.-B.Z.); chenming02@caas.cn (M.C.)

[2]   State Key Laboratory of Crop Stress Biology for Arid Areas, College of Agronomy, Northwest A&F University, Yangling 712100, China; mdh2493@126.com

[3]   Agro-Biotechnology Research Institute, Jilin Academy of Agriculture Sciences, Changchun 130033, China; songxinyuan1980@163.com

[4]   College of Agronomy, Jilin Agricultural University, Changchun 130118, China; majian197916@jlau.edu.cn

*   Correspondence: xuzhaoshi@caas.cn (Z.-S.X.); mayouzhi@caas.cn (Y.-Z.M.)

†   These authors contributed equally to the present work.

Received: 25 September 2019; Accepted: 6 November 2019; Published: 12 November 2019

Abstract: Pentatricopeptide-repeat (PPR) proteins were identified as a type of nucleus coding protein that is composed of multiple tandem repeats. It has been reported that *PPR* genes play an important role in RNA editing, plant growth and development, and abiotic stresses in plants. However, the functions of PPR proteins remain largely unknown in soybean. In this study, 179 DYW subgroup *PPR* genes were identified in soybean genome (*Glycine max* Wm82.a2.v1). Chromosomal location analysis indicated that DYW subgroup *PPR* genes were mapped to all 20 chromosomes. Phylogenetic relationship analysis revealed that DYW subgroup *PPR* genes were categorized into three distinct Clusters (I to III). Gene structure analysis showed that most *PPR* genes were featured by a lack of intron. Gene duplication analysis demonstrated 30 *PPR* genes (15 pairs; ~35.7%) were segmentally duplicated among Cluster I *PPR* genes. Furthermore, we validated the mRNA expression of three genes that were highly up-regulated in soybean drought- and salt-induced transcriptome database and found that the expression levels of *GmPPR4* were induced under salt and drought stresses. Under drought stress condition, *GmPPR4*-overexpressing (*GmPPR4*-OE) plants showed delayed leaf rolling; higher content of proline (Pro); and lower contents of H_2O_2, O_2^- and malondialdehyde (MDA) compared with the empty vector (EV)-control plants. *GmPPR4*-OE plants exhibited increased transcripts of several drought-inducible genes compared with EV-control plants. Our results provided a comprehensive analysis of the DYW subgroup *PPR* genes and an insight for improving the drought tolerance in soybean.

Keywords: Pentatricopeptide-repeat (PPR) proteins; genome-wide analysis; drought responses; hairy root assay; soybean

---

1. Introduction

The pentatricopeptide repeat (PPR) proteins containing tandem 30–40 amino acid sequence motifs, constitute a large gene family in plants [1]. Typically, PPR proteins are classed into two major subfamilies (P and PLS). The P subfamily proteins contain only the canonical P motif (35 amino acid residues); the PLS subfamily proteins are confined to contain a series of PLS triplets (L, long variants of P; S, small variants of P) [2]. L motif and S motif contain 31 and 35 or 36 amino acid residues, respectively, which are considered to be variations of the P motif [2]. Most PPR proteins are highly conserved at the C-terminus and usually have three conserved domains at the C-terminus, E, E+ and DYW domains, respectively [3,4]. Based on these C-terminal domains, the PLS subfamily is divided into four subgroups, PLS subgroup, E subgroup, E+ subgroup, and DYW subgroup [2]. The PPR proteins also contain a distinctive feature that is essentially free of intron. Although members of the PPR family are widely distributed in plants and are large in number, they perform special functions [2].

So far, a considerable amount of PPR proteins have been experimentally identified from various plant species. [5]. For example, 441 and 626 PPR members have been identified in the *A. thaliana* and *Populus trichoccapa* genome, respectively [5–7]. A large number of studies showed that PPR proteins played roles in the growth and development of plants and the formation of organelles [8]. It can participate in the fertility restoration of cytoplasmic male sterility, post-transcriptional processing of RNA and adversity defense [2,7,9–12]. In maize, EMP12 is targeted to the mitochondria and is essential for the splicing of three *nad2* introns and seed development, and the mutant in *EMP12* caused defects complex I [13]. It was reported that the nuclear *OsNPPR1* encoded a PPR protein, which was involved in the Regulation of mitochondrial development and endosperm development. The mutant named *fgr1* exhibited a lower content of starch and delayed seedling growth [14]. Rice *PPS1* encodes a DYW motif-containing PPR protein, which is important for C-to-U RNA editing of *nad3* transcripts. Knock-down and Knock-out of *PPS1* revealed a significant decrease in the editing efficiency of C-to-U at five editing sites of *nad3* transcripts, which resulted in some pleiotropic phenotypes, such as dwarfing, developmental retardation and stunted development in vegetative stages; partial pollen was sterile at the reproductive stage [15]. A chloroplast-localized P-class PPR protein, PpPPR_21, was identified as an essential protein for the accumulation of a stable *psbl-ycf12* mRNA. The knockout *ppr21* mutants displayed decreasing protonemata growth and lower photosynthetic activity [16].

Of particular note, several studies have recently reported roles for *PPR* genes during abiotic stress [17]. For example, *PPR96* from *A. thaliana* was involved in the response to salt, ABA and oxidative stress, and altered translational levels of abscisic stress responsive genes, implying that *PPR96* may take part in the regulation of plant tolerance to abiotics [18]. PGN, an *A. thaliana* mitochondria-localized PPR protein, was reported to play a role in regulating mitochondrial reactive oxygen species balance in abiotic and biotic stress responses. Inactivation of *PGN* displayed hypersensitivity to ABA, glucose, and salinity. Loss of *PGN* resulted in enhanced accumulation of reactive oxygen species (ROS) in seedlings, and the *pgn* mutant notably elevated levels of *ABI4* and *ALTERNATIVE OXIDASE1a* [19]. In rice, *cisc(t)* encoded a PPR protein, which was necessary for chloroplast development in early stages under cold stress [20]. The *A. thaliana ABA overly sensitive 5 (abo5)* mutant showed growth retardation and accumulated increased levers of H_2O_2 in root tips [21]. *AtPPR40* was reported to be involved in salt, ABA and oxidative stress [22].

Previous studies have emphasized the role of *PPR* genes in the growth and development of plants, and several *PPR* genes performed potential roles against abiotic stresses. To date, little information about *PPR* genes relating to abiotic stress response mechanisms in soybean has been reported. Here, a comprehensive genome-wide analysis of the DYW subgroup PPR gene family has been accomplished in soybean. We explored the characteristics of 179 DYW subgroup *PPR* genes, including intron-exon organization, chromosomal location and phylogenetic relationships. The DYW subgroup *PPR* genes were divided into three discrete groups (Cluster I to III). An analysis of a complete set of the Cluster I PPR genes/proteins was performed, including classification, chromosomal location, orthologous relationships, duplication analysis and tissue-specific expression patterns. *GmPPR4* was further investigated owing to significant up-regulation by drought and salt stresses. Our results showed that overexpression of *GmPPR4* improved tolerance to drought stress in soybean. Our study provided an insight into the foundation of the *GmPPR4* gene in abiotic stress responses.

2. Result

2.1. Identification of DYW Subgroup PPR Genes in the Soybean Genome

In our study, 205 proteins were identified in the soybean genome (*Glycine max* Wm82.a2.v1). The candidate members were examined for the required DWY domain, which was unique to DYW subgroup PPR proteins. SMART and Pfam databases were used to verify the presence of the conserved domain in all 205 DYW subgroup PPR proteins, and the results showed that a total of 197 members were identified. Except for the presence of conserved PPR domains, there are great differences in the size and physicochemical properties of the DYW subgroup *PPR* genes (Table S1). The statistical results showed that the amino acid sequence length of the 179 DYW PPR proteins varied from the lowest of 118 to the highest of 1431, the isoelectric point (p*I*) varied from 5.28 to 9.36, and the molecular weight (MW) ranged from 13.7 kDa to 160.3 kDa. The detailed information of the 179 DYW subgroup *PPR* gene sequences is provided in Table S1. For convenience, the 179 DYW subgroup *PPR* genes were named from *GmPPR1* to *GmPPR179* according to the order of their chromosomal locations [23,24].

A considerable amount of research showed that the vast majority of *PPR* genes were found to be lacking or containing several introns in various plant species [5–7]. The intron numbers present within the ORF of each of DYW subgroup *PPR* genes in soybean were decided for analyzing their exon-intron organization. As showed in Figure 1, 59% (105/179) of DYW subgroup *PPR* genes were predicted to lack introns, 17% (30/179) with one intron, 22% (40/179) with two to five introns, and 2% (4/179) with six or more introns.

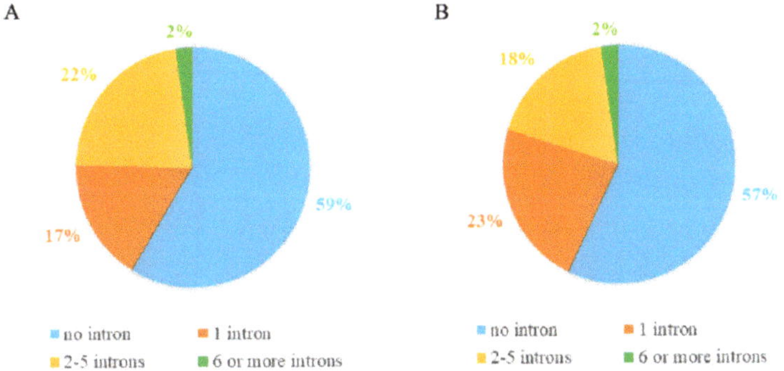

Figure 1. Relative proportions of intron-containing *PPR* genes in soybean. (**A**) Number of introns in the 179 DYW subgroup *PPR genes*. (**B**) Number of introns in the 84 Cluster I *PPR genes*.

2.2. Chromosomal Distribution, Phylogenetic Analysis and Multiple Sequence Alignment

Our study showed 179 DYW subgroup *PPR* genes were distributed widely and unevenly in all the 20 soybean chromosomes (Figure 2). Chromosome 8 of soybean contained the highest number of DYW subgroup *PPR* genes, while fewer numbers of DYW subgroup *PPR* genes were located on chromosome 14 and 19. The detailed position of each DYW subgroup of *PPR* genes on the soybean chromosomes was obtained from Phytozome (Table S1).

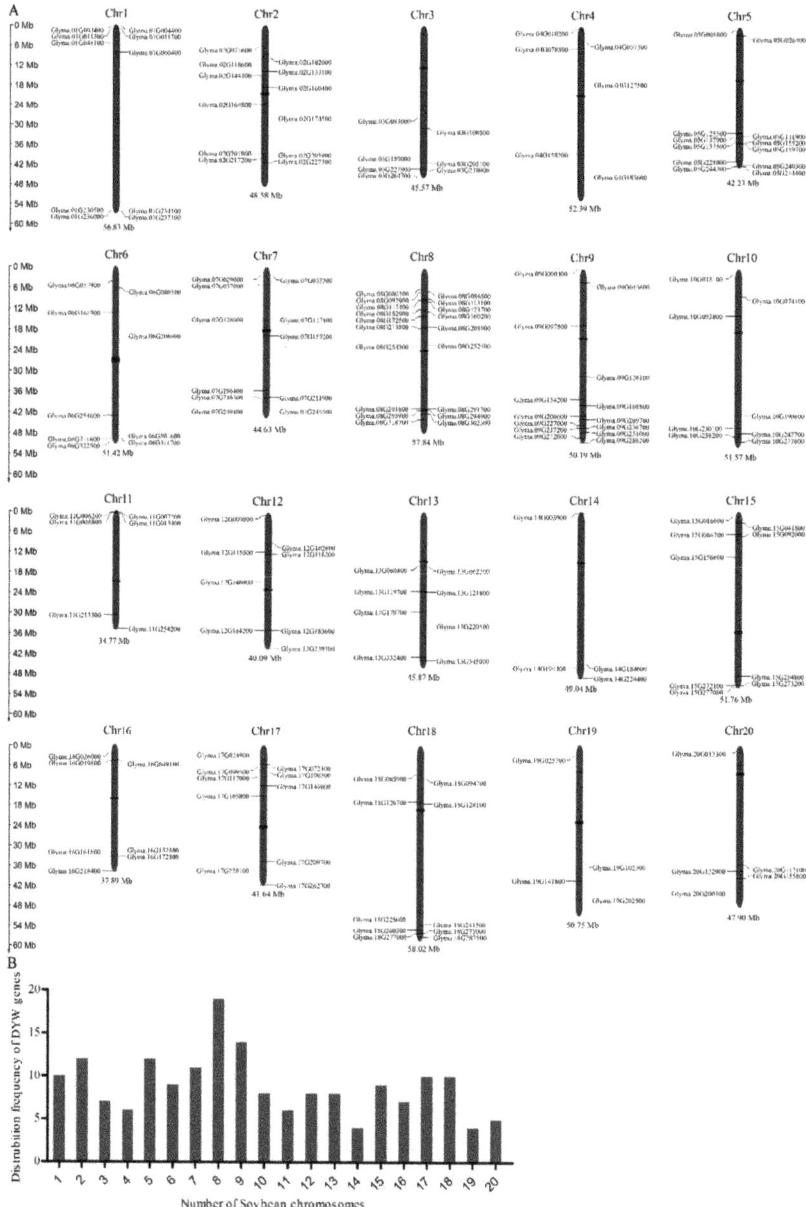

Figure 2. Chromosomal distribution of the 179 DYW subgroup *PPR* genes in soybean. (**A**) The physical location of each member. (**B**) *PPR* genes distribution on 20 soybean chromosomes.

To analyze the evolutionary relationship among DWY subgroup *PPR* genes, the full-length amino acid sequences of the 179 DYW subgroup PPR proteins were aligned and then the phylogenetic tree was conducted using the Maximum likelihood method by MEGA7.0 software (Figure 3). The phylogenetic analysis categorized the DYW subgroup *PPR* genes into three distinct clusters (I to III) comprising of 84, 65 and 30 genes, respectively. In order to isolate drought-inducible DYW subgroup *PPR* genes in soybean, we analyzed the expression patterns of the DYW subgroup *PPR* genes in soybean transcriptome database. We found that three DWY subgroup *PPR* genes were highly up-regulated by drought and salt stresses, including *GmPPR4* (Glyma.01G011700), *GmPPR18* (Glyma.02G174500), and *GmPPR111* (Glyma.11G008800). Interestingly, they all clustered into the Cluster I, implying that Cluster I *PPR* genes may be involved in the abiotic stress of soybean, therefore, we further analyzed the genetic structure and motif of Cluster I *PPR* genes.

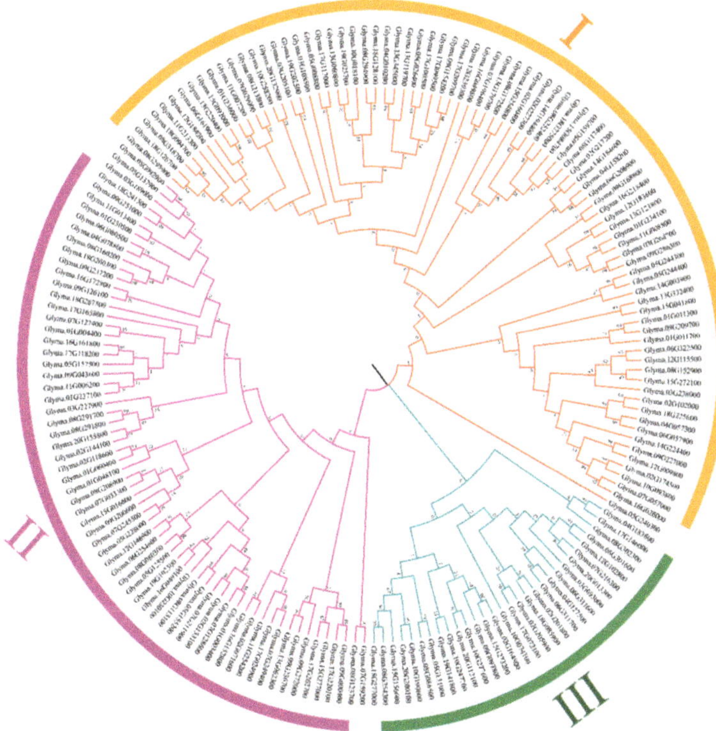

Figure 3. Phylogenetic tree of DYW subgroup *PPR* genes. The complete amino acid sequences of the 179 DYW subgroup PPR proteins were aligned by ClustalW and Maximum-likelihood with MEGA7. Three discrete groups (Cluster I to III) were highlighted in different colors.

To identify conservation of PPR proteins in dicot plant species, six reported DYW proteins of *A. thaliana* and six randomly selected proteins of soybean were used to conduct the multiple sequence alignment, and the C-terminal region is displayed in Figure 4. Results showed that PPR proteins were relatively conserved across *A. thaliana* and soybean, however, only the significant short arrays (no more than four amino acid residues) were absolutely conservative, and 11 members harbored the conservative DYW sequence in C-terminal region that have previously been described as characteristic features of DYW subgroup PPR proteins. In addition, a phylogenic tree was conducted to reveal the relations of 12 selected proteins (Figure S1).

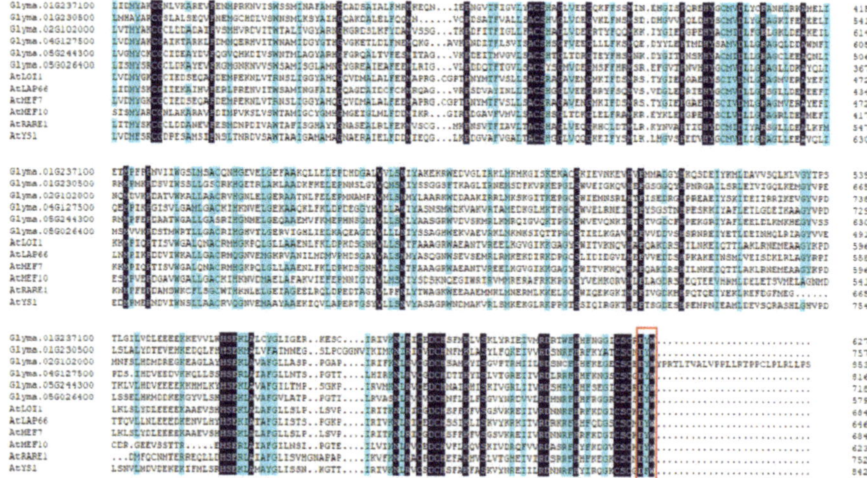

Figure 4. Multiple sequence alignment of 12 DYW subgroup PPR proteins from *A. thaliana* and soybean by DNAMAN. Black or blue shadings represent the 100% or >75% similarity of amino acids. Red rectangle marks highly core short signatures.

2.3. Gene Structure and Motif Composition of Soybean Cluster I PPR Genes

The exon-intron organization of the 84 Cluster I *PPR* genes was conducted to provide valuable information for the structure diversity evolution of the PPR family in soybean. As showed in Figure 5 and Table S1, they almost contained very few introns, which was similar to the previous results (Figure 1). However, the result showed that several genes contained more than 10 introns. Fourteen and 17 introns were found in Glyma.13G220400 and Glyma.15G092000, respectively. Many paralogs exhibited the same numbers and sizes of introns such as gene clusters of Glyma.07G057000, Glyma.16G026000, Glyma.02G217200, and Glyma.14G184600; However, only a few of them showed intron gain/loss phenomenon such as Glyma.03G205100 and Glyma.19G202500. On the whole, the intron distribution and the intron phase are in accordance with the alignment cluster of the Cluster I *PPR* genes.

Also, the Multiple EM for Motif Elicitation (MEME) program was used to search the conserved motifs which were shared with the Cluster I proteins (Figure 6). A total of 10 distinct conserved motifs named 1 to 10, were found. The motifs were defined based on sequence conservation without knowing its structure or function. Motif 1, the most typical DYW domain, comprised of 46 amino acids; the sequence (KNLRVCGDCHSAIKLISKIYNREIIVRDRNRFHHFKBGSCSCGDYW) contains a highly conserved C-terminal region (DYW) (Figure 7). However, this conservative motif is not included in the five proteins, including GmPPR28 (Glyma.03G238000), GmPPR72 (Glyma.08G117400), GmPPR91 (Glyma.09G154200), PPR103 (Glyma.10G093800), and GmPPR165 (Glyma.18G225600), which may be due to the fact that some PPR proteins are truncated at the C-terminus, lacking part, most of or the entire DYW domain. Proteins with higher homology often have the same conserved domain composition, which corresponds to the results of the phylogenetic tree.

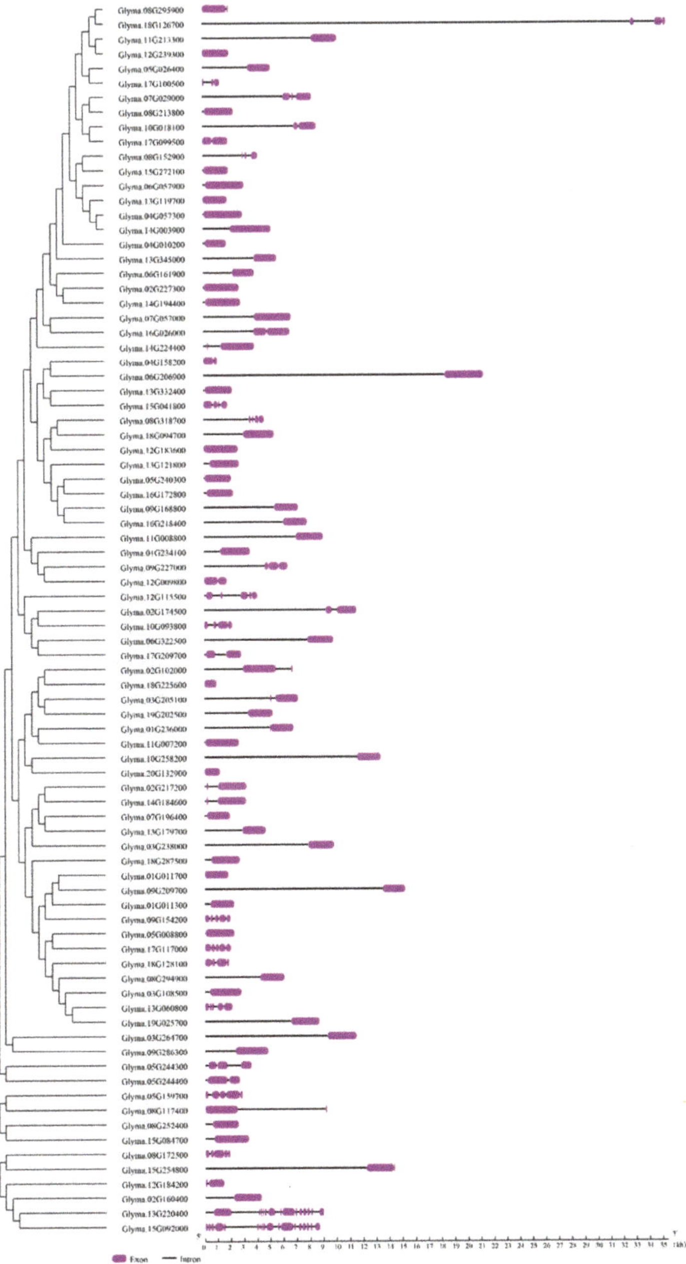

Figure 5. Phylogenetic relationship and structural analysis of the 84 Cluster I *PPR* genes. The phylogenetic tree was constructed via MEGA7.0 software; the different classes of *PPR* genes make up separate clades. The schematic diagram was carried out to represent the gene structure. Introns and exons were indicated by black lines and purple boxes, respectively. The lengths of introns and exons of each gene were displayed proportionally.

Figure 6. Putative motifs of each Cluster I PPR gene by MEME program and TBtools software. Ten putative motifs were indicated in colored boxes. The length of protein can be estimated using the scale at the bottom.

Figure 7. Sequence logo of global multiple alignment of 179 DYW subgroup PPR proteins by MEME program.

2.4. Duplication and Divergence Rate of Soybean Cluster I PPR Genes

Whole genome tandem and segmental duplications play a vital role in multiple copies of genes in a gene family and their subsequent evolution. Among *SiPPR* genes, 190 (95pairs; ~39.01%) were segmentally duplicated [25]. Our research reached a similar conclusion that this proportion is far higher than of genes generally. Among Cluster I *PPR* genes, 30 (15 pairs; ~35.7%) were segmentally duplicated (connected by lines in Figure 8). Among 15 paralog pairs, 13 segmental duplication gene pairs were involved in two chromosomes; only two segmental duplications were intra-chromosomal. In particular, three genes have appeared twice, including *GmPPR3* (Glyma.01G011300), *GmPPR4* (Glyma.01G011700) and *GmPPR94* (Glyma.09G209700). The results indicated that gene segmental duplications might mainly contribute to the expansion of the PPR gene family in soybean.

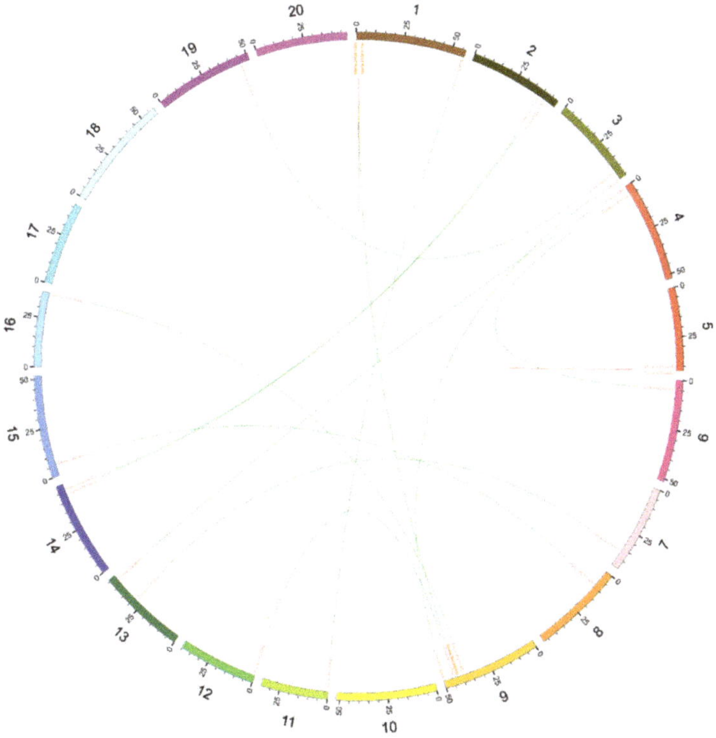

Figure 8. Distribution of segmentally duplicated Cluster I *PPR* genes on soybean chromosomes. Green lines indicate duplicated *PPR* gene pairs, and bold Photozome Locus indicates the gene appears twice.

The ratio of non-synonymous (Ka) versus synonymous (Ks) substitution rates (Ka/Ks) was estimated for 15 segmental duplicated gene pairs to evaluate the role of Darwinian positive selection in duplication and divergence of the Cluster I *PPR* genes (Table S2). The result showed that the Ka/Ks for segmental duplications gene-pairs ranged from 0,14 to 1.03 with an average of 0.61. On the whole, it suggested that the duplication Cluster I *PPR* genes were under purifying selection pressures since their Ka/Ks rations were estimated as <1 except one gene pair with a rate (Ka/Ks) of 1.03.

2.5. Tissue-Specific Expression Pattern of DYW Subgroup PPR Genes

Different members may exhibit significant diversity in expression abundance among different tissues to adapt to different physiological process. To gain insight into the gene expression patterns within the organism in soybean growth and development, we investigated the relative transcript abundance of the 84 DYW subgroup *PPR* genes in six soybean tissues using publicly available RNA-seq data from Phytozome database, including flower, leaves, root, root hairs, seed, and stem, revealing that most DYW subgroup *PPR* genes had a similar pattern of expression in the same tissue. Out of 84 DYW subgroup *PPR* genes, most of them expressed in almost any tissues, while a total of four genes had no expression abundances to be discovered in any tissues, including Glyma.08G318700, Glyma.18G126700, Glyma.18G128100, and Glyma.18G225600. We found that most *PPR* genes showed preferential accumulation in leaves compared to other tissues, which were likely to be responsible for their subcellular localization, and most reported PPR proteins were located in mitochondria or chloroplast. A relatively high expression level was revealed in seeds, suggesting that they have potential function in seed development, which has been reported to be involved in endosperm development [14]. As shown in Figure 9, Glyma.11G008800, Glyma.13G220400 and Glyma.15G192000 showed extremely high expression in roots and root hairs, indicating that the DYW subgroup *PPR* genes may play specific roles in responding to drought stress.

2.6. Cis-Elements Analysis

By examining the expression levels of DYW subgroup *PPR* genes according to soybean drought- and salt-induced transcriptome database, we found that three genes were highly up-regulated to drought and salt stresses. *Cis*-elements presented in the upstream region play major roles in regulating the gene expression at the transcriptional level. To further understand the mechanism of the three genes, a set of 12 important *cis*-elements were identified in their 1.5 Kb 5′ flanking region upstream from the start codons. Various *cis*-elements including ABA-responsive element (ABRE) and MYB-binding site (MBS) were shown in Figure 10, which suggested that the three candidate genes might be involved in responses abiotic stresses.

2.7. Several Candidates Are Involved in Abiotic Stresses

To comprehensively understand the physiological functions of DYW subgroup *PPR* genes, we initially examined the expression patterns of three genes in response to salt and drought stresses by qRT-PCR (Figure 11). Under drought treatment, these three genes exhibited similar expression patterns, reaching a peak at 2 h (~ 4.5-fold, 3-fold and 2.5-fold, respectively), indicating that these genes have a similar biological function in the same environment. Similarly, the expression levels of the three genes were enhanced by salt; the expression peaks of *GmPPR4/18/111* occurred at 4, 8 and 4 h, respectively, which are equivalent to 3-fold, 2.8-fold and 3.2-fold increases, respectively.

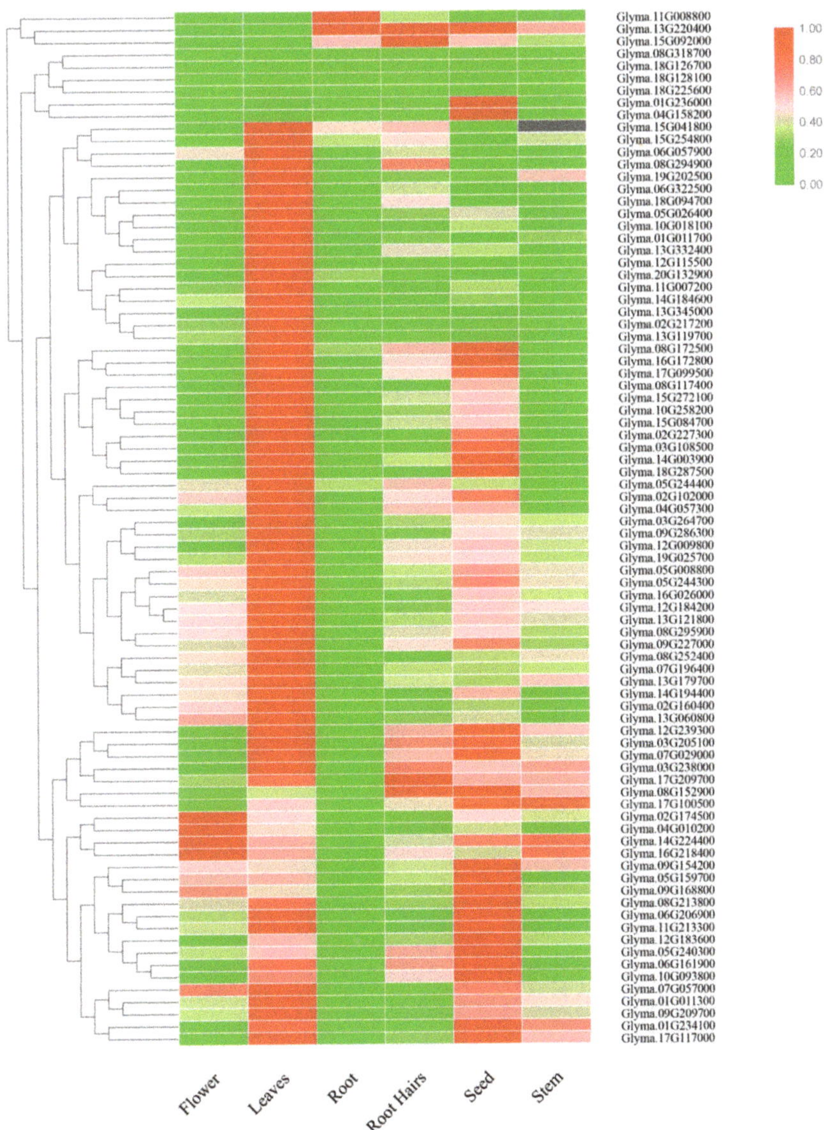

Figure 9. Heat map of expression profiles (in log10-based FPKM) of all DYW subgroup *PPR* genes from six soybean tissues (flower, leaves, root, root hairs, seed, and stem). The expression abundance of each transcript is represented by the color bar: Red, higher expression; and green, lower expression.

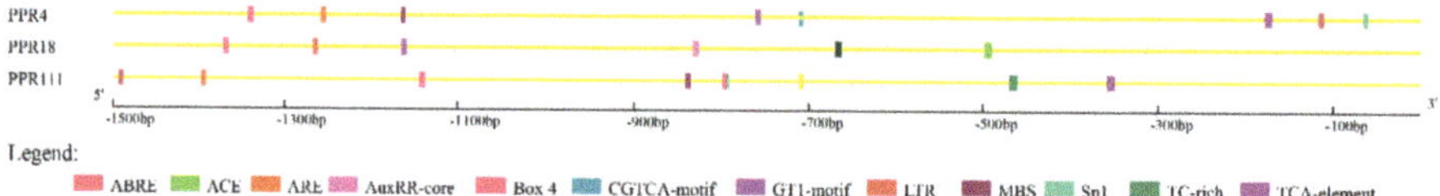

Figure 10. Putative *cis*-element in a 1.5 kb 5′ flanking region upstream from the start codon. Different *cis*-elements were indicated by colored symbols and placed in relative positions on the promoter. The ABA-responsive element (ABRE), light-responsive element (ACE), anaerobic induction element (ARE), auxin responsive element (AuxRR-core), light-responsive element (Box4), MeJA-responsive (CGTCA-motif), light-responsive element (GT1-motif), low temperature responsive element (LTR), MYB-binding site (MBS), light-responsive element (Sp1), salicylic acid responsive element (TCA), and defense and stress responsive element (TC-rich repeat) were analyzed.

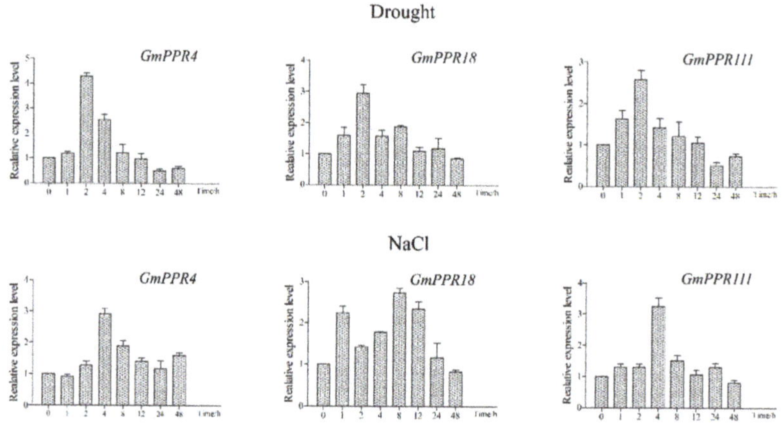

Figure 11. Expression patterns of three selected *PPR* genes under salt and drought treatments by qRT-PCR. *GmPPR4* (Glyma.01G011700), *GmPPR18* (Glyma.02G174500), *GmPPR111* (Glyma.11G008800), respectively. The *actin* gene was used as an internal control. The data are shown as the means ± SD obtained from three biological replicates. ANOVA test demonstrated that there were significant differences (* $p < 0.05$, ** $p < 0.01$).

2.8. GmPPR4 Improved Drought Tolerance in Transgenic Soybean Hairy Roots

Among the three genes, *GmPPR4* (Glyma.01G011300) clearly responded to drought and salt stresses. For this reason, *GmPPR4* was selected for further investigation. To examine the function of *GmPPR4* in vivo, transgenic soybean plants which overexpressed *GmPPR4* (*GmPPR4*-OE) were generated into soybean hairy roots [26]. qRT-PCR analysis showed that *GmPPR4* accumulated in the *GmPPR4*-OE plants. (Figure S2), and about 80% of the roots of transgenic soybean plants were positive. To examine whether *GmPPR4* plays a role in the drought stress tolerance, we compared drought tolerance of *GmPPR4*-OE and WT plants at the vegetable stage; the hairy roots of seedlings which grew in soil were withheld from water for 2 weeks. No significant differences were observed for transgenic plants under normal growth conditions, compared to empty vector (EV)-transformed control hairy roots; drought treatment caused obvious differences in the growth of the EV-control and *GmPPR4*-OE plants; compared with the *GmPPR4*-OE, the EV-control plants showed wilted leaf under drought for 3 days, and seriously dehydrated leaves were observed after 7 days, whereas the *GmPPR4*-OE seedlings showed delayed and less leaf rolling during the drought stress process. (Figure 12A).

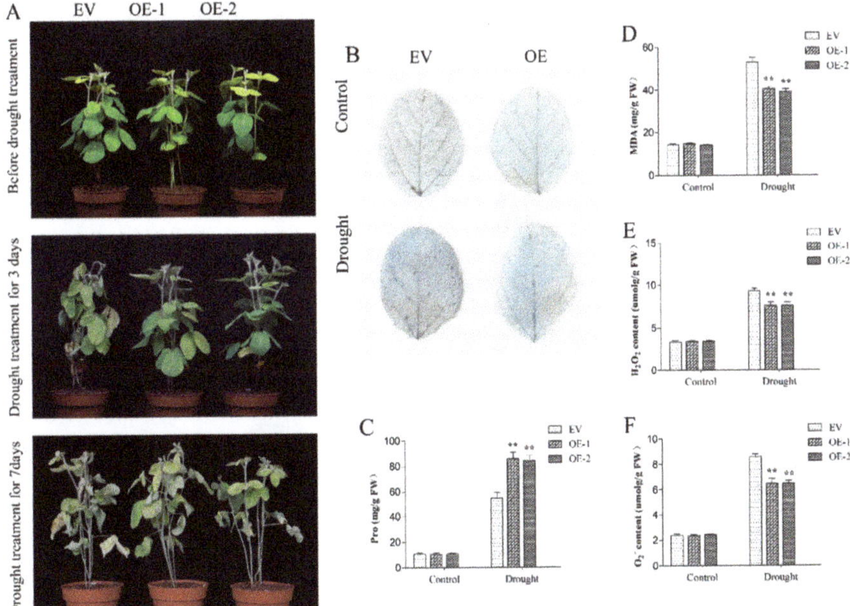

Figure 12. Drought stress analysis of EV-control transgenic plants and *GmPPR4*-OE transgenic plants. (**A**) Phenotypes of *GmPPR4*-OE and EV-control transgenic soybean plants subjected to drought stress without water for 3 days and 7 days. (**B**) Trypan blue staining of soybean plant leaves placed on filter paper for the induction of rapid drought for 3 h, the dead cells can be strained, but living cells cannot. (**C**) The proline content, (**D**) malondialdehyde (MDA) content, (**E**) H_2O_2 content, and (**F**) O_2^- content of *GmPPR4*-OE and EV-control transgenic soybean plants under normal and drought conditions. The data are shown as the means ± SD obtained from three biological replicates. ANOVA test demonstrated that there were significant differences (* $p < 0.05$, ** $p < 0.01$).

To explore the potential physiological mechanism responsible for the improved drought tolerance of the *GmPPR4*-OE seedlings, we compared some stress-related physiological changes in the EV-control and *GmPPR4*-OE plants under both normal growth and drought conditions. We found that proline accumulation in the transgenic plants was much more evident than in EV-control plants (Figure 12C), while the MDA content was decreased due to drought stress (Figure 12D). Furthermore, we also measured the level of H_2O_2 and O_2^- in roots; we found that the drought-treated EV-control roots accumulated much more H_2O_2 and O_2^- than the *GmPPR4*-OE roots (Figure 12E,F).

In addition, we used Trypan blue solution to detect cell activity in EV-control and *GmPPR4*-OE leaves. As shown in Figure 12B, no plant leaves differed under drought stress conditions, however, the color depth of the *GmPPR4*-OE leaves was lower than EV-control leaves under drought treatment, which indicated that the cell membrane integrity and stability in the leaves of the EV-control plants was better than that in the leaves of the *GmPPR4*-OE plants.

NaCl treatment was carried out. However, no obvious differences were observed in EV-control and *GmPPR4*-OE plants under NaCl treatment.

2.9. GmPPR4-OE Plants Exhibited Increased Transcripts of Some Drought-Inducible Genes

A previous study indicated that several genes play an important role in drought stress [27,28], including *DREB2* [29], *DREB3* [30], *MYB84* [31], *bZIP1* [32], *bZIP44* [33], *NAC11* [34], *WRKY13* [35], and *WRKY21* [35,36]. We compared the transcripts of several drought-inducible maker genes between *GmPPR4*-OE and EV-control plants under normal and drought conditions, and we found that drought induced more transcripts of all genes. However, expression differences between *GmPPR4*-OE and EV-control plants under drought treatment occurred in six genes, including *DREB2, DREB3, bZIP1, NAC11, WRKY13,* and *WRKY21*. There was no clear difference in *bZIP44* and *MYB84* expression with drought treatment between *GmPPR4*-OE and EV-control plants (Figure 13).

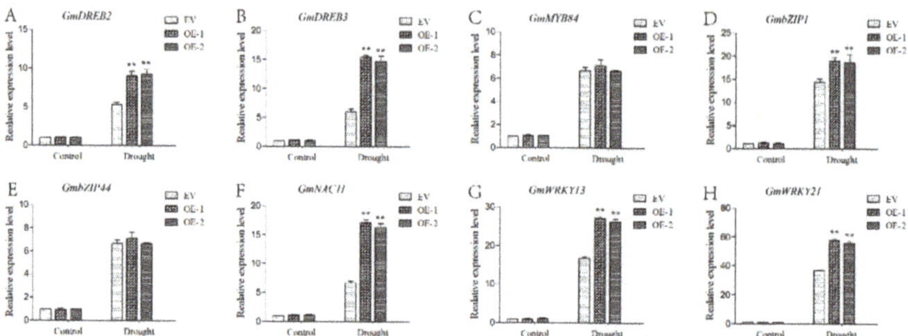

Figure 13. GmPPR4-OE plants exhibited increased transcripts of some drought-inducible genes. The two-week-old soybean seedlings were placed on filter paper for 3 h. qRT-PCR analysis for drought-inducible genes (**A–H**). The actin gene was used as an internal control. The data were shown as the means ± SD obtained from three biological replicates. ANOVA test demonstrated that there were significant differences (* $p < 0.05$, ** $p < 0.01$).

3. Discussion

Soybean is one of the most widely cultivated crops, with a total production of more than 260 million tons in 2010 (FAO data) [37]. Greenhouse and field studies showed that drought stress seriously affects plant development and led to significant reduction in crop yield [38]. Thus, when identifying ideal candidate genes, increasing drought stress resistance is essential for improving soybean yield.

PPR genes comprise a large family, which is ubiquitous to all terrestrial plants. In particular, previous reports indicated that a total of 4000 *PPR* family genes were identified in the spikemoss (*S. moellendorffii*) genome [3]. It is estimated that there were more than 1000 *PPR* genes in soybean. The duplication event was likely to be responsible for the large PPR family size, which has been experimentally verified in many species. Genome-wide identification of *PPR* family genes has been widely carried out in many species that have been sequenced [5–7]. In the present study, we identified 179 DYW subgroup *PPR* genes in the soybean genome (*Glycine max* Wm82.a2.v1). Consistent with previous studies, gene structure analysis revealed that most members of 179 DWY subgroup *PPR* genes were intron-less. Approximately 80% and 65% of *PPR* genes were free of intron in *A. thaliana* and rice, respectively [5,6]. The intron-less nature of the great majority of *PPR* genes may be the result of selection pressures during evolution. However, there are several *PPR* genes that tend to evolve diverse exon-intron organizations with more than 10 introns; we propose that they have a higher probability of evolving new specialized functions and adjusting to their living environment. The intron-less nature may demonstrate a duplication event of *PPR* genes [39]; we presumed that the nature also allows alternative splicing and alteration of splicing pattern in plants. In particular, many coding sequences of PPR proteins with extremely high similarity on the genome were divided only by frameshifts, for example, Glyma.13G332400 and Glyma.15G041800. In our study, *GmPPR3* (Glyma.01G011300), *GmPPR4* (Glyma.01G011700) and *GmPPR94* (Glyma.09G209700) shared more than 95% identity in their coding sequence. In particular, many *PPR* genes have extremely close proximity on the genome, which contributed to the extensive gene-level synteny shared between them [24].

Previous studies have shown that PPR proteins were involved in RNA editing, plant growth and development and abiotic stress. Several PPR proteins have been implicated with trans-splicing of *nad* intron in *A. thaliana*, these include: OTP43 [40], SLO4 [41], PPR19 [42], BIR6 [43], MTL1 [44], ABO5 [21], OTP439 [45], and SG3 [46]. Lots of PPR proteins have a similar function in the development of seed or endosperm, such as EMP5 [47], PPR2263 [48], PPR8522 [49], OGR1 [50], AHG11 [17], and OTP43 [40]. In addition, in rice or *A. thaliana*, six PPR proteins: cisc(t) [20], ABO5 [21], ECB2 [51], PPR40 [22], SVR7 [52], and YS1 [53], were reported to be involved in abiotic stress responses, including ABA, drought, light and cold. In this study, a higher level of H_2O_2 and O_2^- were found in *GmPPPR4*-OE plants compared with EV-control plants under drought stress. In a previous study, the *abo5* mutant accumulated higher H_2O_2 content in roots than wild type [21], and the *ahg11* mutants exhibited higher transcript levels of oxidative stress-responsive genes [17]. The above results suggested that *PPR* genes may play roles in drought stress by regulating the level of H_2O_2.

4. Methods

4.1. Identification of DYW Subgroup PPR Genes in Soybean

DYW subgroup PPR proteins sequences of soybean were downloaded from Phytozome [54]. Predicted proteins from the soybean genome (*Glycine max* Wm82.a2.v1) were scanned using HMMER v3 [55] using the Hidden Markov Model (HMM) corresponding to the Pfam of PPR family (PF14432) [56]. The proteins obtained using the raw DYW HMM, a high-quality protein set (E-value $< 1 \times 10^{-20}$), were aligned and used to construct a soybean-specific DYW HMM profile using hmmbuild from the HMMER v3 suite. This new soybean-specific HMM was used to search for all members in all soybean proteins; in addition, all obtained proteins with an E-value lower than 0.01 were selected. The highly matched sequences were reorganized and merged to remove the redundancy. Then, all protein sequences of putative DYW subgroup *PPR* genes were submitted to the SMART database [57] and Pfam database to confirm the existence of the DYW conserved domain. The sequences lacking a DYW domain were refused in this study. Furthermore, molecular weights and isoelectric points identified that the DYW subgroup PPR proteins were obtained by using tools from EXPASY website [58].

4.2. Chromosomal Location and Phylogenetic Analysis

Positional information of DYW subgroup *PPR* genes on chromosomes of soybean was obtained from the Phytozome database. All DYW subgroup *PPR* genes were mapped to 20 chromosomes in soybean.

Multiple sequence alignment of proteins from soybean was performed via ClustalW with a default parameter. A phylogenetic tree was constructed by using the maximum likelihood with MEGA7 software, with the following parameters: Poisson model, pairwise deletion and 1000 bootstrap replications.

4.3. Gene Structure Analysis and the Sequence of PPR Motif Analysis

The gene structures of *PPR* were illustrated using the online program Gene Structure Display Server [59] by comparing predicted coding sequences with their corresponding genomic DNA sequences.

For sequence analysis of PPR motifs, the MEME online program [60] was used to identify conserved motifs. The scatter diagram was used to display the distribution of PPR motifs at the opposite positions in the PPR proteins by using the TBtools software v0.665 (Guangzhou, China) [61].

4.4. Gene Duplication

All Cluster I PPR proteins were searched for using the BLASTp search (E-value $>1e^{-10}$) with more than 75% sequence similarity being considered to be a pair of tandem repeat genes. Then, the resulting file and GFF3 files of soybean genome (*Glycine max* Wm82.a2.v1) were used to analyze the gene duplication by software MCScanX and visualization using CIRCOS [62].

4.5. Tissue-Specific Expression Patterns of DWY subgroup PPR genes

Transcription databases were obtained from Phytozome database to investigate the tissue expression patterns of the soybean DYW subgroup PPR family; TBtools software was used to conduct a visual hierarchical clustering of the 84 DYW subgroup *PPR* genes under normal conditions. The transcript data are available in Table S3.

4.6. Promoter Sequence Analysis for Potential cis-Elements

For *cis*-elements analysis, 1.5 kb 5′ upstream region sequences were extracted from the Phytozome database. Then, the potential *cis*-elements of promotors for each gene were analyzed using PlantCARE database [63].

4.7. Plant Materials and Treatments

Soybean cultivar Zhonghuang 39 was used to analyze the gene expression pattern. The leaves of 15-day soybean seedlings were collected for RNA extraction and used to further qRT-PCR analysis. For drought stress, the soybean seedlings were placed on filter paper, for NaCl stress, the soybean seedlings were subjected to a 200 mM NaCl solution; samples were collected at 0, 1, 2, 4, 8, 12, 24, and 48 h after treatments. All treated leaf samples were frozen immediately in liquid and then stored at −80 °C for subsequent analysis.

4.8. RNA Extraction and qRT-PCR

Total RNA was extracted from plant samples using the Trizol method as the manufacturer's protocol (TIANGEN, Beijing, China), which was reverse transcribed into cDNA using a PrimeScriptTM RT Reagent Kit (TaKaRa, Shiga, Jappa). All primers used in the study are shown in Table S4 [64].

4.9. Agrobacterium Rhizogenes-Mediated Transformation of Soybean Hairy Roots

Soybean Williams 82, a typical variety, was used to generate *GmPPR4*-OE soybean hairy roots. The coding sequences of *GmPPR4* was ligated into plant transformation vector pCAMBIA3301 with the CaMV 35S promoter. The constructs were transferred into A. rhizogenes strain K599, and *Agrobacteriumrhizo* strain K599 harboring EV-control and *GmPPR4*-OE were injected at the cotyledonary node, as previously described [65]. The injected plants were transferred to the greenhouse with high humidity until hairy roots were generated at the infection site and had grown to about 5 cm long. The original main roots were cut off from 0.5 cm below the infection site. Seedlings were transplanted into nutritious soil and cultured normally in the greenhouse for a week (25 °C 16 h light/8 h dark photoperiod).

4.10. Drought and Salt Stress Assays

For drought treatment, one-week-old seedlings were subjected to dehydration for 15 days. After drought treatment, we re-watered soybean plants for 3 days. At the same time, we carried out salt treatment with 200 mM NaCl solution for 3 days [66].

4.11. Measurement of Proline Content, MDA Content, H_2O_2 Content and O_2^- Content

The content of proline, MDA, H_2O_2 and O_2^- were measured with the corresponding assay kit (Cominbio, Suzhou, China) based on the manufacturer's protocols; all measurements were from four biological replicates.

4.12. Trypan Blue Staining

The leaves separated from the EV-control and *GmPPR4*-OE plants were placed on filter paper for the induction of rapid drought for 3 h, which were used to trypan stain, as previously described [23].

5. Conclusions

This study is the first time identifying the presence of 179 DYW subgroup *PPR* genes in the soybean genome (*Glycine max* Wm82.a2.v1) sequences, which were designated to *GmPPR1* through *GmPPR179* on the basis of their chromosomal location. We conducted a comprehensive and systematic analysis of the DYW subgroup PPR family. Based on the gene expression patterns under drought and salt stresses, we found that three *PPR* genes were highly up-regulated under salt and drought treatment, and *GmPPR4* was selected for validating its role in drought stress tolerance. Compared with the EV-control plants, *GmPPR4*-OE exhibited drought tolerant phenotypes. Our results showed that *GmPPR4* could improve tolerance to drought in soybean.

Supplementary Materials: Supplementary materials can be found at http://www.mdpi.com/1422-0067/20/22/5667/s1. Table S1. DYW-subgroup *PPR* genes in soybean. Detailed genomic information including Phytozome locus, class present, number of introns within ORF, protein, genomic locus (chromosomal location), +/− stand, isoelectric point, and molecular weight (kDa) of the PPR proteins for each *PPR* gene. Table S2. The Ka/Ks ratios for segmentally duplicated the Cluster I PPR proteins. Table S3. Transcript data of 84 DYW subgroup *PPR* genes for heat map. Table S4. The sequences of primers used in the study. Figure S1. The phylogenic tree of 12 selected proteins. Figure S2. qRT-PCR analysis of *GmPPR4* expression in *GmPPR4*-OE and EV-control transgenic hairy roots.

Author Contributions: Z.-S.X. coordinated the project, conceived and designed experiments, and edited the manuscript; H.-G.S. performed experiments and wrote the first draft; B.L., X.-Y.S., and J.M. conducted the bioinformatic work and performed experiments; J.C., Y.-B.Z. and M.C. provided analytical tools and managed reagents; D.-H.M., Z.-S.X., and Y.-Z.M. contributed with valuable discussions. All authors have read and approved the final manuscript.

Funding: This research was financially supported the National Transgenic Key Project of the Chinese Ministry of Agriculture (2018ZX0800909B and 2016ZX08002-002), which were used to design the study and write the manuscript, the National Natural Science Foundation of China (31871624) was used to publish the results.

Acknowledgments: We are grateful to Lijuan Qiu and Shi Sun of the Institute of Crop Science, Chinese Academy of Agricultural Sciences for kindly providing soybean seeds.

Conflicts of Interest: The authors declare no conflict of interest.

Abbreviations

NCBI	National Center for Biotechnology Information
PPR	Pentatricopeptiderepeat
qRT-PCR	Quantitative real-time PCR
ABA	abscisic acid
MDA	malondialdehyde
Pro	proline

References

1. Hammani, K.; Bonnard, G.; Bouchoucha, A.; Gobert, A.; Pinker, F.; Salinas, T.; Giegé, P. Helical repeats modular proteins are major players for organelle gene expression. *Biochimie* **2014**, *100*, 141–150. [CrossRef] [PubMed]

2. Small, I.D.; Peeters, N. The PPR motif—A TPR-Related motif prevalent in plant organellar proteins. *Trends Biochem. Sci.* **2000**, *25*, 45–47. [CrossRef]

3. Cheng, S.; Gutmann, B.; Zhong, X.; Ye, Y.; Fisher, M.F.; Bai, F.; Liu, X. Redefining the structural motifs that determine RNA binding and RNA editing by pentatricopeptide repeat proteins in land plants. *Plant J.* **2016**, *85*, 532–547. [CrossRef] [PubMed]

4. Boussardon, C.; Avon, A.; Kindgren, P.; Bond, C.S.; Challenor, M.; Lurin, C.; Small, I. The cytidine deaminase signature HxE(x)nCxxC of DYW1 binds zinc and is necessary for RNA editing of *ndhD-1*. *New Phytol.* **2014**, *203*, 1090–1095. [CrossRef] [PubMed]

5. Lurin, C.; Andrés, C.; Aubourg, S.; Bellaoui, M.; Bitton, F.; Bruyère, C.; Lecharny, A. Genome-Wide analysis of *Arabidopsis* pentatricopeptide repeat proteins reveals their essential role in organelle biogenesis. *Plant Cell* **2004**, *16*, 2089–2103. [CrossRef] [PubMed]

6. Chen, G.; Zou, Y.; Hu, J.; Ding, Y. Genome-wide analysis of the rice PPR gene family and their expression profiles under different stress treatments. *BMC Genom.* **2018**, *19*, 720. [CrossRef] [PubMed]

7. Xing, H.; Fu, X.; Yang, C.; Tang, X.; Guo, L.; Li, C.; Luo, K. Genome-wide investigation of pentatricopeptide repeat gene family in poplar and their expression analysis in response to biotic and abiotic stresses. *Sci. Rep.* **2018**, *8*, 2817. [CrossRef] [PubMed]

8. Klein, R.R.; Klein, P.E.; Mullet, J.E.; Minx, P.; Rooney, W.L.; Schertz, K.F. Fertility restorer locus *Rf1* of sorghum (*Sorghum bicolor* L.) encodes a pentatricopeptide repeat protein not present in the colinear region of rice chromosome 12. *Theor. Appl. Genet.* **2005**, *111*, 994–1012. [CrossRef] [PubMed]

9. Cui, X.; Wise, R.P.; Schnable, P.S. The *rf2* nuclear restorer gene of male-sterile T-cytoplasm maize. *Science* **1996**, *272*, 1334–1336. [CrossRef] [PubMed]

10. Liu, X.; Yu, F.; Rodermel, S. An *Arabidopsis* pentatricopeptide repeat protein, SUPPRESSOR OF VARIEGATION7, is required for FtsH-mediated chloroplast biogenesis. *Plant Physiol.* **2010**, *154*, 1588–1601. [CrossRef] [PubMed]

11. Ding, Y.H.; Liu, N.Y.; Tang, Z.S.; Liu, J.; Yang, W.C. *Arabidopsis* GLUTAMINE-RICH PROTEIN23 is essential for early embryogenesis and encodes a novel nuclear PPR motif protein that interacts with RNA polymerase II subunit III. *Plant Cell* **2006**, *18*, 815–830. [CrossRef] [PubMed]

12. Banks, J.A.; Nishiyama, T.; Hasebe, M.; Bowman, J.L.; Gribskov, M.; DePamphilis, C.; Ashton, N.W. The Selaginella genome identifies genetic changes associated with the evolution of vascular plants. *Science* **2011**, *332*, 960–963. [CrossRef] [PubMed]

13. Sun, F.; Xiu, Z.; Jiang, R.; Liu, Y.; Zhang, X.; Yang, Y.Z.; Tan, B.C. The mitochondrial pentatricopeptide repeat protein EMP12 is involved in the splicing of three *nad2* introns and seed development in maize. *J. Exp. Bot.* **2018**, *70*, 963–972. [CrossRef] [PubMed]

14. Hao, Y.; Wang, Y.; Wu, M.; Zhu, X.; Teng, Y.; Sun, Y.; Li, J. The nuclear-localized PPR protein OsNPPR1 is important for mitochondrial function and endosperm development in rice. *J. Exp. Bot.* **2019**, *70*, 4705–4720. [CrossRef] [PubMed]

15. Xiao, H.; Zhang, Q.; Qin, X.; Xu, Y.; Ni, C.; Huang, J.; Zhu, Y. Rice *PPS1* encodes a DYW motif-containing pentatricopeptide repeat protein required for five consecutive RNA-editing sites of *nad3* in mitochondria. *New Phytol.* **2018**, *220*, 878–892. [CrossRef] [PubMed]

16. Ebihara, T.; Matsuda, T.; Sugita, C.; Ichinose, M.; Yamamoto, H.; Shikanai, T.; Sugita, M. The P-class pentatricopeptide repeat protein PpPPR_21 is needed for accumulation of the *psbI-ycf12* dicistronic mRNA in Physcomitrella chloroplasts. *Plant J.* **2019**, *97*, 1120–1131. [CrossRef] [PubMed]

17. Murayama, M.; Hayashi, S.; Nishimura, N.; Ishide, M.; Kobayashi, K.; Yagi, Y.; Hirayama, T. Isolation of *Arabidopsisahg11*, a weak ABA hypersensitive mutant defective in *nad4* RNA editing. *J. Exp. Bot.* **2012**, *63*, 5301–5310. [CrossRef] [PubMed]

18. Liu, J.M.; Zhao, J.Y.; Lu, P.P.; Chen, M.; Guo, C.H.; Xu, Z.S.; Ma, Y.Z. The E-subgroup pentatricopeptide repeat protein family in *Arabidopsisthaliana* and confirmation of the responsiveness PPR96 to abiotic stresses. *Front. Plant Sci.* **2016**, *7*, 1825. [CrossRef] [PubMed]

19. Laluk, K.; AbuQamar, S.; Mengiste, T. The *Arabidopsis* mitochondria-localized pentatricopeptide repeat protein PGN functions in defense against necrotrophic fungi and abiotic stress tolerance. *Plant Physiol.* **2011**, *156*, 2053–2068. [CrossRef] [PubMed]

20. Lan, T.; Wang, B.; Ling, Q.; Xu, C.; Tong, Z.; Liang, K.; Wu, W. Fine mapping of *cisc(t)*, a gene for cold-induced seedling chlorosis, and identification of its candidate in rice. *Chin. Sci. Bull.* **2010**, *55*, 3149–3153. [CrossRef]

21. Liu, Y.; He, J.; Chen, Z.; Ren, X.; Hong, X.; Gong, Z. *ABA overly-sensitive 5 (ABO5)*, encoding a pentatricopeptide repeat protein required for cis-splicing of mitochondrial *nad2* intron 3, is involved in the abscisic acid response in *Arabidopsis*. *Plant J.* **2010**, *63*, 749–765. [CrossRef] [PubMed]

22. Zsigmond, L.; Rigó, G.; Szarka, A.; Székely, G.; Ötvös, K.; Darula, Z.; Szabados, L. *Arabidopsis* PPR40 connects abiotic stress responses to mitochondrial electron transport. *Plant Physiol.* **2008**, *146*, 1721–1737. [CrossRef] [PubMed]

23. Du, Y.T.; Zhao, M.J.; Wang, C.T.; Gao, Y.; Wang, Y.X.; Liu, Y.W.; Ma, Y.Z. Identification and characterization of GmMYB118 responses to drought and salt stress. *BMC Plant Biol.* **2018**, *18*, 320. [CrossRef] [PubMed]

24. Mishra, A.K.; Muthamilarasan, M.; Khan, Y.; Parida, S.K.; Prasad, M. Genome-wide investigation and expression analyses of WD40 protein family in the model plant foxtail millet (*Setaria italica* L.). *PLoS ONE* **2014**, *91*, e86852. [CrossRef] [PubMed]

25. Liu, J.M.; Xu, Z.S.; Lu, P.P.; Li, W.W.; Chen, M.; Guo, C.H.; Ma, Y.Z. Genome-wide investigation and expression analyses of the pentatricopeptide repeat protein gene family in foxtail millet. *BMC Genom.* **2016**, *171*, 840. [CrossRef] [PubMed]

26. Gao, Y.; Ma, J.; Zheng, J.C.; Chen, J.; Chen, M.; Zhou, Y.B.; Ma, Y.Z. The Elongation Factor GmEF4 Is Involved in the Response to Drought and Salt Tolerance in Soybean. *Int. J. Mol. Sci.* **2019**, *20*, 3001. [CrossRef] [PubMed]

27. Xu, Z.S.; Chen, M.; Li, L.C.; Ma, Y.Z. Functions of the ERF transcription factor family in plants. *Botany* **2008**, *86*, 969–977. [CrossRef]

28. Xu, Z.S.; Chen, M.; Li, L.C.; Ma, Y.Z. Functions and Application of the AP2/ERF Transcription Factor Family in Crop Improvement, F. *J. Integr. Plant Biol.* **2011**, *53*, 570–585. [CrossRef] [PubMed]

29. Chen, M.; Wang, Q.Y.; Cheng, X.G.; Xu, Z.S.; Li, L.C.; Ye, X.G.; Ma, Y.Z. GmDREB2, a soybean DRE-binding transcription factor, conferred drought and high-salt tolerance in transgenic plants. *Biochem. Biophys. Res. Commun.* **2007**, *353*, 299–305. [CrossRef] [PubMed]

30. Nasreen, S.; Amudha, J.; Pandey, S.S. Isolation and characterization of Soybean DREB 3 transcriptional activator. *J. Appl. Biol. Biotechnol.* **2013**, *1*, 9–12.

31. Wang, N.; Zhang, W.; Qin, M.; Li, S.; Qiao, M.; Liu, Z.; Xiang, F. Drought tolerance conferred in soybean (*Glycine max.* L.) by GmMYB84, a novel R2R3-MYB transcription factor. *Plant Cell Physiol.* **2017**, *58*, 1764–1776. [CrossRef] [PubMed]

32. Gao, S.Q.; Chen, M.; Xu, Z.S.; Zhao, C.P.; Li, L.; Xu, H.J.; Ma, Y.Z. The soybean GmbZIP1 transcription factor enhances multiple abiotic stress tolerances in transgenic plants. *Plant Mol. Biol.* **2011**, *75*, 537–553. [CrossRef] [PubMed]

33. Liao, Y.; Zou, H.F.; Wei, W.; Hao, Y.J.; Tian, A.G.; Huang, J.; Chen, S.Y. Soybean *GmbZIP44*, *GmbZIP62* and *GmbZIP78* genes function as negative regulator of ABA signaling and confer salt and freezing tolerance in transgenic *Arabidopsis*. *Planta* **2008**, *228*, 225–240. [CrossRef] [PubMed]

34. Hao, Y.J.; Wei, W.; Song, Q.X.; Chen, H.W.; Zhang, Y.Q.; Wang, F.; Ma, B. Soybean NAC transcription factors promote abiotic stress tolerance and lateral root formation in transgenic plants. *Plant J.* **2011**, *68*, 302–313. [CrossRef] [PubMed]

35. Zhou, Q.Y.; Tian, A.G.; Zou, H.F.; Xie, Z.M.; Lei, G.; Huang, J.; Chen, S.Y. Soybean WRKY-type transcription factor genes, *GmWRKY13*, *GmWRKY21*, and *GmWRKY54*, confer differential tolerance to abiotic stresses in transgenic *Arabidopsis* plants. *Plant Biotechnol. J.* **2008**, *6*, 486–503. [CrossRef] [PubMed]

36. Vidal, R.O.; Nascimento, L.C.D.; Maurício Costa Mondego, J.; Amarante Guimarães Pereira, G.; Falsarella Carazzolle, M. Identification of SNPs in RNA-seq data of two cultivars of *Glycine max* (soybean) differing in drought resistance. *Genet. Mol. Biol.* **2012**, *35*, 331–334. [CrossRef] [PubMed]

37. Ohyama, T.; Minagawa, R.; Ishikawa, S.; Yamamoto, M.; Hung, N.V.P.; Ohtake, N.; Sueyoshi, K.; Sato, T.; Nagumo, Y.; Takahashi, Y. Soybean seed production and nitrogen nutrition. In *A Comprehensive Survey of International Soybean Research-Genetics, Physiology, Agronomy and Nitrogen Relationships*; Board, J.E., Ed.; InTech: Rijeka, Croatia, 2013; pp. 115–157.

38. Frederick, J.R.; Camp, C.R.; Bauer, P.J. Drought-Stress effects on branch and mainstem seed yield and yield components of determinate soybean. *Crop Sci.* **2001**, *41*, 759–763. [CrossRef]

39. O'Toole, N.; Hattori, M.; Andres, C.; Iida, K.; Lurin, C.; Schmitz-Linneweber, C.; Small, I. On the expansion of the pentatricopeptide repeat gene family in plants. *Mol. Biol. Evol.* **2008**, *25*, 1120–1128. [CrossRef] [PubMed]

40. De Longevialle, A.F.; Meyer, E.H.; Andrés, C.; Taylor, N.L.; Lurin, C.; Millar, A.H.; Small, I.D. The pentatricopeptide repeat gene *OTP43* is required for trans-splicing of the mitochondrial nad1 intron 1 in *Arabidopsis thaliana*. *Plant Cell* **2007**, *19*, 3256–3265. [CrossRef] [PubMed]

41. Weißenberger, S.; Soll, J.; Carrie, C. The PPR protein SLOW GROWTH 4 is involved in editing of nad4 and affects the splicing of nad2 intron 1. *Plant Mol. Biol.* **2017**, *93*, 355–368. [CrossRef] [PubMed]

42. Lee, K.; Han, J.H.; Park, Y.I.; Colas des Francs-Small, C.; Small, I.; Kang, H. The mitochondrial pentatricopeptide repeat protein PPR 19 is involved in the stabilization of NADH dehydrogenase 1 transcripts and is crucial for mitochondrial function and *Arabidopsisthaliana* development. *New Phytol.* **2017**, *215*, 202–216. [CrossRef] [PubMed]

43. Koprivova, A.; des Francs-Small, C.C.; Calder, G.; Mugford, S.T.; Tanz, S.; Lee, B.R.; Kopriva, S. Identification of a pentatricopeptide repeat protein implicated in splicing of intron 1 of mitochondrial nad7 transcripts. *J. Biol. Chem.* **2010**, *285*, 32192–32199. [CrossRef] [PubMed]

44. Haïli, N.; Planchard, N.; Arnal, N.; Quadrado, M.; Vrielynck, N.; Dahan, J.; Mireau, H. The MTL1 pentatricopeptide repeat protein is required for both translation and splicing of the mitochondrial NADH DEHYDROGENASE SUBUNIT7 mRNA in *Arabidopsis*. *Plant Physiol.* **2016**, *170*, 354–366. [CrossRef] [PubMed]

45. Des Francs-Small, C.C.; de Longevialle, A.F.; Li, Y.; Lowe, E.; Tanz, S.; Smith, C.; Small, I. The PPR proteins TANG2 and OTP439 are involved in the splicing of the multipartite nad5 transcript encoding a subunit of mitochondrial complex I. *Plant Physiol.* **2014**, 114. [CrossRef]

46. Hsieh, W.Y.; Liao, J.C.; Chang, C.Y.; Harrison, T.; Boucher, C.; Hsieh, M.H. The SLOW GROWTH3 pentatricopeptide repeat protein is required for the splicing of mitochondrial NADH dehydrogenase subunit7 intron 2 in *Arabidopsis*. *Plant Physiol.* **2015**, *168*, 490–501. [CrossRef] [PubMed]

47. Zheng, P.; He, Q.; Wang, X.; Tu, J.; Zhang, J.; Liu, Y.J. Functional analysis for domains of maize PPR protein EMP5 in RNA editing and plant development in *Arabidopsis*. *Plant Growth Regul.* **2019**, *87*, 19–27. [CrossRef]

48. Sosso, D.; Mbelo, S.; Vernoud, V.; Gendrot, G.; Dedieu, A.; Chambrier, P.; Rogowsky, P.M. PPR2263, a DYW-subgroup pentatricopeptide repeat protein, is required for mitochondrial nad5 and cob transcript editing, mitochondrion biogenesis, and maize growth. *Plant Cell* **2012**, *24*, 676–691. [CrossRef] [PubMed]

49. Sosso, D.; Canut, M.; Gendrot, G.; Dedieu, A.; Chambrier, P.; Barkan, A.; Consonni, G.; Rogowsky, P.M. *PPR8522* encodes a chloroplast-targeted pentatricopeptide repeat protein necessary for maize embryogenesis and vegetative development. *J. Exp. Bot.* **2012**, *63*, 5843–5857. [CrossRef] [PubMed]

50. Kim, S.R.; Yang, J.I.; Moon, S.; Ryu, C.H.; An, K.; Kim, K.M.; An, G. Rice OGR1 encodes a pentatricopeptide repeat–DYW protein and is essential for RNA editing in mitochondria. *Plant J.* **2009**, *59*, 738–749. [CrossRef] [PubMed]

51. Huang, C.; Yu, Q.B.; Li, Z.R.; Ye, L.S.; Xu, L.; Yang, Z.N. Porphobilinogen deaminase HEMC interacts with the PPR-protein AtECB2 for chloroplast RNA editing. *Plant J.* **2017**, *92*, 546–556. [CrossRef] [PubMed]

52. Zoschke, R.; Qu, Y.; Zubo, Y.O.; Börner, T.; Schmitz-Linneweber, C. Mutation of the pentatricopeptide repeat-SMR protein SVR7 impairs accumulation and translation of chloroplast ATP synthase subunits in *Arabidopsis thaliana*. *J. Plant Res.* **2013**, *126*, 403–414. [CrossRef] [PubMed]

53. Zhou, W.; Cheng, Y.; Yap, A.; Chateigner-Boutin, A.L.; Delannoy, E.; Hammani, K.; Huang, J. The *Arabidopsis* gene *YS1* encoding a DYW protein is required for editing of rpoB transcripts and the rapid development of chloroplasts during early growth. *Plant J.* **2009**, *58*, 82–96. [CrossRef] [PubMed]

54. Finn, R.D.; Clements, J.; Eddy, S.R. HMMER web server: Interactive sequence similarity searching. *Nucleic Acids Res.* **2011**, *39*, W29–W37. [CrossRef] [PubMed]

55. Goodstein, D.M.; Shu, S.; Howson, R.; Neupane, R.; Hayes, R.D.; Fazo, J.; Mitros, T.; Dirks, W.; Hellsten, U.; Putnam, N.; et al. Phytozome: A comparative platform for green plant genomics. *Nucleic Acids Res.* **2012**, *40*, D1178–D1186. [CrossRef] [PubMed]

56. Finn, R.D.; Bateman, A.; Clements, J.; Coggill, P.; Eberhardt, R.Y.; Eddy, S.R.; Sonnhammer, E.L. Pfam: The protein families database. *Nucleic Acids Res.* **2013**, *42*, D222–D230. [CrossRef] [PubMed]

57. Letunic, I.; Doerks, T.; Bork, P. SMART 7: Recent updates to the protein domain annotation resource. *Nucleic Acids Res.* **2012**, *40*, D302–D305. [CrossRef] [PubMed]

58. Artimo, P.; Jonnalagedda, M.; Arnold, K.; Baratin, D.; Csardi, G.; de Castro, E.; Duvaud, S.; Flegel, V.; Fortier, A.; Gasteiger, E.; et al. ExPASy: SIB bioinformatics resource portal. *Nucleic Acids Res.* **2012**, *40*, W597–W603. [CrossRef] [PubMed]

59. Guo, A.Y.; Zhu, Q.H.; Chen, X.; Luo, J.C. GSDS: A gene structure display server. *Yi Chuan* **2007**, *29*, 1023–1026. [CrossRef] [PubMed]

60. Bailey, T.L.; Boden, M.; Buske, F.A.; Frith, M. MEME SUITE: Tools for motif discovery and searching. *Nucleic Acids Res.* **2009**, *37*, W202–W208. [CrossRef] [PubMed]

61. Chen, C.; Xia, R.; Chen, H.; He, Y. TBtools, a Toolkit for Biologists integrating various biological data handling tools with a user-friendly interface. *bioRxiv* **2018**, 289660. [CrossRef]

62. Krzywinski, M.; Schein, J.; Birol, I.; Connors, J. Circos: An information aesthetic for comparative genomics. *Genome Res.* **2009**, *19*, 1639–1645. [CrossRef] [PubMed]

63. Lescot, M.; Déhais, P.; Thijs, G.; Marchal, K.; Moreau, Y. PlantCARE, a database of plant cis-acting regulatory elements and a portal to tools for in silico analysis of promoter sequences. *Nucleic Acids Res.* **2002**, *30*, 325–327. [CrossRef] [PubMed]

64. Cui, X.Y.; Gao, Y.; Guo, J.; Yu, T.F.; Zheng, W.J.; Liu, Y.W.; Ma, Y.Z. BES/BZR Transcription Factor TaBZR2 Positively Regulates Drought Responses by Activation of TaGST1. *Plant Physiol.* **2019**, *180*, 605–620. [CrossRef] [PubMed]
65. Wang, F.; Chen, H.W.; Li, Q.T.; Wei, W.; Li, W.; Zhang, W.K.; Man, W.Q. GmWRKY27 interacts with GmMYB174 to reduce expression of GmNAC 29 for stress tolerance in soybean plants. *Plant J.* **2015**, *83*, 224–236. [CrossRef] [PubMed]
66. Xu, Z.S.; Ni, Z.Y.; Liu, L.; Nie, L.N.; Li, L.C.; Chen, M.; Ma, Y.Z. Characterization of the *TaAIDFa* gene encoding a CRT/DRE-binding factor responsive to drought, high-salt, and cold stress in wheat. *Mol. Genet. Genom.* **2008**, *280*, 497–508. [CrossRef] [PubMed]

International Journal of
Molecular Sciences

MDPI

Article

Characterizing the Role of *TaWRKY13* in Salt Tolerance

Shuo Zhou [1,†], Wei-Jun Zheng [2,†], Bao-Hua Liu [3], Jia-Cheng Zheng [4], Fu-Shuang Dong [1], Zhi-Fang Liu [5], Zhi-Yu Wen [1], Fan Yang [1], Hai-Bo Wang [1], Zhao-Shi Xu [6], He Zhao [1,*] and Yong-Wei Liu [1,*]

[1] Institute of Genetics and Physiology, Hebei Academy of Agriculture and Forestry Sciences/Plant Genetic Engineering Center of Hebei Province, Shijiazhuang 050051, China; zhoushuobio@163.com (S.Z.); dongfushuang@126.com (F.-S.D.); wzy1800@126.com (Z.-Y.W.); yangfan21st@163.com (F.Y.); nkywanghb@163.com (H.-B.W.)

[2] College of Agronomy, Northwest A&F University, Yangling 712100, China; zhengweijun@nwafu.edu.cn

[3] Handan Academy of Agricultural Sciences, Handan 056001, China; hm4589@163.com

[4] College of Agronomy, Anhui Science and Technology University, Fengyang, Chuzhou 239000, China; zhengjiachengx2016@126.com

[5] Hebei Seed Station, Shijiazhuang 050031, China; happysky961@163.com

[6] Institute of Crop Sciences, Chinese Academy of Agricultural Sciences (CAAS)/National Key Facility for Crop Gene Resources and Genetic Improvement, Key Laboratory of Biology and Genetic Improvement of Triticeae Crops, Ministry of Agriculture, Beijing 100081, China; xuzhaoshi@caas.cn

* Correspondence: hezhao311@163.com (H.Z.); liuywmail@126.com (Y.-W.L.)

† These authors contributed equally to this work.

Received: 13 October 2019; Accepted: 11 November 2019; Published: 14 November 2019

Abstract: The WRKY transcription factor superfamily is known to participate in plant growth and stress response. However, the role of this family in wheat (*Triticum aestivum* L.) is largely unknown. Here, a salt-induced gene *TaWRKY13* was identified in an RNA-Seq data set from salt-treated wheat. The results of RT-qPCR analysis showed that *TaWRKY13* was significantly induced in NaCl-treated wheat and reached an expression level of about 22-fold of the untreated wheat. Then, a further functional identification was performed in both *Arabidopsis thaliana* and *Oryza sativa* L. Subcellular localization analysis indicated that TaWRKY13 is a nuclear-localized protein. Moreover, various stress-related regulatory elements were predicted in the promoter. Expression pattern analysis revealed that *TaWRKY13* can also be induced by polyethylene glycol (PEG), exogenous abscisic acid (ABA), and cold stress. After NaCl treatment, overexpressed *Arabidopsis* lines of *TaWRKY13* have a longer root and a larger root surface area than the control (Columbia-0). Furthermore, *TaWRKY13* overexpression rice lines exhibited salt tolerance compared with the control, as evidenced by increased proline (Pro) and decreased malondialdehyde (MDA) contents under salt treatment. The roots of overexpression lines were also more developed. These results demonstrate that *TaWRKY13* plays a positive role in salt stress.

Keywords: stress responsive mechanisms; TaWRKY transcription factors; salt tolerance

---

1. Introduction

Unlike animals, plants cannot move when exposed to stress. However, complex signaling network have been established to cope with stress [1]. Under stress, a series of responses are induced to prevent or minimize damage. These are accompanied by many physiological, biochemical and developmental changes [2]. Current research on plant stress response has reached the level of cells and molecules, and combined with genetics, we can explore the stress responsive mechanisms in order to improve plant growth under conditions of stress [3–7].

Many genes are induced by stress; the products of these genes both participate in stress response and regulate the expression of related genes involved in signal transduction pathways in order to avoid or reduce tissue damage [8–10]. Signaling via the hormone, liposome, SnRK2 (sucrose non-fermenting 1-related protein kinase 2) [11], MAPK (mitogen activated protein kinase) [12], ROS signal [13] and stomatal [14] pathways are the main networks by which plants respond to salt and drought stress. Plant adaptation to drought and other stresses depends on both the expression of stress-resistant related genes and the regulation of various signal pathways induced by stress [15]. Products of stress-related genes can be divided into two classes: the first class includes ion channel proteins [16], water channel proteins [17], osmotic regulators (sucrose, proline and betaine), synthases [18] and other products that directly function in stress response, while the products of the second type include proteins involved in stress-related signal transmission and regulators of gene expression, such as protein kinases (PKs) and transcription factors (TFs) [19,20].

Transcription factors play a crucial role in regulating the expression of stress-related genes in plants. When abiotic stress occurs, changes in the activity of transcription factors cause changes in the activity of target genes. Transcription factors involved in plant stress response are widely researched, such as the AP2/EREBP TF family [21], MYC/MYB TF family [22], HSE binding TFs [23], NAC TF family [24], and WRKY TF family [25]. Among them, WRKY TFs are extensively found in higher plants including *Arabidopsis thaliana*, *Oryza sativa*, *Setaria italica*, *Glycine max*, and *Triticum aestivum*, which indicates that WRKY TFs play a significant role in plant stress tolerance [26–30].

Although a large number of studies have shown that WRKY TFs in plants are mainly involved in disease resistance and defense response, some members of the WRKY TFs are involved in abiotic stress response. *TaWRKY1* mediates stomatal movement through an ABA-dependent pathway to improve plant tolerance to drought stress [31]. In addition, *TaWRKY10* acts as a positive regulator under drought, salt, cold, and hydrogen peroxide stress conditions and improves the stress tolerance in transgenic tobacco [32]. In *Arabidopsis*, WRKY proteins are involved in regulating ABA response factors, such as MYB2, DREB1a, DREB2a and Rab18 [33]. The overexpression of *ZmWRKY33* in *Arabidopsis* improved the salt-stress tolerance of transgenic plants [34]. These studies suggested that WRKY TFs play a significant role in plant stress response.

High salt stress is a major obstacle to plant growth and development. High salt conditions lead to increases in reactive oxygen species (ROS), metabolic toxicity, membrane disorganization, the inhibition of photosynthesis, and attenuated nutrient acquisition at different plant growth stages [35]. Recent reports claim that salinity affects about 20% of all irrigated arable land and is an increasing problem in worldwide agriculture (FAO Cereal Supply and Demand Brief. http://www.fao.org/worldfoodsituation/csdb/en/).

Since wheat is rich in thiamine, fat, calcium, niacin, starch, protein, iron, riboflavin, minerals, and vitamin A and can provide abundant energy and protein for humans, wheat is regarded as one of the most important crops in the world [36]. However, wheat production is constrained by environmental conditions, such as drought, salinity, waterlogging, and extremes in temperature. Next in importance to drought stress, salinity affects crop yields worldwide. The improvement of stress tolerance in wheat by biotechnology and transgenic technology could contribute to increased production worldwide. However, the huge wheat genome has slowed progress [37,38]. Although many studies have investigated the roles of WRKY transcription factors in response to various stress conditions, the mechanisms underlying their function need further study. Here, RNA-Seq, real-time fluorescence quantification PCR (RT-qPCR), and several databases were used in a study of *TaWRKY13*. The results demonstrated that overexpression of *TaWRKY13* can improve salt tolerance in *Arabidopsis* and rice.

2. Results

2.1. Identification and Genome Structure Analysis of WRKYs in Triticum aestivum

According to the Plant Transcription Factor Database website (http://planttfdb.cbi.pku.edu.cn/index.php), wheat has 171 TaWRKYs, which are distributed across all chromosomes (1AL, 1BL, 1DL, 2AL, 2AS, 2BS, 2DL, 2DS, 3AL, 3B, 3DL, 4AL, 4AS, 4DS, 5AL, 5BS, 5BL, 5DL, 6AL, 6AS, 6BS, 6DS, 7AL, 7DL). Here, PF03106 was used as a key word to blast WRKYs in wheat on the Phytozome website (https://phytozome.jgi.doe.gov/pz/portal.html). Nucleic acid and amino acid sequences of 100 TaWRKYs that harbor at least one WRKY domain are shown in Supplementary Table S2. Based on the rule that the CDS of TaWRKYs were more than 300 base pairs [30], some TaWRKYs were removed, and then combined with the NCBI database (https://www.ncbi.nlm.nih.gov/pubmed), meaning that 57 TaWRKYs were identified with the annotation gene's name, ID, transcript name and location (Table 1). The location of 57 TaWRKYs on chromosomes was analyzed by using the online website http://mg2c.iask.in/mg2c_v2.0/. From the map, we can see that the locations of TaWRKYs were different on each chromosome; for example, *TaWRKY6, 38, 50, 27, 48,* and *57* were located at the end of chromosome 3B (forward or reverse), while *TaWRKY70* and *TaWRKY 71* were located near the centromere of chromosome 1D. Moreover, the distribution of TaWRKYs on 4A was the combination of both distributions described above (Figure 1). To further explore gene structure differences, a gene structure figure of 56 TaWRKYs is displayed in Figure 2. The TaWRKYs are all different in structure. Most TaWRKYs contain 1 to 5 different exons, which may contain different functional structures, such as zinc finger, leucine, kinase structure, exerting different biological functions. TaWRKY7, 22, 23, 24, 33, 56, and 90 do not harbor introns, only containing exons and/or an upstream structure.

Table 1. Annotation of WRKY transcription factors in *Triticum aestivum*.

Name	ID	Transcript Name	Location
TaWRKY28	31740471	Traes_1BL_9AFA4B870.1	ta_iwgsc_1bl_v1_3809885:1110..4753 reverse
TaWRKY15	31742772	Traes_5BL_E294922A9.2	ta_iwgsc_5bl_v1_10867378:4708..6402 forward
TaWRKY6	31744736	Traes_3B_CDA5ADD75.1	ta_iwgsc_3b_v1_10758590:296..2989 forward
TaWRKY62	31745499	Traes_5DL_C93641E43.1	ta_iwgsc_5dl_v1_4576731:2459..4427 forward
TaWRKY44	31746115	Traes_4AL_2EEECCC4B.1	ta_iwgsc_4al_v2_7093101:3374..6818 forward
TaWRKY74	31747511	Traes_5DL_5C93510D5.1	ta_iwgsc_5dl_v1_4502975:3368..7661 reverse
TaWRKY80	31748920	Traes_6AS_DA75BB1FD.1	ta_iwgsc_6as_v1_4428654:1..1588 reverse
TaWRKY53	31752041	Traes_2DL_F600B5FDF.1	ta_iwgsc_2dl_v1_9719154:1..728 forward
TaWRKY22	31752743	Traes_5AL_6FDB440FB.1	ta_iwgsc_5al_v1_2705439:4..838 forward
TaWRKY45	31765470	Traes_7AL_48C81DE03.1	ta_iwgsc_7al_v1_4556343:539..3297 forward
TaWRKY4	31765472	Traes_7AL_48C81DE031.1	ta_iwgsc_7al_v1_4556343:3718..6476 reverse
TaWRKY35	31766778	Traes_6BL_EEAA2A7E3.1	ta_iwgsc_6bl_v1_4221964:3..2595 forward
TaWRKY79	31767242	Traes_7DL_B09854286.1	ta_iwgsc_7dl_v1_3393496:18..940 reverse
TaWRKY46	31768080	Traes_4DS_FE38A59D0.1	ta_iwgsc_4ds_v1_2280139:4533..7422 reverse
TaWRKY57	31782323	Traes_3B_41047D5E6.2	ta_iwgsc_3b_v1_10527462:3896..6406 reverse
TaWRKY33	31785825	Traes_6DS_8F684013D.1	ta_iwgsc_6ds_v1_1013038:1..419 reverse
TaWRKY12	31787421	Traes_6AL_BA4636569.1	ta_iwgsc_6al_v1_5754118:409..4140 reverse
TaWRKY24	31792629	Traes_4AL_C2A825B6D.1	ta_iwgsc_4al_v2_3841042:1..253 reverse
TaWRKY63	31793891	Traes_3DL_7456F61A3.1	ta_iwgsc_3dl_v1_5877113:2..2892 reverse
TaWRKY68	31798439	Traes_2AL_15A7BB684.1	ta_iwgsc_2al_v1_6374918:10015..11505 forward
TaWRKY50	31799212	Traes_3B_F45FCFE62.1	ta_iwgsc_3b_v1_10625585:4077..6054 forward
TaWRKY58	31811544	Traes_5BL_D3C383CF5.1	ta_iwgsc_5bl_v1_10787947:2038..3881 forward
TaWRKY72	31818595	Traes_2BS_F3097F116.1	ta_iwgsc_2bs_v1_5195103:6587..11319 forward
TaWRKY8	31823877	Traes_5DL_2553A6C33.1	ta_iwgsc_5dl_v1_4566006:8..1007 forward
TaWRKY34	31829399	Traes_2AL_409AB7647.1	ta_iwgsc_2al_v1_6334600:3412..8916 reverse

Table 1. *Cont.*

Name	ID	Transcript Name	Location
TaWRKY9	31836810	Traes_2DS_F6FBC974C.2	ta_iwgsc_2ds_v1_5331381:733..3264 reverse
TaWRKY52	31851405	Traes_3AL_AB2BAE660.1	ta_iwgsc_3al_v1_4270257:1..887 reverse
TaWRKY3	31853252	Traes_2DL_4F9F8F1F0.1	ta_iwgsc_2dl_v1_9906833:634..5055 reverse
TaWRKY51	31854913	Traes_2AL_434E9F101.1	ta_iwgsc_2al_v1_6367445:3985..5752 reverse
TaWRKY27	31865868	Traes_3B_990298FF5.1	ta_iwgsc_3b_v1_10750391:1..2331 reverse
TaWRKY70	31871499	Traes_1DL_DFE1721E0.1	ta_iwgsc_1dl_v1_2268423:4679..7857 forward
TaWRKY41	31872073	Traes_2DS_AD8820C42.1	ta_iwgsc_2ds_v1_5376167:6..1016 reverse
TaWRKY14	31872762	Traes_5BL_B9DD3E76F.1	ta_iwgsc_5bl_v1_10924584:9637..13927 forward
TaWRKY17	31875786	Traes_5BL_8688F70C9.1	ta_iwgsc_5bl_v1_10840877:2232..3827 reverse
TaWRKY56	31876237	Traes_3AL_DED8A29EC.1	ta_iwgsc_3al_v1_382150:704..961 reverse
TaWRKY78	31876678	Traes_2BS_D435A8999.1	ta_iwgsc_2bs_v1_5214231:8279..15893 reverse
TaWRKY32	31888413	Traes_4AS_70DF607CC.1	ta_iwgsc_4as_v2_352920:1884..3594 forward
TaWRKY16	31891223	Traes_4AL_98B1C762B.2	ta_iwgsc_4al_v2_7173949:3935..6881 forward
TaWRKY48	31892659	Traes_3B_B8BF316B8.2	ta_iwgsc_3b_v1_10433739:23..1890 reverse
TaWRKY71	31894510	Traes_1DL_46428511F.1	ta_iwgsc_1dl_v1_2235906:1905..4608 forward
TaWRKY38	31895081	Traes_3B_D6F86ABC3.2	ta_iwgsc_3b_v1_10762199:7310..8822 forward
TaWRKY55	31916438	Traes_6AS_68775100B.1	ta_iwgsc_6as_v1_4413209:7948..9254 forward
TaWRKY76	31917474	Traes_5DL_32D78D06A.1	ta_iwgsc_5dl_v1_4501324:900..1439 reverse
TaWRKY19	31924920	Traes_2BS_380EC4D1E.1	ta_iwgsc_2bs_v1_5227909:9257..13033 reverse
TaWRKY29	31938855	Traes_1AL_4E924201A.1	ta_iwgsc_1al_v2_3969710:4988..6878 reverse
TaWRKY10	31942345	Traes_2DL_362A1F535.1	ta_iwgsc_2dl_v1_9707610:58..446 forward
TaWRKY36	31942939	Traes_3AL_140B829CB.2	ta_iwgsc_3al_v1_4308486:3673..5708 reverse
TaWRKY2	31951792	Traes_5BL_17A712C94.1	ta_iwgsc_5bl_v1_10916210:5033..9621 forward
TaWRKY13	31962353	Traes_2AS_6269D889E.1	ta_iwgsc_2as_v1_5205891:13214..14843 reverse
TaWRKY1	31966248	Traes_5BL_AEF9FE805.1	ta_iwgsc_5bl_v1_10827243:3081..6424 reverse
TaWRKY49	31968771	Traes_3DL_2551BF2C1.1	ta_iwgsc_3dl_v1_6811598:1..1035 reverse
TaWRKY23	31977027	Traes_3AL_4769A72F1.1	ta_iwgsc_3al_v1_805190:2..774 reverse
TaWRKY75	31987126	Traes_1AL_0404BC790.1	ta_iwgsc_1al_v2_3912777:3..1909 forward
TaWRKY5	31988149	Traes_5AL_7164FEAC3.1	ta_iwgsc_5al_v1_2204788:3..342 forward
TaWRKY64	32002393	Traes_4AS_0DA136E0E.1	ta_iwgsc_4as_v2_5962726:2807..4440 forward
TaWRKY90	32002429	Traes_3AL_1B73D2C12.1	ta_iwgsc_3al_v1_4248344:659..1678 forward
TaWRKY61	32024774	Traes_5BS_C46781248.1	ta_iwgsc_5bs_v1_2248873:15934..19692 reverse

Annotations were according to Phytozome (https://phytozome.jgi.doe.gov/pz/portal.html), PlantTFDB (http://planttfdb.cbi.pku.edu.cn/index.php and NCBI (https://www.ncbi.nlm.nih.gov/pubmed).

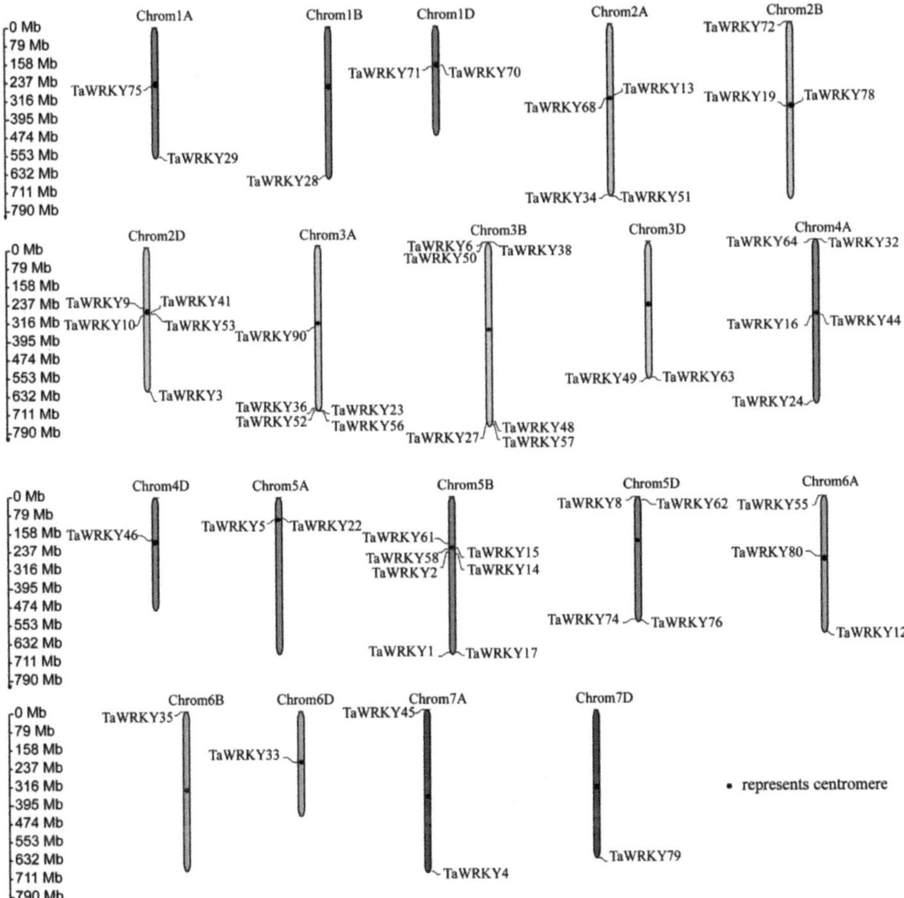

Figure 1. Chromosome location of TaWRKYs listed in Table 1.

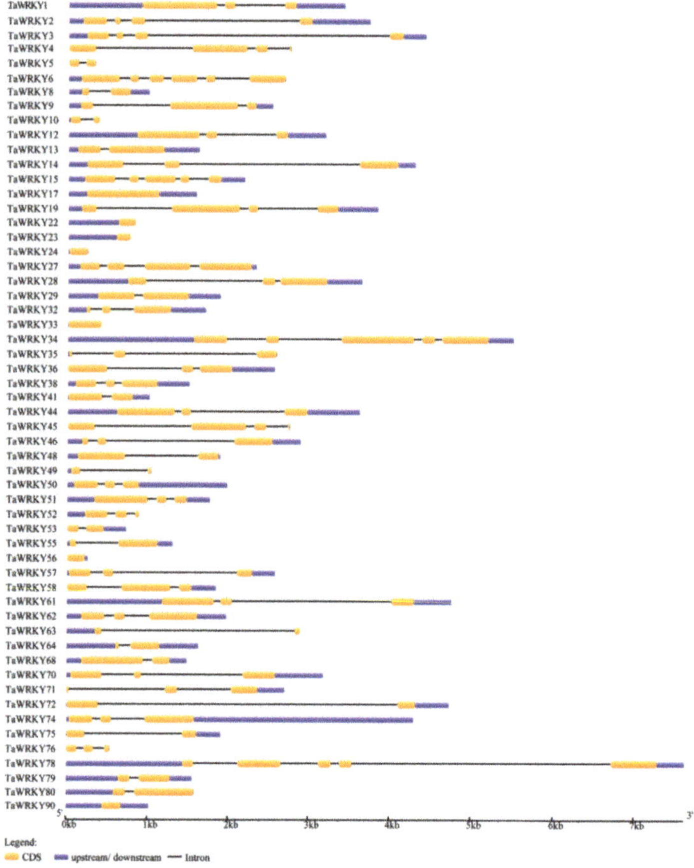

Figure 2. Gene structure analysis of TaWRKYs. Segments in yellow represent CDS, blue indicates upstream/downstream, and black lines represent introns.

2.2. Identification and Biological Analysis of TaWRKY13

To find wheat stress-responsive genes under salt stress, the roots of three-leaf wheat seedlings were immersed in 150 mM NaCl solution for 1 h. Control_Leaf represents the leaf tissue without NaCl treatment, NaCl_Leaf represents the leaf tissue treated as per the above description; each treatment involved two independent replicates which were then sampled for RNA-seq (Supplementary Table S1). Twelve TaWRKYs (TaWRKY4, 9, 12, 13, 15, 22, 29, 33, 34, 44, 53, and 70) were selected based on the rule $\log_2$ (NaCl_Leaf /Control_Leaf) > 2. As shown in Figure 3, *TaWRKY13* gave the highest relative expression in response to salt stress, peaking at more than 20-fold at 1 h. *TaWRKY13* (ID: 31962353, Traes_2AS_6269D889E.1) was selected for further investigation. *TaWRKY13* contained a 975 bp open reading frame (ORF) encoding 324 amino acids; the molecular weight of the protein was 81.02 kDa with pI 4.99 (https://web.expasy.org/protparam/). The predicted amino acid sequence showed that *TaWRKY13* only harbored one WRKY domain with a highly conserved WRKYGQK motif and a CX4-5CX22-23HXH zinc-finger motif.

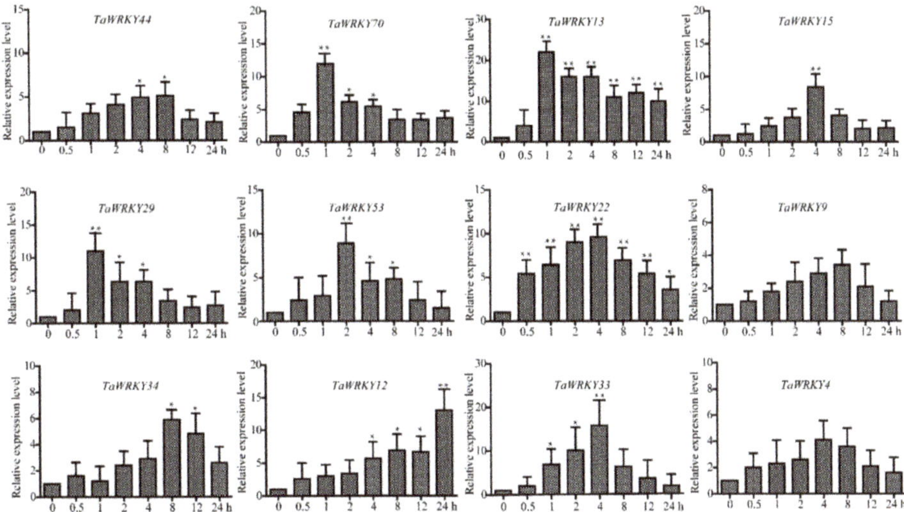

Figure 3. Real-time fluorescence quantification PCR of 12 TaWRKYs under salt treatment. The expression level of TaActin was used as a loading control. The data represent the means ± SD of three biological replications. The ANOVA demonstrated significant differences (* $p < 0.05$, ** $p < 0.01$).

2.3. Phylogenetic Analysis of AtWRKYs, OsWRKYs and TaWRKYs

Phylogenetic analysis is a useful method that can provide some clues to the possible functions of predicted or analyzed target genes. It would be useful to know the homologs of *Triticum aestivum* WRKYs (TaWRKYs), especially *TaWRKY13*, with WRKYs of *Arabidopsis thaliana* (AtWRKYs) and WRKYs of *Oryza sativa* (OsWRKYs) with reference to previous results. A phylogenic tree was constructed by the neighbor-joining method [39] to investigate the evolutionary relationships between AtWRKYs, OsWRKYs and TaWRKYs. There are 398 WRKYs for phylogenetic analysis (90 AtWRKYs, 128 OsWRKYs and 171 TaWRKYs) (Figure 4). According to Figure 4, AtWRKYs, OsWRKYs and TaWRKYs were scattered across different branches of the phylogenic tree, and all WRKYs were divided into three broad categories; among them, there were more WRKYs in groups I and II than in group III. *TaWRKY13* (ID: Traes_2AS_6269D889E.1) and *AtWRKY13* (ID: AT4G39410) were in group II, and *OsWRKY13* (ID: LOC-Os01g546600) belonged to group I. The results of phylogenetic analysis preliminarily indicated that *TaWRKY13* has a closer homology with *AtWRKY13* than *OsWRKY13*.

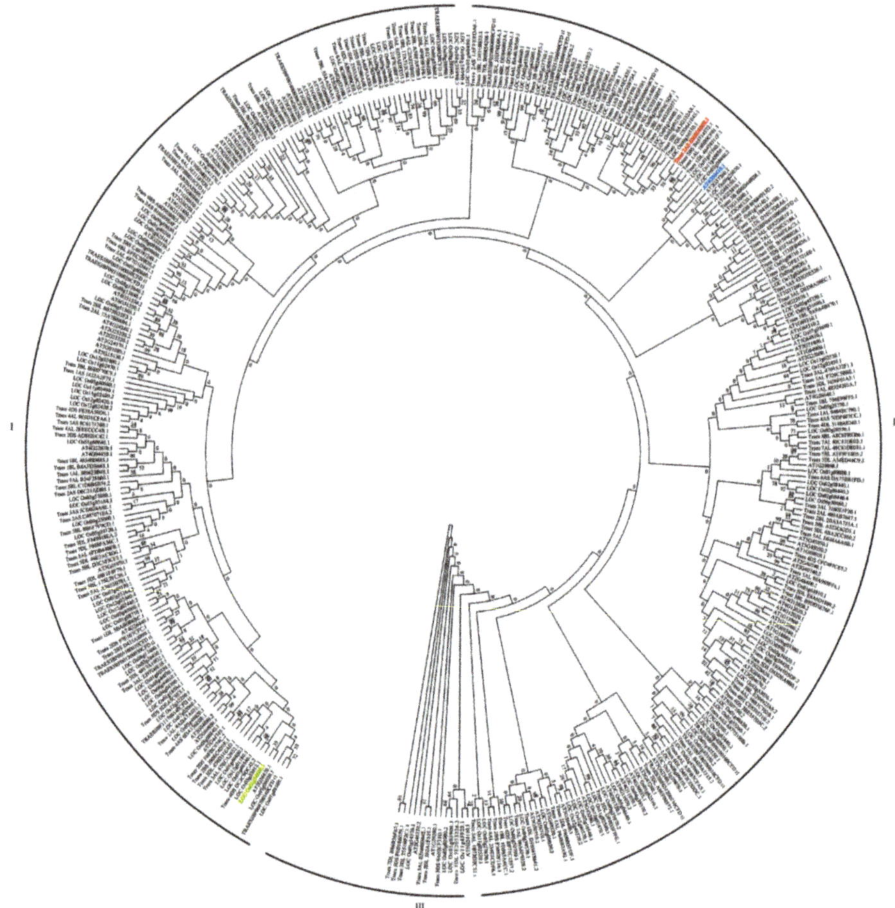

Figure 4. Phylogenetic analysis of AtWRKYs, OsWRKYs and TaWRKYs. The phylogenetic tree was produced using the aligned file with 1000 bootstrap replications in MEGA 6.0. *TaWRKY13, AtWRKY13* and *OsWRKY13* are highlighted in red, blue and yellow, respectively. The numbers at nodes are bootstrap values, and the length of branches represent evolutionary distance. Number of bootstrap replications: 1000.

2.4. TaWRKY13 was Localized in the Nucleus

To investigate the biological activity of *TaWRKY13*, the coding sequence fused to the N-terminus of the green fluorescent protein (GFP) was inserted into wheat mesophyll protoplasts by the PEG-mediated method. As the control, 35S::GFP was transformed [40]. The fluorescence of the control GFP was distributed throughout the cells, whereas the fluorescence of 35S::TaWRKY13-GFP was specifically localized in the nucleus (Figure 5). Thus, TaWRKY13 is a nuclear-located protein.

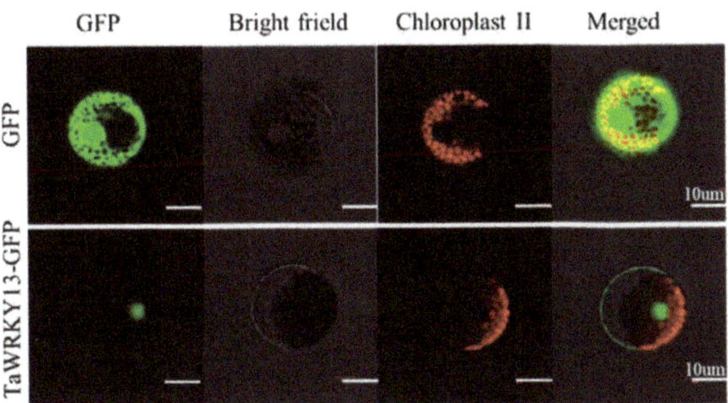

Figure 5. Subcellular localization of TaWRKY13. 35S::GFP and 35S::TaWRKY13-GFP constructs were transformed into wheat mesophyll protoplasts under the control of the Cauliflower Mosaic Virus 35S (CaMV35S) promoter. Wherein, green color represents fluorescence emitted by green fluorescent protein under confocal laser scanning microscope and the red color represents the fluorescence emitted by chloroplasts under confocal laser scanning microscope. Results were observed by a confocal laser scanning microscope (LSM700; CarlZeiss, Oberkochen Germany) after incubation in darkness at 22 °C for 18–20 h. Scale bars, 10 μm.

2.5. Tissue-Specific Expression of TaWRKY13

Studies of genes with a specific expression in different tissues are necessary to understand the regulatory mechanisms of plant growth and development and the relationship between cell type and function. Here, the promoter sequence of *TaWRKY13* was fused to the pCAMBIA1305 vector, which contains a β-glucuronidase (GUS) reporter gene in the N-terminus (Figure 6). The GUS reporter gene can preliminarily determine the tissue specificity of the gene by observing the tissue location with a blue color after staining [41]. qRT-PCR was used to further verify the relative expression level at the molecular level. *TaWRKY13* was expressed in the roots, stems and leaves of T_3 generation transgenic *Arabidopsis* plants under normal and salt-stress conditions, with the relative expression in roots being higher than in leaves and stems. After NaCl treatment, the expression levels in roots, stems and leaves were significantly increased, indicating that *TaWRKY13* might be responsive to salt stress.

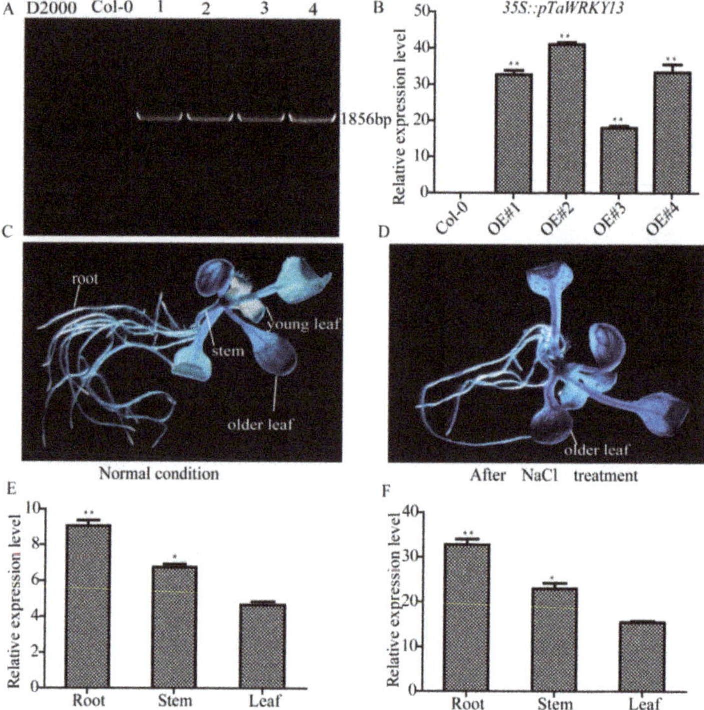

Figure 6. Tissue-specific expression analysis of *TaWRKY13*. (**A**) Identification of homozygous lines by agarose gel electrophoresis. (**B**) Three transgenic lines selected by RT-qPCR. (**C**) β-glucuronidase (GUS) staining of transgenic *Arabidopsis* under normal conditions. (**D**) GUS staining of transgenic *Arabidopsis* after NaCl treatment. (**E**) qRT-PCR for tissue-specific expression analysis of *TaWRKY13* under normal conditions. (**F**) qRT-PCR for tissue-specific expression analysis of *TaWRKY13* after NaCl treatment. All data are means ± SDs of three independent biological replicates. The ANOVA demonstrated significant differences (* $p < 0.05$, ** $p < 0.01$).

2.6. TaWRKY13 Is Involved in Various Stress Responses

WRKY proteins are reported to be involved in various biotic and abiotic stresses [25]. Expression pattern analyses were conducted to determine whether *TaWRKY13* was responsive to abiotic stresses. The results indicated that *TaWRKY13* participated in salt PEG, ABA and cold-stress responses (Figure 7). For PEG treatment, the relative expression level of *TaWRKY13* was rapidly induced at 1 h after the imposition of PEG stress (Figure 7A). After NaCl treatment for 1 h, *TaWRKY13* was highly induced at a maximum level of about 22-fold (Figure 7B). Exogenous ABA and cold stress also significantly affected the expression of *TaWRKY13* (Figure 7C,D). The rapid increase in relative expression levels of *TaWRKY13* following different stress treatments indicated an important role at the initial stages of stress response.

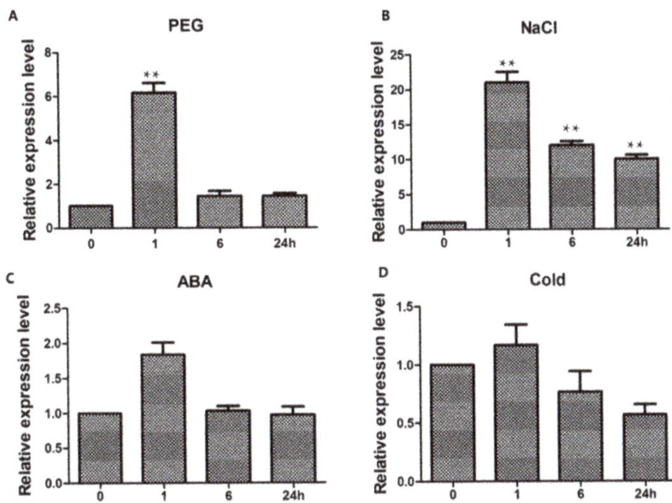

Figure 7. Expression patterns of *TaWRKY13* under (**A**) PEG, (**B**) NaCl, (**C**) exogenous abscisic acid (ABA), and (**D**) cold treatments. The ordinates are relative expression levels (fold) of *TaWRKY13* compared to the non-stressed control. The horizontal ordinate is the treatment time, at 0, 1, 6 and 24 h. The expression level of TaActin as a loading control. All experiments were repeated three times. Error bars represent standard deviations (SDs). All data are means ± SDs of three independent biological replicates. The ANOVA demonstrated significant differences (** $p < 0.01$).

2.7. Stress-Related Regulatory Elements in the Promoter of TaWRKY13

The 1.856 kb promoter region upstream of the *TaWRKY13* ATG start codon was isolated to gain an insight into the regulatory mechanism. We searched for putative cis-acting elements in the promoter regions using the database PLACE (http://www.dna.affrc.go.jp/PLACE/). The results are shown in Table 2. Numerous stress-related regulatory elements were present, including a W-BOX, MYB element and TATA-BOX, which take part in the response to both drought and high-salt stress, as well as low-temperature responsive (LTR), ABA-responsive element (ABRE) and GT1, which mainly participate in salt-stress response. Moreover, there were various light, gibberellin, SA (salicylic acid) and high-temperature responsive elements, indicating that *TaWRKY13* is involved in abiotic stress response and plant hormone-related signal transduction.

Table 2. *Cis*-element analysis of the *TaWRKY13* promotor.

Cis-Element	Target Sequences	Number	Function
W-BOX	TTGAC/TGACT TGACY/CTCAY	27	Drought, high salt responsive elements
MYB	GGATA/WAACCA/TAACARA/ TAACAAA/CCWACC/GNGTTR	20	Drought, high salt responsive elements
LTR	CCGAC/CCGAAA	7	Low-temperature, salt responsive elements
ABRE	ACGTGKC	5	ABA-responsive elements
TATA-BOX	TATATAA	6	Drought, cold, high salt responsive elements
GTI	CAAAAA	3	Salt responsive elements
GATA-BOX	GATA	22	Light, gibberellin responsive elements
WRKY	TAGA	20	Light, salicylic acid responsive elements
HSP70A	SCGAYNR(N)$_{15}$HD	7	High temperature responsive elements

"Number" corresponds to the number of each type of cis-element in the promoter.

2.8. Root System Analysis Indicates That Overexpression Lines Respond to Salt Stress in Arabidopsis

To explore the mechanism of *TaWRKY13* under salt stress, a pCAMBIA1302-*TaWRKY13* (*35S::TaWRKY13*) vector was constructed and transformed into *Arabidopsis* for root length assay [40]. The results of the identification of homozygotes by agarose gel electrophoresis (AGE) and the selection of three transgenic lines (*35S::TaWRKY13#1, #2, #3*) by RT-qPCR are available in Supplementary Figure S1. Seedlings of control (Columbia-0) and three T$_3$ generation overexpression lines were first grown on MS (Murashige & Skoog) medium for one week and then transplanted to MS medium supplemented with various NaCl concentrations (0, 100, 120 mM) for salt treatment. As shown in Figure 8, the overexpression lines have an advantage in terms of the main root length and total surface area compared to Col-0 under NaCl treatment.

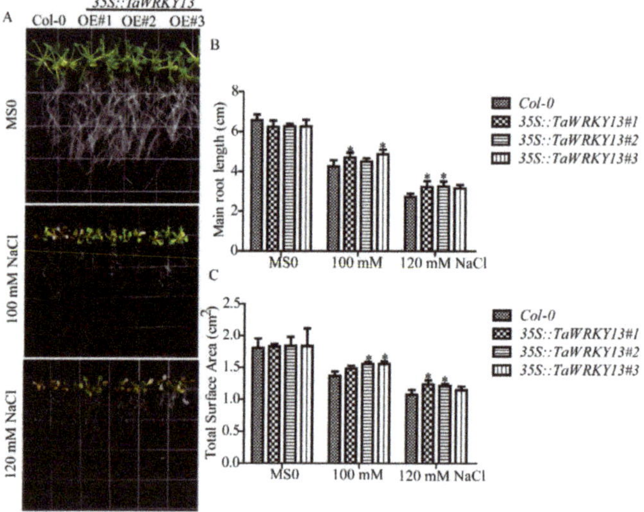

Figure 8. Root length phenotypes of *Arabidopsis* overexpression lines after NaCl treatment. (**A**) Image of the root length phenotype of transgenic lines grown in 0, 100 and 120 mM NaCl. (**B**) Analysis of the main root lengths of transgenic lines under NaCl treatment. (**C**) Analysis of total surface areas of transgenic lines under NaCl treatment. The main root length and total surface area of *Arabidopsis* roots were measured by the WinRHIZO system. All data are means ± SDs of three independent biological replicates. The ANOVA demonstrated significant differences (* $p < 0.05$).

2.9. TaWRKY13 Overexpression Response to Salt Stress in Oryza sativa

Two-week-old T$_3$ rice lines seedlings of the control (Nipponbare) and three overexpression lines (*35S::TaWRKY13#1, #2, #3*) were grown hydroponically in untreated control solution or in the same solution supplemented with 150 mM NaCl to explore the physiological tolerance of *TaWRKY13* overexpression rice lines to salt stress [42]. The verification of homozygotes and the selection of three transgenic lines were conducted by AGE and RT-qPCR, respectively (Figure 9A,B). As shown in Figure 9C, before NaCl treatment, both Nipponbare and the three transgenic lines showed similar growth patterns, with no or little difference in plant height, root length, and proline (Pro) and malondialdehyde (MDA) contents. After 7 days of NaCl treatment, both Nipponbare and the overexpression lines showed leaf shedding (Figure 9D). Compared with the transgenic lines, Nipponbare plants showed evidence of wilting, water loss and yellowing, whereas the transgenics lines showed less severe symptoms. Meanwhile, the overexpression of *TaWRKY13* increased the proline content and decreased MDA content under NaCl treatment (Figure 9E,F). The root length of Nipponbare was significantly lower than for transgenic plants; the surface areas of transgenic plants were higher than for Nipponbare

(Figure 9G,H). These results indicated that the overexpression of *TaWRKY13* enhanced salt tolerance in rice.

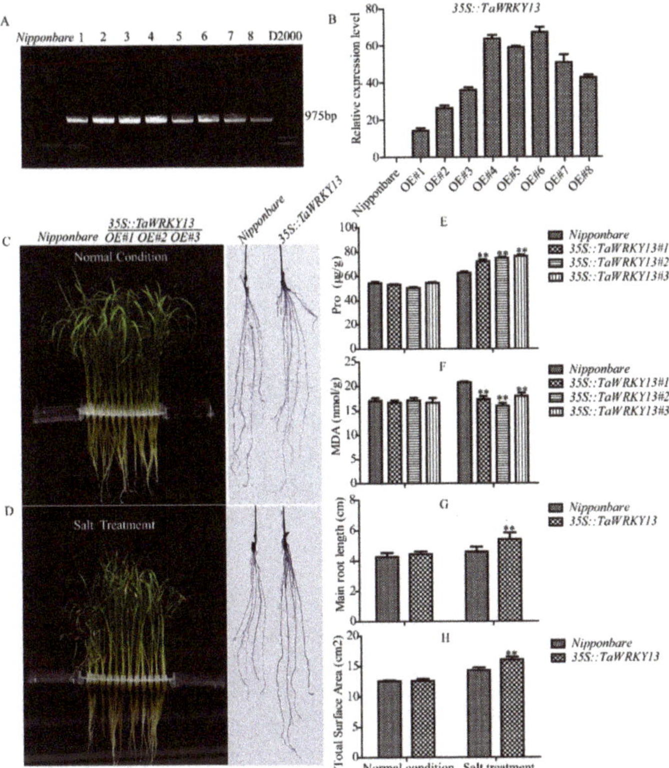

Figure 9. Phenotype identification of *TaWRKY13* transgenic rice under NaCl treatment. (**A**) Confirmation of homozygotes by agarose gel electrphoresis. (**B**) Selection of three transgenic lines by RT-qPCR. (**C**) Rice seedlings and root system diagram of Nipponbare and *35S::TaWRKY13* before treatment. (**D**) Rice seedlings and root system diagram of Nipponbare and *35S::TaWRKY13* after 150 mM NaCl treatment for 7 days. (**E**) Proline contents in Nipponbare and *35S::TaWRKY13* seedlings under normal conditions and NaCl treatment. (**F**) Malondialdehyde (MDA) contents in in Nipponbare and *35S::TaWRKY13* rice seedlings under normal growth conditions and NaCl treatment. (**G**) Root length measurements of Nipponbare and *35S::TaWRKY13* transformants with and without NaCl treatment. (**H**) Total surface areas of Nipponbare and *35S::TaWRKY13* with and without NaCl treatment. Main root lengths and total surface areas were measured by the WinRHIZO system (Hang xin, Guangzhou, China). All data are means ± SDs of three independent biological replicates. The ANOVA demonstrated significant differences (** $p < 0.01$).

3. Discussion

Regarded as one group among many important transcription factors in plants, WRKY TFs are represented by 90 members in *Arabidopsis* and more than 100 in rice [43]. The functions of WRKY TFs have been studied in detail in various plant species since their first discovery.

Since the application of transcriptome sequencing technology, researchers have sequenced the genome of wheat [44,45]. However, owing to the large and complex genome of heterohexaploid wheat, the task has posed many challenges [46]. Recently, transgenic *Arabidopsis* plants of *TaWRKY2* and *TaWRKY19* have shown improved stress tolerance, and the overexpression of *TaWRKY2* and *TaWRKY19* has exhibited salt, osmotic/dehydration and freezing stress tolerance [47]. More than 160 TaWRKYs were characterized according to their sequence alignment, motif type and phylogenetic relationship analysis by Sezer et al. [48]. Although the WRKY genes associated with stress can be identified by transcriptome sequencing and family analysis, functional identification and mechanism analysis in wheat is limited. Salt stress is one of the most serious stresses that cannot be reversed after damage [49].

Here, on the basis of the previous research, combining RNA-Seq, real-time quantitative PCR (RT-qPCR), and the latest wheat database, *TaWRKY13* was isolated from the wheat genome for further study. RNA-Seq was conducted first (Supplementary Table S1); meanwhile, using the wheat database, 57 TaWRKY genes were annotated (Table 1). The results showed that TaWRKYs were differently distributed (number and location) on wheat chromosomes (Figure 1). Studies of the genome structure and the phylogenetic analysis of TaWRKY genes were initially difficult, because the wheat genome was too complex for statistical analysis; there were 171 TaWRKY genes according to the database (https://phytozome.jgi.doe.gov/pz/portal.html). Based on the rule that the CDS of TaWRKYs were more than 300 base pairs, we removed redundant TaWRKY genes and, combined with the NCBI database (https://www.ncbi.nlm.nih.gov/pubmed), 56 TaWRKY genes were selected for the analysis of the gene structure (Figure 2). Major TaWRKY genes harbored different CDS and binding motifs responsible for special function; for example, *TaWRKY1* contained an N-terminal CUT domain and a C-terminal NL domain [30]. To further explore TaWRKY genes that respond to salt stress, 12 TaWRKY genes were chosen for verification by qRT-PCR (Figure 3). All 12 genes were up-regulated under salt stress, and *TaWRKY13* was chosen for further study due to its higher expression level under salt treatment. Phylogenetic analysis demonstrated that TaWRKY genes have different evolutionary relationships and homologies to WRKYs in *Arabidopsis* and rice (Figure 4); compared to *OsWRKY13*, *AtWRKY13* was closer to *TaWRKY13*, possibly indicating similar biological functions [50]. For *OsWRKY13*, the non-conservation of evolution may provide a basis for the subsequent functional identification of *TaWRKY13* in rice, in that the influence of rice itself in *OsWRKY13* was eliminated. Subcellular localization showed that TaWRKY13 is a nuclear protein (Figure 5) which may mainly be involved in nuclear signal transduction [51,52]. Although many cotton (*Gossypium hirsutum*) WRKY genes were expressed at low levels during development, a few *GhWRKYs* expressed highly in specific tissues such as roots, stems, leaves and embryos [53]. Our results showed that *TaWRKY13* was expressed in roots, stems and leaves in transgenic lines, the relative expression level of roots was higher than stems and leaves in transgenic lines, and under salt-stress conditions, the relative expression level was double that of the normal condition (Figure 6).

An increasing number of studies have shown that WRKY TFs play important roles in abiotic stress response; for instance, the overexpression of *GmWRKY21* improved cold tolerance in *Arabidopsis*, because of the regulation of DREB2A and STZ/Zat10. *GmWRKY54* conferred salt and drought tolerance; *GmWRKY13*, which was insensitive to ABA (abscisic acid) but markedly sensitive to salt and mannitol, may function in both lateral root development and the abiotic stress response [54]. Expression pattern analyses revealed that *TaWRKY13* was induced significantly by PEG, salt, low-temperature and ABA (Figure 7). Compared with PEG, low-temperature, and ABA stress, *TaWRKY13* achieved the highest relative expression level under salt treatment, which was in accordance with the following root length assay in *Arabidopsis* and the rice resistance assay. Products of WRKY TFs bind to specific cis-regulatory sequences such as the W-BOX in the promoter to induce the expression of downstream target genes [55]. Many regulatory cis-elements that are responsive to drought (W-BOX, MYB and TATA-BOX), high salt (LTR, ABRE and GT1), SA (salicylic acid, WRKY) and cold were recognized in the *TaWRKY13* promoter, showing that *TaWRKY13* is capable of responding to stress (Table 2). WRKY13 participated in various physiological processes; for example, a weaker stem phenotype, reduced sclerenchyma development, and altered lignin synthesis were observed in an *AtWRKY13* mutant, showing that it functioned in stem development [56]. When *AtWRKY13* was disturbed under short-day conditions, *AtWRKY13* promoted flowering [57]. Furthermore, WRKY13 was also involved in the cross talk between abiotic and biotic stress signaling pathways, and *OsWRKY13* displayed selective binding to different cis-elements to regulate various stress [58]. In this study, a root length assay of overexpression lines was conducted in *Arabidopsis* for an analysis of the stress tolerance of *TaWRKY13*; overexpression lines had longer root lengths and a higher total root area than Col-0 (Figure 8A–C). Additionally, the overexpression of *TaWRKY13* enhanced salt tolerance in transgenic rice (Figure 9). Under NaCl treatment, the transgenic lines of *TaWRKY13* grew vigorously, whereas Nipponbare seedlings were more wilted and yellow (Figure 9D); the transgenic lines also had higher proline (Pro) and reduced malondialdehyde (MDA) contents (Figures 8F and 9E) under NaCl treatment. In addition, the roots of transgenic lines were longer and more developed than Nipponbare (Figure 9G,H). These results all showed that *TaWRKY13* was responsive to salt stress, in agreement with data from other species [54,56,58]. In accordance with the present study, the results suggested that *TaWRKY13* has a potential role in improving salt tolerance in wheat. These results are only preliminary in exploring the putative role of *TaWRKY13* in salt tolerance; more researches about the role and regulation mechanism of *TaWRKY13* are still needed in wheat. For instance, based on the above findings, *TaWRKY13* was transformed into wheat for functional verification and mechanism analysis to further improve the role of *TaWRKY13* in wheat stress tolerance pathways.

4. Materials and Methods

4.1. De Novo Transcriptome Sequencing of Salt-Treated Wheat

Wheat (*Triticum aestivum* L. cultivar Jinhe 9123, from the Institute of Genetics and Physiology, Hebei Academy of Agriculture and Forestry Sciences, Shijiazhuang, China) was cultivated in a 10 cm × 10 cm pot (vermiculite:soil, 1:3) supplemented with Hoagland's liquid medium at 22 °C under a 16 h light/8 h darkness photoperiod for 10 days. When the wheat seedlings were at the three-leaf stage, the pots were immersed in 150 mM NaCl solution and water (control) for 1 h, respectively [30], prior to the sampling of 0.1 g fresh leaf tissue. Samples were submerged immediately in liquid nitrogen and stored at −80 °C for RNA-Seq. The experiment was performed in three independent replications. In Supplementary Table S1, Control_Leaf means a sample without NaCl treatment, and NaCl_Leaf means a sample with salt treatment; each treatment involved three independent replicates, which were then sampled for RNA-Seq. Data are shown in Supplementary Table S1.

4.2. Identification and Annotation of TaWRKY Response to Salt Stress

According to the RNA-Seq data, the rule was adopted that the expression level was up-regulated and $\log_2$(NaCl_treat/Control_treat) > 2 to select TaWRKYs which responded to salt stress. Several databases—NCBI (https://www.ncbi.nlm.nih.gov/pubmed) PlantTFDB (http://planttfdb.cbi.pku.edu.cn/) and Phytozome (https://phytozome.jgi.doe.gov/pz/portal.html)---were used to annotate the gene name, ID, transcript name and localization.

4.3. Structure Analysis and Phylogenetic Analysis of TaWRKYs

According to the information listed in Table 1, the chromosome location of TaWRKYs was analyzed by using the online website http://mg2c.iask.in/mg2c_v2.0/. For gene structure analysis, the data of TaWRKYs that were identified in Section 4.2 were uploaded to GSDS (http://gsds.cbi.pku.edu.cn/) to obtain the map of the TaWRKYs' structure. For phylogenetic analysis, a tree of WRKYs from wheat, rice and *Arabidopsis* was constructed using the neighbor-joining method in MEGA 6.0 with 1000 bootstrap replications [39]. Data for gene structure and the phylogenetic tree analysis were downloaded from PlantTFDB (http://planttfdb.cbi.pku.edu.cn/) and are shown in Supplementary Tables S2 and S3.

4.4. RNA Extraction of Stress Treatments and RT-qPCR Analyses

Wheat seeds were sown as previously described; vermiculite and soil were removed by water after being grown for 10 days, and the fresh leaf tissue of three-leaf-stage wheat seedlings were used for the RNA extraction of different stress treatments. For the identification of TaWRKY responses to salt stress, the seedlings roots were immersed in 150 mM NaCl solution, and 0.1 g of fresh leaf tissue was sampled at different times (0, 0.5, 1, 2, 4, 8, 12 and 24 h). For the expression pattern analyses, the roots of wheat seedlings were immersed in 10% PEG6000, 150 mM NaCl and 100 μmol·L^{-1} ABA solutions. Wheat seedlings for cold treatment were placed in a 10 h light/14 h darkness, 4/2 °C chamber and sampled at different periods (0, 1, 6 and 24 h) [30,59,60]. For specific tissue expression assays, T$_3$ generation transgenic *Arabidopsis* (*35S::pTaWRKY13*) plants were surface-sterilized with 10% Chloros and washed three times with sterile water. Sterilized seeds were sown on MS (Murashige & Skoog) medium, vernalized in darkness for 3–4 days at 4 °C, then grown in a chamber at 22 °C and 75% humidity under a 16 h light/8 h darkness photoperiod for one week. The seedings were transplanted to soil (vermiculite:soil, 1:3), 0.1 g fresh roots, stems and leaves tissue of 10-day-old transgenic *Arabidopsis* seedlings with or without 150 mM NaCl treatment were sampled for RNA the extraction of different tissues [56].

All samples after collection were submerged immediately in liquid nitrogen and stored at −80 °C for RNA extraction using an RNA prep plant kit (TIANGEN, Beijing, China); cDNA was synthesized using a Prime Script First-Strand cDNA Synthesis Kit (TransGen, Beijing, China) following the manufacturer's instructions. RT-qPCR was performed with Super Real PreMix Plus (TransGen, Beijing, China) on an ABI Prism 7500 system (Applied Biosystems, Foster city, CA, USA). Specific primers for TaActin, AtActin and TaWRKY4, 9, 12, 13, 15, 22, 19, 33, 34, 44, 53 and 70 for RT-qPCR are listed in Supplementary Table S4. Three biological replicates were used for RT-qPCR analysis, and the 2$^{-\Delta\Delta Ct}$ method was used for quantification.

4.5. Gene Isolation and Subcellular Localization

The ORF (open reading frame) of *TaWRKY13* was amplified by PCR with specific primers from wheat cDNA (cultivar Jinhe 9123). The PCR product was fused into pZeroBack vector (TIANGEN, Beijing, China) and sequenced for further study. The correct sequencing plasmids were treated as templates, the segment with restriction sites was amplified by specific primers, and the PCR product was inserted into the N-terminus of the green fluorescent protein (GFP) containing the CaMV35S promoter for subcellular localization; the 35S::GFP vector was used as the control. Both 35S::GFP and 35S::TaWRKY13-GFP were transferred into wheat mesophyll protoplasts by the PEG-mediated method [29]. A confocal laser scanning microscope (LSM700; CarlZeiss, Oberkochen, Germany) was used to observe the fluorescence after incubation in darkness at 22 °C for 18–20 h. All primers are listed in Supplementary Table S4.

4.6. Tissue-Specific Expression of TaWRKY13 and GUS Staining

Tissue-specific expression analysis of *TaWRKY13* was conducted by two methods. In the first one, the CDS of *TaWRKY13* was amplified as described in Section 4.5, then cloned into the pCAMBIA1302 vector; then, the infected inflorescence of *Arabidopsis* was determined by the Agrobacterium-mediated method [61], grown as described in Section 4.4, until T$_3$ generation transgenic *Arabidopsis* seeds were obtained. The identification of homozygotes and selection of three transgenic lines were conducted by agarose gel electrophoresis and RT-qPCR, respectively [59]. The transgenic *Arabidopsis* seedlings with or without NaCl (150 mM) treatment were used for RT-qPCR as described in Section 4.4. In the second method, promoter fragments of *TaWRKY13* (*pTaWRKY13*) were obtained from Ensemble Plants (plants.ensembl.org/index.html); the *pTaWRKY13* was amplified by PCR with specific primers from wheat cDNA (Jinhe 9123), and the PCR product was fused into pLB vector (TIANGEN, Beijing, China) and sequenced. The fragment of *TaWRKY13* promoter was cloned to the pCAMBIA1305 vector harboring a β-glucuronidase (GUS) tag, obtaining the T$_3$ generation transgenic *Arabidopsis* seeds as per the previous description. T$_3$ generation transgenic *Arabidopsis* (*35S::pTaWRKY13*) seeds were surface-sterilized, sown on MS medium, vernalized, and grown in a chamber at 22 °C and 75% humidity under a 16 h light/8 h darkness photoperiod for one week as described in Section 4.4. Ten-day-old transgenic *Arabidopsis* seedlings were submerged to 150 mM NaCl solution for 1 h. After salt treatment, the liquid was drained with filter paper, and the plant material was subjected to GUS staining solution supplemented with 5-bromo-4-chloro-3-indolylb-d-glucuronic acid (X-gluc) for 3 h; 70% (*vol/vol*) ethanol was used to remove the chlorophyll following the manufacturer's protocol (Real-Times, Beijing, China) [56]. GUS staining was observed by a Leica microscope (Wetzlar, Germany). Primers are listed in Supplementary Table S4.

4.7. Cis-Acting Elements in the TaWRKY13 Promoter

A 1.856 kb promoter fragment upstream of the ATG start codon of *TaWRKY13* was obtained from the Ensemble Plants website (http://plants.ensembl.org/index.html). Cis-acting elements that respond to various stresses in the promoter region were analyzed by PLACE (http://www.dna.affrc.go.jp/PLACE/) [29].

4.8. Root Growth Assays of TaWRKY13 under Salt Stress in Arabidopsis

T_3 generation transgenic *Arabidopsis* lines were obtained as previously described (Section 4.6). Seeds of Col-0 and transgenic lines (*35S::TaWRKY13#1*, #2, #3) were surface-sterilized, sown on MS medium, vernalized, grown in a chamber at 22 °C and 75% humidity under a 16 h light/8 h darkness photoperiod for one week as described above (Section 4.4). Three ten-day-old *Arabidopsis* seedlings (Col-0 and transgenic lines) were transferred to MS medium containing different concentrations of NaCl (0, 100, 120 mM) for one week [40].

4.9. Generation of Transgenic Rice and Stress Identification of TaWRKY13 to Salt Tolerance

Plant expression vector *pCAMBIA1305-TaWRKY13* was constructed and transformed to competent cells of EHA105 as previously described [30]. Genetic transformation was conducted by Dr Chuan-Yin Wu and colleagues at the Institute of Crop Science, Chinese Academy of Agricultural Sciences using the agrobacterium-mediated method, and Nipponbare was used as the control [62]. The selection of three transgenic lines was made by agarose gel electrophoresis and RT-qPCR, respectively, as previously described (Figure S1). T_3 generation transgenics (*35S::TaWRKY13#1*, #2, #3) and Nipponbare were used for further study. Rice seeds were treated with 0.7% hydrogen peroxide for one day for surface sterilization, breaking dormancy and promoting germination, then replaced with 0.7% hydrogen peroxide with water and germinated at 37 °C for 3 days (changing the water once a day). When seeds showed white buds, bare seeds were transplanted to 96-well plates (24 seeds of Nipponbare and *35S::TaWRKY13#1,#2,#3*, respectively) and placed in a growth chamber at 28 °C and a 16 h light/8 h darkness photoperiod and 70% relative humidity for the hydroponic culture. Seedings were cultured in water for one week, then cultured in water supplemented with Hoagland's hydroponic culture solution. The culture solution was replaced every 5 days, and the pH was set at 5.5 [63]. Three-leaf seedlings were treated. For salt treatment, the 96-well plates growing three-leaf stage seedlings were transferred to YS hydroponic culture solution and a YS hydroponic culture solution supplemented with 150 mM NaCl for several days until phenotypes appeared [62]. For each salt treatment, there were three independent replicates. Primers are listed in Supplementary Table S4.

4.10. Measurements of Proline (Pro) and Malondialdehyde (MDA) Contents

To better understand the function of *TaWRKY13* under salt treatment, proline and MDA contents were measured with Pro and MDA assay kits (Comin, Beijing, China) based on the manufacturer's protocols. Main root lengths and total surface areas of *Arabidopsis* and rice roots were measured by the WinRHIZO system (Hang xin, Guangzhou, China). Measurements were made on all three biological replicates; means ± SD and statistically significant differences were based on the ANOVA (* $p < 0.05$, ** $p < 0.01$).

5. Conclusions

We identified the salt-induced WRKY gene *TaWRKY13* (ID: 31962353) from a wheat RNA-Seq database (https://phytozome.jgi.doe.gov/pz/portal.html) and real-time quantitative PCR (RT-qPCR). TaWRKY13 is a nuclear protein that was expressed in the roots, stems and leaves of transgenic *Arabidopsis. TaWRKY13* was responsive to PEG, salt, cold, and exogenous abscisic acid (ABA) treatment. The overexpression of *TaWRKY13* was responsive to salt stress in both *Arabidopsis* and rice, as evidenced by the promotion of root length and the total root surface area. These results provide a basis for further understanding the functions of *TaWRKY13* in wheat when subjected to salt stress.

Supplementary Materials: Supplementary materials can be found at http://www.mdpi.com/1422-0067/20/22/5712/s1. Supplementary Table S1: RNA-Seq data of salt treated wheat. Supplementary Table S2: CDS and amino acid sequences of 100 TaWRKYs used for genome structure and phylogenetic analysis. Supplementary Table S3: Amino acid sequences of 90 AtWRKYs, 128 OsWRKYs and 171 TaWRKY used for phylogenetic analysis. Supplementary Table S4: Primers used in this study. Supplementary Figure S1: Overexpression rice lines. A: Identification of homozygotes by agarose gel electrophoresis B: Selection of three transgenic lines by RT-qPCR.

Author Contributions: Y.-W.L. coordinated the project, wrote and reviewed the manuscript; H.Z. conceived and designed the experiments and edited the manuscript; S.Z. performed the experiments; W.-J.Z. performed validation and formal analysis; B.-H.L., J.-C.Z. and F.-S.D. conducted the bioinformatics and performed related experiments; Z.-S.X. provided analytical tools and analyzed the data; Supervision, Z.-Y.W.; Project administration, Z.-F.L. and F.Y.; Funding acquisition H.-B.W. All authors have read and approved the final manuscript.

Funding: This research was financially supported by the National Natural Science Foundation of China (31600216 and 31871611), the HAAFS Agriculture Science and Technology Innovation Project (2019-4-8-1), and the Natural Science Foundation of Hebei Province (C2017301066).

Acknowledgments: We thank Li-Na Ning for critically reading the manuscript.

Conflicts of Interest: The authors declare no conflict of interest.

Abbreviations

ROS	Reactive oxygen species
SnRK2	Sucrose non-fermenting 1-related protein kinase 2
MAPK	Mitogen activated protein kinase
PKs	Protein kinases
TFs	Transcription factors
CDS	Coding sequence
ORF	Open reading frame
SA	Salicylic acid
ABA	Abscisic acid
ABRE	ABA-responsive element
Pro	Proline
MDA	Malondialdehyde
PEG	Polyethylene glycol
GFP	Green fluorescent protein
RT-qPCR	Real-time quantitative PCR
LTR	Low-temperature responsive
GUS	β-glucuronidase
X-gluc	5-bromo-4-chloro-3-indolylb-d-glucuronic acid
AGE	Agarose gel electrophoresis
MS	Murashige & Skoog

References

1. Bechtold, U.; Field, B. Molecular mechanisms controlling plant growth during abiotic stress. *J. Exp. Bot.* **2018**, *69*, 2753–2758. [CrossRef] [PubMed]
2. Zhu, J.K. Cell signaling under salt, water and cold stresses. *Curr. Opin. Plant Biol.* **2001**, *4*, 401–406. [CrossRef]

3. Bian, S.M.; Jin, D.H.; Li, R.H.; Xie, X.; Gao, G.I.; Sun, W.K.; Li, Y.J.; Zhai, L.L.; Li, X.Y. Genome-Wide Analysis of CCA1-Like Proteins in Soybean and Functional Characterization of GmMYB138a. *Int. J. Mol. Sci.* **2017**, *18*, 2040. [CrossRef] [PubMed]

4. Cao, D.; Li, Y.Y.; Liu, B.H.; Kong, F.J.; Tran, L.P. Adaptive Mechanisms of Soybean Grown on Salt-Affected Soils. *Land Degrad. Dev.* **2017**, *29*, 1054–1064. [CrossRef]

5. Cao, Z.H.; Zhang, S.Z.; Wang, R.K.; Zhang, R.F.; Hao, Y.J. Genome Wide Analysis of the Apple MYB Transcription Factor Family Allows the Identification of MdoMYB121 Gene Conferring Abiotic Stress Tolerance in Plants. *PLoS ONE* **2013**, *8*, e69955.

6. Degregori, J.; Kowalik, T.; Nevins, J.R. Cellular targets for activation by the E2F1 transcription factor include DNA synthesis- and G1/S-regulatory genes. *Mol. Cell. Biol.* **1995**, *15*, 4215. [CrossRef]

7. Lei, Y.; Lu, L.; Liu, H.Y.; Li, S.; Xing, F.; Chen, L.L. CRISPR-P: A web tool for synthetic single-guide RNA design of CRISPR-system in plants. *Mol. Plant* **2014**, *7*, 1494–1496. [CrossRef]

8. Lang, V.; Palva, E.T. The expression of a Rab-related gene, rab18, is induced by abscisic acid during the cold acclimation process of *Arabidopsis thaliana* (L.) Heynh. *Plant Mol. Biol. Rep.* **1992**, *20*, 951–962. [CrossRef]

9. Levine, A.; Tenhaken, R.; Dixon, R.; Lamb, C. H2O2 from the oxidative burst orchestrates the plant hypersensitive disease resistance response. *Cell* **1994**, *79*, 583–593. [CrossRef]

10. Michael, S.; Markus, H.; Baldwin, I.T.; Emmanuel, G. Jasmonoyl-L-isoleucine coordinates metabolic networks required for anthesis and floral attractant emission in wild tobacco (*Nicotiana attenuata*). *Plant Cell* **2014**, *26*, 3964–3983.

11. Kobayashi, Y.; Murata, M.; Minami, H.; Yamamoto, S.; Kagaya, Y.; Hobo, T.; Yamamoto, A.; Hattori, T. Abscisic acid-activated SNRK2 protein kinases function in the gene-regulation pathway of ABA signal transduction by phosphorylating ABA response element-binding factors. *Plant J. Cell. Mol. Biol.* **2010**, *44*, 939–949. [CrossRef] [PubMed]

12. Wang, D. *Functions of MAPK Signaling Pathways in the Regulation of Toxicity of Environmental Toxicants or Stresses*; Springer: Berlin/Heidelberg, Germany, 2019.

13. Yoshinaga, K.; Arimura, S.I.; Niwa, Y.; Tsutsumi, N.; Uchimiya, H.; Kawai-Yamada, M. Mitochondrial Behaviour in the Early Stages of ROS Stress Leading to Cell Death in *Arabidopsis thaliana*. *Ann. Bot.* **2005**, *96*, 337–342. [CrossRef] [PubMed]

14. Zhao, Z.; Zhang, W.; Stanley, B.A.; Assmann, S.M. Functional proteomics of *Arabidopsis thaliana* guard cells uncovers new stomatal signaling pathways. *Plant Cell* **2008**, *20*, 3210–3226. [CrossRef] [PubMed]

15. Peng, Z.; He, S.; Gong, W.; Sun, J.; Pan, Z.; Xu, F.; Lu, Y.; Du, X. Comprehensive analysis of differentially expressed genes and transcriptional regulation induced by salt stress in two contrasting cotton genotypes. *BMC Genom.* **2014**, *15*, 760. [CrossRef] [PubMed]

16. Runnels, L.W.; Yue, L.; Clapham, D.E. TRP-PLIK, a bifunctional protein with kinase and ion channel activities. *Science* **2001**, *291*, 1043–1047. [CrossRef] [PubMed]

17. Mulders, S.M.; Preston, G.M.; Deen, P.M.; Guggino, W.B.; van Os, C.H.; Agre, P. Water channel properties of major intrinsic protein of lens. *J. Biol. Chem.* **1995**, *270*, 9010–9016. [CrossRef] [PubMed]

18. Yovchev, A.G.; Stone, A.K.; Hucl, P.; Scanlon, M.G.; Nickerson, M.T. Polyethylene glycol as an osmotic regulator in dough with reduced salt content. *J. Cereal Sci.* **2017**, *76*, 193–198. [CrossRef]

19. Hunter, T. Protein kinases and phosphatases: The yin and yang of protein phosphorylation and signaling. *Cell* **1995**, *80*, 225. [CrossRef]

20. Madhunita, B.; Ralf, O. WRKY transcription factors: Jack of many trades in plants. *Plant Signal. Behav.* **2014**, *9*, e27700.

21. Xu, Z.S.; Chen, M.; Li, L.C.; Ma, Y.Z. Functions and application of the AP2/ERF transcription factor family in crop improvement. *J. Integr. Plant Biol.* **2011**, *53*, 570–585. [CrossRef]

22. Abe, H.; Yamaguchishinozaki, K.; Urao, T.; Iwasaki, T.; Hosokawa, D.; Shinozaki, K. Role of *Arabidopsis* MYC and MYB homologs in drought- and abscisic acid-regulated gene expression. *Plant Cell* **1997**, *9*, 1859–1868. [PubMed]

23. Fiorenza, M.T.; Farkas, T.; Dissing, M.; Kolding, D.; Zimarino, V. Complex expression of murine heat shock transcription factors. *Nucleic Acids Res.* **1995**, *23*, 467–474. [CrossRef] [PubMed]

24. Nakashima, K.; Takasaki, H.; Mizoi, J.; Shinozaki, K.; Yamaguchishinozaki, K. NAC transcription factors in plant abiotic stress responses. *BBA Gene Regul. Mech.* **2012**, *1819*, 97–103. [CrossRef] [PubMed]

25. Eulgem, T.; Rushton, P.J.; Robatzek, S.; Somssich, I.E. The WRKY superfamily of plant transcription factors. *Trends Plant Sci.* **2000**, *5*, 199–206. [CrossRef]

26. Chen, J.; Nolan, T.; Ye, H.; Zhang, M.; Tong, H.; Xin, P.; Chu, J.; Chu, C.; Li, Z.; Yin, Y. *Arabidopsis* WRKY46, WRKY54 and WRKY70 Transcription Factors Are Involved in Brassinosteroid-Regulated Plant Growth and Drought Response. *Plant Cell* **2017**, *29*, 1425. [CrossRef] [PubMed]

27. Lee, H.; Cha, J.; Choi, C.; Choi, N.; Ji, H.S.; Park, S.R.; Lee, S.; Hwang, D.J. Rice WRKY11 Plays a Role in Pathogen Defense and Drought Tolerance. *Rice* **2018**, *11*, 5. [CrossRef]

28. Muthamilarasan, M.; Bonthala, V.S.; Khandelwal, R.; Jaishankar, J.; Shweta, S.; Nawaz, K.; Prasad, M. Global analysis of WRKY transcription factor superfamily in Setariaidentifies potential candidates involved in abiotic stress signaling. *Front. Plant Sci.* **2015**, *6*, 910. [CrossRef]

29. Shi, W.Y.; Du, Y.T.; Ma, J.; Min, D.H.; Zhang, X.H. The WRKY transcription factor GmWRKY12 confers drought and salt tolerance in soybean. *Int. J. Mol. Sci.* **2018**, *19*, 4087. [CrossRef]

30. He, G.H.; Xu, J.Y.; Wang, Y.X.; Liu, J.M.; Li, P.S.; Chen, M.; Ma, Y.Z.; Xu, Z.S. Drought-responsive WRKY transcription factor genes TaWRKY1 and TaWRKY33 from wheat confer drought and/or heat resistance in *Arabidopsis*. *BMC Plant Biol.* **2016**, *16*, 116. [CrossRef]

31. Ding, W.; Fang, W.; Shi, S.; Zhao, Y.; Li, X.; Xiao, K. Wheat WRKY Type Transcription Factor Gene TaWRKY1 is Essential in Mediating Drought Tolerance Associated with an ABA-Dependent Pathway. *Plant Mol. Biol. Rep.* **2016**, *34*, 1111–1126. [CrossRef]

32. Chen, W.; Deng, P.; Chen, L.; Wang, X.; Hui, M.; Wei, H.; Yao, N.; Ying, F.; Chai, R.; Yang, G. A Wheat WRKY Transcription Factor TaWRKY10 Confers Tolerance to Multiple Abiotic Stresses in Transgenic Tobacco. *PLoS ONE* **2013**, *8*, e65120.

33. Rushton, D.L.; Tripathi, P.; Rabara, R.C.; Lin, J.; Ringler, P.; Boken, A.K.; Langum, T.J.; Smidt, L.; Boomsma, D.D.; Emme, N.J. WRKY transcription factors: Key components in abscisic acid signalling. *Plant Biotech. J.* **2011**, *10*, 2–11. [CrossRef] [PubMed]

34. Li, H.; Gao, Y.; Xu, H.; Dai, Y.; Deng, D.; Chen, J. ZmWRKY33, a WRKY maize transcription factor conferring enhanced salt stress tolerances in *Arabidopsis*. *Plant Growth Regul.* **2013**, *70*, 207–216. [CrossRef]

35. Hasegawa, P.M.; Bressan, R.A.; Zhu, J.K.; Bohnert, H.J. Plant Cellular and Molecular Responses to High Salinity. *Annu. Rev. Plant Biol.* **2000**, *51*, 463–499. [CrossRef] [PubMed]

36. Munns, R.; James, R.A.; Läuchli, A. Approaches to increasing the salt tolerance of wheat and other cereals. *J. Exp. Bot.* **2006**, *57*, 1025–1043. [CrossRef] [PubMed]

37. Röder, M.S.; Korzun, V.; Wendehake, K.; Plaschke, J.; Tixier, M.; Leroy, P.; Ganal, M.W. A Microsatellite Map of Wheat. *Genetics* **1998**, *149*, 2007.

38. Somers, D.; Isaac, P.; Edwards, K. A high-density microsatellite consensus map for bread wheat (*Triticum aestivum* L.). *Theo. Appl. Genet.* **2004**, *109*, 1105–1114. [CrossRef]

39. Tamura, K.; Stecher, G.; Peterson, D.; Filipski, A.; Kumar, S. MEGA6: Molecular evolutionary genetics analysis version 6.0. *Mol. Biol. Evol.* **2013**, *30*, 2725–2729. [CrossRef]

40. Du, Y.T.; Zhao, M.J.; Wang, C.T.; Gao, Y.; Wang, Y.X.; Liu, Y.W.; Chen, M.; Chen, J.; Zhou, Y.B.; Xu, Z.S.; et al. Identification and characterization of GmMYB118 responses to drought and salt stress. *BMC Plant Biol.* **2018**, *18*, 320. [CrossRef]

41. Martin, T.; Wöhner, R.V.; Hummel, S.; Willmitzer, L.; Frommer, W.B.; Gallagher, S.R. The GUS reporter system as a tool to study plant gene expression. *Gus Protocols* **1992**, 23–43. [CrossRef]

42. James, D.; Borphukan, B.; Fartyal, D.; Ram, B.; Yadav, R.; Singh, J.; Manna, M.; Panditi, V.; Sheri, V.; Achary, M. Concurrent Overexpression of OsGS1;1 and OsGS2 Genes in Transgenic Rice (*Oryza sativa* L.): Impact on Tolerance to Abiotic Stresses. *Front. Plant Sci.* **2018**, *9*, 786. [CrossRef] [PubMed]

43. Ross, C.A.; Liu, Y.; Shen, Q.J. The WRKY Gene Family in Rice (*Oryza sativa*). *J. Integr. Plant Biol.* **2010**, *49*, 827–842. [CrossRef]

44. Garnica, D.P.; Upadhyaya, N.M.; Dodds, P.N.; Rathjen, J.P. Strategies for Wheat Stripe Rust Pathogenicity Identified by Transcriptome Sequencing. *PLoS ONE* **2013**, *8*, e67150. [CrossRef] [PubMed]

45. Pingault, L.; Choulet, F.; Alberti, A.; Glover, N.; Wincker, P.; Feuillet, C.; Paux, E. Deep transcriptome sequencing provides new insights into the structural and functional organization of the wheat genome. *Genome Biol.* **2015**, *16*, 29. [CrossRef]

46. Feng, Y.; Zhao, Y.; Wang, K.; Yong, C.L.; Xiang, W.; Yin, J. Identification of vernalization responsive genes in the winter wheat cultivar Jing841 by transcriptome sequencing. *J. Genet.* **2016**, *95*, 1–8. [CrossRef]

47. Niu, C.F.; Wei, W.; Zhou, Q.Y.; Tian, A.G.; Hao, Y.J.; Zhang, W.K.; Ma, B.; Lin, Q.; Zhang, Z.B.; Zhang, J.S. Wheat WRKY genes TaWRKY2 and TaWRKY19 regulate abiotic stress tolerance in transgenic *Arabidopsis* plants. *Plant Cell Environ.* **2012**, *35*, 1156–1170. [CrossRef]

48. Sezer, O.; Ebru, D.; Turgay, U. Transcriptome-wide identification of bread wheat WRKY transcription factors in response to drought stress. *Mol. Genet. Genom.* **2014**, *289*, 765–781.

49. Kawasaki, S.; Borchert, C.; Deyholos, M.; Hong, W.; Brazille, S.; Kawai, K.; Galbraith, D.; Bohnert, H.J. Gene Expression Profiles during the Initial Phase of Salt Stress in Rice. *Plant Cell* **2001**, *13*, 889–905. [CrossRef]

50. Cavallisforza, L.L.; Edwards, A.W. Phylogenetic analysis. Models and estimation procedures. *Am. J. Hum. Genet.* **1967**, *21*, 550–570.

51. Chen, J.N.; Yin, Y.H. WRKY Transcription Factors are Involved in Brassinosteroid Signaling and Mediate the Crosstalk Between Plant Growth and Drought Tolerance. *Plant Signal. Behav.* **2017**, *12*, e1365212. [CrossRef]

52. Cheng, H. Overexpression of the GmNAC2 Gene, an NAC Transcription Factor, Reduces Abiotic Stress Tolerance in Tobacco. *Plant Mol. Biol. Rep.* **2013**, *31*, 435–442.

53. Ling, L.D.; Xiao, H.Z.; Chao, Y.P.; Mei, Z.S.; Heng, L.W.; Shu, L.F.; Shu, X.Y. Genome-wide analysis of the WRKY gene family in cotton. *Mol. Genet. Genom.* **2014**, *289*, 1103.

54. Zhou, Q.Y.; Tian, A.G.; Zou, H.F.; Xie, Z.M.; Lei, G.; Huang, J.; Wang, C.M.; Wang, H.W.; Zhang, J.S.; Chen, S.Y. Soybean WRKY-type transcription factor genes, GmWRKY13, GmWRKY21, and GmWRKY54, confer differential tolerance to abiotic stresses in transgenic *Arabidopsis* plants. *Plant Biotech. J.* **2010**, *6*, 486–503. [CrossRef] [PubMed]

55. Yu, Y.; Wang, N.; Hu, R.; Xiang, F. Genome-wide identification of soybean WRKY transcription factors in response to salt stress. *Springerplus* **2016**, *5*, 920. [CrossRef]

56. Li, W.; Tian, Z.; Yu, D. WRKY13 acts in stem development in *Arabidopsis thaliana*. *Plant Sci.* **2015**, *236*, 205–213. [CrossRef]

57. Wei, L.; Wang, H.; Yu, D. The *Arabidopsis* WRKY transcription factors WRKY12 and WRKY13 oppositely regulate flowering under short-day conditions. *Mol. Plant* **2016**, *9*, 1492–1503.

58. Jun, X.; Hongtao, C.; Xianghua, L.; Jinghua, X.; Caiguo, X.; Shiping, W. Rice WRKY13 Regulates Cross Talk between Abiotic and Biotic Stress Signaling Pathways by Selective Binding to Different *cis*-Elements. *Plant Phys.* **2013**, *163*, 1868–1882.

59. Cui, X.Y.; Du, Y.T.; Fu, J.D.; Yu, T.F.; Wang, C.T.; Chen, M.; Chen, J.; Ma, Y.Z.; Xu, Z.S. Wheat CBL-interacting protein kinase 23 positively regulates drought stress and ABA responses. *BMC Plant Biol.* **2018**, *18*, 93. [CrossRef]

60. Radkova, M.; Vítámvás, P.; Sasaki, K.; Imai, R. Development- and cold-regulated accumulation of cold shock domain proteins in wheat. *Plant Physiol. Bioch.* **2014**, *77*, 44–48. [CrossRef]

61. Bradley, D.; Ratcliffe, O.; Vincent, C.; Carpenter, R.; Coen, E. Inflorescence commitment and architecture in *Arabidopsis*. *Science* **1997**, *275*, 80–83. [CrossRef]

62. Hiei, Y.; Ohta, S.; Komari, T.; Kumashiro, T. Efficient transformation of rice (*Oryza sativa* L.) mediated by *Agrobacterium* and sequence analysis of the boundaries of the T-DNA. *Plant J. Cell Mol. Biol.* **2010**, *6*, 271–282. [CrossRef] [PubMed]

63. Yoshida, S.; Forno, D.A.; Cock, J.H. *Laboratory Manual for Physiological Studies of Rice*; CABI: Wallingford, UK, 1971.

International Journal of
Molecular Sciences

Article

Genome-Wide Identification and Characterization of *FBA* Gene Family in Polyploid Crop *Brassica napus*

Wei Zhao [1,†], Hongfang Liu [1,†], Liang Zhang [1], Zhiyong Hu [1], Jun Liu [1], Wei Hua [1], Shouming Xu [2,*] and Jing Liu [1,*]

[1] Oil Crops Research Institute of the Chinese Academy of Agricultural Sciences, Key Laboratory of Biology and Genetic Improvement of Oil Crops, Ministry of Agriculture and Rural Affairs, Wuhan 430062, China; zhaowei@caas.cn (W.Z.); wanghw@webmail.hzau.edu.cn (H.L.); zhangliang01@caas.cn (L.Z.); huzhiyong@oilcrops.cn (Z.H.); liujunocr@caas.cn (J.L.); huawei@oilcrops.cn (W.H.)

[2] Henan key laboratory of Plant Stress Biology, School of Life Sciences, Henan University, Kaifeng 475004, China

* Correspondence: xushouming@henu.edu.cn (S.X.); liujing@oilcrops.cn (J.L.)

† These authors contributed equally to this work.

Received: 1 November 2019; Accepted: 14 November 2019; Published: 15 November 2019

Abstract: Fructose-1,6-bisphosphate aldolase (FBA) is a versatile metabolic enzyme involved in multiple important processes of glycolysis, gluconeogenesis, and Calvin cycle. Despite its significance in plant biology, the identity of this gene family in oil crops is lacking. Here, we performed genome-wide identification and characterization of *FBAs* in an allotetraploid species, oilseed rape *Brassica napus*. Twenty-two *BnaFBA* genes were identified and divided into two groups based on integrative analyses of functional domains, phylogenetic relationships, and gene structures. Twelve and ten *B. napus* FBAs (BnaFBAs) were predicted to be localized in the chloroplast and cytoplasm, respectively. Notably, synteny analysis revealed that *Brassica*-specific triplication contributed to the expansion of the *BnaFBA* gene family during the evolution of *B. napus*. Various *cis*-acting regulatory elements pertinent to abiotic and biotic stresses, as well as phytohormone responses, were detected. Intriguingly, each of the *BnaFBA* genes exhibited distinct sequence polymorphisms. Among them, six contained signatures of selection, likely having experienced breeding selection during adaptation and domestication. Importantly, *BnaFBAs* showed diverse expression patterns at different developmental stages and were preferentially highly expressed in photosynthetic tissues. Our data thus provided the foundation for further elucidating the functional roles of individual *BnaFBA* and also potential targets for engineering to improve photosynthetic productivity in *B. napus*.

Keywords: *Brassica napus*; aldolase; Calvin cycle; phytohormones; environmental stresses

1. Introduction

Fructose-1,6-bisphosphate aldolase (FBA, EC4.1.2.13 or aldolase) is an essential metabolism enzyme in the glycolytic pathway [1]. FBA catalyzes the reversible aldol cleavage of fructose-1,6-bisphosphate (FBP) into dihydroxyacetone phosphate (DHAP) and glyceraldehyde-3-phosphate (G3P), two important intermediates for oil biosynthesis [2]. DHAP could be further converted to diacylglycerol (DAG) by multiple enzymatic reactions, and DAG is a key substrate of diacylglycerol acyltransferase (DGAT) for the synthesis of triacylglycerols (TAGs). Meanwhile, G3P could be converted to malonyl CoA (coenzyme A) that is then used to produce fatty acids. Thus, FBA is not only one of the key regulatory enzymes in the glycolysis pathway but also may control the flux of carbohydrates and, therefore, play an important role in the oil yield of oilseeds [3,4].

FBAs could be broadly classified into two classes, namely class-I and class-II, based on their catalytic mechanisms and prevalence among species in evolution [5,6]. Class-I FBAs are usually tetrameric enzymes, forming a Schiff base with the substrate as intermediate, and utilize a lysine residue to generate a nucleophilic enamine from DHAP and are not inhibited by EDTA or affected by potassium ions. Class-II FBAs are found as homodimers and require divalent cations as cofactors to stabilize the DHAP enolate intermediate involved in the aldol condensation reaction and are inhibited by EDTA [7]. FBAs of class-I are found in some bacteria, animals, and plants, while class-II FBAs occur in most bacteria, yeast, and fungi [8–10]. However, FBAs of class-II are also found in wheat and green algae, such as *Euglena gracilis*, *Chlamydomonas mundane*, and *Chlamydomonas rheinhardii* [11–13].

In higher plants, one set of FBA isoenzymes is localized in the cytosolic (cFBA), and another one in the chloroplast/plastid (cpFBA) [14,15]. Both cFBA and cpFBA are encoded by separate nuclear genes that probably evolved from duplication of a common ancestral gene [16]. To date, different members of the *FBA* family genes have been reported in a variety of monocots and eudicots species. For example, in *Arabidopsis thaliana* (*A. thaliana*), eight *FBA* family genes (*AtFBA1–8*) were identified, including three chloroplast/plastid members (*AtFBA1–3*) and five cytosolic members (*AtFBA4–8*) [17]. In tomato, eight *FBA* genes, including five cFBAs and three cpFBAs, were characterized based on the phylogenetic tree, gene structures, and conserved motifs [18,19]. In addition, two homologous genes of cpFBAs were isolated from green leaves of *Nicotiana paniculata* [20]. One *SpFBA* gene was cloned from *Sesuvium portulacastrum* roots [21]. In wheat, 21 genes encoding *Thermus aquaticus* FBA (TaFBA) isoenzymes were identified and categorized into three subgroups of class-I *cpFBAs*, class-I *cFBAs*, and class-II *cpFBAs* [11]. In rice, two *OsFBAs* were reported to localize in the chloroplasts, while the other five *OsFBAs* in the cytoplasm [22]. One chloroplastic *FBA* and one cytosolic *FBA* were detected in the leaves of maize seedlings [23]. Four developmentally up-regulated *FBAs* (*CoFBA1–4*) genes were identified in tea oil tree (*Camellia oleifera*) seeds [24]. However, there is no report on the identification of *FBA* family genes so far in oilseed rape *Brassica napus*.

Recent studies suggest that the *FBA* genes are playing important roles in diverse significant physiological and biochemical processes in plants. The cpFBA is an essential enzyme in the Calvin cycle, in which its activity generates metabolites for starch biosynthesis [25]. For instance, the loss of the *AtFBA6* function resulted in a lower germination rate after abscisic acid (ABA) treatment in *A. thaliana* [17]. The expression levels of *CoFBA1* and *CoFBA3* genes were highly correlated with the amount of tea oil in the seeds of tea oil tree [24]. Reduction of plastid FBA activity inhibits photosynthesis and alters carbon partitioning in potato, whereas increased FBA activity in plastids could promote CO_2 fixation and enhance the growth and photosynthesis in tobacco [26,27]. Inhibition of chloroplastic FBA affects the development of fruit size in tomato [28]. Gibberellin (GA) treatment could increase the levels of cytoplasmic *FBA* in all regions of the roots, resulting in the stimulation of the root growth mediating the energy production in rice [29]. Thus, *FBA* family genes appear to be also associated with the response to phytohormones. Besides, *FBAs* also participate in response to various environmental stimuli in plants, such as salinity, drought, anoxygenic stress, abnormal temperature, light acclimation, and *Rhizoctonia solani Kuhn* infection [21,30–32].

Brassica napus L. (*B. napus*), a relatively recent allotetraploid formed from hybridization between *Brassica rapa* (*B. rapa*) and *Brassica oleracea* (*B. oleracea*), is the second-largest source of vegetable oil crop and cultivated around the world [33,34]. *B. napus* has adapted to diverse climate zones and latitudes by forming three main ecotype groups, namely winter, semi-winter, and spring types [35,36]. Although *FBA* family genes have been studied in several plant species, little is known regarding this gene family in oil crops. Here, we systematically identified *FBA* family genes in *B. napus* and profiled their gene structures, chromosome locations, conserved motifs, *cis*-acting elements in the promoter regions, phylogenetic classifications, and sequence polymorphisms. We further analyzed their expression patterns in different developmental tissues and in response to stresses. Our results provided the foundation for further elucidating the functional roles of *FBA* family genes and potential targets for engineering to improve photosynthetic capacity in *B. napus*.

2. Results

2.1. Identification, Properties, and Genomic Distribution of BnaFBA Genes

To identify each member of the *FBA* gene family in *B. napus*, the glycolytic domain (PF00274) and fructose-bisphosphate aldolase class-II domain (PF01116) from the Pfam database (http://pfam.xfam. org/) were employed as queries to search against *B. napus* var. Darmor-*bzh* protein dataset. The peptides of putative *B. napus* FBAs (BnaFBAs) with the best hit of *A. thaliana* FBAs (AtFBAs) and TaFBAs were further used to predict functional domains by using the Pfam and SMART databases to confirm the presence of the fructose-bisphosphate aldolase domains. As indicated in Table 1, we identified 22 *BnaFBA* genes in *B. napus*. Meanwhile, we found 14 *BraFBAs* and fourteen *BolFBAs* in *B. rapa* var. Z1 and *B. oleracea* var. HDEM, respectively (Table S1).

The transcript length of *BnaFBA* genes varied from 951 bp to 4149 bp. All identified *BnaFBA* genes encoded proteins ranging from 317 (*BnaFBA2e*) to 1382 (*BnaFBA9b*) amino acids (aa), molecular weight (MW) from 34.18 (*BnaFBA2e*) to 148 kDa (*BnaFBA9b*), isoelectric point (pI) from 5.7 (*BnaFBA2a*) to 8.47 (*BnaFBA3a*), and grand average of hydropathy (GRAVY) from −0.263 (*BnaFBA3a*) to 0.076 (*BnaFBA9b*) (Table 1). Twelve BnaFBAs (BnaFBA1a/b/c, BnaFBA2a/b/c/d/e/f, BnaFBA3a/b, and BnaFBA9a) were predicted to be localized in chloroplast, with the other 10 BnaFBAs (BnaFBA5a/b, BnaFBA6, BnaFBA8a/b/c/d/e/f, and BnaFBA9a) being in the cytoplasm (Table 1). Similarly, in *B. rapa* and *B. oleracea*, seven *B. Oleracea* FBAs (BolFBAs) and seven *B. rapa* FBAs (BraFBAs) were predicted to be chloroplast-localized proteins, six BolFBAs and six BraFBAs were cytoplasm-localized ones, whereas BolFBA9 and BraFBA6a were predicted to be localized in plasma membrane and nucleus, respectively (Table S1).

Based on the functional domains contained, the 22 *BnaFBAs* could be divided into two classes—class-I and class-II. All the BnaFBAs of class-I harbored one glycolytic domain (PF00274) across the proteins, and BnaFBA9a and BnaFBA9b of class-II had one fructose-bisphosphate aldolase class-II domain (PF01116). Except for the glycolytic domain (PF00274), some members of class-I FBA, such as BnaFBA5b, as the longest FBA of class-I with 870 aa, also harbored PRK (phosphoribulokinase kinase) (PF00485) and UPRTase (uracil phosphoribosyltransferase) (PF14681) domains in the N terminal. Likewise, in *B. oleracea*, BolFBA5 contained three functional domains of PF00274, PF00485, and PF14681, and BolFBA6b had two tandem copies of glycolytic domains (PF00274). Interestingly, BraFBA6a contained both glycolytic domain (PF00274) and RNA polymerase II-binding domain (PF04818), suggesting that it might be involved in RNA processing in the nucleus (Table S1).

To attain a general view of the distribution of *BnaFBA* genes on the genome of *B. napus*, the 22 *BnaFBAs* were mapped on the corresponding chromosomes according to their physical positions. Of the 22 *BnaFBA* genes, 19 were evenly distributed on the 15 *B. napus* chromosomes, while the three other *BnaFBAs* were assigned to the random chromosomes (two on the An chromosomes and one on the Cn chromosome). The number of *BnaFBAs* per chromosome ranged from one to two. Among them, eleven *BnaFBA* genes were distributed over the seven A subgenome chromosomes of *B. napus*, including A01, A02, A04, A06, A07, A08, and A09, as well as Ann_random. Equally, the other 11 *BnaFBAs* were distributed over eight C subgenome chromosomes of *B. napus*, including C01, C02, C03, C04, C05, C06, C07, and C08, as well as Cnn_random. Most of the *BnaFBAs* were not localized in the terminal regions of the chromosomes, where the gene density was relatively high in *B. napus* (Figure 1).

Table 1. Summary Information on *FBA* Family Genes in *B. Napus*.

Gene Name	Locus Name	Gene Location [1]	Transcript Length (bp)	Protein Length (aa)	MW (kDa)	pI	GRAVY	Domain [2] (Start-End aa)	Homolog of *Arabidopsis*	Classifi-cation	Subcellular Localization
BnaFBA1a	BnaA04g12130D	chrA04:10353159–10354905:−	1173	390	42.2	6.92	−0.194	PF00274 (55–390)	AT2G21330 (*AtFBA1*)	I	Chloroplast
BnaFBA1b	BnaC04g33570D	chrC04:35146154–35147860:+	1173	390	42.16	6.92	−0.19	PF00274 (55–390)	AT2G21330 (*AtFBA1*)	I	Chloroplast
BnaFBA1c	BnaC08g35820D	chrC08:33457869–33459923:+	1197	398	42.81	6.78	−0.155	PF00274 (54–398)	AT2G21330 (*AtFBA1*)	I	Chloroplast
BnaFBA2a	BnaA08g16820D	chrA08:13562506–13564430:+	1152	383	41.3	5.7	−0.179	PF00274 (44–383)	AT4G38970 (*AtFBA2*)	I	Chloroplast
BnaFBA2b	BnaC03g60280D	chrC03:49437707–49439536:−	1152	383	41.33	5.7	−0.185	PF00274 (44–383)	AT4G38970 (*AtFBA2*)	I	Chloroplast
BnaFBA2c	BnaC07g47470D	chrC07:44670819–44672806:+	1182	393	42.58	6.87	−0.196	PF00274 (54–393)	AT4G38970 (*AtFBA2*)	I	Chloroplast
BnaFBA2d	BnaA06g37230D	chrA06:24314148–24316076:+	1182	393	42.64	7.61	−0.208	PF00274 (54–393)	AT4G38970 (*AtFBA2*)	I	Chloroplast
BnaFBA2e	BnaAnng40850D	chrAnn_random:46872917–46874555:+	951	317	34.18	6.77	−0.036	PF00274 (54–306)	AT4G38970 (*AtFBA2*)	I	Chloroplast
BnaFBA2f	BnaC01g00070D	chrC01:20706–22753:−	1200	399	42.95	6.07	−0.137	PF00274 (44–399)	AT4G38970 (*AtFBA2*)	I	Chloroplast
BnaFBA3a	BnaA02g27140D	chrA02:20044496–20046087:+	1170	389	42.3	8.47	−0.263	PF00274 (45–389)	AT2G01140 (*AtFBA3*)	I	Chloroplast
BnaFBA3b	BnaC02g33660D	chrC02:36012928–36014598:−	1170	389	42.28	8.47	−0.257	PF00274 (45–389)	AT2G01140 (*AtFBA3*)	I	Chloroplast
BnaFBA5a	BnaA01g15640D	chrA01:8017642–8018970:+	1077	358	38.23	6.38	−0.072	PF00274 (11–358)	AT4G26530 (*AtFBA5*)	I	Cytoplasm
BnaFBA5b	BnaC01g18640D	chrC01:12962252–12967551:+	2613	870	95.97	6.82	−0.107	PF00485 (52–237); PF14681 (266–458); PF00274 (523–870)	AT4G26530 (*AtFBA5*)	I	Cytoplasm
BnaFBA6	BnaAnng07310D	chrAnn_random:7183202–7186036:+	1032	343	36.91	5.8	−0.189	PF00274 (11–338)	AT2G36460 (*AtFBA6*)	I	Cytoplasm
BnaFBA8a	BnaA09g33290D	chrA09:24581809–24583398:−	1077	358	38.45	6.28	−0.197	PF00274 (11–358)	AT3G52930 (*AtFBA8*)	I	Cytoplasm
BnaFBA8b	BnaC08g24100D	chrC08:26197819–26199427:−	1077	358	38.45	6.28	−0.197	PF00274 (11–358)	AT3G52930 (*AtFBA8*)	I	Cytoplasm
BnaFBA8c	BnaC06g14270D	chrC06:17065486–17066935:−	1077	358	38.42	6.22	−0.177	PF00274 (11–358)	AT3G52930 (*AtFBA8*)	I	Cytoplasm
BnaFBA8d	BnaA07g15900D	chrA07:13618484–13619930:−	1077	358	38.42	6.22	−0.177	PF00274 (11–358)	AT3G52930 (*AtFBA8*)	I	Cytoplasm
BnaFBA8e	BnaA04g05120D	chrA04:3768229–3769737:+	1077	358	38.49	6.28	−0.218	PF00274 (11–358)	AT3G52930 (*AtFBA8*)	I	Cytoplasm

Table 1. *Cont.*

Gene Name	Locus Name	Gene Location [1]	Transcript Length (bp)	Protein Length (aa)	MW (kDa)	pI	GRAVY	Domain [2] (Start-End aa)	Homolog of *Arabidopsis*	Classifi-cation	Subcellular Localization
BnaFBA8f	BnaCnng50220D	chrCnn_random:49773855–49775405:+	1077	358	38.49	6.28	−0.218	PF00274 (11–358)	*AT3G52930* (*AtFBA8*) *AT1G18270*	I	Cytoplasm
BnaFBA9a	BnaA06g12420D	chrA06:6440041–6449829:–	4143	1380	147.66	5.82	0.067	PF01116 (1104–1379)	(ketose-bisphosphate aldolase class-II family protein) *AT1G18270*	II	Chloroplast
BnaFBA9b	BnaC05g14020D	chrC05:8111222–8120981:–	4149	1382	148	5.75	0.076	PF01116 (1106–1381)	(ketose-bisphosphate aldolase class-II family protein)	II	Cytoplasm

[1] Chromosome: start position-end position: strand, (–) means antisense strand of chromosome, (+) means positive-sense strand of chromosome. [2] Glycolytic, fructose-bisphosphate aldolase class-I (PF00274); F_bP_aldolase, fructose-bisphosphate aldolase class-II (PF01116); Uracil phosphoribosyltransferase, UPRTase (PF14681); PRK, phosphoribulokinase / uridine kinase family (PF00485); AP2, apetala 2 (PF00847).

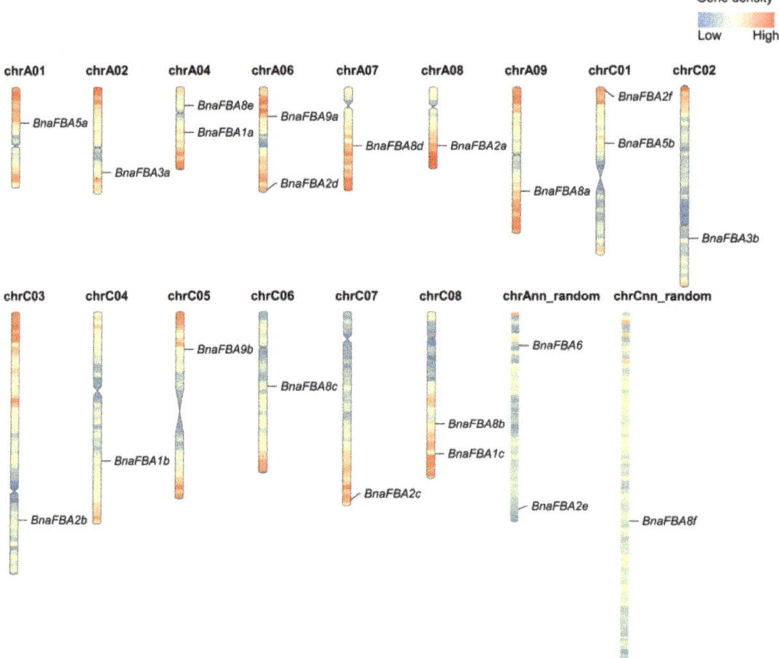

Figure 1. Genomic distributions of *BnaFBA* genes on *B. napus* chromosomes. The *BnaFBAs* were plotted based on the location of genes, length of chromosomes, and positions of centromeres. Heatmap of each chromosome indicated the gene density by the frequency per 1 Mb.

2.2. Phylogenetic and Structure Analysis of BnaFBAs

To explore the molecular evolution of the *FBA* gene family in *B. napus*, a total of 79 *FBA* genes from *B. napus*, *B. rapa*, *B. oleracea*, *A. thaliana*, and wheat, were used to construct an unrooted phylogenetic tree. According to the phylogenetic relationships of these *FBA* genes, they could be divided into two independent classes, consistent with the classification by functional domains they contained (Figure 2A). Furthermore, the *FBA* genes of class-I could be further classified into four subclasses, namely class-Ia, class-Ib, class-Ic, and class-Id. Therefore, in *B. napus*, there were seven *BnaFBA* genes in the class-Ia group, two in class-Ib, nine in class-Ic, two in class-Id, and two in class-II. Consistent with its polyploid origin, except for the FBA4 gene, the genome of *B. napus* maintained homologs of each *FBA* gene derived from the diploid parents, *B. rapa* and *B. oleracea* (Figure 2A).

Based on the gene information of the genome available in the GENOSCOPE database, we performed gene structure analysis by comparing the coding sequence (CDS) of *BnaFBAs*, *BraFBAs*, and *BolFBAs*. In *B. napus*, A subgenome homologs and C subgenome homologs of *FBA* genes that were in the adjacent branches of the phylogenetic tree exhibited the same gene structure (Figure 2B). Within each class of *FBA* genes, the features of exons, such as order, length, and number, were largely conserved except for *FBA6* genes (Figure 2B). Besides, the organization of the introns of *BnaFBA* genes was highly variable. The length of introns varied extensively in different members of the *BnaFBA* gene family, ranging from 30 to 4152 bp, with the number ranging from 2 to 41. Compared to the class-I *BnaFBA* genes, the class-II genes were much longer and had much more exons and introns (Figure 2B).

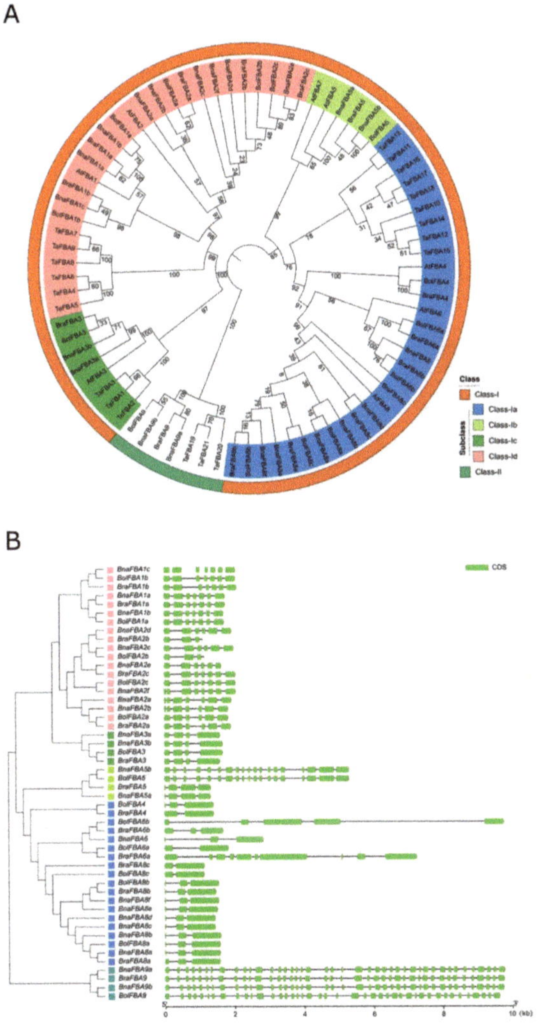

Figure 2. Phylogenetic relationships and gene structure of *BnaFBA* genes. (**A**) Phylogenetic relationships of *BnaFBAs*, *BraFBAs*, *BolFBAs*, *AtFBAs*, and *TaFBAs*. The unrooted tree was generated using MEGA7.1 software by the neighbor-joining method. The numbers next to the branch show the 1000 bootstrap replicates expressed in percentage. The phylogenetic classes of *FBA* genes were marked by corresponding colors that are shown in the color legend at the bottom right. (**B**) The schematic diagrams of the exon-intron organization of *FBA* genes in *B. napus*, *B. rapa,* and *B. oleracea*. The phylogenetic tree of the *FBA* genes is placed at the left, and the color squares represent phylogenetic classes. The green boxes and lines indicate CDS (coding sequence) and introns, respectively. The length of the scale is at the bottom.

To further explore the higher-order structure of the BnaFBA proteins, the three-dimensional (3D) structural models for BnaFBA1a, BnaFBA8a, and BnaFBA9a were generated using SWISS-MODEL. Based on the experimental structure of class-I rabbit muscle aldolase, the SWISS-MODEL analysis results revealed that BnaFBA1a and BnaFBA8a of the class-I group could form tetramers structures, and interfaces A and B were observed in the BnaFBA1a and BnaFBA8a tetramers. Different from BnaFBA1a and BnaFBA8a, BnaFBA9a belonged to class-II group and form dimers based on its relatively high similarity to class-II aldolases of *Thermus aquaticus* (Figure 3A–C). Furthermore, the catalytic residues of D74-K147-K186-R188-E226-E228-K269-S301-R331 and D30-K103-K142-R144-E183-E185-K225-S266-R298 were observed in BnaFBA1a and BnaFBA8a, respectively. Additionally, similar to the class-II FBAs of *Thermus aquaticus*, BnaFBA9a contained active sites of H1181-E1232-H1277-H1309 that also serve as the divalent metal cation binding sites (Figure 3E,F). Multiple sequence alignment results indicated that class-I BnaFBAs, BraFBAs, BolFBAs, and AtFBAs had high conserved catalytic residues (Figure S1). The active sites among class-II BnaFBAs, BraFBAs, and BolFBAs were the same as FBA isozymes in the *Thermus aquaticus* (Figure S2).

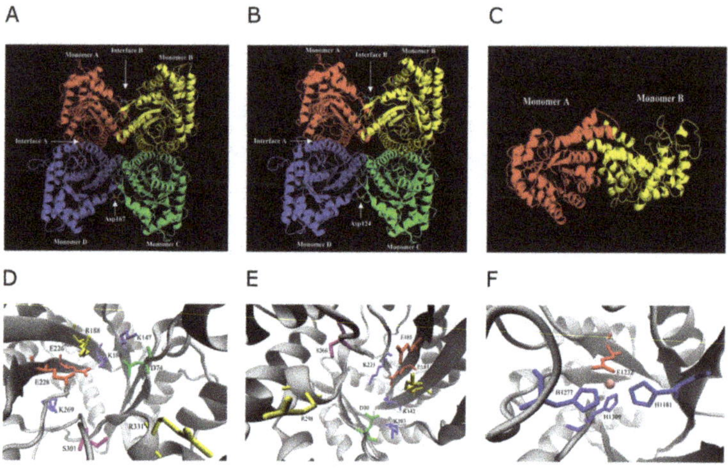

Figure 3. Predicted three-dimensional model of representative *Brassica napus* Fructose-1,6-bisphosphate aldolase (BnaFBA) proteins. (**A**) Tetrameric BnaFBA1a. (**B**) Tetrameric BnaFBA8a. (**C**) Dimeric BnaFBA9a. (**D**) Active site residues of BnaFBA1a. (**E**) Active site residues of BnaFBA8a. (**F**) Active site residues of BnaFBA9a. The sites of the Asp167/124 substitutions are indicated on interface B, and the filled circle represents the divalent metal cation of BnaFBA9a.

2.3. Synteny and Gene Duplication of BnaFBA Genes

A. thaliana is the most prominent model system for plant molecular biology and genetics research, whose structural genes have been identified and functionally characterized. Thus, we traced the orthologous gene pairs between *A. thaliana* and *Brassica* species to investigate the evolutionary history by syntenic gene analysis. A total of 29 orthologous gene pairs were identified between *A. thaliana* and *B. napus*, 20 between *A. thaliana* and *B. rapa*, and 19 between *A. thaliana* and *B. oleracea*. In addition, we also obtained 29, 15, and 15 paralogous gene pairs within *B. napus*, *B. rapa*, and *B. oleracea*, respectively (Figure 4A). The previous study revealed that crucifer (*Brassicaceae*) lineage experienced two whole-genome duplications (named α and β) and one triplication event (γ) shared by most dicots [37]. Moreover, *Brassica* species experienced an extra whole-genome triplication (WGT) event compared with *A. thaliana* [38]. As WGT of the *Brassica* ancestor, *FBA* genes in the *A. thaliana* genome might have triplicated orthologous copies in *B. rapa* and *B. oleracea*. Consequently, some *FBA* genes (i.e., *FBA2* and *FBA8*) existed triple the number of those in *A. thaliana*, while the other genes (i.e., *FBA1*, *FBA3*, *FBA5*, and *FBA6*) had double or equal the number (Figure 4B). The *FBA* genes of *B. napus* were inherited from its diploid ancestors; thus, most of the *BnaFBA* genes were double the number of those in *B. rapa* and *B. oleracea* (i.e., *FBA2*, *FBA3*, *FBA5*, *FBA8*, and *FBA9*). However, both *FBA1* and *FBA6* genes lost one copy in the C subgenome of *B. napus*, while FBA4 lost all the copies in *B. napus* compared to its two ancestors (Figure 4B). Gene duplication analysis with syntenic and phylogenomic approaches using tool DupGen_finder in *B. napus* showed that all *BnaFBA* genes had corresponding duplicate genes. In *B. napus*, a total of 42 gene pairs were derived from whole-genome duplication (WGD), with one gene pair of *BnaFBA2e-BnaFBA2f* being derived from transposed duplication (Table S2).

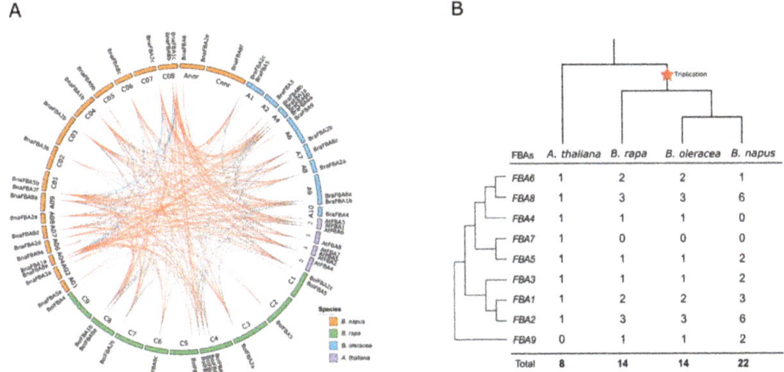

Figure 4. Collinear correlations and copy number variation of the *FBA* family genes in *B. napus, B. rapa, B. oleracea,* and *A. thaliana.* (**A**) Collinear correlations of FBA genes in the *B. napus, B. rapa, B. oleracea,* and *A. thaliana* genomes. The *B. napus, B. rapa, B. oleracea,* and *A. thaliana* chromosomes were colored by corresponding colors that are shown in the color legend at the bottom right. The blue lines represent the collinear correlations of FBA genes within *B. napus,* and the orange lines are for the collinear correlations of *FBA* genes between *B. napus* and the other species, with the grey lines representing the collinear correlations of *FBA* genes among *B. rapa, B. oleracea,* and *A. thaliana.* The figure was created using CIRCOS software. (**B**) Copy number variation of the *FBA* family genes in *B. napus, B. rapa, B. oleracea,* and *A. thaliana.* The phylogenetic tree of *FBA* genes is shown on the left, with the species tree shown at the top. The *Brassica*-specific triplication was indicated on the branches of the trees according to the Plant Genome Duplication Database. The numbers are the copy numbers of each FBA gene in *A. thaliana, B. napus, B. rapa,* and *B. oleracea.*

2.4. Cis-Acting Elements in the Putative Promoter Regions of BnaFBA Genes

As important molecular switches, *cis*-acting elements in the promoter region could provide useful information to understand the function and regulation of the genes during plant development and responses to various stresses. The 1.5 kb genomic DNA sequences identified from upstream of the *BnaFBA* genes were extracted and deployed in *cis*-acting regulatory elements analysis with PlantCARE. Various *cis*-acting regulatory elements existed within the promoter regions of *BnaFBA* genes (Table 2 and Table S3). For example, *BnaFBA* genes contained multiple phytohormone responsive elements, such as ABRE (abscisic acid-responsive element), AuxRE (auxin-responsive element), ERE (ethylene-responsive element), GARE (gibberellin-responsive element), MeJARE (MeJA-responsive element), and SARE (salicylic acid-responsive element). This suggested that the expression of *BnaFBAs* might be induced by different phytohormones. Besides, the *cis*-acting regulatory elements involved in stress-responsive elements, such as ARE (anoxic-responsive element), DRE (damage-responsive element), DIRE (drought-responsive element), DSRE (drought- and stress-responsive element), HSRE (heat stress-responsive element), LTRE (low-temperature-responsive element), and WRE (wound-responsive element), were also found within the promoters of *BnaFBA* genes, suggesting that expression levels of *BnaFBAs* might be also regulated by various environmental factors like drought, heat, and low-temperature. Globally, three phytohormone-related elements (i.e., ABRE, ERE, and MeJARE) and two stress-responsive elements (ARE and HSRE) were detected with high frequency in the promoter regions of *BnaFBA* genes. Notably, each *BnaFBA* had multi-copy LREs (light-responsive elements) ranging from 8 to 26, implying that *BnaFBAs* might play roles in light responses.

Table 2. *Cis*-Acting Elements in the Promoter Region of 22 *BnaFBA* Genes.

Gene	ABRE	ARE	AuxRE	Circadian	DIRE	DRE	DSRE	ERE	GARE	HSRE	LRE	LTRE	MeJARE	SARE	WRE
													Cis-Acting Elements [1]		
BnaFBA1a	13	3	1					1			26	1	2		
BnaFBA1b	11	7			2		1			2	16	2	8	2	
BnaFBA1c	5	3	1				1	1			21		4	1	
BnaFBA2a	4	1	2				1	3	1		16	1	10		
BnaFBA2b	4		2				1	3	2		18		12		
BnaFBA2c	4	1	2			1		1	2		14		4		
BnaFBA2d	4	3	3					1	2		14		4		
BnaFBA2e	4	6						1	1	2	13	1		1	
BnaFBA2f	12	4						1		1	19	2	4		
BnaFBA3a	2			2			1	1	1	1	10	1		2	
BnaFBA3b		3	1		2			1	2	1	9		2		
BnaFBA5a	5	1		1	2		2	1	1	1	13		2		
BnaFBA5b	1	3				1			2	1	10		10		1
BnaFBA6	1				1	3		1	2	1	13		4	1	
BnaFBA8a	7	5				2	1	2		1	16	1	4	1	
BnaFBA8b	1	2				1	1	2	1	1	10		2		
BnaFBA8c	2	9			1			4		2	8	1			1
BnaFBA8d	2	4		1	1		1			5	9	1		2	
BnaFBA8e	4	2	1	1			1	3		1	15		4	1	
BnaFBA8f	3	3	1	1						1	11		8		
BnaFBA9a	15	2	1		1		1	1		1	23		10	1	1
BnaFBA9b	3	3				1	1	1		1	13	1	2		

[1] Abscisic acid-responsive element (ABRE), Anoxic-responsive element (ARE), Auxin-responsive element (AuxRE), Circadian-responsive element (Circadian), Damage-responsive element (DRE), Defense- and stress-responsive element (DSRE), Drought-responsive element (DIRE), Ethylene-responsive element (ERE), Gibberellin-responsive element (GARE), Heat stress-responsive element (HSRE), Light-responsive element (LRE), Low-temperature-responsive element (LTRE), MeJA-responsive element (MeJARE), Salicylic acid-responsive element (SARE), and Wound-responsive element (WRE).

2.5. Natural Variations of BnaFBA Family Genes in B. Napus

Critical sequence polymorphism across the gene and its flanking regions may reflect the evolutionary process of species adapting to different environments. The public resequencing datasets of 991 *B. napus* germplasm accessions covering three main ecotype groups, namely winter, semi-winter, and spring types, were collected for variation analysis of the *BnaFBA* family genes [39]. The polymorphism sites of CDS in *BnaFBA* family genes ranged from two (*BnaFBA2c*) to 130 (*BnaFBA6*) (Table 3). The π and θ_w of nucleotide diversity parameters extended from 0.00057 (*BnaFBA3b*) to 0.04158 (*BnaFBA6*) and from 0.00054 (*BnaFBA3b*) to 0.02255 (*BnaFBA6*), respectively. Some members of *BnaFBA* gene family were conserved, such as *BnaFBA3b* (θ_w = 0.00054), *BnaFBA2c* (θ_w = 0.00092), and *BnaFBA2d* (θ_w = 0.00098), while others had high polymorphism, such as *BnaFBA9b* (θ_w = 0.00511), *BnaFBA5a* (θ_w = 0.00535), *BnaFBA8b* (θ_w = 0.0056), *BnaFBA8e* (θ_w = 0.00598), *BnaFBA3a* (θw = 0.00665), *BnaFBA8a* (θ_w = 0.00706), *BnaFBA1c* (θ_w = 0.00762), *BnaFBA8d* (θ_w = 0.00807), and *BnaFBA6* (θ_w = 0.0225). Besides, the *BnaFBA6* variation ratio reached a peak of 12.60 among the 22 *BnaFBA* genes, whereas *BnaFBA2c* had the lowest variation ratio of 0.17, with only two polymorphic sites (Table 3). Generally, due to the longer length of genes, *BnaFBA9a* and *BnaFBA9b* of class-II had much more variations than *FBA* genes of class-I; however, the variation ratio and nucleotide diversity of the coding regions of *BnaFBA9a* and *BnaFBA9b* showed no difference with the *BnaFBA* genes of class-I except for *BnaFBA6* (Table 3). In addition to CDS regions, a total of 1029 and 814 variations were also identified in the upstream/downstream 1.5 kb regions and intronic regions of *BnaFBA* genes, respectively (Table S4). Notably, *BnaFBA8a*, *BnaFBA8b*, and *BnaFBA6* each harbored one stop-gain mutation that led to premature stop codons, which indicated that these genes exhibited loss-of-function (Figure 5, Table S5). Only a few of the Indels had been detected in *BnaFBAs*. For example, *BnaFBA1c* had three non-frameshift Indels, and *BnaFBA3b* contained only one non-frameshift Indel, with *BnaFBA6* harboring two frameshift deletion/insertion variations (Figure 5). To study the population selection pressure, we conducted neutral testing using Tajima's D. Tajima's D values of all the *BnaFBAs* were positive, with significant Tajima's D values ($p < 0.01$) being observed in *BnaFBA1a*, *BnaFBA1e*, *BnaFBA2d*, and *BnaFBA2e* (Table 3). Particularly, the Tajima's D values of *BnaFBA1b* and *BnaFBA8a* reached extremely significant levels ($p < 0.001$ and $p < 0.0001$). Based on the variation of *BnaFBA* genes, we performed principal component analysis (PCA), in which no significant difference in the polymorphism of *BnaFBA* genes was seen between different ecotype groups of *B. napus* (Figure S3).

Table 3. Summary of Polymorphic Sites of the 22 *BnaFBA* Genes.

Gene	CDS Length (bp)	Sample Size	No. of Polymorphic Sites [1]	Sequence Variation Ratio (%)	Nucleotide Diversity π/bp [2]	θ_w/bp [3]	Tajima's D [4]
BnaFBA1a	1173	401	11	0.94	0.00376	0.00187	2.28214 *
BnaFBA1b	1173	469	7	0.60	0.00286	0.00155	2.56152 **
BnaFBA1c	1197	198	60	5.01	0.01299	0.00762	2.14201 *
BnaFBA2a	1152	163	23	2.00	0.00657	0.00394	1.45126
BnaFBA2b	1152	505	10	0.87	0.00279	0.00211	0.98657
BnaFBA2c	1182	780	2	0.17	0.00113	0.00092	1.89306
BnaFBA2d	1182	727	3	0.25	0.00145	0.00098	1.9818 *
BnaFBA2e	951	289	10	1.05	0.0044	0.00239	2.43882 *
BnaFBA2f	1200	552	21	1.75	0.00509	0.00298	1.90478
BnaFBA3a	1170	249	34	2.91	0.00894	0.00665	0.81257
BnaFBA3b	1170	210	5	0.43	0.00057	0.00054	0.17874
BnaFBA5a	1077	307	21	1.95	0.00783	0.00535	1.49221
BnaFBA5b	2613	228	42	1.61	0.0053	0.00349	1.63186
BnaFBA6	1032	132	130	12.60	0.04158	0.02255	1.99668
BnaFBA8a	1077	379	29	2.69	0.00959	0.00706	1.32819
BnaFBA8b	1077	529	29	2.69	0.01251	0.0056	3.68226 ***
BnaFBA8c	1077	726	5	0.46	0.00241	0.00143	1.43186
BnaFBA8d	1077	251	38	3.53	0.01206	0.00807	1.33435
BnaFBA8e	1077	234	31	2.88	0.0108	0.00598	1.93228
BnaFBA8f	1077	429	12	1.11	0.0033	0.00208	1.81155
BnaFBA9a	4143	100	79	1.91	0.00467	0.00355	0.97966
BnaFBA9b	4149	43	86	2.07	0.00673	0.00511	1.19193

[1] polymorphic includes the SNPs (single nucleotide polymorphisms) and Indels (insertions and deletions). [2] π, average nucleotide differences per site between the two sequences. [3] θ_w, Watterson estimator. [4] Tajima's D, test for neutral selection (*: significant at $p < 0.01$; **: significant at $p < 0.001$; ***: significant at $p < 0.0001$). CDS, coding sequence.

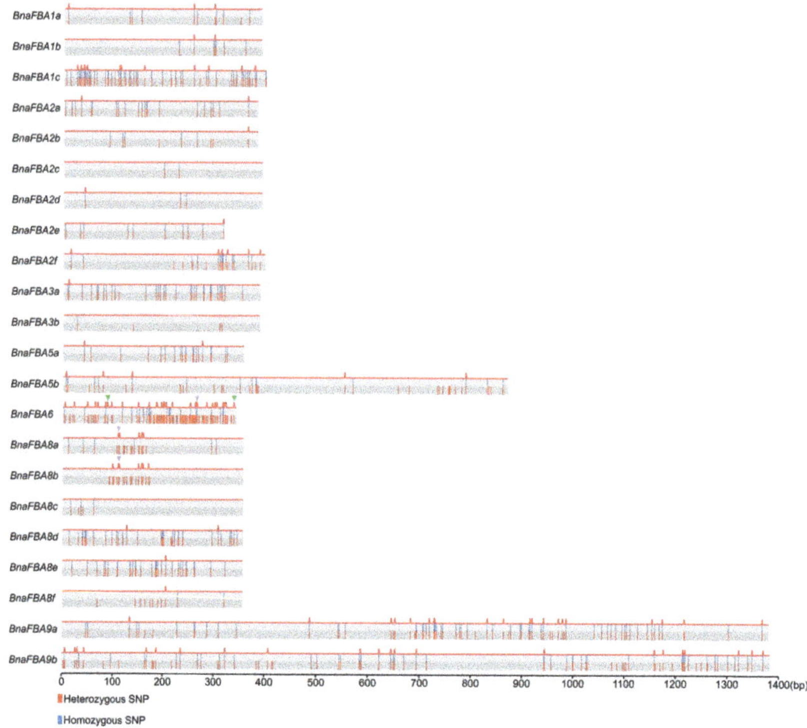

Figure 5. The variation distribution of *BnaFBA* genes. The heterozygous SNPs and homozygous SNPs were marked as red and blue lines, respectively. The red peak on the top of each *FBA* gene represents nonsynonymous variation. The green and purple triangles signify frameshift deletion/insertion and stop-gain variations, respectively. The length of the scale is placed at the bottom.

2.6. Expression Patterns of BnaFBA Genes

The members of the *FBA* gene family are playing important roles in diverse significant physiological and biochemical processes in plants [21,25]. Besides, FBAs also participate in response to various environmental stimuli [30–32]. To identify the function of *BnaFBA* genes under various conditions, the expression levels of *BnaFBAs* were evaluated during growth and development, as well as in response to biotic stresses and phytohormones in *B. napus*.

To explore the expression patterns of the *BnaFBAs* during growth in *B. napus*, we analyzed their expression levels in twelve various tissues (leaf, root, stem, cauline leaf, pistil, stamen, petal, flower bud, axillary bud, silique wall, embryo, and seed) at different developmental stages. As a result, at the trefoil stage and the flowering stage, *BnaFBA* genes showed opposite expression patterns between leaves and roots. For example, *BnaFBA1a/b/c*, *BnaFBA2a/b/c/d/f*, and *BnaFBA5a/b* displayed relatively high expression level in leaves than that in roots. On the contrary, *BnaFBA3a/b* and *BnaFBA8a/b/c/d/e/f* exhibited relatively high expression levels in roots than that in leaves. At the flowering stage, the expression levels of *BnaFBA1a/b/c* and *BnaFBA2a/b/c/d/f* were higher in pistil than that in stamen and petal tissues. At the pod stage, *BnaFBA1a/b/c* and *BnaFBA2a/b/c/d/f* were highly expressed in the silique wall, with their expression levels being increased during the development of pod. *BnaFBA3a*, *BnaFBA3b*, *BnaFBA8c*, and *BnaFBA8d* were highly and constantly expressed in all tissues, particularly showing the highest expression levels in the embryo at 25 days after pollination (DAP). In addition, the expression of *BnaFBA2a/b/c/d/e/f* increased during the development of seeds and reached high levels at 4 weeks after pollination (WAP), then decreased in the following mature stages of seed development (Figure 6A).

BnaFBA9a and *BnaFBA9b* of class-II showed constitutive expression at relatively low levels in all the tissues examined. *BnaFBA6* of class-I exhibited unexpressed in most tissues. Taken together, these results suggested that *BnaFBA* genes displayed diverse spatiotemporal expression patterns during the growth and development of different tissues in *B. napus*.

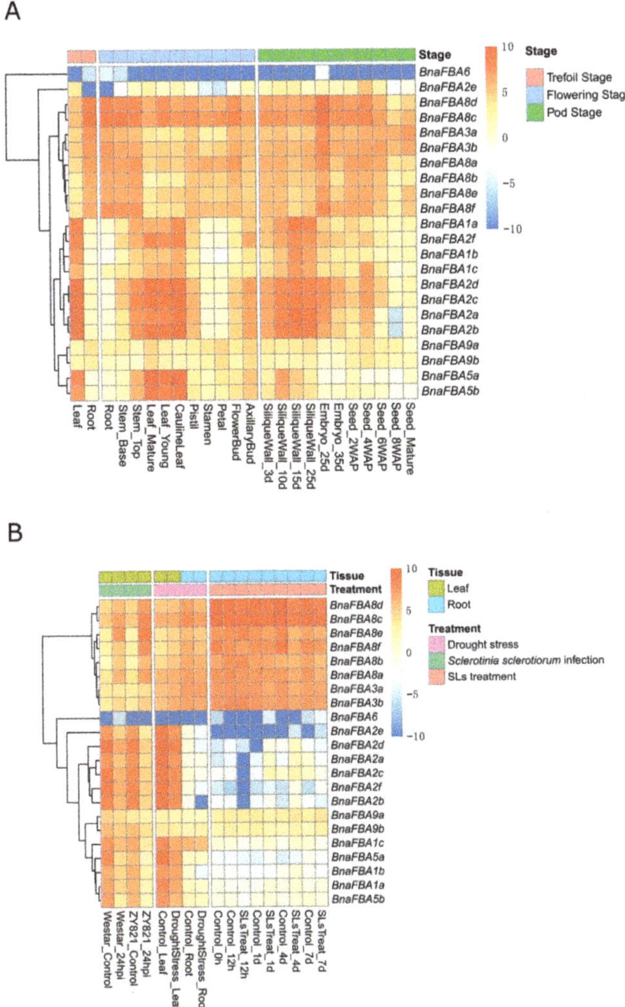

Figure 6. Expression analysis of the 22 *BnaFBA* genes in *B. napus*. (**A**) Expression patterns of the *BnaFBA* genes in various tissues at different developmental stages. The color bar represents log2 expression levels (FPKM, fragments per kilobase of exon per million fragments mapped) of each gene. (**B**) Expression patterns of the *BnaFBA* genes under drought stress, *Sclerotinia sclerotiorum* infection, and strigolactones treatments in *B. napus*. The color scale of heatmap indicates expression values, with blue-white and red representing low and high levels of transcript abundance, respectively. The cluster tree of *BnaFBA* genes, based on the expression level, is shown on the left.

To identify the physiological functions of *BnaFBAs* in response to various environmental stresses and phytohormones, as an illustration, we investigated the expression of *BnaFBA* genes in leaf and root tissues under drought stress, *Sclerotinia sclerotiorum* infection, and strigolactones (SLs) treatments. After

24 h of *Sclerotinia sclerotiorum* infection, *BnaFBA1a/b/c*, *BnaFBA2a/b/c/d/e/f*, *BnaFBA5a/b*, and *BnaFBA9b* were down-regulated, while *BnaFBA3a/b* and *BnaFBA8a/b/c/d/e/f* were up-regulated compared to the control in the leaves of the two *B. napus* cultivars—wester and ZY821 (Figure 6B). Under drought stress, the expression levels of *BnaFBA1a/c*, *BnaFBA2a/b/c/d/e/f*, and *BnaFBA5a/b* decreased in the leaves. Similarly, the expression levels of *BnaFBA1b*, *BnaFBA2b/d/e/f* also decreased in roots compared to the control (Figure 6B). We further examined the expression of *BnaFBAs* under exogenous SLs treatments in *B. napus*. Notably, the members of the *BnaFBA2* group showed diverse expression patterns after SLs treatments in *B. napus*. At 12 h after SLs treatments, the expression levels of *BnaFBA2a/b/c/f* were significantly down-regulated, while the expression of *BnaFBA2d* was slightly decreased relative to the control. At 1 day after SLs treatment, *BnaFBA2a/b/c/d* expression levels were significantly up-regulated compared to the control, whereas *BnaFBA2e/f* showed no difference. At 4 days after SLs treatment, all members of the *BnaFBA2* group exhibited up-regulated expression compared to the control. Then, at 7 days after SLs treatments, the expression levels of *BnaFBA2a/b/c/d* turned to be down-regulated, while *BnaFBA2e/f* expression levels were still up-regulated compared to the control (Figure 6B).

2.7. qRT-PCR Analysis of Selected BnaFBA Genes under Abiotic Stresses

To further validate the functional roles of *BnaFBAs* under abiotic stresses, four *BnaFBA* genes from different clusters were selected for the examination of their expression levels under three stress conditions using quantitative real-time PCR (qRT-PCR) in *B. napus*. These genes included *BnaFBA2a*, *BnaFBA5b*, and *BnaFBA8a* of class-I, and *BnaFBA9a* of class-II. The qRT-PCR analysis was carried out using rapeseed plants exposed to salt, heat, and drought stresses. At 3 days after NaCl treatment, the expression level of *BnaFBA8a* was significantly up-regulated by approximately 11-fold compared with the control, while *BnaFBA2a*, *BnaFBA5b*, and *BnaFBA9a* were significantly down-regulated, particularly, the expression levels of *BnaFBA5b* and *BnaFBA9a* were largely reduced (Figure 7). In addition, at three days after heat and drought stress treatments, the expression level of *BnaFBA5b* was significantly reduced nearly by half, whereas the expression of *BnaFBA8a* was slightly increased compared to the control (Figure 7).

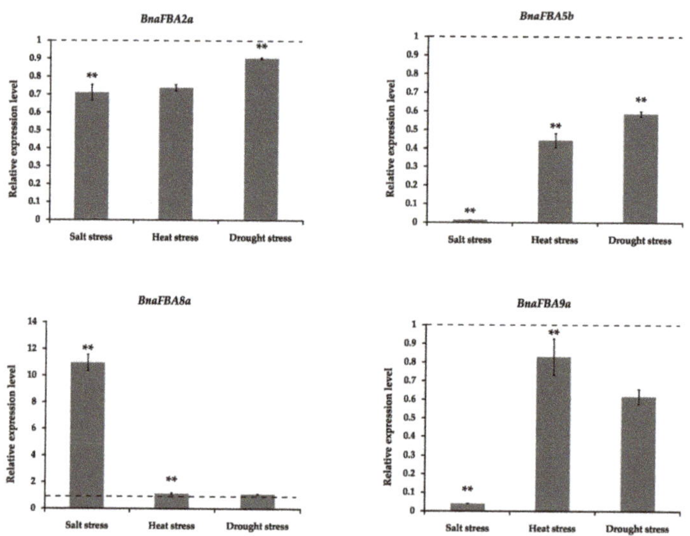

Figure 7. The real-time PCR analysis of the selected four representative *BnaFBA* genes responded to salt, heat, and drought stresses. The dotted lines represent the equivalent levels of expression. Statistically significant differences (Student's *t*-test) are indicated as followed: ** $p < 0.01$.

2.8. Co-Regulatory Networks of BnaFBA Genes

Based on the publically available RNA-seq datasets collected from different tissues, biotic and abiotic stresses, and phytohormone treatments in *B. napus*, we calculated the Pearson correlation coefficients (PCCs) of the expression levels of *BnaFBA* genes and constructed co-regulatory networks. Positive correlations were observed between members of class-Ia and class-Ib, such as *BnaFBA3a/b*, *BnaFBA8a/b/c/d/e/f*, and *BnaFBA6*. Likewise, the *BnaFBA* genes of class-Ic and class-Id also showed positive correlations between each other. Particularly, members of class-Id, including *BnaFBA1a/b/c* and *BnaFBA2a/b/c/d*, exhibited strong positive correlations. *BnaFBA2b* and *BnaFBA2f* showed significant negative correlation with *BnaFBA3a/b* and *BnaFBA8a/b/e/f*. However, they displayed a significant positive correlation with other *BnaFBA* genes of class-Ic and class-Id (Figure 8A). All the significant PCCs (*p*-value ≤ 0.05 and |PCC| > 0.5) of *BnaFBAs* were extracted and used to construct co-regulatory networks delineated by Cytoscape (version 3.1, Seattle, WA, USA) (Figure 8B). The co-regulatory networks of *BnaFBAs* were constituted with 22 nodes and 83 edges. There were four *BnaFBA* gene pairs showing negative correlations (*p*-value ≤ 0.05 and PCC < −0.5), including *BnaFBA2f* and *BnaFBA8a*, *BnaFBA2b* and *BnaFBA8a*, *BnaFBA2f* and *BnaFBA8b*, and *BnaFBA2d* and *BnaFBA8a*. Besides, 79 *BnaFBA* gene pairs showed positive correlations, of which 38 pairs exhibited strong positive correlations (*p*-value ≤ 0.05 and PCC > 0.8). Notably, three gene clusters of *BnaFBAs* were observed, including cluster 1 consisting of *BnaFBA3a/b*, *BnaFBA6*, and *BnaFBA8a/b/c/d/e/f*, cluster 2 consisting of *BnaFBA1a/b/c*, *BnaFBA2a/b/c/d/e/f*, and *BnaFBA5a/b*, and cluster 3 consisting of *BnaFBA9a/b*. Members within each cluster showed strong positive correlations between the expression levels, whereas only a few significant negative correlations existed between cluster 1 and cluster 2. Moreover, cluster 3 was independent among the three clusters (Figure 8B).

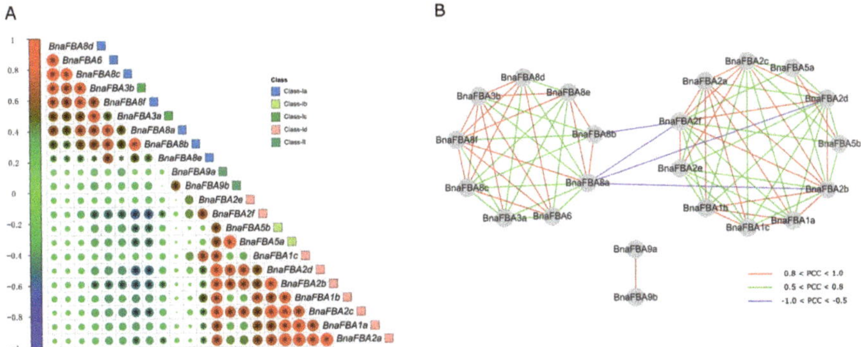

Figure 8. Correlations and co-regulatory networks of *BnaFBA* genes. (**A**) Correlation analysis of *BnaFBA* genes was performed based on the Pearson correlation coefficients (PCCs) of *BnaFBA* gene pairs. Correlations are indicated by the size and color of circles. The left bar represents the correlation values of PCCs. The class information for *BnaFBAs* is indicated by the squares with different colors at the right. Black star signifies the correlation with *p*-value ≤ 0.05. (**B**) The co-regulatory network of *BnaFBA* genes was generated on the basis of the significant PCCs of gene pairs (*p*-value ≤ 0.05). The distinct correlation levels of gene pairs are marked by edge lines with different colors shown at the bottom. The co-regulatory network was illustrated by Cytoscape (version 3.1, Seattle, WA, USA).

3. Discussion

The Calvin cycle is the initial pathway of photosynthetic carbon fixation, which plays a requisite role in the growth and development of plants [40]. Extensive efforts for seeking a breakthrough in the regulation of this cycle have been made to substantially enhance photosynthetic CO_2 capacity and plant productivity. The carboxylation capacity of ribulose bisphosphate carboxylase oxygenase (Rubisco) and the regenerative capacity of ribulose diphosphate (RuBP) are uncovered to be essential for maintaining high photosynthetic CO_2 fixation capacity. Previous studies revealed that three non-regulated enzymes, including fructose–1,6-bisphosphate aldolase (FBA or aldolase), sedoheptulose 1,7-bisphosphatase (SBPase), and transketolase (TK), had significantly higher control coefficient on photosynthesis than the other Calvin cycle enzymes, which indicated that they could limit photosynthetic rate and exert significant control over photosynthetic carbon flux other than Rubisco [41]. Particularly, FBA could catalyze the reversible conversion of DHAP and FBP and also catalyze DHAP and erythrose 4-phosphate (E4P) to sedoheptulose 1,7-bisphosphate (SBP). Thus, FBA is not only one of the key regulatory enzymes in the glycolysis pathway but also may lie in a vital strategic position to determine the carbon partitioning in the Calvin cycle, which made FBA probably an important candidate target of engineering to boost photosynthetic carbon CO_2 fixation capacity [42].

B. napus is the second-largest source of vegetable oil crops and is cultivated around the world [33,34]. *B. napus* seeds contain oil, carbohydrates, and proteins as major storage reserves. In most seeds, glycolysis in plastids supplies carbon for fatty acid (FA) synthesis [43]. FBA, as a key enzyme in the glycolytic metabolism, provides precursors for amino acid and fatty acid synthesis [26]. To our best knowledge, there is not any report in the literature describing *FBA* family genes and their function in *B. napus* so far. In this study, we performed a comprehensive identification and characterization of the *FBA* gene family in *B. napus*. Compared to the number of *FBA* genes identified in other plant species, such as eight in *A. thaliana* [17], seven in rice [22], eight in tomato [19], and 21 in wheat [11], *B. napus* had more *FBA* genes, including 22 *BnaFBA* genes distributed in fifteen *B. napus* chromosomes and two random chromosomes. The *BnaFBAs* can be classified into two classes based on the functional domains contained, class-I and class-II. Enzyme kinetics analysis of aldolase 1 (class-I FBAs) and aldolase 2 (class-II FBAs) in *Escherichia coli* revealed that aldolase 1 and 2 hydrolyze fructose 1,6-bisphosphate by the aldol cleavage reaction [12]. Class-I BnaFBA proteins could form tetramer structures with high conserved catalytic residues of D-K-K-R-E-E-K-S-R that were homologous to those of rabbit FBA isozymes [44]. Class-II BnaFBA could form dimers with the active sites of H-E-H-H that corresponded to those of FBA isozymes in the *Thermus aquaticus* [45,46]. Eleven *BnaFBAs* of class-I and *BnaFBA9a* of class-II were predicted to be localized in chloroplast, while nine class-I *BnaFBAs* and *BnaFBA9b* of class-II were in the cytoplasm. This is consistent with the subcellular localization of their homologs in *A. thaliana* [17] and wheat [11]. Although AT1G18270 in *Arabidopsis* is the homolog of BnaFBA9a and BnaFBA9b, and it is annotated as ketose-bisphosphate aldolase class-II family protein; however, it contains fructose-bisphosphate aldolase class-II (pfam01116) domain and zinc-binding site and is further assigned as fructose-bisphosphate aldolase (NCBI Reference Sequence: NP_001117303.1). More accurately, AT1G18270 is fructose-bisphosphate aldolase class-II family protein. This inaccuracy might result from the homologous protein annotation in bacteria, such as *Variovorax paradoxus B4*. Now, the ketose-bisphosphate aldolase, class-II protein, was further assigned as fructose-1,6-bisphosphate aldolase, class-II in *Variovorax paradoxus B4* (GenBank: AGU50177.1). Therefore, *BnaFBA9a* and *BnaFBA9b* were classified as class-II *FBA* genes in *Brassica napus*.

In contrast to the model plant *A. thaliana*, except for the paleo-polyploidization of alpha (α)-beta (β)-gamma (γ) WGD events, the *Brassica* species, such as *B. rapa* and *B. oleracea*, experienced an extra whole-genome triplication (WGT) event at approximately 15.9 Mya [47–49]. *B. napus* is a relatively new species of *Brassica* genus with a short history of post-Neolithic speciation (~7500 years) and domestication (~700 years) and a recent allotetraploid formed from hybridization between *B. rapa* and *B. oleracea* [35]. Therefore, *B. napus* has experienced five genome duplication events ($3 \times 2 \times 2 \times 3 \times 2$) at times during the evolution process. Gene duplication that resulted from whole-genome duplication (WGD), tandem duplication (TD), proximal duplication, transposed duplication, or dispersed duplication is one of the primary driving forces to the evolution of morphological and physiological diversity in plants [50]. Our results showed that *Brassica* species had an extended *FBA* gene family, containing extra copies of *FBA1*, *FBA2*, *FBA6*, and *FBA8*. Most gene duplications of the *FBA* gene family resulted from WGD (Table S2). Considering the collinear correlations and subgenome, transposed duplication and triplication events in the *FBA* family had contributed to this expansion prior to the speciation of *Brassica* species. WGD is often followed by intensive gene loss to adapt to continuously changing environments. Compared to *A. thaliana* and its two ancestors, *B. napus* species lost some copies of *FBA* genes (i.e., *FBA1, 3, 4, 5,* and *6*). The remaining duplicated or triplicated FBA genes identified might be conducive to the adaptation of *B. napus* to various adverse environments during speciation and domestication.

Genome resequencing provides an effective way to identify a large number of variations, which lay a foundation for further identification and functional validation of candidate genes contributed to important traits in crop plants, such as rice, tomato, and soybean [51–53]. Critical sequence polymorphisms across the gene and its flanking regions may reflect the evolutionary trends and breeding selection effects on the genes. Based on the resequencing of a worldwide collection of 991 *B. napus* germplasm accessions that released recently, we explored the pattern of genetic polymorphisms of the *BnaFBA* family genes. The results showed that the 22 *BnaFBAs* diversified in sequence polymorphism, with the polymorphism sites ranging from two to 130 (Table 3). The *BnaFBA* genes showed different levels of polymorphism. For example, *BnaFBA6, BnaFBA1c,* and *BnaFBA8d* with the highest levels of genetic variation polymorphism might have experienced weak selection pressure during the evolution process, whereas *BnaFBA2c, BnaFBA2d,* and *BnaFBA3b* were highly conserved after strong selection. Particularly, *BnaFBA6* harbored the highest levels of genetic variation, including two missense polymorphism sites with a frequency of 12.12% and 5.72%, respectively (Table S5). Moreover, a one-stop gained site with a frequency of 1.02% was also found in the CDS regions of the *BnaFBA6* gene. These results suggested that the *BnaFBA6* gene might be under the pseudogenization process. Furthermore, *BnaFBA* genes might have experienced balancing selection or population shrinkage. For example, six *FBA* family genes, including *BnaFBA8b, BnaFBA1b, BnaFBA1c, BnaFBA2e, BnaFBA1a,* and *BnaFBA2d,* harbored some selected signals and probably underwent breeding selection in the process of natural selection and domestication in *B. napus.*

Fructose-1,6-bisphosphate aldolase (FBA) is a non-regulated enzyme in the Calvin cycle, whose activity is not regulated by effectors or posttranslational modification but by expressional regulation or protein degradation [42]. Recent studies using transgenic plants with reduced enzyme levels have revealed that FBAs play important roles in regulating carbon flux through the Calvin cycle. Elevated plastid aldolase activity accelerated RuBP regeneration and resulted in increased photosynthetic capacity, growth rate, and biomass yield in tobacco and cyanobacterium [27,54]. Increased activity of FBA in *Anabaena sp.* strain PCC 7120 stimulated photosynthetic yield [55]. Generally, the expression levels of *BnaFBA* family genes were higher in the overground tissues than that in the underground. In *B. napus,* some members of the FBA family, such as *BnaFBA5a, BnaFBA5b, BnaFBA2b, BnaFBA2c,* and *BnaFBA2d,* showed higher expression levels in mature leaves and the top of stems than that in young leaves and the base of stems. Notably, *BnaFBA* genes were preferentially highly expressed in the photosynthetic tissues and stages, particularly leaves and silique wall in *B. napus.* For example, during the development of silique, *BnaFBA5a* and *BnaFBA5b* exhibited the highest expression levels at 10 days after DAP, while *BnaFBA1a, BnaFBA2a,* and *BnaFBA2b* had highest expression levels at 15 days after DAP in the silique wall. Besides, *BnaFBA1a, BnaFBA2a, BnaFBA2b, BnaFBA2c, BnaFBA2d, BnaFBA2e,* and *BnaFBA2f* showed the highest expression levels at four weeks after DAP in the seed tissues when seed fill begins with rapid embryo growth, oil biosynthesis, and protein accumulation [56]. In addition, *BnaFBA3a, BnaFBA3b, BnaFBA8c,* and *BnaFBA8d* also showed the highest expression levels at 25 days in the embryo tissues. Thus, the members of *BnaFBAs* that have a strong correlation with photosynthetic events are potential targets for engineering to improve photosynthetic capacity in the future, which would be further investigated in our next study.

Previous studies revealed *FBA* family genes also involved in plant defense and response to various biotic and abiotic stresses, such as cold and heat stress [17], salt stress [21], drought stress [21], water-deficit stress [30], stress with *Rhizoctonia solani* Kuhn [31], and high light acclimation stress [32]. Here, we found that the members of the *BnaFBA* family showed diverse expression patterns in response to *Sclerotinia sclerotiorum* infection and drought stress in *B. napus.* For example,

BnaFBA1a/b/c, *BnaFBA2a/b/c/d/e/f*, *BnaFBA5a/b*, and *BnaFBA9b* were down-regulated, while *BnaFBA3a/b* and *BnaFBA8a/b/c/d/e/f* were up-regulated compared to the control in the leaves at 24 h after *Sclerotinia sclerotiorum* infection. Under drought stress, the expression levels of 10 *BnaFBA* genes (*BnaFBA1a/c*, *BnaFBA2a/b/c/d/e/f*, and *BnaFBA5a/b*) decreased in the leaves and five *BnaFBA* genes (*BnaFBA1b*, *BnaFBA2b/d/e/f*) lowered in roots. Besides, we applied salt, heat, and drought stresses on *B. napus* seedlings to explore the expression changes of *BnaFBAs*. The expression of the *BnaFBAs* selected was induced by the three stresses. The expression levels of *BnaFBA2a*, *BnaFBA5b*, and *BnaFBA9a* were down-regulated, while *BnaFBA8a* was up-regulated compared to the control after treatments. In addition, *FBA* family genes were reported to play roles in response to phytohormones, such as ABA [17] and GA [29]. Various *cis*-acting regulatory elements related to phytohormone responsive elements were observed in the promoter regions of *BnaFBA* genes in *B. napus*, including ABRE, AuxRE, ERE, GARE, MeJARE, and SARE, indicating that the expression levels of *BnaFBA* family genes might be affected by ABA, IAA (indolylacetic acid), ethylene, GA, MeJA, and SA. Furthermore, our results showed that *BnaFBAs*, such as *BnaFBA2a/b/c/d/e/f*, could be also induced by strigolactones (SLs), a new class of plant hormones playing functional roles in the development of root.

In wheat, some FBA genes showed close correlations in expression patterns and could be classified into different clusters, such as *TaFBA1/2/3*, *TaFBA14/15/17*, and *TaFBA19/20/21* [11]. Similarly, based on the co-regulatory networks of *BnaFBAs*, the 22 *BnaFBA* family genes could be divided into three gene clusters. Strong positive correlations between the expression levels were observed among the members within each gene cluster of *BnaFBAs*. Meanwhile, only a few significant negative correlations appeared between minority members of cluster 1 and cluster 2. Cluster 3, consisting of *BnaFBA9a/b* of class-II, was independent among the three clusters. The results suggested that the *BnaFBA* genes might have functional redundancy within gene clusters and functional diversification among gene clusters in *B. napus*.

In summary, we performed a genome-wide identification of *FBA* family genes in *B. napus*, as well as *B. rapa* and *B. oleracea*, and further investigated their gene structures and phylogenetic relationships. The *cis*-acting regulatory elements in the promoter regions, natural variations, and expression patterns of *BnaFBAs* in different tissues or under treatments were analyzed. Our findings provided useful information regarding *FBA* family genes in *B. napus*. Remarkably, we found that some members of the *FBA* family that underwent breeding selection and were highly expressed in leaves and silique wall probably had positive roles in the promotion of photosynthetic capacity in *B. napus*. These *BnaFBA* genes might be utilized in the development and selection of high-yield *B. napus* cultivars.

4. Materials and Methods

4.1. Identification and Property Analysis of BnaFBA Genes

The genome sequence, annotation, and protein datasets of *B. napus* var. Darmor-*bzh*, *B. rapa* var. Z1, and *B. oleracea* var. HDEM were obtained from the GENOSCOPE database (http://www.genoscope.cns.fr/brassicanapus/ and http://www.genoscope.cns.fr/externe/plants/). The glycolytic domain (PF00274) and fructose-bisphosphate aldolase class-II domain (PF01116) from the Pfam database (http://pfam.xfam.org/) were applied as queries to search against local *B. napus* protein sequence dataset using HMMER (version 3.2.1, HHMI, Chevy Chase MD, USA) with *E*-value setting at 1e-5 [57]. Then, for further confirmation of BnaFBA proteins, the sequences of predicted BnaFBA proteins were searched against all the annotated proteins of *A. thaliana* and wheat (*Triticum aestivum* L.) using BLASTP (version 2.2.26, Bethesda, MD, USA) with *E*-value < 1×10^{-5}. The putative BnaFBAs with best hits of the *A. thaliana* and wheat FBA proteins remained and were further deployed to determine the fructose-bisphosphate aldolase domains by using the Pfam and SMART databases (http://smart.embl-heidelberg.de/). The molecular weight (MW), isoelectric point (pI), and grand average of hydropathy (GRAVY) of each BnaFBA were calculated using the ProtParam tool (http://web.expasy.org/protparam/).

4.2. Cis-Acting Regulatory Elements and Subcellular Localization Analysis

The 1.5 kb promoter sequence upstream from the transcription start site of each *BnaFBA* gene was extracted from the *B. napus* genome sequence and used to predict *cis*-acting regulatory elements by PlantCARE [58]. The subcellular localization of each BnaFBA was predicted by Plant-mPLoc (http://www.csbio.sjtu.edu.cn/bioinf/plant/) [59].

4.3. Structure and Chromosomal Localization Analysis

The gene structures of the *BnaFBA* genes were inferred by aligning their coding sequences to the *B. napus* genomic sequence. Then, a schematic map of the exon-intron structure of each *BnaFBA* gene was drawn by Gene Structure Display Server 2.0 (http://gsds.cbi.pku.edu.cn/) [60]. Three-dimensional (3D) protein models of the BnaFBAs were generated using SWISS-MODEL [61,62]. The physical chromosomal locations of *BnaFBAs* were obtained from the *B. napus* genome annotation information. The graphical representation of the *BnaFBA* genes on chromosomes was plotted using R software with RIdeogram package (https://github.com/TickingClock1992/RIdeogram) [63].

4.4. Multiple Alignments and Phylogenetic Analysis

Multiple alignments of FBA protein sequences from *B. napus*, *B. rapa*, *B. oleracea*, *A. thaliana*, and wheat were performed using MUSCLE (version 3.8, Hinxton, Cambridge, UK) with default parameters [64]. The phylogenetic tree was generated using MEGA7.1 with the neighbor-joining method, and the robustness of each node in the tree was determined using 1000 bootstrap replicates [65].

4.5. Synteny and Duplicate Gene Analysis

The genomic collinearity between all pairwise combinations of *B. napus*, *B. rapa*, *B. oleracea*, and *A. thaliana* genomes was analyzed using MCScanX (version Nov. 11, 2013, Athens, GA, USA) with the default parameters [66]. Then, the syntenic relationships of *BnaFBAs*, *BraFBAs*, *BolFBAs*, and *AtFBAs* were determined according to the genomic collinearity between pairwise genomes. The syntenic map was illustrated by CIRCOS (version 0.69-9, BCGSC, Vancouver, BC, Canada) software [67,68]. The duplicated gene pairs and the modes of gene duplication were identified among *FBA* genes using DupGen_finder (https://github.com/qiao-xin/DupGen_finder) in *B. napus*, *B. rapa*, *B. oleracea*, and *A. thaliana* [69].

4.6. Variations Analysis and Principal Component Analysis

The publically available genome resequencing datasets of 991 *B. napus* germplasm accessions were downloaded from the National Center of Biotechnology Information (NCBI) under SRP155312 [39]. The sequencing reads for each accession were mapped to *B. napus* var. Darmor-*bzh* reference genome using the MEM algorithm of Burrows–Wheeler Aligner (version 0.7.17, Hinxton, Cambridge, UK) [70]. The mapping results were sorted using SAMTOOLS (version 1.1, Hinxton, Cambridge, UK) [71]. The duplicated reads were marked with PICARD (version 2.0.1, Broad Institute, Cambridge, MA, USA) [72]. Variations, including SNPs and InDels, were called using the HaplotypeCaller module in GATK (version 4.1.3.0, https://github.com/broadinstitute/gatk/) for each accession. SNP and InDels annotation was performed using ANNOVAR (version 2018Apr16, Philadelphia, PA, USA) software based on the annotation of *B. napus* var. Darmor-*bzh* genome [73]. Principal component analysis (PCA) was performed using the R software with the ggbiplot package (https://github.com/vqv/ggbiplot).

4.7. Expression Analysis of BnaFBA Genes

The publically available RNA-seq dataset of 12 different tissues (sepal, pistil, stamen, ovule, pericarp, blossomy pistil, wilting pistil, root, flower, leaf, silique wall, and stem) collected from different stages of the growth (BioProject ID: PRJNA394926) [36], RNA-seq dataset of leaf and root tissues under drought stress (BioProject ID: PRJNA256233) [74], RNA-seq dataset of seeds across four phases of the development (BioProject ID: PRJNA311067) [56], RNA-seq dataset of stems and leaves after *Sclerotinia sclerotiorum* infection (BioProject ID: PRJNA321917) [75], and time-series RNA-seq dataset of roots under a synthetic analog of strigolactones (rac-GR24) treatments (BioProject ID: PRJNA484313) [76] were downloaded from the NCBI SRA database, and further used as main sources to perform gene expression profiling of *BnaFBA* genes in *B. napus*. The transcriptome reads were mapped to *B. napus* var. Darmor-*bzh* reference genome using HISAT2 (version 2.1.0, Baltimore, MD, USA) with the default settings [77]. The read counts per gene were generated by featureCounts [78]. Fragments per kilobase of exon per million fragments mapped (FPKM) was used for the quantification of gene expression. The clustered heatmaps were visualized with expression levels (log2) of *BnaFBA* genes by R software using the pheatmap function package (https://cran.r-project.org/web/packages/pheatmap/).

4.8. Plant Materials and Treatments

Rapeseed seeds were germinated on a filter paper saturated with distilled water in darkness at 22 °C for two days. Then, the seedling plants were transferred to a 4 L hydroponic system containing continuously aerated 1/2 Murashige and Skoog (MS) liquid solution (pH 5.8, without agar and sugar) and grown in an incubator under a photosynthetic flux of 160 µmol photons m^{-2} s^{-1} and a humidity of about 50% (16 h light at 25 °C/8 h darkness at 22 °C). The 1/2 MS liquid solution was changed once every two days. After three weeks, the seedlings were transferred to a new 1/2 MS liquid solution (pH 5.8, without agar and sugar) for different stress treatments. For salt, heat, and drought stress treatments, seedlings were exposed to 1/2 MS solution (pH 5.8, without agar and sugar) containing 250-mM NaCl, 40 °C conditions, and 20% (*w/v*) polyethylene glycol (PEG), respectively. Seedlings exposed to 1/2 MS solution at 22 °C were used as controls. Leaf samples were collected 3 days after each treatment. All collected samples were immediately frozen in liquid nitrogen and stored at −70 °C for further analysis.

4.9. RNA Isolation and Quantitative Real-Time Polymerase Chain Reaction (qRT-PCR) Analysis

Total RNAs were extracted from each sample using an RNA extraction kit (Takara, Dalian, China) following the manufacturer's procedure. Two micrograms of total RNA was used to synthesize the first-strand cDNA using the Prime Script RT reagent Kit (Takara, Dalian, China) according to the manufacturer's protocol. Real-time quantitative PCR was performed using 2 mL of cDNA in a 20 mL reaction volume using an SYBR Green PCR kit (GeneCopoeia Inc., Rockville, MD, USA) with ViiATM 7 Dx platform (ABI, Los Angeles, CA, USA). The qRT-PCR reaction condition was as follows: 95 °C for 5 min, 40 cycles at 95 °C for 30 s, 55 °C for 30 s, and 72 °C for 30 s. Gene-specific primers were designed and listed in Supplementary Materials (Table S6). The relative expression levels of these genes were analyzed by the $2^{-\Delta\Delta Ct}$ method. The *BnaTMA7* gene (*BnaC05g11560D*), which exhibited stable expression in different/same tissues under various experimental conditions, was used as an internal control [79]. All qRT-PCR reactions were assayed in triplicates.

4.10. Pearson Correlation Analysis

On the basis of the RNA-seq results, the Pearson correlation coefficients (PCCs) and *p*-value of the expression levels of *BnaFBA* gene pairs were calculated using R software with cor and cor.test function packages, respectively. The correlation heatmap was generated by R software with the corrplot function package. The gene co-regulatory networks were constructed by Cytoscape (version 3.1, Seattle, WA, USA) based on the PCCs of *BnaFBA* gene pairs with a *p*-value ≤ 0.05 [80].

Supplementary Materials: Supplementary materials can be found at http://www.mdpi.com/1422-0067/20/22/5749/s1.

Author Contributions: J.L. (Jing Liu) and W.Z. designed the research and wrote the article. W.Z. and H.L. collected data and performed most of the data analysis. Z.H. and L.Z. performed the experiments. W.H. provided the research facility. J.L. (Jun Liu) participated in the revision of the manuscript. S.X. coordinated the study. All authors read and approved the final manuscript.

Funding: This research was supported by the National Key Basic Research Program of China (2015CB150200), the National Natural Science Foundation of China (Grant No. U1304303 and 31500237), the Ministry of Science and Technology of China (2016YFD0100500), and the Agricultural Science and Technology Innovation Project of Chinese Academy of Agricultural Sciences (CAAS-ASTIP-OCRI).

Conflicts of Interest: The authors declare no conflict of interest.

Abbreviations

ABA	Abscisic acid
IAA	Indolylacetic acid
MeJA	Methyl Jasmonic acid
GA	Gibberellin
SA	Salicylic acid
SLs	Strigolactones
EDTA	Ethylene diamine tetraacetic acid
CDS	Coding sequence
GRAVY	Grand average of hydropathy
FPKM	Fragments per kilobase of exon per million fragments mapped
PCCs	Pearson correlation coefficients
SNP	Single nucleotide polymorphism
Indels	Insertions and deletions
MW	Molecular weight
pI	Isoelectric point
aa	Amino acid
bp	Base pair
kb	Kilobase
Mb	Megabase
kDa	Kilodalton

References

1. Rutter, W.J. Evolution of aldolase. *Fed Proc.* **1964**, *23*, 1248–1257. [PubMed]
2. Berg, I.A.; Kockelkorn, D.; Ramos-Vera, W.H.; Say, R.F.; Zarzycki, J.; Hügler, M.; Alber, B.E.; Fuchs, G. Autotrophic carbon fixation in archaea. *Nat. Rev. Microbiol.* **2010**, *8*, 447. [CrossRef] [PubMed]
3. Vigeolas, H.; Waldeck, P.; Zank, T.; Geigenberger, P. Increasing seed oil content in oil-seed rape (*Brassica napus* L.) by over-expression of a yeast glycerol-3-phosphate dehydrogenase under the control of a seed-specific promoter. *Plant. Biotechnol. J.* **2007**, *5*, 431–441. [CrossRef] [PubMed]
4. Cao, H.; Shockey, J.M.; Klasson, K.T.; Chapital, D.C.; Mason, C.B.; Scheffler, B.E. Developmental regulation of diacylglycerol acyltransferase family gene expression in tung tree tissues. *PloS ONE* **2013**, *8*, e76946. [CrossRef]
5. Nakahara, K.; Yamamoto, H.; Miyake, C.; Yokota, A. Purification and characterization of class-I and class-II fructose-1, 6-bisphosphate aldolases from the cyanobacterium *Synechocystis* sp. PCC 6803. *Plant Cell physiol.* **2003**, *44*, 326–333. [CrossRef]
6. Marsh, J.J.; Lebherz, H.G. Fructose-bisphosphate aldolases: An evolutionary history. *Trends Biochem. Sci.* **1992**, *17*, 110–113. [CrossRef]
7. Gross, W.; Lenze, D.; Nowitzki, U.; Weiske, J.; Schnarrenberger, C. Characterization, cloning, and evolutionary history of the chloroplast and cytosolic class I aldolases of the red alga *Galdieria sulphuraria*. *Gene* **1999**, *230*, 7–14. [CrossRef]
8. Tolan, D.R.; Niclas, J.; Bruce, B.D.; Lebo, R.V. Evolutionary implications of the human aldolase-A,-B,-C, and-pseudogene chromosome locations. *Am. J. Hum. Genet.* **1987**, *41*, 907.
9. Penhoet, E.; Kochman, M.; Valentine, R.; Rutter, W.J. The subunit structure of mammalian fructose diphosphate aldolase. *Biochemistry* **1967**, *6*, 2940–2949. [CrossRef]
10. Henze, K.; Morrison, H.G.; Sogin, M.L.; Müller, M. Sequence and phylogenetic position of a class II aldolase gene in the amitochondriate protist, *Giardia lamblia*. *Gene* **1998**, *222*, 163–168. [CrossRef]

11. Lv, G.; Guo, X.; Xie, L.; Xie, C.; Zhang, X.; Yang, Y.; Xiao, L.; Tang, Y.; Pan, X.; Guo, A.; et al. Molecular characterization, gene evolution, and expression analysis of the fructose-1, 6-bisphosphate aldolase (FBA) gene family in wheat (*Triticum aestivum* L.). *Front. Plant. Sci.* **2017**, *8*, 1030. [CrossRef] [PubMed]

12. Stribling, D.; Perham, R.N. Purification and characterization of two fructose diphosphate aldolases from *Escherichia coli* (Crookes' strain). *Biochem. J.* **1973**, *131*, 833–841. [CrossRef] [PubMed]

13. Guerrini, A.M.; Cremona, T.; Preddie, E.C. The aldolases of *Chlamydomonas reinhardii. Arch. Biochem. Biophys.* **1971**, *146*, 249–255. [CrossRef]

14. Lebherz, H.G.; Rutter, W.J. Distribution of fructose diphosphate aldolase variants in biological systems. *Biochemistry* **1969**, *8*, 109–121. [CrossRef] [PubMed]

15. Bukowiecki, A.C.; Anderson, L.E. Multiple forms of aldolase and triose phosphate isomerase in diverse plant species. *Plant Sci. Lett.* **1974**, *3*, 381–386. [CrossRef]

16. Plaxton, W.C. The organization and regulation of plant glycolysis. *Annu Rev. Plant. Physiol. Plant. Mol. Biol.* **1996**, *47*, 185–214. [CrossRef]

17. Lu, W.; Tang, X.; Huo, Y.; Xu, R.; Qi, S.; Huang, J.; Zheng, C.; Wu, C.A. Identification and characterization of fructose 1, 6-bisphosphate aldolase genes in *Arabidopsis* reveal a gene family with diverse responses to abiotic stresses. *Gene* **2012**, *503*, 65–74. [CrossRef]

18. Zhang, G.; Liu, Y.; Ni, Y.; Meng, Z.; Lu, T.; Li, T. Exogenous calcium alleviates low night temperature stress on the photosynthetic apparatus of tomato leaves. *PLoS ONE* **2014**, *9*, e97322. [CrossRef]

19. Cai, B.; Li, Q.; Xu, Y.; Yang, L.; Bi, H.; Ai, X. Genome-wide analysis of the fructose 1, 6-bisphosphate aldolase (FBA) gene family and functional characterization of *FBA7* in tomato. *Plant Physiol. Biochem.* **2016**, *108*, 251–265. [CrossRef]

20. Yamada, S.; Komori, T.; Hashimoto, A.; Kuwata, S.; Imaseki, H.; Kubo, T. Differential expression of plastidic aldolase genes in *Nicotiana* plants under salt stress. *Plant Sci.* **2000**, *154*, 61–69. [CrossRef]

21. Fan, W.; Zhang, Z.; Zhang, Y. Cloning and molecular characterization of fructose-1, 6-bisphosphate aldolase gene regulated by high-salinity and drought in *Sesuvium portulacastrum. Plant Cell Rep.* **2009**, *28*, 975–984. [CrossRef] [PubMed]

22. Zhang, Y. Functional Analysis of a Fructose-1, 6-diphosphatase Aldolase Gene ALD Y in Rice. Master's Degree Dissertation, Huazhong Agricultural University, Wuhan, China, June 2014.

23. Valenti, V.; Pupillo, P.; Scagliarini, S. Compartmentation of aldolase isoforms in maize leaves. *J. Exp. Bot.* **1987**, *38*, 1228–1237. [CrossRef]

24. Zeng, Y.; Tan, X.; Zhang, L.; Jiang, N.; Cao, H. Identification and expression of fructose-1, 6-bisphosphate aldolase genes and their relations to oil content in developing seeds of tea oil tree (*Camellia oleifera*). *PLoS ONE* **2014**, *9*, e107422. [CrossRef] [PubMed]

25. Sonnewald, U.; Lerchl, J.; Zrenner, R.; Frommer, W. Manipulation of sink-source relations in transgenic plants. *Plant Cell Environ.* **1994**, *17*, 649–658. [CrossRef]

26. Haake, V.; Zrenner, R.; Sonnewald, U.; Stitt, M. A moderate decrease of plastid aldolase activity inhibits photosynthesis, alters the levels of sugars and starch, and inhibits growth of potato plants. *Plant. J.* **1998**, *14*, 147–157. [CrossRef]

27. Uematsu, K.; Suzuki, N.; Iwamae, T.; Inui, M.; Yukawa, H. Increased fructose 1, 6-bisphosphate aldolase in plastids enhances growth and photosynthesis of tobacco plants. *J. Exp. Bot.* **2012**, *63*, 3001–3009. [CrossRef]

28. Obiadalla-Ali, H.; Fernie, A.R.; Lytovchenko, A.; Kossmann, J.; Lloyd, J.R. Inhibition of chloroplastic fructose 1, 6-bisphosphatase in tomato fruits leads to decreased fruit size, but only small changes in carbohydrate metabolism. *Planta* **2004**, *219*, 533–540. [CrossRef]

29. Konishi, H.; Yamane, H.; Maeshima, M.; Komatsu, S. Characterization of fructose-bisphosphate aldolase regulated by gibberellin in roots of rice seedling. *Plant. Mol. Biol.* **2004**, *56*, 839–848. [CrossRef]

30. Khanna, S.M.; Taxak, P.C.; Jain, P.K.; Saini, R.; Srinivasan, R. Glycolytic enzyme activities and gene expression in *Cicer arietinum* exposed to water-deficit stress. *Appl. Biochem. Biotechnol.* **2014**, *173*, 2241–2253. [CrossRef]

31. Mutuku, J.M.; Nose, A. Changes in the contents of metabolites and enzyme activities in rice plants responding to *Rhizoctonia solani* Kuhn infection: Activation of glycolysis and connection to phenylpropanoid pathway. *Plant Cell Physiol.* **2012**, *53*, 1017–1032. [CrossRef]

32. Oelze, M.L.; Muthuramalingam, M.; Vogel, M.O.; Dietz, K.J. The link between transcript regulation and de novo protein synthesis in the retrograde high light acclimation response of *Arabidopsis thaliana. BMC Genom.* **2014**, *15*, 320. [CrossRef] [PubMed]

33. Rana, D.; van den Boogaart, T.; O'Neill, C.M.; Hynes, L.; Bent, E.; Macpherson, L.; Park, J.Y.; Lim, Y.P.; Bancroft, I. Conservation of the microstructure of genome segments in *Brassica napus* and its diploid relatives. *Plant J.* **2004**, *40*, 725–733. [CrossRef] [PubMed]

34. Lu, C.; Napier, J.A.; Clemente, T.E.; Cahoon, E.B. New frontiers in oilseed biotechnology: Meeting the global demand for vegetable oils for food, feed, biofuel, and industrial applications. *Curr. Opin. Biotechnol.* **2011**, *22*, 252–259. [CrossRef] [PubMed]

35. Chalhoub, B.; Denoeud, F.; Liu, S.; Parkin, I.A.; Tang, H.; Wang, X.; Chiquet, J.; Belcram, H.; Tong, C.; Samans, B.; et al. Early allopolyploid evolution in the post-Neolithic *Brassica napus* oilseed genome. *Science* **2014**, *345*, 950–953. [CrossRef] [PubMed]

36. Sun, F.; Fan, G.; Hu, Q.; Zhou, Y.; Guan, M.; Tong, C.; Li, J.; Du, D.; Qi, C.; Jiang, L.; et al. The high-quality genome of *Brassica napus* cultivar 'ZS11' reveals the introgression history in semi-winter morphotype. *Plant J.* **2017**, *92*, 452–468. [CrossRef] [PubMed]

37. Bowers, J.E.; Chapman, B.A.; Rong, J.; Paterson, A.H. Unravelling angiosperm genome evolution by phylogenetic analysis of chromosomal duplication events. *Nature* **2003**, *422*, 433. [CrossRef]

38. Cheng, F.; Wu, J.; Wang, X. Genome triplication drove the diversification of *Brassica* plants. *Hortic Res.* **2014**, *1*, 14024. [CrossRef]

39. Wu, D.; Liang, Z.; Yan, T.; Xu, Y.; Xuan, L.; Tang, J.; Zhou, G.; Lohwasser, U.; Hua, S.; Wang, H.; et al. Whole-Genome Resequencing of a Worldwide Collection of Rapeseed Accessions Reveals the Genetic Basis of Ecotype Divergence. *Mol. Plant* **2019**, *12*, 30–43. [CrossRef]

40. Furbank, R.T.; Taylor, W.C. Regulation of photosynthesis in C3 and C4 plants: A molecular approach. *Plant Cell* **1995**, *7*, 797. [CrossRef]

41. Raines, C.A. The Calvin cycle revisited. *Photosynth. Res.* **2003**, *75*, 1–10. [CrossRef]

42. Graciet, E.; Lebreton, S.; Gontero, B. Emergence of new regulatory mechanisms in the Benson-Calvin pathway via protein-protein interactions: A glyceraldehyde-3-phosphate dehydrogenase/CP12/phosphoribulokinase complex. *J. Exp. Bot.* **2004**, *55*, 1245–1254. [CrossRef] [PubMed]

43. Hay, J.; Schwender, J. Computational analysis of storage synthesis in developing *Brassica napus* L. (oilseed rape) embryos: Flux variability analysis in relation to ^{13}C metabolic flux analysis. *Plant J.* **2011**, *67*, 513–525. [CrossRef] [PubMed]

44. Blom, N.; Sygusch, J. Product binding and role of the C-terminal region in Class I D-fructose 1, 6-bisphosphate aldolase. *Nat. Struct. Biol.* **1997**, *4*, 36–39. [CrossRef] [PubMed]

45. Sauve', V.; Sygusch, J. Crystallization and preliminary X-ray analysis of native and selenomethionine fructose-1,6-bisphosphate aldolase from *Thermus aquaticus*. *Acta Crystallogr.* **2001**, *57*, 310–313.

46. Izard, T.; Sygusch, J. Induced fit movements and metal cofactor selectivity of class II aldolases: Structure of *Thermus aquaticus* fructose-1,6-bisphosphate aldolase. *J. Biol. Chem.* **2004**, *279*, 11825–11833. [CrossRef] [PubMed]

47. Lysak, M.A.; Koch, M.A.; Pecinka, A.; Schubert, I. Chromosome triplication found across the tribe *Brassiceae*. *Genome Res.* **2005**, *15*, 516–525. [CrossRef] [PubMed]

48. Town, C.D.; Cheung, F.; Maiti, R.; Crabtree, J.; Haas, B.J.; Wortman, J.R.; Hine, E.E.; Althoff, R.; Arbogast, T.S.; Tallon, L.J.; et al. Comparative genomics of *Brassica oleracea* and *Arabidopsis thaliana* reveal gene loss, fragmentation, and dispersal after polyploidy. *Plant Cell* **2006**, *18*, 1348–1359. [CrossRef]

49. Liu, S.; Liu, Y.; Yang, X.; Tong, C.; Edwards, D.; Parkin, I.A.; Zhao, M.; Ma, J.; Yu, J.; Huang, S.; et al. The *Brassica oleracea* genome reveals the asymmetrical evolution of polyploid genomes. *Nat. Commun.* **2014**, *5*, 3930. [CrossRef]

50. Paterson, A.H.; Freeling, M.; Tang, H.; Wang, X. Insights from the comparison of plant genome sequences. *Annu. Rev. Plant Biol.* **2010**, *61*, 349–372. [CrossRef]

51. Xu, X.; Liu, X.; Ge, S.; Jensen, J.D.; Hu, F.; Li, X.; Dong, Y.; Gutenkunst, R.N.; Fang, L.; Li, J.; et al. Resequencing 50 accessions of cultivated and wild rice yields markers for identifying agronomically important genes. *Nat. Biotechnol.* **2011**, *30*, 105. [CrossRef]

52. Lin, T.; Zhu, G.; Zhang, J.; Xu, X.; Yu, Q.; Zheng, Z.; Zhang, Z.; Lun, Y.; Li, S.; Wang, X.; et al. Genomic analyses provide insights into the history of tomato breeding. *Nat. Genet.* **2014**, *46*, 1220. [CrossRef] [PubMed]

53. Zhou, Z.; Jiang, Y.; Wang, Z.; Gou, Z.; Lyu, J.; Li, W.; Yu, Y.; Shu, L.; Zhao, Y.; Ma, Y.; et al. Resequencing 302 wild and cultivated accessions identifies genes related to domestication and improvement in soybean. *Nat. Biotechnol.* **2015**, *33*, 408. [CrossRef] [PubMed]

54. Miyagawa, Y.; Tamoi, M.; Shigeoka, S. Overexpression of a cyanobacterial fructose-1,6-/sedoheptulose-1,7-bisphosphatase in tobacco enhances photosynthesis and growth. *Nat. Biotechnol.* **2001**, *19*, 965–969. [CrossRef] [PubMed]

55. Ma, W.; Wei, L.; Long, Z.; Chen, L.; Wang, Q. Increased activity of only an individual non-regulated enzyme fructose-1, 6-bisphosphate aldolase in *Anabaena* sp. strain PCC 7120 stimulates photosynthetic yield. *Acta Physiol. Plant* **2008**, *30*, 897. [CrossRef]

56. Wan, H.; Cui, Y.; Ding, Y.; Mei, J.; Dong, H.; Zhang, W.; Wu, S.; Liang, Y.; Zhang, C.; Li, J.; et al. Time-Series Analyses of Transcriptomes and Proteomes Reveal Molecular Networks Underlying Oil Accumulation in Canola. *Front. Plant Sci.* **2017**, *7*, 2007. [CrossRef]

57. Finn, R.D.; Clements, J.; Eddy, S.R. HMMER web server: Interactive sequence similarity searching. *Nucleic Acids Res.* **2011**, *39*, W29–W37. [CrossRef]

58. Rombauts, S.; Déhais, P.; Van Montagu, M.; Rouzé, P. PlantCARE, a plant *cis*-acting regulatory element database. *Nucleic Acids Res.* **1999**, *27*, 295–296. [CrossRef]

59. Chou, K.C.; Shen, H.B. Cell-PLoc: A package of Web servers for predicting subcellular localization of proteins in various organisms. *Nat. Protoc.* **2008**, *3*, 153. [CrossRef]

60. Kayum, M.A.; Park, J.I.; Nath, U.K.; Saha, G.; Biswas, M.K.; Kim, H.T.; Nou, I.S. Genome-wide characterization and expression profiling of *PDI* family gene reveals function as abiotic and biotic stress tolerance in Chinese cabbage (*Brassica rapa* ssp. *pekinensis*). *BMC Genom.* **2017**, *18*, 885. [CrossRef]

61. Biasini, M.; Bienert, S.; Waterhouse, A.; Arnold, K.; Studer, G.; Schmidt, T.; Kiefer, F.; Gallo Cassarino, T.; Bertoni, M.; Bordoli, L.; et al. SWISS-MODEL: Modelling protein tertiary and quaternary structure using evolutionary information. *Nucleic Acids Res.* **2014**, *42*, W252–W258. [CrossRef]

62. Arnold, K.; Bordoli, L.; Kopp, J.; Schwede, T. The SWISS-MODEL Workspace: A web-based environment for protein structure homology modeling. *Bioinformatics* **2006**, *22*, 195–201. [CrossRef]

63. Hao, Z.; Lv, D.; Ge, Y.; Shi, J.; Weijers, D.; Yu, G.; Chen, J. RIdeogram: Drawing SVG graphics to visualize and map genome-wide data on the idiograms. *PeerJ Prepr.* **2019**, *7*, e27928v1.

64. Edgar, R.C. MUSCLE: Multiple sequence alignment with high accuracy and high throughput. *Nucleic Acids Res.* **2004**, *32*, 1792–1797. [CrossRef]

65. Tamura, K.; Peterson, D.; Peterson, N.; Stecher, G.; Nei, M.; Kumar, S. MEGA5: Molecular evolutionary genetics analysis using maximum likelihood, evolutionary distance, and maximum parsimony methods. *Mol. Biol. Evol.* **2011**, *28*, 2731–2739. [CrossRef]

66. Wang, Y.; Tang, H.; DeBarry, J.D.; Tan, X.; Li, J.; Wang, X.; Lee, T.; Jin, H.; Marler, B.; Guo, H.; et al. MCScanX: A toolkit for detection and evolutionary analysis of gene synteny and collinearity. *Nucleic Acids Res.* **2012**, *40*, e49. [CrossRef]

67. Tang, H.; Wang, X.; Bowers, J.E.; Ming, R.; Alam, M.; Paterson, A.H. Unraveling ancient hexaploidy through multiply-aligned angiosperm gene maps. *Genome Res.* **2008**, *18*, 1944–1954. [CrossRef]

68. Krzywinski, M.; Schein, J.; Birol, I.; Connors, J.; Gascoyne, R.; Horsman, D.; Jones, S.J.; Marra, M.A. Circos: An information aesthetic for comparative genomics. *Genome Res.* **2009**, *19*, 1639–1645. [CrossRef]

69. Qiao, X.; Li, Q.; Yin, H.; Qi, K.; Li, L.; Wang, R.; Zhang, S.; Paterson, A.H. Gene duplication and evolution in recurring polyploidization-diploidization cycles in plants. *Genome Bio.* **2019**, *20*, 38. [CrossRef]

70. Li, H.; Durbin, R. Fast and accurate short read alignment with Burrows-Wheeler transform. *Bioinformatics* **2009**, *25*, 1754–1760. [CrossRef]

71. Li, H.; Handsaker, B.; Wysoker, A.; Fennell, T.; Ruan, J.; Homer, N.; Marth, G.; Abecasis, G.; Durbin, R. Genome Project Data Processing, Subgroup. The sequence alignment/map format and SAMtools. *Bioinformatics* **2009**, *25*, 2078–2079. [CrossRef]

72. Picard Toolkit. Broad Institute, GitHub Repository. Available online: http://broadinstitute.github.io/picard/ (accessed on 5 August 2019).

73. Wang, K.; Li, M.; Hakonarson, H. ANNOVAR: Functional annotation of genetic variants from next-generation sequencing data. *Nucleic Acids Res.* **2010**, *38*, e164. [CrossRef] [PubMed]

74. Liu, C.Q.; Zhang, X.K.; Zhang, K.; An, H.; Hu, K.N.; Wen, J.; Shen, J.S.; Ma, C.Z.; Yi, B.; Tu, J.X.; et al. Comparative Analysis of the *Brassica napus* Root and Leaf Transcript Profiling in Response to Drought Stress. *Int. J. Mol. Sci.* **2015**, *16*, 18752–18777. [CrossRef] [PubMed]

75. Girard, I.J.; Tong, C.; Becker, M.G.; Mao, X.; Huang, J.; de Kievit, T.; Fernando, W.G.D.; Liu, S.; Belmonte, M.F. RNA sequencing of *Brassica napus* reveals cellular redox control of *Sclerotinia* infection. *J. Exp. Bot.* **2017**, *68*, 5079–5091. [CrossRef] [PubMed]

76. Ma, N.; Hu, C.; Wan, L.; Hu, Q.; Xiong, J.L.; Zhang, C.L. Strigolactones Improve Plant Growth, Photosynthesis, and Alleviate Oxidative Stress under Salinity in Rapeseed (*Brassica napus* L.) by Regulating Gene Expression. *Front. Plant Sci.* **2017**, *8*, 1671. [CrossRef] [PubMed]

77. Kim, D.; Langmead, B.; Salzberg, S.L. HISAT: A fast spliced aligner with low memory requirements. *Nat. Methods* **2015**, *12*, 357–360. [CrossRef] [PubMed]

78. Liao, Y.; Smyth, G.K.; Shi, W. featureCounts: An efficient general purpose program for assigning sequence reads to genomic features. *Bioinformatics* **2014**, *30*, 923–930. [CrossRef]

79. Zhu, X.Y.; Huang, C.Q.; Zhang, L.; Liu, H.F.; Yu, J.H.; Hu, Z.Y.; Hua, W. Systematic Analysis of *Hsf* Family Genes in the *Brassica napus* Genome Reveals Novel Responses to Heat, Drought and High CO_2 Stresses. *Front. Plant Sci.* **2017**, *8*, 1174. [CrossRef]

80. Shannon, P.; Markiel, A.; Ozier, O.; Baliga, N.S.; Wang, J.T.; Ramage, D.; Amin, N.; Schwikowski, B.; Ideker, T. Cytoscape: A software environment for integrated models of biomolecular interaction networks. *Genome Res.* **2003**, *13*, 2498–2504. [CrossRef]

 International Journal of
Molecular Sciences

Review

Stay-Green Trait: A Prospective Approach for Yield Potential, and Drought and Heat Stress Adaptation in Globally Important Cereals

Nasrein Mohamed Kamal [1,2,*], Yasir Serag Alnor Gorafi [1,2], Mostafa Abdelrahman [1,3], Eltayb Abdellatef [4] and Hisashi Tsujimoto [1,*]

[1] Arid Land Research Center, Tottori University, 1390 Hamasaka, Tottori 680-0001, Japan; yasirserag@tottori-u.ac.jp (Y.S.A.G.); meettoo2000@tottori-u.ac.jp (M.A.)

[2] Agricultural Research Corporation, Wad-Medani P.O. Box 126, Sudan

[3] Botany Department, Faculty of Science, Aswan University, Aswan 81528, Egypt

[4] Commission for Biotechnology and Genetic Engineering, National Center for Research, Khartoum P.O. Box 6096, Sudan; eltaybfarah@gmail.com

* Correspondence: renokamal@gmail.com (N.M.K.); tsujim@tottori-u.ac.jp (H.T.)

Received: 27 September 2019; Accepted: 13 November 2019; Published: 20 November 2019

Abstract: The yield losses in cereal crops because of abiotic stress and the expected huge losses from climate change indicate our urgent need for useful traits to achieve food security. The stay-green (SG) is a secondary trait that enables crop plants to maintain their green leaves and photosynthesis capacity for a longer time after anthesis, especially under drought and heat stress conditions. Thus, SG plants have longer grain-filling period and subsequently higher yield than non-SG. SG trait was recognized as a superior characteristic for commercially bred cereal selection to overcome the current yield stagnation in alliance with yield adaptability and stability. Breeding for functional SG has contributed in improving crop yields, particularly when it is combined with other useful traits. Thus, elucidating the molecular and physiological mechanisms associated with SG trait is maybe the key to defeating the stagnation in productivity associated with adaptation to environmental stress. This review discusses the recent advances in SG as a crucial trait for genetic improvement of the five major cereal crops, sorghum, wheat, rice, maize, and barley with particular emphasis on the physiological consequences of SG trait. Finally, we provided perspectives on future directions for SG research that addresses present and future global challenges.

Keywords: stay-green; drought stress; heat stress; quantitative trait loci; yield

---

1. Introduction

Sorghum (*Sorghum bicolor* L. Moench), wheat (*Triticum aestivum*), rice (*Oryza sativa*), maize (*Zea mays*), and barley (*Hordeum vulgare* L.) are considered as major staple foods for a large portion of the world population [1] (Figure 1). However, global food security is being haunted by the rapid increase in the world population and drastic changes in the climate [2–4]. For instance, heat and drought are the two most important environmental stresses imposing huge impact on crop growth, development, grain yield, and biomass productivity [4–6] (Figure 2). With the increasing expectations of crop yield losses because of the global climate change and the exponential population growth, there is an urgent need to accelerate plant breeding and mining of novel traits for increased yield potential and better adaptation to abiotic stresses to secure the food availability and meet the future demand for agricultural production [7]. In this context, stay-green (SG) genotype selection can be a principle strategy for increasing crop production to meet the mandate of an expected increase in population, particularly under heat and water-limited conditions.

SG genotypes constitute a potential germplasm source for the genetic improvement of important crops to mitigate heat and drought stresses. SG genotype is characterized by delayed senescence because of chlorophyll (Chl) loss compared with a non-SG standard genotype. Therefore, SG is considered as an important agronomic trait that allows plants to maintain their leaves photosynthetically active and subsequently improved the grain-filling process even under stress conditions [8–14]. SG has two types, functional and non-functional. The functional SG genotypes are agronomically important as they are able to maintain their photosynthetic capacity compared with the non-SG genotypes. The functional SG genotypes delay the onset of senescence (type-A) or initiate the senescence on schedule, but proceeds more slowly (type-B) [15]. In the non-functional/cosmetic SG genotypes the senescence is initiated on a normal time-scale, however, leaf greenness is maintained because of the failure of the Chl degradation pathway with decline in photosynthetic capacity (type-C), or leaf pigment remain because of freezing or drying such as frozen spinach or herbarium specimen (type-D), or the intensely green genotype may have normal ontogenetic photosynthetic capacity but their absolute pigment contents classified it as a SG type-E [15–19]. SG has been used initially as a phenotype descriptor by legume breeders in *Vicia faba* [20], and later on, it was established as a superior characteristic and marketing feature for many commercial grain crops [11].

In addition to the beneficial roles of SG trait in yield improvement and tolerance against drought and heat stresses [11–13,21–25], desirable morphological traits associated with SG trait, including a greater number of grains per ear [22], enhanced resistance to stem lodging [26], and greater tolerance to biotic stress such as spot blotch infection [27,28] have been reported. An increase in leaf area, rate and duration of grain filling and photosynthetic competence, water use efficiency, leaf anatomy, have been found to be a characteristic for the SG trait [24,29] (Figure 3). It has been reported that breeding with a SG phenotype can improve yield under post-flowering drought stress (terminal drought stress), without yield penalties in environments not affected by drought [24,30,31].

Quantitative trait loci (QTL) mapping for SG has been performed for the major five cereal crops, wheat [32,33], maize [34,35], rice [36,37], sorghum [38,39], and barley [22]. In maize, the utilization of SG trait in breeding programs results in significant genetic progress for high grain yield and tolerance to abiotic stress (Figure 4). Understanding the physiology underlying the SG trait will facilitate the identification of functional markers and/or genes for adaptation to limited water and heat stress environments. Therefore, there is a need to increase knowledge of the SG potentiality to increase grain yield under drought and heat stresses in cereals and to explore this trait extensively in breeding programs to harness more advantages of this trait. In this review, we summarize and discuss the recent progress in the application of SG trait as a breeding target under high temperature and water-limited conditions in sorghum, wheat, rice, maize, and barley. This review has aimed to shed light on the main aspects of the SG applications for plant breeding in cereals, with assertion on the physiological consequences of staying green, and its potential use to improve yield under drought and heat stress environments.

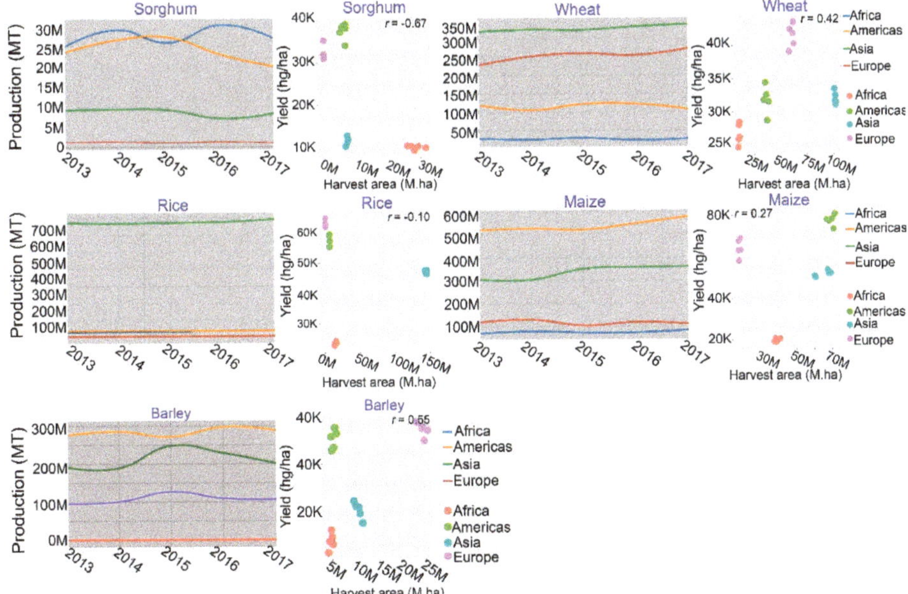

Figure 1. Total production (million tones, MT) and yield (hectogram/hectare, hg/ha) of major cereal crops, including sorghum, wheat, rice, maize and barley. Total production and scatter plot of the relationships between yield (kg/ha) and harvest area (ha) in the five major cereal crops. Correlation coefficient calculated by the Pearson method. The data was obtained from FAOSTAT (http://www.fao. org/faostat/en/#data/QC) database, accessed September 2019 [1].

2. Stay-Green in Sorghum

2.1. Stay-Green QTLs in Sorghum

Drought stress is often a limiting factor for sorghum production and can lead to complete crop failure [40]. In the long history of sorghum breeding for drought adaptation, SG is the best-characterized trait contributing to drought adaptation in sorghum [12,13,21,38,41–43]. Different genotype sources for SG trait have been identified in sorghum, including "B35," "SC56," and "E36-1" with "B35" genotype being the most popular [13,44–46].

QTLs for SG have been identified in the three source lines using several bi-parental populations. Four QTLs for the SG trait has been identified, following analysis on a recombinant inbred line (RIL) population produced from the cross between "B35" (SG line) and "Tx7000" (senescent, post-flowering drought-sensitive) [44,47]. Among these QTLs, *Stg*1 and *Stg*2 QTLs have been mapped to chromosome 3, whereas *Stg*3 and *Stg*4 QTLs have been located on chromosomes 2 and 5, respectively (Figure 5a) [44,45,47].

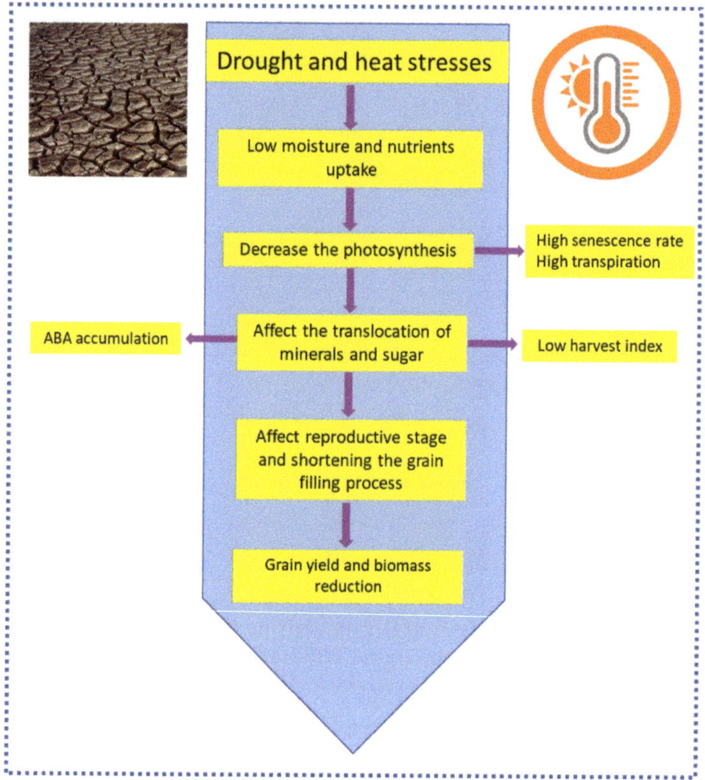

Figure 2. The plant features under heat/drought stress relevant to crop biomass and yield [4–6].

The four SG QTLs combined explained 53.5% of the phenotypic variation within the "B35" × "Tx7000" RIL population [45]. Following these efforts, several QTLs contributing to SG phenotype expression under drought have been validated across different research groups (Figure 5a) [21,38,39,44–50]. However, the four QTLs identified in "B35" are the most stable and significant, and are currently being introgressed in several genetic backgrounds through marker-assisted breeding (MAS) [21,51,52]. Several studies reported a close association between the SG phenotype and plant response to stress, as illustrated in the co-localization of QTLs for SG with QTLs for temperatures and drought stress tolerance [15,44,53,54]. Another example of QTL co-localization was found in studies on RIL populations derived from an original cross between lines with different nodal root angles (narrow vs. wide-angle) [55]. These nodal root angle QTLs have been found to overlap with SG QTLs. Co-localization between SG and root suggested that modified root architecture is likely to be a contributor to the SG trait observed in this population, which will enhance the water extraction capabilities, especially under stress conditions [11,53]. Therefore, selection for SG QTLs can simultaneously lead to the inheritance of stress tolerance features [15,53].

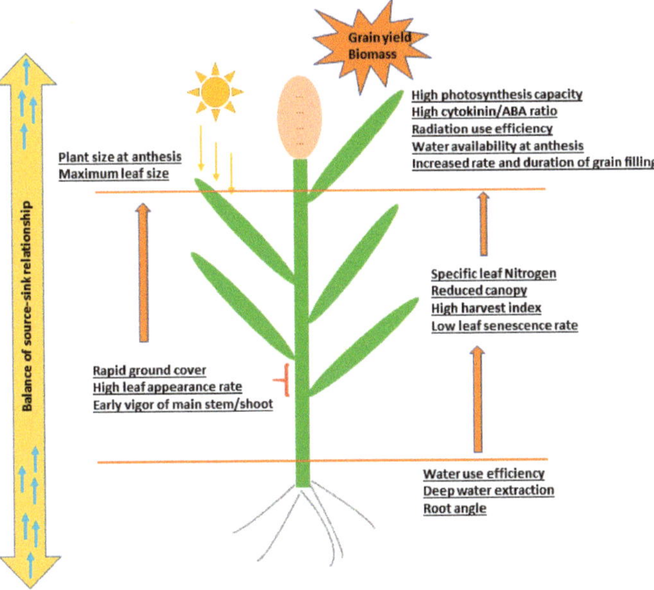

Figure 3. The physiological features of the stay-green plants including photosynthesis, transport of photosynthates and source-sink relationship. These physiological parameters operate to determine grain yield and biomass [11–13,24,29,40,56].

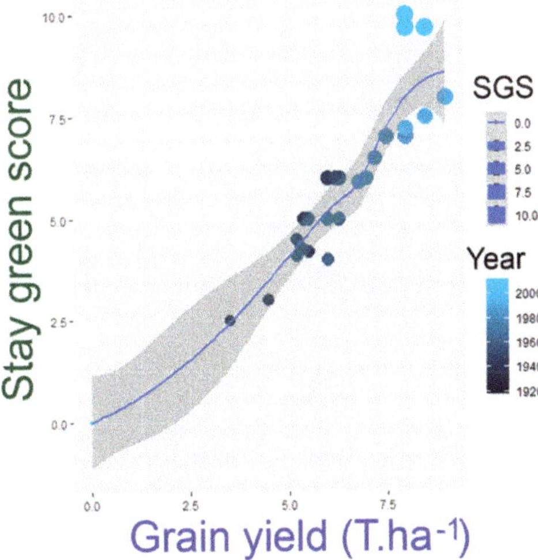

Figure 4. The relationship between the increases in yields and stay-green scores of maize varieties produced since 1930 according to [15], illustrating the contribution of the SG in increasing the yield.

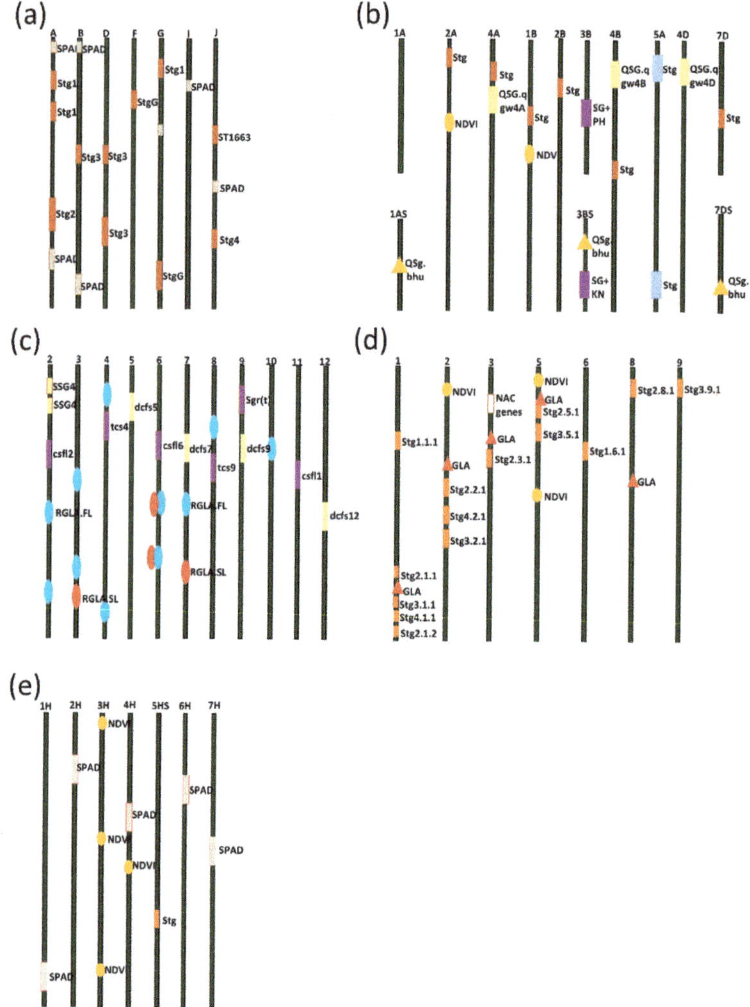

Figure 5. This schematic diagram illustrates the major stay-green quantitative trait loci (QTLs) mapped (based on the previous literature) in the five major cereal crops; sorghum (**a**), wheat (**b**), rice (**c**) maize (**d**), and barley (**e**). In sorghum, the same QTLs are mapped in different populations. In wheat, different stay-green QTLs indicated with different colors are mapped in different populations. Stg denotes stay-green, PH denotes plant height, and KN denotes kernel number. In rice, different colors denote QTLs identified by different groups, RGLA.FL denotes retention green leaf area of flag leaf; RGLA.SL denotes retention green leaf area of the second upper leaf; SSG4 denotes stay-green; dcfs denotes degree of chlorophyll content in flag and second leaves; tcs denotes total cumulative SPAD value of the four upper leaves; csfl denotes cumulative chlorophyll content of flag leaf. In maize, GLA denotes green leaf area, NDVI denotes normalized difference vegetation index.

2.2. The Physiology of Stay-Green in Sorghum

Considerable efforts have been made to understand the physiological mechanism of the SG in sorghum. The earlier studies demonstrated the role of the N uptake by the SG and non-SG sorghum hybrids [41]. SG hybrids grown under terminal drought stress were able to balance between the N demand by the grain and N supply during grain filling. In these hybrids, at anthesis, the leaf N content was correlated with the onset and rate of leaf senescence under terminal drought stress [41]. On the other hand, it is reported that SG loci influences the root architecture and increases the water accessibility during grain filling under water-limited field conditions [57].

Recently it has been reported that adaptation of SG sorghum to drought is a consequence of canopy development, leaf anatomy, root growth, and water uptake and utilization [11–13,53,58]. For example, [21,53] showed that SG QTLs impact water uptake, transpiration efficiency, and grain yield; however, this impact depends on the genetic background and the environment (Figure 2). Furthermore, at flowering stage SG QTLs modify tillering, leaf number, and leaf size, and thus reduce the canopy size. With this small canopy size at flowering, SG reduces pre-anthesis water use (Figures 2 and 3) and increases water availability during grain filling, which in turn increases grain yield under post-flowering water stress [11,12]. Grain yield can be increased with just small increases in water use during grain filling. From simulation studies in sorghum, it is reported that addition of 1 mm transpired water during grain filling could increase the grain yield by about 30 kg ha^{-1} [59]. A recent report by [13] suggested that SG QTLs introgressed from "B35" into Sudanese background "Tabat" can regulate their transpiration rate and water utilization depends on the drought severity. All this information indicates that SG trait increased water availability after anthesis and caused a delay in leaf senescence, which subsequently improved the yield in sorghum. Thus, it is possible to consider that SG phenotype is a result of the interaction between SG loci that largely regulate the plant size, and hence water demand and utilization by the crop, and the environment that regulates water supply by the soil. Based on all these research efforts, SG could be considered as a harmonized system that operates to make available the necessary water for growth and production under terminal drought stress. On the other hand, because SG phenotypes increase water availability, the photosynthesis rate, harvest index, and biomass of the SG introgression lines were better than non-SG phenotypes [13]. These recent investigations illustrated the high productivity of the SG genotypes under the terminal drought stress in sorghum. Interestingly, the positive impact of SG on yield across multiple genetic backgrounds was reported by [60]; and all four SG QTLs increased the grain yield under drought as well as under well-watered conditions without yield penalty under well-watered conditions [13,41,61]. In sorghum, only enhanced water productivity (TE) could simultaneously improve grain and stover yield along with the crop resilience [40] (Figure 3).

Recently, metabolic and transcriptomics studies are able to provide further insights into the molecular and physiological basis of the SG trait by identifying differential expression of specific genes across different varieties or in response to a change in the environment [62–64]. These techniques can reveal more about the processes involved in the SG phenotype, without needing to identify the genes underlying the QTLs. Few studies have used transcriptomic approaches to analyze the change in gene expression in sorghum as a result of abiotic stress such as osmotic stress and abscisic acid treatments [62,65,66]. Johnson [64] compared the transcriptome of "B35" (SG) and "R16" (senescent) plants grown in non-stress conditions. In the "B35" line, 1038 genes were upregulated and 998 genes were downregulated compared with "R16." However, there is no study that utilized the near-isogenic lines of the four SG QTLs and illustrated the differences between the plants under the stress and non-stress conditions. This area of research remains untapped; however, it can provide more detailed information about SG, improve our understanding, and enable more efficient opportunities to deploy SG in breeding and crop improvement. It is anticipated that the genes regulate the SG in sorghum can be modulated in the other major cereals (wheat, maize, and rice) to improve their adaptation to drought wherever water is limited after flowering.

Although sorghum is predominantly grown in the arid and semiarid regions of the world, where heat stress is known to induce significant yield losses, very few reports are handling the heat stress and combined drought-heat stress issues [63]. Recently Tacka [64], from 29 years field-trial data spanning 408 hybrid cultivars, suggested warming scenarios break down, and identified a 33 °C as a temperature threshold, after which, yields start to decline. They suggested that both pre- and post-flowering stages were equally important for overall yields; furthermore, they concluded that the introduction of wider genetic diversity for heat adaptation into the ongoing breeding programs will facilitate sorghum resilience under climate change [64]. As SG trait improved grain yield in sorghum under terminal drought stress, it can also tolerate high temperatures better than non-SG crops, but no report described the SG as adaptation trait for heat and combined heat-drought stresses in sorghum, which remains a future task.

3. Stay-Green in Wheat

3.1. Stay-Green QTLs in Wheat

Compared to sorghum, in wheat, the progress in SG research is relatively small. On the other hand, compared to drought in sorghum, in wheat, the SG has been studied intensively as one of the adaptation traits for heat stress, the major abiotic stress affecting yield in wheat. Kumar [32] identified three QTLs on chromosomes 7DS, 3BS, and 1AS using a recombinant inbred lines (RIL) population between the SG "Chirya 3" and non-SG "Sonalika" under natural field conditions (Figure 5b). He suggested that cultivars with SG characteristics offer a better option for high-temperature and drought environments, and the identified QTLs provide initial information to generate a finer map and to recruit a marker-assisted selection (MAS) strategy. A high-density genetic map consisting of 2575 markers constructed by Shenkui [67] was used to map QTLs of SG and other agronomic traits under four different water regimes. A total of 108 additive QTLs were identified. Twenty-eight QTLs were for Chl content detected on 11 chromosomes, 43 were detected for normalized difference vegetation index (NDVI) on all chromosomes except 5B, 5D, and 7D [67].

Interestingly in many cases, the SG QTLs in wheat found to be co-located with QTLs for other important traits, the thing that can allow the simultaneous selection and improvement. For example, Huang [68] detected SG QTL on chromosome 3B in a similar linkage group where a QTL for plant height is positioned. Also, the SG QTL and a QTL for kernel number per spike were identified in the same region on chromosome 3BS [69] (Figure 5b). Pinto [70] identified a total of 44 loci linked to SG and related traits, spread through the genome. Of these 44 loci, those on chromosomes 1B, 2A, 2B, 4A, 4B, and 7D possessed the strongest and most repeatable effects. Pinto [70] showed that the association of the SG trait and all SG-related traits with stress tolerance is reinforced by results demonstrating that the same genomic regions have an effect on kernel number, yield, grain weight, NDVI, canopy temperature, and also the rate and length of grain-filling. Christopher [24], identified SG QTLs associated with QTLs for seedling root number and *Rht*-height genes. These findings of the co-location of QTL for SG and performance traits confirms the usefulness of SG for productivity enhancement under heat and drought stresses, and suggest avenues for further research to clarify the physiological and genetic mechanisms of SG for better understanding and exploiting SG in wheat.

The moderate and greater effect of each SG QTL is reported in wheat and other cereal crops. Three QTLs identified for SG in wheat explained up to 38.7% of phenotypic variation in a study by [32]. In maize, single SG QTL explained from 3.2 to 12.5% of the phenotypic variance [71]. In sorghum, on the other hand, each of the four key SG QTLs possessed a considerably higher percentage of variation, 10 to 30% [47]. Accordingly, in comparison with other crops, it is possible to speculate that SG in wheat is a function of several genes with relatively small effects. The identification of genetic loci regulating SG in a wheat mapping population offers the tools to enable MAS to accelerate and advance the competence of plant breeding. However, to what extent these markers can be largely applicable is still under examination. As the genetic mechanisms controlling SG and yield-associated traits are very complex, the use of high-density linkage map will enable exploration of novel favorable alleles. In wheat only and not the other species physical deletion maps are useful for physically allocating ESTs and genes to small chromosomal regions for targeted mapping. Sourdille [72] improved the usefulness of deletion stocks for chromosome bin mapping and characterized 84 deletion lines covering the 21 chromosomes of wheat by 725 microsatellites. Several genes and QTLs have been physically mapped on the deletion maps. These deletion stocks could potentially help to elucidate the location and if possible, the cloning of SG genes and other associated traits.

3.2. The Physiology of Stay-Green in Wheat

A positive correlation between yield and SG phenotypes has been recognized under heat and drought stresses, and non-stress conditions [70,73–76]; however, negative effects on yield have rarely been reported [77,78]. Pinto [70] showed that SG trait was positively associated with yield and yield components in bread wheat grown under both heat-stress and non-heat stress conditions. Christopher [73] showed that SG genotypes exhibited higher mean grain mass and showed small variances in water use before anthesis, or better water extraction from depth after anthesis. The deep soil moisture was depleted, indicating that the extraction of deep soil moisture was essential for adaptation of the SG genotypes. However, it is clear from their study that mechanisms other than root traits also exist.

The usage of SG trait alone or in combination with other markers/traits related to water stress has great potential for selecting either for specific or broad water-stress adaptation [24,79]. Christopher [24] reported that SG traits integrating senescence, plus time from anthesis to onset, mid-point, and near completion of senescence were positively correlated with high yield in the severe and mild water-stress environment. He finally suggested that these traits have prospective to surge the rate of progress toward higher yield with better yield stability of wheat in a wide range of environments. The improvement of molecular markers for the selection of these traits would be highly needed and will enable selection in early generations. Although many traits such as canopy temperature depression (CTD) have been suggested in early studies as selection criteria to assess heat tolerance [80,81], published studies on a possible association between the SG trait and CTD in the different crops are scarce.

In durum wheat, a SG mutant has been characterized with increased rate and duration of grain filling, leaf area, and photosynthetic competence [29]. During the grain development, flag leaf SG duration and harvest index showed positive relationships with water use efficiency [82]. Under field conditions, senescence has been quantified by NDVI and linked with yield and response to drought and heat [70,74,83]. Several studies [84,85] demonstrated that a decrease in late-season leaf senescence in wheat accessions was correlated with increased yield. The late-season maintenance of Chl and reduced senescence, slow down the decrease in photosynthetic capacity (i.e., RuBP regeneration and Rubisco activity,) and resulted in a longer photosynthesis duration and higher production potential [29,86]. In a study by [87], delayed flag leaf senescence was associated with late heading date and high grain yield in three water regimes. Bogard [75] showed a negative correlation between anthesis date and the onset of senescence and between leaf senescence and grain yield, which explained by associations between QTLs affecting leaf senescence and QTLs for anthesis date [70,75]. Liang et al. [87] provided evidence that grain yield was sink-limited in three different moisture levels until the final stages of growth, at that time positive relationship between grain yield and light-saturated net carbon assimilation at anthesis and negative relationship between grain yield and flag leaf senescence indicated that sustained photosynthesis contributed to additional grain filling that increased grain yield. Their results implied that delayed leaf senescence and late-season photosynthesis were driven by the size of the reproductive carbon sink, which is greatly controlled by factors affecting the grain numbers.

Under heat and combined heat-drought stresses, SG calculated based on NDVI at physiological maturity and the rate of senescence showed positive and negative relationships with yield, respectively. In addition, canopy temperature at the mid-grain-filling stage and SG variables accounted for ~30% of yield variability in multiple regression analysis, suggesting that SG traits may offer cumulative effects, together with other traits, to improve adaptation under heat stress [88].

Few studies dealt with large-scale phenotyping of SG and early senescence phenotypes under field conditions. Sebastian [89] estimated the onset of senescence of flag leaves in 50 winter wheat cultivars using spectral remote sensing tool as a high-throughput phenotyping tool, and identified the SG and the early senescence phenotypes. Recently, [90] using a handheld color spectrometer successfully converted spectra of the whole canopy into color values measured at the flag leaf level. They confirmed that spectral remote sensing is a suitable method for the high-throughput phenotyping of flag leaf senescence. The work of [24] showed how reliably the SG trait value could be predicted for use in breeding through phenotyping together with environmental simulation and characterization. On the other hand, the selection of functionally SG germplasm from large breeding populations can be achieved easily by complementing the use of NDVI with the new high-throughput phenotyping tools that have the ability to precisely monitor changes in leaf area, greenness and photosynthetic activity (via changes in canopy temperature) [91].

The SG characters are genetically complex with environmental influences that require further exploration, therefore, in the wheat understanding of SG genetic control, and the QTLs expression in different sets of environments would ease the selection for the trait. By measuring late-season photosynthesis or simply select genotypes with higher grain numbers, which is likely to be associated with the ability of the plant to remain photosynthetically active late in the growing season under optimal and stress conditions, wheat breeders might be able to selected genotypes with improved grain yield. The high-throughput phenotyping methods will facilitate the uncovering of senescence mechanisms of cereal plants in huge field trials and helps to better identify the impact of the senescence on grain yield and grain protein content.

4. Stay-Green in Rice

4.1. QTLs for the Stay-Green Trait in Rice

Four QTLs in rice (*Oryza sativa*) (*Csfl12*, *TCS4*, *Csfl6*, *and Csfl9/Tcs9*) were detected in two RILs populations obtained from the combination of "Suweon490" (japonica and synchronized) x "SNU-SG1" (japonica and SG) and "Andabyeo" (India and synchronized) x "*SNU-SG1*" (Figure 5c). Identification of the SG QTLs Csfl6 and Tcs9 in the same positions with the two-grain yield QTLs (Yld6 and Yld9) strengthens the connection between the presence of SG and high productivity in rice [92]. For Chl content, [93] reported six QTLs on five chromosomes using backcross lines, and [94] reported other three QTLs on three chromosomes using a doubled haploid (DH) population obtained from "japonica × indica" hybrid. Jiang [95] analyzed the genetic basis of SG using DH lines obtained from "indica × japonica" hybrid and detected 46 main-effect QTLs in 25 chromosomal regions and 50 digenic interactions concerning 66 loci on 12 chromosomes. Yue [36] identified more than 30 QTLs for flag leaf traits, degree of greenness and SG-related traits, of which 10 QTLs were consistently detected in different years. They reported that region RM255-RM349 on chromosome 4 controlled the three-leaf morphological traits (leaf length, width and area) simultaneously and explained a great part of the variation, which was useful for the genetic improvement of grain yield. The region RM422-RM565 on chromosome 3 was linked with SG traits, although the utilization of this region in breeding needs to be evaluated by constructing near-isogenic lines [36]. Lim [96] identified the main-effect of QTLs for the functional SG traits in the japonica rice SNU-SG1 and isolated candidate genes. They carried out QTL analysis using 131 molecular markers with F_7 RILs from a cross of japonica rice "*SNU-SG1*" and indica rice "Milyang23 (M23)." They identified 18 QTLs for eight traits related to the physiological response of the SG which provide valuable data for breeding high yielding rice.

A recessive *sgr* mutant has been isolated and mapped on chromosome 9. It delays the process of senescence but does not maintain photosynthetic capability [97]. After this, several natural variants or mutants exhibit SG in rice has been reported, such as nyc1 [98], nyc3 [99], SGR [100], and nol [16]. Rice SG mutant retains Chl b and LHCII (light-harvesting Chl-binding protein complexes of PSII) in the light as well as in the darkness [101]. However, Chl *a* as well as other Chl–protein complexes decrease during senescence in this mutant. Jiang [100] isolated another 60Co γ-rays induced rice SG mutant. The gene of this mutant was cloned by a positional cloning strategy and is found to be an allele of the SG rice gene sgr reported by [97].

About 132 rice senescence-associated genes (*SAGs*) allocated on all the 12 chromosomes have been annotated in the leaf senescence database (http://psd.cbi.pku.edu.cn/). These *SAG* genes are classified into five groups: (a) natural, (b) dark-induced senescence, (c) nutrition deficiency-induced senescence, (d) stress-induced senescence, and (f) others [102]. The corresponding mutants of SAGs can be divided into two key categories according to their phenotypes: delayed senescence mutants and premature senescence mutants [103]. However, there are much more premature senescence mutants have been reported in rice compared with the delayed senescence mutants, for instance, the noe1 [104], ospse1 [103], psd128 [105], es1-1 [106], lts [107], rls1 [108], and ps1-D [109] mutants, are involved in different complex regulatory networks of senescence. [100] proposed that SG rice mutant *sgr* involved in regulating or taking part in the activity of pheophorbide an oxygenase (PaO), and then may influence Chl breakdown and degradation of pigment–protein complex. From ethyl methane sulfonate (EMS) mutant bank of rice cultivar Zhong Jian100, an additional five premature leaf senescence mutants (psl15, psl117, psl50, psl89, and psl270) were identified [37]. The influence of these mutations on the agronomic traits as well as the physio-biochemical properties including chloroplast structure, Chl contents, photosynthetic ability, expression profile of ABA, and senescence-related genes, response to darkness and ABA, and the genetic controls of their premature senescence phenotypes were investigated [37]. These results obtained by He [37] provided the basis for the isolation of these premature senescence genes and the elucidation of the senescence mechanism in rice. Zhao [19] from genome wide association analysis (GWAS) of 368 rice accessions reported 25 known genes, among which the pleiotropic candidate gene *OsSG1* accounted for natural variation in Chl content and SG. Further analysis indicated that the significant phenotypic differences between alleles are caused by 20 large-effect, non-synonymous SNPs within six known genes around GWAS signals and three SNPs in the promoter of OsSG1 [19]. Moreover, [19] found all OsFRDL1 and CHR729 haplotypes in wild rice, and OsFRDL1-1 and CHR729-2 haplotypes were prevalent in japonica rice, whereas OsFRDL1-3, OsFRDL1-2, and CHR729-1 haplotypes predominated in Indica rice. They concluded that during domestication of japonica the cultivated areas progressively extended from low to high altitudes along with the variations in light intensity and day length. During this adaptation, new natural mutations for higher SG and Chl were maintained and gradually accrued along with natural elite variation from wild rice [19]. Interestingly, the 368 rice accessions showed no significant correlation between Chl content and SG. The Chl content was higher and SG was stronger in japonica than in Indica [19].

4.2. The Physiology of Stay-Green in Rice

The SG trait in rice cultivars is known as the ability to maintain green leaves and benefits dry matter production in drought-prone areas [110]. Chl degradation and the disassembly of the photosynthetic apparatus were the most remarkable phenomena in leaf senescence, which result in decreases in photosynthetic capacity, energy, and efficiency. Also, there is a considerable decline in electron transport chain for the remaining components in the leaf [111–113]. From their research on rice, [114] reported that the major advances in understanding the origins of SG occurred after the elucidation of the Chl catabolism pathway and the associated genes, which pointed out the functional significance of the photosynthetic and N remobilization phases of leaf development. According to Kusaba [115], continuous biosynthesis of Chl in excess of the activity of the catabolic pathway provide another way to SG. In rice, several studies have been conducted at the molecular level to clarify the leaf senescence process. Several genes involved in leaf senescence have been identified, including hormonal factors, transcription factors, phytochrome B, the Chl degradation genes, and defense-related proteins. Moreover, a comprehensive understanding of leaf senescence was achieved by a time-course gene expression profiling of leaves during the grain-filling period. Rong [116] reported that the overexpression of SG rice like gene (*SGRL*) reduces the level of Chl and Chl-binding protein in leaves, and accelerate their degradation in dark-induced senescence in rice leaves; therefore, they suggested that the *SGRL* protein is involved in Chl degradation. Moreover, the presence of conserved amino acid domain in SGRL and SG rice implies similar biochemical functions [114]. Mao [117] showed that overexpression of *OsNAC2* results in premature senility rice, which indicated the *OsNAC2* role in ABA-induced leaf senescence pathway. The results obtained by [37] have provided principles for further studies on the fine mapping, functional analysis, and isolation of the premature leaf senescence corresponding genes. The senescence-associated genes (*SAGs*), including the six *NACTFs* were identified from the gene expression profiling of the flag leaves from vegetative to senescence stages [37]. On the other hand, mapping of these *SAGs* into cellular processes enabled the identification of the key cellular mechanisms of the shared and differential senescence programs between flag leaf and second leaf. When the changes after panicle removal observed among the differential senescence programs, invariable core senescence programs were distinguished from the variable senescence programs.

Although the contribution of the SG genotype to stable yield production under drought stress has been studied in other crops like sorghum [55], a robust relationship between grain yield increase and leaf greenness has not been reported yet in rice. In contrast, a negative correlation was reported [36,95]. The intra subspecies cross, japonica/japonica has been used by Fu [56] to investigate the inheritance mode and the genetic relationship between the SG traits and the yield and its components. They found that the correlation between the seed-setting rate and SG was higher than that between SG and yield, indicating that SG enhances the yield through the direct improvement of seed setting. In this case, the simultaneous increase of the source (photosynthetic rate) and sink (partitioning to grain) strengths is most likely to be the drive to achieve grain yield [56] (Figure 3). Park and Lee [118] measured Chl content and photosynthesis under light saturation (Pmax) in SNU-SG1 and other two rice varieties. They found that SNU-SG1 maintains high Chl content and photosynthetic capability for longer during the monocarpic senescence, and has improved seed setting rate. Thus, they concluded that SNU-SG1 could be used as a desirable genetic source of functional SG, in breeding programs to increase crop productivity.

5. Stay-Green in Maize

5.1. QTLs for the Stay-Green Trait in Maize

The genetic analysis of complex traits in maize under abiotic stresses has focused mainly on drought tolerance [119–121]. Only a few studies have been conducted to map the SG QTLs in maize. In temperate maize germplasm, [122] mapped three and five QTLs in F_4 progenies and their test crosses, respectively. Zheng [34] mapped 14 QTLs in $F_{2:3}$ progenies, and [70] mapped 14 QTLs in F_2 plants. Zheng [34], reported that the respective QTL contribution to phenotypic variance ranged from 5.40% to 11.49%, with trait synergistic action from Q319. In tropical germplasm, Câmara [123] mapped 20 and 33 QTLs using $F_{2:3}$ progenies from two populations. Additional QTL analyses indicated that multiple intervals of SG QTLs overlapped with yield QTLs. Yang [124] performed a QTL mapping by using 165 $F_{3:4}$ recombinant inbred lines population derived from a cross between a SG inbred line (Zheng58) and a non-SG inbred line (B73) genotyped using 211 polymorphic simple sequence repeat markers. A total of 23 QTLs for Chl content, photosystem II photochemical efficiency, and SG area at maturity stage were mapped on nine chromosomes. The single QTL explained 3.7–13.5% of the phenotypic variance. They validated some important SG QTLs, which were significantly correlated with the plant yield. These studies provide a better insight into the mechanism that regulates leaf SG in maize and contributes to the development of novel elite maize varieties with delayed leaf senescence through MAS. Recently, Zhang [14] in a mapping population derived from the Illinois High Protein 1 (IHP1) and Illinois Low Protein 1 (ILP1) lines, identified a novel QTL controlling functional SG, showing different rates of leaf senescence. They further described the role of NAC7 (transcription factors) in improving functional SG and yield through the regulation of the resource allocation from vegetative source to reproductive organs. These findings of [14] highlight and draw attention to NAC7 as a core target for improving functional SG and yields in maize and other crops. Sekhon [125] studied a SG line able to maintain high grain filling and photosynthetic capacity for six additional critical days compared to a naturally senescent genotype and revealed the genetic architecture of senescence in maize. Furthermore, they placed nine candidate genes represent diverse processes, including sugar-mediated signaling (trps13), transport (*ZmSWEET1b* and *ZmMST4*), and control of sugar uptake (*dek10*) in one model, and elucidated the role of sugar partitioning and sugar signaling in inducing senescence [125].

QTL pyramiding in elite inbreeds, would have enhanced levels of SG in maize. MAS has been successfully used in maize for grain yield, as well as in other crops for other traits [126–128]. For functional marker development and rapid identification of candidate genes or loci, [129] suggested a strategy of combining Meta QTL (MQTLs) analysis and regional association mapping. Several maize orthologs of rice yield-related genes were identified in these MQTL regions [129]. Based on the results of the meta-analysis and regional association mapping, three potential candidate genes GRMZM2G359974, GRMZM2G301884, and GRMZM2G083894) associated with kernel size and weight within three MQTL regions were identified.

5.2. The Physiology of Stay-Green in Maize

Swanckaert [130] characterized two SG types in maize: SG type and normal type. The SG varieties characterized by higher photosynthetic capacity values coincided with higher values for the proxies. Although a higher photosynthetic capacity did not induce the higher accumulation of assimilates in the leaves, the SG trait was characterized as a cosmetic SG. The SG trait influenced N dynamics in the plant since the lower translocation of N from the leaves to the ear resulted in low N concentration in the ear and consequently lower ear dry matter yield. No differences found either in whole-plant N concentration or whole-plant dry matter yield. As the SG trait mainly cause shifts in dry matter partitioning and N balance between vegetative and reproductive tissues, the energy source also shifts from starch (from ear source) to cell wall material (from stover source) [130].

SG maize cultivars have been selected to maintain a high green leaf area during the post silking development phase. Consequently, they are able to sustain higher photosynthetic capacity than non-SG cultivars at a time of high demand for photosynthates [131,132]. This stage is critical because the highest amount of dry matter partitioned to the grain is accumulated after silking [133]. A SG line P3845 showed a delay in leaf senescence correlated with increased levels of Chl, when compared to a non-SG line Hokkou 55, [134]. P3845 showed high levels of cytokinins (trans-zeatin riboside, t-ZR; dihydro zeatin riboside, DHZR; isopentenyl adenosine, iPA) and low level of ABA in the leaves. On the other hand, in roots, P3845 showed increased levels of t-ZR, DHZR, and ABA, but decreased concentrations of iPA [134]. Therefore, [134] concluded that the delayed senescence in P3845 is a result of the higher rate of cytokinin transport from roots to leaves, and the translocation of ABA from roots to shoots may be plugged in the SG cultivar, which results in leaf senescence retardation [134].

In maize, the SG trait used to be evaluated at the leaf level using portable Chl meters, such as the Minolta SPAD. Chl content and imaging spectroscopy were also used to evaluate SG trait in maize [135–137]. The measurement of the NDVI during canopy development stages was proposed as a secondary trait to be included in maize breeding to indicate early vigor and grain yield under drought and well-water conditions [138,139]. The SG trait should be considered in maize breeding programs; it accounts for grain yield and other important related traits of agronomic/economic importance, particularly drought tolerance. However, information on the inheritance of the SG trait in maize is limited. Many reports have shown that maize breeding programs aiming at improving grain yield increased the SG during selection. Thus, newer hybrids [140] and newer populations [141] are more SG than the old ones; that is, the increased level of delayed senescence of newer hybrids contribute to their higher productivity (Figure 4) [15,142]. Moreover, SG maize genotypes had a high tolerance to abiotic stresses (such as drought and high population density) compared to non-GS genotypes.

Cerrudo [139], laid out the basics to utilize high throughput phenotyping and facilitated the identification of climate-adapted germplasm. To identify and select germplasm with high grain yield under drought, heat, and combined drought-heat stresses, starting at anthesis, they used an airplane mounted multispectral camera to estimate the area under the curve (AUC) for vegetation indices to measure and compute secondary traits. Among the secondary traits, NDVI was found to be the best secondary trait to breed for high grain yield and extended SG under drought, heat, and combined drought-heat stresses [139]. Furthermore [139] found that the prediction accuracy of the secondary traits like NDVI was better than the prediction accuracy of grain yield under stress and non-stress conditions. NDVI is a highly heritable trait with moderate and consistent correlation with grain yield under well-watered conditions [138,143].

Most of SG genes have been identified and functionally characterized in maize and rice (Table 1), work in sorghum and wheat is still behind, although in sorghum the trait extensively studied and sorghum genome was sequenced. Genes were identified, but the SG related function still un-known.

6. Stay-Green in Barley

6.1. QTLs for the Stay-Green Trait in Barley

Unlike other cereal crops, SG research in barley is very limited. Emeberi [54] in multiple barley populations identified nine QTLs related to SG. Of these nine QTLs, only that on the short arm of barley chromosome 5H showed a consistency under different environments in all populations; however, its expression possessed high GxE interaction. On the other hand, the presence of only one consistent QTL suggests that the leaf senescence/Chl loss during the maturity stage is controlled by simple genetic factors [54]. Sallam [144] found that QTLs for leaf rolling and leaf chlorophyll content on chromosome two, four, and five are syntenic between barley and wheat. Obsa [145] evaluated three interconnected doubled haploid populations in drought-prone environments and detected 18 QTLs for drought adaptation. Among these 18 QTLs, four and two QTLs were detected for SG related traits NDVI and SPAD, respectively. Fox [146] evaluated the SG expression under terminal heat and drought stresses in 100 barley lines from a ND24260 × Flagship doubled haploid population. They detected ten SG QTLs on chromosomes 3H, 4H, 5H, 6H, and 7H. Out of these ten QTLs, six QTLs were associated with terminal heat-stress and four with terminal drought stress. These QTLs did not co-localized with previously reported barley stress-response QTL and therefore considered as novel QTLs. However, the two heat-stress QTLs mapped to bPb-5529 on chromosome 5H, are near to QTLs reported for root/shoot ratio and root length [146]. After field validation these identified QTLs could be good candidates for MAS targeting improvement of barley abiotic stress tolerance.

6.2. The Physiology of Stay-Green in Barley

Although drought-tolerant barley has been reported by Gonzalez [147] very few reports discussed the SG trait. SG reported frequently for leaf greenness while other organs contribution was detected. CO_2 estimates indicate that the spikes' contribution to grain yield can reach up to 70% depending on the conditions in wheat and barley grown under stress [148]. Vaezi1 [149], found that among 11 barley genotypes evaluated under drought stress, the highest yielding genotype possess SG characteristics. Therefore, they suggested that potential grain yield can be improved by increasing plant photosynthetic capacity and assimilates production during the later phase of grain filling. Seiler [150] studied a number of barley lines showing senescence or SG phenotype and demonstrated the superior yield performance of the SG lines under drought conditions. They found that the reason of the difference between the SG and senescing lines in their assimilation capacity under drought stress is that the ABA synthesis levels in senescing lines are greater than that of SG lines under short and long-term drought stress. Based on this finding they suggested that a greater ABA flux metabolism in the senescing lines negatively affected assimilation and water use efficiency [150]. Shirdelmoghanloo [151] studied 157 barley genotypes under heat stress in two environments comprising three sowing dates and detected genetic variation for grain growth components, grain plumpness, and SG traits. Their results showed a significant positive correlation between the SG and the grain filling duration which suggests the role of the SG in stabilization of the grain filling duration in barley under heat stress. Moreover, they demonstrated the possibility of developing heat-tolerant barley genotypes through appropriate focus on grain filling rate and SG traits in breeding programs [151].

7. Stay-Green and Grain or End-Use Quality

Although, grain quality or end-use quality in cereals is an important aspect and very hot topic in the research community, knowledge on the effects of "SG" expression on grain or end-use quality under normal or stress conditions is extremely limited. On the other hand, as it has been reported that SG trait is capable of protecting the cereal yields under stress conditions through organization and stabilization of grain development, it is possible to speculate that it might have a positive impact on the grain or end-use quality. In sorghum, SG plants showed increased resistance to pest and disease invasion, better quality forages for animals, high chl content, and extended pigment source for food industry, as well as the attractive ornamental period [44]. In study of 50 winter wheat genotypes, a negative correlation between the onset of leaf senescence and grain yield ($r^2 = 0.81$) and a positive correlation with grain protein content ($r^2 = 0.48$) was observed [89]. It is concluded that the SG phenotypes were characterized by higher N uptake during grain filling and longer maintenance of greenness. In addition, the use of photosynthetic glucose for the synthesis of amino acids rather than for starch decreased wheat yield and increased grain protein content [89]. NDVI measurements using a drone to estimate the SG phenology have recently been demonstrated to improve wheat grain quality and yield predictions [152] In rice, it has been speculated that delaying the senescence at the terminal stage of maturity may lead to increased yield and improved grain quality [95] In barley, Gous [153] examined the effect of SG expression on starch biosynthesis in grains of Flagship (a cultivar without "SG"-like characteristics) and ND24260 (SG"-like cultivar) under mild and severe drought stress conditions at anthesis. In this study Flagship possessed higher grain amylose and long amylopectin branches under the mild drought stress, suggesting that drought stress affects starch biosynthesis in grain, probably because of early termination of grain filling. In contrast, ND24260 did not possess any changes in starch molecular structure under the different drought levels. As long as changes in starch molecular structure can affect starch properties, such as enzymatic degradation rates, and hence its nutritional value, the ND24260 has a greater potential to maintain starch biosynthesis and hence better grain quality under drought conditions. These results make the "SG"-like traits potentially useful to ensure food quality and quantity [153].

Table 1. List of stay-green-related genes and their functional characterization in the four major cereal crops (Rice, Wheat, Maize, and Sorghum).

Genes	Function	Cereals				References
		Rice	Wheat	Maize	Sorghum	
Stay-Green Rice like (*SGRL*)	Affect Chlorophyll (Chl) degradation during natural and dark-induced leaf senescence	√	-	-	0	[116,154]
Glucuronic acid substitution of xylan1 (*GUX1*)	Required for substitution of the xylan backbone with 4-O-methylglucuronic acid [Me]GlcA	-	0	√	0	[125,155–158]
β-Glucosidase (*BGLU42*)	The exact role remains to be determined	√	-	√	0	[125]
Trehalose-6-phospate synthase13(*trps13*)	Sugar-mediated signaling in stay-green	-	-	√	0	[125]
Monosaccharide transporter (*MST4*)	Associated with senescence and nitrogen use efficiency, transport of Sucrose out of leaf cells by sugar transporters to alternative sinks	√	-	√	-	[125,159]
Sugar will Eventually be exported transporters (*SWEET*)	Transport of Sucrose out of leaf cells by sugar transporters to alternative sinks	√	-	√	-	[125,160]
Cell wall invertase (*incw4*)	Sinks hydrolyzation	-	-	√	0	[125,161]
Mitochondrial pentatricopeptide repeat protein (*dek10*)	Activation sugar-sinks	-	-	√	-	[125,162,163]
Sb09g004170 and *Sb09g022580*	DRGs (Associated with *Stg1* QTL)	-	-	-	√	[164]
NAC-transcription factor 9(*nactf9*)	Appears to be one of the master regulators of stay-green, acting in conjunction with ZmIRX15-L, ZmGUX1, mlg3	-	-	√	-	[125]

√, -, 0: Identified and characterized, not identified yet, identified, however, their contribution to the stay-green trait not yet known.

8. Conclusions and Future Prospective

This review discussed the recent progress made in the research of the SG as an important trait to combat abiotic stresses in the major cereals. The review discussed the identification of SG and SG-related traits QTLs and the SG genes. The five cereal crops differ in their chromosome numbers and genome sizes; however, they have been diverged from common ancestor 60 million years ago [165] and there is a degree of synteny between their genomes [166] and [22]. Considering the synteny between cereals genomes, we suggest that comparative mapping approaches using the huge genomic information became available recently would elucidate inheritance, physiology, and expression of SG, and would generate massive genetic information regarding the SG including identification of the genes leading to full understanding of the SG mechanism. Interestingly this can be done without extensive phenotyping.

Despite the knowledge of genomic regions conferring the SG trait, it is surprising that knowledge about the physiological mechanisms of the SG is still relatively limited. Early explanations focused mainly on the role of SG in the maintenance of photosynthetic activity. However, the SG trait has been suggested to be involved (regulates) in (1) the plant N/C balance and in particular to increase the capacity of C capture and N mobilization during the post-anthesis period; (2) increased water availability during the post-anthesis period. However, the improvements in water uptake and water use efficiency because of SG phenotype are still unexplained, which could be accounted for either a deeper soil extraction because of improved root architecture or water-saving traits operating at early stages. This could be a useful future study to elucidate the water relations in the SG plants for efficient cereal breeding under drought. The strong association between SG QTLs in major cereal crops and other useful agronomic traits such as grain yield improvement, sunlight interception, and conversion and biomass allocation, especially under drought and heat stress conditions, would also provide opportunities to use both phenotypic and molecular markers of SG trait to accelerate the breeding of new cereal varieties.

The studies reported here support the use of leaf and canopy photosynthesis, as a target trait to breed high-yielding cultivars. High yielding genotypes adopt different strategies to achieve high production, and this can explain the complexity of grain yield formation under favorite and stress conditions. This review provides evidence that there is a need to phenotype photosynthetic capacity-related traits, leaf anatomy, canopy development at vegetative, pre, and post-anthesis stages. As the SG mechanisms become more evident and as DNA-sequencing offers intensive genome coverage, the possibility is that the future of manipulating the SG trait will be about manipulating its physiological components. Thus, this review has contributed to increasing knowledge on the understanding of the physiological mechanisms associated with SG trait and photosynthetic efficiency in cereals as a prospective approach for high yield under stress conditions. Maybe the key to breaking the plateau of productivity associated with adaptation to stress conditions particularly heat and drought stresses. More exploration needs to be done extensively in breeding programs to benefit from this trait like gene identifications, grain quality, and deep physiological analysis, including metabolomic, transcriptomic and ionomic studies. This trait in combination with other useful traits may provide the solution against the major environmental problems (heat and drought). Information about QTLs for SG trait in major cereal crops would also provide opportunities to use this trait in breeding programs.

At last, taking into account the knowledge accumulated about the SG in the five cereal crops and the synteny between them we can describe the SG as plant mechanism manage the canopy size, water uptake and utilization, nitrogen and carbon dynamics, leaf senescence, photosynthesis capacity, and finally assimilates partitioning. All these comprise very complicated and interacted biochemical processes through hormonal balance and other bath ways. Based on this, it is obvious that SG is a very complicated trait and to be more precise SG should be nominated as a system and not just a trait.

Author Contributions: N.M.K. conceived the presented idea, drafted and edited the manuscript. Y.S.A.G. was involved in planning, revising, and edited the paper with input from all authors. M.A. revised and edited the manuscript, and E.A. made the SG genes table and worked on the manuscript. H.T. supervised the work.

Int. J. Mol. Sci. **2019**, *20*, 5837

Funding: This work was supported by the Science and Technology Research Partnership for Sustainable Development in collaboration with the Japan Science and Technology Agency (JPMJSA1805), the Japan International Cooperation Agency (JICA).

Conflicts of Interest: The authors declare no conflicts of interest.

Abbreviations

SG	Stay-green
Chl	Chlorophyll
NDVI	Normalized difference vegetation index
CTD	Canopy temperature depression
MAS	Marker-assisted selection

References

1. FAO Statistics 2019. Available online: http://www.fao.org/faostat/en/#data (accessed on 20 November 2019).
2. Lesk, C.; Rowhani, P.; Ramankutty, N. Influence of extreme weather disasters on global crop production. *Nature* **2016**, *529*, 84–87. [CrossRef] [PubMed]
3. Daryanto, S.; Wang, L.; Jacinthe, P.A. Global synthesis of drought effects on maize and wheat production. *PLoS ONE* **2016**, *11*, e0156362. [CrossRef] [PubMed]
4. Viola, D.; Daniel, K.Y.; Tan, I.D. Impact of High Temperature and Drought Stresses on Chickpea Production. *Agronomy* **2018**, *8*, 145.
5. Barnabas, B.; Jäger, K.; Fehér, A. The effect of drought and heat stress on reproductive processes in cereals. *Plant Cell Environ.* **2008**, *31*, 11–38. [CrossRef] [PubMed]
6. Abdelrahman, M.; Burritt, D.J.; Gupta, A.; Tsujimoto, H.; Tran, L. Heat stress effects on source–sink relationships and metabolome dynamics in wheat. *J. Exp. Bot.* 2019. [CrossRef] [PubMed]
7. Abdelrahman, M.; El-Sayed, M.; Jogaiah, S.; Burritt, D.J.; Tran, L.S.P. The "STAY-GREEN" trait and phytohormone signaling networks in plants under heat stress. *Plant Cell Rep.* **2017**, *36*, 1009–1025. [CrossRef] [PubMed]
8. Rosenow, D.T. Breeding for Resistance to Root and Stalk Rots in Texas. In *Sorghum Root and Stalk Rots, A Critical Review*; ICRISTAT: Patancheru, AP, India, 1983; pp. 209–217.
9. Peingao, L. Structural and biochemical mechanism responsible for the stay-green phenotype in common wheat. *Chi. Sci. Bull.* **2013**, *51*, 2595–2603.
10. Gregersen, N.; Bross, P.; Vang, S.; Christensen, J.H. Protein misfolding and human disease. *Genomics Hum. Genet.* **2006**, *7*, 103–124. [CrossRef]
11. Borrell, A.K.; Mullet, J.E.; George-Jaeggli, B.; van Oosterom, E.J.; Hammer, G.L.; Klein, P.E.; Jordan, D.R. Drought adaptation of stay-green cereals associated with canopy development, leaf anatomy, root growth and water uptake. *J. Exp. Bot.* **2014**, *65*, 6251–6263. [CrossRef]
12. Jaegglia, B.G.; Mortlockb, M.Y.; Borrell, A. Bigger is not always better: Reducing leaf area helps stay-green sorghum use soil water more slowly. *Environ. Exp. Bot.* **2017**, *138*, 119–129. [CrossRef]
13. Kamal, N.M.; Gorafi, Y.S.A.; Tsujimoto, H.; Ghanim, A.M.A. Stay-green QTLs response in adaptation to post-flowering drought depends on the drought severity. *BioMed Res. Int.* **2018**, *2018*, 7082095. [CrossRef] [PubMed]
14. Zhang, J.; Fengler, K.A.; John, L.; Hemert, V.; Gupta, R.; Mongar, N.; Sun, J.; Allen, W.B.; Wang, Y.; Weers, B.; et al. Identification and characterization of a novel stay-green QTL that increases yield in maize. *Plant Biot. J.* **2019**, *17*, 2272–2285. [CrossRef] [PubMed]
15. Thomas, H.; Ougham, H. The stay-green trait. *J. Exp. Bot.* **2014**, *65*, 3889–3900. [CrossRef]
16. Sato, Y.; Morita, R.; Katsuma, S.; Nishimura, M.; Tanaka, A.; Kusaba, M. Two short-chain dehydrogenase/reductases, Non-Yellow Coloring 1 and NYC1-LIKE, are required for Chl b and light-harvesting complex II degradation during senescence in rice. *Plant J.* **2009**, *57*, 120–131. [CrossRef] [PubMed]
17. Schelbert, S.; Aubry, S.; Burla, B.; Agne, B.; Kessler, F.; Krupinska, K.; Hörtensteiner, S. Pheophytin pheophorbide hydrolase (pheophytinase) is involved in Chl breakdown during leaf senescence in Arabidopsis. *Plant Cell* **2009**, *21*, 767–785. [CrossRef] [PubMed]

18. Shimoda, Y.; Ito, H.; Tanaka, A. Arabidopsis stay-green, Mendel's green cotyledon gene, encodes magnesium-dechelatase. *Plant Cell* **2016**, *28*, 2147–2160. [CrossRef] [PubMed]

19. Zhao, Y.; Chenggen, Q.; Wang, X.; Chen, Y.; Jinqiang, D.; Jiang, C.; Sun, X.; Chen, H.; Li, J.; Piao, W.; et al. New alleles for Chl content and stay-green traits revealed by a genome wide association study in rice (*Oryza sativa*). *Sci. Rep.* **2019**, *9*, 2541.

20. Sjödin, J. Induced morphological variation in Vicia faba L. *Hereditas* **1971**, *67*, 155–179. [CrossRef]

21. Kassahun, B.; Bidinger, F.; Hash, C.; Kuruvinashetti, M. Stay-green expression in early generation sorghum [Sorghum bicolor (L.) Moench] QTL introgression lines. *Euphytica* **2010**, *172*, 351–362. [CrossRef]

22. Luche, H.S.; Silva, J.A.G.; Nörnberg, R.; Silveira, F.S.; Baretta, D.; Groli, E.L.; Maia, L.C.; Oliveira, A.C. Stay-green: A potentiality in plant breeding. *Ciência Rural* **2015**, *45*, 1755–1760. [CrossRef]

23. Fukao, T.; Yeung, E.; Bailey-Serres, J. The Submergence tolerance gene SUB1A delays leaf senescence under prolonged darkness through hormonal regulation in rice. *Plant Physiol.* **2012**, *160*, 1795–1807. [CrossRef] [PubMed]

24. Christopher, J.T.; Christopher, M.J.; Borrell, A.K.; Fletcher, S.; Chenu, K. Stay-green traits to improve wheat adaptation in well-watered and water-limited environments. *J. Exp. Bot.* **2016**, *67*, 5159–5172. [CrossRef] [PubMed]

25. Antonietta, M.; Acciaresi, H.A.; Guiamet, J.J. Responses to N deficiency in stay green and non-stay green Argentinean hybrids of maize. *J. Agron. Crop Sci.* **2015**, *202*, 231–242. [CrossRef]

26. Adeyanju, A.; Yu, J.M.; Little, C.; Rooney, W.; Klein, P.; Burke, J.; Tesso, T. Sorghum RILs segregating for stay-green QTL and leaf dhurrin content show differential reaction to stalk rot diseases. *Crop Sci.* **2016**, *56*, 2895–2903. [CrossRef]

27. Distelefeld, A.; Avni, R.; Fishcher, A.M. Senescence, nutrient remobilization, and yield in wheat and barley. *J. Exp. Bot.* **2014**, *65*, 3783–3798. [CrossRef]

28. Joshi, A.K.; Kumari, M.; Singh, V.P.; Reddy, C.M.; Kumar, S.; Rane, J.; Chand, R. Stay green trait: Variation, inheritance and its association with spot blotch resistance in spring wheat (*Triticum aestivum* L.). *Euphytica* **2007**, *153*, 59–71. [CrossRef]

29. Spano, G.; Di, F.N.; Perrotta, C.; Platani, C.; Ronga, G.; Lawlor, D.W.; Napier, J.A.; Shewry, P.R. Physiological characterization of stay green mutants in durum wheat. *J. Exp. Bot.* **2003**, *54*, 1415–1420. [CrossRef]

30. Veyradier, M.; Christopher, J.; Chenu, K. Quantifying the potential yield benefit of root traits. In *Proceedings of the 7th International Conference on Functional-Structural Plant Models, Saariselkä, Finland, 9–14 June 2013*; Sievänen, R., Nikinmaa, E., Godin, C., Lintunen, A., Nygren, P., Eds.; Finnish Society of Forest Science: Vantaa, Finland; MELTA: Helsinki, Finland; pp. 317–319.

31. Kumar, R.R.; Goswami, S.; Shamim, M. Biochemical defense response: Characterizing the plasticity of source and sink in spring wheat under terminal heat stress. *Front. Plant Sci.* **2017**, *8*, 1603. [CrossRef]

32. Kumar, U.; Joshi, A.K.; Kumari, M.; Paliwal, R.; Kumar, S.; Röder, M.S. Identification of QTLs for stay green trait in wheat (*Triticum aestivum* L.) in the 'Chirya 3' 3 'Sonalika' population. *Euphytica* **2010**, *174*, 437–445. [CrossRef]

33. Vijayalakshmi, K.; Fritz, A.K.; Paulsen, G.M.; Bai, G.H.; Pandravada, S.; Gill, B.S. Modelling and mapping QTL for senescence-related traits in winter wheat under high temperature. *Mol. Breed.* **2010**, *26*, 163–175. [CrossRef]

34. Zheng, H.J.; Wu, A.Z.; Zheng, C.C.; Wang, Y.F.; Cai, R.; Shen, X.F.; Xu, R.R.; Liu, P.; Kong, L.J.; Dong, S.T. QTL mapping of maize (*Zea mays*) stay-green traits and their relationship to yield. *Plant Breed.* **2009**, *128*, 54–62. [CrossRef]

35. Ribeiro, T.; Alves da Silva, D.; Fátima Esteves, J.A.; Azevedo, C.V.G.; Gonçalves João, G.R.; Carbonell, S.A.M.; Chiorato, A.F. Evaluation of common bean genotypes for drought tolerance. *Bragantia* **2019**, *78*. [CrossRef]

36. Yue, B.; Wei-Ya, X.; Li-Jun, L.; Yong-Zhong, X. QTL analysis for flag leaf characteristics and their relationships with yield and yield traits in rice. *Acta Genetica Sin.* **2006**, *33*, 824–832. [CrossRef]

37. He, Y.; Li, L.; Zhang, Z.; Wu, J. Identification and comparative analysis of premature, senescence leaf mutants in rice (Oryza sativa L.). *Int. J. Mol. Sci.* **2018**, *19*, 140. [CrossRef] [PubMed]

38. Harris, K.; Subudhi, P.K.; Borrell, A.; Jordan, D.; Rosenow, D.; Nguyen, H.; Klein, P.; Klein, R.; Mullet, J. Sorghum stay-green QTL individually reduce post-flowering drought-induced leaf senescence. *J. Exp. Bot.* **2007**, *58*, 327–338. [CrossRef] [PubMed]

39. Tao, Y.Z.; Henzell, R.G.; Jordan, D.R.; Butler, D.G.; Kelly, A.M.; McIntyre, C.L. Identification of genomic regions associated with stay green in sorghum by testing RILs in multiple environments. *Theor. Appl. Gen.* **2000**, *100*, 1225–1232. [CrossRef]

40. Kholová, J.; Tharanya, M.; Sivasakthi, K.; Srikanth, M.; Rekha, B.; Hammer, G.L.; McLean, G.; Deshpande, S.; Hash, C.T.; Craufurd, P.; et al. Modelling the effect of plant water use traits on yield and stay-green expression in sorghum. *Funct. Plant Biol.* **2014**, *41*, 1019–1034. [CrossRef]

41. Borrell, A.K.; Hammer, G.L.; Douglas, A.C.L. Does maintaining green leaf area in sorghum improve yield under drought? I. Leaf growth and senescence. *Crop Sci.* **2000**, *40*, 1026–1037. [CrossRef]

42. Borrell, A.K.; Hammer, G.L.; van Oosterom, E. Stay-green: A consequence of the balance between supply and demand for nitrogen during grain filling? *Ann. Appl. Biol.* **2001**, *138*, 91–95. [CrossRef]

43. Jordan, D.R.; Tao, Y.; Godwin, I.D.; Henzell, R.G.; Cooper, M.; McIntyre, C.L. Prediction of hybrid performance in grain sorghum using RFLP markers. *Theor. App. Genet.* **2003**, *106*, 559–567. [CrossRef]

44. Xu, W.W.; Subudhi, P.K.; Crasta, O.R.; Rosenow, D.T.; Mullet, J.E.; Nguyen, H.T. Molecular mapping of QTLs conferring stay-green in grain sorghum (Sorghum bicolor L. Moench). *Genome* **2000**, *43*, 461–469. [CrossRef] [PubMed]

45. Subudhi, P.K.; Rosenow, D.T.; Nguyen, H.T. Quantitative trait loci for the stay green trait in sorghum (*Sorghum bicolor* L. Moench): Consistency across genetic backgrounds and environments. *Theor. App. Genet.* **2000**, *101*, 733–741. [CrossRef]

46. Kebede, H.; Subudhi, P.K.; Rosenow, D.T.; Nguyen, H.T. Quantitative trait loci influencing drought tolerance in grain sorghum (Sorghum bicolor L. Moench). *Theor. App. Genet.* **2001**, *103*, 266–276. [CrossRef]

47. Sanchez, A.C.; Subudhi, P.K.; Rosenow, D.T.; Nguyen, H.T. Mapping QTLs associated with drought resistance in sorghum (Sorghum bicolor L. Moench). *Plant Mol. Biol.* **2002**, *48*, 713–726. [CrossRef]

48. Hash, C.T.; Bhasker, A.G.; Lindup, S. Opportunities for marker assisted selection (MAS) to improve the feed quality of crop residues in pearl millet and sorghum. *Field Crops Res.* **2003**, *84*, 79–88. [CrossRef]

49. Haussmann, B.I.G.; Mahalakshmi, V.; Reddy, B.V.S.; Seetharama, N.; Hash, C.T.; Geiger, H.H. QTL mapping of stay-green in two sorghum recombinant inbred populations. *Theor. Appl. Genet.* **2002**, *106*, 133–142. [CrossRef]

50. Reddy, N.R.R.; Ragimasalawada, M.; Sabbavarapu, M.M.; Nadoor, S.; Patil, J.V. Detection and validation of stay-green QTL in post-rainy sorghum involving widely adapted cultivar, M35-1 and a popular stay-green genotype B35. *BMC Genom.* **2014**, *15*, 909. [CrossRef]

51. Kamal, N.M.; Gorafi, Y.S.A.; Ghanim, A.M.A. Performance of sorghum stay-green introgression lines under post-flowering drought. *Int. J. Plant Res.* **2017**, *7*, 65–74.

52. Ngugi, K.W.; Kimani, D.K. Introgression of stay-green trait into a Kenyan farmer preferred sorghum variety. *Afr. Crop Sci. J.* **2010**, *18*, 141–146.

53. Vadez, V.; Deshpande, S.P.; Kholova, J.; Hammer, G.L.; Borrell, A.K.; Talwar, H.S.; Hash, C.T. Stay-green quantitative trait loci's effects on water extraction, transpiration efficiency and seed yield depend on recipient parent background. *Funct. Plant Biol.* **2011**, *38*, 553–566. [CrossRef]

54. Emebiri, L.C. QTL dissection of the loss of green colour during post anthesis grain maturation in two-rowed barley. *Theor. Appl. Genet.* **2013**, *126*, 1873–1884. [CrossRef] [PubMed]

55. Mace, E.; Singh, V.; van Oosterom, E.; Hammer, G.; Hunt, C.; Jordan, D. QTL for nodal root angle in sorghum (Sorghum bicolor L. Moench) co-locate with QTL for traits associated with drought adaptation. *Theor. Appl. Genet.* **2012**, *124*, 97–109. [CrossRef] [PubMed]

56. Fu, J.; Yan, F.; Lee, W. Physiological characteristics of a functional stay-green rice "SNU-SG1" during grain-filling period. *J. Crop Sci. Biotechnol.* **2009**, *12*, 47–52. [CrossRef]

57. Manschadi, A.M.; Christopher, J.; Devoil, P.; Hammer, G.L. The role of root architectural traits in adaptation of wheat to water-limited environments. *Funct. Plant Biol.* **2006**, *33*, 823–837. [CrossRef]

58. Borrell, A.K.; van Oosterom, E.J.; Mullet, J.E.; George-Jaeggli, B.; Jordan, D.R.; Klein, P.E.; Hammer, G.L. Stay-green alleles enhance grain yield in sorghum under drought by modifying canopy development and enhancing water uptake. *New Phytol.* **2014**, *203*, 817–830. [CrossRef] [PubMed]

59. Hammer, G.; Cooper, M.; Tardieu, F.; Weich, S.; Walsh, B.; van Eeuwijk, F.; Chapman, S.; Podlich, D. Models for navigating biological complexity in breeding improved crop plants. *Trends Plant Sci.* **2006**, *11*, 587–593. [CrossRef] [PubMed]

60. Jordan, D.R.; Hunt, C.H.; Cruickshank, A.W.; Borrell, A.K.; Henzell, R.G. The relationship between the stay-green trait and grain yield in elite sorghum hybrids grown in a range of environments. *Crop Sci.* **2012**, *52*, 1153–1161. [CrossRef]

61. Borrell, A.K.; Hammer, G.L.; Henzel, R.G. Does maintaining green leaf area in sorghum improve yield under drought? II. Dry matter production and yield. *Crop Sci.* **2000**, *40*, 1037–1048. [CrossRef]

62. Johnson, S.M.; Cummins, I.; Lim, F.L.; Slabas, A.R.; Knight, M.R. Transcriptomic analysis comparing stay-green and senescent Sorghum bicolor lines identifies a role for proline biosynthesis in the stay-green trait. *J. Exp. Bot.* **2015**, *66*, 7061–7073. [CrossRef]

63. Nanjundaswamy, A.; Vadlani, P.V.; Prasad, P.V.V. Evaluation of drought and heat stressed grain sorghum (Sorghum bicolor) for ethanol production. *Ind. Crops Prod.* **2011**, *33*, 779–782.

64. Tacka, J.; Lingenfelserb, J.; Krishna, J.S.V. Disaggregating sorghum yield reductions under warming scenarios exposes narrow genetic diversity in US breeding programs. *PNAS* **2017**, *114*, 9296–9301. [CrossRef] [PubMed]

65. Buchanan, C.D.; Lim, S.; Salzman, R.A.; Kagiampakis, I.; Morishige, D.T.; Weers, B.D.; Klein, R.R.; Pratt, L.H.; Cordonnier-Pratt, M.M.; Klein, P.; et al. Sorghum bicolor's transcriptome response to dehydration, high salinity and ABA. *Plant Mol. Biol.* **2005**, *58*, 699–720. [CrossRef] [PubMed]

66. Dugas, D.V.; Monaco, M.K.; Olesen, A.; Klein, R.R.; Kumari, S.; Ware, D.; Klein, P.E. Functional annotation of the transcriptome of Sorghum bicolor in response to osmotic stress and abscisic acid. *BMC Genom.* **2011**, *12*, 21. [CrossRef] [PubMed]

67. Shi, S.; Azam, F.; Li, H.; Chang, X.; Li, B.; Jing, R. Mapping QTL for stay-green and agronomic traits in wheat under diverse water regimes. *Euphytica* **2017**, *2017*, 213–246. [CrossRef]

68. Huang, X.Q.; Kempf, H.; Ganal, M.W.; Röder, M.S. Advanced backcross QTL analysis in progenies derived from a cross between a German elite winter wheat variety and a synthetic wheat (Triticum aestivum L.). *Theor. Appl. Genet.* **2004**, *109*, 933–943. [CrossRef]

69. Marza, F.; Bai, G.H.; Carver, B.F.; Zhou, W.C. Quantitative trait loci for yield and related traits in the wheat population Ning7840 9 Clark. *Theor. Appl. Genet.* **2006**, *112*, 688–698. [CrossRef]

70. Pinto, R.S.; Reynolds, M.P.; Mathews, K.L.; McIntyre, C.L.; Olivares-Villegas, J.J.; Chapman, S.C. Heat and drought adaptive QTL in a wheat population designed to minimize confounding agronomic effects. *Theor. Appl. Genet.* **2010**, *121*, 1001–1021. [CrossRef]

71. Wang, A.-Y.; Li, Y.; Zhang, C.-Q. QTL mapping for stay-green in maize (Zea mays). *Can. J. Plant Sci.* **2012**, *92*, 249–256. [CrossRef]

72. Sourdille, P.; Singh, S.; Cadalen, T.; Brown-Guedira, G.L.; Gay, G.; Qi, L.; Gill, B.S.; Dufour, P.; Murigneux, A.; Bernard, M. Microsatellite based deletion bin system for the establishment of genetic-physical map relationships in wheat (Triticum aestivum L.). *Funct. Integr. Genomics* **2004**, *4*, 12–25. [CrossRef]

73. Christopher, J.T.; Manschadi, A.M.; Hammer, G.L.; Borrell, A.K. Developmental and physiological traits associated with high yield and stay-green phenotype in wheat. *Aust. J. Agric. Res.* **2008**, *59*, 354–364. [CrossRef]

74. Lopes, M.S.; Reynolds, M.P. Stay-green in spring wheat can be determined by spectral reflectance measurements (normalized difference vegetation index) independently from phenology. *J. Exp. Bot.* **2012**, *63*, 3789–3798. [CrossRef] [PubMed]

75. Bogard, M.; Jourdan, M.; Allard, V.; Martre, P.; Perretant, M.R.; Ravel, C.; Heumez, E.; Orford, S.; Snape, J.; Griffiths, S. Anthesis date mainly explained correlations between post-anthesis leaf senescence, grain yield, and grain protein concentration in a winter wheat population segregating for flowering time QTLs. *J. Exp. Bot.* **2011**, *62*, 3621–3636. [CrossRef] [PubMed]

76. Pask, A.; Pietragalla, J. Leaf area, green crop area and senescence. In *Physiological Breeding II: A Field Guide to Wheat Phenotyping*; Pask, A., Pietragalla, J., Mullan, D., Reynolds, M., Eds.; International Maize and Wheat Improvement Center (CIMMYT): Mexico City, Mexico, 2012; pp. 58–62.

77. Kichey, T.; Hirel, B.; Heumez, E.; Dubois, F.; Le Gouis, J. In winter wheat (Triticum aestivum L.), post-anthesis nitrogen uptake and remobilization to the grain correlates with agronomic traits and nitrogen physiological markers. *Field Crops Res.* **2007**, *102*, 22–32. [CrossRef]

78. Derkx, A.; Orford, S.; Griffiths, S.; Foulkes, J.; Hawkesford, M.J.I. dentification of differentially senescing mutants of wheat and impacts on yield, biomass and nitrogen partitioning. *J. Integ. Plant Biol.* **2012**, *54*, 555–566. [CrossRef]

79. Christopher, J.T.; Veyradier, M.; Borrell, A.K.; Harvey, G.; Fletche, S.; Chenu, K. Phenotyping novel stay-green traits to capture genetic variation in senescence dynamics. *Funct. Plant Biol.* **2014**, *41*, 1035–1048. [CrossRef]

80. Reynolds, M.P.; Balota, M.; Delgado, M.I.B.; Amani, I.; Fischer, R.A. Physiological and morphological traits associated with spring wheat yield under hot irrigated conditions. *Aust. J. Plan Physiol.* **1994**, *21*, 717–730. [CrossRef]

81. Reynolds, M.P.; Singh, R.P.; Ibrahim, A.; Ageeb, O.A.A.; Larque Saavedra, A.; Quick, J.S. Evaluating Physiological traits to complement empirical selection for wheat in warm environments. *Euphytica* **1998**, *100*, 84–95. [CrossRef]

82. Gorny, A.G.; Garczynski, S. Genotypic and nutrition-dependent variation in water use efficiency and photosynthetic activity of leaves in winter wheat (Triticum aestivum L.). *J. Appl. Genet.* **2002**, *43*, 145–160.

83. Montazeaud, G.; Karatogma, H.; Ozturk, I.; Roumet, P.; Ecarnot, M.; Crossa, J.; Lopes, M.S. Predicting wheat maturity and stay-green parameters by modeling spectral reflectance measurements and their contribution to grain yield under rainfed conditions. *Field Crop Res.* **2016**, *196*, 191–198. [CrossRef]

84. Gizaw, S.A.; Garland-Campbell, K.; Carter, A.H. Evaluation of agronomic traits and spectral reflectance in Pacific Northwest wheat under rain-fed and irrigated conditions. *Field Crop Res.* **2016**, *196*, 168–179. [CrossRef]

85. Foulkes, M.J.; Sylvester-Bradley, R.; Weightman, R.; Snape, J.W. Identifying physiological traits associated with improved drought resistance in winter wheat. *Field Crops Res.* **2007**, *103*, 11–24. [CrossRef]

86. Fischer, R.A. The importance of grain or kernel number in wheat: A reply to Sinclair and Jamieson. *Field Crops Res.* **2008**, *105*, 15–21. [CrossRef]

87. Liang, X.; Liu, Y.; Chen, J.; Adams, C. Late-season photosynthetic rate and senescence were associated with grain yield in winter wheat of diverse origins. *J. Agron. Crop Sci.* **2018**, *204*, 1–12. [CrossRef]

88. Lopes, M.S.; Reynolds, M.P.; Manes, Y.; Singh, R.P.; Crossa, J.; Braun, H.J. Genetic yield gains and changes in associated traits of CIMMYT spring bread wheat in a "Historic" set representing 30 years of breeding. *Crop Sci.* **2012**, *52*, 1123–1131. [CrossRef]

89. Kipp, S.; Mistele, B.; Schmidhalter, U. Identification of stay-green and early senescence phenotypes in high-yielding winter wheat, and their relationship to grain yield and grain protein concentration using high-throughput phenotyping techniques. *Funct. Plant Biol.* **2013**, *41*, 227–235. [CrossRef]

90. João, P.; Pennacchi, D.; Elizabete, C.D.; Andralojc, P.J.; Feuerhelm, D.; Powers, S.J.; Martin Parry, A.J. Dissecting wheat grain yield drivers in a mapping population in the UK. *Agronomy* **2018**, *8*, 94.

91. Rebetzke, G.J.; Jimenez-Berni, J.A.; Bovill, W.D.; Deery, D.M.; James, R.A. High-throughput phenotyping technologies allow accurate selection of stay-green. *J. Exp. Bot.* **2016**, *67*, 4919–4924. [CrossRef]

92. Fu, J.D.; Yan, Y.F.; Kim, M.Y.; Lee, S.H.; Lee, B.W. Population-specific quantitative trait loci mapping for functional stay-green trait in rice (Oryza sativa L.). *Genome* **2011**, *5*, 235. [CrossRef]

93. Ishimaru, K.; Yano, M.; Aoki, N.; Ono, K.; Hirose, T. Toward the mapping of physiological and agronomic characters on a rice function map: QTL analysis and comparison between QTLs and expressed sequence tags. *Theor. Appl. Genet.* **2004**, *102*, 793–800. [CrossRef]

94. Teng, S.; Qian, Q.; Zeng, D.; Kunihiro, Y.; Fujimoto, K.; Huang, D.; Zhu, L. QTL analysis of leaf photosynthetic rate and related physiological traits in rice (Oryza sativa L.). *Euphytica* **2004**, *135*, 1–7. [CrossRef]

95. Jiang, G.H.; He, Y.Q.; Xu, C.G.; Li, X.H.; Zhang, Q. The genetic basis of stay-green in rice analyzed in a population of doubled haploid lines derived from an indica by japonica cross. *Theor. Appl. Genet.* **2004**, *108*, 688–698. [CrossRef] [PubMed]

96. Lim, J.H.; Yang, H.J.; Jung, K.H.; Yoo, S.C.; Paek, N.C. Quantitative trait locus mapping and candidate gene analysis for plant architecture traits using whole genome re-sequencing in rice. *Mol. Cells* **2014**, *37*, 149–160. [CrossRef] [PubMed]

97. Cha, K.W.; Koh, H.J.; Lee, Y.J.; Lee, B.M.; Nam, Y.W.; Paek, N.C. Isolation, characterization, and mapping of the stay-green mutant in rice. *Theor. Appl. Genet.* **2002**, *104*, 526–532. [CrossRef]

98. Kusaba, M.; Ito, H.; Morita, R.; Iida, S.; Sato, Y.; Fujimoto, M.; Kawasaki, S.; Tanaka, R.; Hirochika, H.; Nishimura, M.; et al. Rice non-yellow coloring 1 is involved in light-harvesting complex II and grana degradation during leaf senescence. *Plant Cell* **2007**, *19*, 1362–1375. [CrossRef] [PubMed]

99. Morita, R.; Sato, Y.; Yu, M.; Nishimura, M.; Kusaba, M. Defect in non-yellow coloring 3, an alpha/beta hydrolase-fold family protein, causes a stay-green phenotype during leaf senescence in rice. *Plant J.* **2009**, *59*, 940–952. [CrossRef] [PubMed]

100. Jiang, H.; Li, M.; Liang, N.; Yan, H.; Wei, Y.; Xu, X.; Liu, J.; Xu, Z.; Chen, F.; Wu, G. Molecular cloning and function analysis of the stay green gene in rice. *Plant J.* **2007**, *52*, 197–209. [CrossRef]
101. Tanaka, A.; Tanaka, R. Chlorophyll metabolism. *Curr. Opin. Plant Biol.* **2006**, *9*, 248–255. [CrossRef]
102. Liu, X.; Li, Z.; Jiang, Z.; Zhao, Y.; Peng, J.; Jin, J.; Guo, H.; Luo, J. LSD: A leaf senescence database. *Nucl. Acid. Res.* **2011**, *39*, D1103–D1107. [CrossRef]
103. Wu, H.; Wang, B.; Chen, Y.; Liu, Y.; Chen, L. Characterization and fine mapping of the rice premature senescence mutant ospse1. *Theor. Appl. Genet.* **2013**, *126*, 1897–1907. [CrossRef]
104. Lin, A.; Wang, Y.; Tang, J.; Xue, P.; Li, C.; Liu, L.; Hu, B.; Yang, F.; Loake, G.J.; Chu, C. Nitric oxide and protein S-nitrosylation are integral to hydrogen peroxide-induced leaf cell death in rice. *Plant Physiol.* **2012**, *158*, 451–464. [CrossRef]
105. Huang, Q.; Shi, Y.; Zhang, X.; Song, L.; Feng, B.; Wang, H.; Xu, X.; Li, X.; Guo, D.; Wu, J. Single base substitution in OsCDC48 is responsible for premature senescence and death phenotype in rice. *J. Integr. Plant Biol.* **2016**, *58*, 12–28. [CrossRef] [PubMed]
106. Rao, Y.; Yang, Y.; Xu, J.; Li, X.; Leng, Y.; Dai, L.; Huang, L.; Shao, G.; Ren, D.; Hu, J.; et al. EARLY SENESCENCE1 encodes a SCAR-LIKE PROTEIN2 that affects water loss in rice. *Plant Physiol.* **2015**, *169*, 1225–1239. [CrossRef] [PubMed]
107. Wu, L.; Ren, D.; Hu, S.; Li, G.; Dong, G.; Jiang, L.; Hu, X.; Ye, W.; Cui, Y.; Zhu, L.; et al. Down-regulation of a nicotinate phosphoribosyl transferase gene, OsNaPRT1, leads to withered leaf tips. *Plant Physiol.* **2016**, *171*, 1085–1098. [PubMed]
108. Jiao, B.; Wang, J.; Zhu, X.; Zeng, L.; Li, Q.; He, Z. A novel protein RLS1 with NB-ARM domains is involved in chloroplast degradation during leaf senescence in rice. *Mol. Plant* **2012**, *5*, 205–217. [CrossRef]
109. Liang, C.; Wang, Y.; Zhu, Y.; Tang, J.; Hu, B.; Liu, L.; Ou, S.; Wu, H.; Sun, X.; Chu, J.; et al. OsNAP connects abscisic acid and leaf senescence by fine-tuning abscisic acid biosynthesis and directly targeting senescence-associated genes in rice. *PNAS* **2014**, *111*, 10013–10018. [CrossRef]
110. Hoang, T.B.; Kobata, T. Stay-green in Rice (*Oryza sativa* L.) of drought prone areas in desiccated soils. *Plant Prod. Sci.* **2009**, *12*, 397–408. [CrossRef]
111. Adams, W.W.; Winter, K.; Schreiber, U.; Schramel, P. Photosynthesis and Chl fluorescence characteristics in relationship to changes in pigment and element composition of leaves of Platanus occidentalis L. during autumnal leaf senescence. *Plant Physiol.* **1990**, *92*, 1184–1190. [CrossRef]
112. Weng, X.Y.; Xu, H.X.; Jiang, D.A. Characteristics of gas exchange, Chl fluorescence and expression of key enzymes in photosynthesis during leaf senescence in rice plants. *J. Integr. Plant Biol.* **2005**, *47*, 560–566. [CrossRef]
113. Zhang, C.J.; Chen, G.X.; Gao, X.X.; Chu, C.J. Photosynthetic decline in flag leaves of two field-grown spring wheat cultivars with different senescence properties. *S. Afr. J. Bot.* **2006**, *72*, 15–23. [CrossRef]
114. Hörtensteiner, S.; Kräutler, B. Chl breakdown in higher plants. *Biochim. Biophys. Acta* **2011**, *1807*, 977–988. [CrossRef]
115. Kusaba, M.; Tanaka, A.; Tanaka, R. Stay-green plants: What do they tell us about the molecular mechanism of leaf senescence. *Photosynth Res.* **2013**, *117*, 221–234. [CrossRef] [PubMed]
116. Rong, H.; Tang, Y.; Zhang, H.; Wu, P.; Chen, Y.; Li, M.; Wu, G.; Jiang, H. The Stay-green rice like (SGRL) gene regulates Chl degradation in rice. *J. Plant Physiol.* **2013**, *170*, 1367–1373. [CrossRef] [PubMed]
117. Mao, C.; Lu, S.; Lv, B.; Zhang, B.; Shen, J.; He, J.; Luo, L.; Xi, D.; Chen, X.; Ming, F. A rice NAC transcription factor promotes leaf senescence via ABA biosynthesis. *Plant Physiol.* **2017**, *174*, 1747–1763. [CrossRef] [PubMed]
118. Park, J.H.; Lee, B.W. Photosynthetic characteristics of rice cultivars with depending on leaf senescence during grain filling. *Korean J. Crop Sci.* **2003**, *48*, 216–223.
119. Agrama, H.A.S.; Moussa, M.E. Mapping QTLs in breeding for drought tolerance in maize (Zea mays L.). *Euphytica* **1996**, *91*, 89–97. [CrossRef]
120. Ribaut, J.M.; Jiang, C.; Gonzalez-de-Leon, D.; Edmeades, G.O.; Hoisington, D.A. Identification of quantitative trait loci under drought conditions in tropical maize: Part 2. Yield components and marker-assisted selection strategies. *Theor. Appl. Genet.* **1997**, *94*, 887–896. [CrossRef]
121. Tuberosa, R.; Salvi, S.; Sanguineti, M.C.; Landi, P.; Maccaferri, M.; Conti, S. Mapping QTLs regulating morpho-physiological traits and yield: Case studies, shortcomings and perspectives in drought-stressed maize. *Ann. Bot.* **2002**, *89*, 941–963. [CrossRef]

122. Beavis, W.D.; Smith, O.S.; Grant, D.; Fincher, R. Identification of quantitative trait loci using a small sample of top crossed and F4 progeny from maize. *Crop Sci.* **1994**, *34*, 882–896. [CrossRef]

123. Câmara, T.M.M. Mapping QTL for Traits Related to Drought Stress Tolerance in Tropical Maize (In Portuguese with English Abstract). Ph.D. Thesis, Agriculture College Luiz de Queiroz, University of São Paulo, Piracicaba, SP, Brazil, 2006.

124. Yang, Z.; Li, X.; Zhang, N.; Wang, X.; Zhang, Y.; Ding, Y.; Kuai, B.; Huang, X. Mapping and validation of the quantitative trait loci for leaf stay-green-associated parameters in maize. *Plant Breed.* **2017**, *136*, 188–196. [CrossRef]

125. Sekhon, R.D.; Saski, C.; Kumar, R.; Flinn, B.S.; Luo, F.; Beissinger, T.M.; Ackerman, A.J.; Breitzman, M.M.W.; Bridges, W.C.; de Leon, N.; et al. Integrated genome-scale analysis identifies novel genes and networks underlying senescence in maize. *Plant Cell* **2019**, *31*, 1968–1989. [CrossRef]

126. Benchimol, L.L.; de Souza, C.L., Jr.; de Souza, A.P. Microsatellite-assisted backcross selection in maize. *Genet. Mol. Biol.* **2005**, *28*, 789–797. [CrossRef]

127. Neereja, C.N.; Maghirang-Rodriguez, R.; Pamplona, A.; Heuer, S.; Collard, B.C.Y.; Sptiningsih, E.M.; Vergara, G.; Sanchez, D.; Xu, K.; Ismail, A.M.; et al. A marker-assisted backcross approach for developing submergence-tolerant rice cultivars. *Theor. Appl. Genet.* **2007**, *115*, 767–776. [CrossRef] [PubMed]

128. Garzón, L.N.; Ligarreto, G.A.; Blair, M.W. Molecular marker-assisted backcrossing of anthracnose resistance into Andean climbing beans Phaseolus vulgaris L. *Crop Sci.* **2008**, *48*, 562–570. [CrossRef]

129. Chen, L.; An, Y.; Li, Y.; Li, C.; Shi, Y.; Song, Y.; Zhang, D.; Wang, T.; Li, Y. Candidate loci for yield-related traits in maize revealed by a combination of meta QTL analysis and regional association mapping. *Front. Plant Sci.* **2017**, *8*, 2190. [CrossRef] [PubMed]

130. Swanckaert, J.; Pannecocque, J.; Vanwaes, J.; Steppee, K.; Vanlabeke, M.-C.; Rehul, D. Stay-green characterization in Belgian forage maize. *J. Agric. Sci.* **2017**, *155*, 766–776. [CrossRef]

131. Hay, R.K.M.; Porter, J.R. *The Physiology of Crop Yield*; Blackwell Publishing: Oxford, UK, 2006; ISBN 9781405108591.

132. Pennisi, E. Plant genetics: The blue revolution, drop by drop, gene by gene. *Science* **2008**, *320*, 171–173. [CrossRef]

133. Ludlow, M.M.; Muchow, R.C. A critical evaluation of traits for improving crop yields in water-limited environments. *Adv. Agron.* **1990**, *43*, 107–153.

134. He, Y.; Liu, Y.; Cao, W.; Huai, M.; Xu, B.; Huang, B. Effects of salicylic acid on heat tolerance associated with antioxidant metabolism in Kentucky Bluegrass. *Am. J. Crop Sci.* **2005**, *45*, 988–995. [CrossRef]

135. Cai, H.; Chu, Q.; Yuan, L.; Liu, J.; Chen, X.; Chen, F.; Mi, G.; Zhang, F. Identification of quantitative trait loci for leaf area and Chl content in maize (Zea mays L.) under low nitrogen and low phosphorus supply. *Mol. Breed.* **2012**, *30*, 251–266. [CrossRef]

136. Cai, H.; Chu, Q.; Yuan, L.; Liu, J.; Chen, X.; Chen, F.; Mi, G.; Zhang, F. Identification of QTL for plant height, ear height and grain yield in maize (Zea mays L.) in response to nitrogen and phosphorus supply. *Plant Breed.* **2012**, *131*, 502–510. [CrossRef]

137. Li, L.; Zhang, Q.; Huang, D.A. Review of imaging techniques for plant phenotyping. *Sensors* **2014**, *14*, 20078–20111. [CrossRef] [PubMed]

138. Trachsel, S.; Sun, D.; San Vicente, F.M.; Zheng, H.; Atlin, G.N.; Suarez, E.A. Identification of QTL for early vigor and stay-green conferring tolerance to drought in two connected advanced backcross populations in tropical maize (Zea mays L.). *PLoS ONE* **2016**, *11*, e0149636.

139. Cerrudo, D.; Cao, S.; Yuan, Y.; Martinez, C.; Suarez, E.A.; Babu, R.; Zhang, X.; Trachsel, S. Genomic selection outperforms marker assisted selection for grain yield and physiological traits in a maize doubled haploid population across water treatments. *Front. Plant Sci.* **2018**, *9*, 366. [CrossRef] [PubMed]

140. Duvick, D.N.; Smith, J.S.C.; Cooper, M. Long-term selection in a commercial hybrid maize breeding program. *Plant Breed Rev.* **2004**, *24*, 109–151.

141. Crosbie, T.M.; Pearce, R.B.; Mock, J.J. Selection for high CO2 exchange rate among inbred lines of maize. *Crop Sci.* **1981**, *21*, 629–631. [CrossRef]

142. Valentinuz, O.R.; Tollenaar, M. Effect of genotype, nitrogen, plant density, and row spacing on the area-per leaf profile in maize. *Agron. J.* **2006**, *98*, 94–99. [CrossRef]

143. Cairns, J.E.; Sanchez, C.; Vargas, M.; Ordoñez, R.; Araus, J.L. Dissecting maize productivity: Ideotypes associated with grain yield under drought stress and well-watered conditions. *J. Integr. Plant Biol.* **2012**, *54*, 1007–1020. [CrossRef]

144. Sallam, A.; Alqudah, A.M.; Dawood, M.F.A.; Baenziger, P.S.; Börner, A. Drought Stress Tolerance in Wheat and Barley: Advances in Physiology, Breeding and Genetics Research. *Int. J. Mol. Sci.* **2019**, *20*, 3137. [CrossRef]

145. Obsa, B.T.; Eglinton, J.; Coventry, S.; March, T.; Langridge, P.; Fleury, D. Genetic analysis of developmental and adaptive traits in three doubled haploid populations of barley (*Hordeum vulgare* L.). *Theor. Appl. Genet.* **2016**, *129*, 1139–1151. [CrossRef]

146. Fox, G.P. Discovery of QTL for stay-green and heat-stress in barley (*Hordeum vulgare*) grown under simulated abiotic stress conditions. *Euphytica* **2015**, *207*, 2.

147. Gonzalez, A.; Martin, I.; Ayerbe, L. Response of barley genotypes to terminal soil moisture stress: Phenology, growth, and yield. *Aust. J. Agric. Res.* **2007**, *58*, 29–37. [CrossRef]

148. Maydup, M.L.; Antonietta, M.; Buiamet, J.J.; Graciano, C.; Lopez, J.R.; Tambussi, E.A. The contribution of ear photosynthesis to grain filling in bread wheat (*Triticum aestivum* L.). *Field Crops Res.* **2010**, *119*, 48–58. [CrossRef]

149. Vaezi1, B.; Bavei1, V.; Shiran, B. Screening of barley genotypes for drought tolerance by agro-physiological traits in field condition. *Afr. J. Agric. Res.* **2010**, *5*, 881–892.

150. Seiler, C.; Harshavardhan, V.T.; Reddy, P.S.; Hensel, G.; Kumlehn, J.; Eschen-Lippold, L.; Rajesh, K.; Korzun, V.; Wobus, U.; Lee, J.; et al. Abscisic acid flux alterations result in differential ABA signaling responses and impact assimilation efficiency in barley under terminal drought stress. *Plant Physiol.* **2014**, *164*, 1677–1696. [CrossRef] [PubMed]

151. Shirdelmoghanloo, H.; Paynter, B.; Chen, K.; D'Antuono, M.; Balfour, C.A.; Angesa, T.; Westcott, S.; Li, C. Grain plumpness in barley under grain filling heat stress: Association with grain growth components and stay-green. In Proceedings of the 19th Australian Barley Technical Symposium, Perth, WA, Australia, 9–12 September 2019.

152. Magney, T.S.; Eitel, J.U.H.; Huggins, D.R.; Viering, L.A. Proximal NDVI derived phenology improves in-season predictions of wheat quality and quantity. *Agric. Forest Meteorol.* **2015**, *217*, 46–60. [CrossRef]

153. Gous, P.W.; Hasjim, J.; Franckowiak, J.; Fox, G.P.; Gilbert, R. Barley genotype expressing "SG"-like characteristics maintains starch quality of the grain during water stress condition. *J. Cereal Sci.* **2013**, *58*, 414–419. [CrossRef]

154. Zhou, C.; Han, L.; Pislariu, C.; Nakashima, J.; Fu, C.; Jiang, Q.; Quan, L.; Blancaflor, E.B.; Tang, Y.; Bouton, J.H. From model to crop: Functional analysis of a Stay-green gene in the model legume Medicago truncatula and effective use of the gene for alfalfa improvement. *Plant Physiol.* **2011**, *157*, 1483–1496. [CrossRef]

155. Patro, L.; Mohapatra, P.K.; Biswal, U.C.; Biswal, B. Dehydration induced loss of photosynthesis in Arabidopsis leaves during senescence is accompanied by the reversible enhancement in the activity of cell wall β-glucosidase. *J. Photochem. Photobiol.* **2014**, *137*, 49–54. [CrossRef]

156. Bromley, J.R.; Busse-Wicher, M.; Tryfona, T.; Mortimer, J.C.; Zhang, Z.; Brown, D.M.; Dupree, P. GUX1 and GUX2 glucuronyl transferases decorate distinct domains of glucuronoxylan with different substitution patterns. *Plant J.* **2013**, *74*, 423–434. [CrossRef]

157. Rennie, E.A.; Hansen, S.F.; Baidoo, E.E.; Hadi, M.Z.; Keasling, J.D.; Scheller, H.V. Three members of the Arabidopsis glycosyltransferase family 8 are xylan glucuronosyl transferases. *Plant Physiol.* **2012**, *159*, 1408–1417. [CrossRef]

158. Zeng, W.; Chatterjee, M.; Faik, A. UDP-Xylose-stimulated glucuronyl transferase activity in wheat microsomal membranes: Characterization and role in glucuronol (arabino) xylan biosynthesis. *Plant Physiol.* **2008**, *147*, 78–91. [CrossRef] [PubMed]

159. Wang, Y.; Xiao, Y.; Zhang, Y.; Chai, C.; Gang, W.; Wei, X.; Xu, H.; Wang, M.; Ouwerkerk, P.B.F.; Zhu, Z. Molecular cloning, functional characterization and expression analysis of a novel monosaccharide transporter gene OsMST6 from rice (Oryza sativa L.). *Planta* **2008**, *228*, 525–535. [CrossRef] [PubMed]

160. Zhou, Y.; Liu, L.; Huang, W.; Yuan, M.; Zhou, F.; Li, X. Overexpression of OsSWEET5 in rice causes growth retardation and precocious senescence. *PLoS ONE* **2014**, *9*, e94210. [CrossRef] [PubMed]

Int. J. Mol. Sci. **2019**, *20*, 5837

161. Kim, J.Y.A.; Mahé, A.; Guy, S.; Brangeona, J.; Rochea, O.; Chourey, P.S.; Priou, J.L. Characterization of two members of the maize gene family, Incw3andIncw4, encoding cell-wall invertases. *Gene* **2000**, *245*, 89–102. [CrossRef]

162. Gal, C.; Moore, K.M.; Paszkiewicz, K.; Kent, N.A.; Whitehall, S.K. The impact of the HIRA histone chaperone upon global nucleosome architecture. *Cell Cycle* **2015**, *14*, 123–134. [CrossRef] [PubMed]

163. Qi, W.; Yang, Y.; Feng, X.; Zhang, M.; Song, R. Mitochondrial function and maize kernel development requires Dek2, a pentatricopeptide repeat protein involved in nad1 mRNA splicing. *Genetics* **2017**, *205*, 239–249. [CrossRef]

164. Abdi Woldesemayat, A.; Modise, D.M.; Gemeildien, J.; Ndimba, B.K.; Christoffels, A. Cross-species multiple environmental stress responses: An integrated approach to identify candidate genes for multiple stress tolerance in sorghum (*Sorghum bicolor* (L.) Moench) and related model species. *PLoS ONE* **2018**, *13*, e0192678.

165. Moore, G. Cereal genome evolution: Pastoral pursuits with Lego' genomes. *Curr. Opin. Genet. Dev.* **1995**, *5*, 717–724. [CrossRef]

166. Foote, T.; Roberts, M.; Kurata, N.; Sasaki, T.; Moore, G. Detailed comparative mapping of cereal chromosome regions corresponding to the *Phl* locus in wheat. *Genetics* **1997**, *147*, 801–807.

International Journal of
Molecular Sciences

Article

Functional Analysis of the Soybean *GmCDPK3* Gene Responding to Drought and Salt Stresses

Dan Wang [1,2,†], Yuan-Xia Liu [1,†], Qian Yu [1], Shu-Ping Zhao [2], Juan-Ying Zhao [2], Jing-Na Ru [2], Xin-You Cao [3], Zheng-Wu Fang [4], Jun Chen [2], Yong-Bin Zhou [2], Ming Chen [2], You-Zhi Ma [2], Zhao-Shi Xu [2,*] and Jin-Hao Lan [1,*]

1   College of Agronomy, Qingdao Agricultural University, Qingdao 266109, China; wangdanyx@stu.qau.edu.cn (D.W.); yuanxialiu@163.com (Y.-X.L.); qianstyle123@sina.com (Q.Y.)
2   Key Laboratory of Biology and Genetic Improvement of Triticeae Crops, Ministry of Agriculture, Institute of Crop Sciences, Chinese Academy of Agricultural Sciences (CAAS)/National Key Facility for Crop Gene Resources and Genetic Improvement, Beijing 100081, China; zhaoshuping001@163.com (S.-P.Z.); zjy0502@yeah.net (J.-Y.Z.); rujingna1993@163.com (J.-N.R.); chenjun01@caas.cn (J.C.); zhouyongbin@caas.cn (Y.-B.Z.); chenming02@caas.cn (M.C.); mayouzhi@caas.cn (Y.-Z.M.)
3   National Engineering Laboratory for Wheat and Maize/Key Laboratory of Wheat Biology and Genetic Improvement, Crop Research Institute, Shandong Academy of Agricultural Sciences, Jinan 250100, China; cxytvs@163.com
4   College of Agronomy, College of Agriculture, Yangtze University, Jingzhou 434025, China; fangzhengwu88@163.com
*   Correspondence: xuzhaoshi@caas.cn (Z.-S.X.); jinhao2005@163.com (J.-H.L.)
†   These authors contributed equally to this work.

Received: 22 October 2019; Accepted: 20 November 2019; Published: 25 November 2019

Abstract: Plants have a series of response mechanisms to adapt when they are subjected to external stress. Calcium-dependent protein kinases (CDPKs) in plants function against a variety of abiotic stresses. We screened 17 CDPKs from drought- and salt-induced soybean transcriptome sequences. The phylogenetic tree divided CDPKs of rice, *Arabidopsis* and soybean into five groups (I–V). *Cis*-acting element analysis showed that the 17 CDPKs contained some elements associated with drought and salt stresses. Quantitative real-time PCR (qRT-PCR) analysis indicated that the 17 CDPKs were responsive after different degrees of induction under drought and salt stresses. *GmCDPK3* was selected as a further research target due to its high relative expression. The subcellular localization experiment showed that GmCDPK3 was located on the membrane of *Arabidopsis* mesophyll protoplasts. Overexpression of *GmCDPK3* improved drought and salt resistance in *Arabidopsis*. In the soybean hairy roots experiment, the leaves of *GmCDPK3* hairy roots with RNA interference (*GmCDPK3*-RNAi) soybean lines were more wilted than those of *GmCDPK3* overexpression (*GmCDPK3*-OE) soybean lines after drought and salt stresses. The trypan blue staining experiment further confirmed that cell membrane damage of *GmCDPK3*-RNAi soybean leaves was more severe than in *GmCDPK3*-OE soybean lines. In addition, proline (Pro) and chlorophyll contents were increased and malondialdehyde (MDA) content was decreased in *GmCDPK3*-OE soybean lines. On the contrary, *GmCDPK3*-RNAi soybean lines had decreased Pro and chlorophyll content and increased MDA. The results indicate that *GmCDPK3* is essential in resisting drought and salt stresses.

Keywords: calcium-dependent protein kinase; abiotic stresses; responsive mechanism; soybean hairy root; *Arabidopsis*

---

1. Introduction

Plants inevitably experience a variety of abiotic stresses during their growth and development process, such as drought, salt, and extreme temperature [1]. To adapt to certain environmental changes, plants evolved regulatory pathways in response to different stress signal stimuli [2]. As signals are transmitted through the cell, the plant will regulate the expression of some genes and produce new proteins, which will also change noticeably at the morphological, physiological, and biochemical levels [3].

Calcium ions (Ca^{2+}) are ubiquitous and are the most important messenger molecules in higher plant cells. Ca^{2+} is involved in regulating growth and development of plants and functions directly in the biotic and abiotic stress response of plants [4]. When plants are exposed to different external stimuli, Ca^{2+} acts as the second messenger in the cell, and its concentration in the cytoplasm undergoes transient and complex changes, producing Ca^{2+} oscillations, which are transmitted to the calcium-binding protein via calcium receptors, and are then transmitted and amplified by calcium-binding proteins. The signal goes downstream, causing changes in the expression of the corresponding gene, thereby regulating the response of the plant to adverse stimuli [4,5].

Calcium-dependent protein kinases (CDPKs) are a class of typical Ser/Thr protein kinases [6]. CDPKs are widely distributed in some plants, algae, and some lower animals, but not in bacteria, fungi, or higher animals [7,8]. CDPKs are essential in Ca^{2+}-mediated signal transduction in plant response to stresses. Unlike other calcium binding proteins, CDPKs are a single peptide chain. CDPKs possess four specific domains: the N-terminal domain, the ATP-binding kinase domain, the autoinhibitory junction domain, and the C-terminus domain [9,10]. The C-terminus domain of CDPKs binds with Ca^{2+}, resulting in conformational changes of CDPK proteins and causing activation of the kinase domain and autoimmune changes. This change results in important functions of CDPKs in plant responses to abiotic stresses [11].

CDPKs are widely present in various organs of plants and are expressed in roots, stems, leaves, flowers, fruits, and seeds [12–15]. CDPKs are mainly involved in physiological activities such as plant drought resistance, salt resistance, disease resistance, hormone signal transduction, photoperiod regulation, and nutrient metabolism [16]. In the past 10 years, research on plant CDPKs developed rapidly, and a large number of CDPKs were found in many plants. A total of 34 CDPKs have been found in *Arabidopsis* [10], 31 CDPKs in rice [11], 40 CDPKs in maize and 20 CDPKs in wheat [17]. In addition, CDPKs have been increasingly studied in horticultural crops, and 19 CDPKs have been identified in cucumber [13], 29 CDPKs in tomato [18], 31 CDPKs in pepper [19], 19 CDPKs in grape [20], and 29 CDPKs in millet [16].

Soybean (*Glycine max*) is one of the most important crops in the world, providing oil and protein to humans and livestock. There are currently 50 CDPKs identified from soybean [21]. The *GmCDPKs* transcript levels change after wounding, exhibiting specific expression patterns after being simulated by *Spodoptera exigua* feeding or soybean aphid (*Aphis glycines*) herbivory, and are largely independent of the phytohormones jasmonic acid and salicylic acid [15]. Based on previous reports, this study screened GmCDPKs with high expression levels in soybean transcriptome sequences from our laboratory following drought and salt stresses, and screened for 17 GmCDPKs responding to drought and salt. The genes were further verified by qRT-PCR analysis, and *GmCDPK3* was selected for further research. The results showed that in transgenic *Arabidopsis*, *GmCDPK3* gene was resistant to drought and salt environmental stress, and this characteristic was verified in soybean root experiment. The discovery could help develop plants that are resistant to drought and salt.

2. Results

2.1. Phylogenetic Tree Inference in CDPKs

All amino acid sequences were used to create phylogenetic tress in MEGA X 10.0 (Figure 1). The adjacent junction is the complete sequence of CDPKs in *Arabidopsis*, rice, and soybean, and the phylogenetic tree was divided into five groups (I–V). For 17 screened genes, 6 were in group I, 3 in group II, 2 in group III, 5 in group IV, and 1 in group V.

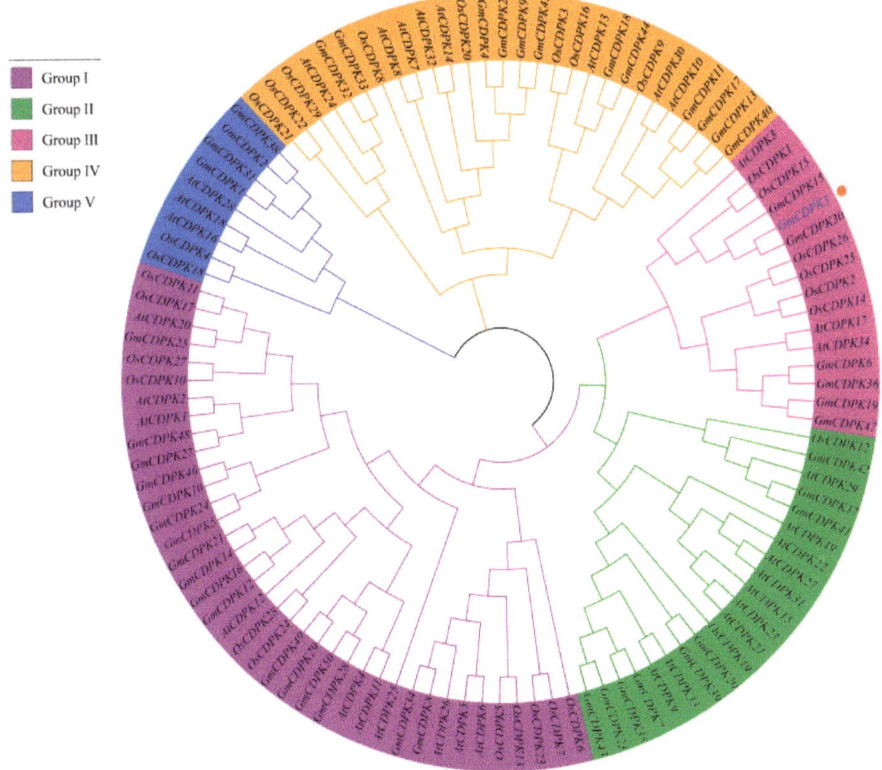

Figure 1. Phylogenetic relationships of CDPKs among *Arabidopsis*, rice, and soybean. The phylogenetic tree was generated by comparison of the CDPK amino acid sequences in MEGA 10.0. The neighbor-joining (NJ) method was used and the bootstrap value was set to 1000. Frequency values (%) higher than 50 are displayed. All the genes were divided into five groups and *GmCDPK3* is classified as group III.

2.2. Analysis of Gene and Protein Structure of 17 Selected GmCDPK

A large number of CDPKs were found in the soybean drought and salt transcriptome. To further analyze their roles in abiotic stress, we selected 17 GmCDPKs and queried basic information in the Phytozome, Pfam, and SMART databases (Table S1) [22]. The chromosomal locations of the 17 GmCDPK genes (Figure 2) on the 12 chromosomes. As shown in figure, three CDPK genes were mapped on chromosomal 2 and 5; tow CDPK genes on chromosomal 10; one CDPK gene was mapped on chromosomal 1, 3, 4, 6, 11, 14, 16, 18, and 19. To characterize the 17 GmCDPKs, we used the gene structure display server (GSDS) website (http://gsds.cbi.pku.edu.cn/) to analyze genomic sequences by submitting coding DNA. The results showed that 17 GmCDPKs contained an S-TKc protein domain and four EF hand-shaped structures (Figure 3a). The genes had exon-intron structure (Figure 3c). In addition, the 17 GmCDPKs contained nine motifs, in which *GmCDPK1/13/15* had the same structure, and the structures of other genes were similar (Figure 3b). Results showed that CDPKs tend to have a close genetic relationship with similar structures, suggesting that they evolved from the same pattern.

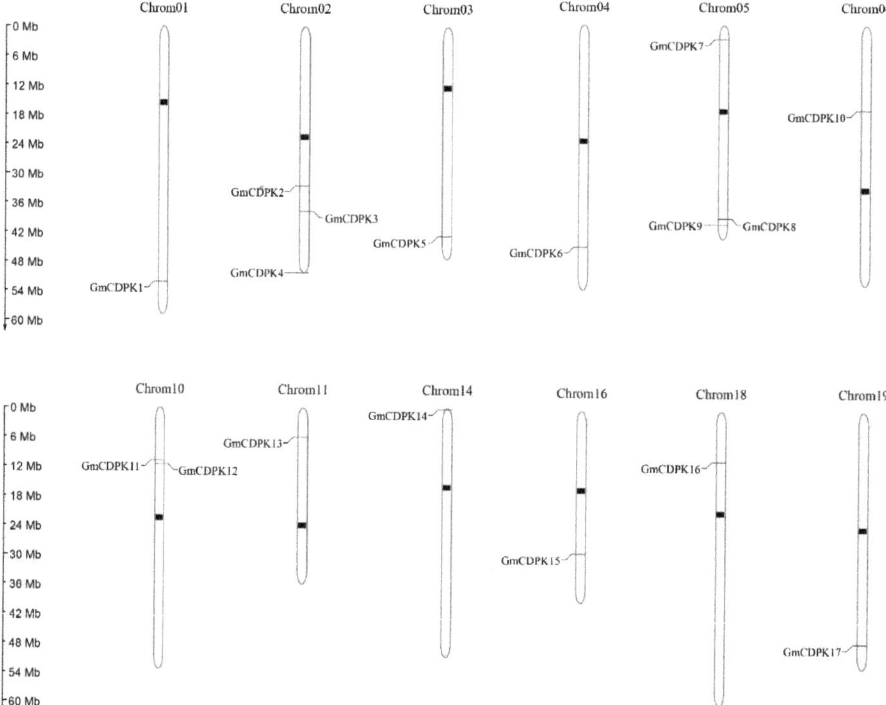

Figure 2. Chromosomal locations of the 17 GmCDPK genes. Using Map Gene 2 Chromosomal to depicted chromosomal location. The location of the gene is marked with short lines. black mark is the centromere of the chromosome. There are 17 GmCDPK genes that are distributed on 12 chromosomes.

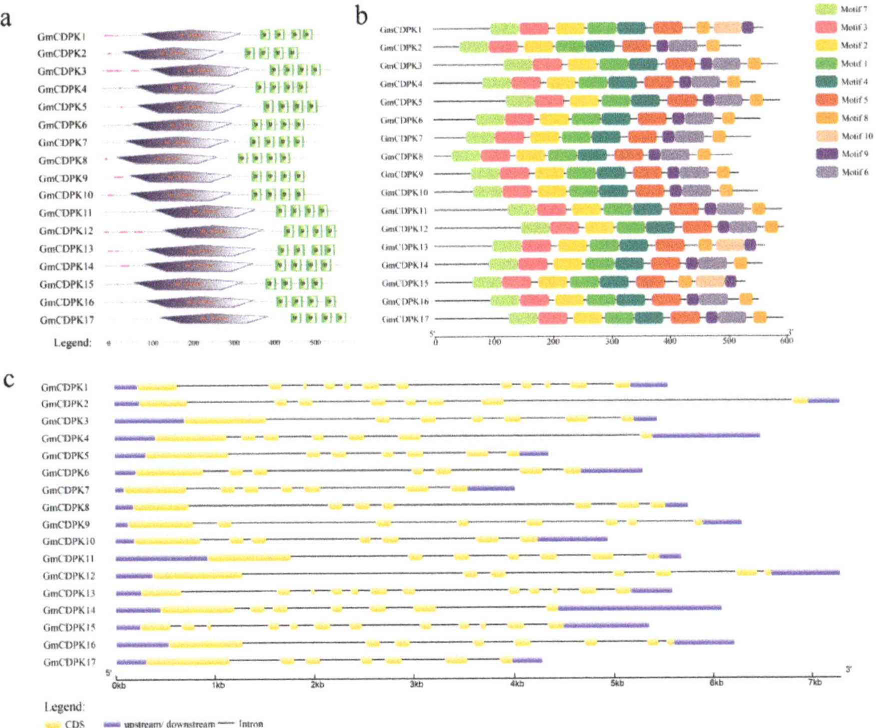

Figure 3. Bioinformatics analysis of 17 GmCDPK proteins and genes. The gene and protein sequences of 17GmCDPK were analyzed with bioanalysis tools. (**a**) The domain of the GmCDPK protein; (**b**) GmCDPKs protein motif; and (**c**) intron exon structure of GmCDPKs.

2.3. GmCDPK Protein Tertiary Structure Homology Modeling

To visualize the conserved Ser/Thr protein kinase domain and EF hand structure on the GmCDPK, protein structure models of these 17 GmCDPKs were constructed (Figure S1). The simulated three-dimensional structure shows that in addition to the unique coiled Ser/Thr protein kinase domain, there are four EF-hand structures, consisting of four helical region-bubble region-coil regions with Ca^{2+} binding in the middle of the hand structure region.

2.4. Expression of 17 GmCDPKs in Different Tissues and at Different Developmental Stages

Gene registration numbers were submitted to SoyBase (http://soybase.org/soyseq/) [23], and quantitatively predicted tissue expression data was obtained for 14 tissues (Figure S2) (details are in Table S2). The results showed that *GmCDPK6/7/8/9/10* were highly expressed in roots and basal tissues; *GmCDPK14* was highly expressed at all times in soybean; and most of the other genes were slightly expressed or had no expression under stress-free conditions. These results indicated that these genes are essential in the soybean response to stresses.

2.5. Promoter Regions of 17 GmCDPKs Contain Various Stress Response Elements

The 2000 bp sequence before the start codon ATG was cloned in these 17 GmCDPK promoters. To investigate the mechanism of response to abiotic stress, the plant *cis*-acting elements of the 17 GmCDPK promoter regions were submitted to PLACE (http://bioinformatics.psb.ugent.be/webtools/plantcare/html/) (Table S3). Many regulatory factors that respond to drought and salt stresses have been identified, including ABRE, DRE, ERE, GT-1, LTRF, MYB, and W-box elements. All of the genes except *GmCDPK12* have an ERE *Cis*-acting element, of which *GmCDPK9/13* is the most abundant. *GmCDPK3* contains seven ABRE, four EREs, two GT-1s, one LTRE, two MYBs, and four W-box *Cis*-acting elements. This information indicates that 17 GmCDPKs may be involved in abiotic stress responses.

2.6. Candidate Genes Involved in Drought and Salt Stresses

To gain insight into the potential functions of these 17 GmCDPKs in plants subjected to abiotic stresses, we detected the expression patterns under drought and salt stresses by qRT-PCR. After the drought treatment, *GmCDPK1/17* reached the highest expression at the 2nd hour, *GmCDPK2/3/15* had the highest expression at the fourth hour, and *GmCDPK8* and *GmCDPK13* had the highest expression at the first and eighth hours, respectively (Figure 4a). After the salt treatment, the expression of *GmCDPK3/8/11/12/13* reached the highest level at the eighth hour. Furthermore, the transcription levels of other genes slowly increased, but the changes were not significant (Figure 4b). These results indicate that the transcription levels of most of the GmCDPKs are affected by drought and salt stresses. *GmCDPK3* apparently responds to a variety of these stresses, so *GmCDPK3* was selected for further study.

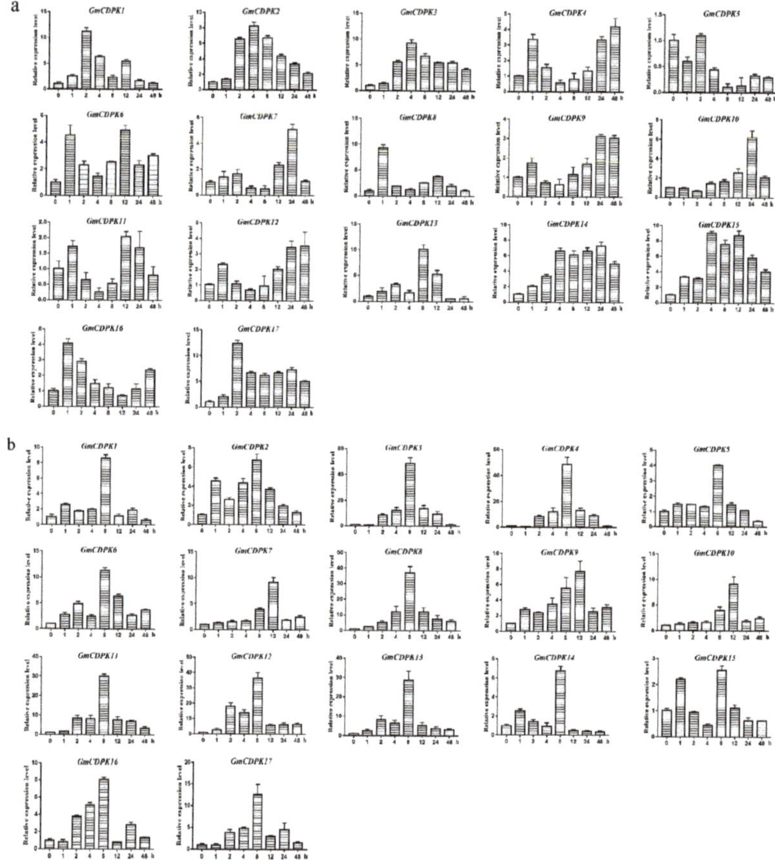

Figure 4. Expression patterns of the 17 GmCDPKs under drought and salt treatment. Using qRT-PCR to measure expression levels of 17 GmCDPKs under drought and salt stresses. (**a**) Drought treatment was applied for 0, 1, 2, 4, 8, 12, 24, and 48 h. (**b**) Expression levels of 17 GmCDPKs under 250 mM NaCl treatment for 0, 1, 2, 4, 8, 12, 24, and 48 h. The result is the mean ± SD of three experiments.

2.7. GmCDPK3 Localized on the Cell Membrane

To determine the subcellular localization of *GmCDPK3*, the *GmCDPK3* gene sequence was fused with the N-terminal end of the GFP reporter gene, and connected to the 16318 hGFP expression vector under the control of cauliflower mosaic virus (CaMV). The constructed vector was transferred to the protoplast of *Arabidopsis* and the location of GFP expression was observed. We found that GmCDPK3-GFP fusion protein was mainly located in the cell membrane (Figure 5).

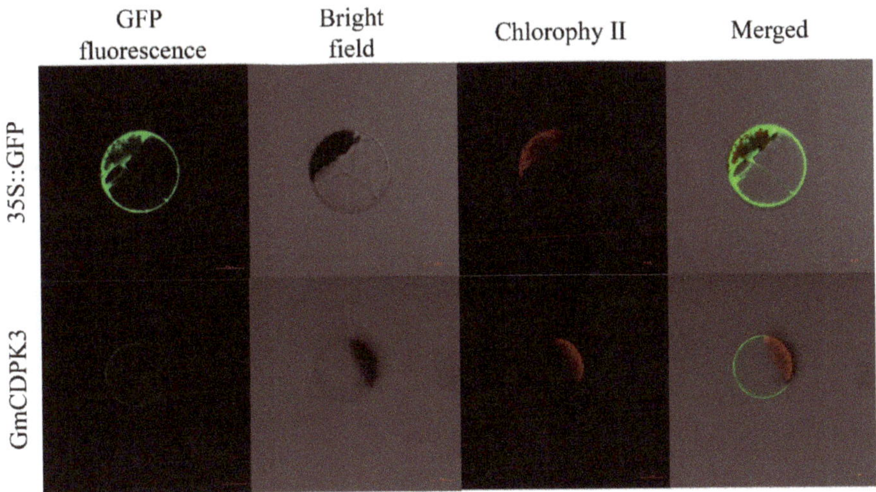

Figure 5. Subcellular localization of *GmCDPK3* overexpression plants in *Arabidopsis*. Plasmid labeled with GFP was transferred into the protoplast of *Arabidopsis* and subcellular localization image of GmCDPK3-GFP was observed under a laser scanning confocal microscope, and the image showed that it was expressed in the cell membrane. Bar = 20 μm.

2.8. GmCDPK3 Conferred Drought Tolerance in Transgenic Arabidopsis

Genes have potential functions in enhancing abiotic stress tolerance, particularly in overexpression plants [24–28]. To further study the biological function of *GmCDPK3*, overexpression (OE, in lines 5, 14, and 15) lines were treated with polyethylene glycol (PEG6000). To determine germination rate, seeds of OE and WT were sown on 1/2 MS0 medium containing different concentrations of PEG6000, and at 0, 12, 24, 36, 48, 60, and 72 h, and the number of germinations was recorded. The germination rate on 1/2 MS0 medium did not differ, however, in the medium containing PEG6000, the germination of WT seeds was inhibited to a greater extent than that of OE seeds (Figure 6a). Under normal conditions, the germination rates of WT and OE ranged from 96–100% at 84 h (Figure 6b). Under 9% PEG6000 treatment, the germination rate of OE was 92.79–95.92%, higher than WT (85.71%) (Figure 6c). Under 12% PEG6000 treatment, the germination rate of OE (83.67–96.94%) was still higher than that of WT (73.47%) (Figure 6d).

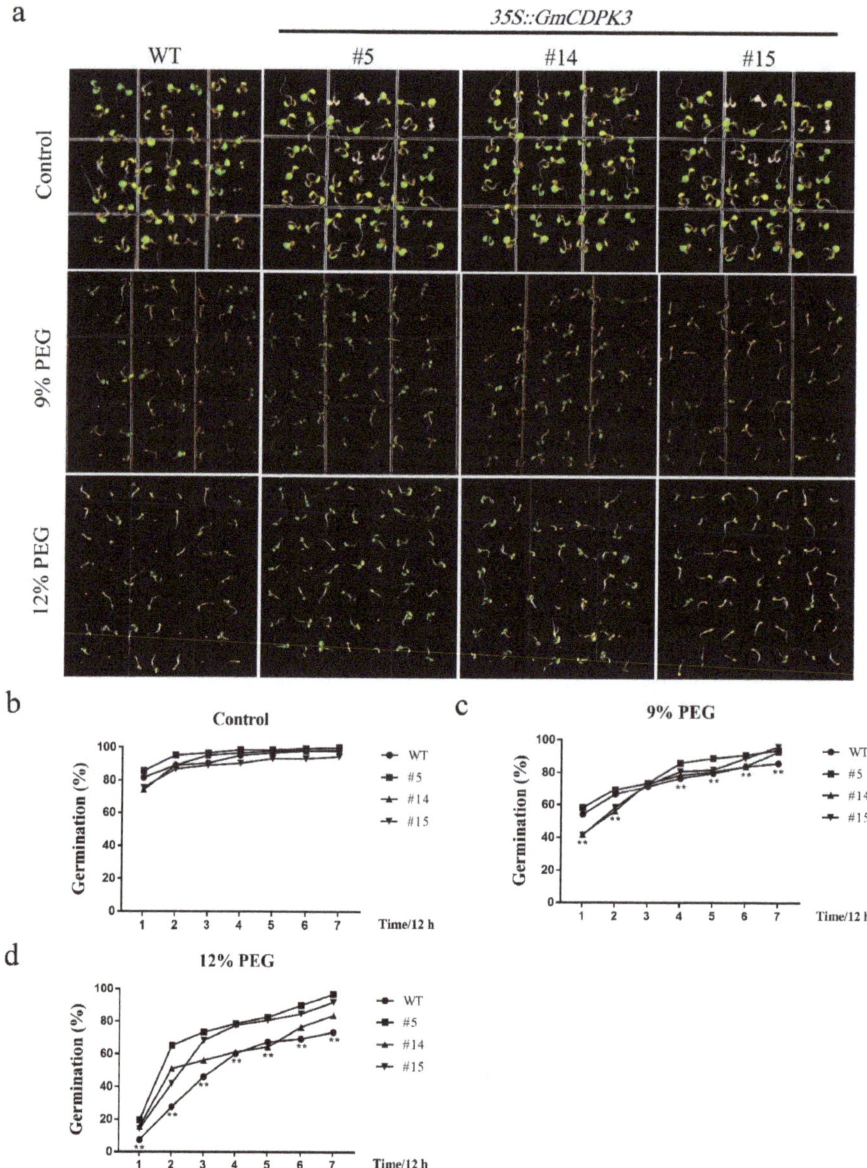

Figure 6. Germination rate of the OE and WT under different concentrations of PEG6000. After disinfection of *Arabidopsis* seeds, select the same strong seeds and sow them on the corresponding medium. Record the number of germinations every 12 h., and each experiment was repeated three times. (**a**) Germination rate phenotypes treated with different concentrations of PEG6000. (**b**) Untreated control germination rate statistics. (**c**) Germination rate statistics under 9% PEG treatment. (**d**) Germination rate statistics under 12% PEG treatment. Data are mean ± SD of three experiments ($n = 64$). The ANOVA test showed a significant difference (** $p < 0.01$).

Six-day-old *Arabidopsis* seedlings were transferred to 1/2 MS0 medium containing different concentrations of PEG6000 for a week (Figure 7a). Under normal conditions and 6% PEG6000, the root length phenotype of OE was similar to WT, with no differences among lines (Figure 7b, c). Under treatment of 9% and 12% PEG6000, the root length of the OE was significantly longer than WT (Figure 7d, e). With the increase of PEG6000 concentration, the root length became shorter.

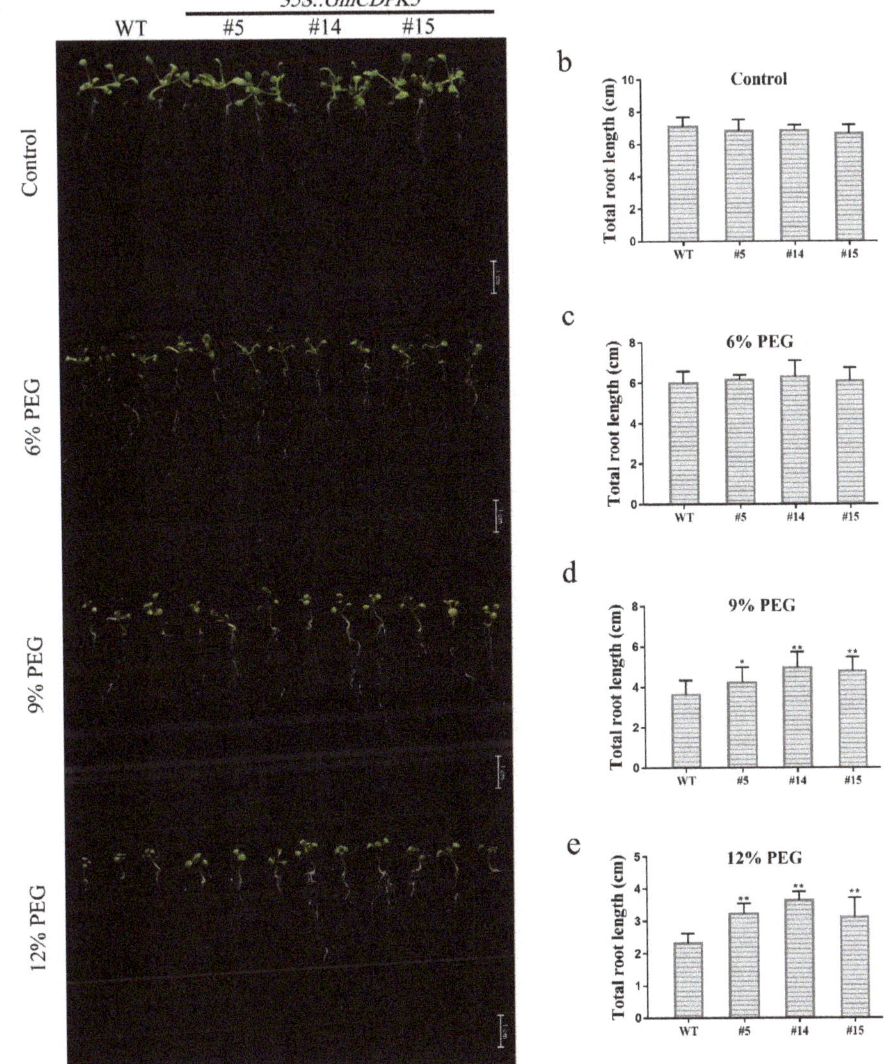

Figure 7. Root length phenotype of the OE and WT under different concentrations of PEG6000. Seedlings with consistent root length and growth state were selected to grow in Ms0 medium for about three days and transferred to another medium to observe root length. Each experiment was repeated three times. (**a**) Root growth phenotypes of OE and WT under different concentrations of PEG6000. (**b**) Untreated control total root length statistics. (**c**) Total root length statistics under 6% PEG treatment. (**d**) Total root length statistics under 9% PEG treatment. (**e**) Total root length statistics under 12% PEG treatment. Data are mean ± SD of three experiments ($n = 30$). The ANOVA test showed a significant difference (* $p < 0.05$, ** $p < 0.01$).

In the seedling treatment, 3-week-old OE and WT seedings were subjected to drought for 14 days, and then rehydration for 3 days, and the survival rates were recorded separately. The survival rate of OE after 3 days of fluid replacement was 90.05–91.63%, which was significantly higher than the WT (50.50%) (Figure 10a,c).

2.9. Salt Tolerance of GmCDPK3 in Arabidopsis

To elucidate the role of *GmCDPK3* in plant growth and development under salt treatment conditions, salt tolerance experiments of OE and WT were performed. For the germination assay, seeds of OE and WT were cultured on 1/2 MS0 medium containing different concentrations of NaCl, and the germination rate was measured (Figure 8a). OE and WT had similar germination rates on 1/2 MS0 medium. Under the treatment of 100 mM NaCl, the germination rate of OE seeds was 82.31–85.76%, which was higher than that of WT seeds (72.11%). Under the treatment of 125 mM NaCl, the germination rate of OE seeds was 65.31–86.73%, which was higher than that of WT seeds (61.22%). When treated with 150 mM NaCl, the germination rate of OE seeds (72.45–82.65%) was significantly higher than that of WT seeds (60.20%) (Figure 8b–e). The germination of OE and WT seeds was inhibited under NaCl treatment.

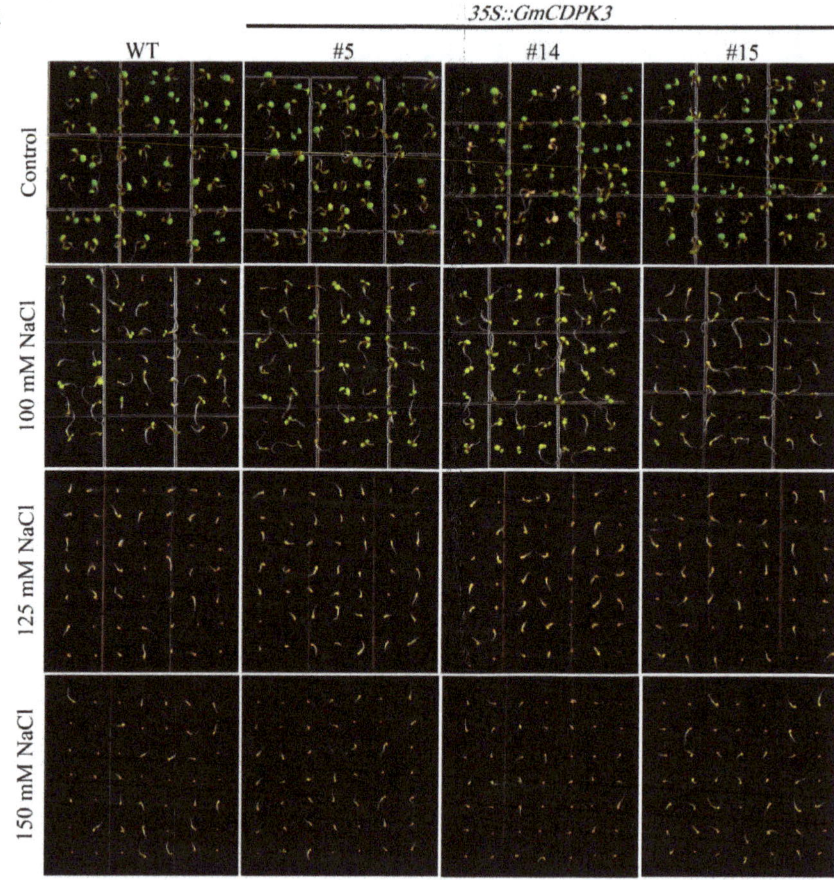

Figure 8. *Cont.*

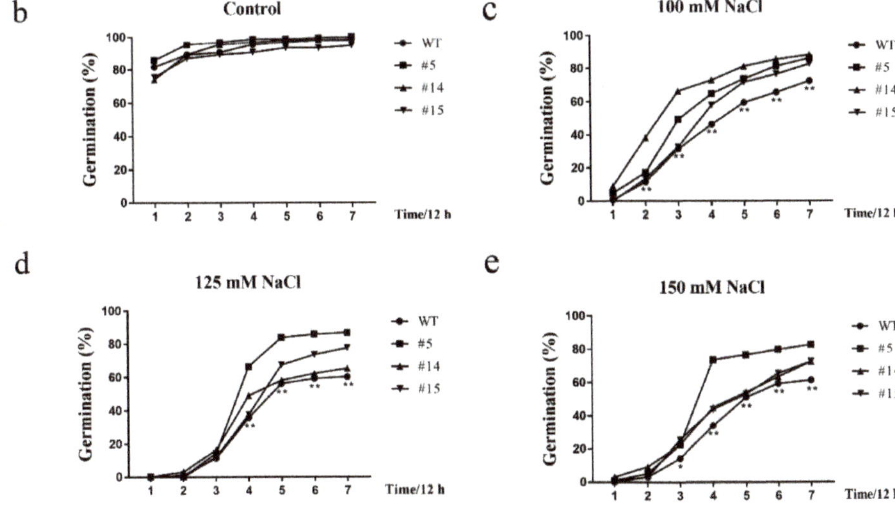

Figure 8. Germination rate of OE and WT under different concentrations of NaCl. After disinfection of *Arabidopsis* seeds, select the same strong seeds and sow them on the corresponding medium. Record the number of germinations every 12 h., and each experiment was repeated three times. (**a**) Germination rate phenotypes treated with different concentrations of NaCl. (**b**) Untreated control germination rate statistics. (**c**) Germination rate statistics under 100 mM NaCl treatment. (**d**) Germination rate statistics under 125 mM NaCl treatment. (**e**) Germination rate statistics under 150 mM NaCl treatment. Data are mean ± SD of three experiments (n = 64). The ANOVA test showed a significant difference (* $p < 0.05$, ** $p < 0.01$).

For root length analysis, OE and WT seeds were grown on 1/2 MS0 medium at 22 °C for a week, then transferred to 1/2 MS0 medium containing different concentrations of NaCl and grown for 7 days (Figure 9a). Under normal conditions, OE and WT had similar root length. The OE roots were significantly longer than WT under 100 mM NaCl treatment. Under the treatment of 125 mM and 150 mM NaCl, the root length of the OE was significantly longer than WT (Figure 9b–e). With the increase of NaCl concentration, the taproot length became shorter and the number of lateral roots increased in OE plants.

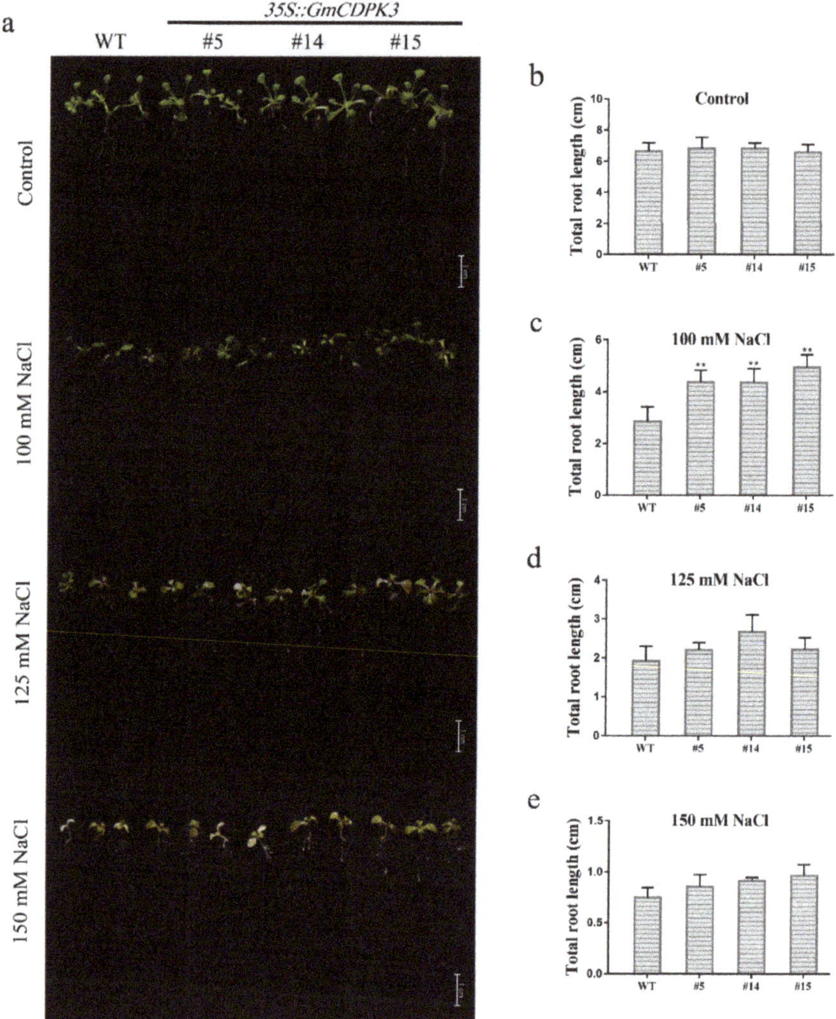

Figure 9. Root length phenotypes of OE and WT under different concentrations of NaCl. Seedlings with consistent root length and growth state were selected to grow in Ms0 medium for about three days and transferred to other medium to observe root length. Each experiment was repeated three times. (**a**) Root growth phenotypes of OE and WT under different concentrations of NaCl. (**b**) Untreated control total root length statistics. (**c**) Total root length statistics under 100 mM NaCl treatment. (**d**) Total root length statistics under 125 mM NaCl treatment. (**e**) Total root length statistics under 150 mM NaCl treatment. Data are mean ± SD of three experiments ($n = 30$). The ANOVA test showed a significant difference (** $p < 0.01$).

In the seedling treatment, 3-week-old OE and WT seedlings were cultured for 14 days at 250 mM NaCl, and the survival rate of the OE was 91.56–92.36%, which was significantly higher than WT (74.94%) (Figure 10b,d). The results indicate that *GmCDPK3* can be used to increase the tolerance of transgenic plants to salt stress.

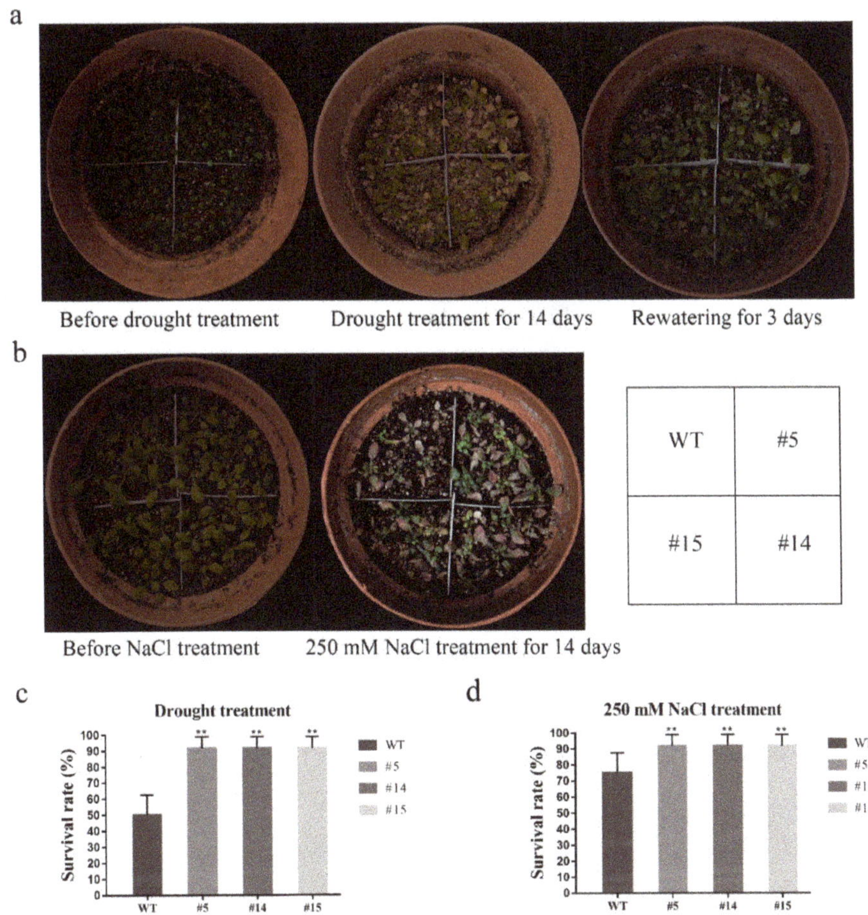

Figure 10. OE and WT phenotypes at the seedling stage under drought and salt treatments. The seedlings of *Arabidopsis* with the same growth condition were transferred to the soil, after growing for two weeks, they were treated with drought and salt. (**a**) Drought-tolerant phenotype of OE and WT in the absence of water and after rehydration. (**b**) Salt-tolerant phenotype of the OE and WT in 250 mM NaCl treatment. (**c**) Monitoring the survival rate of OE and WT under water stress 3 days after rehydration. (**d**) Monitoring the survival rate of OE and WT under 250 mM NaCl treatment for 14 days. Data are means ± SD of three experiments ($n = 24$). The ANOVA test showed a significant difference (** $p < 0.01$).

2.10. Positive Effect of GmCDPK3 in Transgenic Soybean Hairy Roots Under Drought and Salt Treatment

GmCDPK3 has a positive effect on drought and salt stresses of transgenic soybean hairy roots. To further investigate the biological function of *GmCDPK3* under drought and salt treatment conditions, we used *Agrobacterium rhizogenes*-mediated transformation of soybean hairy roots to produce *GmCDPK3*-OE (*GmCDPK3* overexpression) and Gm*CDPK3*-RNAi (*GmCDPK3* with RNA interference) soybean hairy roots. In the drought treatment, the hairy roots of the seedlings were dried for 7 days first and then rehydrated for 3 days. Under drought stress, the leaf wilting degree of *GmCDPK3*-RNAi was higher than that of EV (empty-vector) and *GmCDPK3*-OE (Figure 11a). Under salt stress, the phenotype of the plants was observed to be consistent with the drought treatment, in which the *GmCDPK3*-RNAi plants showed more pronounced leaf drooping, yellowing, and wilting (Figure 12a).

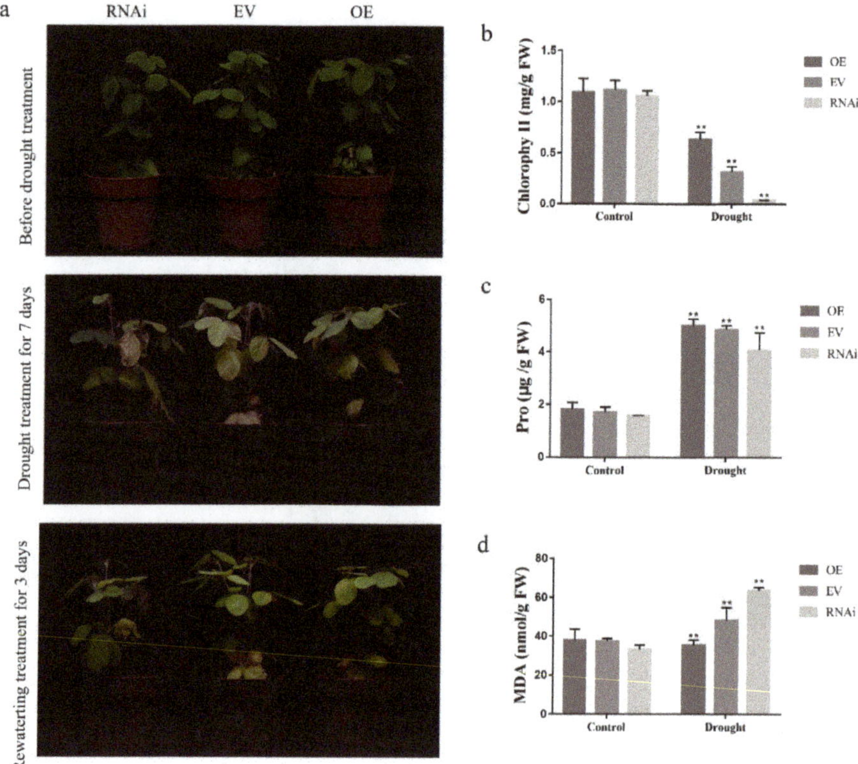

Figure 11. Phenotypic analysis of *GmCDPK3*-OE, *GmCDPK3*-RNAi, and EV under drought stress. Soybean plants with the new root hair were transplanted and grown for another week without watering and drought treatment three times per group. (**a**) Phenotypes of *GmCDPK3*-OE and *GmCDPK3*-RNAi under normal conditions and drought treatments. (**b**) Changes in chlorophyll content in *GmCDPK3*-OE, *GmCDPK3*-RNAi and EV under drought stress. (**c**) Changes in Pro content in *GmCDPK3*-OE, *GmCDPK3*-RNAi, and EV under drought stress. (**d**) Changes in MDA content in *GmCDPK3*-OE, *GmCDPK3*-RNAi, and EV under drought stress (** $p < 0.01$).

To investigate the potential physiological mechanisms by which *GmCDPK3*-OE enhances plant tolerance, we determined proline (Pro), malondialdehyde (MDA) and chlorophyll content in *GmCDPK3*-OE, EV, and *GmCDPK3*-RNAi plants under normal and stress conditions. The results showed that under drought conditions, the MDA content of the *GmCDPK3*-RNAi (63.76 nmol/g) was higher than that of EV (48.41 nmol/g); MDA content (35.42 nmol/g) of *GmCDPK3-OE* was lower than EV (48.41 nmol/g). The Pro and chlorophyll content of *GmCDPK3*-OE (5.02 µg/g and 0.64 mg/g, respectively) were higher than EV (4.88 µg/g and 0.31 mg/g, respectively) and *GmCDPK3*-RNAi (4.1 µg/g and 0.04 mg/g, respectively) (Figure 11b,c). Under salt treatment conditions, the MDA content of the *GmCDPK3*-RNAi (54.57 nmol/g) was significantly higher than EV (41.44 nmol/g) and *GmCDPK3*-OE (31.90 nmol/g). The Pro and chlorophyll content of *GmCDPK3*-RNAi (1.43 µg/g and 0.55 mg/g, respectively) were significantly lower than EV (1.91 µg/g and 0.77 mg/g, respectively) and *GmCDPK3*-OE (4.42 µg/g and 0.83 mg/g, respectively) (Figure 12b,c) (Table S4).

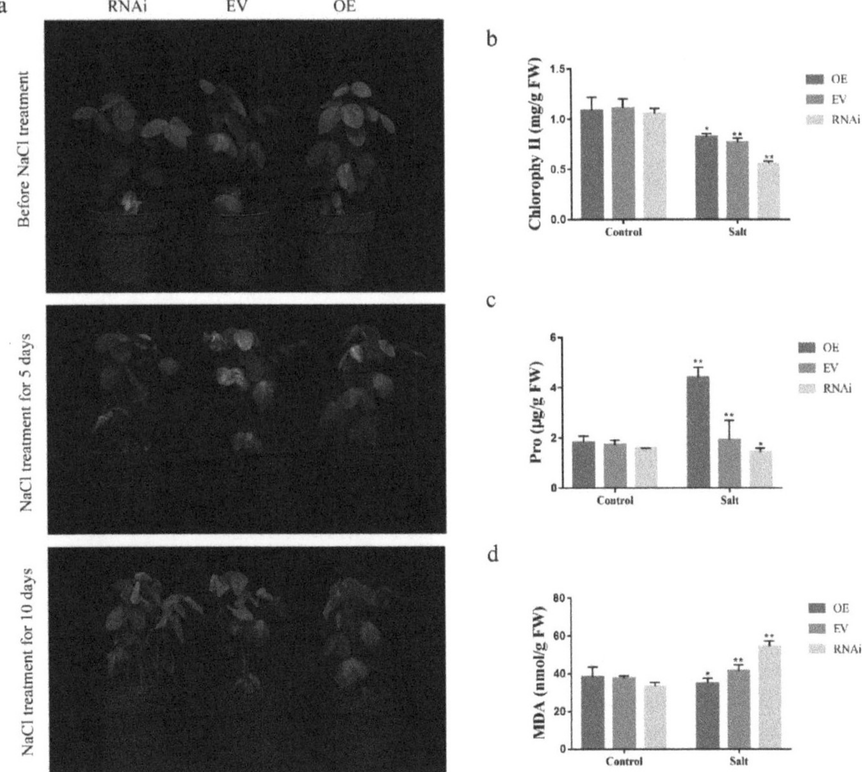

Figure 12. Phenotypic analysis of *GmCDPK3*-OE, *GmCDPK3*-RNAi, and EV under salt stress. Soybean plants with the new root hair were transplanted and grown for another week and 250 mM NaCl treatment three times per group. (**a**) Phenotypes of *GmCDPK3*-OE and *GmCDPK3*-RNAi under normal conditions and salt treatments. (**b**) Changes in chlorophyll content in *GmCDPK3-OE*, *GmCDPK3*-RNAi, and EV under salt stress. (**c**) Changes in Pro content in *GmCDPK3-OE*, *GmCDPK3*-RNAi, and EV under salt stress. (**d**) Changes in MDA content in *GmCDPK3*-OE, *GmCDPK3*-RNAi, and EV under salt stress (* $p < 0.05$ ** $p < 0.01$).

When cells are damaged or die, trypan blue penetrates the denatured cell membrane and binds to the disintegrated DNA to color the cell, while living cells prevent the dye from entering the cell. Staining of the soybean leaves after salt and drought treatment was observed. Compared with the control, the leaves of the treated group had different degrees of cell membrane damage, and the *GmCDPK3*-OE had the least damage, followed by EV and *GmCDPK3*-RNAi (Figure S5).

3. Discussion

In this study, we analyzed 17 GmCDPK basic molecular characterization and identified the *GmCDPK3* gene from soybean. Subcellular localization results suggested that *GmCDPK3* functions in cell membranes. We obtained transgenic *Arabidopsis* seedlings and soybean hairy roots for studying the functions of *GmCDPK3*. Our results indicated that overexpression of *GmCDPK3* improved plants tolerance to drought and salt stresses compared with control, after PEG6000 and NaCl treatments.

The regulation of CDPKs in the response to plant stresses has been widely reported. Overexpression of *SiCDPK24* in *Arabidopsis* enhanced drought resistance [16]. *OsCPK24* overexpressing plants have stronger resistance to low temperature by increasing amino acid content and increasing the GSH/GSSG (reduced glutathione/oxidized glutathione) ratio [23]. Overexpression of rice *OsCPK12* can inhibit the accumulation of intracellular ROS, thereby enhancing the tolerance of transgenic rice to salt stress [24]. *AtCPK10* functions in abscisic acid- (ABA) and Ca^{2+}-mediated stomatal regulation in response to drought stress [25]. *AtCPK32*, through phosphorylating ABA-induced transcription factor ABF4, is involved in ABA/stress responses [26]. *OsCPK21* phosphorylates *OsGF14e* to facilitate the response to ABA and salt stress [27]. In our research, qRT-PCR analysis indicated that most of 17 GmCDPKs responded highly to drought or salt treatment. We hypothesize that soybean CDPKs may be play important roles in enhancing tolerance to drought or salt stresses.

Drought, salinity, and other abiotic stresses are important factors that restrict plant growth and affect plant morphology [28]. There are many microorganisms living around the root system of plants. Changing the composition of plant-related flora in the root region and selecting a combination that is adapted to abiotic stress can improve plant resistance to stress sources, promote plant health and drought tolerance. An important factor in alleviating plant drought stress is *mycorrhizal fungi* (PGPR), which promotes plant growth [29]. Studies have shown that promoting the growth of plant *rhizobacteria* can improve the drought resistance of crops [30]. In addition, plants have evolved various physiological and biochemical mechanisms to adapt to environmental changes [28]. Under drought stress, the water loss rate of the drought-sensitive mutant dsm2 was faster than that of the wild type; the photosynthetic rate, biomass and grain yield of the mutant were significantly reduced, while the malondialdehyde level and stomatal aperture increased. *DSM2* gene plays an important role in the regulation of the rice lutein cycle and ABA synthesis; that is, it provides the ability for the establishment of a drought tolerance mechanism in rice [31]. GmNHX1, located in the vacuole membrane, can enhance the salt tolerance of plants by maintaining a high K^+/Na^+ ratio and inducing the expression of SKOR, SOS1, and AKT1. Many physiological indicators are often used to verify plant tolerance during growth [32]. It has been proven that proline accumulation under adverse conditions can resist the damage abiotic stress causes plants [33,34]. The higher the content of proline, the stronger the resistance of plants. Malondialdehyde is produced by peroxidation of membrane lipids in tissues or organs when plants are aging or injured in adverse conditions [35]. The higher the malondialdehyde content, the more damaged the plant is. Chlorophyll content can reflect the plant is being hurt by adversity stress level. Research has shown that PSII of salinity–alkalinity stress is the most sensitive; the chloroplast is one of the most important organelles in plant response to salt stress [36]. According to the data obtained in our experiment, it is certain that overexpression of *GmCDPK3* gene can protect plants from adverse conditions.

Roots are one of the main vegetative organs of plants. They absorb water and minerals dissolved in the soil and transport them to the aerial parts of plants [37]. In the case of drought and high salt conditions, in order to improve plant vigor, the roots face the problem of controlling ion migration into or out of the cell membrane to maintain ion balance. In the root length experiment of *Arabidopsis*, it was observed that the root phenotype of *Arabidopsis* gradually changed with the increase of PEG6000 concentration (Figure 7). In the WT, the primary roots gradually shortened, the lateral roots decreased, the length became shorter and the root hairs became sparse. However, in overexpressing lines, with the increase of PEG6000 concentration, *Arabidopsis* showed short roots, increased the number of lateral roots and density of root hair. Under salt treatment conditions, the roots of WT changed in the same way as the drought treatment with the increase of salt concentration (Figure 9). In the overexpressed lines, with the increase of salt concentration, the same primary roots of *Arabidopsis* shortened and the number of lateral roots increased. The number of root hairs increased, but the overall change was not as obvious as the phenotype during drought treatment. Therefore, we hypothesize that CDPK regulates plant adaptation to the stress environment by regulating the multiple pathways of Ca^{2+} in *Arabidopsis* and may have direct or indirect regulatory functions on plant root growth and development.

4. Materials and Methods

4.1. Phylogenetic Tree Analysis and Gene Source

CDPK genes and protein sequences of *Arabidopsis*, rice, and soybean were obtained from TAIR (http://arabidopsis.org), RGAP (http://rice.plantbiology.msu.edu), and Phytozome (http://megasoftware.net). The phylogenetic tree was inferred using the neighbor-joining method in MEGA 10.0. Based on previously published information [22], combined with the existing soybean drought and salt transcriptome data in the laboratory to screen out all CDPK genes, and selected 17 genes with a multiple log2 Fold Change greater than 2 for the following experiment.

4.2. Sequence Analysis of 17 Drought and Salt-Tolerant GmCDPKs

Gene sequences of 17 GmCDPKs were analyzed at the transcriptome level, and the domain and tertiary structure of 17 GmCDPKs were separately analyzed at the protein level. To depict the chromosomal locations of the 17 GmCDPK genes Map Gene 2 Chromosomal (http://mg2c.iask.in/mg2c_v2.0/) was used. The tertiary structure was predicted using SWISS-MODEL (https://www.swissmodel.expasy.org), and evaluated in SAVES (https://servicesn.mbi.ucla.edu/SAVES/) [38]. The SMART (http://smart.embl-heidelberg.de/) online tool and ExPASy Proteomics Sever (http://expasy.org/) were used to analyze the functional domain of GmCDPKs [39]. The motifs of GmCDPKs were analyzed with the online tool MAMA (https://www.ebi.ac.uk/Tools) [40]. An exon–intron structure map was created using GSDS online tools (http://gsds.cbi.pku.edu.cn/). Expression of 17 GmCDPKs at different tissue and developmental stages was analyzed by SoyBase (https://www.soybase.org/soyseq/) [41].

4.3. Promoter Analysis of GmCDPKs

The 2000 bp region upstream of the ATG start codon of the 17 GmCDPKs was submitted to PLACE (http://bioinformatics.psb.ugent.be/webtools/plantcare/html/) to identify the cis-acting elements and calculate the number of each element [42].

4.4. Planting of Plant Materials

Soybean seeds (Our Laboratory supply, Tiefeng 8) were planted in pots until the seedlings grew to 10 cm and had two new leaves, and then the seedlings were subjected to drought and salt stresses. Drought stress involved putting soybean seedlings on filter paper for rapid drought for 0, 1, 2, 4, 8, 12, 24, and 48 h; salt treatment involved transferring seedlings to 250 mM NaCl with the same timing, samples were separately taken and immediately immersed in liquid nitrogen and stored at −80 °C for RNA extraction [43].

The seeds of *Arabidopsis* (Col-0) were sterilized and planted on 1/2 MS0 medium, after vernalization at 4 °C for 3 days. Plates containing seeds were placed at 22 °C and in light conditions of 40 μmol/m^2/s^1 with a photoperiod of 16h light/8h dark in the growth chamber the seedlings were then used in further experiments [43,44].

4.5. RNA Extraction and qRT-PCR

Plant samples stored at −80 °C were run according to the method provided in the Plant Total RNA Extraction Kit (TIANGEN). qRT-PCR was performed using the experimental method provided by the PrimeScript^{TM} RT Kit (Takara, Shiga, Japan). The primers were designed by Primer Premier 5.0, in which the soybean actin gene was used as a control. An ABI Prism 7500 real-time PCR system (Applied Biosystems, Foster City, CA, USA) was used to perform qRT-PCR [26]. The resulting data was analyzed using the $2^{-\Delta\Delta CT}$ method [45].

4.6. Subcellular Localization of GmCDPK3

The full-length cDNA sequence of *GmCDPK3* was ligated to the N-terminus of the hGFP gene carrying the CaMV 35S promoter. The recombinant plasmid of *GmCDPK3*-GFP was transformed into *Arabidopsis* protoplasts using a PEG4000-mediated method [46]. It was placed in the dark for more than 12 h and the GFP signal was detected by laser scanning confocal microscopy (Zeiss LSM 700, Oberkochen, Germany) [44].

4.7. Drought and Salt Stress Assays of Transgenic Arabidopsis Plants

The full-length cDNA sequence of *GmCDPK3* was transformed into a pCAMBIA1302 plant transformation vector to obtain OE, and then transformed into *Agrobacterium tumefaciens* (GV3101) after sequencing. The gene of interest was transferred into WT *Arabidopsis* (Col-0) plants using the floral dip method mediated by *Agrobacterium* [47]. The seeds were sterilized with 70% alcohol and 0.1% sodium hypochlorite. After vernalization for 3 days at 4 °C, the plates containing the seeds were transferred to a growth chamber [46]. The qRT-PCR analysis of *GmCDPK3* gene expression was conducted in 3-week-old *Arabidopsis* seedlings (Figure S3).

For germination rate experiments, sterilized seeds from WT and OE plants (lines 5, 14, and 15) were sown in various concentrations of PEG6000 (0, 6, 9, and 12%) or NaCl (0, 100, 125, and 150 mM). Each concentration was repeated three times on 1/2 MS0 medium. The plate was housed in a growth chamber maintained at 22 °C, with illumination intensity of 40 μmol/m^2/s^1 and photoperiod of 16h-light/8h-dark [48,49]. The number of germinating seeds was counted every 12 h and at least 80 seeds per genotype were measured, and each experiment was repeated three times.

For root growth experiments, sterilized WT and OE seeds were seeded on 1/2 MS0 medium. After 5 days, the seedlings were transferred to medium containing different concentrations of PEG6000 (0, 6, 9, and 12%) or NaCl (0, 100, 125, and 150 mM) for one week, and each concentration was repeated three times. Photographs were taken one week later and the root length was assessed by the Epson Expression 11000XL Root Scan Analyzer (Epson, Nagano, Japan) [44]. To test drought and salt response at later developmental stages, 3-week-old seedlings were treated with drought or 250 mM NaCl for 14 days. The experiment was repeated three times for each concentration and treatment. In the end, the survival rate was calculated, and seeds were photographed.

4.8. Vector Construction of GmCDPK3

After amplification of the CDS of *GmCDPK3*, the restriction site sequences (NcoI and BsTEII) and gene-specific primers were ligated to the end of *GmCDPK3*. The PCR products and the pCAMBIA3301 vector were digested with NcoI and BsTEII (ThermoFisher Scientific, USA), respectively, and the product was ligated to obtain pCAMBIA3301-*GmCDPK3* [44,50]. The RNAi sequence was selected on the sense strand encoding the mRNA, the sense strand and the antisense strand of the selected interference sequence were inserted before and after the sequence, and the non-coding sequence (153 bp) from maize was inserted in the middle to form a large hairpin structure. The structure was clipped to cause the gene to be disturbed without expression. After selecting the sequence, it was submitted to the AUGCT company (Beijing, China) for synthesis and testing. The RNAi sequence and the pCAMBIA3301 vector were digested with NcoI and BsTEII, respectively, and the product was ligated to obtain pCAMBIA3301-*GmCDPK3i*.

4.9. Transformation of Soybean Hairy Roots

Soybean seeds (Williams 82) were subjected to *Agrobacterium tumefaciens*-mediated transformation of the hairy root [51]. The seeds were planted in pots for 9–14 days, and after the seedlings grew to 10 cm, the hairy root strains carrying the genes of interest were inoculated 2 mm below the soybean cotyledons, and the soil was overturned until the cotyledons were exposed. The newly grown roots were treated after bacterial infection, and the old roots were removed and cultured for 4 days, then subjected to drought and 250 mM NaCl treatment for 16 and 7 days, respectively [52]. Due to the uncertainty of inoculation, the experiment must be expanded to ensure that each treatment is repeated at least three times. The qRT-PCR analysis of *GmCDPK3* expression in *GmCDPK3*-OE, EV-control (3301), and *GmCDPK3*-RNAi transgenic soybean hairy roots before processing (Figure S4).

4.10. Trypan Blue Staining

The isolated at least 5 leaves from the same position of soybean seedlings were treated with drought for 7 days, and 250 mM NaCl for 3 days. The leaves were completely immersed in a 0.5% trypan) solution for 12 h, then immersed in 75% ethanol for decolorization until the leaves turned white and were photographed to visualize leaf staining [44].

4.11. Determination of Pro, MDA, and Chlorophyll Contents

The leaves were tested for Pro, MDA, and chlorophyll content after 1–3 days of drought or 250 mM NaCl stress [52]. Untreated leaves were used as controls, all measurements were repeated three times and statistical analysis was performed using the ANOVA test. Pro, MDA, and chlorophyll contents were measured in accordance with the instructions provided in the corresponding kit produced by Suzhou Comin Biotechnology Co., Ltd (Suzhou, China).

5. Conclusions

We screened GmCDPK genes induced by abiotic stresses from transcriptome, and our results demonsrated that *GmCDPK3* can increase plant resistance to drought and salt stresses.

Supplementary Materials: Supplementary materials can be found at http://www.mdpi.com/1422-0067/20/23/5909/s1.

Author Contributions: Z.-S.X. conceived of and designed experiments, and edited the manuscript; D.W. completed the experiment and wrote the first draft of the manuscript; Y.-X.L., X.-Y.C., and Z.-W.F. conducted bioinformatics analysis; Q.Y., J.-Y.Z., and J.-N.R. provided analysis tools and analyzed the data; S.-P.Z., Y.-B.Z., M.C., and J.C. collected samples and processed images; J.-H.L., and Y.-Z.M. made suggestions concerning the experiment. All authors read and approved the final manuscript.

Funding: This research was financially supported by the National Natural Science Foundation of China (31701414 and 31871624) and the National Transgenic Key Project of the Ministry of Agriculture of China (2018ZX0800909B and 2016ZX08002-002).

Acknowledgments: We are grateful to Li-Juan Qiu and Shi Sun of the Institute of Crop Science, Chinese Academy of Agricultural Sciences for kindly providing soybean seeds.

Conflicts of Interest: The authors declare no conflict of interest. The funders had no role in the design of the study; in the collection, analyses, or interpretation of data; in the writing of the manuscript, or in the decision to publish the results.

Abbreviations

A. rhizogenes	*Agrobacterium rhizogenes*
ABA	Abscisic acid
CDPK	Calcium-dependent protein kinase
qRT-PCR	Quantitative real-time PCR
pI	Isoelectric points
GFP	Green fluorescent protein
MDA	Malondialdehyde
PRO	Proline
PEG	Polyethylene glycol

References

1. Bohnert, H.J.; Nelson, D.E.; Jensen, R.G. Adaptations to environmental stresses. *Plant. Cell.* **1995**, *7*, 1099–1111. [CrossRef] [PubMed]

2. Bray, E.A. Plant responses to water deficit. *Trends Plant. Sci.* **1997**, *2*, 48–54. [CrossRef]

3. Zhu, J.K.; Hasegawa, P.M.; Bressan, R.A.; Bohnert, H.J. Molecular aspects of osmotic stress in plants. *Crit. Rev. Plant. Sci.* **1997**, *16*, 253–277. [CrossRef]

4. Ghosh, A.; Greenberg, M.E. Calcium signaling in neurons: Molecular mechanisms and cellular consequences. *Science* **1995**, *268*, 239–247. [CrossRef] [PubMed]

5. Rudd, J.J.; Franklin Tong, V.E. Unravelling response-specificity in Ca^{2+} signaling pathways in plant cells. *New Phytol.* **2001**, *151*, 7–33. [CrossRef]

6. Hamel, L.P.; Sheen, J.; Seguin, A. Ancient signals: Comparative genomics of green plant CDPKs. *Trends Plant. Sci.* **2014**, *19*, 79–89. [CrossRef]

7. Harper, J.F.; Harmon, A. Plants, symbiosis and parasites: A calcium signalling connection. *Nat. Rev. Mol. Cell Biol.* **2005**, *6*, 555–566. [CrossRef]

8. Valmonte, G.R.; Arthur, K.; Higgins, C.M.; MacDiarmid, R.M. Calcium-dependent protein kinases inplants: Evolution, expression and function. *Plant. Cell Physiol.* **2014**, *55*, 551–569. [CrossRef]

9. Roberts, D.M.; Harmon, A.C. Calcium-modulated proteins: Targets of intracellular calcium signals in higher plants. *Annu. Rev. Plant. Physiol. Plant. Mol. Biol.* **1992**, *43*, 375–414. [CrossRef]

10. Wu, Z.G.; Wu, S.J.; Wang, Y.C.; Zheng, L.L. Advances in studies of calcium-dependent protein kinase (CDPK) in plants. *Acta Prataculturae Sinica.* **2018**, *27*, 204–214.

11. Cheng, S.H.; Willmann, M.R.; Chen, H.C.; Sheen, J. Calcium signaling through protein kinases. The *Arabidopsis* calcium-dependent protein kinase gene family. *Plant Physiol.* **2002**, *129*, 469–485. [CrossRef] [PubMed]

12. Ray, S.; Agarwal, P.; Arora, R.; Kapoor, S.; Tyagi, A.K. Expression analysis of calcium-dependent proteinkinase gene family during reproductive development and abiotic stress conditions in rice (*Oryza sativa* L. ssp. *indica*). *Mol. Genet. Genom.* **2007**, *278*, 493–505. [CrossRef] [PubMed]

13. Xu, X.W.; Liu, M.; Lu, L.; He, M.; Qu, W.Q.; Xu, Q.; Qi, X.H.; Chen, X.H. Genome-wide analysis and expression of the calcium-dependent protein kinase gene family in cucumber. *Mol. Genet. Genom.* **2015**, *290*, 1403–1414. [CrossRef] [PubMed]

14. Li, A.L.; Zhu, Y.F.; Tan, X.M.; Wang, X.; Wei, B.; Guo, H.Z.; Zhang, Z.L.; Chen, X.B.; Zhao, G.Y.; Kong, X.Y.; et al. Evolutionary and functional study of the CDPK gene family in wheat (*Triticumaestivum* L.). *Plant Mol. Biol.* **2008**, *66*, 429–443. [CrossRef] [PubMed]

15. Liu, H.L.; Che, Z.J.; Zeng, X.R.; Zhou, X.Q.; Hélder, M.S.; Wang, H.; Yu, D.Y. Genome-wide analysis of calcium-dependent protein kinases and their expression patterns in response to herbivore and wounding stresses in soybean. *Funct. Integr. Genom.* **2016**, *16*, 481–493. [CrossRef] [PubMed]

16. Yu, T.F.; Zhao, W.Y.; Fu, J.D.; Liu, Y.W.; Chen, M.; Zhou, Y.B.; Ma, Y.Z.; Xu, Z.S.; Xi, Y.J. Genome-Wide Analysis of CDPK Family in Foxtail Millet and Determination of SiCDPK24 Functions in Drought Stress. *Front. Plant. Sci.* **2018**, *9*, 651. [CrossRef] [PubMed]

17. Asano, T.; Tanaka, N.; Yang, G.X.; Hayashi, N.; Komatsu, S. Genome-wide identification of the rice calcium-dependent protein kinase and its closely related kinase gene families: Comprehensive analysis of the CDPKs gene family in rice. *Plant. Cell Physiol.* **2005**, *46*, 356–366. [CrossRef]

18. Wang, J.P.; Xu, Y.P.; Munyampundu, J.P.; Liu, T.Y.; Cai, X.Z. Calcium-dependent protein kinase (CDPK) and CDPK-related kinase (CRK) gene families in tomato: Genome-wide identification and functional analyses in disease resistance. *Mol. Genet. Genom.* **2016**, *291*, 661–676. [CrossRef]

19. Cai, H.Y.; Cheng, J.B.; Yan, Y.; Xiao, Z.L.; Li, J.Z.; Mou, S.L.; Qiu, A.L.; Lai, Y.; Guan, D.Y.; He, S.L. Genome-wide identification and expression analysis of calcium-dependent protein kinase and its closely related kinase genes in Capsicum annuum. *Front. Plant. Sci.* **2015**, *6*, 737. [CrossRef]

20. Zhang, K.; Han, Y.T.; Zhao, F.L.; Hu, Y.; Gao, Y.R.; Ma, Y.F.; Zheng, Y.; Wang, Y.J.; Wen, Y.Q. Genome-wide identification and expression analysis of the CDPK gene family in grape, *Vitis* spp. *BMC Plant. Biol.* **2015**, *15*, 164. [CrossRef]

21. Hettenhausen, C.; Sun, G.L.; He, Y.B.; Zhuang, H.F.; Sun, T.; Qi, J.F.; Wu, J.Q. Genome-wide identification of calcium-dependent protein kinases in soybean and analyses of their transcriptional responses to insect herbivory and drought stress. *Sci. Rep.* **2016**, *6*, 18973. [CrossRef] [PubMed]

22. Liu, Y.; Xu, C.; Zhu, Y.; Zhang, L.; Zhou, F.; Chen, H.; Lin, Y. The calcium-dependent kinase OsCPK24 functions in cold stress responses in rice. *BMC Genom. Biol.* **2018**, *60*, 173–188. [CrossRef]

23. Bian, S.M.; Jin, D.H.; Li, R.H.; Xie, X.; Gao, G.; Sun, W.; Li, Y.; Zhai, L.; Li, X. Genome-Wide Analysis of CCA1-Like Proteins in Soybean and Functional Characterization of GmMYB138a. *Int. J. Mol. Sci.* **2017**, *18*, 2040. [CrossRef] [PubMed]

24. Asano, T.; Hayashi, N.; Kobayashi, M.; Aoki, N.; Mitsuhara, I.; Ichikawa, H.; Komatus, S.; Hirochika, H.; Kikuchi, S.; Ohsuqi, R. A rice calcium-dependent protein kinase OsCPK12 oppositely modulates salt-stress tolerance and blast disease resistance. *Plant. J.* **2012**, *69*, 26–36. [CrossRef] [PubMed]

25. Zou, J.J.; Wei, F.J.; Wang, C.; Wu, J.J.; Ratnasekera, D.; Liu, W.X.; Wu, W.H. *Arabidopsis* calcium-dependent protein kinase CPK10 functions in abscisic susceptibility to early blight pathogen in potato via reactive oxygen species burst. *New Phytol.* **2012**, *196*, 223–237.

26. Choi, H.I.; Park, H.J.; Park, J.H.; Kim, S.; Im, M.Y.; Seo, H.H.; Kim, Y.W.; Hwang, I.; Kim, S.Y. Arabidopsis calcium-dependent protein kinase AtCPK32 interacts with ABF4, a transcriptional regulator of abscisic acid-responsive gene expression, and modulates its activity. *Plant. Physiol.* **2005**, *139*, 1750–1761. [CrossRef] [PubMed]

27. Chen, Y.; Zhou, X.; Chang, S.; Chu, Z.; Wang, H.; Han, S.; Wang, Y. Calcium-dependent protein kinase 21 phosphorylates 14–3-3 proteins in response to ABA signaling and salt stress in rice. *Biochem. Biophys. Res. Commun.* **2017**, *493*, 1450–1456. [CrossRef] [PubMed]

28. Boyer, J.S. Plant productivity and environment. *Science* **1982**, *218*, 443–448. [CrossRef]

29. Liu, F.; Ma, H.; Peng, L.; Du, Z.; Ma, B.; Liu, X. Effect of the inoculation of plant growth-promoting rhizobacteria on the photosynthetic characteristics of *Sambucus williamsii* Hance container seedlings under drought stress. *AMB Expr.* **2019**, *9*, 2–9. [CrossRef]

30. Vurukonda, S.S.; Vardharajula, S.; Shrivastava, M.; SkZ, A. Enhancement of drought stress tolerance in crops by plant growth promoting rhizobacteria. *Microbiol. Res.* **2016**, *184*, 13–24. [CrossRef]

31. Du, H.; Wang, N.L.; Cui, F.; Li, X.H.; Xiao, J.H.; Xiong, L.Z. Characterization of the β-Carotene Hydroxylase Gene DSM2 Conferring Drought and Oxidative Stress Resistance by Increasing Xanthophylls and Abscisic Acid Synthesis in Rice. *Plant. Physiol.* **2010**, *154*, 1304–1318. [CrossRef] [PubMed]

32. Sun, T.J.; Fan, L.; Yang, J.; Cao, R.Z.; Yang, C.Y.; Zhang, J.; Wang, D.M. A Glycine max sodium/hydrogen exchanger enhances salt tolerance through maintaining higher Na$^+$ efflux rate and K$^+$/Na$^+$ ratio in *Arabidopsis*. *BMC Plant. Biol.* **2019**, *19*, 469. [CrossRef] [PubMed]

33. Szabados, L.; Savour, A. Proline: A multifunctional amino acid. *Trends Plant. Sci.* **2009**, *15*, 89–97. [CrossRef] [PubMed]

34. Lehmann, S.; Funk, D.; Szabados, L.; Rentsch, D. Proline metabolism and transport in plant development. *Amino Acids* **2010**, *39*, 949–962. [CrossRef]

35. Dimitrios, T. Assessment of lipid peroxidation by measuring malondialdehyde (MDA) and relatives in biological samples: Analytical and biological challenges. *Anal. Biochem.* **2017**, *524*, 13–30.

36. Baker, N.R. A possible role for photosystem II in environmental perturbations of photosynthesis. *Physiol. Plant.* **1991**, *81*, 563–570. [CrossRef]

37. Kolb, E.; Legué, V.; Bogeat-Triboulot, M.B. Physical root-soil interactions. *Phys. Biol.* **2017**, *16*, 065004. [CrossRef]

38. Waterhouse, A.; Bertoni, M.; Bienert, S.; Studer, G.; Tauriello, G.; Gumienny, R.; Heer, F.T.; Beer, T.A.P.; Rempfer, C.; Bordoli, L.; et al. SWISS-MODEL: Homology modelling of protein structures and complexes. *Nucleic Acids Res.* **2018**, *46*, W296–W303. [CrossRef]

39. Xu, Z.S.; Chen, M.; Li, L.C.; Ma, Y.Z. Functions of the ERF transcription factor family in plants. *Botany* **2008**, *86*, 969–977. [CrossRef]

40. Xu, Z.S.; Chen, M.; Li, L.C.; Ma, Y.Z. Functions and application of the AP2/ERF transcription factor family in crop improvement. *Integr. Plant. Biol.* **2011**, *53*, 570–585. [CrossRef]

41. Liu, P.; Xu, Z.S.; Pan-Pan, L.; Hu, D.; Chen, M.; Li, L.C.; Ma, Y.Z. A wheat PI4K gene whose product possesses threonine autophophorylation activity confers tolerance to drought and salt in *Arabidopsis*. *Exp. Bot.* **2013**, *64*, 2915–2927. [CrossRef] [PubMed]

42. Flowers, T.J. Improving crop salt tolerance. *Exp. Bot.* **2004**, *55*, 307–319. [CrossRef] [PubMed]

43. Du, Y.T.; Zhao, M.J.; Wang, C.T.; Gao, Y.; Wang, Y.X.; Liu, Y.W.; Chen, M.; Chen, J.; Zhou, Y.B.; Xu, Z.S.; et al. Identification and characterization of GmMYB118 responses to drought and salt stress. *BMC Plant. Biol.* **2018**, *18*, 320. [CrossRef] [PubMed]

44. Riechmann, J.L.; Heard, J.; Martin, G.; Reuber, L.; Jiang, C.; Keddie, J.; Adam, L.; Pineda, O.; Ratcliffe, O.J.; Samaha, R.R.; et al. *Arabidopsis* transcription factors: Genome-wide comparative analysis among eukaryotes. *Science* **2000**, *290*, 2105–2110. [CrossRef]

45. Le, D.T.; Nishiyama, R.; Watanabe, Y.; Mochida, K.; Yamaguchi-Shinozaki, K.; Shinozaki, K.; Tran, L.S. Genome-wide expression profiling of soybean two component system genes in soybean root and shoot tissues under dehydration stress. *DNA Res.* **2011**, *18*, 17–29. [CrossRef]

46. He, G.H.; Xu, J.Y.; Wang, Y.X.; Liu, J.M.; Li, P.S.; Chen, M.; Ma, Y.Z.; Xu, Z.S. Drought responsive WRKY transcription factor genes TaWRKY1 and TaWRKY33 from wheat confer drought and/or heat resistance in Arabidopsis. *BMC Plant. Biol.* **2016**, *16*, 116. [CrossRef]

47. Clough, S.J.; Bent, A.F. Floral dip: A simplified method for agrobacterium mediated transformation of *Arabidopsis thaliana. Plant. J.* **1998**, *16*, 735–743. [CrossRef]

48. Bu, Q.; Lv, T.; Shen, H.; Luong, P.; Wang, J.; Wang, Z.; Huang, Z.; Xiao, L.; Engineer, C.; Kim, T.H.; et al. Regulation of drought tolerance by the F-box protein MAX2 in *Arabidopsis. Plant. Physiol.* **2014**, *164*, 424–439. [CrossRef]

49. Feng, C.Z.; Chen, Y.; Wang, C.; Kong, Y.H.; Wu, W.H.; Chen, Y.F. *Arabidopsis* RAV1 transcription factor, phosphorylated by SnRK2 kinases, regulates the expression of ABI3, ABI4, and ABI5 during seed germination and early seedling development. *Plant. J.* **2014**, *80*, 654–668. [CrossRef]

50. Kereszt, A.; Li, D.X.; Indrasumunar, A.; Nguyen, C.D.; Nontachaiyapoom, S.; Kinkema, M.; Gresshoff, P.M. Agrobacterium *rhizogenes*-mediated transformation of soybean to study root biology. *Nat. Protoc.* **2007**, *2*, 948–952. [CrossRef]

51. Wang, F.; Chen, H.W.; Li, Q.T.; Wei, W.; Li, W.; Zhang, W.K.; Ma, B.; Bi, Y.D.; Lai, Y.C.; Liu, X.L.; et al. GmWRKY27 interacts with GmMYB174 to reduce expression of GmNAC29 for stress tolerance in soybean plants. *Plant. J.* **2015**, *83*, 224–236. [CrossRef] [PubMed]

52. Zhao, S.P.; Xu, Z.S.; Zheng, W.J.; Zhao, W.; Wang, Y.X.; Yu, T.F.; Chen, M.; Zhou, Y.B.; Min, D.H.; Ma, Y.Z.; et al. Genome-wide analysis of the RAV family in soybean and functional identification of GmRAV-03 involvement in salt and drought stresses and exogenous ABA treatment. *Front. Plant. Sci.* **2017**, *8*, 905. [CrossRef] [PubMed]

International Journal of *Molecular Sciences*

Article

Bioinformatics Analysis of the Lipoxygenase Gene Family in Radish (*Raphanus sativus*) and Functional Characterization in Response to Abiotic and Biotic Stresses

Jinglei Wang [1], Tianhua Hu [1], Wuhong Wang [1], Haijiao Hu [1], Qingzhen Wei [1], Xiaochun Wei [2] and Chonglai Bao [1,*]

[1] Institute of Vegetables Research, Zhejiang Academy of Agricultural Sciences, Hangzhou 310021, China; syauwjl@163.com (J.W.); hutianh@126.com (T.H.); hongge5@163.com (W.W.); huhj0571@126.com (H.H.); weiqz@mail.zaas.ac.cn (Q.W.)

[2] Institute of Horticulture, Henan Academy of Agricultural Sciences, Zhengzhou 450002, China; jweixiaochun@126.com

* Correspondence: baocl@mail.zaas.ac.cn; Tel.: +86-571-8640-9722

Received: 27 September 2019; Accepted: 28 November 2019; Published: 3 December 2019

Abstract: Lipoxygenases (LOXs) are non-heme iron-containing dioxygenases involved in many developmental and stress-responsive processes in plants. However, little is known about the radish LOX gene family members and their functions in response to biotic and abiotic stresses. In this study, we completed a genome-wide analysis and expression profiling of *RsLOX* genes under abiotic and biotic stress conditions. We identified 11 *RsLOX* genes, which encoded conserved domains, and classified them in 9-LOX and 13-LOX categories according to their phylogenetic relationships. The characteristic structural features of 9-LOX and 13-LOX genes and the encoded protein domains as well as their evolution are presented herein. A qRT-PCR analysis of *RsLOX* expression levels in the roots under simulated drought, salinity, heat, and cold stresses, as well as in response to a *Plasmodiophora brassicae* infection, revealed three tandem-clustered *RsLOX* genes that are involved in responses to various environmental stresses via the jasmonic acid pathway. Our findings provide insights into the evolution and potential biological roles of RsLOXs related to the adaptation of radish to stress conditions.

Keywords: radish; lipoxygenase (LOX); gene family; abiotic stress; tandem duplication

1. Introduction

Lipoxygenases (EC 1.13.11.12; LOXs), which are ubiquitously distributed in plants, animals, and fungi, are important monomeric, non-heme iron-containing dioxygenases [1] that catalyze the oxygenation of polyunsaturated fatty acids to produce fatty acid hydroperoxides [2]. A typical LOX comprises the following two major domains: a C-terminal LOX domain and an N-terminal polycystin-1, lipoxygenase, alpha-toxin (PLAT) domain [3]. The LOX domain contains five conserved histidine (His) residues that form a fragment [His-(X)4-His-(X)4-His-(X)17-His-(X)8-His] involved in the binding of an iron atom [4]. The PLAT domain, which contains an eight-stranded antiparallel β-barrel, can bind to procolipase to mediate membrane associations [5,6]. Depending on whether the oxidation of polyunsaturated fatty acids occurs at carbon atom 9 or at carbon atom 13, the plant LOXs are grouped in the 9-LOX or 13-LOX categories [2]. Additionally, LOXs may also be categorized as type I or type II based on their structural and sequence similarities. The type I LOX sequences are highly similar and do not encode a transit peptide, whereas the type II LOX sequences are only moderately similar and encode an N-terminal chloroplast transit peptide [7].

A previous study revealed that 9S-hydroperoxyoctadecadienoic acid (9-HPOD) and 13S-hydroperoxyoctadecadienoic acid (13-HPOD), which are the products generated by 9-LOX and 13-LOX, respectively, can be further converted to various oxylipins by a variety of enzymes [8]. In plants, LOXs and the derived oxylipins are reportedly involved in several biological events, including seed germination [9]; tuber development [10]; fruit ripening [11]; sex determination [12]; and, most importantly, immune responses [13–15].

The oxylipins derived from 13-HPOD, such as jasmonic acid (JA) and green leaf volatiles, are important for responses to abiotic and biotic stresses [16,17]. In maize, the 9-HPOD-derived products include death acids, such as 10-oxo-11-phytodienoic acid and 10-oxo-11-phytoenoic acid, which accumulate to inhibit infections by fungi and herbivorous insects when plants are infected with the pathogen responsible for southern leaf blight (*Cochliobolus heterostrophus*) [18,19]. Moreover, LOX activities have been detected during responses to biotic and abiotic stresses in diverse plant species. For example, six LOXs were identified in *A. thaliana* [20], with the *AtLOX1* and *AtLOX5* expression induced by stress-related hormones influencing immune responses [21,22] and *AtLOX2* and *AtLOX6* expression contributing to JA biosynthesis [23,24]. Among solanaceous plant species, tomato *SlLOXD* [15], tobacco *NaLOX3* [25,26], and potato *StLOXH3* [27] are involved in JA biosynthesis, whereas *NaLOX2* and *StLOXH1* mediate the biosynthesis of green leaf volatiles [28,29]. Furthermore, pepper *CaLOX1* contributes to the scavenging of reactive oxygen species and the activation of defense-related genes in responses to abiotic stresses [30]. In persimmon, *DkLOX3* plays a crucial role in the regulation of senescence and the tolerance to salt stress [31].

Several studies have investigated LOX genes using bioinformatic tools to clarify their functions in various physiological and molecular events in diverse plant species, including *Arabidopsis thaliana* [32], rice [20], cotton [33], maize [34], and cucumber [35]. Radish (*Raphanus sativus* L., $2n = 2x = 18$), which belongs to the family Brassicaceae and a relative of *Brassica rapa* and *Brassica oleracea*, is a root vegetable cultivated worldwide. The growth and economic value of *R. sativus* are affected by a variety of biotic and abiotic stresses, such as extreme weather, saline–alkaline conditions, and diseases due to insect infestations. Because of the importance of the LOX gene family in responses to biotic and abiotic stresses, a comprehensive investigation of the *R. sativus* LOX genes is needed.

The present study involved a thorough analysis of the *R. sativus* LOX gene family. Specifically, 11 LOX genes were identified in the *R. sativus* genome based on bioinformatics resources and then characterized regarding the encoded enzyme sequences, chromosomal distribution, phylogenetic relationships, classifications, structure, encoded conserved motifs, synteny, tandem duplications, as well as expression patterns in various organs and in response to abiotic and biotic stresses. The results of our study may form the basis of future investigations aimed at more thoroughly characterizing the LOX genes in *R. sativus* and related species.

2. Results

2.1. Identification and Characterization of LOX Gene Family Members

We applied a bioinformatics-based approach to identify 11, 14, 11, and six LOX genes in *R. sativus*, *B. rapa*, *B. oleracea*, and *A. thaliana*, respectively (Table S1). The *A. thaliana* LOX genes identified in this study were identical to those described in a previous report [20], confirming the reliability of our results. The *RsLOX* genes were distributed on only five of the nine radish chromosomes (R01, R02, R05, R07, and R08). Additionally, R07 contained five LOX genes, which was the most of any of the radish chromosomes, followed by R01 and R02, both of which included two LOX genes (Figure 1).

The physical and chemical properties of all 11 *RsLOX* genes were analyzed (Table S2). The encoded RsLOX enzymes comprised 676 to 920 amino acids, with a predicted molecular weight of 77.01 to 103.32 kDa. Among the 11 RsLOXs, RsLOX10 was the shortest and had the lowest molecular weight. The predicted aromaticity ranged from 0.092 to 0.104, whereas the isoelectric point ranged from 5.17 to 9.11. These differences mostly result from variation in the nonconserved regions' amino acid sequences. A comparison of the RsLOX sequences revealed a sequence identity of 18.7–91.3% and a sequence similarity of 27.4–95.6% at the amino acid level (Figure 2).

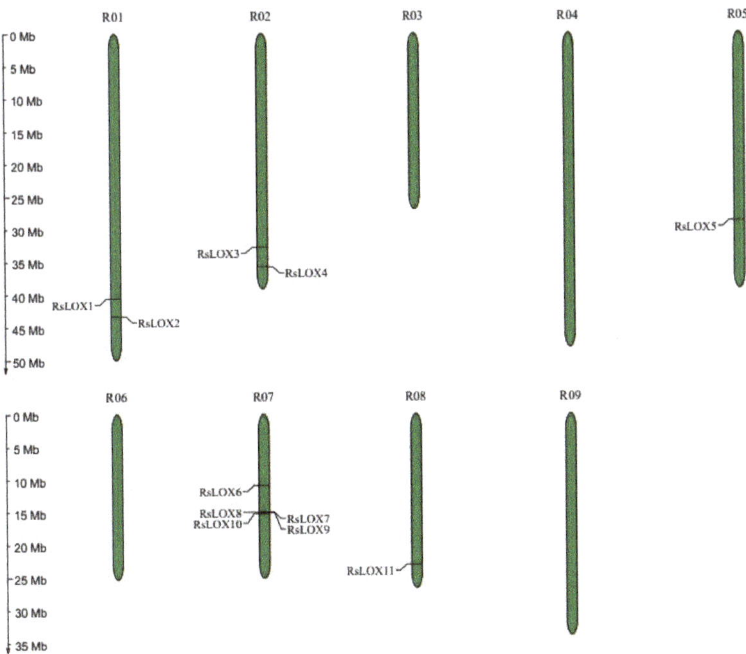

Figure 1. Distribution of *RsLOX* genes on *R. sativus* chromosomes. The line on the green bars indicates the location of LOX genes on chromosomes. The left values corresponding to the scales indicate chromosomes physical distance.

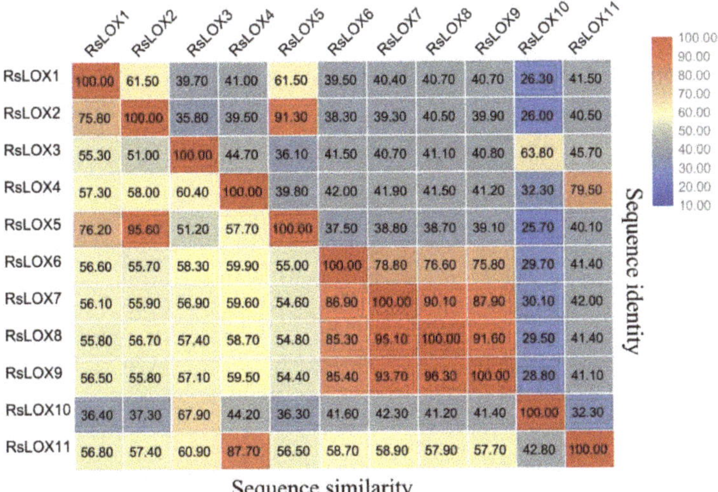

Figure 2. *R. sativus* LOX amino acid sequence identities and similarities (%).

2.2. Phylogenetic Relationships among LOX Family Members from Diverse Plant Species

To determine the evolutionary relationships of the *R. sativus* LOX family members to other known LOX families, we constructed a phylogenetic tree based on 123 LOX amino acid sequences from 17 species (Figure 3 and Table S3). The tree categorized the LOXs as 9-LOX and 13-LOX enzymes. Of the 11 identified *R. sativus* LOXs, three (RsLOX1, RsLOX2, and RsLOX5) were characterized as 9-LOX enzymes, whereas the other eight RsLOXs were designated as 13-LOX enzymes. The RsLOXs were grouped with the LOXs from related species (i.e., *B. rapa* and *B. oleracea*). Furthermore, the 13-LOX enzymes were further classified as type I and II. However, the analyzed Brassica species lacked 13-LOX type I enzymes, which is consistent with previous report [36].

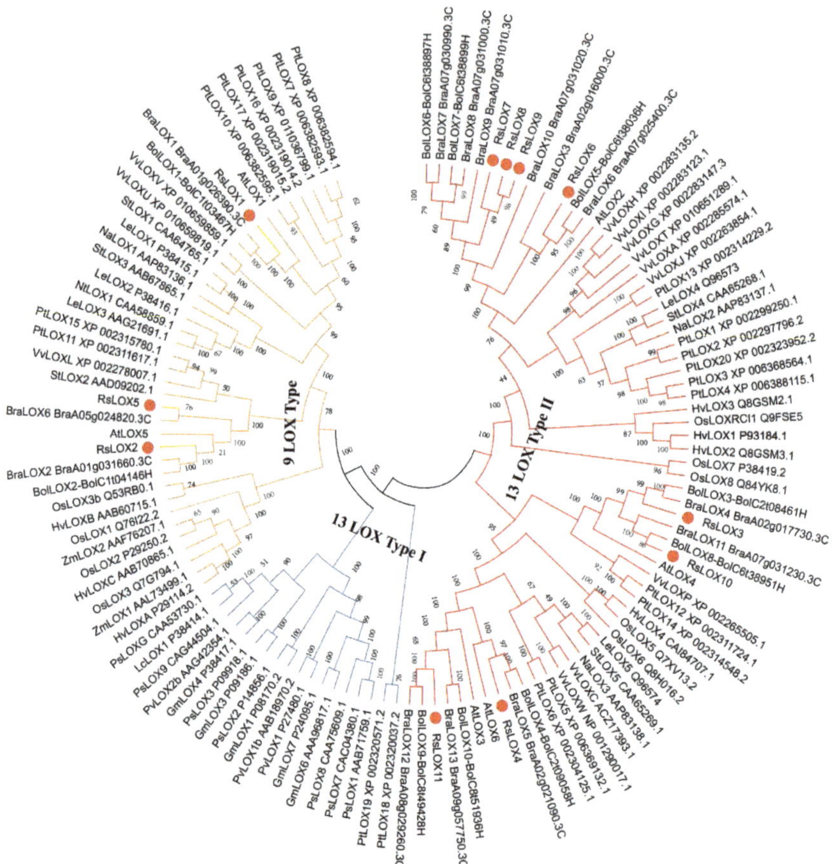

Figure 3. Phylogenetic relationships among LOXs from *R. sativus* and other species. The 9-LOX branches are presented in orange. The 13-LOX type I and type II branches are presented in blue and red, respectively. The identified radish LOXs are highlighted with a red circle. Species abbreviations are as provided in Table S3.

2.3. Conserved Domain and Structural Analyses of R. sativus LOXs

To further clarify the motifs and structural variations among RsLOXs, a separate phylogenetic tree with RsLOX protein sequences was constructed. Additionally, *RsLOX* exon–intron organizations and the encoded conserved motifs were compared. The phylogenetic relationships of the RsLOXs were identical to those of the RsLOXs in the phylogenetic tree based on sequences from 17 species (Figure 4A).

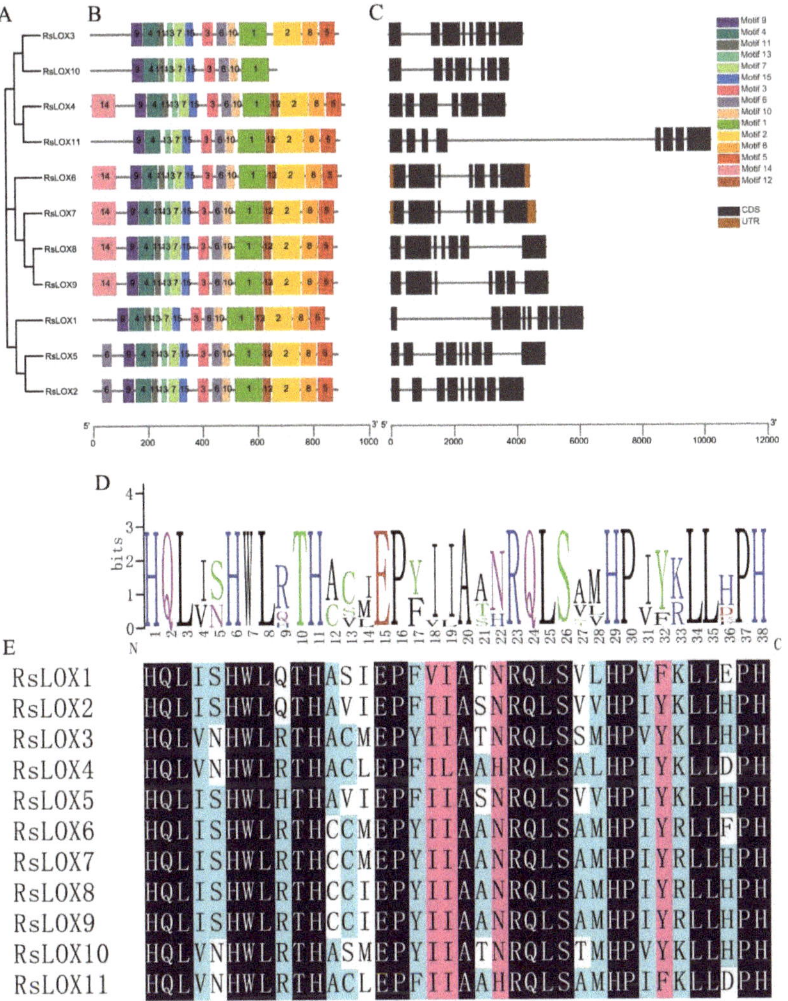

Figure 4. Phylogenetic, motif, structural, and conserved domain analyses of RsLOXs. (**A**) Phylogenetic tree comprising RsLOX proteins. (**B**) Schematic representation of the conserved RsLOX motifs. (**C**) Exon–intron structures in *R. sativus* LOX genes. (**D**) Sequences logo of a 38-residue RsLOX motif. (**E**) Alignment of a 38-residue conserved motif. The dark color indicates the amino acids totally matching, while the light indicates the part matching.

Most of the *RsLOX* genes shared similar exon–intron structures regarding the number and length of exons. The genes encoding 9-LOX enzymes (*RsLOX1*, *RsLOX2*, and *RsLOX5*) comprised seven to eight introns, whereas the genes for the 13-LOX enzymes had five to seven introns. Moreover, *RsLOX4* had the fewest introns (five), whereas RsLOX2 and RsLOX5 contained the most introns (eight) (Figure 4C).

To gain insight into potential functions and diversifcation of the *LOX* genes in *R. sativus*, we analyzed the predicted distinct motifs. As expected, most of closely related family members shared common motifs, suggesting they are functionally similar. Upon closer inspection, we determined that RsLOX10 lacks some of the conserved motifs present in other RsLOXs. Moreover, 12 of the 15 identified motifs (motifs 1–10, 12, and 15) were present in all analyzed LOXs (Figure 4B). Motif 1, which comprised 38 amino acid residues, was highly similar to the aforementioned LOX domain [His-(X)4-His-(X)4-His-(X)17-His-(X)8-His]. This motif was conserved in all 11 RsLOXs identified in this study, suggesting it may affect enzyme stability and activity (Figure 4D,E). An analysis of subcellular localization indicated that the 13-LOX type II enzymes included a chloroplast transit peptide, whereas RsLOX1, RsLOX2, and RsLOX5 (i.e., 9-LOX enzymes) did not (Table S4).

2.4. Tandem Duplications and Synteny of LOX Genes

Whole genome and tandem duplications (TD) provide sources of primitive genetic material for genome complexity and evolutionary novelty [37]. We investigated the syntenic and tandem relationship of *LOX* genes between *A. thaliana* and *R. sativus* to trace the evolutionary history of *LOX* genes. Of the six *AtLOX* genes, four have a syntenic relationship with seven *RsLOX* genes. An analysis with the SynOrths program suggested all of these syntenic relationships formed via whole genome triplication (WGT) or segmental duplication (SD) events. Syntenic relationships were also detected among the genes of the other analyzed species (Table S5). The *AtLOX1* gene had a syntenic relationship with only one *B. rapa* gene. Additionally, *AtLOX2* lacked a syntenic gene in *R. sativus*, *B. oleracea*, and *B. rapa*. In contrast, *AtLOX3–6* had syntenic genes in all three of these species (Figure 5).

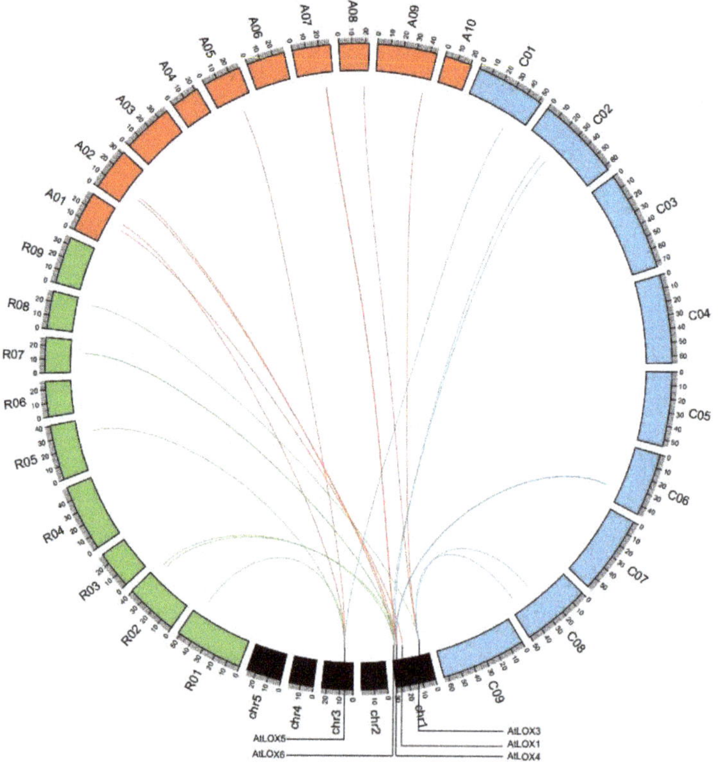

Figure 5. Syntenic relationships among *LOX* genes of *R. sativus*, *B. oleracea*, *B. rapa*, and *Arabidopsis* was visualized in a circos plot. The chromosomes of *Arabidopsis*, *R. sativus*, *B. rapa*, and *B. oleracea* were shaded with black, green, orange, and blue colors, respectively.

Interestingly, of the six *AtLOX* genes, *AtLOX4* had the most syntenic copies in *R. sativus*, *B. oleracea*, and *B. rapa*. Furthermore, the *AtLOX4* syntenic genes Rsa10033815, BraA07g030990.3C, and BolC6t38897H were tandemly duplicated. These observations implied that WGT and TD events substantially contributed to the expansion of the LOX gene families in *R. sativus*, *B. oleracea*, and *B. rapa* (Table S6).

2.5. Expression Profiles of R. sativus LOX Genes in Various Tissues

To clarify the *RsLOX* expression patterns, we analyzed previously reported transcriptome data for six tissues (flowers, siliques, leaves, stems, callus, and roots) (Table S7) [38]. A heat map representing *RsLOX* expression levels indicated these genes were expressed in several tissues, with some genes expressed mainly in particular tissues. A hierarchical cluster analysis classified the *RsLOX* genes into two main groups based on their expression patterns (Figure 6). However, these groups were not consistent with the 9-LOX and 13-LOX categorizations. The *RsLOX1* and *RsLOX2* genes, which encode 9-LOX enzymes, were similarly expressed and were clustered together, implying they may have similar functions. Duplicated genes may undergo a non-functionalization, neofunctionalization, or subfunctionalization [39]. The tandemly duplicated genes (*RsLOX7*, *RsLOX8*, and *RsLOX9*) were differentially expressed, implying they are functionally diverse.

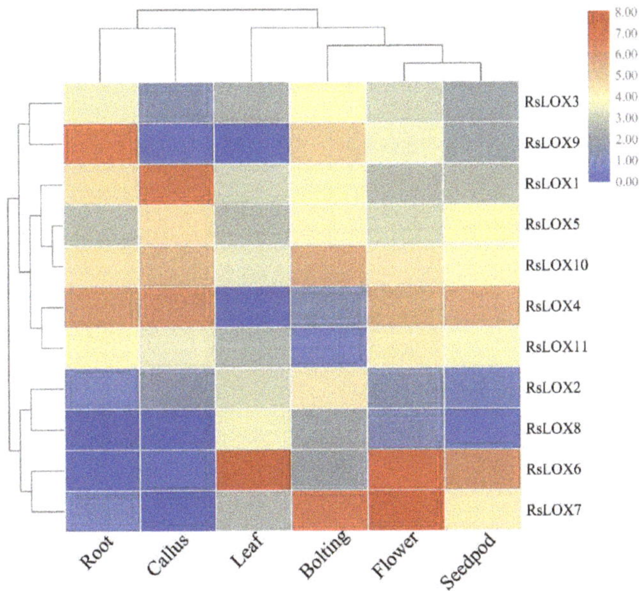

Figure 6. Hierarchical clustering of *R. sativus* LOX gene expression profiles in various tissues. The FPKM values were log-transformed and visualized in a heat map.

2.6. Analysis of RsLOX Expression in Response to Abiotic and Biotic Stresses

A quantitative real-time polymerase chain reaction (qRT-PCR) assay was performed to examine *RsLOX* expression levels in root tissue under high and low temperature, salinity, and simulated drought (polyethylene glycol (PEG) treatment) conditions (Figure 7). The qRT-PCR results revealed that the expression levels of the tandemly duplicated genes (*RsLOX7*, *RsLOX8*, and *RsLOX9*) were considerably upregulated by the high and low temperatures and the PEG treatment. Additionally, *RsLOX11* expression was upregulated approximately 5-fold in response to the high-temperature and PEG treatments.

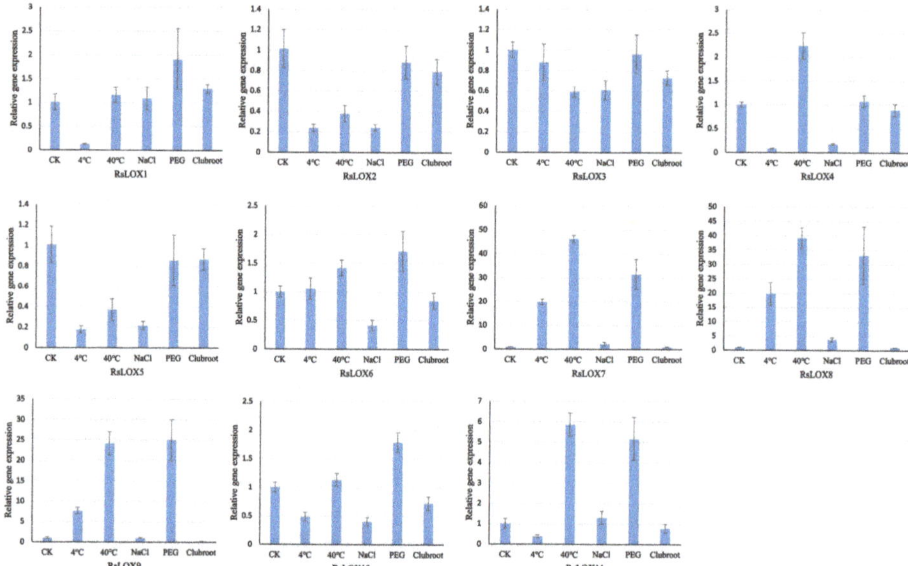

Figure 7. Quantitative real-time PCR analysis of *RsLOX* expression levels in the roots under various abiotic and biotic stress conditions. The presented gene expression levels are relative to the expression of the reference gene. Data are presented as the mean ± standard deviation of three independent experiments. CK: control treatment.

Regarding the low-temperature treatment, the expression levels of all analyzed genes, except for *RsLOX7*, *RsLOX8*, and *RsLOX9*, were downregulated or unchanged. Similar to the effects of the low-temperature treatment, the exposure to NaCl increased the *RsLOX7* and *RsLOX8* expression levels by 2.2-fold and 3.7-fold, respectively, but decreased or did not affect the expression levels of the other *RsLOX* genes. Moreover, the expression levels of all *RsLOX* genes were not affected by the *Plasmodiophora brassicae* infection.

3. Discussion

With the increasing availability of genomic data, the LOX gene family has been studied in several plant species regarding their potential functions in development and stress responses. However, there has been relatively little research into the LOX gene families of *R. sativus* and Brassica species. We identified and characterized 11 *R. sativus* LOX genes in terms of their phylogenetic relationships, gene and protein structures, and expression profiles. We revealed diverse gene structures, conserved motifs, and differential expression patterns in various tissues and in response to abiotic stresses (cold, heat, simulated drought, and salinity). Our findings suggest the *RsLOX* genes have similar sequences and functions, but fulfill slightly different tissue-specific roles related to plant development and stress responses. Thus, the data presented herein provide insights into the *RsLOX* family and may be useful for functionally characterizing the *RsLOX* genes.

In this study, the *RsLOX* genes were divided into two groups (13-LOX and 9-LOX) according to phylogenetic relationships, which was consistent with the results of many previous studies [20,32–35]. Furthermore, the 9-LOX enzymes (RsLOX1, RsLOX2, and RsLOX5) have highly similar sequences (>75%), whereas, except for the enzymes encoded by the tandemly duplicated genes (*RsLOX7*, *RsLOX8*, and *RsLOX9*), the 13-LOX type II enzyme sequences are only moderately similar (Figure 1). Additionally, the 13-LOX type II RsLOXs contain a chloroplast transit peptide. Our findings are consistent with the results of previous studies that indicated the type I LOX sequences are highly similar and lack a transit peptide, whereas the type II LOX sequences are moderately similar and include an N-terminal chloroplast transit peptide [2,7]. Therefore, the classifications of the RsLOXs based on the phylogenetic tree and analyzed sequence characteristics were consistent, which confirmed the validity of our methodology and predictions. The analysis with the MEME suite and the examination of exon–intron structures revealed the similarity in the gene structures and motif compositions of the RsLOXs within sub-branches, suggesting these RsLOXs are functionally similar [40].

Gene duplication events (e.g., WGT, SD, and TD) have led to the expansion and increased functional diversity of specific gene families [41,42]. Previous studies concluded that a WGT event occurred in the common ancestor of *R. sativus*, *B. oleracea*, and *B. rapa* following its divergence from *A. thaliana* approximately 14–20 million years ago [43,44]. In the current study, we determined that four of the six *AtLOX* genes have syntenic copies in *R. sativus*, which is consistent with the evolution of these species. However, differences in the number of retained syntenic genes suggest there may have been some variability in the gene loss events during evolution. Functionally redundant genes are always lost in the diploidization process occurring after paleopolyploidy events [45]. In contrast, most of the *AtLOX4* syntenic copies, as well as the tandemly duplicated genes, are present in *R. sativus*, *B. oleracea*, and *B. rapa*, suggesting the associated tandem duplication occurred in the common ancestor of these three species. The gene dosage effects hypothesis suggests that the increase in the *AtLOX4* copy number reflects the considerable need for the encoded enzyme in multiple complex physiological and biological processes [46]. Our findings may be useful for elucidating the expansion of the *LOX* gene family.

Gene expression patterns can provide important clues regarding gene functions. Previous research produced evidence that LOXs play fundamental roles in plant development and stress responses. In the current study, *RsLOX* expression patterns were observed that reproductive organs were clustered together. Under stress conditions, the expression levels of the 9-LOX genes (*RsLOX1*, *RsLOX2*, and *RsLOX5*) were downregulated, which is similar to pepper 9-LOX gene expression patterns in response to JA and thrips [47]. These results are consistent with the fact that 9-LOX enzymes mediate insect resistance, protein storage, and tuber development [2,18]. The 13-LOX enzymes reportedly contribute to abiotic stress responses [16,17]. The data presented herein indicated that the expression levels of *RsLOX7*, *RsLOX8*, *RsLOX9*, and *RsLOX11* (i.e., 13-LOX genes) are considerably affected by several stresses. Moreover, RsLOX7, RsLOX8, and RsLOX9 clustered with AtLOX2 in the phylogenetic tree. In *A. thaliana*, AtLOX2 is required for wound-induced JA accumulation, but whether it is involved in responses to other stresses remains unclear [23]. JA is important for regulating plant responses to various abiotic stresses, including cold and drought conditions [48–50]. Therefore, it is reasonable to speculate that *RsLOX7*, *RsLOX8*, and *RsLOX9* are the main *RsLOXs* in the JA biosynthetic pathway, and the increased production of JA is directly involved in responses to various environmental stresses. We also observed that the *RsLOXs* are unaffected by NaCl and *P. brassicae* treatments, possibly because of an insufficient processing time or concentration. Alternatively, the *RsLOXs* may simply be unresponsive to these two stresses. The predicted *RsLOX* functions will need to be experimentally verified in future studies.

4. Materials and Methods

4.1. Sequence Acquisition and Identification of LOX Genes

The *A. thaliana* genome assemblies were downloaded from the TAIR10 database [51]. Genome resources for *R. sativus* [52] and *B. rapa* [53] were downloaded from the BRAD database (http://brassicadb.org/brad/) [54], and *B. oleracea* HDEM genome data [55] was download from Genoscope database. The LOX domain model (PF00305) was downloaded from the Pfam database 32.0 [56]. Additionally, Hmmsearch, from the HMMER suite (version 3.1) [57], was used to search for the LOX domain (PF00305) in the proteins of each species. The putative LOX proteins among the initially screened proteins were filtered with an E-value cutoff of 1×10^{-5} and sequence coverage of the

Pfam domain models of at least 60%. The detected proteins were then examined for the presence of the LOX and PLAT domains with the InterProScan website. The Biopython module Bio.SeqUtils.ProtParam was used to predict the molecular weight, isoelectric point, and other physical and chemical properties of the *R. sativus* LOX proteins. The MG2C website was used to visualize the location of *RsLOX* genes on nine pseudomolecular chromosomes [58].

4.2. Multiple Sequence Alignment and Phylogenetic Tree Construction

The global alignment tool Needle of the EMBOSS software suite [59] was used for the pairwise alignment of RsLOX proteins to determine the sequence identity and similarity. The LOX proteins from 17 species were included in a phylogenetic analysis. The amino acid sequences for the *R. sativus*, *B. oleracea*, *B. rapa*, and *A. thaliana* LOX proteins were identified in this study, whereas the sequences for the other 13 species were obtained from published articles [36,60]. The MUSCLE program [61] was used to align the complete LOX amino acid sequences. A phylogenetic tree was constructed with the MEGA-X program [62]. Specifically, the neighbor-joining method with the Jones–Taylor–Thornton model was used. The phylogenetic tree was constructed with 500 bootstrap replicates used to assess the statistical support for each tree node. Additionally, uniform rates and homogeneous lineages were applied and a partial deletion with a site coverage cutoff of 70% was used for treating gaps. Another phylogenetic tree comprising RsLOX proteins was constructed using the same methods.

4.3. Analyses of Conserved Motifs, Gene Structures, and Subcellular Localizations

To identify the conserved motifs encoded by LOX genes, the following parameters of the MEME suite [63] were applied to search for motifs: the maximum number to be found was set to 15 and the motif window length was set to 8–100 bp. The TBtools program [64] was used to analyze gene structures and for visualizing gene structures and the encoded conserved domains. Sequence logos for the conserved LOX domain in *R. sativus* proteins were generated with WebLogo 3 [65]. Additionally, the conserved 38-amino acid sequences were aligned with the DNAMAN program (version 9) (Lynnon Biosoft Company, Quebec, QC, Canada). The subcellular localizations of RsLOXs were predicted with the online ChloroP 1.1 Server [66].

4.4. Analyses of Tandem Duplications and Synteny

To clarify the evolution of the LOX genes in *R. sativus* and related species, we analyzed the tandem duplications and syntenic relationships of LOX genes in *R. sativus*, *A. thaliana*, *B. oleracea*, and *B. rapa*. Tandem duplications were defined as gene copies that were separated by 10 or fewer genes and within 200 kb. The SynOrths program was used to identify syntenic orthologs based on the sequence similarity and the collinearity of the flanking genes [67], and the circos software was used to visualize the syntenic relationships of LOXs among these species [68].

4.5. Transcriptional Profile Analysis

The *RsLOX* expression profiles were analyzed based on published RNA sequencing data for six *R. sativus* tissues (flowers, siliques, leaves, stems, callus, and roots) [38]. The reads were aligned to the *R. sativus* "XYB-36-2" genome [52] with TopHat2 [69]. The transcript abundance for each gene was calculated according to the fragments per kilobase of transcript per million mapped reads (FPKM) values with the Cufflinks program [70].

4.6. Plant Materials and Stress Treatments

Seeds of radish variety 'BXC02' were incubated at 25 °C for 2 days in darkness. Germinated seeds were sown in plastic pots and incubated in a growth chamber with a 16-h day (24 °C)/8-h night (21 °C) cycle. Seedlings at the six true-leaf stage were used for the subsequent experiments. For the heat and cold stress treatments, the seedlings were incubated at 40 °C and 6 °C for 24 h, respectively. The control seedlings were subjected to a 24 °C (day)/21 °C (night) cycle. For the simulated drought and salinity stress treatments, the seedlings were subjected to 20% PEG 6000 and 300 mM NaCl, respectively, for 24 h with a 24 °C (day)/21 °C (night) cycle. Regarding the *P. brassicae* infection, a 5-mL aliquot of a 3×10^8/mL resting spore suspension was injected in the soil near the roots of seedlings at the four-leaf stage. The roots were harvested and analyzed 7 days after the inoculation. The control seedlings were treated with sterile water. The taproots of five seedlings were collected as an independent biological replicate, and each treatment was completed with three replicates. The samples were stored at −80 °C prior to the subsequent RNA isolation step.

4.7. RNA Extraction and qRT-PCR Analysis

Total RNA was extracted from all samples with the Trizol reagent (Invitrogen, Carlsbad, CA, USA). A DNase I treatment was used to eliminate any contaminating genomic DNA. The quality of the RNA samples was checked with the NanoDrop 2000 spectrophotometer (ThermoFisher Scientific, Beijing, China) and 1% denaturing agarose gels. The RNA was used as the template for the first-strand cDNA synthesis with PrimeScript reverse transcriptase (TaKaRa Biotechnology, Dalian, China). The Primer Premier 5.0 program was used to design gene-specific primers (Table S8). Additionally, the *RsGAPDH* gene was used as an internal control for normalizing the gene expression data [71]. The *RsLOX* expression levels were analyzed in a qRT-PCR assay, which was completed with the SYBR Green qPCR kit (TaKaRa Biotechnology, Dalian, China) and the Stratagene Mx3000P thermocycler (Agilent, Santa Clara, CA, USA). The PCR program was as follows: 95 °C for 5 min then 40 cycles of 95 °C for 15 s and 60 °C for 30 s. The relative LOX gene expression levels were calculated with the $2^{-\Delta\Delta Ct}$ method [72]. The analysis included three biological replicates, each with three technical replicates.

5. Conclusions

In this study, we comprehensively characterized the radish LOX gene family via a systematic approach comprising analyses of phylogeny, gene structure, motif composition, evolution, and gene expression patterns in response to various abiotic and biotic stresses. Our results revealed the conservation and diversity in the sequence characteristics and functions among *RsLOX* genes. Moreover, a qRT-PCR analysis of *RsLOX* expression indicated that three tandem-clustered *RsLOX* genes are involved in responses to various environmental stresses through the jasmonic acid pathway. This comprehensive analysis provides a foundation for the functional characterization of LOX genes under stress conditions in radish and potentially in other species, including cotton and other Brassicaceae species.

Supplementary Materials: Supplementary materials can be found at http://www.mdpi.com/1422-0067/20/23/6095/s1. Table S1: Protein LOX and their alignment information in radish and related species. Table S2: Characteristics of protein RsLOX. Table S3: Plant LOX protein sequences used for phylogenetic tree construction. Table S4: Subcellular localization analysis of radish LOXs. Table S5: Syntenic relationships of LOX between *A. thaliana* and radish, *B. oleracea*, *B. rapa*. Table S6: Tandem duplicated LOX genes in radish, *B. oleracea* and *B. rapa*. Table S7: FPKM of LOX genes in different radish tissues. Table S8: RT-PCR primers of *RsLOXs*.

Author Contributions: Conceptualization, C.B.; methodology, J.W.; software, J.W.; validation, J.W. and C.B.; formal analysis, J.W.; investigation, T.H.; resources, T.H. and X.W.; data curation, W.W.; writing—original draft preparation, J.W.; writing—review and editing, J.W.; visualization, Q.W.; supervision, C.B. and T.H.; project administration, H.H.; funding acquisition, C.B.

Funding: This work was supported by grant from the National Technology System of Commodity Vegetable Industry-Radish (CARS-23-A-14).

Conflicts of Interest: The authors declare no conflicts of interest.

Abbreviations

LOXs	Lipoxygenases
PLAT	Polycystin-1, lipoxygenase, alpha-toxin
His	Histidine
9-HPOD	9s-hydroperoxyoctadecadienoic acid
13-HPOD	13s-hydroperoxyoctadecadienoic acid
JA	Jasmonic acid
WGT	Genome triplication
SD	Segmental duplication
TD	Tandem duplication
PEG	Polyethylene glycol
qRT-PCR	Quantitative real-time polymerase chain reaction
FPKM	Fragments per kilobase of transcript per million mapped reads

References

1. Hildebrand, D.F. Lipoxygenases. *Physiol. Plant.* **1989**, *76*, 249–253. [CrossRef]
2. Feussner, I.; Wasternack, C. The lipoxygenase pathway. *Plant Biol.* **2003**, *53*, 275–297. [CrossRef] [PubMed]
3. Teng, L.; Han, W.; Fan, X.; Xu, D.; Zhang, X.; Dittami, S.M.; Ye, N. Evolution and expansion of the prokaryote-like lipoxygenase family in the brown alga *Saccharina japonica*. *Front. Plant Sci.* **2017**, *28*, 8. [CrossRef] [PubMed]
4. Minor, W.; Steczko, J.; Stec, B.; Otwinowski, Z.; Bolin, J.T.; Walter, R.; Axelrod, B. Crystal structure of soybean lipoxygenase L-1 at 1.4 A resolution. *Biochemistry* **1996**, *35*, 10687–10701. [CrossRef] [PubMed]
5. Bateman, A.; Sandford, R. The PLAT domain: A new piece in the PKD1 puzzle. *Curr. Biol.* **1999**, *9*, R588–R590. [CrossRef]
6. Tomchick, D.R.; Phan, P.; Cymborowski, M.; Minor, W.; Holman, T.R. Structural and functional characterization of second-coordination sphere mutants of soybean lipoxygenase-1. *Biochemistry* **2001**, *40*, 7509–7517. [CrossRef]
7. Brash, A.R. Lipoxygenases: Occurrence, functions, catalysis, and acquisition of substrate. *J. Biol. Chem.* **1999**, *274*, 23679–23682. [CrossRef]
8. Porta, H.; Rochasosa, M. Plant lipoxygenases. physiological and molecular features. *Plant Physiol.* **2002**, *130*, 15–21. [CrossRef]
9. Bailly, C.; Bogatekleszczynska, R.; Come, D.; Corbineau, F. Changes in activities of antioxidant enzymes and lipoxygenase during growth of sunflower seedlings from seeds of different vigour. *Seed Sci. Res.* **2002**, *12*, 47–55. [CrossRef]
10. Kolomiets, M.V.; Hannapel, D.J.; Chen, H.; Tymeson, M.; Gladon, R.J. Lipoxygenase is involved in the control of potato tuber development. *Plant Cell* **2001**, *13*, 613–626. [CrossRef]
11. Barry, C.S.; Giovannoni, J.J. Ethylene and fruit ripening. *J. Plant Growth Regul.* **2007**, *26*, 143–159. [CrossRef]
12. Chuck, G. Molecular mechanisms of sex determination in monoecious and dioecious plants. *Adv. Bot. Res.* **2010**, *54*, 53–83.
13. Blee, E. Impact of phyto-oxylipins in plant defense. *Trends Plant Sci.* **2002**, *7*, 315–322. [CrossRef]
14. Moran, P.J.; Thompson, G.A. Molecular responses to aphid feeding in *Arabidopsis* in relation to plant defense pathways. *Plant Physiol.* **2001**, *125*, 1074–1085. [CrossRef]
15. Yan, L.; Zhai, Q.; Wei, J.; Li, S.; Wang, B.; Huang, T.; Du, M.; Sun, J.; Kang, L.; Li, C.B. Role of tomato lipoxygenase D in wound-induced jasmonate biosynthesis and plant immunity to insect herbivores. *PLoS Genet.* **2013**, *9*. [CrossRef]
16. Browse, J. Jasmonate passes muster: A receptor and targets for the defense hormone. *Annu. Rev. Plant Biol.* **2009**, *60*, 183–205. [CrossRef]
17. Hassan, M.N.; Zainal, Z.; Ismail, I. Green leaf volatiles: Biosynthesis, biological functions and their applications in biotechnology. *Plant Biotechnol. J.* **2015**, *13*, 727–739. [CrossRef]

18. Christensen, S.A.; Huffaker, A.; Kaplan, F.; Sims, J.; Ziemann, S.; Doehlemann, G.; Ji, L.; Schmitz, R.J.; Kolomiets, M.V.; Alborn, H.T. Maize death acids, 9-lipoxygenase–derived cyclopente(a)nones, display activity as cytotoxic phytoalexins and transcriptional mediators. *Proc. Natl. Acad. Sci. USA* **2015**, *112*, 11407–11412. [CrossRef]

19. Christensen, S.A.; Huffaker, A.; Hunter, C.T.; Alborn, H.T.; Schmelz, E.A. A maize death acid, 10-oxo-11-phytoenoic acid, is the predominant cyclopentenone signal present during multiple stress and developmental conditions. *Plant Signal. Behav.* **2016**, *11*, e1120395. [CrossRef]

20. Umate, P. Genome-wide analysis of lipoxygenase gene family in *Arabidopsis* and rice. *Plant Signal. Behav.* **2011**, *6*, 335–338. [CrossRef]

21. Melan, M.A.; Dong, X.; Endara, M.E.; Davis, K.; Ausubel, F.M.; Peterman, T.K. An *Arabidopsis thaliana* lipoxygenase gene can be induced by pathogens, abscisic acid, and methyl jasmonate. *Plant Physiol.* **1993**, *101*, 441–450. [CrossRef]

22. Vellosillo, T.; Martinez, M.; Lopez, M.A.; Vicente, J.; Cascon, T.; Dolan, L.; Hamberg, M.; Castresana, C. Oxylipins produced by the 9-lipoxygenase pathway in *Arabidopsis* regulate lateral root development and defense responses through a specific signaling cascade. *Plant Cell* **2007**, *19*, 831–846. [CrossRef]

23. Bell, E.; Creelman, R.A.; Mullet, J.E. A chloroplast lipoxygenase is required for wound-induced jasmonic acid accumulation in *Arabidopsis*. *Proc. Natl. Acad. Sci. USA* **1995**, *92*, 8675–8679. [CrossRef]

24. Grebner, W.; Stingl, N.; Oenel, A.; Mueller, M.J.; Berger, S. Lipoxygenase6-dependent oxylipin synthesis in roots is required for abiotic and biotic stress resistance of *Arabidopsis*. *Plant Physiol.* **2013**, *161*, 2159–2170. [CrossRef]

25. Rayko, H.; Baldwin, I.T. Antisense LOX expression increases herbivore performance by decreasing defense responses and inhibiting growth-related transcriptional reorganization in *Nicotiana attenuata*. *Plant J.* **2010**, *36*, 794–807.

26. Kessler, A.; Halitschke, R.; Baldwin, I.T. Silencing the jasmonate cascade: Induced plant defenses and insect populations. *Science* **2004**, *305*, 665–668. [CrossRef]

27. Royo, J.; Vancanneyt, G.; Pérez, A.G.; Sanz, C.; Störmann, K.; Rosahl, S.; Sánchez-Serrano, J.J. Characterization of three potato lipoxygenases with distinct enzymatic activities and different organ-specific and wound-regulated expression patterns. *J. Biol. Chem.* **1996**, *271*, 21012. [CrossRef]

28. Allmann, S.; Halitschke, R.; Schuurink, R.C.; Baldwin, I.T. Oxylipin channelling in *Nicotiana attenuata*: Lipoxygenase 2 supplies substrates for green leaf volatile production. *Plant Cell Environ.* **2010**, *33*, 2028–2040. [CrossRef]

29. José, L.; Joaquín, R.; Guy, V.; Carlos, S.; Helena, S.; Gareth, G.; Sánchez-Serrano, J.J. Lipoxygenase H1 gene silencing reveals a specific role in supplying fatty acid hydroperoxides for aliphatic aldehyde production. *J. Biol. Chem.* **2002**, *277*, 416–423.

30. Lim, C.W.; Han, S.; Hwang, I.S.; Kim, D.S.; Hwang, B.K.; Lee, S.C. The pepper lipoxygenase CaLOX1 plays a role in osmotic, drought and high salinity stress response. *Plant Cell Physiol.* **2015**, *56*, 930–942. [CrossRef]

31. Hou, Y.; Meng, K.; Han, Y.; Ban, Q.; Wang, B.; Suo, J.; Lv, J.; Rao, J. The persimmon 9-lipoxygenase gene *DkLOX3* plays positive roles in both promoting senescence and enhancing tolerance to abiotic stress. *Front. Plant Sci.* **2015**, *6*, 1073. [CrossRef] [PubMed]

32. Bannenberg, G.; Martinez, M.; Hamberg, M.; Castresana, C. Diversity of the enzymatic activity in the lipoxygenase gene family of *Arabidopsis thaliana*. *Lipids* **2009**, *44*, 85–95. [CrossRef]

33. Shaban, M.; Ahmed, M.M.; Sun, H.; Ullah, A.; Zhu, L. Genome-wide identification of lipoxygenase gene family in cotton and functional characterization in response to abiotic stresses. *BMC Genom.* **2018**, *19*, 599. [CrossRef] [PubMed]

34. Ogunola, O.; Hawkins, L.K.; Mylroie, E.; Kolomiets, M.V.; Borrego, E.J.; Tang, J.D.; Williams, W.P.; Warburton, M.L. Characterization of the maize lipoxygenase gene family in relation to aflatoxin accumulation resistance. *PLoS ONE* **2017**, *12*. [CrossRef]

35. Liu, S.Q.; Liu, X.H.; Jiang, L.W. Genome-wide identification, phylogeny and expression analysis of the lipoxygenase gene family in cucumber. *Gen. Mol. Res.* **2011**, *10*, 2613–2636. [CrossRef]

36. Chen, Z.; Chen, X.; Yan, H.; Li, W.; Li, Y.; Cai, R.; Xiang, Y. The lipoxygenase gene family in poplar: Identifcation, classifcation, and expression in response to MeJA treatment. *PLoS ONE* **2015**, *10*, 1–23. [CrossRef]

37. Yu, J.; Wang, L.; Guo, H.; Liao, B.; King, G.; Zhang, X. Genome evolutionary dynamics followed by diversifying selection explains the complexity of the *Sesamum indicum* genome. *BMC Genom.* **2015**, *18*, 257. [CrossRef]

38. Wang, J.; Qiu, Y.; Wang, X.; Yue, Z.; Yang, X.; Chen, X.; Zhang, X.; Shen, D.; Wang, H.; Song, J. Insights into the species-specific metabolic engineering of glucosinolates in radish (*Raphanus sativus* L.) based on comparative genomic analysis. *Sci. Rep.* **2017**, *7*, 16040. [CrossRef]

39. Force, A.; Lynch, M.; Pickett, F.B.; Amores, A.; Yan, Y.; Postlethwait, J.H. Preservation of duplicate genes by complementary, degenerative mutations. *Genetics* **1999**, *151*, 1531–1545.

40. Paterson, A.H.; Chapman, B.; Kissinger, J.C.; Bowers, J.E.; Feltus, F.A.; Estill, J.C. Many gene and domain families have convergent fates following independent whole-genome duplication events in *Arabidopsis*, Oryza, Saccharomyces and Tetraodon. *Trends Genet.* **2006**, *22*, 597–602. [CrossRef]

41. Wang, Y.; Wang, X.; Tang, H.; Tan, X.; Ficklin, S.P.; Feltus, F.A.; Paterson, A.H. Modes of gene duplication contribute differently to genetic novelty and redundancy, but show parallels across divergent angiosperms. *PLoS ONE* **2011**, *6*. [CrossRef] [PubMed]

42. Wang, Y. Locally duplicated ohnologs evolve faster than nonlocally duplicated ohnologs in *Arabidopsis* and Rice. *Genome Biol. Evol.* **2013**, *5*, 362–369. [CrossRef]

43. Beilstein, M.A.; Alshehbaz, I.A.; Kellogg, E.A. Brassicaceae phylogeny and trichome evolution. *Am. J. Bot.* **2006**, *93*, 607–619. [CrossRef]

44. Lysak, M.A.; Koch, M.A.; Pecinka, A.; Schubert, I. Chromosome triplication found across the tribe Brassiceae. *Genome Res.* **2005**, *15*, 516–525. [CrossRef]

45. Tang, H.; Bowers, J.E.; Wang, X.; Ming, R.R.; Alam, M.; Paterson, A.H. Synteny and collinearity in plant genomes. *Science* **2008**, *320*, 486–488. [CrossRef]

46. Birchler, J.A.; Veitia, R.A. The Gene Balance Hypothesis: From classical genetics to modern genomics. *Plant Cell* **2007**, *19*, 395–402. [CrossRef]

47. Sarde, S.J.; Kumar, A.; Remme, R.N.; Dicke, M. Genome-wide identification, classification and expression of lipoxygenase gene family in pepper. *Plant Mol. Biol.* **2018**, *98*, 375–387. [CrossRef]

48. Mahouachi, J.; Argamasilla, R.; Gomezcadenas, A. Influence of exogenous glycine betaine and abscisic acid on papaya in responses to water-deficit stress. *J. Plant Growth Regul.* **2012**, *31*, 1–10. [CrossRef]

49. Zhao, J.; Li, S.; Jiang, T.; Liu, Z.; Zhang, W.; Jian, G.; Qi, F. Chilling stress–the key predisposing factor for causing *Alternaria alternata* infection and leading to cotton (*Gossypium hirsutum* L.) leaf senescence. *PLoS ONE* **2012**, *7*, e36126. [CrossRef]

50. Creelman, R.A.; Mullet, J.E. Jasmonic acid distribution and action in plants: Regulation during development and response to biotic and abiotic stress. *Proc. Natl. Acad. Sci. USA* **1995**, *92*, 4114–4119. [CrossRef]

51. Huala, E.; Dickerman, A.W.; Garciahernandez, M.; Weems, D.; Reiser, L.; Lafond, F.; Hanley, D.; Kiphart, D.; Zhuang, M.; Huang, W. The *Arabidopsis* Information Resource (TAIR): A comprehensive database and web-based information retrieval, analysis, and visualization system for a model plant. *Nucleic Acids Res.* **2001**, *29*, 102–105. [CrossRef]

52. Xiaohui, Z.; Zhen, Y.; Shiyong, M.; Yang, Q.; Xinhua, Y.; Xiaohua, C.; Feng, C.; Zhangyan, W.; Yuyan, S.; Yi, J. A de novo genome of a chinese radish cultivar. *Hortic. Plant J.* **2015**, *1*, 155–164.

53. Wang, X.; Wang, H.; Wang, J.; Sun, R.; Wu, J.; Liu, S.; Bai, Y.; Mun, J.; Bancroft, I.; Cheng, F. The genome of the mesopolyploid crop species *Brassica rapa*. *Nat. Genet.* **2011**, *43*, 1035–1039. [CrossRef] [PubMed]

54. Cheng, F.; Liu, S.; Wu, J.; Fang, L.; Sun, S.; Liu, B.; Li, P.; Hua, W.; Wang, X. BRAD, the genetics and genomics database for Brassica plants. *BMC Plant Biol.* **2011**, *11*, 136. [CrossRef] [PubMed]

55. Belser, C.; Istace, B.; Denis, E.; Dubarry, M.; Baurens, F.; Falentin, C.; Genete, M.; Berrabah, W.; Chevre, A.; Delourme, R. Chromosome-scale assemblies of plant genomes using nanopore long reads and optical maps. *Nat. Plants* **2018**, *4*, 879–887. [CrossRef]

56. Elgebali, S.; Mistry, J.; Bateman, A.; Eddy, S.R.; Luciani, A.; Potter, S.C.; Qureshi, M.; Richardson, L.; Salazar, G.A.; Smart, A. The Pfam protein families database in 2019. *Nucleic Acids Res.* **2019**, *47*, D427–D432. [CrossRef]

57. Finn, R.D.; Clements, J.; Eddy, S.R. HMMER web server: Interactive sequence similarity searching. *Nucleic Acids Res.* **2011**, *39*, 29–37. [CrossRef]

58. Jiangtao, C.; Yingzhen, K.; Qian, W.; Yuhe, S.; Daping, G.; Jing, L.; Guanshan, L. MapGene2Chrom, a tool to draw gene physical map based on Perl and SVG languages. *Hereditas* **2015**, *37*, 91.

59. Rice, P.M.; Longden, I.; Bleasby, A.J. EMBOSS: The european molecular biology open software suite. *Trends Genet.* **2000**, *16*, 276–277. [CrossRef]

60. Zhu, J.; Wang, X.; Guo, L.; Xu, Q.; Zhao, S.; Li, F.; Yan, X.; Liu, S.; Wei, C. Characterization and alternative splicing profiles of the lipoxygenase gene family in tea plant (*Camellia sinensis*). *Plant Cell Physiol.* **2018**, *59*, 1765–1781. [CrossRef]

61. Edgar, R.C. MUSCLE: Multiple sequence alignment with high accuracy and high throughput. *Nucleic Acids Res.* **2004**, *32*, 1792–1797. [CrossRef] [PubMed]

62. Kumar, S.; Stecher, G.; Li, M.; Knyaz, C.; Tamura, K. MEGA X: Molecular evolutionary genetics analysis across computing platforms. *Mol. Biol. Evol.* **2018**, *35*, 1547–1549. [CrossRef] [PubMed]

63. Bailey, T.L.; Boden, M.; Buske, F.A.; Frith, M.C.; Grant, C.E.; Clementi, L.; Ren, J.; Li, W.W.; Noble, W.S. MEME Suite: Tools for motif discovery and searching. *Nucleic Acids Res.* **2009**, *37*, 202–208. [CrossRef] [PubMed]

64. Chen, C.; Xia, R.; Chen, H.; He, Y. TBtools, a Toolkit for Biologists integrating various HTS-data handling tools with a user-friendly interface. *bioRxiv* **2018**. [CrossRef]

65. Crooks, G.E.; Hon, G.C.; Chandonia, J.M.; Brenner, S.E. WebLogo: A sequence logo generator. *Genome Res.* **2004**, *14*, 1188–1190. [CrossRef]

66. Emanuelsson, O.; Nielsen, H.; Von Heijne, G. ChloroP, a neural network-based method for predicting chloroplast transit peptides and their cleavage sites. *Protein Sci.* **1999**, *8*, 978–984. [CrossRef]

67. Cheng, F.P.; Wu, J.; Fang, L.; Wang, X. Syntenic gene analysis between Brassica rapa and other Brassicaceae species. *Front. Plant Sci.* **2012**, *3*, 198. [CrossRef]

68. Krzywinski, M.; Schein, J.E.; Birol, I.; Connors, J.M.; Gascoyne, R.D.; Horsman, D.; Jones, S.J.M.; Marra, M.A. Circos: An information aesthetic for comparative genomics. *Genome Res.* **2009**, *19*, 1639–1645. [CrossRef]

69. Kim, D.; Pertea, G.; Trapnell, C.; Pimentel, H.; Kelley, R.; Salzberg, S.L. TopHat2: Accurate alignment of transcriptomes in the presence of insertions, deletions and gene fusions. *Genome Biol.* **2013**, *14*, R36. [CrossRef]

70. Trapnell, C.; Roberts, A.; Goff, L.; Pertea, G.; Kim, D.; Kelley, D.R.; Pimentel, H.; Salzberg, S.L.; Rinn, J.L.; Pachter, L. Differential gene and transcript expression analysis of RNA-seq experiments with TopHat and Cufflinks. *Nat. Protoc.* **2012**, *7*, 562. [CrossRef]

71. Duan, M.; Wang, J.; Zhang, X.; Yang, H.; Wang, H.; Qiu, Y.; Song, J.; Guo, Y.; Li, X. Identification of optimal reference genes for expression analysis in radish (*Raphanus sativus* L.) and its relatives based on expression stability. *Front. Plant Sci.* **2017**, *8*, 1605. [CrossRef]

72. Livak, K.J.; Schmittgen, T.D. Analysis of relative gene expression data using real-time quantitative PCR and the 2(-Delta Delta C(T)) Method. *Methods* **2001**, *25*, 402–408. [CrossRef]

MDPI
St. Alban-Anlage 66
4052 Basel
Switzerland
Tel. +41 61 683 77 34
Fax +41 61 302 89 18
www.mdpi.com

International Journal of Molecular Sciences Editorial Office
E-mail: ijms@mdpi.com
www.mdpi.com/journal/ijms

9 783039 361144